MASTERING
AUTOCAD 14
FOR MECHANICAL ENGINEERS

D1737394

Mastering AUTOCAD® 14 FOR MECHANICAL ENGINEERS

George Omura
Steven Keith

SYBEX®

San Francisco • Paris • Düsseldorf • Soest

Associate Publisher: Amy Romanoff
Acquisitions Manager: Kristine Plachy
Acquisitions & Developmental Editor: Melanie Spiller
Editor: Nancy Conner
Project Editor: Brenda Frink
Technical Editors: Bill Hill and Steve Wells
Book Designers: Patrick Dintino and Catalin Dulfu
Graphic Illustrator: Andrew Benzie
Electronic Publishing Specialist: Bob Bihlmayer
Production Coordinator: Charles Mathews
Indexer: Matthew Spence
Companion CD: Molly Sharp and John D. Wright
Cover Designer: Design Site
Cover Photographer: Mark Johann
Cover Photo Art Direction: Ingalls + Associates

Screen reproductions produced with Collage Complete.
Collage Complete is a trademark of Inner Media Inc.
SYBEX is a registered trademark of SYBEX Inc.
Mastering is a trademark of SYBEX Inc.

TRADEMARKS: SYBEX has attempted throughout this book to distinguish proprietary trademarks from descriptive terms by following the capitalization style used by the manufacturer. Netscape Communications, the Netscape Communications logo, Netscape, and Netscape Navigator are trademarks of Netscape Communications Corporation.
CD Interface music from GIRA Sound AURIA Music Library ©℗GIRA Sound 1996.
The author and publisher have made their best efforts to prepare this book, and the content is based upon final release software whenever possible. Portions of the manuscript may be based upon pre-release versions supplied by software manufacturer(s). The author and the publisher make no representation or warranties of any kind with regard to the completeness or accuracy of the contents herein and accept no liability of any kind including but not limited to performance, merchantability, fitness for any particular purpose, or any losses or damages of any kind caused or alleged to be caused directly or indirectly from this book.
Photographs and illustrations used in this book have been downloaded from publicly accessible file archives and are used in this book for news reportage purposes only to demonstrate the variety of graphics resources available via electronic access. Text and images available over the Internet may be subject to copyright and other rights owned by third parties. Online availability of text and images does not imply that they may be reused without the permission of rights holders, although the Copyright Act does permit certain unauthorized reuse as fair use under 17 U.S.C. Section 107.

An earlier version of this book was published under the title *Mastering AutoCAD 14* ©1997 SYBEX Inc.

Copyright ©1998 SYBEX Inc., 1151 Marina Village Parkway, Alameda, CA 94501. World rights reserved. No part of this publication may be stored in a retrieval system, transmitted, or reproduced in any way, including but not limited to photocopy, photograph, magnetic or other record, without the prior agreement and written permission of the publisher.

Library of Congress Card Number: 97-61716

ISBN: 0-7821-2108-X

Manufactured in the United States of America

10 9 8 7 6 5 4 3 2 1

*George Omura:
To my family and
my teachers*

*Steven Keith:
To Jill; without her
this would not have
been possible*

Acknowledgments

Book production is a complicated process, so I'm always amazed at how quickly *Mastering AutoCAD* evolves from manuscript to finished product. In many ways, it's quite magical; but behind the magic, there are many hardworking people giving their best effort. I'd like to thank those people who helped bring this book to you.

Heartfelt thanks go to the editorial and production teams at Sybex for their efforts in getting this book to press on an incredible schedule. Developmental Editor Melanie Spiller got things going and offered many great suggestions. Project Editor Brenda Frink and Editor Nancy Conner (along with her assistant Tamsen Conner) made the frantic schedule bearable with humor and encouragement. Steven Wells and Bill Hill, Technical Editors, provided helpful suggestions as they carefully reviewed the book. Bob Bihlmayer, Electronic Publishing Specialist, created the pages you see before you, and Charles Mathews, Production Coordinator, tracked the book's progress through the production cycle. Finally, Molly Sharp and Dale Wright compiled our CD and made it easy and fun to use.

Thanks also go to the people at Autodesk for their support. Jim Quanci was always willing to give a helping hand. Kathy Koepke cheerfully provided the materials we needed and was always quick to respond to my questions.

I also wish to thank the many contributors to this book. First, a big thanks to Mike Gunderloy for his great work on ActiveX Automation in Chapter 20. Many thanks to my friend and colleague Robert B. Callori for his work on the *AutoCAD Instant Reference* on the CD-ROM. And thanks to Paul Richardson and Christine Merredith of Technical Publications for their work on the appendices and the *ABCs of AutoLISP*, which is also on the CD.

And finally, a great big thanks to my wife and sons, who are always behind my work 100 percent.

—George Omura

In addition to all those George acknowledged, I owe thanks to my employer, the Westcor Division of Vicor Corporation, for allowing me a light workload during the production of this book and for providing the hardware and software necessary to remain current in an ever-changing world.

My wife and family deserve hugs and thanks for their encouragement and support of this work.

—Steven Keith

Contents at a Glance

		Introduction	.xxiii
PART I		**THE BASICS**	**1**
	1	This Is AutoCAD	3
	2	Creating Your First Drawing	37
	3	Learning the Tools of the Trade	87
	4	Organizing Your Work	131
PART II		**BUILDING ON THE BASICS**	**190**
	5	Editing for Productivity	193
	6	Enhancing Your Drawing Skills	243
	7	Adding Text to Drawings	263
	8	Using Dimensions	309
	9	Advanced Productivity Tools	361
	10	Drawing Curves and Solid Fills	415
PART III		**MODELING AND IMAGING IN 3D**	**462**
	11	Introducing 3D	465
	12	Mastering 3D Solids	523
	13	Using 3D Surfaces	581
PART IV		**PRINTING AND PLOTTING AS AN EXPERT**	**640**
	14	Printing and Plotting	643
	15	3D Rendering in AutoCAD	681
	16	Working with Existing Drawings and Raster Images	727

PART V	CUSTOMIZATION: TAKING AUTOCAD TO THE LIMIT	756
17	Storing and Linking Data with Graphics	759
18	Getting and Exchanging Data from Drawings	811
19	Introduction to Customization	847
20	Using ActiveX Automation with AutoCAD	903
21	Integrating AutoCAD into Your Projects and Organization	941

APPENDICES		993
A	Hardware and Software Tips	993
B	Installing and Setting Up AutoCAD	1005
C	What's on the Companion CD-ROM	1031
D	System and Dimension Variables	1045
	Index	1087

Table of Contents

Introduction . . . . *xxiii*

PART I • THE BASICS

1   This Is AutoCAD   3

Taking a Guided Tour 6
    The AutoCAD Window 9
    The Pull-Down Menus 13
    The Toolbars 19
Working with AutoCAD 26
    Opening an Existing File 27
    Getting a Closer Look 28
    Saving a File As You Work 32
    Making Changes 33
    Closing AutoCAD 35
If You Want to Experiment... 35

2   Creating Your First Drawing   37

Getting to Know the Draw Toolbar 40
Starting Your First Drawing 44
Specifying Distances with Coordinates 49
    Specifying Polar Coordinates 49
    Specifying Relative Cartesian Coordinates 51
Interpreting the Cursor Modes and Understanding Prompts 54
    Choosing Command Options 55
Selecting Objects 61
    Selecting Objects in AutoCAD 61
    Selecting Objects Before the Command: Noun/Verb 70
    Restrictions on Noun/Verb Object Selection .74
Editing with Grips 75
    Stretching Lines Using Grips 75
    Moving and Rotating with Grips 78
Getting Help 81
Displaying Data in a Text Window 84
If You Want to Experiment... 85

3 Learning the Tools of the Trade — 87

- Setting Up a Work Area 90
- Specifying Units 90
  - Fine-Tuning the Measurement System 94
  - Setting Up the Drawing Limits 96
  - Understanding Scale Factors 98
- Using the AutoCAD Modes as Drafting Tools 100
  - Using the Grid Mode as a Background Grid 100
  - Using the Snap Mode 102
  - Using Grid and Snap Together 103
  - Using the Coordinate Readout as Your Scale 104
- Exploring the Drawing Process 106
  - Locating an Object in Reference to Others 106
  - Getting a Closer Look 107
  - Modifying an Object 109
- Planning and Laying Out a Drawing 118
  - Making a Preliminary Sketch 124
  - Using the Layout 126
  - Putting On the Finishing Touches 127
- If You Want to Experiment... 129

4 Organizing Your Work — 131

- Creating a Symbol 134
- Inserting a Symbol 136
  - Using an Existing Drawing as a Symbol 140
  - Unblocking and Modifying a Block 147
  - Saving a Block as a Drawing File 150
  - Replacing Existing Files with Blocks 151
  - Other Uses for Blocks 151
  - Grouping Objects 152
- Organizing Information with Layers 158
  - Creating and Assigning Layers 159
  - Working On Layers 167
  - Controlling Layer Visibility 169
  - Finding the Layers You Want 171
  - Assigning Linetypes to Layers 174
- Keeping Track of Blocks and Layers 181
  - Using the Log File Feature 182
- Finding Files on Your Hard Disk 184
- Inserting Symbols with Drag and Drop 187
- If You Want to Experiment... 188

PART II • BUILDING ON THE BASICS

5   Editing for Productivity — 193

- Creating and Using Templates 196
  - Creating a Template 196
  - Using a Template 198
- Copying an Object Multiple Times 199
  - Making Polar Arrays 200
  - Making Row and Column Copies 204
- Developing Your Drawing 213
  - Importing Settings 213
  - Using and Editing Lines 217
  - Finding Distances Along Arcs 228
  - Changing the Length of Objects 231
- Drawing Parallel Lines 232
  - Customizing Multilines 234
  - Joining and Editing Multilines 237
- Eliminating Blocks, Layers, Linetypes, Shapes, and Styles 239
  - Selectively Removing Unused Elements 239
  - Removing All Unused Elements 240
- If You Want to Experiment… 241

6   Enhancing Your Drawing Skills — 243

- Assembling the Parts 246
- Taking Control of the AutoCAD Display 252
  - Understanding Regeneration and Redraw 253
  - Exploring Other Ways of Controlling AutoCAD's Display 254
  - Freezing Layers to Control Regeneration Time 261
  - Block Visibility with Freeze and Thaw 261
  - Taking Control of Regens 261
- If You Want to Experiment… 262

7   Adding Text to Drawings — 263

- Adding Text to a Drawing 266
- Understanding Text Formatting in AutoCAD 269
  - Adjusting the Text Height and Font 269
  - Adding Color, Stacked Fractions, and Special Symbols 272
  - Adjusting the Width of the Text Boundary Window 274
  - Adjusting the Text Alignment 275
  - Editing Existing Text 278
  - Understanding Text and Scale 279
- Organizing Text by Styles 280
  - Creating a Style 281
  - Using a Type Style 282
  - Setting the Current Default Style 283
  - Understanding the Text Style Dialog Box Options 283
  - Renaming a Text Style 285

What Do the Fonts Look Like?	287
Adding Special Characters	289
Adding Simple Text Objects	291
Editing Single-Line Text Objects	293
Justifying Single-Line Text Objects	295
Using Special Characters with Single-Line Text Objects	297
Checking Spelling	299
Choosing a Dictionary	301
Substituting Fonts	302
Accelerating Zooms and Regens with Qtext	305
Bonus Text Editing Utilities	306
If You Want to Experiment...	307

8 Using Dimensions 309

Creating a Dimension Style	312
Setting the Dimension Unit Style	313
Setting the Height for Dimension Text	315
Setting the Location of Dimension Text	316
Choosing an Arrow Style and Setting the Dimension Scale	317
Drawing Linear Dimensions	319
Finding the Dimension Toolbar	319
Placing Horizontal and Vertical Dimensions	320
Continuing a Dimension	321
Editing Dimensions	327
Appending Data to Dimension Text	328
Locating the Definition Points	331
Making Minor Adjustments to Dimensions Using Grips	332
Changing Style Settings of Individual Dimensions	333
Editing Dimensions and Other Objects Together	338
Dimensioning Nonorthogonal Objects	342
Dimensioning Nonorthogonal Linear Distances	342
Dimensioning Angles	345
Dimensioning Radii, Diameters, and Arcs	347
Adding a Note with an Arrow	350
Using Multiline Text with Leaders	351
Skewing Dimension Lines	352
Applying Ordinate Dimensions	354
Adding Tolerance Notation	356
If You Want to Experiment...	359

9 Advanced Productivity Tools 361

Editing More Efficiently	364
Editing an Existing Drawing	364
Singling Out Proximate Objects	369
Using External References (Xrefs)	371
Attaching a Drawing as an External Reference	372
Other Differences between External References and Blocks	375
Other External Reference Options	376

Clipping Xref Views and Improving Performance 378
Importing Named Elements from External References 382
Using Tiled Viewports 385
Understanding Model Space and Paper Space 388
Getting Back to Full-Screen Model Space 395
Working with Paper Space Viewports 396
Scaling Views in Paper Space 398
Setting Layers in Individual Viewports 400
Linetype Scales and Paper Space 403
Dimensioning in Paper Space 404
Other Uses for Paper Space 405
Advanced Tools: Selection Filter and Calculator 405
Filtering Selections 405
Finding Geometry with the Calculator 409
If You Want to Experiment... 414

10 Drawing Curves and Solid Fills   415

Introducing Polylines 417
Drawing a Polyline 417
Polyline Options 420
Editing Polylines 421
Smoothing Polylines 425
Editing Vertices 428
Creating a Polyline Spline Curve 438
Using True Spline Curves 441
Drawing a Spline 441
Fine-Tuning Spline Curves 444
Marking Divisions on a Curve 450
Dividing Objects into Segments of Equal Length 450
Dividing Objects into Specified Lengths 452
Sketching with AutoCAD 454
Freehand Sketching with AutoCAD 454
Filling In Solid Areas 455
Drawing Solid Filled Areas 455
Drawing Filled Circles 456
Toggling Solid Fills On and Off 458
If You Want to Experiment... 459

PART III • MODELING AND IMAGING IN 3D

11 Introducing 3D   465

Creating a 3D Drawing 468
Preparing to Draw Solid Models 469
Drawing a Solid Cylinder 469
Turning a Polyline into a Solid 470
Removing the Volume of One Solid From Another 471
Using the Hide Tool to Verify Your Design 472

Editing a Solid Model with the Chamfer Tool 474
Adding the Volumes of Two Solids 475
Using the Box Primitive 477
The UCS and the WCS 478
Managing the UCS Icon's Appearance 479
The User Coordinate System 479
Coordinate Systems for UCS and WCS 480
Viewing Your Model 495
Viewing a 3D Drawing 496
Finding Isometric and Orthogonal Views 496
Using a Dialog Box to Select 3D Views 498
More Sculpturing of 3D Solids 500
Using the Fillet Tool to Edit a Solid Model 502
Making a Solid Model from a 2D Drawing 506
Aligning Two 3D Objects 508
If You Want to Experiment... 522

12 Mastering 3D Solids   523

Putting Two Solids Together 526
Using the Fillet Tool with Solids 528
Using The Revolve Tool to Make New Parts 536
Making Slides to Improve Communication 546
Mslide and Vslide 546
More Primitives 548
Wedge 549
Cone 550
Editing Solids with Slice 551
The Section Tool 553
The Region: A 2D Object for Boolean Operations 555
Interference Checking: Looking for the Not-So-Obvious 559
Enhancing the 2D Drawing Process 561
The Setup View Tools 565
Managing the New Layers 571
Finding the Mass Properties 573
Taking Advantage of Rapid Prototyping 575
Data Translation and AutoCAD 575
Talking to the Shop 577
3D and Solid Modeling Design Tips 578
If You Want to Experiment... 578

13 Using 3D Surfaces   581

Creating a Surface Model 584
Changing a 2D Polyline into a 3D Model 586
Creating a 3D Object 588
Getting the 3D Results You Want 589
Making Horizontal Surfaces Opaque 590
Setting Layers Carefully 591

Drawing 3D Surfaces 592
  Using Point Filters 592
  Laying Out a 3D Form Object 593
  Adding a 3D Face 596
  Hiding Unwanted Surface Edges 600
  Using Pre-Defined 3D Surface Shapes 602
Creating Complex 3D Surfaces 603
  Creating Curved 3D Surfaces 603
  Adjusting the Settings That Control Meshes 610
Other Surface Drawing Tools 611
  Using Two Objects to Define a Surface 612
  Extruding an Object along a Straight Line 615
  Extruding a Circular Surface 617
Editing a Mesh 619
  Other Mesh Editing Options 621
Moving Objects in 3D Space 622
  Aligning Objects in 3D Space 622
  Rotating an Object in 3D 624
Viewing Your Model in Perspective 624
  Setting Up Your Perspective View 626
  Adjusting Distances 627
  Adjusting the Camera and Target Positions 629
  Changing Your Point of View 631
  Using the Zoom Option as a Telephoto Lens 634
  Twisting the Camera 635
  Using Clip Planes to Hide Parts of Your View 636
If You Want to Experiment... 638

PART IV • PRINTING AND PLOTTING AS AN EXPERT

14 Printing and Plotting   643

Plotting the Plan 646
Selecting an Output Device 648
  Understanding Your Plotter's Limits 652
  Knowing Your Plotter's Origins 653
Selecting a Paper Size and Orientation 654
Controlling the Appearance of Output 655
  Designating Hidden Lines, Fills, and Text Quality 655
  Determining What to Print 658
Controlling Scale and Location 664
  Specifying Drawing Scale 664
  Setting the Output's Origin and Rotation 665
Adjusting Pen Parameters and Plotter Optimization 667
  Working with Pen Assignments and Line Weights 667
  Optimizing Plotter Speed 671

Other Plot Controls 672
   Previewing a Plot 672
   Saving Your Settings 673
Batch Plotting 675
   Other Extended Batch Plot Utility Options 676
Sending Your Drawings to a Service Bureau 679
If You Want to Experiment... 679

15   3D Rendering in AutoCAD   681

Things to Do Before You Start 684
Creating a Quick Study Rendering 684
   Simulating the Sunlight Angle 686
   Adding Shadows 691
   Adding Materials 696
   Adjusting the Materials' Appearance 699
Adding a Background 702
Effects with Lighting 704
   Spotlights 705
   Controlling Lights with Scenes 707
Adding Reflections and Detail with Ray Tracing 712
   Getting a Sharp, Accurate Shadow with Ray Tracing 713
Creating and Adjusting Texture Maps 714
Rendering Output Options 719
   Rendering to the Render Window 719
   Rendering Directly to a File 721
Improving Your Image and Editing 722
Smoothing Out the Rough Edges 724
If You Want to Experiment... 725

16   Working with Existing Drawings and Raster Images   727

Tracing, Scaling, and Scanning Drawings 730
   Tracing a Drawing with a Digitizing Tablet 731
   Cleaning Up a Traced Drawing 735
   Scaling a Drawing 738
   Scanning a Drawing 739
Importing Raster Images 740
   Controlling Object Visibility and Overlap with Raster Images 742
   Reordering a Raster Image 744
   Clipping a Raster Image 745
   Adjusting Brightness, Contrast, and Strength .748
   Turning off the Frame, Adjusting Overall Quality and Transparency 750
Importing PostScript Files 751
If You Want to Experiment... 752

PART V • CUSTOMIZATION: TAKING AUTOCAD TO THE LIMIT

17 Storing and Linking Data with Graphics   759

Creating Attributes 762
   Linking Attributes to Parts 762
   Adding Attributes to Blocks 763
   Changing Attribute Specifications 765
   Inserting Blocks Containing Attributes 768
   Using a Dialog Box to Answer Attribute Prompts 772
Editing Attributes 772
   Editing Attributes One at a Time 773
   Editing Several Attributes in Succession 773
   Making Minor Changes to an Attribute's Appearance 774
   Making Global Changes to Attributes 776
   Making Invisible Attributes Visible 778
   Redefining Blocks Containing Attributes 779
Extracting and Exporting Attribute Information .781
   Determining What to Extract 781
   Extracting Block Information Using Attributes 783
   Performing the Extraction 785
   Using Extracted Attribute Data with Other Programs 788
Accessing External Databases 789
   Setting Up ASE to Locate Database Files 790
   Loading the External Database Toolbar 793
   Opening a Database from AutoCAD 794
   Finding a Record in the Database 795
   Adding a Row to a Database Table 797
Linking Objects to a Database 798
   Creating a Link 799
   Adding Labels with Links 800
   Updating Rows and Labels 802
   Finding and Selecting Graphics through the Database 803
   Deleting a Link 804
   Filtering Selections and Exporting Links .805
   Using SQL Statements 807
   Where to Go from Here 808
If You Want to Experiment... 808

18 Getting and Exchanging Data from Drawings   811

Getting Information about a Drawing 814
   Finding the Area or Location of an Object 814
   Finding the Area of Complex Shapes 817
   Determining the Drawing's Status 820
   Keeping Track of Time 823
   Getting Information from System Variables 824
   Keeping a Log of Your Activity 824

Capturing and Saving Text Data from the AutoCAD Text Window 827
Recovering Corrupted Files 828
Exchanging CAD Data with Other Programs 828
Using the DXF File Format 829
Using AutoCAD Drawings in Desktop Publishing 833
Exporting Raster Files 833
Exporting Vector Files 836
Combining Data from Different Sources 839
Editing Links 843
Options for Embedding Data 844
Using the Clipboard to Export AutoCAD Drawings 845
If You Want to Experiment... 846

19 Introduction to Customization 847

Enhancements Straight from the Source 850
Tools for Managing Layers 850
Tools for Editing Text 853
Bonus Standard Tools 859
Tools on the Bonus Pull-Down Menu 870
Utilities Available from Other Sources 876
Putting AutoLISP to Work 877
Loading and Running an AutoLISP Program 878
Working with the Load AutoLISP, ADS, and ARX Files Dialog Box 879
Loading AutoLISP Programs Automatically 879
Creating Keyboard Macros with AutoLISP 881
Using Third-Party Software 884
Custom-Tailoring AutoCAD 884
Third-Party Product Information on the World Wide Web 885
Autodesk's Own Offerings 885
Getting the Latest Information from Online Services 886
Posting and Accessing Drawings on the World Wide Web 886
Creating a Web-Compatible Drawing 887
Adding a DWF file to a Web Page 889
Viewing Your Web Page 891
Opening, Inserting, and Saving DWG Files over the Web 896
If You Want to Experiment... 900

20 Using ActiveX Automation with AutoCAD 903

What Is ActiveX Automation? 906
Integrating Applications 906
Clients and Servers 907
Automation Objects 908
Automation and AutoCAD 909
An Automation Sample 910
The AutoCAD Object Model 913
The Object Hierarchy 914
Application 915
Preferences 918

Document 919
Block 921
Entity Properties 922
Dictionary 923
DimStyle 924
Group and SelectionSet 924
Layer 924
LineType 926
RegisteredApplication 926
TextStyle 927
UserCoordinateSystem 927
View 928
Viewport 928
Plot 929
Utility 930
Automation Techniques 931
Creating a Drawing 932
Drawing a Line 933
User Interaction 935
Setting Active Properties 936
The AutoCAD VBA Preview 938
If You Want to Experiment... 939

21 Integrating AutoCAD into Your Projects and Organization 941

Customizing Toolbars 944
    Taking a Closer Look at the Toolbars Dialog Box 944
    Creating Your Own Toolbar 945
    Customizing Buttons 949
Adding Your Own Pull-Down Menu 956
    Creating Your First Pull-Down Menu 956
    Loading a Menu 957
    How the Pull-Down Menu Works 959
Creating Custom Linetypes and Hatch Patterns 966
    Viewing Available Linetypes 966
    Creating a New Linetype 969
    Creating Complex Linetypes 971
    Creating Hatch Patterns 974
Supporting Your System and Working in Groups 979
    Getting Outside Support 979
    Choosing In-House Experts 980
    Acclimatizing the Staff 981
    Learning the System 982
    Making AutoCAD Use Easier 982
    Managing an AutoCAD Project 983
Establishing Company Standards 984
    Establishing Layering Conventions 984
Maintaining Files 985
    Backing Up Files 986
    Labeling Hard Copies 987

Using Networks with AutoCAD 987
Keeping Records 988
Understanding What AutoCAD Can Do for You 989
    Seeing the Hidden Advantages 989
    Taking a Project Off AutoCAD 990

APPENDICES

A   Hardware and Software Tips   993

The Graphics Display 994
Pointing Devices 995
    The Digitizing Tablet 995
Output Devices 996
    Printers 996
    Plotters 997
Fine-Tuning PostScript File Export 997
Memory and AutoCAD Performance 999
    AutoCAD and Your Hard Disk 1000
    What to Do for Out of RAM
        and Out of Page Space Errors 1000
When Things Go Wrong 1001
    Difficulty Starting Up or Opening a File 1001
    Restoring Corrupted Files 1001
    Troubleshooting 1002

B   Installing and Setting Up AutoCAD   1005

Before Installing AutoCAD 1006
    Installing the AutoCAD Software 1006
    The Program Files 1008
Configuring AutoCAD 1009
    Files 1009
    Performance 1012
    Compatibility 1014
    General 1016
    Display 1018
    Pointer 1019
    Printer 1020
    Profiles 1021
    What Happened to the AutoCAD .ini File? 1022
    Configuring Your Digitizing Tablet 1022
    Configuring the Tablet Menu Area 1022
    Turning on the Noun/Verb Option 1025
    Turning on the Grips Feature 1026
Setting Up AutoCAD to Use ODBC 1027
    Create a Reference Database File 1027

C What's on the Companion CD-ROM　　1031

- The Mastering AutoCAD Bonus Software 1032
  - The Figures, Eye2eye, and SI Mechanical Utilities 1033
- Opening the Installation Program 1033
- Installing and Using the Sample Drawing Files 1034
- Installing the Bonus Add-On Packages 1035
- Eye2eye 1035
  - Using Eye2eye 1036
  - Using Eye2eye for Model Construction 1039
  - Setting the Eyes 1040
  - Controlling the Eyes 1040
- SWLite 1040
  - Installing SWLite 1041
- The ActiveX Automation Samples 1041
- The AutoCAD Instant Reference 1042
- The ABCs of AutoLISP 1042
- The Whip2 Netscape Communicator Plug-in 1043

D System and Dimension Variables　　1045

- Setting System Variables 1046
- Setting Dimension Variables 1069
  - Controlling Associative Dimensioning 1074
  - Storing Dimension Styles through the Command Line 1075
  - Restoring a Dimension Style from the Command Line 1075
  - Notes on Metric Dimensioning 1076
- A Closer Look at the Dimension Styles Dialog Box 1076
  - The Geometry Dialog Box 1077
  - The Format Dialog Box 1080
  - The Annotation Dialog Box 1081
  - Importing Dimension Styles from Other Drawings 1084
  - Drawing Blocks for Your Own Dimension Arrows and Tick Marks 1084

Index *1087*

Introduction

Welcome to *Mastering AutoCAD 14 for Mechanical Engineers*. As many readers have already discovered, *Mastering AutoCAD* offers a unique blend of tutorial and source book that offers everything you need to get started and stay ahead with AutoCAD.

How to Use This Book

Rather than just showing you how each command works, *Mastering AutoCAD 14 for Mechanical Engineers* shows you AutoCAD in the context of a meaningful activity. You will learn how to use commands while working on an actual project and progressing toward a goal. It also provides a foundation on which you can build your own methods for using AutoCAD, and become an AutoCAD expert yourself. For this reason, we haven't covered every single command or every permutation of a command response. The *AutoCAD Instant Reference*, which we've included on the companion CD-ROM, will fill that purpose nicely. This online resource will help you quickly locate the commands you need. You should think of *Mastering AutoCAD 14 for Mechanical Engineers* as a way to get a detailed look at AutoCAD as it is used on a real project. As you follow the exercises, we encourage you to explore AutoCAD on your own, applying the techniques you learn to your own work.

If you are not an experienced user, you may want to read *Mastering AutoCAD 14 for Mechanical Engineers* as a tutorial. You'll find that each chapter builds on the skills and information you learned in the previous one. To help you navigate, the exercises are shown in numbered steps. This book can also be used as a ready reference for your day-to-day problems and questions about commands. Optional exercises at the end of each chapter will help you review what you have learned and look at different ways to apply these skills.

Getting Information Fast

We've also included plenty of Notes, Tips, and Warnings. *Notes* supplement the main text; *Tips* are designed to make practice easier; and *Warnings* steer you away from pitfalls. Also, in each chapter you will find more extensive tips and discussions in the form of specially screened *sidebars*. Together the Notes, Tips, Warnings, and sidebars provide a wealth of information gathered over years of using AutoCAD on a variety of projects in different environments. You may want to browse through the book, looking at the projects and sidebars, to get an idea of how the information they contain might be useful to you.

Another quick reference you'll find yourself turning to often is Appendix D. This appendix contains tables of all the system settings and comments on their use.

What to Expect

Mastering AutoCAD 14 for Mechanical Engineers is divided into five parts, each representing a milestone in your progress toward becoming an expert AutoCAD user. Here is a description of those parts and what each will show you.

Part I   The Basics

As with any major endeavor, you must begin by tackling small, manageable tasks. In this first part, you will become familiar with the way AutoCAD looks and feels. Chapter 1, *This Is AutoCAD*, shows you how to get around in AutoCAD. In Chapter 2, *Creating Your First Drawing*, you will learn how to start and exit the program and how to respond to AutoCAD command prompts. Chapter 3, *Learning the Tools of the Trade*, tells you how to set up a work area, edit objects, and lay out a drawing. In Chapter 4, *Organizing Your Work*, you will explore some tools unique to CAD: symbols, blocks, and layers. As you are introduced to AutoCAD, you will also get a chance to make some drawings that you can use later in the book and perhaps even in future projects of your own.

Part II   Building on the Basics

Once you have the basics down, you will begin to explore some of AutoCAD's more subtle qualities. Chapter 5, *Editing for Productivity*, tells you how to re-use drawing setup information and parts of an existing drawing. In Chapter 6, *Enhancing Your Drawing Skills*, you will learn how to assemble and edit a large drawing file. In Chapter 7, *Adding Text to Drawings*, you will learn how to annotate your drawing and edit your notes. Chapter 8, *Using Dimensions*, gives you practice in using automatic dimensioning, another unique CAD capability. Chapter 9, *Advanced Productivity Tools*, adds to and refines your editing toolbox. In Chapter 10, *Drawing Curves and Solid Fills*, you will get an in-depth look at some special drawing objects, such as spline and fitted curves. Along the way, we will be giving you tips on editing and problems you may encounter as you begin to use AutoCAD for more complex tasks.

Part III Modeling and Imaging in 3D

While 2D drafting is AutoCAD's workhorse application, AutoCAD's 3D capabilities give you a chance to expand your ideas and look at them in a new light. Chapter 11, *Introducing 3D*, explores solid modeling and the User Coordinate System. Chapter 12, *Mastering 3D Solids*, extends your knowledge of solids and shows you how to create a 2D drawing almost automatically from your 3D model. You'll also learn how to use AutoCAD to calculate the mass properties of a solid object. Chapter 13, *Using 3D Surfaces*, sorts out some of the mysteries of creating and managing surfaces.

Part IV Printing and Plotting as an Expert

The section demonstrates the wide variety of hardcopy (paper) output available from AutoCAD. Chapter 14, *Printing and Plotting*, shows you how to get your drawing onto paper. In Chapter 15, *3D Rendering in AutoCAD*, you'll see how you can use the AutoCAD rendering tool to produce lifelike views of your 3D drawings. Chapter 16, *Working with Existing Drawings and Raster Images*, will teach techniques for transferring paper drawings to AutoCAD.

Part V Customization: Taking AutoCAD to the Limit

In the last part of the book, you will learn how you can take full control of AutoCAD. In Chapter 17, *Storing and Linking Data with Graphics*, you'll learn how to attach information to drawing objects and how to link your drawing to database files. Chapter 18, *Getting and Exchanging Data from Drawings*, will give you practice getting information about a drawing, and you'll learn how AutoCAD can interact with other applications such as spreadsheets and desktop publishing programs. Chapter 19, *Introduction to Customization*, gives you a gentle introduction to the world of AutoCAD customization. You'll learn how to load and use existing utilities that come with AutoCAD and find out how you can publish high-resolution drawings on the World Wide Web. Chapter 20, *Using ActiveX Automation With AutoCAD*, shows you how you can tap the power of automation to add new functions to AutoCAD and link AutoCAD to other applications. Chapter 21, *Integrating AutoCAD into Your Projects and Organization*, shows you how you can adapt AutoCAD to your own work style. Customizing menus, linetypes, and screens are only three of the many topics.

The Appendices

Finally, this book has four appendices. Appendix A, *Hardware and Software Tips*, offers information on hardware related to AutoCAD. It also provides tips on improving AutoCAD's performance and on troubleshooting. Appendix B, *Installing and Setting Up AutoCAD*, contains an installation and configuration tutorial. If AutoCAD is not already installed on your system, you should follow this tutorial before starting Chapter 1. Appendix C, *What's on the Companion CD-ROM*, describes the utilities available on the companion CD-ROM. Appendix D, *System and Dimension Variables*, will illuminate the references to the system variables scattered throughout the book. Appendix D also discusses the many dimension settings and system features AutoCAD has to offer.

The Minimum System Requirements

This book assumes you have an IBM-compatible Pentium computer that will run AutoCAD and support a mouse. Your computer should have at least one CD-ROM drive, and a hard disk with 100MB or more free space after AutoCAD is installed (about 70MB for AutoCAD to work with and another 30MB available for drawing files). In addition to these requirements, you should also have enough free disk space to allow for a Windows virtual memory page file of at least 60MB. Consult your Windows manual or Appendix A of this book for more on virtual memory.

AutoCAD Release 14 runs best on systems with at least 32MB of RAM, although you can get by with 16MB for Windows 95 and with 24MB for Windows NT. Your computer should also have a high-resolution monitor and a color display card. The current standard is the Super Video Graphics Array (SVGA) display. This is quite adequate for most AutoCAD work. The computer should also have at least one serial port. If you have only one, you may want to consider having another one installed. We also assume you are using a mouse and have the use of a printer or a plotter. Most computers come equipped with a sound card, though you won't necessarily need one to use this book.

If you want a more detailed explanation of hardware options with AutoCAD, see Appendix A. You will find a general description of the available hardware options and their significance to AutoCAD.

Doing Things in Style

Much care has been taken to see that the stylistic conventions in this book—the use of upper- or lowercase letters, italic or boldface type, and so on—will be the ones most likely to help you learn AutoCAD. On the whole, their effect should be subliminal. However, you may find it useful to be conscious of the rules that we have followed:

1. Pull-down selections are shown by a series of menu options separated by the ➣ symbol (for example, choose File ➣ New).
2. Keyboard entries are shown in boldface (for example, enter **rotate.↵**). AutoCAD is not case-sensitive, so keyboard entries are shown in lower case to make typing easier.
3. Command-line prompts are also shown in a different font (for example, Select object:).

For most functions, we describe how to select options from toolbars and the menu bar. In addition, where applicable, we include related keyboard shortcuts and command names. By providing command names, we have provided continuity for those readers already familiar with earlier releases of AutoCAD.

All This, and Software Too

Finally, we have included a CD-ROM containing a wealth of utilities, symbols libraries, and sample programs that can greatly enhance your use of AutoCAD. We have also included two online books: the *AutoCAD Instant Reference* and the *ABCs of AutoLISP*. These easy-to-use online references complement *Mastering AutoCAD 14 for Mechanical Engineers* and will prove invaluable for quick command searches and customization tips. Appendix C gives you detailed information about the CD-ROM, but here's a brief rundown of what's available. Check it out!

Software You Can Use Right Away

Si-Mech, a mechanical add-on to AutoCAD, offers many hardware symbols, formats, and quick drawing tools that can save hundreds of drafting and design hours per year. It is a simple, straightforward add-on to AutoCAD that won't take you months to master.

Eye2eye is a utility that makes perspective viewing of your 3D work a simple matter of moving camera and target objects. This utility lets you easily fine-tune your perspective views so you can use them to create rendered images with AutoCAD 14's enhanced rendering tools.

Online Resources

If you just need to quickly find information about a command, the online version of the *AutoCAD Instant Reference* is here to help you. It is a comprehensive guidebook that walks you through every feature and command of AutoCAD Release 14. *Mastering AutoCAD* and the *AutoCAD Instant Reference* have always been a great combination. We have included an electronic version of this best-selling reference so that you can have the best AutoCAD resources in one place.

And if you want in-depth coverage of AutoLISP, AutoCAD's macro programming language, you can delve into the *ABCs of AutoLISP*. This book is an online AutoLISP reference and tutorial. AutoCAD users and developers alike have found the original *ABCs of AutoLISP* book an indispensable resource in their customization efforts. Now in its new HTML format, it's even easier to use.

Drawing Files for the Exercises

We have also included drawing files from the exercises in this book. These are provided so that you can pick up an exercise anywhere in the book, without having to work through the book from front to back. You can also use these sample files to repeat exercises or just to explore how files are organized and put together.

New Features of Release 14

AutoCAD Release 14 offers a higher level of speed, accuracy, and ease of use. It has always provided drawing accuracy to 16 decimal places. With this kind of accuracy, you can create a computer model of the earth and include details down to submicron levels. It also means that no matter how often you edit an AutoCAD drawing, its dimensions will remain true. And AutoCAD Release 14 has greatly improved its overall speed. The interface is more consistent than in earlier releases, so learning and using AutoCAD is easier than ever.

Other new features include

- Improved layer and display controls
- Solid fills for irregular shapes
- Improved hatch patterns that require much less memory
- New raster image tools
- Expanded keyboard shortcuts
- External reference and raster image clipping to show just the portion of a drawing you want

- Full rendering capabilities, including Ray Tracing
- Simplified configuration
- Internet tools to allow full Web and FTP access for reading and posting drawings
- Support for ActiveX automation
- Full TrueType support for improved text quality
- A more consistent, easier-to-use interface

Finally, perhaps the most important feature is what AutoCAD doesn't offer. You will not see a Release 14 version for DOS, Macintosh, SGI, or UNIX. By eliminating these other platforms and concentrating on Windows 95 and NT, Autodesk is able to produce a leaner, meaner AutoCAD. It uses less memory than its previous Windows version and is faster than the previous DOS version. In many ways, this is the AutoCAD you've been waiting for.

The AutoCAD Package

This book assumes you are using AutoCAD Release 14. If you are using an earlier version of AutoCAD, you will want to refer to *Mastering AutoCAD Release 13 for Windows 95 and NT* or its appropriate predecessor. You can find out about other books on AutoCAD published by Sybex by visiting the Sybex Web site, at http://www.sybex.com.

With AutoCAD Release 14, you receive a set of manuals in both hardcopy and electronic formats. They are

- The *AutoCAD Command Reference*
- The *AutoCAD User's Guide*
- The *Installation Guide*
- The *Customization Guide*

In addition, the AutoCAD package contains the AutoCAD Learning Assistant. This is a CD-ROM-based multimedia training and reference tool designed for those users who are upgrading from earlier versions of AutoCAD. It offers animated video clips, tips, and tutorials on a variety of topics. You'll need a sound card to take full advantage of the Learning Assistant.

You'll probably want to read the installation guide for Windows first, and then browse through the *Command Reference* and *User's Guide* to get a feel for the kind of information available there. You may want to save the *Customization Guide* for when you've become more familiar with AutoCAD.

AutoCAD comes on a CD-ROM and offers several levels of installation. This book assumes that you will use the full installation, which includes the Internet and Bonus utilities. You'll also want to install the ActiveX Automation software, included on the AutoCAD CD-ROM, if you plan to explore this new feature.

The Digitizer Template

If you intend to use a digitizer tablet in place of a mouse, Autodesk also provides you with a digitizer template. Commands can be selected directly from the template by pointing at the command on the template and pressing the pick button. Each command is shown clearly by name and a simple icon. Commands are grouped on the template by the type of operation the command performs. Before you can use the digitizer template, you must configure the digitizer. See Appendix A for a more detailed description of digitizing tablets and Appendix B for instructions on configuring the digitizer.

We won't specifically discuss using the digitizer for selecting commands because the process is straightforward. If you are using a digitizer, you can use its puck like a mouse for all of the exercises in this book.

We hope that *Mastering AutoCAD 14 for Mechanical Engineers* will be of benefit to you and that, once you have completed the tutorials, you will continue to use the book as a reference. If you have comments, criticisms, or ideas about how the book can be improved, you can write to us, or send e-mail to the addresses below. And thanks for choosing *Mastering AutoCAD 14*.

George Omura
P.O. Box 6357
Albany, CA 94706-0357
Gomura@sirius.com

Stephen Keith
San Jose, California
Steven.Keith@home.com

PART 1

The Basics

LEARN TO:

- **Perform Basic AutoCAD Tasks**
- **Use AutoCAD Tools**
- **Create an AutoCAD Drawing**
- **Organize Your Work**

Chapter 1

This Is AutoCAD

FEATURING

Navigating the AutoCAD Window

Starting AutoCAD

Opening an Existing File

Enlarging Your View

Saving a File

Erasing an Object

Exiting AutoCAD

Chapter 1

This Is AutoCAD

Over the last few years, AutoCAD has evolved from a DOS-based, command line-driven program, to a full-fledged Windows95/NT application. With AutoCAD Release 14, Autodesk is making a complete break from the DOS world. Neither DOS nor UNIX is supported with AutoCAD Release 14.

By concentrating on a single operating system, Autodesk is able to create a more efficient, faster AutoCAD. As an added benefit, Release 14 offers smaller file sizes than Release 13 did and reduced memory requirements. Its speed matches—and in many cases exceeds—those found in earlier DOS versions of AutoCAD.

If you're a DOS AutoCAD user who has been waiting for a faster Windows-based AutoCAD, your wait is over. AutoCAD Release 14 offers the speed you demand with the convenience of a Windows multitasking environment. You'll also find that AutoCAD makes great use of the Windows environment. For example, you can use Windows' OLE features to paste documents directly into AutoCAD from Excel, Windows Paint, or any other programs that support OLE as a server application. And, as in Releases 12 and 13, you can export AutoCAD drawings directly to other OLE clients. This means no more messy conversions and reworking to get spreadsheet, database, text, or other data into

AutoCAD. It also means that if you want to include a photograph in your AutoCAD drawing, all you have to do is cut and paste. Text-based data can also be cut and pasted, saving you time in transferring data, such as layer or block names.

NOTE

OLE stands for *Object Linking and Embedding*—a Windows feature that lets different applications share documents. See Chapter 18 for a more detailed discussion of OLE.

With Windows, you have the freedom to arrange AutoCAD's screen by clicking and dragging its components. AutoCAD 14 offers many time-saving tools not found in the older DOS-based version, such as a drop-down list for layer settings and line types, and toolbars for easy access to all of AutoCAD's commands. There's even an expanded help system, with online tutorials and full documentation.

If you are new to AutoCAD, this is the version you may have been waiting for. Even though they have added many new features, the programmers at Autodesk have managed to make AutoCAD easier to use than previous releases. AutoCAD's interface has been trimmed down and is more consistent than prior versions. Release 14 is fully compliant with the Windows interface guidelines and is specially designed to conform to the Microsoft Office standard interface. So if you are familiar with the Microsoft Office suite of programs, you will feel right at home with AutoCAD Release 14.

In this first chapter, you will look at many of AutoCAD's basic operations, such as opening and closing files, getting a close-up look at part of a drawing, and making changes to a drawing.

Taking a Guided Tour

First, in this chapter, you will get a chance to familiarize yourself with the AutoCAD screen and how you communicate with AutoCAD. Along the way, you will also get a feel for how to work with this book. Don't worry about understanding or remembering everything that you see in this chapter. You will get plenty of opportunities to probe the finer details of the program as you work through the later chapters. If you are already familiar with earlier versions of AutoCAD, you might want to read through this chapter anyway, to get acquainted with new features and the graphical interface. To help you remember the material, you will find a brief exercise at the end of each chapter. For now, just enjoy your first excursion into AutoCAD.

TAKING A GUIDED TOUR   7

> If you are new to computers or need a little help with Windows, consider purchasing *Mastering Windows 95* by Bob Cowart or *The ABCs of Windows 95* by Sharon Crawford, both published by Sybex Inc.

If you have already installed AutoCAD, and you are ready to jump in and take a look, then proceed with the following steps to launch the program:

1. Click on the Start button in the lower-left corner of the Windows 95 or NT 4.0 screen. Then choose Program ➢ AutoCAD R14 ➢ AutoCAD R14. You can also double-click on the AutoCAD R14 icon on your Windows desktop.

2. You will see an opening greeting, called a screen splash, telling you which version of AutoCAD you are using, who the program is registered to, and the AutoCAD dealer's name and phone number should you need help.

3. Next, you will see the Start Up dialog box. This dialog box is a convenient tool for setting up new drawings. You'll learn more about this tool in later chapters. For now, click on Cancel in the Start Up dialog box.

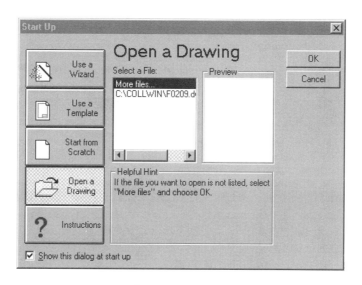

Message to Veteran AutoCAD Users

Autodesk is committed to the Windows operating environment. The result is a graphical user interface that is easier on the AutoCAD neophyte, but perhaps a bit foreign to a veteran AutoCAD user.

If you've been using AutoCAD for a while, and you prefer the older interface, you can still enter AutoCAD commands through the keyboard, and you can still mold AutoCAD's interface into one that is more familiar to you.

You can, for example, restore the side menu that appears in the DOS version of AutoCAD. Here's how it's done:

1. Select Tools ➣ Preferences.
2. At the Preferences dialog box, click on the Display tab.
3. Click on the check box labeled Display AutoCAD Screen Menu in Drawing Window.
4. Finally, click on OK. The side menu will appear.

A word of caution: If you are accustomed to pressing Ctrl+c to cancel an operation, you must now retrain yourself to press the Esc (Escape) key. Ctrl+c now conforms to the Windows standard, making this key combination a shortcut for saving marked items to the Clipboard. Similarly, instead of using F1 to view the full text window, you must use F2. F1 is most commonly reserved for the Help function in Windows applications.

If you prefer entering commands through the keyboard, you'll also want to know about some changes to specific commands in Release 14. Several commands that usually invoke dialog boxes can be used through the Command window prompt. Here is a list of those commands:

Bhatch	Boundary	Group	Hatchedit
Layer	Linetype	Mtext	Pan

When you enter these commands through the keyboard, you will normally see a dialog box. In the case of Pan, you will see a Realtime Pan hand graphic. To utilize these commands from the command line prompt, add a minus sign (–) to the beginning of the command name. For example, to use the Layer command in the older command-line method, enter **–layer** at the command prompt. To use the old Pan command, enter **–pan** at the command prompt.

Even if you don't care to enter commands through the keyboard, knowing how to use the minus sign can help you create custom macros. See Chapter 21 for more on AutoCAD customization.

The AutoCAD Window

The AutoCAD program window is divided into five parts:

- Pull-down menu bar
- Docked and floating toolbars
- Drawing area
- Command window
- Status bar

A sixth, hidden component, the Aerial View window, displays your entire drawing and lets you select close-up views of parts of your drawing.

Figure 1.1 shows a typical layout of the AutoCAD program window. Along the top is the *menu bar*, and at the bottom are the *Command window* and the *status line*. Just below the menu bar and to the left of the window are the *toolbars*. The rest of the screen is occupied by the *drawing area*.

Many of the elements within the AutoCAD window can be easily moved and reshaped. Figure 1.2 demonstrates how different AutoCAD can look after some simple rearranging of window components. Toolbars can be moved from their default locations to any location on the screen. When they are in their default location, they are in their *docked* position. When they are moved to a location where they are free-floating, they are *floating*.

The menu bar at the top of the drawing area (shown in Figure 1.3) offers pull-down menus from which you select commands in a typical Windows fashion. The toolbars offer a variety of commands through tool buttons and drop-down lists. For example, the *layer* name or number you are presently working on is displayed in a drop-down list in the Object Properties toolbar. The layer name is preceded by tools that inform you of the status of the layer. The tools and lists on the toolbar are plentiful, and you'll learn more about all of them later in this chapter and as you work through this book.

A *layer* is like an overlay that allows you to separate different types of information. AutoCAD allows an unlimited number of layers. On new drawings, the default layer is 0. You'll get a detailed look at layers and the meaning of the Layer tools in Chapter 4.

10 CHAPTER 1 • THIS IS AUTOCAD

FIGURE 1.1

A typical arrangement of the elements of the AutoCAD window

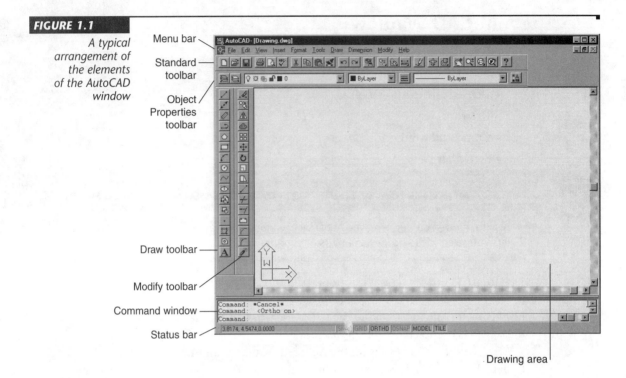

FIGURE 1.2

An alternative arrangement of the elements of the AutoCAD window

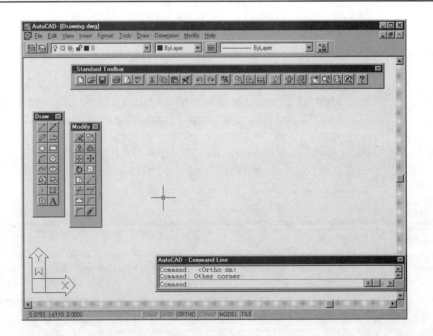

FIGURE 1.3

The components of the menu bar and toolbar

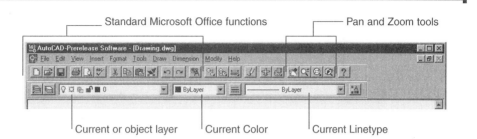

The Draw and Modify toolbars (Figure 1.4) offer commands that create new objects and edit existing ones. These are just two of many toolbars available to you.

FIGURE 1.4

The Draw and Modify toolbars as they appear when they are floating

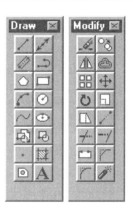

The drawing area—your workspace—occupies most of the screen. Everything you draw appears in this area. As you move your mouse around, you will see crosshairs appear to move within the drawing area. This is your drawing cursor, which lets you point to locations in the drawing area. At the bottom of the drawing area, the status bar (see Figure 1.5) gives you information at a glance about the drawing. For example, the coordinate readout toward the far left of the status line tells you the location of your cursor. The Command window can be moved and resized in a manner similar to toolbars. By default, the Command window is in its docked position, shown here. Let's practice using the coordinate readout and drawing cursor.

FIGURE 1.5

The status bar and Command window

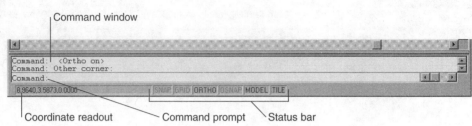

Picking Points

1. Move the cursor around in the drawing area. As you move, note how the coordinate readout changes to tell you the cursor's location. It shows the coordinates in an X,Y format.

2. Now place the cursor in the middle of the drawing area, and press and immediately release the left mouse button. You have just picked a point. Move the cursor, and a rectangle follows. This is a *selection window*; you'll learn more about this window in Chapter 2.

3. Move the cursor a bit in any direction; then press and let go of the left mouse button again. Notice that the rectangle disappears.

4. Try picking several more points in the drawing area.

Terminology to Remember: The operation you performed in steps 2 and 3—placing the cursor on a specific point and pressing the left mouse button—is referred to as *clicking on* a point. From now on, we will use the expression "click on" to describe the process of placing the cursor on an item or in an area and pressing the left mouse button.

UCS Icon

In the lower-left corner of the drawing area, you see a thick, L-shaped arrow outline. This is the *User Coordinate System* (UCS) icon, which tells you your orientation in the drawing. This icon becomes helpful as you start to work with complex 2D drawings and 3D models. The *X* and *Y* inside the icon indicate the X- and Y-axes of your drawing. The *W* tells you that you are in what is called the *World Coordinate System*. We will discuss this icon in detail in Chapter 11. For now, you can use it as a reference to tell you the direction of the axes.

NOTE

If you can't find the UCS icon... The UCS icon can be turned on and off, so if you are on someone else's system and you don't see the icon, don't panic. It also changes shape depending on whether you are in Paper Space or Model Space mode! If you don't see the icon or it doesn't look like it does in this chapter, see *Switching to Paper Space* in Chapter 9 for more on Paper Space and Model Space. Also see Chapter 11 for more on the UCS icon.

The Command Window

At the bottom of the screen, just above the status bar, is a small horizontal window, which is the *Command window*. Here AutoCAD displays responses to your input. It shows three lines of text. The bottom shows the current messages while the top two show messages that have scrolled by, or in some cases, components of the current message that do not fit in a single line. Right now, the bottom line shows the word Command: (see Figure 1.5). This tells you that AutoCAD is waiting for your instructions. As you click on a point in the drawing area, you'll see the message Other corner. At the same time, the cursor starts to draw a selection window that disappears when you click on another point.

As a new user, you should pay special attention to messages displayed in the Command window because this is how AutoCAD communicates with you. Besides giving you messages, the Command window records your activity in AutoCAD. You can use the scroll bar to the right of the Command window to review previous messages. You can also enlarge the window for a better view. (We'll discuss this in more detail in Chapter 2.)

NOTE

As you become more familiar with AutoCAD, you may find you don't need to rely on the Command window as much. For new users, however, it can be quite helpful in understanding what steps to take as you work.

Now let's look at AutoCAD's window components in detail.

The Pull-Down Menus

Like many Windows programs, the pull-down menus available on the menu bar offer an easy-to-understand way to access the general controls and settings for

AutoCAD. Within these menus you'll find the commands and functions that are the heart of AutoCAD. By clicking on menu items, you can cut and paste items to and from AutoCAD, change the settings that make AutoCAD work the way you want it to, set up the measurement system you want to use, access the Help system, and much more.

> To close a pull-down menu without selecting anything, press the Esc (Escape) key. You can also click on any other part of the AutoCAD window or on another pull-down menu.

The pull-down menu options perform four basic functions:
- Display additional menu choices
- Display a dialog box that contains settings you can change
- Issue a command that requires keyboard or drawing input
- Offer an expanded set of the same tools found in the Draw and Modify toolbars

As you point to commands and options in the menus or toolbars, AutoCAD provides additional help for you in the form of brief descriptions of each menu option, which appear in the status bar.

Here's an exercise to let you practice with the pull-down menus and get acquainted with AutoCAD's interface:

1. Click on View in the menu bar. The list of items that appears contains the commands and settings that let you control how AutoCAD displays your drawings. Don't worry if you don't understand these items; you'll get to know them in later chapters.

2. Move the highlight cursor slowly down the list of menu items. As you highlight each item, notice that a description of it appears in the status line at the bottom of the AutoCAD window. These descriptions help you choose the menu option you need.

3. Some of the menu items have triangular pointers to their right. This means the command has additional choices. For instance, highlight the Zoom item, and you'll see another set of options appear to the right of the menu.

TAKING A GUIDED TOUR

NOTE

If you look carefully at the command descriptions in the status bar, you'll see an odd word at the end. This is the keyboard command equivalent to the highlighted option in the menu or toolbar. You can actually type in these keyboard commands to start the tool or menu item that you are pointing to. You don't have to memorize these command names, but knowing them will be helpful to you later if you want to customize AutoCAD.

This second set of options is called a *cascading menu*. Whenever you see a pull-down menu item with the triangular pointer, you know that this item opens a cascading menu offering a more detailed set of options.

You might have noticed that other pull-down menu options are followed by an ellipsis (...). This indicates that the option brings up a dialog box, as the following exercise demonstrates.

NOTE

If you prefer, you can click and drag the highlight cursor over the pull-down menu to select an option.

1. Now move the highlight cursor to the Tools option in the menu bar.
2. Click on the Preferences... item. The Preferences dialog box appears.

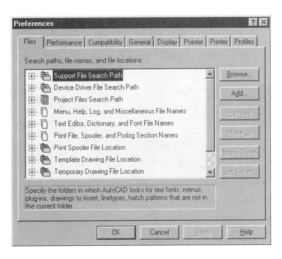

NOTE If you're familiar with the Windows 95 Explorer, you should feel at home with the Files tab of the Preferences dialog box. The plus sign to the left of the items in the list expands the option to display more detail.

This dialog box contains several "pages," indicated by the tabs across the top, that contain settings for controlling what AutoCAD shows you on its screens, where you want it to look for special files, and other "housekeeping" settings. You needn't worry about what these options mean at this point. Appendix B describes this dialog box in more detail.

3. In the Preferences dialog box, click on the tab labeled General. The options change to reveal new options.

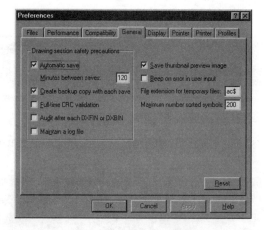

In the upper-left corner of the dialog box, you'll see a check box labeled Automatic Save, with the Minutes between Saves setting set at every 120 minutes. This setting controls how frequently AutoCAD performs an automatic save.

4. Change the 120 to 20, then click on OK. You have just changed AutoCAD's Automatic Save feature to automatically save files every 20 minutes instead of every two hours. (Let this be a reminder to give your eyes a rest!)

The third type of item you'll find on pull-down menus is a command that directly executes an AutoCAD operation. Let's try an exercise to explore these commands:

1. Click on the Draw option on the menu bar, then click on the Rectangle command. Notice that the Command window now shows the comment

Chamfer/Elevation/Fillet/Thickness/Width/<First corner>:

AutoCAD is asking you to select the first corner for the rectangle.

2. Click on a point roughly in the lower-left corner of the drawing area, as shown in Figure 1.6. Now as you move your mouse, you'll see a rectangle follow the cursor with one corner fixed at the position you just selected. You'll also see the following message in the Command window:

```
Other corner:
```

3. Click on another point anywhere in the upper-right region of the drawing area. A rectangle appears (see Figure 1.7). You'll learn more about the different cursor shapes and what they mean in Chapter 2.

At this point you've seen how most of AutoCAD's commands work: You'll find that dialog boxes are offered when you want to change settings, while many drawing and editing functions present messages in the Command window. Also, be aware that many of the pull-down menu items are duplicated in the toolbars that you will explore next.

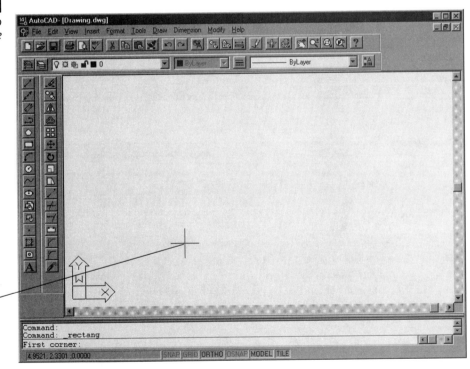

FIGURE 1.6
Selecting points to define a rectangle

Click on a point roughly here.

FIGURE 1.7

Once you've selected your first point of a rectangle, the cursor disappears and you see a rectangle follow the motion of your mouse.

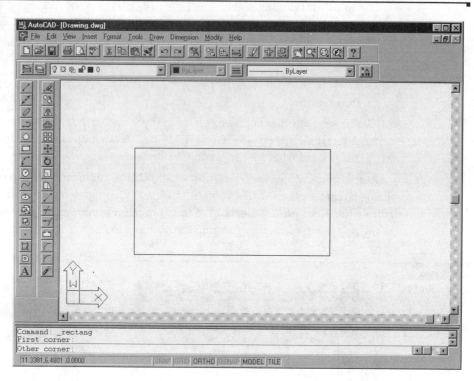

Communicating with AutoCAD

AutoCAD is the perfect servant: It does everything you tell it to, and no more. You communicate with AutoCAD using the pull-down menus and the tools on toolbars. These devices invoke AutoCAD commands. A command is a single-word instruction you give to AutoCAD telling it to do something, such as draw a line (the Line button on the Draw toolbar) or erase an object (the Erase button on the Modify toolbar). Whenever you invoke a command, by typing it in or by selecting a menu or toolbar item, AutoCAD responds by presenting messages to you in the Command window or by displaying a dialog box.

Continued

CONTINUED

The messages in the Command window often tell you what to do next, or they offer a list of options. A single command will often present several messages, which you answer to complete the command. These messages serve as an aid to new users who need a little help. If you ever get lost while using a command or forget what you are supposed to do, look at the Command window for clues. As you become more comfortable with AutoCAD, you will find that you won't need to refer to these messages as frequently.

A dialog box is like a form you fill out on the computer screen. It lets you adjust settings or make selections from a set of options pertaining to a command. You'll get a chance to work more with commands and dialog boxes later in this chapter.

The Toolbars

While the pull-down menus offer a full range of easy-to-understand options, they require some effort to navigate. The toolbars, on the other hand, offer quick, single-click access to the most commonly used AutoCAD features.

The tools in the toolbars perform three types of actions, just like the pull-down menu commands: They display further options, open dialog boxes, and issue commands that require keyboard or cursor input.

The Toolbar Tool Tips

AutoCAD's toolbars contain tools that represent commands. To help you understand each tool, a *tool tip* appears just below the arrow cursor when you rest the cursor on a tool. Each tool tip helps you identify the tool with its function. A tool tip appears when you follow these steps:

1. Move the arrow cursor onto one of the toolbar tools and leave it there for a second or two. Notice that the command's name appears nearby—this is the tool tip. In the status bar, a brief description of the button's purpose appears. See Figure 1.8.
2. Move the cursor across the toolbar. As you do, notice that the tool tips and status bar descriptions change to describe each tool. The keyboard command equivalent of the tool is also shown in the status bar at the end of the description.

FIGURE 1.8

Tool tips show you the function of each tool in the toolbar. AutoCAD also displays a description of the tool in the status bar.

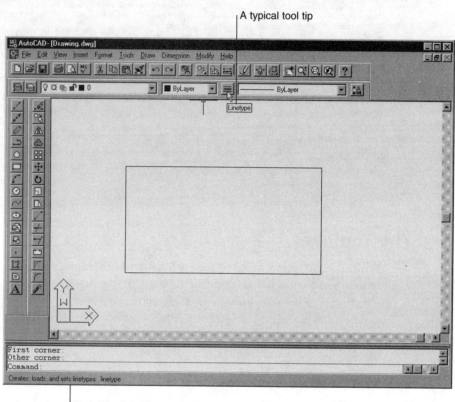

Flyouts Most toolbar tools start a command as soon as you click on them, but other tools will display a set of additional tools (similar to the cascading menus) that are related to the tool you have selected. This set of additional tools is called a toolbar *flyout*. If you've used other Windows graphics programs, chances are you've seen flyouts. Look closely at the tools just below the Help pull-down menu option on your screen or in Figure 1.8. You'll be able to identify which toolbar tools have flyouts; they'll have a small right-pointing arrow in the lower-right corner of the tool.

Remember: When an instruction says "click on," you should lightly press the left mouse button until you hear a click, then immediately let it go. Don't hold it down.

Let's see how a flyout works:

1. Move the cursor to the Zoom Window button in the Standard toolbar. Click and hold the left mouse button to display the flyout. Don't release the mouse button.

Flyout

2. Still holding down the left mouse button, move the cursor over the flyout; notice that the tool tips appear here as well. Also, notice the description in the status bar.
3. Move the cursor to the Zoom Window tool at the top of the flyout and release the mouse button.
4. You won't need to use this tool yet, so press Esc to cancel this tool.

As you can see from this exercise, you get a lot of feedback from AutoCAD!

Moving the Toolbars

One unique characteristic of AutoCAD's toolbars is their mobility. They can either be floating anywhere on the Windows screen or in a *docked* position. This means the toolbar is placed against the top and side borders of the AutoCAD window, so that the toolbar occupies a minimal amount of space. If you want to, you can move the toolbar to any location on your desktop, thus turning it into a *floating toolbar*.

Later in this section you'll find descriptions of all AutoCAD's toolbars, but first try the following exercise to move the Object Properties toolbar away from its current position in the AutoCAD window:

1. Move the arrow cursor so that it points to the border of the Object Properties toolbar.

2. Press and hold down the left mouse button. Notice that a gray rectangle appears by the cursor.

3. Still holding down the mouse button, move the mouse downward. The gray box follows the cursor.

When the gray box is over the drawing area, release the mouse button and the Object Properties toolbar—now a floating toolbar—moves to its new location.

Terminology to Remember: The action you perform in steps 2 and 3 of this exercise—holding down the mouse/pick button while simultaneously moving the mouse—is called *click and drag*. (If you have used other Windows applications, you already know this.) From now on, we will use "click and drag" to describe this type of action.

You can now move the Object Properties toolbar to any location on the screen that suits you. You can also change the shape of the toolbar; try this:

4. Place the cursor on the bottom-edge border of the Object Properties toolbar. The cursor becomes a double-headed arrow.

5. Click and drag the border downward. The gray rectangle jumps to a new, taller rectangle as you move the cursor.

6. When the gray rectangle changes to the shape you want, release the mouse button to reform the toolbar.

7. To move the toolbar back into its docked position, place the arrow cursor on (point to) the toolbar's title bar and slowly click and drag the toolbar so the cursor is in position in the upper-left corner of the AutoCAD window. Notice how the gray outline of the toolbar changes as it approaches its docked position.

8. When the outline of the object properties toolbar is near its docked position, release the mouse button. The toolbar moves back into its previous position in the AutoCAD window.

You can move and reshape any of AutoCAD's toolbars to place them out of the way, yet still have them at the ready to give you quick access to commands. You can also put them away altogether when you don't need them, and bring them back at will, as shown in these next steps:

9. Click and drag the Draw toolbar from its position at the left of the AutoCAD window to a point near the center of the drawing area.

10. Click on the Close button in the upper-right corner of the Draw floating toolbar. This is the small square button with the X on it. The toolbar disappears.

NOTE

Terminology to remember: When we ask you to select an option from the menu bar, we will use the notation *Menu* ➢ *Option*. For cascading menus, we will use the notation *Menu* ➢ *Option* ➢ *Option;* the second ➢ *Option* is in a cascading menu. In either case, the selected menu option issues a command that performs the function being discussed. As mentioned earlier, the actual command name appears in the status bar when you point to a menu option or toolbar tool.

11. To recover the toolbar, click on View and then on Toolbars (View ➤ Toolbars). The Toolbars dialog box appears.

You can also right-click on any toolbar to open the Toolbars dialog box.

12. Locate Draw in the list of toolbars shown in this dialog box, then click in the check box next to the Draw item so that an X appears in the box. The Draw toolbar reappears.

13. Click on the Close button in the upper-right corner of the dialog box.

14. Click and drag the Draw toolbar back to its docked position in the far-left side of the AutoCAD window.

AutoCAD will remember your toolbar arrangement between sessions. When you exit and then reopen AutoCAD later, the AutoCAD window will appear just as you left it.

If you are using a shared computer, you will find the screen arrangement to be the one left by the last user. You may want to use your own screen arrangement, called a profile. This option is available as a tab in the Preferences dialog box.

You may have noticed several other toolbars listed in the Toolbars dialog box that don't appear in the AutoCAD window. To keep the screen from becoming cluttered, many of the toolbars are not placed on the screen. The toolbars you'll be using most are displayed first; others that are less frequently used are kept out of sight until they are needed. Here are brief descriptions of all the toolbars available from the Toolbars dialog box:

Draw: Commands for creating common objects, including lines, arcs, circles, curves, ellipses, and text. This toolbar appears in the AutoCAD window by default. Many of these commands are duplicated in the Draw pull-down menu.

Modify: Commands for editing existing objects. You can Move, Copy, Rotate, Erase, Trim, Extend, and so on. Many of these commands are duplicated in the Modify pull-down menu.

Modify II: Commands for editing special, complex objects such as polylines, multilines, 3D solids, and hatches.

Dimension: Commands that help you dimension your drawings. See Chapter 8. Many of these commands are duplicated in the Dimension pull-down menu.

Solids: Commands for creating 3D solids. See Chapters 11 and 12.

Surfaces: Commands for creating 3D surfaces. See Chapter 13.

References: Commands that control cross-referencing of drawings.

Render: Commands to operate AutoCAD's rendering feature. See Chapter 15.

External Database: Commands for linking objects to external databases. See Chapter 17.

Select Objects: Tools for modifying the method used to select objects on the screen. See Chapter 2.

Object Snap: Tools to help you select specific points on objects, such as endpoints and midpoints. See Chapter 3.

UCS: Tools for setting up a plane on which to work. This is most useful for 3D modeling, but it can be helpful in 2D drafting, as well. See Chapter 11.

Object Properties: Commands for manipulating the properties of objects. This toolbar is normally docked below the menu bar.

Standard Toolbar: The most frequently used commands for view control, file management, and editing. This toolbar is normally docked below the menu bar.

Inquiry: Commands for finding distances, point coordinates, object properties, mass properties, and areas.

Insert: Commands for importing other drawings, raster images, and OLE objects.

Viewpoint: Tools for viewing 3D models.

Zoom: Commands that allow you to navigate your drawing.

You'll get a chance to work with all of the toolbars as you work through this book. Or, if you plan to use the book as a reference rather than working through it as a chapter-by-chapter tutorial, any exercise you try will tell you which toolbar to use for performing a specific operation.

Menus versus the Keyboard

Throughout this book, you will be told to select commands and command options from the pull-down menus and toolbars. For new and experienced users alike, menus and toolbars offer an easy-to-remember method for accessing commands. If you are familiar with the DOS version of AutoCAD, you still have the option of entering commands directly through the keyboard. Most of the commands you know and love still work as they did from the keyboard.

Another method for accessing commands is to use accelerator keys, which are special keystrokes that open and activate pull-down menu options. You might have noticed that the commands in the menu bar and the items in the pull-down menus each have an underlined character. By pressing the Alt key followed by the key corresponding to the underlined character, you activate that command or option, without having to engage the mouse. For example, to issue File ➢ Open, press Alt, then F, then finally O (Alt+F+O).

Many tools and commands have keyboard shortcuts; one- or two-letter abbreviations of a command name. As you become more proficient with AutoCAD, you may find these shortcuts helpful. As you work through this book, we'll point out the shortcuts for your reference.

Finally, if you are feeling adventurous, you can create your own accelerator keys and keyboard shortcuts for executing commands by adding them to the AutoCAD support files. We'll discuss customization of the menus, toolbars, and keyboard shortcuts in Chapter 21.

Working with AutoCAD

Now that you've been introduced to the AutoCAD window, let's try using a few of AutoCAD's commands. First, you'll open a sample file and make a few simple

WORKING WITH AUTOCAD   27

modifications to it. In the process, you'll get familiar with some common methods of operation in AutoCAD.

Opening an Existing File

In this exercise, you will get a chance to see and use a typical Select File dialog box. To start with, you will open an existing file.

1. From the menu bar, choose File ➢ Open. A message appears asking you if you want to save the changes you've made to the current drawing. Click on No.
2. Next, a Select File dialog box appears. This is a typical Windows file dialog box, with an added twist; notice the large Preview box on the right. It allows you to preview a drawing before you open it, thereby saving time while searching for files.

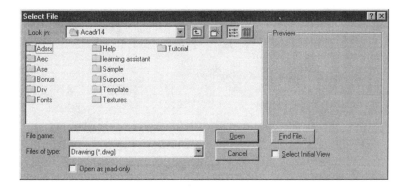

3. In the Select File dialog box, go to the Directories list and locate the directory named Sample (you may need to scroll down the list to find it). Point to it and then double-click (press the mouse/pick button twice in rapid succession). (If you're having trouble opening files with a double-click, here's another way to do it until you are more proficient with the mouse: Click on the file once to highlight it, and then click on the OK button.) The file list on the left changes to show the contents of the Sample directory.

NOTE

The Nozzle3D drawing is included on the companion CD-ROM. If you cannot find this file, be sure you have installed the sample drawings from the CD. See Appendix C for installation instructions.

4. Move the arrow to the file named Nozzle3d, and click on it. Notice that the name now appears in the File Name input box above the file list. Also, the Preview box now shows a thumbnail image of the file.

5. Click on the OK button at the bottom of the dialog box. AutoCAD opens the Nozzle3d file, as shown in Figure 1.9.

FIGURE 1.9

The Nozzle drawing. In the early days, this drawing became the unofficial symbol of AutoCAD, frequently appearing in ads for AutoCAD third-party products.

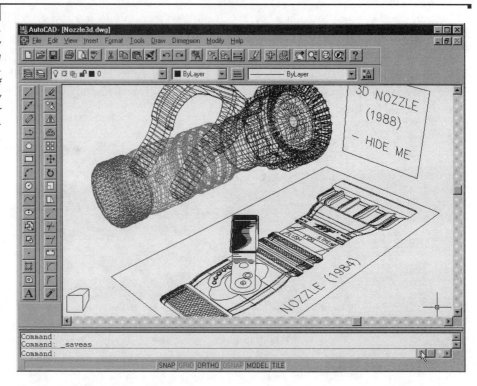

The Nozzle3d file opens to display the entire drawing. Also, the AutoCAD window's title bar displays the name of the drawing. This offers easy identification of the file. This particular file contains both a 2D and 3D model of a fire hose nozzle. The opening view is actually a 3D view.

Getting a Closer Look

One of the most frequently used commands is the Zoom command. Zoom lets you get a closer look at a part of your drawing. It offers a variety of ways to control your view. Now you'll enlarge a portion of the Nozzle drawing to get a more detailed look. To tell AutoCAD what area you wish to enlarge, you will use what is called a *window*.

1. Choose View ➢ 3D Viewpoint ➢ Plan View ➢ World UCS. Your view changes to display a two-dimensional view looking down on the drawing.
2. Click on the Zoom Window button on the Standard toolbar.

You can also Choose View ➢ Zoom ➢ Window from the menu bar.

3. The Command window displays `First corner:`. Look at the top image of Figure 1.10. Move the crosshair cursor to a location similar to the one shown in the figure; then press the left button on the mouse. Move the cursor, and you see the rectangle appear, one corner fixed on the point you just picked, while the other corner follows the cursor.
4. The Command window now displays `First corner: Other corner:`. Position the other corner of the window so it encloses the handle of the nozzle, as shown in the figure, and press the mouse/pick button. The handle enlarges to fill the screen (bottom image of Figure 1.10).

FIGURE 1.10

Placing the Zoom window around the nozzle handle. After clicking on the Zoom Window button in the Standard toolbar, select the two points shown in this figure.

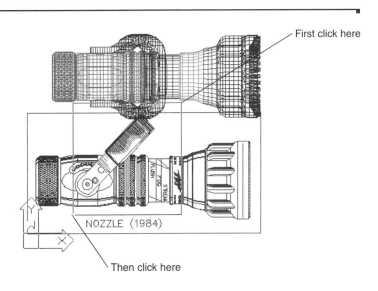

In this exercise, you used the Window option of the Zoom command to define an area to enlarge for your close-up view. You saw how AutoCAD prompts you to indicate first one corner of the window, and then the other. These messages are helpful for first-time users of AutoCAD. You will be using the Window options frequently—not just to define views, but also to select objects for editing.

You will notice that tiny crosses appear where you picked points. These are called *blips*—markers that show where you've selected points. They do not become a permanent part of your drawing, nor do they print onto hard copy output. You can turn off blips if you don't like them with the Blipmode command.

Getting a close-up view of your drawing is crucial to working accurately with a drawing, but you'll often want to return to a previous view to get the overall picture. To do so, click on the Zoom Previous button on the Standard toolbar.

Do this now, and the previous view—one showing the entire nozzle—returns to the screen. You can also get there by choosing View ➢ Zoom ➢ Previous.
You can quickly enlarge or reduce your view using the Zoom Realtime button on the Standard toolbar.

You can also zoom in and out using the Zoom In and Zoom Out buttons in the Zoom Window flyout of the Standard toolbar. The Zoom In button shows a magnifying glass with a plus sign; Zoom Out shows a minus sign.

1. Click on the Zoom Realtime button on the Standard toolbar.

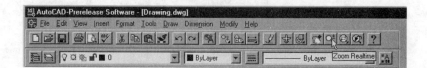

The cursor changes into a magnifying glass.

2. Place the Zoom Realtime cursor slightly above the center of the drawing area, and then click and drag downward. Your view zooms out to show more of the drawing.
3. While still holding the left mouse button, move the cursor upward. Your view zooms in to enlarge your view. When you have a view similar to the one shown in Figure 1.11, release the mouse button. (Don't worry if you don't get the *exact* same view as the figure. This is just for practice.)

FIGURE 1.11

The final view you want to achieve in step 3 of the exercise

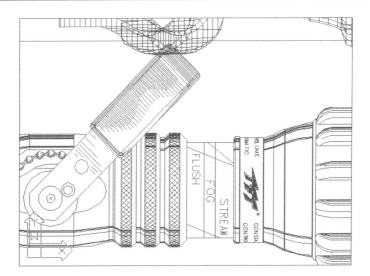

4. You are still in Zoom Realtime mode. Click and drag the mouse again to see how you can further adjust your view. To exit, you can select another command besides a Zoom or Pan command, press the Esc key, or right-click on your mouse.

5. Go ahead and right-click now. A pop-up menu appears.

This menu lets you select other display-related options.

6. Click on Exit from the pop-up menu to exit the Zoom Realtime command.

As you can see from this exercise, you have a wide range of options for viewing your drawing, just by using a few buttons. In fact, these three buttons, along with the scroll bars at the right side and bottom of the AutoCAD window, are all you need to control the display of your 2D drawings.

The Aerial View Window

The *Aerial View* window is an optional AutoCAD display tool. It gives you an overall view of your drawing, no matter how much magnification you may be using for the drawing editor. Aerial View also makes it easier to get around in a large-scale drawing. You'll find that this feature is best suited to more complex drawings that cover great areas, such as site plans, topographical maps, or city planning documents.

We won't discuss this view much here, as it can be a bit confusing for the first-time AutoCAD user. However, as you become more comfortable with AutoCAD, you may want to try it out.

Saving a File As You Work

It is a good idea to periodically save your file as you work on it. You can save it under its original name (with File ➤ Save) or under a different name (with File ➤ Save As), thereby creating a new file.

By default, AutoCAD automatically saves your work at 120-minute intervals under the name AUTO.SV$; this is known as the *Autosave* feature. Using system variables, you can change the name of the autosaved file and control the time between autosaves. See the sidebar, *Using AutoCAD's Automatic Save Feature*, in Chapter 3 for details.

Let's first try the Save command. This quickly saves the drawing in its current state without exiting the program.

Choose File ➢ Save. You will notice some disk activity while AutoCAD saves the file to the hard disk. As an alternative to picking File ➢ Save from the menus, you can type Alt+F+S. This is the accelerator key, also called the *hotkey*, for the File ➢ Save command.

Now try the Save As command. This command brings up a dialog box that allows you to save the current file under a new name.

1. Choose File ➢ Save As or type **saveas**↵ at the command prompt. The Select File dialog box appears. Note that the current file name, Nozzle3d, is highlighted in the File Name input box at the bottom of the dialog box.

2. Type **myfirst**↵. As you type, the name Nozzle3d disappears from the input box and is replaced by Myfirst. You don't need to enter the .dwg file name extension. AutoCAD adds it to the file name automatically when it saves the file.

3. Click on the Save button. The dialog box disappears, and you will notice some disk activity.

You now have a copy of the Nozzle file under the name Myfirst.DWG, and the name of the file displayed in the AutoCAD window's title bar has changed to Myfirst. From now on, when you use the File ➢ Save option, your drawing will be saved under its new name. Saving files under a different name can be useful when you are creating alternatives or when you just want to save one of several ideas you are trying out.

If you are working with a monitor that is on the small side, you may want to consider closing the Draw and Modify toolbars. The Draw and Modify pull-down menus offer the same commands, so you won't loose any functionality by closing these toolbars. If you really want to maximize your drawing area, you can also turn off the scroll bars and reduce the Command window to a single line. See Appendix B for details on how to do this.

Making Changes

You will be making frequent changes to your drawings. In fact, one of the AutoCAD's chief advantages is the ease with which you can make changes. The following exercise shows you a typical sequence of operations involved in making a change to a drawing.

1. On the Modify toolbar, click on the Erase tool (the one with a pencil eraser touching paper). This activates the Erase command. You can also choose Modify ➢ Erase from the menu bar.

Notice that the cursor has turned into a small square; this square is called the *pickbox*. You also see `Select object:` in the command prompt area. This message helps remind new users what to do.

2. Place the pickbox on the diagonal pattern of the nozzle handle (see Figure 1.12) and click on it. The 2D image of the nozzle becomes highlighted. The pickbox and the `Select object:` prompt remain, telling you that you can continue to select objects.

3. Now press ↵. The nozzle and the rectangle disappear. You have just erased a part of the drawing.

In this exercise, you first issued the Erase command, and then selected an object by clicking on it using a pickbox. The pickbox tells you that you must select items on the screen. Once you've done that, press ↵ to move on to the next step. This sequence of steps is common to many of the commands you will work with in AutoCAD.

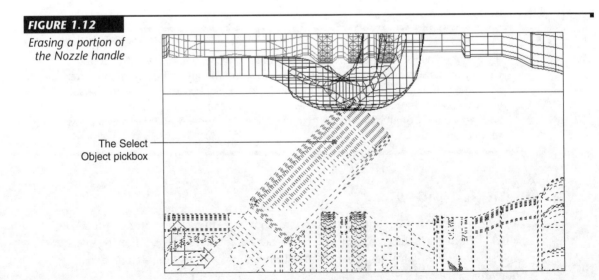

FIGURE 1.12

Erasing a portion of the Nozzle handle

The Select Object pickbox

Closing AutoCAD

When you are done with your work on one drawing, you can open another drawing, temporarily leave AutoCAD, or close AutoCAD entirely. To close a file and exit Auto-CAD, use the Exit option on the File menu.

1. Choose File ➤ Exit, which is the last item in the menu. A dialog box appears, asking you if you want to "Save Changes to Myfirst?" and offering three buttons labeled Yes, No, and Cancel.

2. Click on the No button. AutoCAD exits the Nozzle drawing and closes without saving your changes.

Whenever you attempt to exit a drawing that has been changed, you will get this same inquiry box. This request for confirmation is a safety feature that lets you change your mind and save your changes before you exit AutoCAD. In the previous exercise, you discarded the changes you made, so the Nozzle drawing reverts back to its state before you erased the handle.

If you only want to exit AutoCAD temporarily, you can minimize it so it appears as a button on the Windows 95 or NT 4.0 toolbar. You do this by clicking on the Minimize button in the upper-right corner of the AutoCAD window; this is the button with the underline sign. Alternatively, you can use the Alt+Tab key combination to switch to another program.

If You Want to Experiment...

Try opening and closing some of the sample drawing files.

1. Start AutoCAD by choosing Start ➤ Programs ➤ AutoCAD R14 ➤ AutoCAD R14.
2. Click on File ➤ Open.
3. Use the dialog box to open the Myfirst file again. Notice that the drawing appears on the screen with the handle enlarged. This is the view you had on screen when you used the Save command in the earlier exercise.
4. Erase the handle, as you did in the earlier exercise.

5. Click on File ➤ Open again. This time, open the Dhouse file from the companion CD-ROM. Notice that you get the Save Changes inquiry box you saw when you used the Exit option earlier. File ➤ Open acts just like Exit, but instead of exiting AutoCAD altogether, it closes the current file and then opens a different one.

6. Click on the No button. The 3D Dhouse drawing opens.

7. Click on File ➤ Exit. Notice that you exit AutoCAD without getting the Save Changes dialog box. This is because you didn't make any changes to the Dhouse file.

Chapter 2

Creating Your First Drawing

FEATURING

Understanding the AutoCAD Interface

Drawing Lines

Giving Distance and Direction

Drawing Arcs

Selecting Objects for Editing

Specifying Point Locations for Moves and Copies

Using Grips

Selecting Specific Geometric Elements

Getting Help

Chapter 2

Creating Your First Drawing

In this chapter we'll examine some of AutoCAD's basic functions and practice with the drawing editor by building a simple drawing to use in later exercises. We'll discuss giving input to AutoCAD, interpreting prompts, and getting help when you need it. We'll also cover the use of coordinate systems to give AutoCAD exact measurements for objects. You'll see how to select objects you've drawn, and how to specify base points for moving and copying.

If you're not a beginning AutoCAD user, you might want to move on to the more complex material in Chapter 3. You can use the files supplied on the companion CD-ROM to this book to continue the tutorials at that point.

Getting to Know the Draw Toolbar

Your first task in learning how to draw in AutoCAD is to try drawing a line. But before you begin drawing, take a moment to familiarize yourself with the toolbar you'll be using more than any other to create objects with AutoCAD: the Draw toolbar.

The AutoCAD Mechanical Toolbar

The Mechanical toolbar is slightly different from other toolbars. The top four tools are quite special; each one changes the focus of the rest of the toolbar. The topmost tool is the Draw tool, the second is the Modify tool, the third is the Edit tool, and the fourth is the Dimension or Annotation tool. If you click on the Modify tool, all of the tools below the Dimension tool change to a set of modification tools. If you click on the Edit tool, again the tools change. Experienced Windows users will see that this style of toolbar means that instead of having all four toolbars occupying valuable screen space, you will only have one.

1. Start AutoCAD just as you did in the first chapter, by clicking on Start ➢ Programs ➢ AutoCAD R14 ➢ AutoCAD R14.
2. In the AutoCAD window, move the arrow cursor to the top icon in the Draw toolbar, and rest it there so that the tool tip appears.
3. Slowly move the arrow cursor downward over the other tools in the Draw toolbar, and read each tool tip.

NOTE

Moving the arrow cursor onto an element on the screen is also referred to as "pointing to" that element.

GETTING TO KNOW THE DRAW TOOLBAR | 41

In most cases, you'll be able to guess what each tool does by looking at its icon. The icon with an arc, for instance, indicates that the tool draws arcs; the one with the ellipse shows that the tool draws ellipses; and so on. The tool tip gives you the name of the tool for further clarification. For further help, you can look to the status bar at the bottom of the AutoCAD window. For example, if you point to the Arc icon just below the Rectangle icon, the status bar reads Creates an Arc. It also shows you the actual AutoCAD command name: Arc. This name is what you would type in the Command window to invoke the Arc tool. You would also use this word if you are writing a macro or creating your own custom tools.

Figure 2.1 and Table 2.1 will aid you in navigating the two main toolbars, Draw and Modify, and you'll get experience with many of AutoCAD's tools as you work through this book.

FIGURE 2.1

The Draw and Modify toolbars. The options available from each toolbar tool are listed by number in Table 2.1.

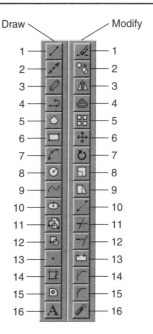

TABLE 2.1: THE OPTIONS THAT APPEAR ON THE DRAW AND MODIFY TOOLBARS AND FLYOUTS

Draw Toolbar

1	Line
2	Construction Line (Xline)
3	Multiline (Mline)
4	Polyline (Pline)
5	Polygon
6	Rectangle
7	Arc
8	Circle
9	Spline
10	Ellipse
11	Insert Block
12	Make Block
13	Point
14	Hatch
15	Region
16	Multiline Text

Modify Toolbar

1	Erase
2	Copy Object
3	Mirror
4	Offset
5	Array
6	Move
7	Rotate
8	Scale
9	Stretch
10	Lengthen
11	Trim
12	Extend
13	Break
14	Chamfer
15	Fillet
16	Explode

As you saw in Chapter 1, clicking on a tool issues a command. Some tools allow clicking *and dragging*, which opens a flyout. A flyout offers further options for that tool. You can identify flyout tools by a small triangle located in the lower-right corner of the tool.

1. Click and drag the Distance tool on the Standard toolbar. A flyout appears with a set of tools. As you can see, there are a number of additional tools for gathering information about your drawing.

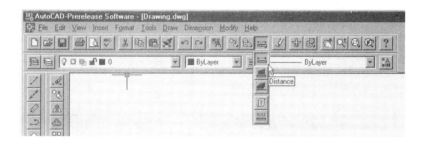

2. Move the cursor down the flyout to the second to last tool, until the tool tip reads "List", and then let go of the mouse button. Notice that the icon representing the Distance tool now changes and becomes the icon from the flyout that represents List. By releasing the mouse you've also issued the List command. This command lists the properties of an object.

3. Press Esc twice to exit the List command.

If you find you are working a lot with one particular flyout, you can easily turn that flyout into a custom toolbar, so that all the flyout options are readily available with a single click. See Chapter 21 for details on how to do this.

By making the most recently selected option on a flyout the default option for the toolbar tool, AutoCAD gives you quick access to frequently used commands. A word of caution, however: This feature can confuse the first-time AutoCAD user. Also, the grouping of options on the flyout menus is not always self-explanatory—even to a veteran AutoCAD user.

NOTE

Release 13 offered flyout menus on nearly every toolbar. Release 14 has reduced the number of flyouts so that on a typical screen they only appear on the Standard toolbar.

Working with Toolbars

As you work through the exercises, this book will show you graphics of the tools to choose, along with the toolbar or flyout that contains the tool. Don't be alarmed, however, if the toolbars you see in the examples don't look exactly like those on your screen. To save page space, we have horizontally oriented the toolbars and flyouts for the illustrations; the ones on your screen may be oriented vertically, like the Draw and Modify toolbars to the left of the AutoCAD Window. Although the shape of your toolbars and flyouts may differ from the ones you see in this book, the contents are the same. So when you see a graphic showing a tool, focus on the tool icon itself with its tool tip name, along with the name of the toolbar in which it is shown.

Starting Your First Drawing

In Chapter 1, you looked at a preexisting sample drawing. This time you will begin to draw on your own, by creating a part that will be used in later exercises. First, though, you must learn how to tell AutoCAD what you want and let AutoCAD help you. As you will see, the command line is also a prompt line which gives AutoCAD the opportunity to ask you the appropriate question in the current context of the command.

1. Choose File ➢ New.

2. When the Create New Drawing dialog box appears, click on the Start from Scratch button, click on English from the Select Default Settings list box, and then click on OK. This creates a new drawing using the standard AutoCAD default settings for the English (inches) system of measurement. You'll get a chance to use other options in this dialog box later. For now, you'll explore some basic AutoCAD options.

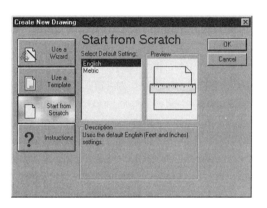

3. To give your new file a unique name, choose File ➢ Save As.
4. At the Save Drawing As dialog box, type **Base**. As you type, the name appears in the File Name input box.

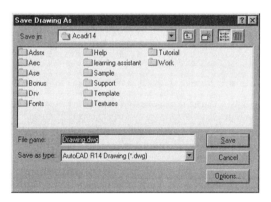

NOTE

You can use up to 255 characters and spaces to name files in Windows 95 and NT. If you are going to save R14 drawings in any previous version, you may want to use the 11-character file-naming convention for compatibility with other file systems. In all cases, a dot (.) and the final three characters are reserved for the .dwg file type.

5. Double-click on the Sample folder shown in the main file list of the dialog box. By doing this, you open the Samples subdirectory.

6. Click on Save. You now have a file called Base.dwg, located in the Samples subdirectory of your AutoCADR14 directory. Of course, your drawing doesn't contain anything yet. You'll take care of that next.

The new file shows a drawing area roughly 16 inches wide by 9 inches high. To check this for yourself, move the crosshair cursor to the upper-right corner of the screen, and observe the value shown in the coordinate readout. This is the standard AutoCAD default drawing area for new drawings.

To begin a drawing, follow these steps:

1. Click on the Draw tool on the Express toolbar, then click on the Line tool or type **L**↵. You've just issued the Line command. AutoCAD responds in two ways. First, you see the message

    From point:

    in the command prompt, asking you to select a point to begin your line. Also, the cursor has changed its appearance; it no longer has a square in the crosshairs. This is a clue telling you to pick a point to start a line.

NOTE

You can also type **line**↵ in the Command window to start the Line command.

2. Using the left mouse button, select a point on the screen near the center. As you select the point, AutoCAD changes the prompt to

    To point:

Now as you move the mouse around, you will notice a line with one end fixed on the point you just selected, and the other end following the cursor (see the top image of Figure 2.2). This action is called *rubber-banding*.

FIGURE 2.2
Two rubber-banding lines

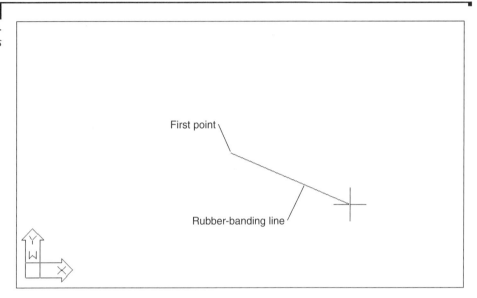

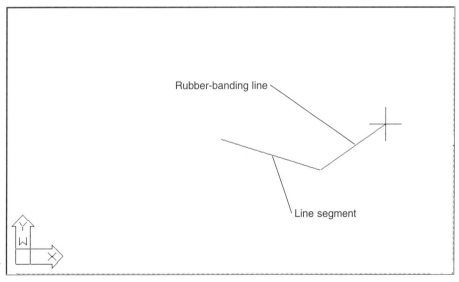

Now continue with the Line command:

3. Move the cursor to a point to the right of the first point you selected, and press the left mouse button again. The first rubber-banding line is now fixed between the two points you selected, and a second rubber-banding line appears (see the bottom image of Figure 2.2).

4. If the line you drew isn't the exact length you want, you can back up during the Line command and change it. To do this, click on Undo on the Standard toolbar, or type **u**↵ from the keyboard. Now the line you drew previously will rubber-band as if you hadn't selected the second point to fix its length.

You've just drawn, and then undrawn, a line of an arbitrary length. The Line command is still active. There are two things that tell you that you are in the middle of a command, as mentioned above. If you don't see the word Command in the bottom of the Command window, you know a command is still active. Also, the cursor will be the plain crosshairs, without the little box at its intersection.

NOTE

From now on, we will refer to the crosshair cursor without the small box as the *Point Selection mode* of the cursor. If you look ahead to Figure 2.8, you'll see all the different modes of the drawing cursor.

Getting Out of Trouble

Beginners and experts alike are bound to make a few mistakes. Before you get too far into the tutorial, here are some powerful, yet easy-to-use tools to help you recover from accidents:

Backspace [←]   If you make a typing error, you can use the Backspace key to back up to your error, and then retype your command or response. Backspace is located in the upper-right corner of the main keyboard area.

Continued

> **CONTINUED**
>
> **Escape [Esc]** This is perhaps the single most important key on your keyboard for getting out of trouble. When you need to quickly exit a command or dialog box without making changes, just press the Escape key in the upper-left corner of your keyboard. Press it twice if you want to cancel a selection set of objects or to make absolutely sure you've canceled a command.
>
> *Tip*: Use the Escape key before editing with grips or issuing commands through the keyboard. You can also press Esc twice to clear grip selections.
>
> **U↵** If you accidentally change something in the drawing and want to reverse that change, click on the Undo tool in the Standard toolbar (the left-pointing curved arrow). Or type **u↵** at the Command prompt. Each time you do this, AutoCAD will undo one operation at a time, in reverse order—so the last command performed will be undone first, then the next to last, and so on. The prompt will display the name of the command being undone, and the drawing will revert to its state prior to that command. If you need to, you can undo everything back to the beginning of an editing session.
>
> **Redo** If you accidentally undo one too many commands, you can redo the last undone command by clicking on the Redo tool (the right-pointing curved arrow) in the Standard toolbar. Or type **redo↵**. Unfortunately, Redo only restores one command, and it can only be invoked immediately after an Undo.

Specifying Distances with Coordinates

Next, you will continue with the Line command to draw a *Plan view* (an overhead view) of a base. The base will be 9 units long and 6.5 units wide. To specify these exact distances in AutoCAD, you can use either *relative polar coordinates*, *Cartesian coordinates*, or *Absolute coordinates*.

Specifying Polar Coordinates

To enter the exact distance of nine units to the right of the last point you selected, do the following:

1. Type **@9<0**. As you type, the letters appear in the command prompt.
2. Press ↵. A line appears, starting from the first point you picked and ending 9 units to the right of it (see Figure 2.3). You have just entered a relative polar coordinate.

FIGURE 2.3
A line 9 units long. Notice that the rubber-banding line now starts from the last point selected. This tells you that you can continue to add more line segments.

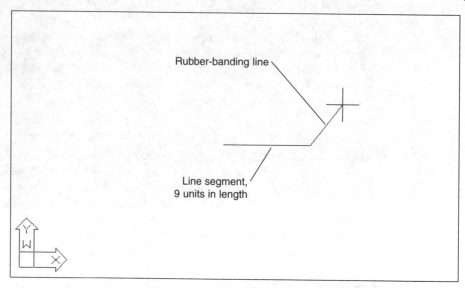

The at sign (@) you entered tells AutoCAD that the distance you are specifying is from the last point you selected. The 9 is the distance, and the less-than symbol (<, which looks a little like an angle symbol) tells AutoCAD that you are designating the angle at which the line is to be drawn. The last part is the value for the angle, which in this case is 0. This is how to use *polar coordinates* to communicate distances and direction to AutoCAD.

NOTE

If you are accustomed to a different method for describing directions, you can set AutoCAD to use a vertical direction or downward direction as 0°. See Chapter 3 for details.

Angles are given based on the system shown in Figure 2.4, where 0° is a horizontal direction from left to right, 90° is straight up, 180° is horizontal from right to left, and so on. You can specify degrees, minutes, and seconds of arc if you want to be that exact. We'll discuss angle formats in more detail in Chapter 3.

FIGURE 2.4

AutoCAD's default system for specifying angles

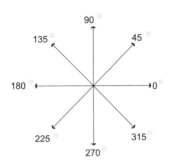

Specifying Relative Cartesian Coordinates

For the next line segment, let's try another method of specifying exact distances.

1. Enter **@0,6.5**↵. A line appears above the endpoint of the last line.

Once again, the @ tells AutoCAD that the distance you specify is from the last point picked. But in this example, you give the distance in X and Y values. The X distance, 0, is given first, followed by a comma, and then the Y distance, 6.5. This is how to specify distances in relative Cartesian coordinates.

2. Enter **@-9,0**↵. The result is a drawing that looks like Figure 2.5.

FIGURE 2.5

Three sides of the base drawn using the Line tool. Points are specified using either relative Cartesian or polar coordinates.

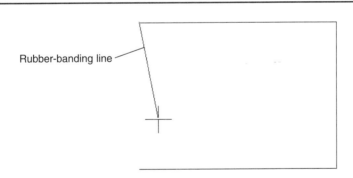

The distance you entered in step 2 was also in X,Y values, but here you used a negative value to specify the X distance. Positive values in the Cartesian coordinate system are to the right of and above the origin (see Figure 2.6). (You may remember this from your high school geometry class!) If you want to draw a line from right to left, you must designate a negative value.

FIGURE 2.6

Positive and negative Cartesian coordinate directions

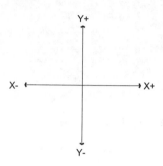

If you have trouble remembering which of the two angle symbols is used to specify a polar coordinate, the correct one is on the same key as the comma. Use a comma for Cartesian coordinates and a shifted comma (<) for polar coordinates.

To finish drawing a series of lines without closing them, you can press Esc, ↵, or the spacebar.

3. Now type **c**↵. This *c* stands for Close. It closes a sequence of line segments. A line connecting the first and last points of a sequence of lines is drawn (see Figure 2.7), and the Line command terminates. The rubber-banding line also disappears, telling you that AutoCAD has finished drawing line segments. You can also use the rubber-banding line to indicate direction while simultaneously

entering the distance through the keyboard. See the sidebar *A Fast Way to Enter Distances*.

FIGURE 2.7

Distance and direction input for the base

A Fast Way to Enter Distances

A third method for entering distances is to simply point in a direction with a rubber-banding line, and then enter the distance through the keyboard. For example, to draw a line 9 units long from left to right, click on the Line tool from the Draw toolbar, click on a start point, and then move the cursor so the rubber-banding line points to the right at some arbitrary distance. While holding the cursor in the direction you want, type **9**↵. The rubber-banding line becomes a fixed line 9 units long.

Using this method, called the Direct Distance method, along with the Ortho mode described in Chapter 3, can be a fast way to draw objects of specific lengths. Use the standard Cartesian or polar coordinate methods when you need to enter exact distances at angles other than those that are exactly horizontal or vertical.

> ### Cleaning Up the Screen
>
> On some systems, the AutoCAD Blipmode setting may be turned on. This will cause tiny cross-shaped markers, called *blips*, to appear where you've selected points. These blips can be helpful to keep track of the points you've selected on the screen. You can also type in **r** and hit Enter (**r↵**).
>
> Blips aren't actually part of your drawing and will not print. Still, they can be annoying. To clear the screen of blips, click on the Redraw tool in the Standard toolbar (it's the one that looks like a pencil point drawing an arc), or type **r↵**. The screen quickly redraws the objects, clearing the screen of the blips. You can also choose View ➢ Redraw View to accomplish the same thing. As you will see in Chapter 6, Redraw can also clear up other display problems.
>
> Another command, Regen, does the same thing as Redraw, but also updates the drawing display database—which means it takes a bit longer to restore the drawing. In general, you will want to avoid Regen, though at times using it is unavoidable. You will examine Regen in Chapter 6.
>
> To turn Blipmode on and off, type **blipmode.↵** at the Command prompt, and then enter **on.↵** or **off.↵**.

Interpreting the Cursor Modes and Understanding Prompts

The key to working with AutoCAD successfully is in understanding the way it interacts with you. In this section you will become familiar with some of the ways AutoCAD prompts you for input. Understanding the format of the messages in the Command window and recognizing other events on the screen will help you learn the program more easily.

As the Command window aids you with messages, the cursor also gives you clues about what to do. Figure 2.8 illustrates the various modes of the cursor and gives a brief description of the role of each mode. Take a moment to study this figure.

The Standard cursor tells you that AutoCAD is waiting for instructions. You can also edit objects using grips when you see this cursor. The Point Selection cursor appears whenever AutoCAD expects point input. It can also appear in conjunction with a rubber-banding line. You can either click on a point or enter a coordinate through the keyboard. The Object Selection cursor tells you that you must select objects—either by clicking on them or by using any of the object-selection options available. The object snap (Osnap) marker appears along with the Point Selection cursor when you invoke an Osnap. Osnaps let you select specific points on an object, such as endpoints or midpoints.

FIGURE 2.8

The drawing cursor's modes

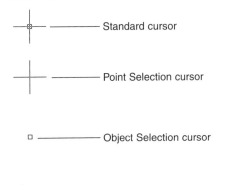

Standard cursor

Point Selection cursor

Object Selection cursor

Osnap marker with tool tip

TIP

If you are an experienced AutoCAD user and would prefer to use the older style crosshair cursor that crosses the entire screen, you can use the Pointer tab of the Preferences dialog box (Tools ➤ Preferences) to set the cursor size. Set the Percent of screen size option near the bottom of the dialog box to 100. The cursor will then appear as it did in prior versions of AutoCAD. As the option implies, you can set the cursor size to any percentage of the screen you want. The default is 5 percent.

Choosing Command Options

Many commands in AutoCAD offer several options, which are often presented to you in the Command window in the form of a prompt. Here, we'll use the Arc command to illustrate the format of AutoCAD's prompts.

The part we are drawing has a square hole with rounded corners. You will draw the hole as an arc followed by a line, then another arc, then another line, and so on until you close the figure. Figure 2.9 shows how this will look when you are finished.

FIGURE 2.9
Example of the base

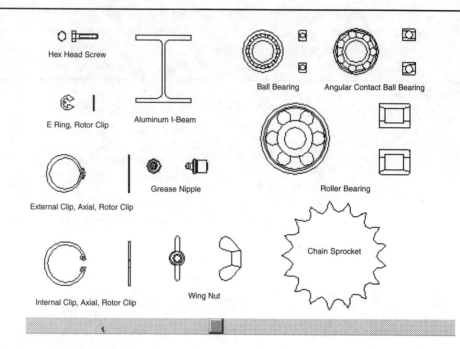

The *default* is the option AutoCAD assumes you intend to use unless you tell it otherwise.

Next, you'll draw the arc for the hole:

1. Click on the Arc tool in the Draw option of the Express toolbar. The prompt Center/<Start point>: appears, and the cursor changes to Point Selection mode.

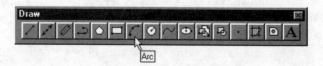

Let's examine this Center/<Start point>: prompt. It contains two options. The *default* option always appears between angle brackets (< >), and all other options are separated by slashes (/). If you choose to take the Start point: default, you can input a point by clicking on a location on the screen or by entering a coordinate.

2. Type **c**↵ to select the Center option. The prompt Center: appears. Notice that you only had to type in the *c* and not the whole word *Center*.

NOTE

When you see a set of options in the Command window prompt, note their capitalization. If you choose to respond to prompts using the keyboard, these capitalized letters are all you need to enter to select that option. In some cases, the first two letters are capitalized to differentiate two options that begin with the same letter, such as LAyer and LType.

3. Now pick a point representing the center of the arc near the middle of the base (see the top image of Figure 2.10). The prompt Start point: appears.
4. Type **@.125<90**↵. The prompt Angle/Length of chord/<End point>: appears.
5. Move the mouse and you will see a temporary arc originating from a point .125 units above the center point you selected and rotating about that center, as in the center image of Figure 2.10.

As the prompt indicates, you now have three options. You can enter an angle, a length of chord, or the endpoint of the arc. The default, indicated by <End point> in the prompt, is to pick the arc's endpoint. Again, the cursor is in Point Selection mode, telling you it is waiting for point input. To select this default option, you only need to pick a point on the screen indicating where you want the endpoint.

6. Type **a**↵ to select the Angle option. At the Included Angle: prompt, type **90**↵ to draw a quarter round arc. The arc is now fixed in place, as in the bottom image of Figure 2.10.
7. To continue with the hole, click on the line tool from the Draw toolbar. At the From point: prompt, type ↵ to use the last point which was at the end of the arc.

The line obtained its starting point in the same way that it did when you drew the large rectangle for the base. AutoCAD was expecting a value for the From point. When you did not give AutoCAD one, it used the last one in memory.

8. The width of the recess is .75 units between the centers. At the `Length of line:` prompt, type **.75**↵ and then ↵ again to end the Line command.

NOTE

You did not have to tell AutoCAD what direction to go when you gave it the length. The line obtained its vector, or direction, from the direction of the arc at the end where the line began.

9. Another arc must be drawn at the end of this line. Click on the Arc tool again. At the `arc Center/<Start point>:` prompt, type ↵ to use the endpoint of the last line. At the `End point:` prompt, type **@.375,-.375**↵ to create an arc .375 units across and 90° around.

10. Draw a line from the end of the last arc. Type **L**↵ to begin the Line command and then ↵ again to begin the line at the end of the last arc.

11. The length of the recess is 1.75 units between the centers. At the `Length of line:` prompt, type **1.75**↵ and then ↵ again to end the Line command.

12. Another arc must be drawn at the end of this line. Click on the Arc tool again. At the `arc Center/<Start point>:` prompt, type ↵ to use the endpoint of the last line. At the `End point:` prompt, type **@.375,.375**↵ to create an arc .375 units across and 90° around.

13. To continue the figure, click on the Line tool and press ↵ to begin the line at the end of the last arc. The width of the recess is .75 units between the centers. At the `Length of line:` prompt, type **.75**↵ and then press ↵ again to end the Line command.

14. Another arc must be drawn at the end of this line. Click on the Arc tool again. At the `arc Center/<Start point>:` prompt, type ↵ to use the endpoint of the last line. At the `End point:` prompt type **@-.125,.125**↵ to create an arc 125 units across and 90° around.

15. Finally, you will draw the line that closes the slot. Click on the Line tool again. At the `From point:` prompt, type ↵ to start the line at the end of the arc. At the `Length of line:` prompt, type **2.25**↵↵ to draw a line 2.25 units long back to the beginning of the first arc and end the command.

INTERPRETING THE CURSOR MODES AND UNDERSTANDING PROMPTS | 59

FIGURE 2.10

Using the Arc command

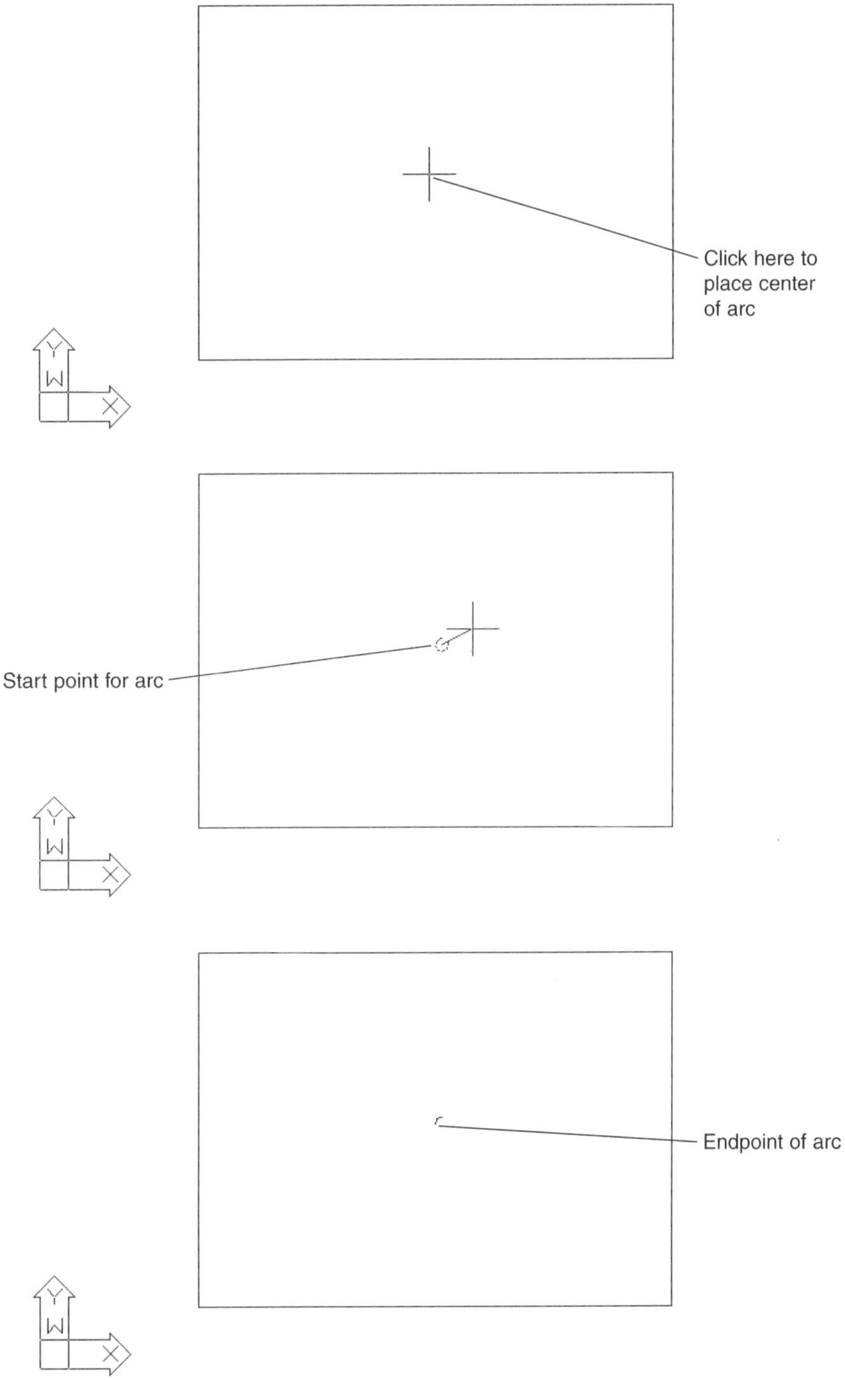

Click here to place center of arc

Start point for arc

Endpoint of arc

PART
I

The Basics

This exercise has given you some practice working with AutoCAD's Command window prompts and entering keyboard commands—a skill you will need when you start to use some of the more advanced AutoCAD functions.

As you can see, AutoCAD has a distinct structure in its prompt messages. You first issue a command, which in turn offers options in the form of a prompt. Depending on the option you select, you will get another set of options, or you will be prompted to take some action, such as picking a point, selecting objects, or entering a value.

As shown in Figure 2.11, the sequence is something like a tree. As you work through the exercises, you will become intimately familiar with this routine. Once you understand the workings of the toolbars, Command window prompts, and dialog boxes, you can almost teach yourself the rest of the program!

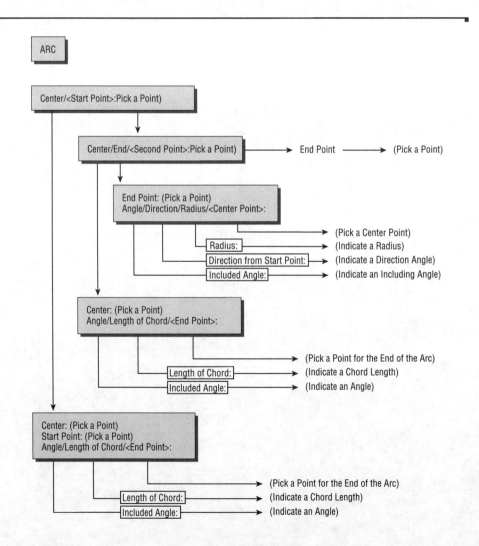

FIGURE 2.11

A typical command structure, using the Arc command as an example. You will see different messages, depending on the options you choose as you progress through the command. This figure shows the various pathways to creating an arc.

Selecting Objects

AutoCAD provides many options for selecting objects. This section has two parts: The first part deals with object-selection methods unique to AutoCAD; the second part deals with the more common selection method used in most popular graphic programs, the *Noun/Verb* method. Because these two methods play a major role in working with AutoCAD, it's a good idea to familiarize yourself with them early on.

If you need to select objects by their characteristics rather than by their location, Chapter 9 describes the Object Selection Filter tool. This feature lets you easily select a set of objects based on their properties, including object type, color, layer assignment, and so on.

Selecting Objects in AutoCAD

Many AutoCAD commands prompt you to Select objects:. Along with this prompt, the cursor will change from crosshairs to a small square (look back at Figure 2.8). Whenever you see this object selection prompt and the square cursor, you have several options while making your selection. Often, as you select objects on the screen, you will change your mind about a selection or accidentally pick an object you do not want. Let's take a look at most of the selection options available in AutoCAD and learn what to do when you make the wrong selection.

1. Choose Move from the Modify toolbar.

2. At the Select objects: prompt, click on the two horizontal lines that compose the hole. As you saw in Chapter 1, whenever AutoCAD wants you to select objects, the cursor turns into the small square pickbox. This tells you that you are in *Object Selection mode*. When you pick an object, it is *highlighted,* as shown in Figure 2.12.

Highlighting means that an object changes from a solid image to one composed of dots. When you see an object highlighted on the screen, you know that you have chosen that object to be acted upon by whatever command you are currently using.

FIGURE 2.12

Selecting the lines of the hole and seeing them highlighted

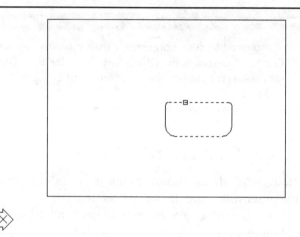

3. After making your selections, you may decide to deselect some items. Click on Undo in the Standard toolbar, or enter **u**↵ from the keyboard.

Notice that one line is no longer highlighted. The Undo option deselects objects, one at a time, in reverse order of selection.

4. There is another way to deselect objects: Hold down the Shift key and click on the remaining highlighted line. It reverts to a solid line, showing you that it is no longer selected for editing.

By now you have deselected both lines. Let's try using another method for selecting groups of objects.

5. Another option for selecting objects is to *window* them. Type **w**↵. The cursor changes to a Point Selection cursor, and the prompt changes to

   First corner:

6. Click on a point below and to the left of the hole. As you move your cursor across the screen, the window appears and stretches across the drawing area.

7. Once the window completely encloses the arc and both adjacent lines, but not the other arcs, click on this location. One arc and two lines will be highlighted. This window selects only objects that are completely enclosed by the window, as shown in Figure 2.13.

NOTE

You might remember that you used a window with the Zoom command in Chapter 1. The Window option under the Zoom command does not select objects. Rather, it defines an area of the drawing you want to enlarge. Remember that the Window option works differently under the Zoom command than it does for other editing commands.

If you are using a mouse you're not familiar with, it's quite easy to accidentally click the right mouse button, which is the button that triggers the ↵ action, when you really wanted to click the left mouse button, and vice versa. If you click the wrong button, you'll get the wrong results. On a two- or a three-button mouse, the right button acts like the ↵ key. On a three-button mouse, the middle button usually pops the Osnap menu.

8. You can add the other arcs and the rest of the objects by simply by clicking on them. Now that you have selected the entire hole, press ↵. *It is important to remember to press* ↵ *as soon as you have finished selecting the objects you want to edit.* Pressing ↵ tells AutoCAD when you have finished selecting objects. A new prompt, Base point or displacement:, appears. The cursor changes to its Point Selection mode.

FIGURE 2.13

Selecting parts of the hole within a window

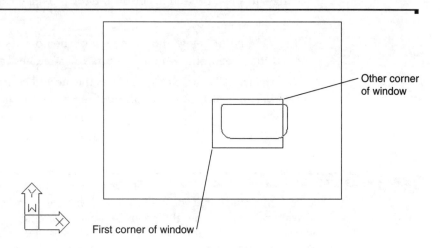

Now you have seen how the selection process works in AutoCAD—but we've left you in the middle of the Move command. In the next section, we'll discuss the prompt that's now on your screen and see how to input base points and displacement distances.

Providing Base Points

When you move or copy objects, AutoCAD prompts you for a *base point*, which is a difficult concept to grasp. AutoCAD must be told specifically *from* where and *to* where the move occurs. The base point is the exact location from which you determine the distance and direction of the move. Once the base point is determined, you can tell AutoCAD where to move the object in relation to that point.

SELECTING OBJECTS | 65

1. To select a base point, hold down the Shift key and press the right mouse button. A menu pops up on the screen. This is the *Object Snap* (*Osnap*) menu.

When right-clicking the mouse, make sure the cursor is within the AutoCAD drawing area. Otherwise, you will not get the results described in this book.

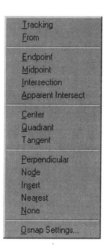

2. Choose Center from the Osnap menu. The Osnap menu disappears.
3. Move the cursor to the lower-left arc of the recess. Notice that as you approach the arc, a small O-shaped graphic appears on the corner. This is called an Osnap marker.
4. After the O-shaped marker appears, hold the mouse motionless for a second or two. A tool tip appears, telling you the current Osnap point AutoCAD has selected.
5. Now press the left mouse button to select the center indicated by the Osnap marker. Whenever you see the Osnap marker at the point you wish to select, you don't have to point exactly at the location with your cursor. Just left-click and the exact Osnap point is selected (see Figure 2.14). In this case, you selected the exact center of the arc.

 **NOTE**

AutoCAD can use Osnap to locate the center of an arc, ellipse, or circle. A common error made by those new to AutoCAD is to place the cursor at the center of the object. AutoCAD does not identify this space as part of the object. You must place the cursor on the circumference of the circle or ellipse or along the length of the arc.

FIGURE 2.14

Using the Osnap cursor

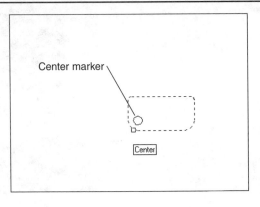

6. At the Second point of displacement: prompt, hold down the Shift key and press the right mouse button again. You'll use the From Osnap this time, but instead of clicking on the option with the mouse, type **f**↵.

7. You are going to locate the arc center of the recess, 3 AutoCAD units to the right of and 2.75 units down from the upper-left corner of the part. At the From point: _from Base point: prompt, once again hold down the Shift key and press the right mouse button. You'll use the Endpoint Osnap this time; type **e**↵.

8. Now move the cursor anywhere along the top line or the left side line. A box-shaped marker appears. The end of the line that AutoCAD chooses depends on which side of the line's midpoint you are over. Move the cursor until the box marker appears at the upper-left corner and press the left mouse button. You did not have to place the cursor exactly at the corner, because you had already told AutoCAD to go to the endpoint.

9. You will now tell AutoCAD how far from the corner of the plate you want to place the center of the arc circle. At the <Offset>: prompt type **@3,-2.75**↵ (see Figure 2.15).

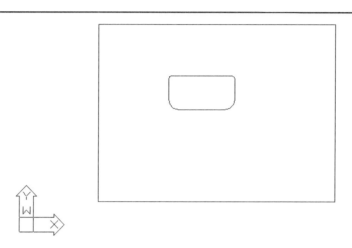

FIGURE 2.15

The hole in its new position using the From, Center, and Endpoint Osnaps

As you can see, the Osnap options allow you to select specific points on an object. You used From, Center, and Endpoint in this exercise, but other options are available. We will look at some of the other options in Chapter 3. You may also have noticed that the Osnap marker is different for each of the options you used. You'll learn more about Osnaps in Chapter 3. Now let's continue with our look at point selection.

If you want to specify an exact distance and direction by typing in a value, you can select any point on the screen as a base point. Or you can just type @ followed by ↵ at the Base point: prompt; then enter the second point's location in relative coordinates. Remember that @ means the last point selected. In the next exercise, you'll move the entire hole an exact distance of 1 unit in a 45° angle.

1. Click on the Move tool from the Modify toolbar.
2. Type **p**↵. The set of objects you selected in the previous command is highlighted. *P* selects the previously selected set of objects.
3. Now press ↵ to tell AutoCAD you have finished your selection. The cursor changes to Point Selection mode.
4. At the Base point or displacement: prompt, pick a point on the screen between the hole and the left side of the screen (see Figure 2.16).
5. Move the cursor around slowly and notice that the hole moves as if the base point you selected were attached to the hole. The hole moves with the cursor, at

a fixed distance from it. This demonstrates how the base point relates to the objects you select.

6. Now type **@1.25<270**↵. The hole will move to a new location on the screen at a distance of 1.25 units from its previous location and at an angle of 270°.

If AutoCAD is waiting for a command, you can repeat the last command used by pressing the right mouse button, the spacebar, or the ↵ key.

FIGURE 2.16

The highlighted hole and the base point just left of the hole. Note that the base point does not need to be on the object you are moving.

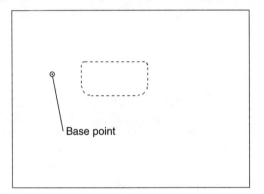

This exercise illustrates that the base point does not have to be on the object you are manipulating. The base point can be virtually anywhere on your drawing. You also saw how you can reselect objects that were selected previously, without having to duplicate the selection process.

In performing a Move command, you will often pick both the From: and To: points with your mouse. A common error made by those new to AutoCAD is to attempt to pick the To: point with the right mouse button. If you do this, AutoCAD treats the From: point as a displacement, and typically places the object far beyond your currently displayed view. Your moved object seems to disappear. Remember that the left mouse button is used for picking points, and the right mouse button acts like the ↵ key.

Other Selection Options

There are several other selection options you haven't tried yet. This sidebar describes these other options. You'll see how these options work in exercises later in this book. Or if you are adventurous, try them out now on your own. To use these options, type their keyboard abbreviations (shown in brackets in the following list) at any Select objects: prompt.

All [all↵] selects all the objects in a drawing except those in frozen or locked layers (see Chapter 4 for more on layers).

Crossing [c↵] is similar to the Select Window option but will select anything that crosses through the window you define.

Crossing Polygon [cp↵] acts exactly like WPolygon (see below) but, like the Select Crossing option, will select anything that crosses through a polygon boundary.

Fence [f↵] selects objects that are crossed over by a temporary line called a fence. The operation is like crossing out the objects you want to select with a line. When you invoke this option, you can then pick points, as when you are drawing a series of line segments. When you are done drawing the fence, press ↵, then go on to select other objects, or press ↵ again to finish your selection.

Last [l↵] selects the last object you input.

Multiple [m↵] lets you select several objects first, before AutoCAD highlights them. In a very large file, picking objects individually can cause AutoCAD to pause after each pick, while it locates and highlights each object. The Multiple option can speed things up by letting you first pick all the objects quickly, and then highlight them all by pressing ↵. This has no menu equivalent.

Previous [p↵] selects the last object or set of objects that was edited or changed.

Window [w↵] forces a standard selection window. This option is useful when your drawing area is too crowded to use the Autoselect feature to place a window around a set of objects (see the Auto entry in this sidebar). It prevents you from accidentally selecting an object with a single pick when you are placing your window.

Continued

> **CONTINUED**
>
> **Window Polygon [wp↵]** lets you select objects by enclosing them in an irregularly shaped polygon boundary. When you use this option, you see the prompt `First polygon point:`. You then pick points to define the polygon boundary. As you pick points, the prompt `Undo/<Endpoint of line>:` appears. Select as many points as you need to define the boundary. You can Undo boundary line segments as you go by clicking on the Undo tool on the Standard toolbar or by pressing **u**. With the boundary defined, press ↵. The bounded objects are highlighted and the `Select objects:` prompt returns, allowing you to use more selection options.
>
> The following two selection options are also available, but are seldom used. They are intended for use in creating custom menu options or custom toolbar tools.
>
> **Auto [au↵]** forces the standard automatic window or crossing window when a point is picked and no object is found (see *Using Autoselect* later in this chapter). A standard window is produced when the two window corners are picked from left to right. A crossing window is produced when the two corners are picked from right to left. Once this option is selected, it remains active for the duration of the current command. Auto is intended for use on systems where the Autoselect feature has been turned off.
>
> **Single [si↵]** forces the current command to select only a single object. If you use this option, you can pick a single object; then the current command will act on that object as though you had pressed ↵ immediately after selecting the object. This has no menu equivalent.
>
> If you need more control over the selection of objects, you will find the Add/Remove selection mode setting useful. This setting lets you deselect a set of objects within a set of objects you've already selected. Both settings must be used in Object Selection mode.
>
> **Remove [r↵]** removes objects from the selection set. Or, if you only need to deselect a single object, hold Shift and click on the object.
>
> **Add [a ↵]** continues to add more objects to the selection set.

Selecting Objects Before the Command: Noun/Verb

Nearly all graphics programs today have tacitly acknowledged the *Noun/Verb* method for selecting objects. This method requires you to select objects *before* you issue a command to edit them. The next set of exercises shows you how to use the Noun/Verb method in AutoCAD.

SELECTING OBJECTS | 71

This chapter presents the standard AutoCAD method of object selection. AutoCAD also offers selection methods with which you may be more familiar. Refer to the section on the Object Selection Settings dialog box in Appendix B to learn how you can control object selection methods.

You have seen that when AutoCAD is waiting for a command, it displays the crosshair cursor with the small square. This square is actually a pickbox superimposed on the cursor. It tells you that you can select objects, even while the command prompt appears at the bottom of the screen and no command is currently active. The square momentarily disappears when you are in a command that asks you to select points. From now on, we'll refer to this crosshair cursor with the small box as the *Standard cursor.*

Now try moving objects by first selecting them and then using the Move command.

1. First, press Esc twice to make sure AutoCAD isn't in the middle of a command you may have accidentally issued. Then click on the arc at the lower-right end of the hole. The arc is highlighted, and you may also see squares appear at its endpoints and midpoint. These squares are called *grips*. You may know them as workpoints from other graphics programs. You'll get a chance to work with them a bit later.

2. Choose Move from the Modify toolbar. The cursor changes to Point Selection mode.

3. At the `Base point or displacement:` prompt, pick any point on the screen. The prompt `Second point of displacement:` appears.

4. Type **@.5<0**↵. The arc moves to a new location .5 units to the right.

If you find that this exercise does not work as described here, chances are the Noun/Verb setting has been turned off on your copy of AutoCAD. Refer to Appendix B to find out how to activate this setting.

In this exercise, you picked the arc *before* issuing the Move command. Then, when you clicked on the Move tool, you didn't see the object selection prompt. Instead, AutoCAD assumed you wanted to move the arc you had selected and went directly to the base-point prompt.

Using Autoselect

Next you will move the rest of the hole in the same direction by using the Autoselect feature.

1. Pick a point just below and to the left of the rest of the hole objects. Be sure not to pick the arc you just moved. Now a window appears that you can drag across the screen as you move the cursor. If you move the cursor to the left of the last point selected, the window appears dotted (see the top image of Figure 2.17). If you move the cursor to the right of that point, it appears solid (see the bottom image of Figure 2.17).

2. Now pick a point above and to the right of the rest of the objects that make up the hole, so that they are completely enclosed by the window. Do not include the arc that you moved before. The objects are highlighted (and again, you may see small squares appear at the line's endpoints and midpoints).

3. Click on the Move tool again. Just as in the last exercise, the base-point prompt appears.

4. Pick any point on the screen; then enter **@.5<0↵**. The hole is reassembled.

The two different windows you have just seen—the solid one and the dotted one—represent a *standard window* and a *crossing window*. If you use a standard window, anything that is completely contained within the window will be selected. If you use a crossing window, anything that crosses through the window or is contained within it will be selected. These two kinds of window start automatically when you click on any blank portion of the drawing area with a Standard cursor or Point Selection cursor: hence the name Autoselect.

FIGURE 2.17

The dotted window (top image) indicates a crossing window; the solid window (bottom image) indicates a standard window.

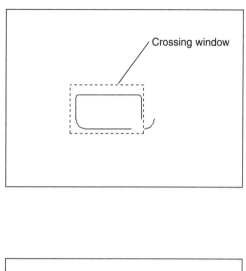

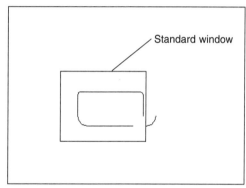

Next, you will select objects with an automatic crossing window.

1. Pick a point below and to the right of the hole. As you move the cursor to the left, the crossing (dotted) window appears.
2. Select the next point so that the window encloses the right side arcs and one line and part of the two horizontal lines (see Figure 2.18). Three lines and two arcs are highlighted.
3. Click on the Move tool.
4. Pick any point on the screen; then enter **@.5<180**↵. The highlighted parts of the hole move back to their original location.

FIGURE 2.18

Parts of the hole selected by a crossing window

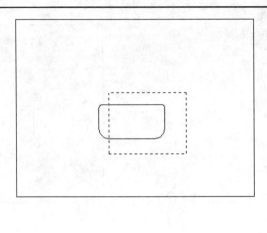

You'll find that in most cases, the Autoselect standard and crossing windows are all you need when selecting objects. They will really save you time, so you'll want to get familiar with these features.

Before we continue, use File ➤ Save to save the Base file. You won't want to save the changes you make in the next section, so saving now will store the current condition of the file on your hard disk for safekeeping.

Restrictions on Noun/Verb Object Selection

If you prefer to work with the Noun/Verb selection feature, you should know that its use is limited to the following subset of AutoCAD commands, listed here in no particular order.

Array	Mirror	Wblock	Block
Dview	Move	Erase	Explode
Change	Rotate	Chprop	Hatch
Scale	Copy	List	Stretch

For all other modifying or construction-oriented commands the Noun/Verb selection method is inappropriate, because for those commands you must select more than one set of objects. But you do not need to remember this list. You'll know if a command accepts the Noun/Verb selection method right away. Commands that don't accept the Noun/Verb selection will clear the selection and display a `Select object:` prompt.

NOTE

If you want to take a break, now is a good time to do it. If you wish, you can exit AutoCAD and return to this point in the tutorial later. When you return, start AutoCAD and open the Base file.

Editing with Grips

WARNING

If you did not see small squares appear on the base in the previous exercise, your version of AutoCAD may have the Grips feature turned off. Before continuing with this section, refer to the information on grips in Appendix B.

Earlier, when you selected the base, little squares appeared at the endpoints and midpoints of the lines and arcs. These squares are called *grips*. Grips can be used to make direct changes to the shape of objects, or to quickly move and copy them.

So far, you have seen how operations in AutoCAD have a discrete beginning and ending. For example, to draw an arc, you first issue the Arc command and then go through a series of operations, including answering prompts and picking points. When you are done, you have an arc and AutoCAD is ready for the next command.

The Grips feature, on the other hand, plays by a different set of rules. Grips offer a small yet powerful set of editing functions that don't conform to the lockstep command/prompt/input routine you have seen so far. As you work through the following exercises, it will be helpful to think of the Grips feature as a "subset" of the standard method of operation within AutoCAD.

To practice using the Grips feature, you'll make some temporary modifications to the Base drawing.

Stretching Lines Using Grips

In this exercise, you'll stretch one corner of the base by grabbing the grip points of two lines.

1. Press Esc to make sure AutoCAD has your attention and you're not in the middle of a command. Click on a point below and to the left of the base to start a selection window.
2. Click above and to the right of the base to select it.
3. Place the cursor on the lower-left corner grip of the rectangle, *but don't press the pick button yet.* Notice that the cursor jumps to the grip point.
4. Move the cursor to another grip point. Notice again how the cursor jumps to it. When the cursor is placed on a grip, the cursor moves to the exact center of the grip point. This means, for example, that if the cursor is placed on an endpoint grip, it is on the exact endpoint of the object.
5. Move the cursor to the upper-left corner grip of the rectangle and click on it. The grip becomes a solid color, and is now a *hot grip*. The prompt displays the following message:

STRETCH
<Stretch to point>/Base point/Copy/Undo/eXit:

This prompt tells you that the Stretch mode is active. Notice the options shown in the prompt. As you move the cursor, the corner follows and the lines of the rectangle stretch (see Figure 2.19).

When you select a grip by clicking on it, it turns a solid color and is known as a *hot grip*. You can control the size and color of grips using the Grips dialog box (see Appendix B).

6. Move the cursor upward and click on a point. The rectangle deforms, with the corner placed at your pick point (see Figure 2.19).

When you click on the corner grip point, AutoCAD selects the overlapping grips of two lines. When you stretch the corner away from its original location, the endpoints of both lines follow.

EDITING WITH GRIPS   77

FIGURE 2.19

Stretching lines using hot grips. The top image shows the rectangle's corner being stretched upward. The bottom image shows the new location of the corner at the top of the arc.

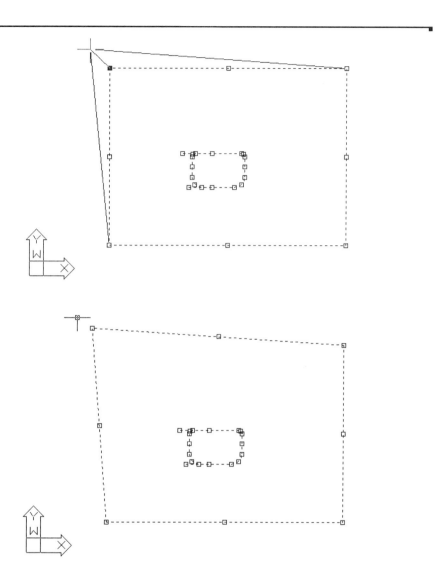

Here you saw that a command called **STRETCH** is issued simply by clicking on a grip point. As you will see, a handful of other hot grip commands are also available.

1. Notice that the grips are still active. Click on the grip point that you moved before to make it a hot grip again.

2. Right-click on the mouse. A pop-up menu of Grip Edit options appears.

3. Select Base Point from the menu, and then click on a point to the right of the hot grip. Now as you move the cursor, the hot grip moves relative to the cursor.

4. Right-click again, and then select the Copy option from the pop-up menu and enter **@1<-30**↵. Instead of moving the hot grip and changing the lines, copies of the two lines are made, with their endpoints one unit below and to the right of the first set of endpoints.

5. Pick another point just below the last. More copies are made.

6. Press ↵ or enter **x**↵ to exit the Stretch mode. You can also right-click again and select Exit from the pop-up menu.

In this exercise, you saw that you can select a base point other than the hot grip. You also saw how you can specify relative coordinates to move or copy a hot grip. Finally, with grips selected on an object, right-clicking the mouse opens a pop-up menu showing Grip Edit options.

Moving and Rotating with Grips

As you've just seen, the Grips feature offers an alternate method of editing your drawings. You've already seen how you can stretch endpoints, but there is much more that you can do with grips. The next exercise demonstrates some other options. You will start by undoing the modifications you made in the last exercise.

1. Click on the Undo tool in the Standard toolbar, or type **u**↵. The copies of the stretched lines disappear.
2. Press ↵ again. The deformed base snaps back to its original form.

Pressing ↵ at the command prompt causes AutoCAD to repeat the last command entered—in this case, Undo.

3. Select the entire base by first clicking on a blank area below and to the right of the base.
4. Move the cursor to a location above and to the left of the rectangular portion of base, and click. Since you went from right to left, you created a crossing window. Recall that this selects anything enclosed in or crossing through the window.
5. Click on the lower-left grip of the base to turn it into a hot grip. Just like before, as you move your cursor, the corner stretches.
6. Right-click, then at the Grip Edit pop-up menu, select Move. The Command window shows

    ```
    **MOVE**
    <Move to point>/Base point/Copy/Undo/eXit:
    ```

    Now as you move the cursor, the entire base moves with it.
7. Position the base near the center of the screen and click there. The base moves to the center of the screen. Notice that the command prompt returns, yet the base remains highlighted, telling you that it is still selected for the next operation.
8. Click on the lower-left grip again, and right-click on the mouse. This time, select Rotate from the pop-up menu. The Command window shows

    ```
    **ROTATE**
    <Rotation angle>/Base point/copy/Undo/Reference/eXit:
    ```

    As you move the cursor, the base rotates about the grip point.
9. Position the cursor so that the base rotates approximately 90° (see Figure 2.20). Then, while holding down the Shift key, press the mouse/pick button. A copy of the base appears in the new rotated position, leaving the original base in place.
10. Click on the Undo tool on the Standard toolbar, or type **u**↵ to eliminate the rotated copy of the base. Press the Esc key until the object highlighting and grips are gone.

11. Select the entire base again using a crossing window and click on the lower-left grip again.
12. Right-click on the mouse. This time select Copy. Type **@2.5<90**↵ to create another slot 2.5 units above the first slot. Your drawing should look like the one in Figure 2.20.
13. Press ↵ to exit the Grip Edit mode.

You've seen how the Move command is duplicated in a modified way as a hot grip command. Other hot grip commands (****Stretch****, ****Rotate****, ****Scale****, and ****Mirror****) also have similar counterparts in the standard set of AutoCAD commands.

After you've completed any operation using grips, the objects are still highlighted with their grips still active. To clear the grip selection, press Esc twice.

FIGURE 2.20

Rotating and copying the base using a hot grip. Notice that more than one object is affected by the grip edit, even though only one grip is "hot."

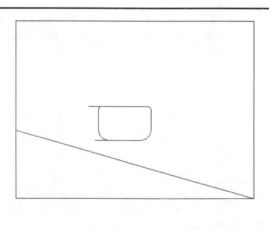

In this exercise, you saw how hot grip options appear in a pop-up list. Several other options are available in that list, including Exit, Base Point, Copy, and Undo. You can also make adjustments to an object's properties using the Properties option.

Many of these Grip Edit options are also available by pressing the spacebar or ↵ while a grip is selected. With each press, the next option becomes active. The commands then repeat if you continue to press ↵. The Shift key acts as a shortcut to the Copy option. You only have to use it once; then each time you click on a point, a copy is made.

A Quick Summary of the Grips Feature

The exercises in this chapter using hot grips include only a few of the Grips options. You'll get a chance to use other hot grip commands in later chapters. Meanwhile, here is a summary of Grips.

- Clicking on endpoint grips causes those endpoints to stretch.
- Clicking on midpoint grips of lines causes the entire line to move.
- If two objects meet end to end and you click on their overlapping grips, both grips are selected simultaneously.
- You can select multiple grips by holding down the Shift key and clicking on the desired grips.
- When a hot grip is selected, the Stretch, Move, Rotate, Scale, and Mirror commands are available to you by right-clicking on the mouse.
- OR you can cycle through the Stretch, Move, Rotate, Scale, and Mirror commands by pressing ↵ while a hot grip is selected.
- All the hot grip commands allow you to make copies of the selected objects by using the Copy option or by holding down the Shift key while selecting points.
- All the hot grip commands allow you to select a base point other than the originally selected hot grip.

Getting Help

Eventually, you will find yourself somewhere without documentation and you will have a question about an AutoCAD feature. AutoCAD provides an online Help facility that will give you information on nearly any topic related to AutoCAD. Here's how to find help:

1. Click on Help on the menu bar and choose AutoCAD Help Topics. A Help window appears.
2. If it isn't already selected, click on the Contents tab. This window shows a table of contents. There are two more tabs labeled Index and Find, and each offers assistance in finding specific topics.

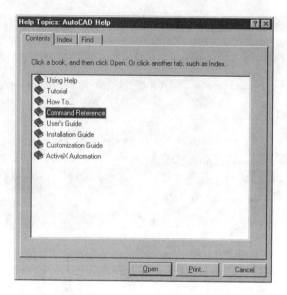

You can also press F1 to open the AutoCAD Help window.

3. Scan down the screen until you see the topic named Command Reference, and double-click on it. The list expands to show more topics.

4. Double-click on the item labeled Commands. The Help window expands to show a list of command names.

5. At the top of the list is a set of alphabet buttons. In the main window, you see a list of commands, beginning with 3D. You can click on the alphabetical button to go to a listing of commands that start with a specific letter. For now, scroll down the list and click on the word Copy shown in green. A detailed description of the Copy command appears.

6. Click on the Help Topics button at the top of the window. The Help Topics dialog box appears.

7. Click on the Find tab. If this is the first time you've selected the Find tab, you will see the Find Setup Wizard. This dialog box offers options for the search database that Find uses to locate specific words.

8. Accept the default option by pressing the Next button at the bottom of the dialog box. Find options appear with a list of topics in alphabetical order. You can enter a word to search for in the drop-down list at the top of the dialog, or you can choose a topic in the list box.

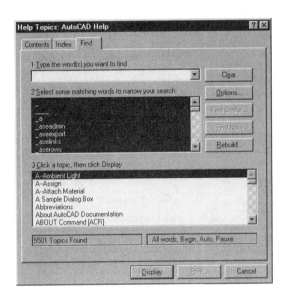

9. Type the word **Change**. The list box immediately goes to the word Change in the list.
10. Click on the word CHANGE in capital letters. Notice that the list box at the bottom of the dialog box changes to show some options.
11. Double-click on Change Command [ACR] in the list. A description of the Change command appears.

AutoCAD also provides *context-sensitive help* to give you information related to the command you are currently using. To see how this works, try the following:

1. Close or minimize the Help window and return to the AutoCAD window.
2. Click on the Move tool in the Modify toolbar to start the Move command.
3. Press the F1 function key, or choose Help ➤ AutoCAD Help Topics. The Help window appears, with a description of the Move command.

4. Click on the Close button or press Esc.
5. Press Esc to exit the Move command.

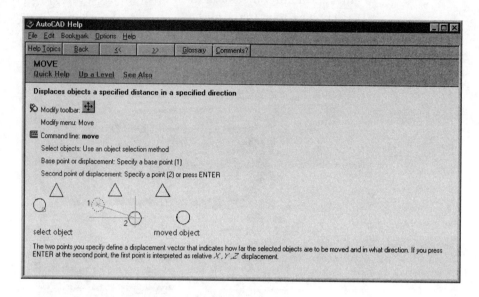

If you are already familiar with the basics of AutoCAD, you might want to install the AutoCAD Learning Assistant. This tool offers quick tips and brief tutorials on a wide variety of topics, including working in collaborative groups and making the most of the Windows environment. The Learning Assistant is on its own CD as part of the AutoCAD R14 package.

Displaying Data in a Text Window

Some commands produce information that requires a Text window. This frequently happens when you are trying to get information about your drawing. The following exercise shows how you can get an enlarged view of messages in a Text window.

1. Click on the List tool in the Standard toolbar (it's the icon that looks like a piece of paper with writing on it). This tool offers information about objects in your drawing.

2. At the `Select objects:` prompt, click on one of the arcs and press ↵. Information about the arc is displayed the AutoCAD Text window (see Figure 2.21). Toward the bottom is the list of the arc's properties. Don't worry if you don't understand this listing. As you work through this book, you'll learn what the different properties of an object mean.

3. Press F2. The AutoCAD Text window closes.

The F2 function key offers a quick way to switch between the drawing editor and the Text window.

The Text window not only shows you information about objects, it also displays a history of the command activity for your AutoCAD session. This can be helpful for remembering data you may have entered earlier in a session, or to help recall an object's property that you have listed earlier. The scroll bar to the right of the Text window lets you scroll to earlier events. You can even set the number of lines AutoCAD will retain in this Text window using the Preferences dialog box, or you can have AutoCAD record the Text window information in a text file.

4. Now you are done with the base drawing, so choose File ➤ Exit.

5. At the Save Changes? dialog box, click on the No button. (You've already saved this file in the condition you want it, so you do not need to save it again.)

FIGURE 2.21

The AutoCAD text screen showing the data displayed by the List tool

```
Command:
Command: list
1 found
                      ARC        Layer: 0
                                 Space: Model space
                       Handle = 42
          center point, X=    7.1444   Y=    5.0424   Z=    0.0000
          radius     0.3750
            start angle    180
              end angle    270
           length     0.5890
```

If You Want to Experiment...

Try drawing the latch shown in Figure 2.22.

1. Start AutoCAD, open a new file, and name it **Latch**.

2. When you get to the drawing editor, use the Line command to draw the straight portions of the latch. Start a line as indicated in the figure; then enter relative coordinates from the keyboard. For example, for the first line segment, enter **@4<180↵** to draw a line segment four units long from right to left.

3. Draw an arc for the curved part. To do this, click on the Arc tool on the Draw toolbar.

4. Use the Endpoint Osnap to pick the endpoint indicated in the figure to start your arc.

5. Type **e**↵ to issue the End option of the Arc command.

6. Using the Endpoint Osnap again, click on the endpoint above where you started your line. A rubber-banding line and a temporary arc appear.

7. Type **d**↵ to issue the Direction option of the Arc command.

8. Position your cursor so that the ghosted arc looks like the one in the figure, and then press the mouse/pick button to draw the arc.

FIGURE 2.22
Try drawing this latch. Dimensions are provided for your reference.

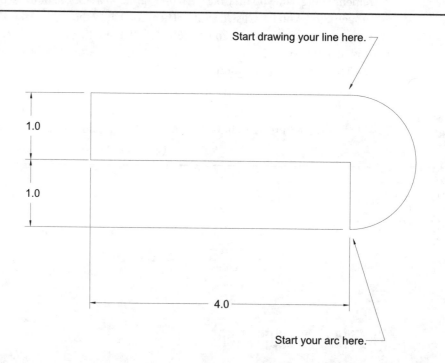

Chapter 3

Learning the Tools of the Trade

FEATURING

Setting Up a Work Area

Designating the Measurement System

Understanding Scale in AutoCAD

Using Snaps, Grids, and the Coordinate Readout

Enlarging and Reducing Your View of the Drawing

Trimming and Making Parallel Copies of Objects

Laying Out a Drawing with Lines

Learning the Tools of the Trade

So far we have covered the most basic information you need to understand the workings of AutoCAD. Now you will put your knowledge to work. In this tutorial, which begins here and continues through Chapter 12, you will draw a number of parts. The tutorial illustrates how to use AutoCAD commands and will give you a solid understanding of the basic AutoCAD package. With these fundamentals, you can use AutoCAD to its fullest potential, regardless of the kinds of drawings you intend to create or the enhancement products you may use in the future.

In this chapter you will start drawing a Plan view of the machined bracket shown in Figure 3.1. In the process, you will learn how to use AutoCAD's basic tools.

FIGURE 3.1
The bracket with dimensions

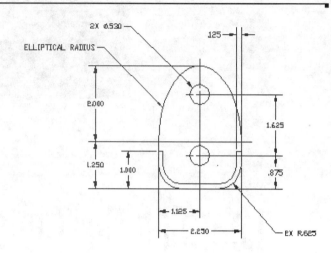

Setting Up a Work Area

Before beginning most drawings, you will want to set up your work area. To do this you must determine the *measurement system* you want to use. The default work area is roughly 9"×6" at full scale, given a decimal measurement system where 1 unit equals 1 inch. These are appropriate settings for your drawing, so you don't have to do very much setting up. Often, however, you will be doing drawings of various sizes and scales. For example, you may want to create a drawing in a measurement system where you can specify feet, inches, and fractions of inches at 1"=1' scale, and print the drawing on an 8½"×11" sheet of paper. In this section, you will learn how to set up a drawing the way you want.

Specifying Units

Start by creating a new file called DTO100.

1. Start up AutoCAD; then pick File ➤ New.

2. In the Create New Drawing dialog box, click on the Start from Scratch button and select English from the Select Default Setting list.
3. Click on OK to open the new file.
4. Choose File ➢ Save As.
5. At the Save Drawing As dialog box, enter **DTO100** for the file name.
6. Check to make sure you are saving the drawing in the Samples subdirectory or the directory you have chosen to store your exercise files, then click on Save.

NOTE

Although you could start drawing in the AutoCAD window immediately after starting up AutoCAD and then save the file later under the name DT0100, use File ➢ New for this exercise—in case you are using a system that has an altered default setup for new files.

The first thing you will want to tell AutoCAD is the *unit style* you intend to use. So far, you've been using the default, which is decimal. You can think of decimal as either inches or millimeters. In this style, whole units represent inches or millimeters, and decimal units are decimal inches or millimeters. If you want to be able to enter distances in feet, then you must change the unit style to one that accepts feet as input. This is done through the Units Control dialog box.

NOTE

If you are a civil engineer, you will want to know that the Engineering unit style allows you to enter feet and decimal feet for distances. For example, the equivalent of 12'-6" would be 12.5'. Earlier versions of AutoCAD did not have this feature, so engineers had to resort to using the Decimal unit style and feet as the base unit instead of inches. This caused problems when architectural drawings were combined with civil drawings: The scales of each type of drawing did not match. Even though this ability to input distances in feet and decimal feet has existed since Release 12, old habits die hard. If you use the Engineering unit style, you will insure that your drawings will conform to the scale of drawings created by your architectural colleagues. And you will have the ability to enter decimal feet.

1. Choose Formats ➢ Units. Or type **un**↵. The Units Control dialog box appears. Let's look at a few of the options available.

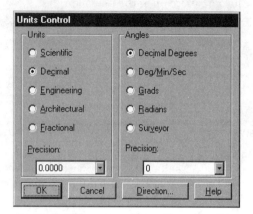

The Units Control settings can also be controlled using several system variables. To set the unit style, you can type `'lunits`↵ at the command prompt. (The apostrophe lets you enter this command while in the middle of other commands.) At the `New value for Lunits <2>:` prompt, enter **2** for Decimal. See Appendix D for other settings.

2. Notice the unit styles listed in the Units group.
3. Click on the down-pointing arrow in the Precision drop-down list at the bottom of the Units group. Notice the options available. The default is four-place decimal precision. AutoCAD uses this value to display settings values in the coordinate readout on the status bar and in dialog boxes. For now, you can set the smallest unit AutoCAD will display in this drawing to three decimal places by selecting the three decimal option.

Units do not control the precision AutoCAD is using to create and maintain geometry or the accuracy of the database. The AutoCAD database is always accurate to slightly more than 16 decimal places. The inaccuracy is due to truncation of values in 64-bit calculations.

4. Close the drop-down list and then click on the Direction button at the bottom of the dialog box. The Direction Control dialog box appears. This dialog box lets you set the direction for the 0° angle and the direction for positive degrees. For now, don't change these settings—you'll read more about them in a moment.
5. Click on the OK button.

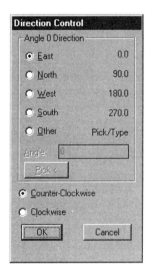

6. Click on OK in the main Units Control dialog box to return to the drawing.

NOTE

Remember that the status bar displays a description of the tool or pull-down menu option, including the command name. If you prefer entering commands through the keyboard, look at the tool description in the status bar during the exercises. The command name will be listed last. You can type in the command name to issue the command instead of clicking on the tool during any of these exercises. Command names are also useful when you want to create your own custom macros. You'll get a chance to create some macros in Chapter 19.

You picked Decimal measurement units for this tutorial, but your own work may require a different unit style. You saw the unit styles available in the Units Control dialog box. Table 3.1 shows examples of how the distance 15.5 is entered in each of these styles.

TABLE 3.1: MEASUREMENT SYSTEMS AVAILABLE IN AUTOCAD

Measurement System	AutoCAD's Display of Measurement
Scientific	1.55E+01 (inches)
Decimal	15.5000 (inches)
Engineering	1'-3.5" (input as 1'3.5")
Architectural	1'-3½" (input as 1'3-1/2")
Metric	15.5000 (mm, cm, or meters)
Fractional	15½" (input as 15-1/2")

In the previous exercise, you needed to change only one setting. Let's take a look at the other Units Control settings in more detail. As you read, you may want to refer to the illustration of the Units Control dialog box.

Fine-Tuning the Measurement System

Most of the time, you will be concerned only with the Units and Angles setting groups of the Units Control dialog box. But as you saw from the last exercise, you can control many other settings related to the input and display of units.

To find the distance between two points, click on Distance on the Standard toolbar, or type **di**↵ and then click on the two points (Di is the shortcut for entering **dist**↵). But if you find that this command doesn't give you an accurate distance measurement, examine the Units/Precision option in the Units Control dialog box. If it is set too high, the value returned by Dist may be rounded to a value greater than your tolerances allow, even though the distance is drawn accurately.

The Precision drop-down list in the Units group lets you specify the smallest unit value that you want AutoCAD to display in the status line and in the prompts. If you choose a measurement system that uses fractions, the Precision list will include fractional units. This setting can also be controlled with the Luprec system variable.

The Angles group lets you set the style for displaying angles. You have a choice of five angle styles: decimal degrees, degrees/minutes/seconds, grads, radians, and surveyor's units. In the Angles group's Precision drop-down list, you can determine the degree of accuracy you want AutoCAD to display for angles. These settings can also be controlled with the Aunits and Auprec system variables.

NOTE

You can find out more about system variables in Appendix D.

The Direction Control dialog box lets you set the direction of the 0° base angle. The default base angle (and the one used throughout this book) is a direction from left to right. However, there may be times when you will want to designate another direction as the 0° base angle. You can also tell AutoCAD which direction is positive, either clockwise or counterclockwise. In this book we use the default, which is counterclockwise. These settings can also be controlled with the Angbase and Angdir system variables (see Appendix D for more on the AutoCAD system variables).

Things to Watch Out for When Entering Distances

When you are using Engineering, Architectural, or Fractional units, there are two points you should be aware of:

- Hyphens are used only to distinguish fractions from whole inches.
- You cannot use spaces while giving a dimension. For example, you can specify eight feet, four and one-half inches as 8'4-1/2" or 8'4.5, but not as 8'4 1/2".

These idiosyncrasies are a source of confusion to many engineers and architects new to AutoCAD, because the program often displays dimensions with fractions in the standard format but does not allow you to enter dimensions that way.

When inputting distances and angles in unusual situations, here are some tips:

- When entering distances in inches and feet, you can omit the inch (") sign. If you are using the Engineering unit style, you can enter decimal feet and forego the inch sign entirely.
- You can enter fractional distances and angles in any format you like, regardless of the current unit system. For example, you can enter a distance as **@1/2<1.5708r** even if your current unit system is set for decimal units and decimal degrees (1.5708r is the radian approximation of 90°).
- If you have your angle units set to degrees, grads, or radians, you do not need to specify g, r, or d after the angle. You *do* have to specify g, r, or d, however, if you want to use these units when they are not the current default angle system.

Continued

> **CONTINUED**
>
> - If your current angle system is set to something other than degrees, but you want to input angles in degrees, you can use a double less-than symbol (<<) in place of the single less-than symbol (<) to override the current angle system of measure. The << also assumes the base angle of 0° to be a direction from left to right and the positive direction to be counterclockwise.
> - If your current angle system uses a different base angle and direction (other than left to right for 0° and a counterclockwise direction for positive angles), and you want to specify an angle in the standard base direction, you can use a triple less-than symbol (<<<) to indicate the angle.
> - You can specify a denominator of any size when specifying fractions. However, you should be aware that the value you have set for the maximum number of digits to the right of decimal points (under the Units setting) will restrict the actual fractional value AutoCAD will use. For example, if your units are set for a maximum of 2 digits of decimals and you give a fractional value of 5/32, AutoCAD will round it out to 3/16 or 0.16.
> - You are allowed to enter decimal feet for distances in the Architectural unit style.

Setting Up the Drawing Limits

One of the big advantages in using AutoCAD is that you can draw at full scale; you aren't limited to the edges of a piece of paper the way you are in manual drawing. But you still have to consider what will happen when you want a printout of your drawing. If you're not careful, you may create a drawing that won't fit on the paper size you want at the scale you want. When you start a new drawing, it helps to limit your drawing area to one that can be scaled down to fit on a standard sheet size. While this is not absolutely necessary with AutoCAD, the limits will give you a frame of reference between your work in AutoCAD and the final printed output.

In order to set up the drawing work area, you need to understand how standard sheet sizes translate into full-scale drawing sizes. Table 3.2 lists widths and heights of drawing areas in inches, according to scales and final printout sizes. The scales are listed in the far left column; the output sheet sizes are listed across the top.

Let's take an example: To find the area needed in AutoCAD for your bracket if it is shown sitting on a 6'×3' table with the long side of the table oriented horizontally, look across from the scale ⅛"=1" to the column that reads 8½"×11" at the top. You'll find the value 68×132. This means the table's drawing area of 72"×36" needs to fit within an area 68"×88" in AutoCAD in order to fit a printout of a ⅛"=1" scale drawing on an 8½"×11" sheet of paper. You will want the drawing area to be oriented horizontally, so that the 11" will be in the X axis and the 8.5" will be in the Y axis.

TABLE 3.2: WORK AREA IN DRAWING UNITS BY SCALE AND PLOTTED SHEET SIZE

Scale	A 8½"×11"	B 11"×17"	C 17"×22"	D 22"×34"	E 34"×44"	F 28"×40"
50"=1"	.17×.22	.22×.34	.34×.44	.44×.68	.68×.88	.56×.80
20"=1"	.42×.55	.55×.85	.85×1.1	1.1×1.7	1.7×2.2	1.4×2.0
10"=1"	.85×1.1	1.1×1.7	1.7×2.2	2.2×3.4	3.4×4.4	2.8×4.0
4"=1"	2.12×2.75	2.75×4.25	4.25×5.5	5.5×8.5	8.5×11	7.0×10.0
2"=1"	4.25×5.5	5.5×8.5	8.5×11	11×17	17×22	14×20
¾"=1"	11×14	14×22	22×29	29×45	45×58	37×53
½"=1"	17×22	22×34	34×44	44×68	68×88	56×80
⅜"=1"	22×29	29×45	45×58	58×90	90×117	74×106
¼"=1"	34×44	44×68	68×88	88×136	136×176	112×160
⅛"=1"	68×88	88×136	136×176	176×272	272×352	224×320
1"=1'	102×132	132×204	204×264	264×408	408×528	336×480
¾"=1'	136×176	176×272	272×352	352×544	544×704	448×640
½"=1'	204×264	264×408	408×528	528×816	816×1056	672×960
¼"=1'	408×528	528×816	816×1056	1056×1632	1632×2112	1344×1920
⅛"=1'	816×1056	1056×1632	1632×2112	2112×3264	3264×4224	2688×3840
¹⁄₁₆"=1'	1632×2112	2112×3264	3264×4224	4224×6528	6528×8448	5376×7680
¹⁄₃₂"=1'	3264×4224	4224×6528	6528×8448	8448×13056	13056×16896	10752×15360
1"=10'	1020×1320	1320×2040	2040×2640	2640×4080	4080×5280	3360×4800
1"=20'	2040×2640	2640×4080	4080×5280	5280×8160	8160×10560	6720×9600
1"=30'	3060×3960	3960×6120	6120×7920	7920×12240	12240×15840	10080×14400
1"=40'	4080×5280	5280×8160	8160×10560	10560×16320	16320×21120	13440×19200
1"=50'	5100×6600	6600×10200	10200×13200	13200×20400	20400×26400	16800×24000
1"=60'	6120×7920	7920×12240	12240×15840	15840×24480	24480×31680	20160×28800

Now that you know the area you need, you can use the Limits command to set up the area.

1. Choose Format ➤ Drawing Limits.
2. At the ON/OFF/<Lower left corner> <0'-0",0'-0">: prompt, specify the lower-left corner of your work area. Press ↵ to accept the default.

3. At the Upper right corner <12.000,9.000>: prompt, specify the upper-right corner of your work area. (The default is shown in brackets.) Enter ↵ to accept the default.

4. Next, choose View ➣ Zoom ➣ All. You can also click on the Zoom All tool from the Zoom Window flyout on the Standard toolbar or type **z**↵**a**↵. Though it appears that nothing has changed, your drawing area is now set to a size that will allow you to draw your bracket at full scale.

You can toggle through the different coordinate readout modes by pressing F6 or by double-clicking on the coordinate readout of the status bar. For more on the coordinate readout modes, see Chapter 1 and *Using the Coordinate Readout as Your Scale* later in this chapter.

5. Move the cursor to the upper-right corner of the drawing area and watch the coordinate readout. You will see that now the upper-right corner has a Y coordinate of approximately 12.000, 9.000 ". The X coordinate will vary depending on the proportion of your AutoCAD window.

In step 5 above, the coordinate readout shows you your drawing area, but there are no visual clues to tell you where you are or what distances you are dealing with. To help you get your bearings, you can use the Grid mode, which you will learn about shortly. But first, let's take a closer look at scale factors and how they work.

Understanding Scale Factors

When you draft manually, you work on the final drawing directly with pen or pencil. With a CAD program, you are a few steps removed from the actual finished product. Because of this, you need to have a deeper understanding of your drawing scale and how it is derived. In particular, you will want to understand *scale factors*.

For example, one of the more common uses of scale factors is in translating text size in your CAD drawing to the final plotted text size. When you draw manually, you simply draw your notes at the size you want. In a CAD drawing, you need to translate the desired final text size to the drawing scale.

When you start adding text and dimensions to your drawing (Chapters 7 and 8), you will have to specify a text height. The scale factor will help you determine the appropriate text height for a particular drawing scale. For example, you may want your text to appear ⅛" high in your final plot. But if you drew your text to ⅛" in your drawing, it would be multiplied by the plot scale when plotted. If your drawing was to be plotted at ⅛"=1" scale, the text would be ¹⁄₆₄" high. If the plot scale was 4"=1", your text would be ½" high. The text has to be scaled to a size that, when scaled at plot time, will appear ⅛" high. So for a ¼"=1" scale drawing you would multiply the ⅛" text height by a scale factor of 4 to get .5". Your text should be .5" high in the CAD drawing in order to appear ⅛" high in the final plot.

All the drawing sizes in Table 3.2 were derived by using scale factors. Table 3.3 shows scale factors as they relate to standard drawing scales. These scale factors are the values by which you multiply the desired final printout size to get the equivalent full-scale size. For example, if you have a sheet size of 11"×17", and you want to know the equivalent full-scale size for a ¼"=1"-scale drawing, you multiply the sheet measurements by 4. In this way, 11" becomes 44" (48×11) and 17" becomes 86". Your work area must be 44" by 86" if you intend to have a final output of 11"×17" at ¼"=1".

If you get the message **Outside limits, it means you have selected a point outside the area defined by the limits of your drawing *and* the Limits command's limits-checking feature is on. (Some third-party programs may use the limits-checking feature.) If you must select a point outside the limits, issue the Limits command and then enter **off**↵ at the ON/OFF <Lower left corner>... prompt to turn off the limits-checking feature.

The scale factor for fractional inch scales is derived by multiplying the denominator of the scale by 12, and then dividing by the numerator. For example, the scale factor for ¼"=1'-0" is (4×12)/1, or 48/1. For whole-foot scales like 1"=10', multiply the feet side of the equation by 12. Metric scales require simple decimal conversions.

If you are using the metric system, the drawing scale can be used directly as the scale factor. For example, a drawing scale of 1:10 would have a scale factor of 10; a drawing scale of 1:50 would have a scale factor of 50; and so on.

TABLE 3.3: WORK AREA IN METRIC UNITS (MILLIMETERS) BY SCALE AND PLOTTED SHEET SIZE					
Scale	A0 or F 841mm× 1189mm (33.11"× 46.81")	A or D 594mm× 841mm (23.39"× 33.11")	A2 or C 420mm× 594mm (16.54"× 3.39")	A3 or B 297mm× 420mm (11.70"× 16.54")	A4 or A 210mm× 297mm (8.27"× 11.70")
1:2	1682mm× 2378mm	1188mm× 1682mm	840mm× 1188mm	594mm× 840mm	420mm× 594mm
1:5	4205mm× 5945mm	2970mm× 4205mm	2100mm× 2970mm	1485mm× 2100mm	1050mm× 1485mm
1:10	8410mm× 11890mm	5940mm× 8410mm	4200mm× 5940mm	2970mm× 4200mm	2100mm× 2970mm

You will be using scale factors to specify text height and dimension settings, so understanding them now will pay off later.

Using the AutoCAD Modes as Drafting Tools

After you have set up your work area, you can begin the Plan view of the bracket. We will use this example to show you some of AutoCAD's drawing aids. These tools might be compared to a background grid (the *Grid mode*), scale (the *coordinate readout*), and a T square and triangle (the *Ortho mode*). These drawing modes can be indispensable tools when used properly. The Drawing Aids dialog box helps you visualize the modes in an organized manner and simplifies their management.

Using the Grid Mode as a Background Grid

Using the *Grid mode* is like having a grid under your drawing to help you with layout. In AutoCAD, the Grid mode also lets you see the limits of your drawing and helps you visually determine the distances you are working with in any given view. In this section, you will learn how to control the grid's appearance. The F7 key toggles the Grid mode on and off; you can also double-click on the Grid button in the status bar. Start by setting the grid spacing.

USING THE AUTOCAD MODES AS DRAFTING TOOLS | **101**

1. Choose Tools ➤ Drawing Aids or type **rm↵** to display the Drawing Aids dialog box, showing all the mode settings. You see four button groups: Modes, Snap, Grid, and Isometric Snap/Grid.

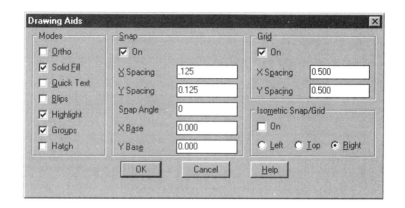

2. Let's start with the Grid group. Notice that the X Spacing input box contains a value of 0.500.

3. Double-click on the X Spacing input box. The 0.500 highlights. You can now type in a new value for this setting.

You can use the Gridunit system variable to set the grid spacing. Enter **'gridunit↵**, and at the New value for GRIDUNIT <0'0",0'0">: prompt, enter **12,12↵**. Note that the Gridunit value must be entered as an X,Y coordinate.

4. Enter **.125↵**. Notice that the Y Spacing input box automatically changes to 0.1250" (the .125 you entered remains 12 until the next time you open this dialog box). AutoCAD assumes you want the X and Y grid spacing to be the same, unless you specifically ask for a different Y setting.

If you want to change an entry in an input box, you can double-click on it to highlight the whole entry, and then replace the entry by simply typing in a new one. If you just want to change a part of the entry, use you mouse to highlight the part you want to change, or click on the input box and then use the cursor keys to move the vertical bar cursor to the exact character you want to change. You can use the Delete or Backspace keys to delete characters.

5. Click on the On check box near the top of the Grid group. This setting makes the grid visible.

6. Click on OK. The grid now appears as an array of dots with a 0.125" spacing in your drawing area. They will not print or plot with your drawing.

With the grid at a 0.125-unit spacing, you can see your work area more clearly. Since the grid will appear only within the drawing limits, you are better able to see your work area. In the next section, you'll see how the Snap mode works.

7. Press F7, or double-click on the word GRID in the status bar (you can also hold down the Ctrl key and press g). The grid disappears.

8. Press F7 again to turn the grid back on.

If your view is such that the grid spacing appears quite small, AutoCAD will not display the grid, in order to preserve the readability of the drawing. If this situation occurs, you will see the message Grid too dense to display in the command window.

Using the Snap Mode

The *Snap mode* has no equivalent in hand drafting. This mode forces the cursor to step a specific distance. It is useful if you want to maintain accuracy while entering distances with the cursor. The F9 key toggles the Snap mode on and off. Or, just like the Grid mode, there is a Snap button in the status bar that you can double-click. Follow these steps to access the Snap mode.

1. Choose Tools ➢ Drawing Aids or type **rm**↵. The Drawing Aids dialog box appears again.

2. In the Snap group of the dialog box, double-click on the X Spacing input box and enter **.125**↵. As with the Grid setting, AutoCAD assumes you want the X and Y snap spacing to be the same, unless you specify a different Y setting.

3. Click on the On check box so a check mark appears.

4. Click on OK, and start moving the cursor around. Notice how the cursor seems to move in "steps" rather than in a smooth motion. Also notice that the word Snap in the status bar is solid black, indicating that the Snap mode is on. If this is difficult to see, type **zoom**↵**2x**↵ to zoom in closer. When you are done type **zoom**↵**p**↵ to return to your previous view. Zoom will be discussed a little later.

5. Press F9 or double-click on the word SNAP in the status bar (you can also hold down the Ctrl key and press b); then move the cursor slowly around the drawing area. The Snap mode is now off.

6. Press F9 again to turn the Snap mode back on.

You can use the Snapunit system variable to set the snap spacing. Enter `'snapunit↵`, then at the `New value for SNAPUNIT <0'0",0'0">:` prompt, enter **.125,.125**↵. Note that the Snapunit value must be entered as an X,Y coordinate.

Take a moment to look at the Drawing Aids dialog box. The other options in the Snap group allow you to set the snap origin point (X Base and Y Base), rotate the cursor to an angle other than its current 0–90° (Snap angle), and set the horizontal snap spacing to a value different from the vertical spacing (X Spacing and Y Spacing). You can also adjust other settings, such as the grid/snap orientation that allows isometric-style drawings (Isometric Snap/Grid).

Using Grid and Snap Together

You can set the grid spacing to be the same as the snap setting, allowing you to see every snap point. Let's take a look at how Grid and Snap work together.

1. Open the Drawing Aids dialog box.

2. Double-click on the X Spacing input box in the Grid group, and enter **0**↵.

3. Click on OK. Now the grid spacing has changed to reflect the .125" snap spacing. Move the cursor, and watch it snap to the grid points.

4. Open the Drawing Aids dialog box again.

5. Double-click on the X Spacing input box in the Snap group, and enter **.05**↵.

6. Click on OK. The grid automatically changes to conform to the new snap setting. When the grid spacing is set to 0, the grid then aligns with the snap points. At this density, the grid is overwhelming.

7. Open the Drawing Aids dialog box again.

8. Double-click on the X Spacing input box in the Grid group, and enter **.125**↵.

9. Click on OK. The grid spacing is now at 12 again, which is a more reasonable spacing for the current drawing scale.

With the snap spacing set to 0.05, it is difficult to tell if Snap is turned on based on the behavior of the cursor, but the coordinate readout in the status bar gives you a clue. As you move your cursor, the coordinates appear as rounded numbers with no fractional distances less than the snap spacing. Next, you'll look at other ways the coordinate readout helps you.

Using the Coordinate Readout as Your Scale

As you move the cursor over the drawing area, the coordinate readout dynamically displays its position in absolute Cartesian coordinates. This allows you to find a position on your drawing by locating it in reference to the drawing origin—0,0—which is in the lower-left corner of the sheet. You can also set the coordinate readout to display relative coordinates. Throughout these exercises, coordinates will be provided to enable you to select points using the dynamic coordinate readout. (If you want to review the discussion of AutoCAD's coordinate display, see Chapter 1.)

1. Click on the Line tool on the Draw toolbar or type **L**↵. You could also select Draw ➤ Line from the menu bar.

2. Using your coordinate readout for guidance, start your line at the coordinate 2.2500,2.0000.

3. Press F6 until you see the relative polar coordinates appear in the coordinate readout at the bottom of the AutoCAD window. You can also double-click on the coordinate readout. Polar coordinates allow you to see your current location in reference to the last point selected. This is helpful when you are using a command that requires distance and direction input.

4. Move the cursor until the coordinate readout lists 2.250<0,0,0.000, and pick this point. As you move the cursor around, the rubber-banding line follows it at any angle.

5. You can also force the line to be orthogonal. Press F8, or double-click on the ORTHO label in the status bar (you can also hold down the Ctrl key and press o to toggle on the Ortho mode), and move the cursor around. Now the rubber-banding line will only move vertically or horizontally.

NOTE

The Ortho mode (also available under Modes in the Drawing Aids dialog box) is analogous to the T square and triangle. Note that the word ORTHO appears on the status bar to tell you that the Ortho mode is on.

6. Move the cursor down until the coordinate readout lists 0'- 1.2500 "< 270 ,0.000 and click on this point.

7. Continue drawing the other two sides of the rectangle by using the coordinate readout. You should have a drawing that looks like Figure 3.2.

FIGURE 3.2

The rectangle

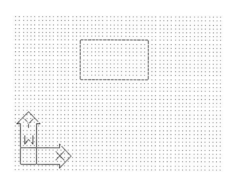

In steps 2 and 4, the coordinate readout showed some extra values. The 0.0000 that you see at the end of the coordinate readout listing indicates the Z value of the coordinate. This extra coordinate is significant only when you are doing 3D modeling, so for the time being, you can ignore it.

While this exercise tells you to use the Line tool to draw the bracket, you could also use the Rectangle tool. The Rectangle tool creates what is known as a *polyline*, which is a set of line or arc segments that acts like a single object. You'll learn more about polylines in Chapter 10.

NOTE

If you'd like to know more about the additional Z coordinate listing in the coordinate readout, see *Spherical and Cylindrical Coordinate Formats* in Chapter 13.

By using the Snap mode in conjunction with the coordinate readout, you can measure distances as you draw lines. This is similar to the way you would draw using a scale. Be aware that the smallest distance the coordinate readout will register depends on the area you have displayed in your drawing area. For example, if you are displaying an area the size of a football field, the smallest distance you can indicate with your cursor may be 6". On the other hand, if your view shows an area of only one square inch, you can indicate distances as small as 1/1000" using your cursor.

Exploring the Drawing Process

In this section, you will look at some of the more common commands and use them to complete a simple drawing. As you draw, watch the prompts and notice how your responses affect them. Also note how you use existing drawing elements as reference points.

While drawing with AutoCAD, you create gross geometric forms to determine the basic shapes of objects, and then modify the shapes to fill in detail. In essence, you alternately lay out, create, and edit objects to build your drawing.

AutoCAD offers many drawing tools. These include lines, arcs, circles, text, polylines, points, 3D faces, ellipses, spline curves, solids, Multiline text, and others. All drawings are built on these objects. In addition, there are five different 3D meshes, which are three-dimensional surfaces composed of 3D faces. You are familiar with lines and arcs; these, along with circles, are the most commonly used objects. As you progress through this book, you will be introduced to the other objects and how they are used.

Locating an Object in Reference to Others

To continue drawing the bracket, you will use an ellipse.

1. Click on the Ellipse tool in the Draw toolbar or type **el**↵. You can also choose Draw ➤ Ellipse ➤ Axis, End.

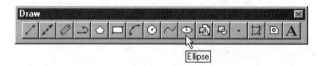

2. At the Arc/Center/<Axis endpoint 1>: prompt, pick the midpoint of the bottom horizontal line of the rectangle. Do this by bringing up the Osnap pop-up menu and selecting Midpoint; then move the cursor toward the bottom line. (Remember, to bring up the Osnap menu, press Shift and click the right mouse button.) When you see the Midpoint Osnap marker appear on the line, press the left mouse button.

EXPLORING THE DRAWING PROCESS

3. At the Axis endpoint 2: prompt, move the cursor up until the coordinate readout lists 4.000 < 90 ,0.000.
4. Pick this as the second axis endpoint.
5. At the <Other axis distance>/Rotation: prompt, move the cursor horizontally from the center of the ellipse until the coordinate readout lists 1.1258< 0 ,0.000.
6. Pick this as the axis distance defining the width of the ellipse. Your drawing should look like Figure 3.3.

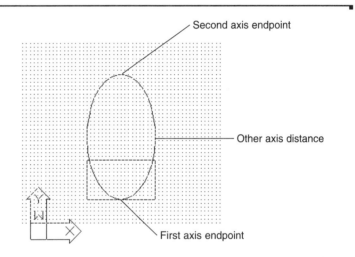

FIGURE 3.3
The ellipse added to the rectangle

NOTE

As you work with AutoCAD, you will eventually run into NURBS. NURBS stands for Non-Uniform Rational B-Splines—a fancy term meaning that curved objects are based on accurate mathematical models. When you trim the ellipse in a later exercise, it becomes a NURBS curve, not a segmented polyline as in earlier versions of AutoCAD. You'll learn more about polylines and NURBS curves in Chapter 10.

Getting a Closer Look

During the drawing process, you will want to enlarge areas of a drawing to edit its objects more easily. In Chapter 1, you already saw how the Zoom command is used for this purpose.

1. Click on the Zoom Window tool on the Standard toolbar or type **z↵w↵**. You can also Choose View ➢ Zoom ➢ Window.

2. At the First corner: prompt, pick a point below and to the left of your drawing at coordinate 1.5000,0.6250.
3. At the Other corner: prompt, pick a point above and to the right of the drawing at coordinate 5.1250,4.8750, so that the bracket is completely enclosed by the view window. To obtain this view, use the Zoom Window tool. You can also use the Zoom Realtime tool in conjunction with the Pan Realtime tool. The bracket enlarges to fill more of the screen (see Figure 3.4).

To issue the Zoom Realtime tool from the keyboard, type **z↵↵**.

FIGURE 3.4

A close-up of the details of the drawing

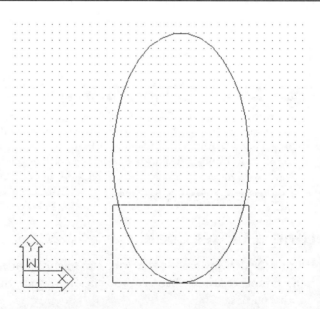

Modifying an Object

Now let's see how editing commands are used to construct an object. To define the bottom area of the bracket, let's create copies of the three lines defining the thickness of the bottom area .125" toward the center.

   1. Click on the Copy Object tool in the Modify toolbar or type **co**↵. You can also select Modify ➤ Copy from the menu bar.

You can also use the Grip edit tools to make the copy. See Chapter 2 for more on Grip editing.

   2. At the Select object: prompt, pick the bottom horizontal line. The line is highlighted. Press ↵ to confirm your selection.

   3. At the <Base point or displacement>/Multiple: prompt, pick a base point near the line.

   4. Type **@.125<90**↵ to tell AutoCAD to use a relative coordinate location of .125 units distance and 90° direction from the base point. Your drawing should look like top of Figure 3.5.

   5. Continue copying the other two sides of the rectangle by using the coordinate location method or the coordinate readout method. You should have a drawing that looks like bottom of Figure 3.5.

You will have noticed that the Copy command acts exactly like the Move command you used in Chapter 2, except that Copy does not alter the position of the objects you select.

FIGURE 3.5
The copied lines

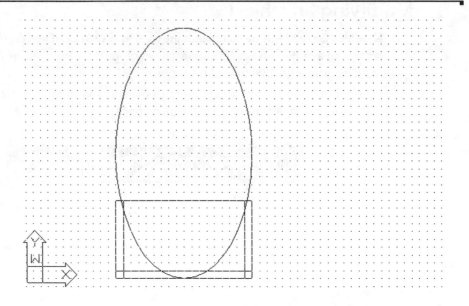

Trimming an Object

Now you must move the ellipse and delete the part that is not needed. You will use the Trim command to trim off parts of the ellipse.

1. Click on the Move tool in the Modify toolbar.
2. At the `Select objects:` prompt, click anywhere on the ellipse, then press ↵ to complete the selection set and go on to the next part of the command.
3. At the `Base point or displacement:` prompt, pick a point near the line.
4. At the `Second point of displacement:` prompt, type **@.75<270**↵.
5. Turn the Snap mode off by pressing F9 or double-clicking on the word SNAP in the status bar. Snap may be a hindrance at this point in your editing session because it may keep you from picking the points you want. Snap mode forces the cursor to move to points at a given interval, so you will have difficulty selecting a point that doesn't fall exactly at one of those intervals.

EXPLORING THE DRAWING PROCESS | **111**

6. Click on the Trim tool in the Modify toolbar.

You will see this prompt:

```
Select cutting edges: (Projmode=UCS, Edgemode=No extend)
Select objects:
```

7. Click on the top line of the rectangle—the one that crosses through the ellipse—and press ↵ to finish your selection.
8. At the <Select object to trim>/Project/Edge/Undo: prompt, pick the bottom portion of the ellipse below the line. This trims the ellipse back to the line.
9. Press ↵ to exit the Trim command.

Selecting Close or Overlapping Objects

At times, you will want to select an object that is in close proximity to or lying underneath another object, and AutoCAD won't obey your mouse click. It's frustrating when you click on the object you want to select, but AutoCAD selects the one next to it instead. To help you make your selections in these situations, AutoCAD provides Object Selection Cycling. To use it, hold down the Ctrl key while simultaneously clicking on the object you want to select. If the wrong object is highlighted, press the left mouse button again (you do not need to hold down the Ctrl key for the second time), and the next object in close proximity will be highlighted. If several objects are overlapping or close together, just continue to press the left mouse button until the correct object is highlighted. When the object you want is finally highlighted, press ↵ and continue with further selections.

In step 6 of the foregoing exercise, the Trim command produces two messages in the prompt. The first message, Select cutting edges..., tells you that you must first select objects to define *the edge to which you wish to trim an object*. In step 8, you are again prompted to select objects, this time to select the *objects to trim*. Trim is one of a handful of AutoCAD commands that asks you to select two sets of objects: The first set defines a boundary, and the second is the set of objects you want to edit. The two sets of objects are not mutually exclusive. You can, for example, select the cutting edge objects as objects to trim. The next exercise shows how this works.

1. Start the Trim tool again by clicking on it in the Modify toolbar.
2. At the Select cutting edges... Select objects: prompt, click on the bottom and left lines that you copied earlier. You are going to trim these lines to create a neat corner.
3. Press ↵ to finish your selection and move to the next step.
4. At the <Select object to trim>/Project/Edge/Undo: prompt, click on the portion of the horizontal line that extends beyond the vertical line. The line trims back. Do the same with the portion of the line that extends below the horizontal line.

NOTE

These Trim options—Project, Edge, and Undo—are described in *The Trim Options* section below.

5. Press ↵ to finish your selection and move to the next step. Press ↵ to begin the command again, but you will not explicitly select new objects.
6. At the <Select object to trim>/Project/Edge/Undo: prompt, press ↵. This null entry option sets all possible objects to be cutting edges, so you can trim a selected object back to the first object AutoCAD finds that can be used as a cutting edge.
7. Click on the end of the horizontal line that extends beyond the right vertical line. The right side of the line trims back to meet the vertical line. Click on the end of the right vertical line. Your drawing should look like Figure 3.6.
8. Press ↵ to exit the Trim command.

Here you saw how the objects are used both as trim objects and as the objects to be trimmed. Next you will move the top horizontal line of the rectangle down .25 units and use it to trim the inside vertical lines (and the horizontal line between the inside vertical lines) to become the top of the bracket ledge.

FIGURE 3.6

Trimming the ellipse and the line

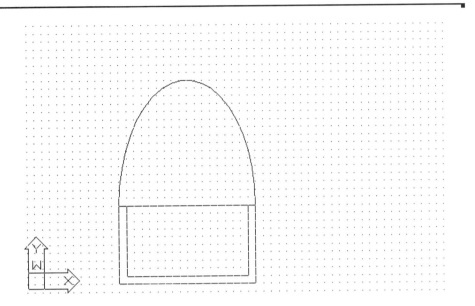

1. Select the Move tool from the Modify toolbar.
2. At the `Select objects:` prompt, select the top horizontal line and press ↵ twice to complete the selection set.
3. At the `Base point or displacement:` prompt, pick a point near the line and at the `Second point of displacement:` prompt, type **@.25<270**↵. Your drawing should look like the top of Figure 3.7.
4. Select the Trim tool from the Modify toolbar.
5. At the `Select cutting edges: (Projmode = UCS, Edgemode = No extend)` prompt, press ↵ to use the Autoselect method.
6. Select the horizontal line anywhere between the two interior vertical lines to trim between them. Select the end of the left vertical line to trim back to the horizontal line. Do the same with the right vertical line. Your drawing should look like the bottom of Figure 3.7.
7. Press ↵ to exit the Trim command.

FIGURE 3.7

Trimming the top of the bracket ledge

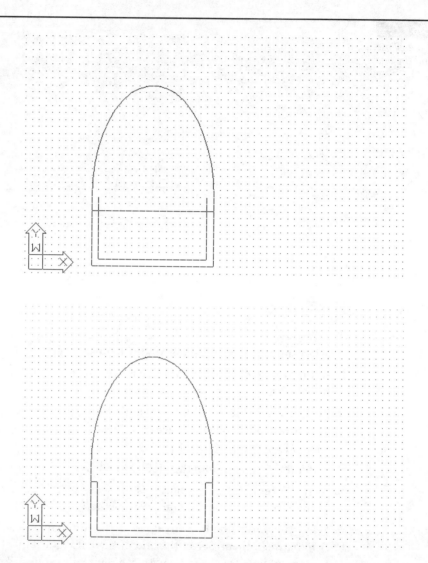

Next you will draw one mounting hole and make a copy of it for the other hole.

1. Select the Circle command from the Draw toolbar.
2. At the 3P/2P/TTR/<Center point>: prompt, Shift+right-click to select the From Osnap. At the Base point: prompt, Shift+right-click to select the Midpoint Osnap. Pick a point along the bottom horizontal line.
3. At the <Offset>: prompt, type **@.875<90**↵ to start the circle .875 units from the bottom middle of the part.

4. At the Diameter/<Radius>: prompt, type **d**↵ for diameter and at the Diameter: prompt, type **.530**↵.
5. Select the Copy tool from the Modify toolbar. At the Select objects: prompt, type **L**↵ to use the last object created, and press ↵ to complete the selection set.
6. At the <Base point or displacement>/Multiple: prompt, select a point near the circle. At the Second point of displacement: prompt, type **@1.625<90**↵.

Your drawing looks nearly complete, except for the arcs called *fillets* at the bottom of the part. It should look like Figure 3.8. Let's talk about the Trim options, and then fillet those corners.

FIGURE 3.8

The bracket with holes added

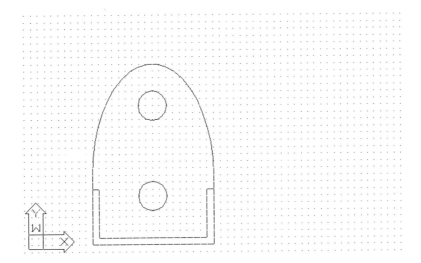

The Trim Options

AutoCAD offers four options for the Trim command: Edge, Project, Undo, and ↵ for edge inference. As described in the following paragraphs, these options give you a higher degree of control over how objects are trimmed.

Edge [E] allows you to trim an object to an apparent intersection, even if the cutting-edge object does not intersect the object to be trimmed (see the top of Figure 3.9). Edge offers two options: Extend and No Extend. These can also be set using the Edgemode system variable.

Project [P] is useful when working on 3D drawings. It controls how Auto-CAD trims objects that are not coplanar. Project offers three options: None, UCS, and View. None causes Trim to ignore objects that are on different planes, so that only coplanar objects will be trimmed. If you choose UCS, the Trim command trims objects based on a plan view of the current UCS and then disregards whether the objects are coplanar or not (see the middle of Figure 3.9). View is similar to UCS but uses the current view's "line of sight" to determine how non-coplanar objects are trimmed (see the bottom of Figure 3.9).

Undo [U] causes the last trimmed object to revert to its original length.

Enter [↵] causes AutoCAD to use the first cutting-edge object that intersects the object you have selected. AutoCAD begins its search for an eligible cutting edge from the point you select on the object. If you have selected an object that is crossed by two eligible cutting edges on either side of the point you selected, AutoCAD will trim the object between the two cutting edges. This process is called *edge inference*.

FIGURE 3.9

The Trim tool's options

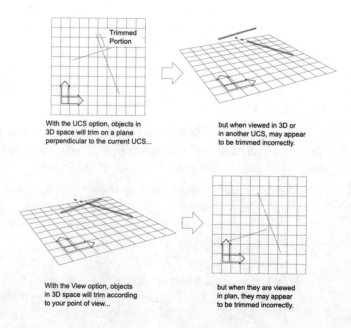

The inside of the detail still has some sharp corners. To round out these corners, you can use the versatile Fillet command (on the Modify toolbar). Fillet allows you to join lines and arcs end to end, and it can add a radius where they join, so there is a smooth transition from arc to arc or line to line. Fillet can join two lines that do not intersect, and it can trim two crossing lines back to their point of intersection.

1. Click on the Fillet tool from the Modify toolbar or type **f**↵. You can also choose Modify ➢ Fillet from the menu bar.

2. At the `Polyline/Radius/Trim/<Select first object>:` prompt, enter **r**↵.
3. At the `Enter fillet radius <0.5000:` prompt, press ↵. By accepting the default, you are telling AutoCAD that you want a .5 unit radius for your fillet. (If the default does not show 0.500, enter **.5**↵.) Look at the drawing of the bracket at the beginning of the exercise. The fillet radius is .625 minus the ledge thickness of .125.
4. Press ↵ to invoke the Fillet command again; this time, pick two inside corner lines. The fillet arc joins and trims the two lines.
5. Press ↵ again and fillet the other inside corner.
6. Press ↵ to invoke the Fillet command again. At the `Polyline/Radius/Trim/<Select first object>:` prompt, enter **r**↵. At the `Enter fillet radius <0'1/2">:` prompt, enter **.625**↵.
7. Press ↵ to invoke the Fillet command again; this time, pick two outside corner lines. The fillet arc joins and trims the two lines.
8. Press ↵ again to fillet the other outside corner Your drawing should look like Figure 3.10.

FIGURE 3.10

The bracket with corners filleted

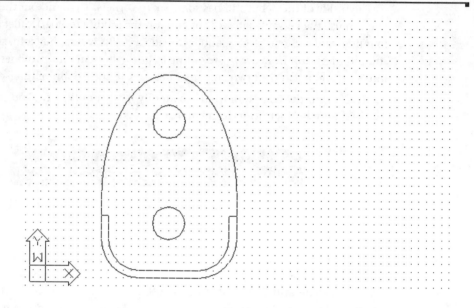

9. Save the DTO100 file.

You've just seen one way to construct these details. However, there are many ways to construct objects. For example, you could have just created an ellipse arc, and you could have used the Grips feature to move the endpoints of the line to meet at the intersection. As you become familiar with AutoCAD, you will start to develop your own ways of working, using the tools best suited to your style.

Planning and Laying Out a Drawing

As a designer, you will often want to take a different look at a design. Let's create another bracket which will look like the drawing in Figure 3.11. You will note that much of the original design is still the same. We will use the parts that are the same and only create the new geometry. Let's look at how to draw the new version. This will help you get a feel for the kind of planning you can do to use AutoCAD effectively. You'll also get a chance to use some of the keyboard shortcuts built into AutoCAD. First, though, go back to the previous view of your drawing, and arrange some more room to work.

PLANNING AND LAYING OUT A DRAWING | 119

FIGURE 3.11

The alternate bracket design

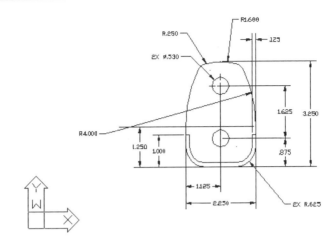

1. Return to your previous view; the one shown in Figure 3.8. A quick way to do this is to click on the Zoom Previous tool on the Standard toolbar or choose View ➢ Zoom ➢ Previous. Your view will return to the one you had before the last Zoom command (Figure 3.12).

FIGURE 3.12

The view of the bracket after using the Zoom Previous tool. You can also obtain this view using the Zoom All tool from the Zoom Window flyout.

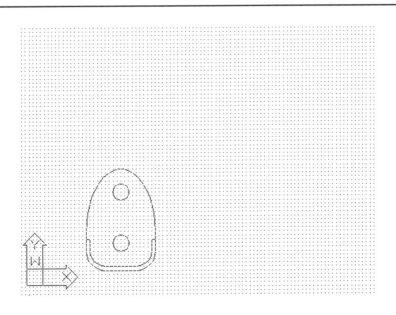

You'll begin the new design by making a copy of the current design.

2. Turn Snap mode on by pressing F9.

3. Type **co**↵. At the `Select objects:` prompt, window the entire bracket or type **all**↵. You can see that all of the bracket is highlighted. Press ↵ again to complete the selection set.

4. At the `Base Point or Displacement:` prompt, pick a point near the object.

5. At the `Second point of displacement:` prompt, check the status bar to see that Ortho is on (double-click the box if it is not on), and drag the cursor to the right. Type **4** and then right-click to create a copy of the bracket four units to the right of the original.

Instead of right-clicking during the direct distance entry method, you can press ↵ or the spacebar. The difference between the two designs seems to be the replacement of the ellipse with arcs. You will now draw one of the large arcs. The arc begins tangent to the side, one unit up from the bottom of the bracket. You do not know exactly how long to draw the arc, so you will draw it longer than it needs to be and trim it back.

1. In the Draw toolbar, click on the Arc tool or type **a**↵. See Figure 3.13 for other Arc options available from the menu bar. This figure shows each pull-down menu option name with a graphic above it depicting the arc, and numbers indicating the sequence of points to select. For example, if you want to know how the Draw ➤ Arc ➤ Start, Center, End option works, you can look to the graphic at the bottom-right corner of the figure. It shows the point-selection sequence for drawing an arc using that option: 1 for the start point, 2 for the center point, then 3 for the end of the arc.

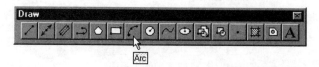

2. At the `Center/<Start point>:` prompt, pick the endpoint of the right side of the ellipse.

3. At the `Center/End/<Second point>:` prompt, type **c**↵ and at the `Center:` prompt, type **@4<180**↵.

4. At the `Angle/Length of chord/<End point>:` prompt, type **a**↵. At the `Included angle:` prompt, type **40**↵ to draw an arc beginning at and tangent to the end of the ellipse and ending at 40°.

5. Type **a**↵, and at the `ARC Center/<Start point>:` prompt, Shift+right-click to select the endpoint. Then select the left endpoint of the ellipse.

FIGURE 3.13

In the Draw ➤ Arc cascading menu, there are some additional options for drawing arcs. These options provide "canned" responses to the Arc command so that you only have to select the appropriate points as indicated by the pull-down menu option name.

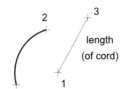
Center, start, length

Center, start, angle

Start, end, Dir

Center, start, end

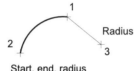
Start, end, radius

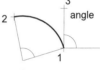
Start, end, angle

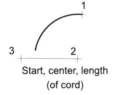
Start, center, length (of cord)

Start, center, angle

3-point

Start, center, end

6. At the Center/End/<Second point>: prompt, type **c**↵, and at the Center: prompt, type **@4<0**↵.

7. At the Angle/Length of chord/<End point>: prompt, type **a**↵. At the Included angle: prompt, type **–40**↵ to draw an arc beginning at and tangent to the end of the ellipse and ending at 40°. The minus sign is required because AutoCAD draws arcs in the counterclockwise direction unless you change this direction in the Units Control dialog box. The minus sign tells AutoCAD to draw the arc in the clockwise direction. Your drawing should look like Figure 3.14.

FIGURE 3.14

The bracket with both arcs

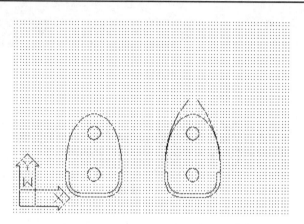

You can use the top of the ellipse to locate another circle. The circle, after it is trimmed, will be the new arc at the top of the bracket. You will need to zoom in a bit to see what is happening next and to select the right points.

1. Type **z**↵.

2. At the First corner: prompt, pick a point below and to the left of your drawing at coordinate 6.000,0.62500. At the Other corner: prompt, pick a point above and to the right of the drawing at coordinate 9.2500,4.8750, so that the bracket is completely enclosed by the view window. To obtain this view, use the Zoom Window tool.

3. Type **c**↵ and at the CIRCLE 3P/2P/TTR/<Center point>: prompt, type **2p**↵ to draw a circle using two points on the diameter.

4. At the First point on diameter: prompt, type **qua**↵ to use the Quadrant object snap.

5. At the of: prompt, select a point near the top of the ellipse. You will see the quadrant marker appear when you are near the top of the ellipse.

6. At the Second point on diameter: prompt, type **@3.376<270**↵. The radius is 1.688, so the diameter is 3.376 and the direction is straight down. Your drawing should look like Figure 3.15.

7. Save the DTO100 file.

FIGURE 3.15
The second bracket with both arcs and the circle

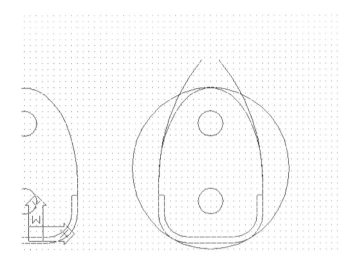

You no longer need the ellipse, as it has served its purpose of providing object snap points for the new arcs. Erase the ellipse.

1. Type **e**↵ to begin the Erase command.

2. Select the ellipse and press ↵ to complete the command.

Trim the two arcs and the circle to each other.

1. Type **tr**↵ to begin the Trim command; select the two arcs and the circle as cutting edges.

2. Pick the circle outside of the arcs. The circle is trimmed back to the arcs.

3. Pick the upper ends of the arcs one at a time. The arcs are trimmed back to what was the circle.

Some of the keyboard shortcuts for tools or commands you've used in this chapter are CO (Copy), E (Erase), EL (Ellipse), F (Fillet), M (Move), O (Offset), and TR (Trim). Remember that keyboard shortcuts, like keyboard commands, can only be entered when the command prompt is visible in the Command window.

Making a Preliminary Sketch

The following exercise will show you how planning ahead can make AutoCAD work more efficiently. When drawing a complex object, you will often have to do some layout before you do the actual drawing. This is similar to drawing an accurate pencil sketch using construction lines that you later trace over to produce a finished drawing. The advantage of doing this layout in AutoCAD is that your drawing doesn't lose any accuracy between the sketch and the final product. Also, AutoCAD allows you to use the geometry of your sketch to aid you in drawing. While planning your drawing, think about what you want to draw, and then decide what drawing elements will help you create that object.

You will use the Offset command to establish reference lines to help you draw a circle tangent to the vertical and horizontal arcs.

Setting Up a Layout

The Offset tool on the Modify toolbar allows you to make parallel copies of a set of objects, such as the arcs detail. When an arc is offset, the radius is either increased or decreased by the offset distance. The centers of the old arc and the new arc are the same point, and the angle subtended by the old arc and the new arc remains the same. When a line is offset, the new line is parallel to and the same length as the original. In Figure 3.16 you can see examples of objects when offset both inside and outside of the original. Offset is different from the Copy command; Offset allows only one object to be copied at a time, but it can remember the distance you specify. The Offset option does not work with all types of objects. Only lines, Xlines, rays, arcs, circles, ellipses, splines, and 2D polylines can be offset.

FIGURE 3.16

Examples of the Offset command on a line, a circle, an arc, a polyline, a polyline rectangle, an ellipse, and a spline

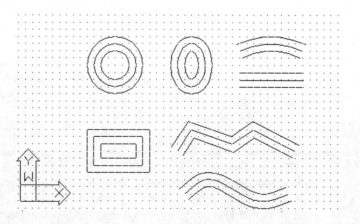

PLANNING AND LAYING OUT A DRAWING

Let's continue with your drawing of the bracket.

1. Click on the Offset tool in the Modify toolbar or type **o**↵. You can also select Modify ➢ Offset from the menu bar.

2. At the Offset distance or Through <Through>: prompt, enter **.25**↵. This enters the distance of .25 as the offset distance, which is the radius of the round.
3. At the Select object to offset: prompt, click on the right side arc.
4. At the Side to offset? prompt, pick a point near and to the left of the right side arc. A copy of the arc appears. You don't have to be exact about where you pick the side to offset; AutoCAD only wants to know on which side of the object you want to make the offset copy.
5. The prompt Select an object to offset: appears again. Click on the top arc and pick a point below the arc to create a new arc. You will have a drawing that looks like Figure 3.17.
6. When you are done, exit the Offset command by pressing ↵.

FIGURE 3.17

The layout with offset arcs

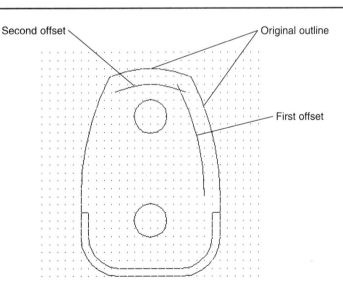

Using the Layout

You have created the intersection of two arcs which you will use to locate the center of a circle tangent to both arcs. You will draw the circle, then trim it and both arcs to create the rounded outside edge.

1. Choose Draw ➢ Circle ➢ Center ➢ Radius or type **c**↵. At the `Circle 3P/2P/TTR/<Center point>:` prompt, type **int**↵ to use the Intersection Osnap.

2. Move the cursor near the intersection of the two inside arcs until you see the intersection marker. Pick this point. At the `Diameter/<Radius>:` prompt, enter **.25**↵ (the radius of the rounded edge is .25 units).

3. Type **z**↵ to use the Zoom Window tool. The null entry default for zoom is Realtime; however, for every rule there is an exception, and if you just pick a point in the edit area or type a 2D point, you can then drag the zoom window. When you pick or type the other corner, AutoCAD will zoom into that window. Drag the cursor close around the circle and arcs. Your view should look like Figure 3.18.

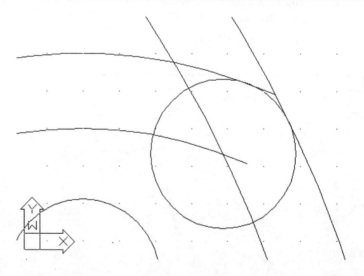

FIGURE 3.18

Using the Zoom Window tool for a closer look at the upper corner of the new bracket

4. Type **e**↵ for Erase; pick the two inside arcs that we have used for our layout and press ↵.

5. Type **tr**↵ for Trim and pick both arcs and the circle at the `Select cutting edges: (Projmode = UCS, Edgemode = No extend)` prompt.

PLANNING AND LAYING OUT A DRAWING | 127

6. See Figure 3.19 to pick the portions of the arcs and circle to be trimmed. First, pick the circle near the bottom; second, pick the horizontal arc near but not on the end; and third, pick a point near the end of the vertical arc.

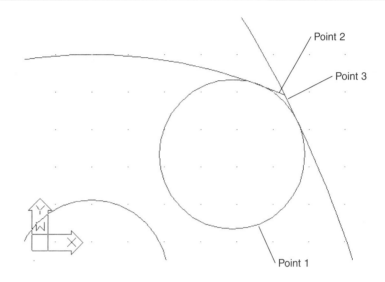

FIGURE 3.19

Pick these points to trim the arcs and the circle

Putting On the Finishing Touches

The process of using the Offset tool to create guide lines is very common in AutoCAD. Many times you will have to choose between using the Offset tool or the Copy tool. Grips, as you will learn in Chapter 9, also give you copy options. You may be asking yourself why there are so many options to do the same thing. Part of the answer is that this is legacy data. These commands reflect how commands were designed in prior releases. Third-party add-on software, as well as customization software, still depends on some of these commands to function as they functioned in the past. The other part of the answer is that different people work in different ways. A new or infrequent user will access the commands differently from a full-time user, and there are many specialties in engineering—each with a slightly different way of creating documents. AutoCAD is used by millions of people around the globe to do all sorts of work, and Autodesk has tried to accommodate as many of these differences as possible.

You will now use the Fillet tool to do the same job we just did with all of the layout and trimming. One way of solving a problem with AutoCAD isn't necessarily better than any other, but some methods may be faster than others.

1. Type **f↵** and at the `(TRIM mode) Current fillet radius = 0.6250 Polyline/Radius/Trim/<Select first object>:` prompt, type **r↵**.

2. At the `Enter fillet radius <0.6250>:` prompt, type **.25↵** for the radius of the round.

3. Press ↵ again to issue the fillet command again. At the `(TRIM mode) Current fillet radius = 0.2500 Polyline/Radius/Trim/<Select first object>:` prompt, pick the top and the left side arcs. AutoCAD draws the new arc tangent to both existing arcs at a radius of .25 units. Your drawing is now complete and should look like Figure 3.20.

4. Save your drawing and exit AutoCAD.

FIGURE 3.20

The complete bracket after filleting

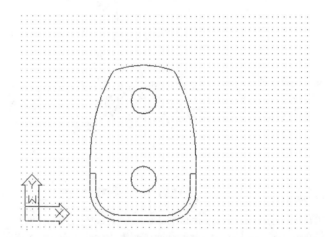

Using AutoCAD's Automatic Save Feature

As you work with AutoCAD, you may notice that AutoCAD periodically saves your work for you. Your file is saved not as its current file name, but as a file called `Auto.SV$`. The default time interval between automatic saves is 120 minutes. You can change this interval by doing the following:

1. Enter **savetime**↵ at the command prompt.

2. At the `New value for SAVETIME < 120 >:` prompt, enter the desired interval in minutes. Or, to disable the Automatic Save feature entirely, enter 0 at the prompt.

If You Want to Experiment...

As you draw, you will notice that you alternate between creating objects, and then copying and editing them. This is where the difference between hand-drafting and CAD really begins to show.

Try drawing the part shown in Figure 3.21. The figure shows you what to do, step by step. Notice how you are applying the concepts of layout and editing to this drawing.

FIGURE 3.21

Drawing a section view of a wide flange beam. Notice how objects are alternately created and edited instead of simply drawing each line segment of the wide flange.

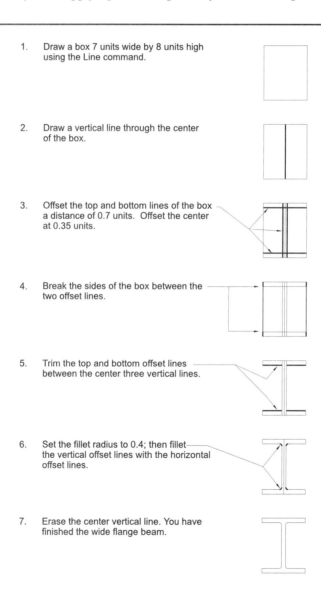

1. Draw a box 7 units wide by 8 units high using the Line command.

2. Draw a vertical line through the center of the box.

3. Offset the top and bottom lines of the box a distance of 0.7 units. Offset the center at 0.35 units.

4. Break the sides of the box between the two offset lines.

5. Trim the top and bottom offset lines between the center three vertical lines.

6. Set the fillet radius to 0.4; then fillet the vertical offset lines with the horizontal offset lines.

7. Erase the center vertical line. You have finished the wide flange beam.

Chapter 4

Organizing Your Work

FEATURING

Creating Symbols Using Blocks

Placing a Block

Restoring Erased Objects

Breaking Down a Block into its Component Objects

Creating Layers and Assigning Objects to Them

Controlling Color

Controlling Layers

Using Linetypes

Inserting Symbols with Drag and Drop

Chapter 4

Organizing Your Work

Drawing the two brackets in Chapter 3 may have taken what seemed an inordinate amount of time. As you continue to use AutoCAD, however, you will learn to draw objects more quickly. You will also need to draw fewer of them because you can save drawings as symbols to be used like rubber stamps, duplicating drawing information instantaneously wherever it is needed. This will save you a lot of time when you're composing drawings.

To make effective use of AutoCAD, you should begin a *symbols library* of drawings you use frequently. A mechanical designer might have a library of symbols for fasteners, cams, valves, or other kinds of parts for his or her application. An electrical engineer might have a symbols library of capacitors, resistors, switches, and the like. And a circuit designer will have yet another unique set of frequently used symbols. On this book's companion CD-ROM, you'll find a variety of ready-to-use symbols libraries. Check them out—you're likely to find some you can use.

In Chapter 3 you drew two objects. In this chapter, you will see how to create symbols from those drawings. You will also learn about layers and how you can use them to organize information.

Creating a Symbol

To save a drawing as a symbol, use the Block tool. In word processors, the term *block* refers to a group of words or sentences selected for moving, saving, or deletion. A block of text can be copied elsewhere within the same file, to other files, or to a separate file on disk for future use. AutoCAD uses blocks in a similar fashion. Within a file, you can turn parts of your drawing into blocks that can be saved and recalled at any time. You can also use entire existing files as blocks.

1. Start AutoCAD and open the existing DT0100 file. Use the one you created in Chapter 3, or open 4-DT0100.dwg on the companion CD-ROM. The drawing appears just as you left it in the last session.

2. In the Draw toolbar, click on the Make Block tool or type **b**↵, the keyboard shortcut for the Make Block tool.

The Block Definition dialog box appears.

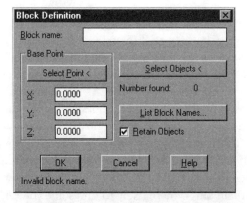

3. Type **bkt_v1** into the Block name input box.
4. In the Base Point button group of the dialog box, click on Select Point. This option enables you to select a base point for the block using your cursor. (The insertion base point of a block is similar to the base point you used as a handle

on an object in Chapter 2.) When you've selected this option, the dialog box will temporarily disappear.

NOTE

Notice that the Block Definition dialog box gives you the option to specify the X, Y, and Z coordinates for the base point, instead of selecting a point.

5. Using the Midpoint Osnap, pick the midpoint of the bottom of left bracket (the bracket that you made with the ellipse) as the base point. If you cannot see all of the bracket on the left, you can use the scroll bar at the bottom of the edit area to scroll the part into view. As you click on the arrow at either end of the scroll bar, your view changes. Remember that you learned how to use an Osnap in Chapter 3; press Shift and right-click to pop the menu and pick the Midpoint option or type **m**↵. All you need to do is point to the midpoint of a line to display the Midpoint Osnap marker, then left-click on your mouse. Once you've selected a point, the Block Definition dialog box will reappear.

6. Next, select the actual objects that you want as part of the block. Click on the Select Objects button. Once again, the dialog box will momentarily disappear. You now see the familiar object-selection prompt in the Command window. Click on a point below and to the left of the left bracket. Then window the entire left bracket; it will be highlighted.

WARNING

Make sure you use the Select Objects option in the Block Definition dialog box to select the objects you want to turn into a block. AutoCAD will let you create a block that contains no objects. This can cause some confusion and frustration, even for an experienced user.

7. Press ↵ to confirm your selection. The dialog box appears again. Make sure the Retain Objects option is checked, and then click on OK. The first bracket drawing is now a block with the name bkt_v1.

8. Repeat the blocking process for the bracket, but this time use the top circle of the bracket as the insertion base point and give the block the name **bkt_v2**. Use the Center Osnap to set the base point at the center of the top circle in the bracket on the right.

NOTE You can press ↵ or right-click on the mouse to start the Make Block tool again.

When you turn an object into a block, it is stored within the drawing file, ready to be recalled at any time. The block remains part of the drawing file even when you end the editing session. When you open the file again, the block will be available for your use. A block acts like a single object, even though it is really made up of several objects. A block can only be modified by unblocking it using the Explode tool in the Modify toolbar. You can then edit it and turn it back into a block. We will look at the block-editing process later in this chapter.

Restoring Objects that Have Been Removed by the Make Block Tool or the Block Command

In prior versions of AutoCAD, the Block command was the only command available to create blocks. This is a command-line version of the Make Block tool and is still available to those users who are more comfortable entering commands via the keyboard. When you use the Block command, the objects you turn into a block will automatically disappear. Release 14's new Make Block tool (Bmake) gives you the option of removing or maintaining the block's source objects by way of the Retain Objects option shown above in step 7.

If you use the Block command, or if for some reason you leave the Retain Objects option unchecked in the Block Definition dialog box, the source objects you select for the block will disappear. You can restore the source using the Oops command. Oops can also be used in any situation where you want to restore an object you accidentally erased. To use it, simply type **oops**↵ at the command prompt. The source objects reappear in their former condition and location, not as a block.

Inserting a Symbol

A block can be recalled at any time, and as many times as you want, as long as you are in the file where it was created. In the following exercise you'll draw a support rail for the bracket.

INSERTING A SYMBOL   137

1. First, delete the bracket drawings. Click on the Erase tool in the Modify toolbar; then enter **all**↵↵. This erases the entire visible contents of the drawing. (It has no effect on the blocks you created previously.)

2. Turn Grid and Snap off. Look in the status bar at the bottom of the screen. If the words are black (instead of gray), double-click in the box until the words become gray.

TIP

If you're in a hurry, enter **insert**↵ at the command prompt, then enter **bkt_1**↵, and then go to step 6.

3. Draw a rectangle 60 × 5. Orient the rectangle so that the long sides go from left to right and the lower-left corner is at coordinate 2.000,3.000. If you draw the rectangle using the Rectangle tool from the Draw toolbar, make sure you explode it using the Explode tool from the Modify toolbar. This is important for later exercises. Use Zoom All (**z**↵**a**↵) to see all of the rectangle. Your drawing should now look like Figure 4.1.

FIGURE 4.1
The outline of the support rail

NOTE

The Insert Block tool can also be found in the Insert toolbar. The Insert toolbar can be opened using the Toolbars dialog box. To open the Toolbars dialog box, right-click on any open toolbar or choose View ➢ Toolbars.

4. In the Draw toolbar, click on the Insert Block tool or type **i**↵.

The Insert dialog box appears.

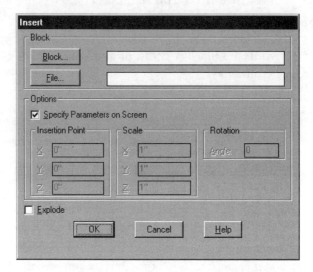

5. Click on the Block button at the top of the dialog box. The Defined Blocks dialog box appears, with a list of the available blocks in the current drawing.

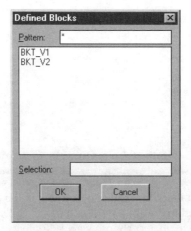

6. Double-click on the block named BKT_V1. The Insert dialog box returns with BKT_V1 in the input box next to the Block button.

7. Click on OK and you see a preview image of the bracket attached to the cursor. The lower-left corner you picked for the bracket's base point is now on the cursor intersection.

INSERTING A SYMBOL 139

8. At the Insertion point: prompt, pick a point above the left end of the bar. Once you've picked the insertion point, notice that as you move your cursor, the preview image of the bracket appears distorted.

9. At the X scale factor <1> / Corner / XYZ: prompt, press ↵ to accept the default, 1. At the Y scale factor (default=X): prompt, press ↵ to accept (default=X). This means you are accepting that the X scale equals the Y scale, which in turn equals 1.

NOTE

The X scale factor and Y scale factor prompts let you stretch the block in one direction or another. You can even specify a negative value to mirror the block. The default on these prompts is always 1, which inserts the block or file at the same size as it was created.

10. At the Rotation angle <0>: prompt, press ↵ to accept the default of 0. You should have a drawing that looks like the image of Figure 4.2.

FIGURE 4.2

The bar with the bracket inserted

As you moved the cursor in step 8, the bracket became distorted. This demonstrates how the X and Y scale factors can affect the item being inserted. Also, in step 11, you can see the bracket rotate as you move the cursor. You can pick a point to fix the block in place, or you can enter a rotation value. The default 0° angle inserts the block or file with the orientation at which it was created.

Using an Existing Drawing as a Symbol

Now you need a bolt head to bolt the bracket to the bar. We have created a bolt head symbol in another drawing and saved in on the CD. This is just an AutoCAD drawing exactly like the ones that you have been creating and saving. You can bring the bolt head into this drawing file and use it as a block.

1. In the Draw toolbar, click on the Insert Block tool or type **i**↵.
2. In the Insert dialog box, click on the File button just below the Block button. The Select Drawing File dialog box appears.
3. Locate the `Chapter_4_bolt` file in the Hardware subdirectory of your CD and double-click on it. You can insert a file as a symbol from any drive on your system or your network.

You can also browse your hard disk by looking at thumbnail views of the drawing files in a directory. See *Finding Files on Your Hard Disk* later in this chapter.

4. When you return to the Insert dialog box, click on OK. As you move the cursor around, you will notice the bolt appears centered on the cursor intersection, as in Figure 4.3.

FIGURE 4.3

The bolt head

5. Pick a point to the right of the bracket and above the bar to insert the bolt.
6. If you use the default setting for the X scale of the inserted block, the bolt will be inserted at the size it was drawn. However, as mentioned earlier, you can specify a smaller or larger size for an inserted object.
7. Press ↵ three times to accept the default Y = X and the rotation angle of 0°.

FIGURE 4.4

The bar with the bracket and bolt head drawing being inserted

The bracket is to mounted to the bar with the bolt. You must draw a circle representing a hole in the bar and place it 3 units from the left of the bar and 3 units from the bottom of the bar. You will need to use object snaps to do this. In this exercise, you'll set up some of the Osnap tools to be available automatically whenever AutoCAD expects a point selection.

1. Choose Tools ➢ Object Snap Settings or type **os**↵. This opens the Osnap Settings dialog box.

2. Make sure the Running Osnap tab is selected, and then click on the check boxes labeled Endpoint, Midpoint, and Intersection so that an X appears in each box; then click on OK.

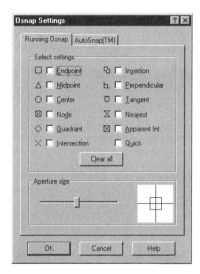

NOTE Take a look at the graphic symbols next to each of the Osnap options in the Osnap Settings dialog box. These are the Osnap markers that appear in your drawing as you select Osnap points. Each Osnap option has its own marker symbol.

Understanding the AutoSnap Settings in the Osnap Settings Dialog Box

When you select the AutoSnap tab of the Osnap Settings dialog box, you will see a set of options pertaining to the new Release 14 AutoSnap feature. AutoSnap looks at the location of your cursor during Osnap selections and locates the Osnap point nearest your cursor. AutoSnap then displays a graphic called a *marker* showing you the Osnap point it has found. If it is the one you want, simply left-click your mouse to select it.

The AutoSnap settings allow you to control its various features. The following is a listing of each setting and its purpose.

Marker   turns the graphic marker on or off.
Magnet   causes the Osnap cursor to "jump to" inferred Osnap points.
Snap Tip   turns the Osnap tool tip on or off.
Display aperture box   turns the old-style Osnap cursor box on or off.
Marker size   controls the size of the graphic marker.
Marker color   controls the color of the graphic marker.

If you prefer the old method of using the Osnaps, you can turn off Marker, Magnet, and Snap Tip, and then turn on Display Aperture Box. The Osnaps will then work as they did prior to Release 14. If you have problems seeing the graphic marker, you may want to change its color using the Marker Color control.

You've just set up the Endpoint, Midpoint, and Intersection Osnaps to be on by default. This is called a *Running Osnap*: AutoCAD will automatically select the nearest Osnap point without your intervention. Now let's see how Running Osnaps work.

TIP When you see an Osnap marker on an object, you can have AutoCAD move to the next Osnap point on the object by pressing the Tab key. If you have several Running Osnap modes turned on (such as Endpoint, Midpoint, and Intersection), pressing the Tab key will cycle through those Osnap points on the object. This feature can be especially useful in a crowded area of a drawing.

The Osnap Options

In the previous exercise, you made several of the Osnap settings automatic so that they were available without your having to select them from the Osnap pop-up menu. Another way to invoke the Osnap options is by typing in their keyboard equivalents while selecting points.

Here is a summary of all the available Osnap options, including their keyboard shortcuts. You've already used many of these options in this chapter and in Chapter 3. Pay special attention here to those options you haven't yet used in the exercises but which you may find useful to your style of work. The full name of each option is followed by its keyboard shortcut name in brackets. To use these options, you can enter either the full name or abbreviation at any point prompt. You can also pick these options from the pop-up menu obtained by pressing Shift and clicking on the right mouse button.

Apparent Intersection [apint] selects the apparent intersection of two objects. This is useful when you want to select the intersection of two objects that do not actually intersect on the same plane. You will be prompted to select the two objects.

Center [cen] selects the center of an arc or circle. You must click on the arc or circle itself, not its apparent center.

Endpoint [endp] selects all the endpoints of lines, polylines, arcs, curves, and 3Dface vertices.

Continued

CONTINUED

From [fro] selects a point relative to a picked point. For example, you can select a point that is 2 units to the left and 4 units to the right of a circle's center.

Insert [ins] selects the insertion point of text, blocks, Xrefs, and overlays.

Intersection [int] selects the intersection of objects.

Midpoint [mid] selects the midpoint of a line or arc. In the case of a polyline, it selects the midpoint of the polyline segment.

Nearest [nea] selects a point on an object nearest the pick point.

Node [nod] selects a point object.

None [non] temporarily turns off running Osnaps.

Perpendicular [per] selects a position on an object that is perpendicular to the last point selected. Normally, this option is not valid for the first point selected in a string of points.

Quadpoint [qua] selects the nearest cardinal (north, south, east, or west) point on an arc or circle.

Quick [qui] improves the speed at which AutoCAD selects geometry, by sacrificing accuracy. Use Quick in conjunction with one of the other Osnap options. For example, to speed up the selection of an intersection, you would enter **quick,int**↵ at a point prompt, and then select the intersection of two objects.

Tangent [tan] selects a point on an arc or circle that represents the tangent from the last point selected. Like the Perpendicular option, Tangent is not valid for the first point in a string of points.

Sometimes you'll want to have one or more of these Osnap options available as the default selection. You can set an Osnap to be on at all times (called a Running Osnap). Choose Tools ➢ Object Snap Settings from the menu bar.

1. Click on View ➢ Zoom ➢ Window and window the area at the left end of the bar. Notice that as you get near any of the objects, their markers appear. Be careful not to pick any of these objects because your zoom window will snap to them and you will not get the results that you want. Your drawing should look like Figure 4.5.

2. Click on the Circle tool from the Draw toolbar. At the 3P/2P/TTR/<Center point>: prompt, move the cursor toward the lower-left corner of the bar. The square endpoint maker appears. If you continue to move the cursor, you will see the center and midpoint markers appear as you approach these points along their respective objects. Return to the lower-left corner of the bar, and when the endpoint marker appears, left-click to accept this point.

FIGURE 4.5

The enlarged bar

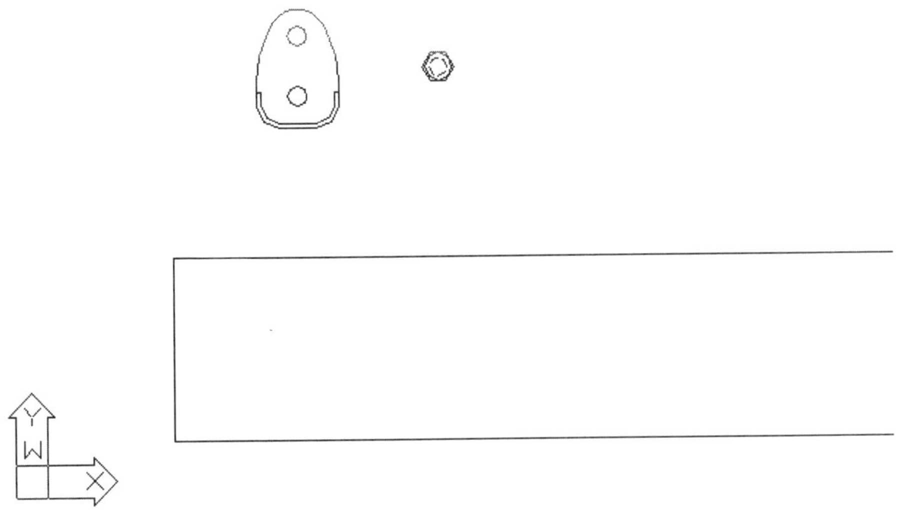

3. At the Diameter/<Radius>: prompt, type **d**↵ (so that you input a diameter instead of a radius) and enter **.5**↵ at the Diameter prompt.
4. Click on Move from the Modify toolbar or type **m**↵.
5. To pick the circle you just created, at the Select objects: prompt click on the circle.
6. At the Base point: prompt, pick any point near the circle and at the Second point: prompt, enter **@3,3**↵ to move the circle 3 units positive X and 3 units positive Y. Your drawing should look like Figure 4.6.

FIGURE 4.6

The circle positioned on the bar

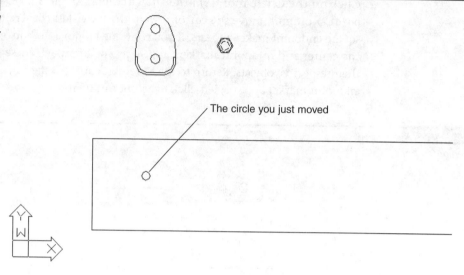

Now you will move the bracket into position on the circle.

1. Click on the Move tool from the Modify toolbar. At the `Select objects:` prompt, pick anywhere on the bracket block. Notice that the entire block is highlighted, not just the single object that you selected. The block is now a single object. Press ↵ to complete the selection set.

2. At the `Base point or displacement:` prompt, move the cursor to the top circle until the round Osnap marker appears. Left-click to accept this location.

3. At the `Second point of displacement:` prompt, move the cursor which drags the bracket. Notice how the Magnet function and the Osnap markers are working. Move the cursor until the center marker of the new circle appears, and left-click to accept this object snap. The bracket moves into position on the bar.

4. Use the same procedure to move the bolt head into position on the top hole of the bracket.

5. Click on the Copy tool from the Modify toolbar. At the `Select objects:` prompt, pick the bracket. Do not use a window, because this would also acquire the bolt and circle. At the `Select objects Other corner: 1 found Select objects:` prompt, press ↵.

6. At the `<Base point or displacement>/Multiple:` prompt, pick a point near the bracket.

7. At the `Second point of displacement:` prompt, type **@9<0↵** to create a copy 9 units to the right of the original. Your drawing should look like Figure 4.7.

FIGURE 4.7

The bar and bracket assembly thus far

Unblocking and Modifying a Block

To modify a block, you break it down into its components, edit them, and then turn them back into a block. This is called *redefining* a block. If you redefine a block that has been inserted into a drawing, each occurrence of that block within the current file will change to reflect the new block definition. You can use this block redefinition feature to make rapid changes to a design.

If the Regenauto setting is turned off, you will have to issue a Regen command to see changes made to redefined blocks. See Chapter 6 for more on Regenauto.

To separate a block into its components, use the Explode command.

1. Click on Explode from the Modify toolbar. You can also type **x↵** to start the Explode command.

2. Click on the right-hand bracket and press ↵ to confirm your selection.

TIP You can simultaneously insert and explode a block by clicking on the Explode check box in the lower-left corner of the Insert dialog box.

Now you can edit the individual objects that make up the bracket. In this case, you will move the lower hole in the bracket up .25 units. Then you will turn the individual objects back into a block, which will update the other copy of the block on the left.

3. From the Modify toolbar select Move, and at the `Select objects:` prompt pick the lower circle on the exploded block; press ↵ to complete the selection set. At the `Base point or displacement:` prompt, pick near the circle and at the `Second point of displacement:` prompt, type **@.25<90**↵.

You are now ready to redefine the block.

4. On the Draw toolbar, select Make Block or type **b**↵.
5. In the Block Definition dialog box, enter **bkt_v1** for the block name.
6. Click on the Select Point button and pick the center of the upper circle.
7. Click on the Select Objects button and select the components of the exploded bracket. Press ↵ when you've finished making your selection.
8. Now click on OK. You will see a Warning that says, "A Block with this name already exists in the drawing. Do you want to redefine it?" You don't want to accidentally redefine an existing block. In this case, you know you want to redefine the bracket, so click on the Redefine button to proceed.

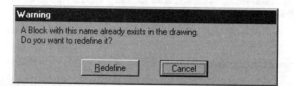

Two things happened when you redefined the block (see Figure 4.8). First, the new base point caused the other copy of the block to move. The new base point is different, but AutoCAD remembers the insertion point relative to the original base point. Second, the lower hole in the copy has moved up. You can check the alignment of the holes after you have moved the displaced copy into the correct position.

9. From the Modify toolbar, click on the Move tool and use the Running Osnaps to select the top circle as the base point and the circle on the bar as the second point of displacement. Erase the old bracket still remaining in the drawing.

10. Now insert the bracket block again, using the Block button in the Insert dialog box. This time however, use the From Osnap from the pop-up menu (Shift+right-click) and select the top hole on the bracket block. At the `Offset distance:` prompt, type **@9<0** to place the new block 9 units to the right. Press ↵ until you have accepted the rest of the defaults.

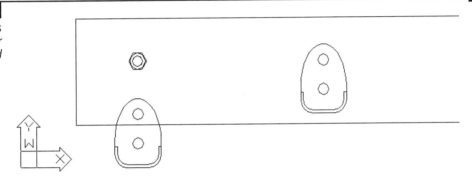

FIGURE 4.8

The bar and blocks positioned after they were redefined

In step 8, you received a Warning that you were about to redefine the existing block. But you had inserted the bracket as a file, not as a block. Whenever you insert a drawing file using the Insert Block tool, the drawing automatically becomes a block in the current drawing. When you redefine a block, however, you do not affect the drawing file you imported. AutoCAD only changes the block within the current file. Next you'll see how you can update an external file with a redefined block.

You may nave noticed that both the Select Objects button and the Select Point button are both followed by the less-than symbol (<). Whenever you see this symbol, you know that by selecting that option, AutoCAD will temporarily close the dialog box and allow you to select objects, pick points, or perform other operations that require a clear view of the drawing area.

Saving a Block as a Drawing File

You've seen that, with very little effort, you can create a symbol that can be placed anywhere in a file. Suppose you want to use this symbol in other files. When you create a block using the Block command, the block exists within the current file only until you specifically instruct AutoCAD to save it as a drawing file on disk. For an existing drawing that has been brought in and modified, such as the bracket, the drawing file on disk associated with that bracket is not automatically updated. To update the bracket file, you must take an extra step and use the Export option on the File menu. Let's see how this works.

Start by turning the two bracket blocks into individual files on disk.

1. Click on File ➤ Export or type **exp**↵. The Export Data dialog box opens. This dialog box is a simple file dialog box.

If you prefer, you can skip step 2, and then in step 3 enter the full file name, including the .dwg extension, as in **bkt_v1.dwg.**

2. Open the List Files of Type drop-down list and select Block (*.dwg).
3. Double-click on the File Name input box and enter **bkt_v1**.
4. Click on the Save button. The dialog box closes.
5. At the Block name: prompt, enter the name of the block you wish to save on disk as the bracket file—in this case, also **bkt_v1**. The first bracket block is now saved as a file.

AutoCAD gives you the option to save a block's file under the same name as the original block or under a different name. Usually you will want to use the same name, which you can do by entering an equals sign (=) after the prompt.

Normally, AutoCAD will save a preview image with a file. This allows you to preview a drawing file prior to opening it. Preview images are not included with files that are exported with the File ➤ Export option or the Wblock command, which is discussed in the next section.

Replacing Existing Files with Blocks

The *Wblock* command does the same thing as File ➤ Export, but output is limited to AutoCAD DWG files. (Veteran AutoCAD users will want to note that Wblock is now incorporated into the File ➤ Export option.) Let's try using the Wblock command this time to save the second bracket block.

1. Issue the Wblock command by typing **wblock**↵ or use the keyboard shortcut by typing **w**↵.
2. At the Create Drawing File dialog box, enter the file name **bkt_v2**.
3. You'll see a prompt asking for a block name. Enter **bkt_v2**↵.
4. Save the current drawing.

In this exercise, you typed the Wblock command at the command prompt instead of using File ➤ Export. The results are the same, regardless of which method you use. If you are in a hurry, the Wblock command is a quick way to save part of your drawing as a file. The File ➤ Export option may be easier for new users to remember.

Other Uses for Blocks

So far, you have used the Make Block tool to create symbols, and the Export and Wblock commands to save those symbols to disk. As you can see, symbols can be created and saved at any time while you are drawing. You have made the bracket symbols into drawing files that you can see when you check the contents of your current directory.

However, creating symbols is not the only use for Insert Block, Block, Export, and Wblock. You can use them in any situation that requires grouping objects (although you may prefer to use the more flexible Object Group command discussed in the next section). You can also use blocks to stretch a set of objects along one axis. Export and Wblock also allow you to save a part of a drawing to disk. You will see instances of these other uses of Block, Export, and Wblock throughout Chapters 5–8.

Make Block, Export, and Wblock are extremely versatile and, if used judiciously, can boost your productivity and simplify your work. Planning your drawings helps you to determine which elements will work best as blocks and to recognize situations where other methods of organization would be more suitable.

> ### An Alternative to Blocks
>
> Another way to create symbols is by creating shapes. Shapes are special objects made up of lines, arcs, and circles. They can regenerate faster than blocks, and they take up less file space. Unfortunately, shapes are considerably more difficult to create and less flexible to use than blocks.
>
> You create shapes by using a coding system developed by Autodesk. The codes define the sizes and orientations of lines, arcs, and circles. First sketch your shape, convert it into the code, and then copy that code into a DOS text file. We won't get into detail on this subject, so if you want to know more about shapes, look in Chapter 3 of your *AutoCAD Customization Manual*.

Another way of using symbols is to use AutoCAD's external reference capabilities. External referenced files, otherwise know as *Xrefs*, are files inserted into a drawing in a way similar to blocks—the difference is that Xrefs do not actually become part of the drawing's database. Instead, they are loaded along with the current file at start-up time. It is as if AutoCAD opens several drawings at once: the main file you specify when you start AutoCAD and any Xrefs associated with the main file.

By keeping Xrefs independent of the current file, you make sure that any changes made to the Xrefs will automatically appear in the current file. You don't have to update the Xrefs as you must for blocks. For example, if you used the External Reference option on the Reference toolbar (to be discussed in Chapter 9) to insert the bolt drawing, and you later made changes to the bolt, the next time you opened the bracket file, you would see the new version of the bolt.

Xrefs are especially useful in workgroup environments, where several people are working on the same project. One person might be updating several files that have been inserted into a variety of other files. Before Xrefs were available, everyone in the workgroup would have had to be notified of the changes and update all the affected blocks in all the drawings that contained them. With Xrefs, the updating is automatic. There are many other features unique to these files, discussed in more detail in Chapter 9.

Grouping Objects

Blocks are an extremely useful tool, but for some situations, they are too restricting. At times, you will want to group objects together so that the objects are connected but can still be edited individually. A group can be copied, then each object can be edited without the group losing its identity. The following exercises demonstrate how this works.

INSERTING A SYMBOL | 153

1. Make a copy of the right bracket block 9 units to the right of the current one and then explode the block.
2. Use the Zoom command to enlarge just the view of the new bracket, as shown in Figure 4.9.

FIGURE 4.9

The bracket after the zoom

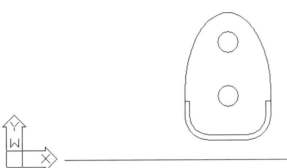

3. Choose Tools ➢ Object Group, or use the keyboard shortcut **g**↵. The Object Grouping dialog box appears.

4. Type **bkt_1_mod**. As you type, your entry appears in the Group Name input box.

5. Click on New in the Create Group button group, near the center of the dialog box. The dialog box temporarily disappears to allow you to select objects for your new group.

6. At the `Select objects:` prompt, window the entire bracket and press ↵. The Object Grouping dialog box returns. Notice that the name BKT_1_MOD appears in the Group Name list box at the top of the dialog box.

7. Click on OK. You have just created a group.

Now, whenever you want to select the bracket, you can click on any part of it and the entire group will be selected. At the same time, you will still be able to modify individual parts of the group—the holes, ellipse, and so on—without losing the grouping of objects.

Modifying Members of a Group

Next, you will remove a notch from the left side of the ellipse. Figure 4.10 is a sketch of the proposed modification.

FIGURE 4.10

A sketch of the modified bracket

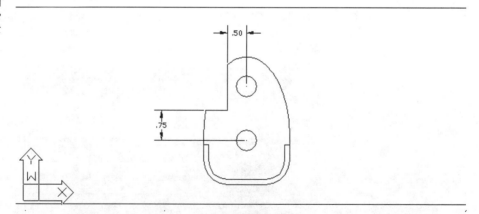

This exercise shows you how to complete your drawing to reflect the design requirements of the sketch.

1. Draw a rectangle in the area of the bracket. The rectangle should be approximately 2 units wide and 3 units high.

2. Move the rectangle so that the lower-right corner is at the center of the lower bracket mounting hole.

3. Move the rectangle again from the current position to @-.5,.75, as shown in the sketch in Figure 4.11.

FIGURE 4.11

The rectangle placed and ready to be used to edit the group

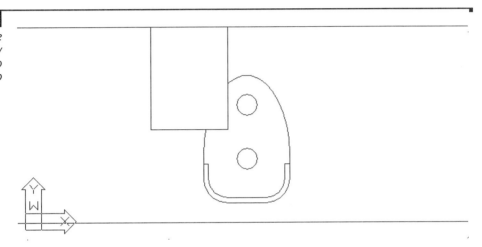

Now that you've got a copy of the bracket as a group and a sketch of the notch, you need to edit the bracket group. If you had used a block for the bracket, you would have had to explode the bracket before you could edit it. Groups, however, let you make changes without undoing their grouping.

1. Press Ctrl+a. This temporarily turns off groupings. You'll see the message <Group off> in the Command window.

2. Using the Trim tool, remove the appropriate parts of the rectangle and the ellipse, as shown in Figure 4.12.

3. Press Ctrl+a again to turn groupings back on. You'll see the message <Group on> in the Command window.

4. To check your bracket, click on one of the objects to see if all of its components are highlighted together. The new parts of the notch and a part of the ellipse are not part of the bracket group. The ellipse arc is now two ellipse arcs, and only one is part of the group.

5. To add the notch and ellipse objects to the group, issue the group command again by pressing Ctrl+a.

6. At the top of the dialog box, pick the bracket group Bkt_V1-Mod and click on the Add button in the Change Group box.

7. The dialog box disappears, and the prompt says Select objects to add. Pick the two lines that make up the notch, and press ↵ to complete the selection set. When the dialog box reappears, click on OK.

8. Save the current drawing.

You can also use the Pickstyle system variable to control groupings. See Appendix D for more on Pickstyle.

FIGURE 4.12

The bracket with the notch removed

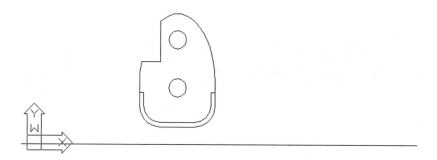

Working with the Object Grouping Dialog Box

Each group has a unique name, and you can also write a brief description of the group in the Object Grouping dialog box. When you copy a group, AutoCAD assigns an arbitrary name to the newly created group. Copies of groups are considered unnamed, but these still can be listed in the Object Grouping dialog box by clicking the Unnamed check box. You can use the Rename button in the Object Grouping dialog box to rename unnamed groups appropriately.

Objects within a group are not bound solely to that group. One object can be a member of several groups, and you can have nested groups.

Here are descriptions of the options available in the Object Grouping dialog box.

Group Identification   Use this button group to identify your groups, using unique elements that help you remember what each group is for.

> The **Group Name**   input box lets you create a new group by naming it first.
>
> The **Description**   input box lets you include a brief description of the group.
>
> The **Find Name**   button lets you find the name of a group by temporarily closing the dialog box so you can click on a group.
>
> The **Highlight**   button highlights a group that has been selected from the group list. This helps you locate a group in a crowded drawing.
>
> The **Include Unnamed**   check box determines whether unnamed groups are included in the Group Name list. Check this box to display the names of copies of groups for processing by this dialog box.

Create Group   Here's where you control how a group is created.

> **New**   lets you create a new group. It temporarily closes the dialog box so you can select objects for grouping. To use this button, you must have either entered a group name or checked the Unnamed check box.
>
> The **Selectable**   check box lets you control whether the group you will create is selectable or not. See the description of the Selectable button in the Change Group just below.
>
> The **Unnamed**   check box lets you create a new group without naming it.

Change Group   These buttons are available only when a group name is highlighted in the Group Name list at the top of the dialog box.

> **Remove**   lets you remove objects from a group.
>
> **Add**   lets you add objects to a group. While you are using this option, grouping is temporarily turned off to allow you to select objects from other groups.
>
> **Rename**   lets you rename a group.
>
> **Reorder**   lets you change the order of objects in a group.
>
> **Description**   lets you modify the description of a group.
>
> **Explode**   separates a group into its individual components.
>
> **Selectable**   turns individual groupings on and off. When a group is selectable, it is selectable as a group. When a group is not selectable, the individual objects in a group can be selected, but not the group.

Organizing Information with Layers

Another tool for organization is the *layer*. Layers are like overlays on which you keep various types of information (see Figure 4.13). Each layer is perfectly registered to the original so that if you are doing a layout of an assembly and wish to create another layer to draw a part, there will be no loss in accuracy. It's also a good idea to keep notes and reference information for the drawing, as well as the drawing's dimensions, on separate layers. As your drawing becomes more complex, the various layers can be turned on and off to allow easier display and modification. If you are making working drawings, you will want to use layers to help you with plotting. A color can be assigned to each layer, as well as a linetype. Most plotters, whether they are of the modern raster type or older pen plotters, use the pen color to assign line weight (width). If you have created a green layer called Dimension, placed all of your dimensions on the Dimension layer, and assigned a .015-wide pen to the color green, the plotter will draw all of the green objects, your dimensions, with a .015-wide pen.

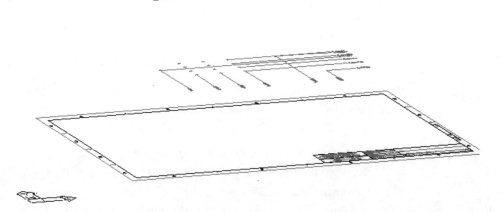

FIGURE 4.13

A comparison of layers and overlays

AutoCAD allows an unlimited number of layers, and you can name each layer anything you want.

ORGANIZING INFORMATION WITH LAYERS | 159

Creating and Assigning Layers

To continue with your bar and bracket drawing, you will create some new layers.

1. Open the Dto100 file you saved earlier in this chapter. If you didn't create one, use the file 4b-dto100.dwg from the companion CD-ROM.

2. To display the Layer & Linetype Properties dialog box, click on the Layers tool in the Object Properties toolbar, or choose Format ➢ Layers from the menu bar, or type **la**↵ to use the keyboard shortcut.

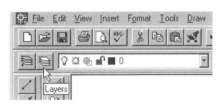

The Layer & Linetype Properties dialog box shows you at a glance the status of your layers. Right now, you only have one layer, but as your work expands, so will the number of layers. You will then find this dialog box indispensable.

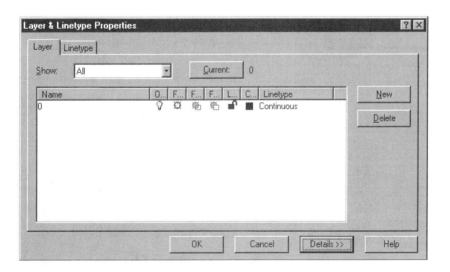

3. Click on the New button in the upper-right corner of the dialog box. A new layer named Layer1 appears in the list box. Notice that the name is highlighted. This tells you that by typing, you can change the default name to something better suited to your needs.

4. Type **Hidden**; as you type, your entry replaces the Layer1 name in the list box.

You can use up to 32 characters, no spaces, to name layers. However, if you are going to use the drawing as an external reference, you should limit the name to eight or fewer characters, because the name of the file will be added to the layer name when the drawing is attached. If the resultant layer name is greater than 32 characters, the attachment won't work. Most AutoCAD users have learned to abbreviate the layer names for this reason.

5. Click on the Details button near the bottom of the dialog box. Additional layer options will appear.

6. With the Hidden layer name highlighted, click on the downward pointing arrow to the right of the Color drop-down list. You see a listing of colors that you can assign to the Hidden layer.

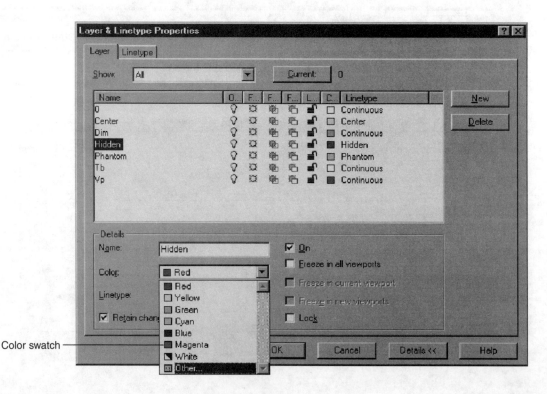

Color swatch

ORGANIZING INFORMATION WITH LAYERS

At first glance, this list may seem a bit limited, but you actually have a choice of 256 colors.

NOTE
Alhough it isn't readily apparent, all the colors in the Color drop-down list, except for the first seven, are designated by numbers. So when you select a color after the seventh color, the color's number—rather than its name—appears in the Color input box at the bottom of the dialog box.

7. Click on Other from the list, or click on the black color swatch in the Hidden layer listing, next to the word Continuous.

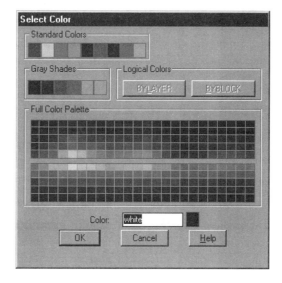

8. In the top row of Standard Colors, click on the red square and then on OK. Notice that the color swatch in the Hidden layer listing is now red. You could have chosen red from the Colors drop-down list, but you selected red from the Select Color dialog box so that you would be aware that many other colors are available.

9. When the Layer & Linetype Properties dialog box returns, click on OK to close it.

The Release 14 Layer & Linetype Properties Dialog Box

The Release 14 Layer & Linetype Properties dialog box conforms to the Windows interface standard. You'll notice that the bar at the top of the layer list box offers several buttons for the various layer properties. Just as you can adjust Windows Explorer, you can adjust the width of each column in the list of layers by clicking and dragging either side of the column head buttons. You can also sort the layer list based on a property simply by clicking on the property name at the top of the list. And, just as with other Windows list boxes, you can Shift+click on names to select a block of layer names, or press Control and click on individual names to select multiples that do not appear together. These features will become helpful as your list of layers expands.

Controlling Layers through the Layer Command

You have seen how the Layer & Linetype Properties dialog box makes it easy to view and edit layer information, and how layer colors can be easily selected from an on-screen toolbar. But layers can also be controlled through the command prompt.

1. First press Esc to make sure any current command is canceled.

2. At the command prompt, enter **–layer**↵. Make sure you include the minus sign in front of the word *layer*. The following prompt appears:

   `?/Make/Set/New/ON/OFF/Color/Ltype/Freeze/Thaw/Lock/Unlock:`

   You'll learn about many of the options in this prompt as you work through this chapter.

3. Enter **n**↵ to select the New option.

4. At the `New layer name(s):` prompt, enter **dim**↵. The ?/Make/Set/New... prompt appears again.

5. Enter **c**↵.

6. At the `Color:` prompt, enter **green**↵. Or you can enter **3**↵, the numeric equivalent of the color green in AutoCAD.

7. At the `Layer Names for color 3 (green) <0>:` prompt, enter **dim**↵. The ?/Make/Set/New... prompt appears again.

8. Press ↵ to exit the Layer command.

Each method of controlling layers has its own advantages: The Layer & Linetype Properties dialog box offers more information about your layers at a glance. On the other hand, the Layer command offers a quick way to control and create layers if you're in a hurry. Also, if you intend to write custom macros, you will want to know how to use the Layer command as opposed to the dialog box, because dialog boxes cannot be controlled through custom toolbar buttons or scripts.

Another advantage to using the keyboard commands is that you can recall previously entered keystrokes by using the ↑ and ↓ cursor keys. For example, to recall the layer named Dim you entered in step 4 above, press the ↓ key until DIM appears in the prompt. This can save time when you are performing repetitive operations such as creating multiple layers. This feature does not work with tools selected from the toolbars.

Assigning Layers to Objects

When you create an object, that object is assigned to the current layer. Until now, only one layer has existed, layer 0—which contains all the objects you've drawn so far. Now that you've created some new layers, you can reassign objects to them using the Properties tool on the Object Properties toolbar. The bar that you have been working on is a channel with the flanges facing away. Let's draw the lines representing the flanges using the Offset tool and then assign them to the Hidden layer.

1. From the Modify toolbar click on the Offset tool. At the `Offset distance or Through <Through>:` prompt, type **.19**↵. Pick the top horizontal line of the bar at the `Object to offset:` prompt, and then pick any point below the line at the side to offset. A new line appears. The command is still working, so pick the bottom line at the `Object to offset:` prompt and choose a point above the line for the side to offset.

2. Click on Properties from the Object Properties toolbar.

3. At the `Select objects:` prompt, click on the two new lines representing the flanges.

4. Press ↵ to confirm your selection. The Change Properties dialog box appears.

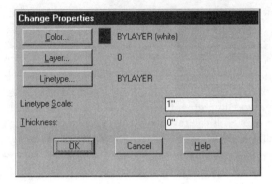

This dialog box allows you to change the layer assignment, color, linetype, and thickness of an object. You'll learn about linetypes later in this chapter.

5. Click on the Layer button. Next you will see the Select Layer dialog box, listing all the existing layers, including the ones you just created.

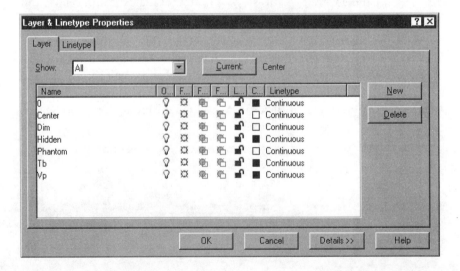

6. Double-click on the Hidden layer. You will return to the Change Properties dialog box.
7. Click on OK to close the dialog box.

The flanges are now on the new layer, Hidden, and the lines are changed to red. Layers are more easily distinguished from one another when you use colors to set them apart.

Next, you will practice the commands you learned in this section by creating some new layers.

1. Bring up the Layer & Linetype Properties dialog box (use Format ➢ Layers or click on the Layers button in the Object Properties toolbar). Create a new layer called **Center** and give it the color yellow.

You can change the name of a layer by clicking on it in the Layer & Linetype Properties dialog box; once it is highlighted, click on it again so that a box surrounds the name. You can then rename the layer. This works in the same way as renaming a file or folder in Windows 95.

Within a block, you can change the color assignment and linetype of only the objects that are on layer 0. See the sidebar *Controlling Colors and Linetypes of Blocked Objects* in this chapter.

2. Use the Layer & Linetype Properties dialog box to create three more layers for phantom lines called **Phantom** (color cyan), for a title block called **Tb** (color white), and for viewports called **Vp** (color magenta), as shown in Table 4.1. Remember that you can open the Select Color dialog box by clicking on the color swatch of the layer listing.

TABLE 4.1: CREATE THESE LAYERS AND SET THEIR COLORS AS INDICATED

Layer Name	Layer Color (number)
Vp	Magenta (6)
Tb	White (7)
Phantom	Cyan (4)

In step 2 above, you used a dialog box that offered several options for modifying the block. When you click on the Properties tool on the Object Properties toolbar, the dialog box displayed will depend on whether you have selected one object or several. With only one object selected, AutoCAD presents options that apply specifically to that object. With several objects selected, you'll see a more limited set of options because AutoCAD can change only the properties that are common to all the objects selected.

NOTE

The Properties button on the Object Properties toolbar issues one of two commands, based on how many objects you selected. Dchprop is the command that opens the Change Properties dialog box; Dmodify opens the Modify Properties dialog box.

Controlling Colors and Linetypes of Blocked Objects

Layer 0 has special importance to blocks. When objects assigned to layer 0 are used as parts of a block, those objects take on the characteristics of the layer on which the block is inserted. On the other hand, if those objects are on a layer other than 0, they will maintain their original layer characteristics even if you insert or change that block to another layer. For example, suppose the bracket is drawn on the Hidden layer, instead of on layer 0. If you turn the bracket into a block and insert it on the Vp layer, the objects the bracket is composed of will maintain their assignment to the Hidden layer, although the bracket block is assigned to the Vp layer.

It might help to think of the Block function as a clear plastic bag that holds together the objects that make up the bracket. The objects inside the bag maintain their assignment to the Hidden layer even while the bag itself is assigned to the Vp layer.

AutoCAD also allows you to have more than one color or linetype on a layer. You can use the Color and Linetype buttons in the Change Properties dialog box (the Object Properties button on the Standard toolbar) to alter the color or linetype of an object on layer 0, for example. That object then maintains its assigned color and linetype—no matter what its layer assignment. Likewise, objects specifically assigned a color or linetype will not be affected by their inclusion in blocks.

Working on Layers

So far you have created layers and assigned objects to those layers. However, the current layer is still 0, and every new object you draw will be on layer 0. Here's how to change the current layer.

   1. Click on the arrow button next to the layer name on the Object Properties toolbar. A drop-down list opens, showing you all the layers available in the drawing.

Notice the icons that appear next to the layer names; these control the status of the layer. You'll learn how to work with these icons later in this chapter. Also notice the box directly to the left of each layer name. This shows you the color of the layer.

   2. Click on the Phantom layer name. The drop-down list closes, and the name Phantom appears in the toolbar's layer name box. Phantom is now the current layer.

NOTE

You can also use the Layer command to reset the current layer. To do this here, enter **–layer**↵ (be sure to include the minus sign) at the command prompt, and at the ?/Make... prompt, enter **s**↵ for set. At the New current layer: prompt, enter **phantom** and then press ↵ twice to exit the Layer command.

   3. Zoom in to the notched bracket, and draw a 5" line, starting at the center of the top mounting hole.

4. Draw a similar line from the bottom mounting hole. Your drawing should look like Figure 4.14.

Because you assigned the color cyan to the Phantom layer, the two lines you just drew are cyan. This gives you immediate feedback about what layer you are on as you draw.

FIGURE 4.14
Bar and bracket with objects on layers

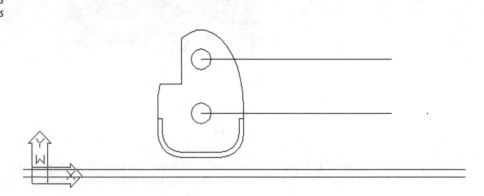

Next, you'll change the layer property of the new line to the Center layer. But instead of using the Properties tool, as you've done in earlier exercises, you'll use a shortcut method.

1. Click on the new cyan-colored line to highlight it. Notice that the layer listing in the Object Properties toolbar changes to Phantom. Whenever you select an object to expose its grips, the Layer, Color, and Linetype listings in the Object Properties toolbar will change to reflect those properties of the selected object.

2. Click on the Layer name in the Object Properties toolbar. The layer drop-down list will open.

3. Click on the Center layer name. The list closes and the line you selected changes to yellow, showing you that it is now on the Center layer. Also, notice that the Color list in the Object Properties toolbar has changed to reflect the new color for the line.

4. Press Esc twice to clear the grip selection. Notice that the layer returns to Phantom, the current layer.

In this exercise, you saw that by selecting an object with no command active, the object's properties are immediately displayed in the Object Properties toolbar in the Layer, Color, and Linetype boxes. Using this method, you can also change an object's color and linetype independent of its layer. Just as with the Properties tool, you can select multiple objects and change their layers through the layer drop-down list.

Controlling Layer Visibility

When you are working on developing a new design, it is often nice to have hidden lines, dimensions, and so on available but off. A word of caution before you get carried away with layer visibility: Objects on layers that are not displayed are not selected—unless they are part of a group or block that is visible. If you have the Hidden layer turned off and you move the rest of the view to a better location, you will find that the hidden lines did not move.

1. Open the Layer & Linetype Properties dialog box by clicking on the Layers tool in the Object Properties toolbar.

2. Click on the Hidden layer in the layer list.

3. Click on the lightbulb icon in the layer list next to the Hidden layer name. You can also click on the check box labeled On in the Details section of the dialog box, so that no check appears there. In either case, the lightbulb icon changes from yellow to gray to indicate that the layer is off.

By momentarily placing the cursor on an icon in the layer, you will get a tool tip giving you a brief description of the icon's purpose.

4. Click on OK to exit the dialog box. When you return to the drawing, the red hidden lines disappear, because you have made these invisible by turning off their layer (see Figure 4.15).

FIGURE 4.15
Bar with the hidden lines turned off

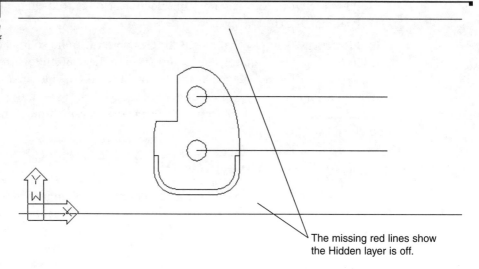

You can also control layer visibility using the layer drop-down list on the Object Properties toolbar.

1. On the Object Properties toolbar, click on the layer drop-down list.
2. Find the Hidden layer and notice that its lightbulb icon is gray. This tells you that the layer is off and not visible.
3. Click on the lightbulb icon to make it yellow.
4. Now click on the drawing area to close the layer list, and the hidden lines reappear.

Figure 4.16 explains the role of the other icons in the layer drop-down list.

FIGURE 4.16
The drop-down layer list icons

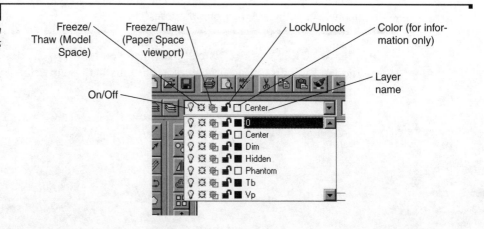

Finding the Layers You Want

With only a handful of layers, it's fairly easy to find the layer you want to turn off. This becomes much more difficult, however, when the number of layers exceeds 20 or 30. The Layer & Linetype Properties dialog box offers some useful tools to help you find the layers you want fast.

Now suppose you have several layers whose names begin with C, such as C-flange, C-grill, and C-seal, and you want find those layers quickly. You can click on the Name button at the top of the layer list to sort the layer names in alphabetical order. Click on the Name button again to reverse the order. To select those layers for processing, click on the first layer name that starts with C; then scroll down the list until you find the last layer of the group and Shift+click on it. All the layers between those layers will be selected. You can also select multiple layer names by holding down the Ctrl key while clicking on each name. If you make a mistake and want to deselect an item, Ctrl+click on it again.

The Color and Linetype buttons at the top of the list let you control what layers appear in the list by color or linetype assignments. Other buttons will sort the list by status: On/Off, Freeze/Thawed, Locked/Unlocked, and so forth. See the *Other Layer Options* sidebar later in this chapter.

NOTE

To delete all the objects on a layer, you can set the current layer to the one you want to edit, and then freeze or lock all the others. Click on Erase in the Modify toolbar and then type **a**↵↵. The *a* issues the All selection option.

You can turn off a set of layers with a single click on a lightbulb icon. You can freeze/thaw, lock/unlock or change the color of a group of layers in a similar manner by clicking on the appropriate layer property. For example, if you had clicked on a color swatch of one of the selected layers, the Select Color dialog box would appear, allowing you to set the color for all the selected layers.

Other Layer Options

You may have noticed the *Freeze* and *Thaw* buttons in the Layer & Linetype Properties dialog box. These options are similar to the On and Off buttons—however, Freeze not only makes layers invisible, it also tells AutoCAD to ignore the contents of those layers

Continued

> **CONTINUED**
>
> when you use the All response to the Select object prompt. Freezing layers can also save time when you issue a command that regenerates a complex drawing. This is because AutoCAD ignores objects on frozen layers during a regen. You will get firsthand experience with Freeze and Thaw in Chapter 6.
>
> Another pair of Layer & Linetype Properties options, *Lock* and *Unlock*, offers a function similar to Freeze and Thaw. If you lock a layer, you will be able to view and snap to objects on that layer, but you won't be able to edit those objects. This feature is useful when you are working on a crowded drawing and you don't want to accidentally edit portions of it. You can lock all the layers except those you intend to edit, and then proceed to work without fear of making accidental changes.

Taming an Unwieldy List of Layers

You may eventually end up with a fairly long list of layers. Managing such a list can become a nightmare, but AutoCAD provides the Layer Filter dialog box to help you locate and isolate only those layers you need to work with.

To use layer filters, click on the Show drop-down list near the top of the Layer & Linetype Properties dialog box. This drop-down list contains options described in Table 4.2. See *Using External References* in Chapter 9 for information on Xref-dependent layers.

TABLE 4.2: THE FILTER OPTIONS

Filter Options	What It Filters
All	All layers regardless of their status
All in use	All layers that have objects assigned to them
All unused	All layers that do not have objects assigned to them
All Xref dependent	All layers that contain Xref objects
All not Xref dependent	All layers that do not contain Xref objects
All that pass filter	All layers that conform to filter criteria set in the Set Filter dialog box
Set Filter Dialog…	Opens the Set Layer Filters dialog box

If you click on the Set Filter Dialog option at the bottom of the list, you open the Set Layer Filters dialog box. Here you can filter out the layers you want to show in the layer list by indicating the layers' characteristics.

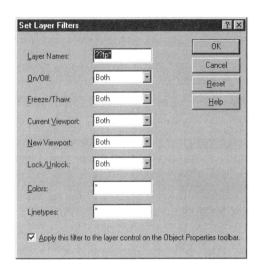

Now suppose you have a drawing whose layer names are set up to help you easily identify groups of layers associated with each part in an assembly layout:

v1-case   v1-cover   v1-chassis   v2 case

v2-cover   v1-panel   v1-screw

If you want to isolate only those layers that have to do with covers, regardless of their assembly, enter **??cover*** in the Layer Names input box, and then click on OK. Now only the layers whose names contain the letters *cover* as their third through sixth characters will appear in the list of layers. You can then easily turn all these layers off, change their color assignment, or change other settings quickly, without having to wade through other layers you don't want to touch.

NOTE

In the ??cover* example, the question marks (??) tell AutoCAD that the first two characters in the layer name can be anything. The "cover" tells AutoCAD that the layer name must contain the word *cover* in these five places of the name. The asterisk (*) at the end tells AutoCAD that the remaining characters can be anything. The question marks and asterisk are known as *wildcard characters*. They are commonly used filtering tools for DOS and Unix operating systems.

The other two input boxes near the bottom, Colors and Linetypes, let you control what layers appear in the list by color or linetype assignments. The five drop-down lists let you filter layers by status: On/Off, Freeze/Thawed, Locked/Unlocked, and so forth. See the *Other Layer Options* sidebar earlier in this chapter.

As the number of layers in a drawing grows, you will find layer filters to be an indispensable tool. But bear in mind that the successful use of the layer filters depends on a careful layer-naming convention.

Assigning Linetypes to Layers

You may have been wondering how to make the lines on the various layers, which seem to be named after linetypes, look like those linetypes. A hidden line, for example, should be made up of a dashed line, and a center line should be a line broken by a dashed line. Drafters use different linetypes to differentiate drafting lines from object lines and other kinds of information. AutoCAD lets you use both color and linetype to indicate the difference between lines. Many people work on a relatively small monitor (relative to a full-sized piece of paper) and need all the help that they can get. Color plotting is often not an option for many reasons. A company may not have a color plotter, or color plots may not be allowed by document control because the company doesn't have a large document color copier. As mentioned earlier, color is still valuable because your plotter still uses color to determine line weight.

You can set a layer to have not only a color assignment but also a linetype assignment. AutoCAD comes with several linetypes, as shown in Figure 4.17. From the top of Figure 4.17 downward, we can see Standard linetypes, then ISO and Complex linetypes, and then a series of lines that can be used to illustrate gas and water lines in civil work, or batt insulation in a wall cavity. ISO linetypes are designed to be used with specific plotted line widths and linetype scales. For example, if you are using a pen width of .5mm, the linetype scale of the drawing should be set to .5 as well (see Chapter 14 for more information on plotting and linetype scale). You can also create your own linetypes (see Chapter 21).

Be aware that linetypes that contain text, such as the gas sample, will use the current text height and font to determine the size and appearance of the text displayed in the line. A text height of 0 (zero) will display the text properly in most cases. See Chapter 7 for more on text styles.

AutoCAD stores linetype descriptions in an external file named ACAD.LIN. You can edit this file in a word processor to create new linetypes or to modify existing ones. You will see how this is done in Chapter 21.

ORGANIZING INFORMATION WITH LAYERS

FIGURE 4.17

Standard AutoCAD linetypes

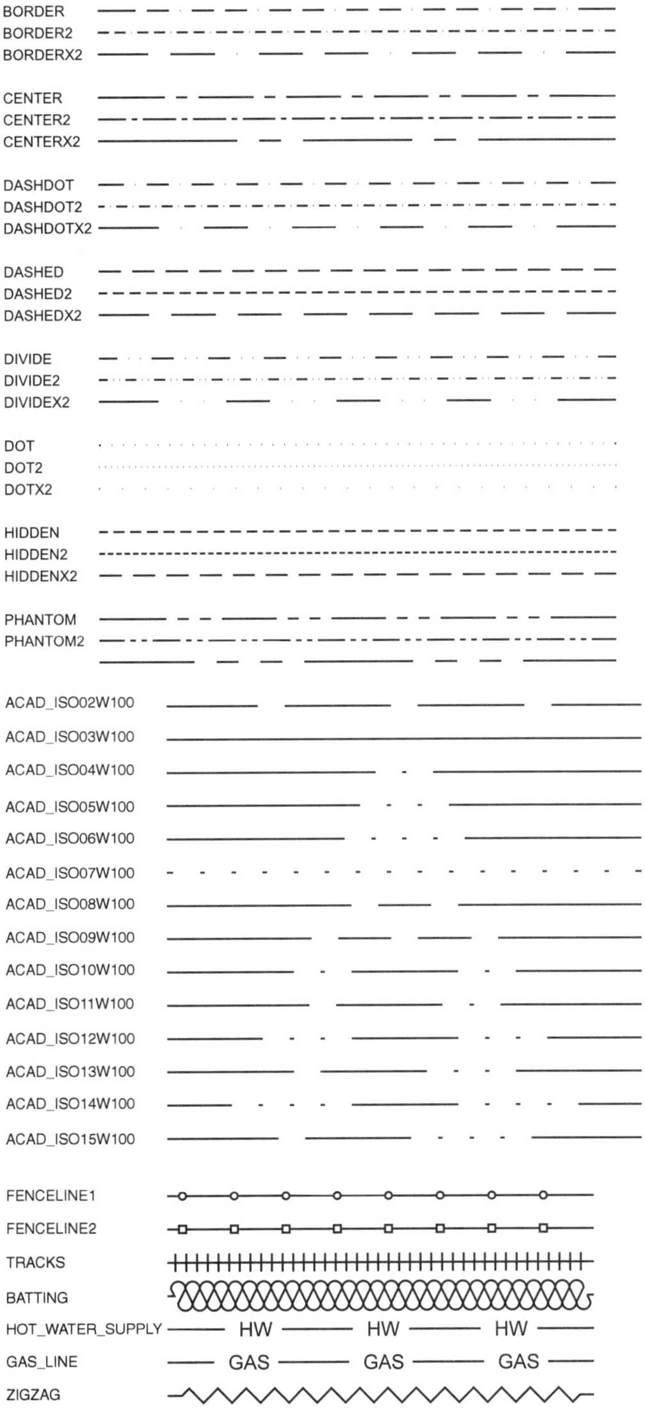

Adding a Linetype to a Drawing

To see how linetypes work, change the linetype of the Hidden layer to a hidden linetype.

1. Open the Layers & Linetype Properties dialog box.

If you are in a hurry, you can simultaneously load a linetype and assign it to a layer by using the Layer command. In this exercise, you would enter **–layer**↵ at the command prompt, then enter **L**↵, **hidden**↵, **hidden**, and then ↵ to exit the Layer command.

2. Click on the word Continuous at the far right of the Hidden layer listing (under the Linetype column).

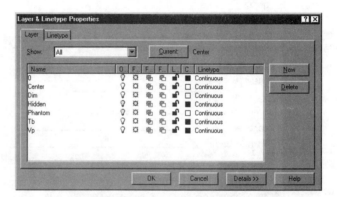

The Select Linetype dialog box appears.

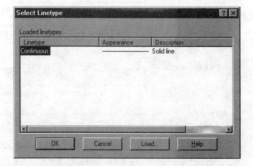

This dialog box offers a list of linetypes to choose from. In a new file, such as the drawing file you have been working on, only one linetype is available by default. You must load any additional linetype you may want to use.

3. Click on the Load button at the bottom of the dialog box. The Load or Reload Linetypes dialog box appears.

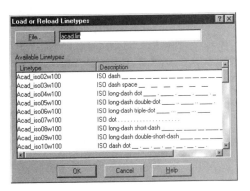

Notice that the list of linetype names is similar to the layer list. You can sort the names alphabetically or by description by clicking on the Linetype or Description buttons at the top of the list.

4. In the Available Linetypes list, scroll down to locate the Hidden linetype, click on it, and then click on OK.

5. Notice that the Hidden linetype is now added to the linetypes available in the Select Linetype dialog box.

6. Click on Hidden to highlight it; then click on OK. Now Hidden appears in the layer listing under Linetype in the Layer & Linetype Properties dialog box.

7. With the Hidden layer still highlighted, click on the Current button to make the Hidden layer current.

8. Click on OK to exit the Layer & Linetype Properties dialog box. The red lines for the flanges have become dashed lines.

Controlling Linetype Scale

Although you have designated that this line is to be a hidden line, it may appear to be solid. Zoom in to a small part of the line, and you'll see that the line is indeed as you specified.

Since you are working at a scale of 1"=1", you may adjust the scale of your linetypes to see them more easily on the screen. You will have to return the linetypes to your plot scale before you print the drawing. This, too, is accomplished in the Layer & Linetype Properties dialog box.

1. Click on the Linetype button in the Object Properties toolbar, or select Format ➢ Linetype. The Layer & Linetype Properties dialog box appears. This is the same dialog box you see when you click on the Layers tool, but in this case the Linetypes tab is visible.

You can also use the Ltscale system variable to set the linetype scale. Type **ltscale**↵, and at the LTSCALE New scale factor <1.0000>: prompt, enter **12**↵.

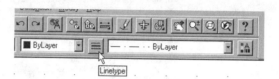

You may notice that the Linetype tab of the Layer & Linetype Properties dialog box also contains the Load and Delete button options that you saw in step 4 of the previous exercise. These offer a way to directly load or delete a linetype without having to go through a particular layer's linetype setting.

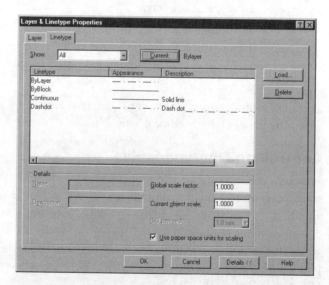

2. Double-click on the Global scale factor input box, and then type **.5**. This is the scale conversion factor for a 1"=.5" scale (see Table 3.3).
3. Click on OK. The drawing regenerates, and the red lines are displayed in the linetype and at the scale you designated. Your drawing looks like Figure 4.18.
4. Choose File ➢ Save to record your work up to now.

FIGURE 4.18

The hidden linetype at a smaller scale

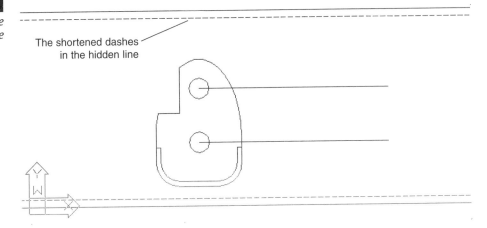

If you change the linetype of a layer or object but the object remains a continuous line, check the Ltscale system variable. It should be set to your drawing scale factor. If this doesn't work, set the Viewres system variable to a higher value (see Chapter 6). Also, linetype scales act differently depending on whether you are in Model Space or Paper Space. See Chapter 9 for more on Model and Paper Space.

Remember that if you assign a linetype to a layer, everything you draw on that layer will be of that linetype. This includes arcs, polylines, circles, and traces. As explained in the *Setting Individual Colors, Linetypes, and Linetype Scales* sidebar that follows, you can also assign different colors and linetypes to individual objects, rather than relying on their layer assignment to define color and linetype. However, you may want to avoid assigning colors and linetypes directly to objects until you have some experience with AutoCAD and a good grasp of your drawing's organization.

In the last exercise, you changed the global linetype scale setting. This affects all non-continuous linetypes within the current drawing. You can also change the linetype scale of individual objects, using the Properties button on the Object Properties toolbar. Or you can set a default linetype scale for all new objects, with the Current button option in the Linetype tab of the Layer & Linetype Properties dialog box.

When individual objects are assigned a linetype scale, they are still affected by the global linetype scale set by the Ltscale system variable. For example, say you assign a linetype scale of 2 to the hidden lines in the previous example. This scale would then be multiplied by the global linetype scale of 12, for a final linetype scale of 48.

The default Linetype Scale setting for individual objects can also be set using the Celtscale system variable. Once this is set, it will affect only newly created objects. You must use the Properties tool to change the linetype scale of individual existing objects.

If the objects you draw appear in a different linetype from that of the layer they are on, check the default linetype, using the Linetype tab of the Layer & Linetype Properties dialog box. Click on Format ➢ Linetype. Select ByLayer in the Linetype list, then click on the Current button. Also, check the linetype scale of the object itself, using the Properties button. A different linetype scale can make a line appear to have an assigned linetype that may not be what you expect. See the *Setting Individual Colors, Linetypes, and Linetype Scales* sidebar below.

The display of linetypes also depends on whether you are in Paper Space or Model Space. If your efforts to control linetype scale have no effect on your linetype's visibility, you may be in Paper Space. See Chapter 9 for more information on how to control linetype scale while in Paper Space.

Setting Individual Colors, Linetypes, and Linetype Scales

If you prefer, you can set up AutoCAD to assign specific colors and linetypes to objects, instead of having objects take on the color and linetype settings of the layer on which they reside. Normally, objects are given a default color and linetype called Bylayer, which means each object takes on the color or linetype of its assigned layer. (You've probably noticed the word Bylayer in the Object Properties toolbar and in various dialog boxes.)

Continued

> **CONTINUED**
>
> Use the Properties tool on the Object Properties toolbar to change the color or linetype of existing objects. This tool opens a dialog box that lets you set the properties of individual objects. For new objects, use the Color tool on the Object Properties toolbar to set the current default color to red (for example), instead of Bylayer. The Color tool opens the Select Color dialog box, where you select your color from a toolbar. Then everything you draw will be red, regardless of the current layer color.
>
> For linetypes, you can use the Linetype drop-down list in the Object Properties toolbar to select a default linetype for all new objects. The list only shows linetypes that have already been loaded into the drawing, so you must have loaded a linetype before you can select it.
>
> Another possible color and linetype assignment is Byblock, which is also set with the Properties button. Byblock makes everything you draw white, until you turn your drawing into a block and then insert the block on a layer with an assigned color. The objects then take on the color of that layer. This behavior is similar to that of objects drawn on layer 0. The Byblock linetype works similarly to the Byblock color.
>
> Finally, if you want to set the linetype scale for each individual object, instead of relying on the global linetype scale (the Ltscale system variable), you can use the Properties button to modify the linetype scale of individual objects. Or you can use the Object Creation Modes dialog box (via the Object Creation button in the Object Properties toolbar) to set the linetype scale to be applied to new objects. In place of using the Properties button, you can set the Celtscale system variable to the linetype scale you want for new objects.
>
> As we mentioned earlier, you should stay away from assigning colors and linetypes to individual objects until you are comfortable with AutoCAD; and even then, use color and linetype assignments carefully. Other users who work on your drawing may have difficulty understanding your drawing's organization if you assign color and linetype properties indiscriminately.

Keeping Track of Blocks and Layers

The Insert dialog box and the Layer & Linetype Properties dialog box let you view the blocks and layers available in your drawing, by listing them in a window. The Layer & Linetype Properties dialog box also includes information on the status of layers. However, you may forget the layer on which an object resides. The List button on the Object Properties toolbar enables you to get information about individual objects, as well as blocks.

1. Click and hold the Distance tool so that a flyout appears.
2. Drag the pointer down to the List tool and select it.

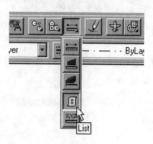

If you just want to check quickly what layer an object is on, click on the object. Its layer will appear in the layer list of the Object Properties toolbar. You can also click on the Properties tool in the Object Properties toolbar, and then click on the object in question. You will get a dialog box showing you the basic properties of the object, including its layer setting.

3. At the object-selection prompt, click on the hidden line and then press ↵. The AutoCAD Text window appears.
4. In the Text window, you will see a listing that shows not only the layer that the line is on, but its space, insertion point, name, color, linetype, rotation angle, and scale.

The Space property you see listed for the line designates whether the object resides in Model Space or Paper Space. You'll learn more about these spaces in Chapters 9 and 12.

Using the Log File Feature

Eventually, you will want a permanent record of block and layer listings. This is especially true if you work on drawing files that are being used by others and your company does not have a layer standard. Here's a way to get a permanent record of the layers and blocks within a drawing, using the Log File option in the Preferences dialog box.

KEEPING TRACK OF BLOCKS AND LAYERS

1. Minimize the Text window (click on the Minimize button in the upper-right corner of the Text window).
2. Click on Tools ➢ Preferences or type **pr**↵. The Preferences dialog box appears.
3. Click on the General tab at the top of the dialog box.

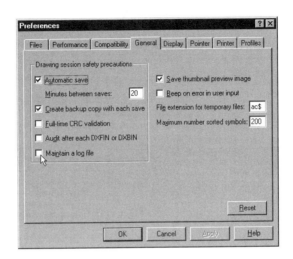

4. Click on the Maintain a log file check box in the bottom-left side of the dialog box. A check mark appears in the check box.
5. Click on OK. Type **–layer**↵ (don't forget the minus sign) at the command prompt, and then **?**↵↵. The AutoCAD Text window appears, and a listing of all the layers scrolls into view.
6. Press F2 to return to the AutoCAD drawing screen. Then click on Tools ➢ Preferences to reopen the Preferences dialog box.
7. Deselect the Maintain a log file check box, and then click on OK.
8. Use the Windows Notepad to open the AutoCAD log file, ACAD.LOG, located in the \ACADR14\ directory. You will see that the layer listing is recorded there.

With the Log File feature, you can record virtually anything that appears in the command prompt. You can even record an entire AutoCAD session. The log file can also be helpful in constructing script files to automate tasks. To have a hard copy of the log file, just print it from an application such as Windows Notepad or your favorite word processor.

If you wish, you can arrange to keep the ACAD.LOG file in a directory other than the default AutoCAD subdirectory. This setting is also in the Preferences dialog box under the Files tab. Locate the Menu, Help, Log, and miscellaneous File Names listing in the

Search path, file names, and file locations list box. Click on the plus sign next to this listing. The listing expands to show the different types of support files available under this listing. Click on the plus sign next to the Log File listing. You will see the location of the ACAD.LOG file.

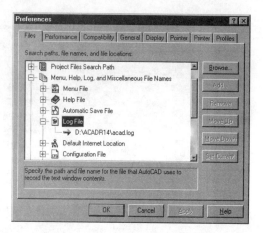

See Appendix B for more on the AutoCAD Preferences dialog box settings.

You can double-click on the ACAD.LOG file location listing to open a Select File dialog box and specify a different location and file name for your log file. This dialog box is a typical Windows file dialog box.

Once you've settled on a disk location for the log file, use the File Manager to associate the log file with the Windows Notepad or Write application. Then click and drag the file to the AutoCAD program group. This gives you quick access to your log file by simply double-clicking on its icon in the AutoCAD program group.

Finding Files on Your Hard Disk

As your library of symbols and files grows, you might begin to have difficulty keeping track of them. Fortunately, AutoCAD offers a utility that lets you quickly locate a file anywhere in your computer. The Find File utility searches your hard disk for specific files. You can have it search one drive or several, or you can limit the search to one directory. You can limit the search to specific file names or use DOS wildcards to search for files with similar names.

The following exercise steps you through a sample Find File task.

1. Click on File ➣ Open. In the Select File dialog box, click on Find File to display the Browse/Search dialog box.

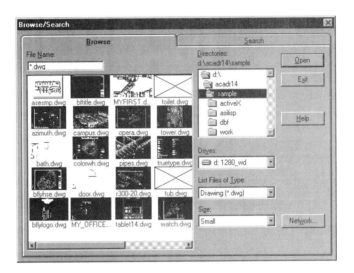

Find File can also be accessed using the Find File button in any AutoCAD file dialog box, including the File option of the Insert Block tool. Find File can help you access and maintain your symbols library.

The Browse/Search dialog box has two tabs: Browse and Search. On the Browse page are all the drawings in the current directory, displayed as thumbnail views so you can identify them easily. You can open a file by double-clicking on its thumbnail view, or by entering its name in the File Name input box at the top.

The Size drop-down list on the Browse page of the Browse/Search dialog box lets you choose the size of the thumbnail views shown in the list box—small, medium, or large. You can scroll through the views using the scroll bars at the bottom of the list box.

2. Click on the Search tab to open the page of Search functions. Use the Search Pattern input box to enter the name of the file for which you wish to search. The default is *.dwg, which will cause Find File to search for all AutoCAD drawing files. Several other input boxes help you set a variety of other search criteria,

such as the date stamp of the drawing, the type of drawing, and the drive and path to be searched. For now, leave these settings as they are.

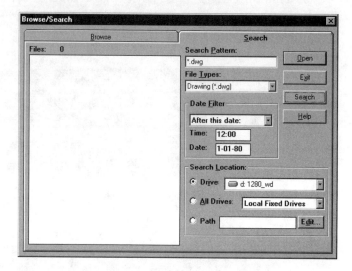

3. Click on the Search button. In a few seconds, a listing of files that meet the criteria specified in the input boxes appears in the Files list on the left, along with thumbnail views of each file. You can click on a file name in the list, and then click on the Open button to open the file in the drawing editor.

4. When you're ready, click on Exit to exit the Browse/Search dialog box, and then click on Cancel to exit the Open File dialog box.

NOTE

In the Browse/Search dialog box, a drawing from a pre-Release 13 version of AutoCAD will be represented as a box with an X through it.

In this exercise you performed a search using the default settings. These settings caused AutoCAD to search for files with the .dwg file name extension, created after 12:00 midnight on January 1, 1980, in the \ACADWIN directory.

Here are descriptions of the items in the Browse/Search dialog box:

Search Pattern lets you give specific file-name search criteria using DOS wildcard characters.

File Types lets you select from a set of standard file types.

Time and Date let you specify a cutoff time and date. Files created before the specified time and date are ignored.

Drives lets you specify the drives to search.

All Drives lets you search all the drives on your computer.

Path lets you specify a path to search.

The **Search** button begins the search process.

The **Open** button opens the file highlighted in the file list, after a search is performed.

The **Help** button provides information on the use of Browse/Search.

The **Edit** button, next to the Path box, opens another dialog box, displaying a directory tree from which you can select a search path.

Inserting Symbols with Drag and Drop

If you prefer to manage your symbols library using Windows Explorer or another third-party file manager, you'll appreciate AutoCAD's support for *drag and drop*. With this feature, you can click on and drag a file from Windows Explorer into the AutoCAD window. You can also drag and drop from the Windows Find File or Folder utilities. AutoCAD will automatically start the Insert command to insert the file. Drag and drop also works with a variety of other AutoCAD support files. AutoCAD also supports drag and drop for other types of data from applications that support Microsoft's ActiveX technology. Table 4.3 shows a list of files with which you can use drag and drop, and the functions associated with them.

TABLE 4.3: AUTOCAD SUPPORT FOR DRAG AND DROP

File Type	Command Issued	Function Performed When File Is Dropped
.dxf	Dxfin	Imports .dxf files
.dwg	Insert	Imports or plots drawing files
.txt	Dtext	Imports texts via Dtext
.lin	Linetype	Loads linetypes
.mnu, .mnx	Menu	Loads menus
.ps	Psin	Imports PostScript files
.psb, .shp, .shx	Style	Loads fonts or shapes
.scr	Script	Runs script
.lsp	(Load..)	Loads AutoLISP routine
.exe, .exp	(Xload..)	Loads ADS application

You can also drag and drop from folder shortcuts placed on your Desktop or even from a Web site.

If You Want to Experiment...

You might want to experiment with creating some other types of symbols. You might also start thinking about a layering system that suits your particular needs.

Open a new file called Mytemp. In it, create layers named 1 through 8 and assign each layer the color that corresponds to its number. For example, give layer 1 the color 1 (red), layer 2 the color 2 (yellow), and so on. Draw each part shown in Figure 4.19, and turn each part into a file on disk using the Export (File ➢ Export) or the Wblock command. When specifying a file name, use the name indicated for each part in the figure. For the insertion point, use the points indicated in the figure. Use the Osnap modes (see Chapter 2) to select the insertion points.

FIGURE 4.19

A typical set of symbols

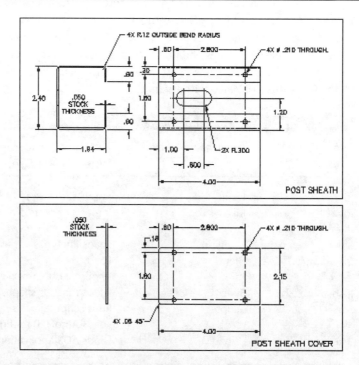

IF YOU WANT TO EXPERIMENT... 189

When you are done creating the parts, exit the file using File ➤ Exit, and then open a new file. Set up the drawing as an engineering drawing with a scale of ¼"=1" on an 11"×17" sheet. Create the drawing in Figure 4.20 using the Insert Block command to place your newly created parts.

FIGURE 4.20

Draw this part using the symbols you create.

1. Set the snap mode to .125 and be sure it is on. Set Ltscale to .25. Draw the figure at right using the dimensions shown as a guide.

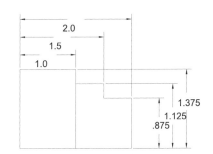

2. Insert the HEXHD drawing at the location shown in the figure at right. Enter .25 for a scale value and when you are asked for a rotation angle, visually orient it as shown.

3. Insert the HIDETH file at the same point and scale as the HEXHD file, and then explode it.

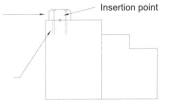

4. Insert the HIDECAP file as shown at right. Enter .25 for the scale and rotate it so it looks like this figure.

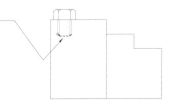

5. Do the same sequence of steps to add the screw shown at right. This time, use a scale factor of .125 when you insert the ROUNDHD file. When you insert the HIDETH file, enter a value of 1 for the X scale factor and .125 for the Y scale factor.

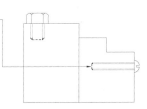

PART II

Building on the Basics

LEARN TO:

- **Create and Use Templates**
- **Draw Parallel Lines and Construction Lines**
- **Take Control of the AutoCAD Display**
- **Add Labels and Notes to Your Drawings**
- **Use Linear and Ordinate Dimensions**
- **Work in Paper Space**
- **Attach External References**
- **Draw and Edit Polylines and Spline Curves**

Chapter 5

Editing for Productivity

FEATURING

Using Existing Drawings as Prototypes for New Drawings

Making Polar and Rectangular Arrays

Using Construction Lines

Marking Regular Intervals along a Line or Arc

Drawing Parallel Lines

Removing Unused Elements Such As Blocks and Layers

Understanding Methods for Constructing New Drawings

Chapter 5

Editing for Productivity

There are at least five commands devoted to duplicating objects, ten if you include the Grips options. Why so many? If you're an experienced drafter, you know that technical drawing is often tedious. So AutoCAD offers a variety of ways to re-use existing geometry, thereby automating much of the repetitious work usually associated with manual drafting.

In this chapter, as you add more detail to your drawing, you will explore some of the ways to exploit existing files and objects while constructing your drawing. For example, you will use existing files as prototypes for new files, eliminating the need to set up layers, scales, and sheet sizes for similar drawings. With AutoCAD you can also duplicate objects in multiple arrays. You have already seen how to use the Osnap overrides on objects to locate points for drawing complex forms. We will look at other ways of using lines to aid your drawing.

And, because you will begin to use Zoom more in the exercises of this chapter, we will review this command as we go along. We'll also introduce you to the Pan command—another tool to help you get around in your drawing.

You're already familiar with many of the commands. So, rather than going through every step of the drawing process, we will sometimes ask you to copy the drawing from a figure, using notes and dimensions as guides and putting objects on the indicated layers. If you have trouble remembering a command you've already learned, just go back and review the appropriate section of the book.

Creating and Using Templates

If you are familiar with the Microsoft Office suite, you are probably familiar with templates. A template is a blank file that is already set up for a specific application. For example, you might want to have letters set up in a way that is different from a report or invoice. You can have a template for each type of document, each template set up for the needs of that document. That way, you don't have to spend time reformatting each new document you create.

Similarly, AutoCAD offers templates, which are drawing files that contain custom settings designed for a particular function. Out of the box, AutoCAD offers templates for ISO, ANSI, DIN, and JIS standard drawing formats. But you aren't limited to these "canned" templates. You can create your own templates set up for your particular style and method of drawing.

If you find that you use a particular drawing setup frequently, you can turn one or more of your typical drawings into a template. For example, you may want to create a set of drawings with the same scale and sheet size as an existing drawing. By turning a typical drawing into a template, you can save a lot of setup time for subsequent drawings.

Creating a Template

The following exercise guides you through creating and using a template drawing for a version of the bar and bracket using the second bracket that you drew and a new, thinner bar. Because the new version will use the same layers, settings, scale, and sheet size as the current version drawing, you can use the DT0100 file as a prototype.

1. Start AutoCAD in the usual way.
2. Click on File ➢ Open.

3. At the Select File dialog box, locate the Bracket file you created in the last chapter. You can also use the file 5-DT0100.dwg from the companion CD-ROM.
4. Click on the Erase button in the Modify toolbar; then type **all**↵↵. This will erase all the objects that make up the current drawing.
5. Choose File ➢ Save As; then at the Save Drawing As dialog box, open the Save as type drop-down list and select Drawing Template File (*.dwt). The file list window will change to display the current template files in the \Template folder.

NOTE

When you choose the Drawing Template File option in the Save Drawing As dialog box, AutoCAD automatically opens the folder containing the template files. The standard installation creates the folder named Template to contain the template files. If you wish to place your templates in a different folder, you can change the default template location using the Preferences dialog box (select Tools ≻ Preferences). Use the Files tab and then double-click on Template Drawing File Location in the list. Double-click on the folder name that appears just below Template Drawing File Location; then select a new location from the Browse for Folder dialog box that appears.

6. Double-click on the File name input box and enter the name **versions**.

7. Click on OK. The Template Description dialog box appears.

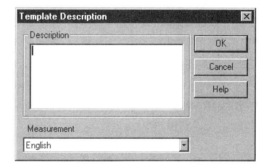

8. Enter the following description:

Bar and bracket template for multiple versions

9. Click on Save. You have just created a template.

While AutoCAD allows long names, only 8-character names appear correctly in the Use a Template list box that you will see in the next exercise. The Template Description dialog box lets you supply more descriptive information regarding your template.

Notice that the current drawing is now the template file you just saved. As with other Windows programs, the File ➢ Save As option makes the saved file current. This also shows that you can edit template files just as you would regular drawing files.

Using a Template

Now let's see how a template is used. You'll use the template you just created as the basis for a new drawing you will work on in this chapter.

1. Choose File ➢ New.
2. At the Create New Drawing dialog box, click on the Use a Template button. A list box called Select a Template appears, along with a file Preview window.

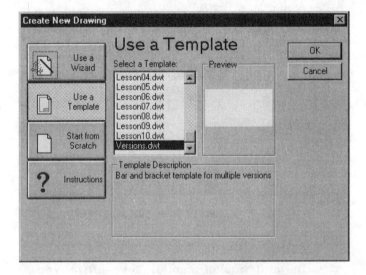

3. Click on versions.dwt from the Use a Template list box. The file is displayed in the Preview window. Since it is a blank file, nothing is displayed in the Preview box. Also notice that the description you entered earlier in the Template Description dialog box appears in the Template Description area below the list.

The More Files option at the top of the Use a Template list box lets you browse your hard drive to locate other files for use as a template.

4. Click on OK. It may not be obvious, but your new file is set up with the same units and drawing limits as the previous drawing. It also contains the two versions of the bracket blocks.

5. Now you need to give your new file a name. Choose File ➤ Save As. Then at the Save Drawing As dialog box, enter **Bar** for the file name and select the appropriate folder in which to save your new Bar file.

NOTE

Before you named it, the file name was the AutoCAD default of Drawing—different from Release 13, which used Unnamed. You will find unnamed.dwg files on many hard disks of Release 13 users who inadvertently saved a file they were working on before they had named it. You will probably do the same thing once or twice. You can easily rename the file from Windows Explorer or use the Save As option in the File menu to save your work under the correct file name.

6. Click on Save to create the Bar file and close the dialog box.

You've created and used your own template file. Later, when you have established a comfortable working relationship with AutoCAD, you can create a set of templates that are custom made to your particular needs.

However, you don't need to create a template every time you want to re-use settings from another file. You can use an existing file as the basis or prototype for a new file without creating a template. Open the prototype file and then use File ➤ Save As to create a new version of the file under a new name. You can then edit the new version without affecting the original file.

Copying an Object Multiple Times

Now let's explore the tools that let you quickly duplicate objects. In the next exercise, you will begin to draw a new bar. This exercise introduces the Array command, which enables you to draw rectangular and circular arrangements of objects.

NOTE

An array can be in a circular pattern called a *polar array,* or in a matrix of columns and rows called a *rectangular array.*

Making Polar Arrays

You do not need a circular array to complete the current drawing, but you will draw and discard a circular array to see how this command works. Later, you will use the Array command in Chapter 12. Here, you will set an Undo marker so that you can back up to this point in the drawing process. AutoCAD either temporarily or permanently keeps a log of the commands that you issue. (We will discuss the Log feature, which is the permanent record of the temporary log, in Chapter 18.) The ability to undo to a certain point provides the AutoCAD user with the opportunity to iterate a design.

1. Type **undo**↵. No toolbar tool exists for this version of Undo. At the Auto/Control/BEgin/End/Mark /Back/<Number>: prompt, type **m**↵ to mark this point in the command sequence of this file. Take a look at the *Undo* sidebar to see what the rest of the functions can do for you.

Undo

The *Undo* feature allows you to undo parts of your editing session. This can be useful if you accidentally execute a command that destroys part or all of your drawing. Undo also allows you to control how much of a drawing is undone.

To Reverse Commands

Command Line: **undo**↵

Menu: Edit ➢ Undo

Auto/Control/BEgin/End/Mark/Back/<default number>: Enter an option or the number of commands to undo.

Options

Auto makes AutoCAD view menu macros as a single command. If Auto is on, the effect of macros issued from a menu will be undone regardless of the number of commands the macro contains.

Control allows you to turn off the Undo feature to save disk space or to limit the Undo feature to single commands. You are prompted for All, None, or One. *All* fully enables the Undo feature, *None* disables Undo, and *One* restricts the Undo feature to a single command.

Continued

CONTINUED

Begin, End allow you to mark a group of commands to be undone together. *Begin* marks the beginning of a sequence of operations. All edits after that point become part of the same group. *End* terminates the group.

Mark, Back allows you to experiment safely with a drawing by first marking a point in your editing session to which you can return. Once a mark has been issued, you can proceed with your experimental drawing addition. Then, you can use *Back* to undo all the commands back to the point at which the mark was issued.

NOTE

Many commands offer an Undo option. The Undo option under a main command will act like the single Undo command and will not offer the options described here.

2. Set the current layer to 0, and toggle the Snap mode on.

NOTE

Since you used the DT0100 file as a template, the Running Osnaps for Endpoint, Midpoint, and Intersection are already turned on and available in this new file.

3. Click on Circle from the Draw toolbar or type **c**↵.

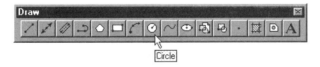

4. At the 3P/2P/TTR/<Center point>: prompt, pick a point at coordinate 14,15.

5. At the Diameter/<Radius>: prompt, enter **3**↵. The circle appears.

If your circle does not appear on your screen, use Zoom Extents (type **z**↵**e**↵, use the Zoom tool located on the Standard toolbar, or choose View ➤ Zoom ➤ Extents). Now you're ready to use the Array command to draw a polar array. You will first draw one line, and then use Array to create the copies.

1. Draw a 4"-long line starting from the coordinate 15.5,15 and ending to the right of that point.
2. Zoom into the circle and line to get a better view. Your drawing should look like Figure 5.1.

FIGURE 5.1

A close-up of the circle and line

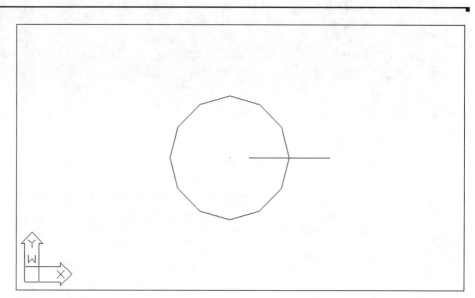

3. Click on Array from the Modify toolbar or type **ar**↵.

4. At the Select objects: prompt, enter **L**↵. This highlights the line you just drew.
5. Press ↵ to confirm your selection.
6. At the Rectangular or Polar array (<R>/P): prompt, type **p**↵ to use the Polar Array option.

7. At the Specify Center point of array: prompt, pick the center of the circle. You can either use the snap point at the center of the circle at coordinate 4,4, or use the Center Osnap.

Remember, to access Osnaps other than those set up as Running Osnaps, Shift+right-click, then select the Osnap from the pop-up menu.

If you use the Center Osnap, you must place the cursor on the circle, not on the circle's center point.

8. At the Number of items: prompt, enter **8**↵. This tells AutoCAD you want seven copies plus the original.

9. At the Angle to fill (+=ccw,-=cw) <360>: prompt, press ↵ to accept the default. The default value of 360 tells AutoCAD to copy the objects so that they are spaced evenly over a 360° arc. (If you had instead entered 180, the lines would be evenly spaced over a 180° arc, filling only half the circle.)

If you want to copy in a clockwise (CW) direction, you must enter a minus sign (–) before the number of degrees.

10. At the Rotate objects as they are copied? <Y>: prompt, press ↵ again to accept the default. The line copies around the center of the circle, rotating as it copies. Your drawing will look like Figure 5.2.

In step 10, you could have the line maintain its horizontal orientation as it is copied around, by entering **n**↵. But since you want it to rotate about the array center, accept the default, which is Yes.

FIGURE 5.2
The completed array

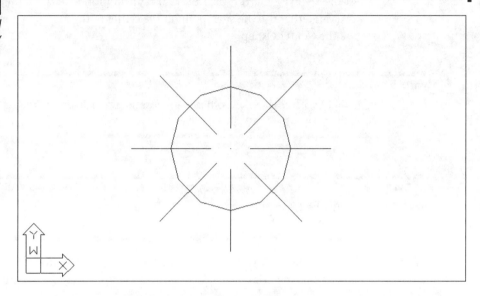

11. Now use the Undo command to return to the point before you created the polar array. Type **undo**↵. At the `Auto/Control/BEgin/End/Mark/Back/<Number>:` prompt, type **b**↵. You see the response on the command line `ARRAY LINE CIRCLE Mark encountered`, which means that AutoCAD searched back through the command log, found the mark that you placed, and undid all of the commands back to that point.

Making Row and Column Copies

You only drew one hole in the bar in the previous exercise, but to mount the bracket properly you needed two holes per bracket. You need to draw another rectangle that represents the new, thinner bar and two concentric circles that represent a tapped mounting hole. The larger circle will be on the Hidden layer. You will then insert the second version of the bracket to find the distance between the mounting holes (you could also open and review the bracket drawing that you created in Chapter 3). You will finish by creating a rectangular pattern of tapped holes along the bar.

1. Set the current layer to 0, and toggle the Snap mode on.

> Since you used the previous exercise as a template, the Running Osnaps for Endpoint, Midpoint, and Intersection are already turned on and available in this new file.

2. Draw a rectangle 60 units long, and approximately 3.375 units high from point 2,3.

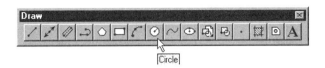

3. Draw a circle using the From Osnap at a point 3 units from the left end of the bar and 1 unit up from the bottom of the bar.
4. At the `Diameter/<Radius>:` prompt, type **d**↵ to select Diameter, then enter **.422**↵. The circle represents the minor diameter of the thread.
5. From the Object Properties toolbar, choose the layer drop-down list and select the Hidden layer to make it current.
6. Using the Center Osnap, draw another circle concentric with the first, and with a diameter of .5. The circle appears in red and consists of a dashed line.
7. Change the current layer to 0 using the layer drop-down list in the Object Properties toolbar.
8. Insert the block bkt_v2 near, but not on, the top left end of the bar. Your drawing should look like Figure 5.3.

FIGURE 5.3

The bar, bracket, and circle

Using the Distance Tool

You will next use the Distance tool to find the distance between the bracket mounting holes. This information is required to create the array of mounting holes along the length of the bar.

> ### The Distance Tool
>
> The Distance tool measures the distance and vector between any two points that you select. The information appears on the command line as
>
> ```
> Distance = 1.6250, Angle in XY Plane = 90, Angle from XY Plane = 0
> Delta X = 0.0000, Delta Y = 1.6250, Delta Z = 0.0000
> ```
>
> The accuracy is determined by the units setting. The positive or negative sign of delta values are returned by the order in which you pick the points. For example, if you pick two points from left to right and from bottom to top, the values will be positive. The angles are also determined by the order of point selection.

9. From the Standard toolbar, choose the Distance tool. At the `First point:` prompt, pick the lower circle in the bracket. At the `Second point:` prompt, pick the upper circle. The response looks like this:

   ```
   Distance = 1.6250, Angle in XY Plane = 90, Angle from XY Plane = 0
   Delta X = 0.0000, Delta Y = 1.6250, Delta Z = 0.0000
   ```

Using the Array Command

Now you're ready to use the Array command to draw the mounting holes.

1. Click on Array from the Modify toolbar or type **ar**↵.

2. At the `Select objects:` prompt, select the two circles and hit ↵. The prompt states `2 found Select objects:`. Press ↵ to finish selecting objects and begin the next step of the command. The circles do not remain highlighted.

3. At the `Rectangular or Polar array (<R>/P):` prompt, type **r**↵ to use the Rectangular Array option.

4. The next prompt is `Number of rows (--) <1>:`. The series of dashes indicates that a row is horizontal. You need two rows for the two mounting holes per bracket, so you should type **2**↵. The next prompt is `Number of columns (|||) <1>:`. The design calls for seven brackets equally spaced on this rail; type **7**↵.

5. From the distance inquiry, you determined the distance between the bracket mounting holes is 1.625. So at the Unit cell or distance between rows (--): prompt, enter **1.625**↵.

6. You'll have to do a little math to determine the input for the Distance between columns (|||): prompt. If the hole pattern is to be centered on the 60-unit bar and the holes are 3 units in from each end, then you must divide the remaining 54 units by 6 (the number of spaces between 7 evenly spaced units). This gives a distance of 9 units. Enter **9**↵. Your drawing should look like Figure 5.4.

FIGURE 5.4
The new bar with the hole pattern

You can also type **array**↵ to start the Array command. Then, after confirming your selection of objects in step 2, type **r**↵.

Unlike prior versions of AutoCAD, the Array command now "remembers" whether you last used the Polar or Rectangular Array option and offers that option as the default.

AutoCAD usually draws a rectangular array from bottom to top, and from left to right. You can reverse the direction of the array by giving negative values for the distance between columns and rows.

At times you may want to do a rectangular array at an angle. To accomplish this you must first set the Snap Angle setting in the Drawing Aids dialog box (Tools ≻ Drawing Aids) to the desired angle. Then proceed with the Array command. Another method is to set the UCS to the desired angle. See Chapter 11 for more information.

You can also use the cursor to graphically indicate an *array cell* (see Figure 5.5). An array cell is a rectangle defining the distance between rows and columns. You may want to use this option when an object is available to use as a reference from which to determine column and row distances. For example, you may have drawn a cross-hatch pattern, as on a calendar, within which you want to array an object. You would use the intersections of the hatch lines as references to define the array cell, which would be one square in the hatch pattern.

FIGURE 5.5

An array cell

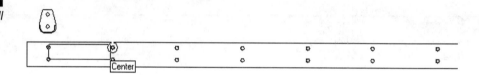

Fine-Tuning Your View

Notice that most of the bracket does not appear on your screen. To move the view over, you can use the Pan command. Pan is similar to Zoom in that it changes your view of the drawing; however, Pan does not alter the magnification of the view the way Zoom does. Rather, Pan maintains the current magnification while moving your view across the drawing, just as you would pan a camera across a landscape.

Pan is especially helpful when you have magnified an area to do some editing, and you need to get to part of the drawing that is near your current view.

If you're not too fussy about the amount you want to zoom out, you can choose View ➢ Zoom ➢ Out to quickly reduce your view.

You have been looking at an overall view of the bar. Zoom in on the bracket and holes at the left end of the bar.

1. From the Standard toolbar, select the Zoom Window tool or type **z↵**. Pick a point slightly below and to the left of the leftmost set of holes. For the second point, pick a point slightly above and to the right of the bracket.

The Zoom Flyout

There are a few tools on the Standard toolbar that offer more than one option. The Zoom tool, located on the Standard toolbar to the right of the Zoom Realtime tool, is one of these. You might have to look closely to see a small, southeast-pointing, black triangle on the tool. Only a few tools have this triangle, which indicates that a flyout toolbar can be accessed by depressing this tool.

Move your cursor to the Zoom tool and press the left mouse button, holding it down. (If you just click the tool button, you will select the tool itself.) When you hold down the left mouse button, the flyout toolbar is displayed, and if you drag the mouse, still pressing the left button, you will see tool tips appear to identify various tools. These are called *flyout tools*. To select a flyout tool, release the mouse button over the tool that you want.

The last flyout tool that you used will become the default tool, shown in place of the original tool on the Standard toolbar. This *last used* feature is designed to make the most recently used tool easy to use again. This feature also has the disconcerting side-effect of making the toolbar look different each time you use a different flyout tool. If you don't see Zoom Window as the default tool on the Standard toolbar, click on whatever Zoom tool appears to the right of the Zoom Realtime tool and hold down the left mouse button until you see the flyout tools. Continue to hold down the left mouse button, drag the cursor down until you highlight the Zoom Window tool, then release the mouse button. If you are not sure which tool is which, keep pressing the left mouse button and hover the cursor over the tool button a moment, until the tool tips appear.

To activate the Pan command, follow these steps:

1. Click on the Pan Realtime tool in the Standard toolbar or type **p**↵.

A small, hand-shaped cursor appears in place of the AutoCAD cursor.

2. Place the hand cursor in the center of the drawing area, then click and drag it downward and to the left. The view follows the motion of your mouse.

3. Continue to drag the view until it looks similar to Figure 5.6; then let go of the mouse.

FIGURE 5.6

The panned view of the bar

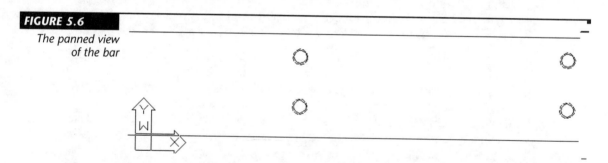

4. Now right-click on the mouse. A pop-up menu appears.

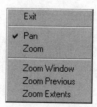

NOTE

The Pan/Zoom pop-up menu also appears when you right-click on your mouse during the Zoom Realtime command.

5. Select Zoom from the list. The cursor changes to the Zoom Realtime cursor.

6. Now place the cursor close to the top of the screen and click and drag the cursor downward to zoom out until your view looks like Figure 5.7. You may need to click and drag the Zoom cursor a second time to achieve this view.

7. Right-click again, then choose Exit from the pop-up menu.

NOTE

To exit the Pan Realtime or Zoom Realtime command without opening the pop-up menu, press the Esc key.

FIGURE 5.7
The final view of the bar holes and bracket

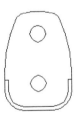

This exercise showed you how you can fine-tune your view by easily switching between Pan Realtime and Zoom Realtime. Once you get the hang of these two tools working together, you'll be able to quickly access the best view for your needs. Each of the other options in the pop-up menu—Zoom Window, Zoom Previous, and Zoom Extents—performs the same function as the tool with the same name on the Standard toolbar and View pull-down menu.

While we're on the subject of display tools, don't forget the scroll bars to the right and bottom of the AutoCAD drawing area. They work like any other Windows scroll bar, offering a simple way to move up, down, left, or right in your current view.

If the scroll bars do not appear in AutoCAD, go to the Display tab of the Preferences dialog box (Tools ➢ Preferences) and make sure the Display Scroll Bars in Drawing Window option is checked. If you prefer to turn the scroll bars off, make sure this option is unchecked.

This would be a very good time to save your work. You could take a break before going on to the next exercise.

Making Random Multiple Copies

The Draw ➢ Array command is useful when you want to make multiple copies in a regular pattern. But what if you need to make copies in a random pattern? You have two alternatives for accomplishing this: the Copy command's Multiple option and the Grips Move option.

To use the Copy command to make random multiple copies:

1. Click on Copy Object from the Modify toolbar or type **co**↵.
2. At the `Select objects:` prompt, select the objects you want to copy and press ↵ to confirm your selections.
3. At the `<Base point or displacement>/Multiple:` prompt, enter **m**↵ to select the Multiple option.
4. At the `Base point:` prompt, select a base point as usual.
5. At the `Second point:` prompt, select a point for the copy. You will be prompted again for a second point, allowing you to make yet another copy of your object.

Continued

CONTINUED

   6. Continue to select points for more copies as desired.
   7. Press ↵ to exit the Copy command when you are done.

When you use the Grips feature to make random multiple copies, you get an added level of functionality because you can also rotate, mirror, and stretch copies by using the pop-up menu (right-click while a grip is selected). Of course, you must have the Grips feature turned on; it is usually on by default, but you may find yourself on a system that has it turned off for some reason.

   1. Press Esc twice to make sure you are not in the middle of a command; then select the objects you want to copy.
   2. Click on a grip point as your base point.
   3. Right-click your mouse and select Move.
   4. Right-click again and select Copy.
   5. Click on the location for the copy. Notice that the rubber-banding line persists and you still see the selected objects follow the cursor.
   6. If desired, click on other locations for more copies.

Finally, you can make square arrayed copies using grips by doing steps 1–3 above, but instead of step 4, Shift+click on a copy location. Continue to hold down the Shift key and select points. The copies will snap to the angle and distance you indicate with the first Shift-select point. Release the Shift key and you can make multiple random copies.

Developing Your Drawing

When using AutoCAD, you first create the most basic forms of your drawing; then you refine them. In this section you will create two views of the original bracket—top and side—that demonstrate this process in more detail.

You will also further examine how to use existing files as blocks. In Chapter 4, you inserted a file into another file. There is no limit to the size or number of files you can insert. As you may already have guessed, you can also *nest* files and blocks; that is, you can insert blocks or files into other blocks or files.

Importing Settings

In this exercise, you will use the Versions template. Then, you will import the bracket, thereby importing the layers and blocks contained in the file.

As you go through this exercise, observe how the drawings begin to evolve from simple forms to complex, assembled forms.

1. First, open a new file using the Versions template. If you skipped creating the template, use the template file named Versions from the companion CD-ROM.
2. Use the Block Insert tool to insert the file DTO100. This file is also available from the companion CD-ROM.
3. The inserted block comprises both brackets. You only need the bracket on the left for this exercise. Use the Explode tool from the Modify toolbar to explode the block.
4. Erase the right-hand bracket. Use a window to surround the bracket, which activates the grips. Then select the Erase tool from the Modify toolbar.
5. Save the new file as **bkt_v1**.

By using the Versions template to start this drawing, you are using the layer and Running Osnaps that you have set in that template.

The New File Setup Wizard

You can use the New File Setup Wizard to start a new drawing. A Wizard is an aid designed to assist users by prompting them for information. You will find Wizards in many Windows 95 programs. The following tutorial demonstrates the use of the Wizard when creating a new drawing. For experienced users, the Setup Wizard is a combination of the File command and the New option, with settings Limits for and Units. In prior releases of AutoCAD, it was necessary to set these system variables in each new drawing or create a template file which already had these system variables set. AutoCAD would then inherit these settings at the start of a new drawing. The difference between the quick and the advanced versions of the Setup Wizard is that the advanced version gives access to more unit-setting options. The drawback to using this feature is that the Wizard does not retain any of your settings, so the next time that you use the Wizard you will find that the settings have reverted to the original values (the defaults) that shipped with AutoCAD.

1. Choose File ➢ New.

Continued

CONTINUED

2. At the Create New Drawing dialog box, click on Use a Wizard. You'll see two options in the Select a Wizard list box. Choose Quick Setup and then click on OK. The Quick Setup dialog box appears. Notice the two tabs labeled Step 1: Units and Step 2: Area.

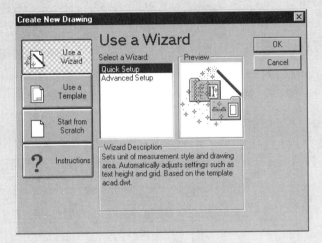

3. The Decimal units radio button is the default.

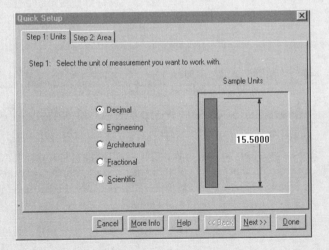

Continued

CONTINUED

4. Click on the Step 2: Area tab.

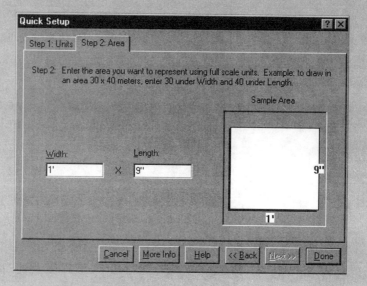

5. Enter **34** in the Width input box and **22** in the length input box. These are the appropriate dimensions for a "D" size ANSI drawing engineering sheet.
6. Click on Done.

Importing Settings from External Reference Files

As explained in the Chapter 4 sidebar, *An Alternative to Blocks,* you can use the External Reference (Xref) Attach option to use another file as a background or Xref file. Xref files are similar to blocks, except that they do not actually become part of the current drawing's database; neither do the settings from the cross-referenced file automatically become part of the current drawing.

Continued

DEVELOPING YOUR DRAWING

> **CONTINUED**
>
> If you want to import layers, linetypes, text styles, and so forth from a Xref file, you must use the Xbind command, which you will learn more about as you work through this book. Xbind allows you to attach dimension style settings (discussed in Chapter 8 and in Appendix D), layers, linetypes (discussed in Chapter 4), or text styles (discussed in Chapter 7) from a cross-referenced file to the current file.
>
> You can also use Xbind to turn a cross-referenced file into an ordinary block, thereby importing all the new settings contained in that file.
>
> See Chapter 9 for a more detailed description of how to use the Xref and Xbind commands.

Using and Editing Lines

You will draw lines in most of your work, so it is important to know how to manipulate lines to your best advantage. In this section, you will look at some of the more common ways to use and edit these fundamental drawing objects. The following exercises show you the process of drawing lines, rather than just how individual commands work.

Roughing In the Line Work

The bracket you inserted in the last section has only one view. In the next exercise, you will draw the side view. The dimensions for this view are shown in Figure 5.8.

FIGURE 5.8

The bracket with a side view and dimensions

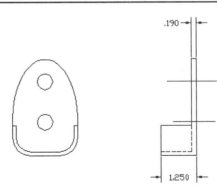

> **NOTE**
>
> You may notice that some of the arcs in your bracket drawing are not smooth. Don't be alarmed; this is how AutoCAD displays arcs and circles in enlarged views. The arcs will be smooth when they are plotted. If you want to see them now as they actually are stored in the file, you can regenerate the drawing by typing **regen.↵** at the command prompt. We will look more closely at regeneration in Chapter 6.

1. Select 0 from the Layer drop-down list in the Object Properties toolbar to make 0 the current layer.
2. Choose Draw ➤ Line or type **L↵**.
3. Shift+right-click to open the Osnap menu; then select From. This option lets you select a point relative to another point.
4. At the From: prompt, open the Osnap menu again and using the Endpoint Osnap, click on the lower-right corner of the bracket (see Figure 5.9). Nothing appears in the drawing area yet.

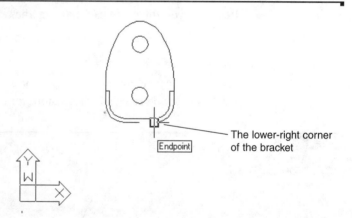

FIGURE 5.9

This point is the lower-right corner of the bracket.

5. Type **@5<0↵**. Now a line starts 5" to the right of the corner of the bracket.
6. To draw the line up, use a point filter. At the To point: prompt, Shift+right-click to pop the Osnap menu. Select Point filters ➤ .Y (AutoCAD will only use the Y value of the next point that you pick). At the .Y of: prompt, move the cursor until you see the triangle marker for the top of the bracket and click the left

DEVELOPING YOUR DRAWING | 219

mouse button. To complete the command, move the cursor to any point above the top right of the bracket and hit ↵. Your drawing should look like Figure 5.10.

FIGURE 5.10

The line and Osnap locations using a point filter

The point filters of the Osnap menu can be very valuable in saving the experienced AutoCAD user a good deal of layout time. You can use the Shift+right-click method to pop up the Osnap menu, or at the To or From point: prompt you can type a period (.) followed by the filter that you would like to use. You can use the X, Y, Z component—or combinations XY, XZ, or YZ components of—the next point that you pick. AutoCAD will then prompt you for a point in the missing direction.

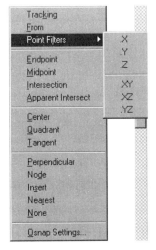

WARNING If you do not select the correct point when using a point filter, the option is discontinued and you must select the filter again. If you find the point filter too difficult to use, try to draw a line from the midpoint of the bottom the bracket to the quadrant of the ellipse. Then move the line 5 units to the right.

In the preceding exercise, the From Osnap allowed you to specify a point in space relative to the lower edge of the bracket. The point filter lets you specify the second point of the line in space relative to the upper edge.

1. Use the Offset or the Copy command to make a copy of the new line .19 units to the left.

TIP The Perpendicular Osnap override can also be used to draw a line perpendicular to a non-orthogonal line—one at a 45° angle, for instance.

2. Draw a line connecting these two lines at the top.

If you find the object snaps and point filters somewhat difficult to understand at first, you can use some of the more classic drafting procedures. You can create layout lines first, and then trim away the parts of the lines that you don't need. Use the Xline command to create some guidelines.

3. On the Draw toolbar, click on the Construction Line or type **xl**↵. At the XLINE Hor/Ver/Ang/Bisect/Offset/<From point>: prompt, move the cursor to the top of the ledge area until the endpoint box marker appears, and pick this point. Check the status bar to see that Ortho mode is on, then at the Through point: prompt, move the cursor to the right and pick any point. You'll see a construction line that extends from one side of your screen to the other. Hit ↵ to end the selection set.

DEVELOPING YOUR DRAWING — 221

The Xline Options

There is more to the Xline command than you have seen in the exercises of this chapter. Here is a list of the Xline options and their uses:

Hor draws horizontal Xlines as you click on points.

Ver draws vertical Xlines as you click on points.

Angle draws Xlines at a specified angle as you pick points.

Bisect draws Xlines bisecting an angle or a location between two points.

4. Use the Endpoint Osnap to pick the left endpoint of the lower horizontal line in the side view. Then pick a point above or below the point to complete the construction line. Hit ↵ to end the selection set.

5. Next, draw a vertical construction line from the left end of the lower horizontal line in the side view. Click on the Construction Line tool on the Draw toolbar and use it to draw the vertical construction line. Your figure should look like Figure 5.11.

FIGURE 5.11

The construction lines added to the bracket

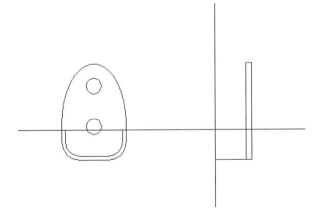

Cleaning Up the Line Work

You've drawn some of the lines, approximating their endpoint locations. Next you will use the Fillet command to join lines exactly end to end.

1. Click on Fillet on the Modify toolbar.

2. Type **r↵0↵** to set the fillet radius to 0; then press ↵ to repeat the Fillet command.

There is another command, the *Chamfer* command, that performs a similar function to Fillet. Unlike Fillet, the Chamfer command allows you to join two lines with an intermediate beveled line, rather than an arc. Chamfer can be set to join two lines at a corner in exactly the same manner as Fillet.

3. Fillet the two lines by picking the vertical and horizontal lines as indicated in Figure 5.12. Notice that these points lie on the portion of the line you want to keep. Your drawing will look like Figure 5.13.

FIGURE 5.12

Pick these points to fillet the first corner.

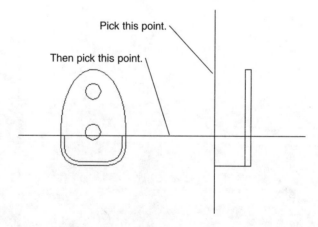

FIGURE 5.13
Your drawing after the first fillet

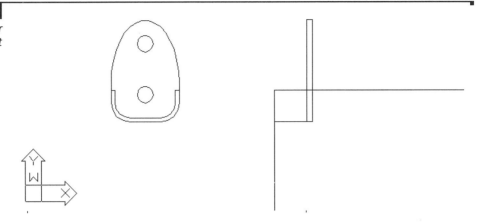

WARNING

If you are a veteran AutoCAD user, you should note that the default value for the Fillet command is now .5 instead of 0.

4. Fillet the bottom of the bracket with the left edge, as shown in Figure 5.14. Make sure the points you pick on the lines are on the side of the line you want to keep, not on the side you want trimmed.

FIGURE 5.14
Pick these points to fillet your second corner.

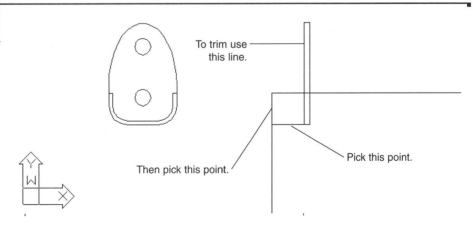

5. Use the inside vertical line to trim the rest of the construction line as shown in Figure 5.14. Your drawing should look like Figure 5.15.

FIGURE 5.15

The construction lines after you've cleaned them up

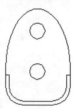

You can select two lines at once for the Fillet operation by using a crossing window. Type **c↵** at the Select first object: prompt. The two endpoints closest to the fillet location will be trimmed.

Where you select the lines will affect how the lines are joined. As you select objects for Fillet, the side of the line you click on is the side that remains when the lines are joined. Figure 5.16 illustrates how Fillet works and shows what the fillet options do.

If you select two parallel lines during the Fillet command, the two lines will be joined with an arc.

Next you will draw the hidden lines. The left vertical line, which is the thickness of the back of the bracket below the ledge, is a hidden line. You will break the vertical line into two segments so that one segment can be an object line and the other segment can be a hidden line.

1. Click on the Break tool in the Modify toolbar, then select the vertical line anywhere along its length.

2. Type **f↵** to use the First Point option; then select the intersection of the upper horizontal line. (Use Shift+right-click to pop the Osnap menu or type **int↵** and move the mouse close to the intersection until you see the X intersection marker. You could also pick one of the lines and AutoCAD will ask you to pick the intersecting line.)

DEVELOPING YOUR DRAWING | 225

FIGURE 5.16

Where you click on the object to select it determines what part of an object gets filleted.

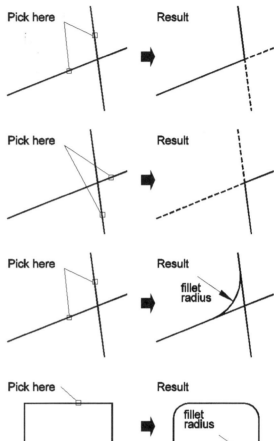

These first two examples show how the pick location affects the way lines are filleted.

This example shows what happens when you set a fillet radius to greater than 0 and the Fillet Trim option is turned off.

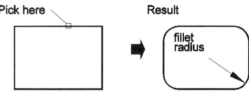

This example shows how Fillet affects polylines. You can fillet polylines using the Polyline and Radius options of the Fillet command.

PART II
Building on the Basics

3. At the Second point: prompt, type **@↵**. The @ symbol forces AutoCAD to use the last point again. You use the same procedure to remove a section of line, except that in that case you would pick two different points on the line or arc and AutoCAD removes the line or arc between the points. For more details, see the following sidebar, *Using the Break Command*.

4. Click on the lower portion of the left vertical line to turn on the Grips feature. Then from the layer drop-down list, select the Hidden layer. Press Esc twice to clear the grips.

5. Click on the layer drop-down list again and click on the Hidden layer to make it current.

Using the Break Command

When you drew the side view of the bracket, you used the Break command with the F option to accurately break the line into two parts. You can also break a line without the F option—with a little less accuracy. By not using the F option, the point at which you select the object is used as the first break point. If you're in a hurry, you can dispense with the F option and simply place a gap in an approximate location. You can then use other tools later to adjust the gap.

In addition, you can use locations on other objects to select the first and second points of a break. For example, you may want to align an opening with another opening some distance away. Once you've selected the line to break, you can then use the F option and select two points on the existing opening to define the first and second break points. The break points will align in an orthogonal direction to the selected points.

Using Construction Lines as Tools

Now you will draw the hidden line, which represents the inside of the ledge, in the side view. Start by drawing a ray.

1. Choose Draw ➢ Ray from the menu bar.

2. At the From point: prompt, start the ray from the endpoint of the horizontal line that represents the inside thickness of the ledge in the front view. If the Ortho mode is on, turn it off. As you move your mouse, you'll see the ray follow its direction (see the top image of Figure 5.17).

3. Type **@1<0**↵. The ray is fixed in a 0° angle. The Ray command persists, allowing you to add more lines.

4. Press ↵ to exit the Ray command.

NOTE A *ray* is a line that starts from a point you select and continues off to some infinite distance. For this reason, you weren't required to enter a distance value in step 3.

FIGURE 5.17
A ray used to project the ledge thickness into the side view

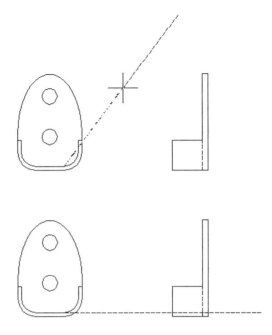

5. Zoom in to see the detail at the bottom of the side view.

6. Trim the ray and the vertical line to look like Figure 5.18.

FIGURE 5.18
The bracket with the ray and vertical line trimmed

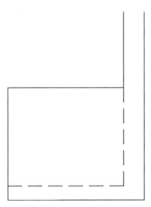

7. Click on View ➤ Zoom ➤ All to view the entire drawing. It will look like Figure 5.19.

8. Save your work.

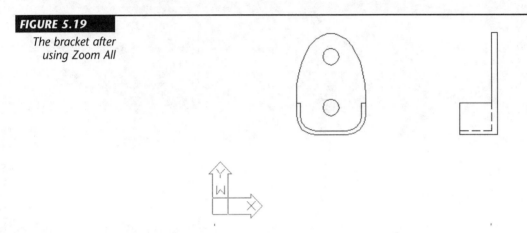

FIGURE 5.19
The bracket after using Zoom All

In this exercise, you used the Ray command to accurately position a construction line which became part of the drawing. This shows that you can freely use rays and construction lines (Xlines) as objects to help construct your drawing.

Finding Distances Along Arcs

You've seen how you can use lines to help locate objects and geometry in your drawing. But if you need to find distances along a curved object such as an arc, lines don't always help. Following are two ways of finding exact distances on arcs. Try these exercises when you're not working through the main tutorial.

Finding a Point a Particular Distance from Another Point

At times you'll need to find the location of a point on an arc that lies at a known distance from another point on the arc. The distance could be described as a chord of the arc, but how do you find the exact chord location? To find a chord along an arc, do the following:

1. Click on Circle on the Draw toolbar.

2. Use the Endpoint Osnap to click on the endpoint of the arc.

3. At the End of diameter/<radius>: prompt, enter the length of the chord distance you wish to locate along the arc.

DEVELOPING YOUR DRAWING

The point where the circle intersects the arc is the endpoint of the chord distance from the endpoint of the arc (see Figure 5.20). You can then use the Intersect Osnap override to select the circle and arc intersection.

FIGURE 5.20

Finding a chord distance along an arc using a circle

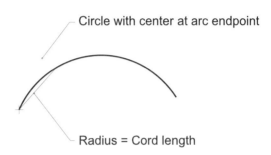

Finding an Exact Distance Along an Arc

To find an exact distance along an arc or curve (nonlinear), or to mark off specific distance increments along an arc or curve, do the following:

1. Choose Format ➢ Point Style from the menu bar to open the Point Style dialog box.

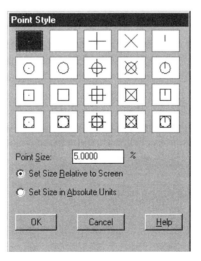

You can also set the point style by setting the Pdmode system variable to 3. See Appendix D for more on Pdmode.

2. At the Point Style dialog box, click on the icon that looks like an X in the top row. Also be sure the Set Size Relative to Screen radio button is selected. Then click on OK.

3. Choose Draw ➢ Point ➢ Measure from the menu bar.

Another command called Divide (Draw ➢ Point ➢ Divide) marks off a line, arc, or curve into equal divisions, as opposed to divisions of a length you specify. You would use Divide to divide an object into 12 equal segments, for example. Aside from this difference in function, Divide works in exactly the same way as Measure.

4. At the `Select object to measure:` prompt, click on the arc near the end from which you wish to find the distance.

5. At the `<Segment length>/Block:` prompt, enter the distance you are interested in. A series of Xs appears on the arc, marking off the specified distance along the arc. You can select the exact location of the Xs using the Node Osnap override (Figure 5.21).

The Block option of the Measure command allows you to specify a block to be inserted at the specified segment length, in place of the Xs on the arc. You have the option of aligning the block with the arc as it is inserted. (This is similar to the Polar Array's Rotate Objects As They Are Copied option.)

The Measure command also works on Bezier curves. You'll get a more detailed look at the Measure command in Chapter 10.

As you work with AutoCAD, you'll find that constructing temporary geometry, such as the circle and points in the two foregoing examples, will help you solve problems in new ways. Don't hesitate to experiment! Remember, you've always got the Save and Undo commands to help you recover from mistakes.

FIGURE 5.21

Finding an exact distance along an arc using points and the Measure command

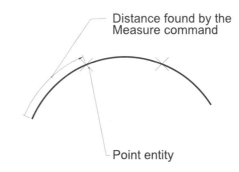

Changing the Length of Objects

Suppose, after finding the length of an arc, you realize you need to lengthen the arc by a specific amount. The Modify ➢ Lengthen command lets you lengthen or shorten arcs, lines, splines, and elliptical arcs. Here's how to lengthen an arc:

1. Click on Lengthen from the Modify toolbar.

2. At the DElta/Percent/Total/DYnamic/<Select object>: prompt, type **t**↵.
3. At the Angle/<Enter total length (1.0000)>: prompt, enter the length you want for the arc, then press ↵.
4. At the <Select object to change>/Undo: prompt, click on the arc you wish to change. Be sure to click on the point nearest the end you want to lengthen. The arc increases in length to the size you specified.

Lengthen will also shorten an object if the object is currently longer than the value you enter.

In this short example, we have demonstrated how to change an object to a specific length. You can use other criteria to change an object's length, using these options available in the Lengthen command:

DElta lets you lengthen or shorten an object by a specific length. To specify an angle rather than a length, use the Angle suboption.

Percent lets you increase or decrease the length of an object by a percentage of its current length.

Total lets you specify the total length or angle of an object.

DYnamic lets you graphically change the length of an object using your cursor.

Drawing Parallel Lines

Multilines are generally not used in mechanical engineering. Civil and architectural engineers will find them convenient when working on projects. You will first do your schematic layout using simple lines for walls. Then, as the design requirements begin to take shape, you can start to add more detailed information about the walls; for example, indicating wall materials or locations for insulation. AutoCAD provides *Multilines* (the Mline command), which are double lines that can be used to represent walls. Mline can also be customized to display solid fills, center lines, and additional linetypes shown in Figure 5.22. You can save your custom Multilines as Mline styles, which are in turn saved in special files for easy access from other drawings.

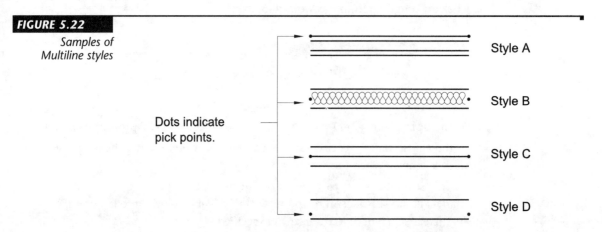

FIGURE 5.22
Samples of Multiline styles

The following exercise shows you how you might build information into your drawings by using Multilines to indicate wall types.

1. Click Multiline from the Draw menu or type **ml**↵. You'll see two lines in the prompt area:

```
Justification = Top, Scale = 1.00, Style = Standard
Justification/Scale/STyle/<From point>:
```

The first line of the prompt gives you the current settings for Mline. You'll learn more about those settings in the section following this exercise.

2. At the Justification/Scale/Style/<From Point>: prompt, type **s**↵ (for Scale).
3. At the Set Mline Scale <0.00>: prompt, type **1**↵.
4. Pick a point to start the double line.
5. Continue to select points to draw more double-line segments, or type **c**↵ to close the series of lines.

Let's take a look at the meaning of the Mline settings included in the prompts you saw in steps 1 and 2 above.

Justification controls how far off center the double lines are drawn. The default sets the double lines equidistant from the points you pick. By changing the justification value to be greater than or less than 0, you can have AutoCAD draw double lines off center from the pick points.

Scale lets you set the width of the double line.

Close closes a sequence of double lines, much as the Line command's Close option does.

Style lets you select a style for Multilines. You can control the number of lines in the Multiline, as well as the linetypes used for each line in the Multiline style, by using the Mledit command (discussed in the next section).

Customizing Multilines

In Chapter 4 you learned how to make a line appear dashed or dotted using linetypes. In a similar way, you can control the appearance of Multilines using the Multiline Styles dialog box. This dialog box allows you to

- Set the number of lines that appear in the Multiline
- Control the color of each line
- Control the linetype of each line
- Apply a fill between the outermost lines of a Multiline
- Control whether and how ends of Multilines are closed

To access the Multiline Styles dialog box, choose Format ➢ Multiline Style from the menu bar, or type **mlstyle**↵ at the command prompt. You will see the Multiline Styles dialog box.

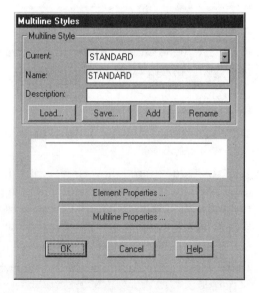

Once you have drawn a Multiline in a particular style, you cannot modify the style settings for that style in the Element Properties and Multiline Properties dialog boxes described later in this section. The Multiline Styles dialog box only allows you to set up a style before it is used in a drawing.

At the top of the dialog box, you see a group of buttons and input boxes that allow you to select the Multiline Style you want to work with. The Current drop-down list offers you a selection of existing styles. In the Name input box you can name a new style you are creating, or rename an existing style. The Description input box lets you attach a description to an Mline style for easy identification. Use the Add and Save buttons to create and save Multiline styles as files so they can be accessed by any AutoCAD drawing, and with the Load button you can retrieve a saved style for use in the current drawing. Rename lets you change the name of an Mline style (the default style in a new drawing is called Standard).

In the lower half of the Multiline Styles dialog box are two buttons—Element Properties and Multiline Properties—that allow you to make adjustments to the Mline style currently indicated at the top of the dialog box. This Mline is also previewed in the middle of the dialog box.

Element Properties

In the Element Properties dialog box, you can control the properties of the individual elements of a newly created style, including the number of lines that appear in the Mline, their color, and the distance they appear from your pick points.

The Element Properties settings are not available for existing Mline styles.

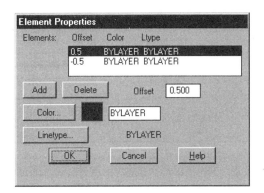

For example, click on the Add button, and another line is added to your Multiline. The offset distance of the new line appears in the list box. The default value for new lines is 0.0, which places the line at the center of the standard Multiline. To delete a line, highlight its offset value in the list box and click on the Delete button. To change the amount of offset, highlight the current value in the Offset input box and enter a new value.

To change the color and linetype of individual lines, use the Color and Linetype buttons, which open the Color and Select Linetype dialog boxes, both of which you have already worked with. In Figure 5.23 you see some examples of Multilines and their corresponding Element Properties settings.

You can easily indicate insulation in a drawing by adding a third center line (offset of 0.0), and giving that center line a Batting line. This linetype draws an S-shaped pattern typically used to represent fiberglass batt insulation. To see how other patterns can be created, see the sections on linetype customization in Chapter 21.

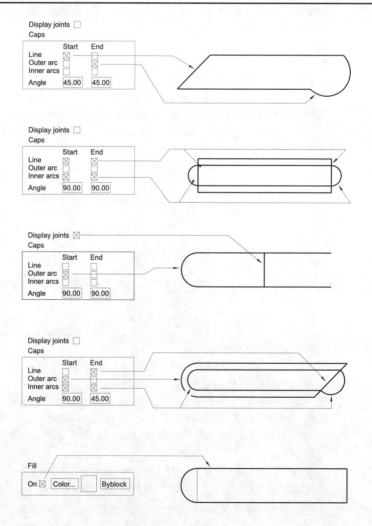

FIGURE 5.23

Samples of Mline styles you can create

Multiline Properties

The Multiline Properties dialog box lets you control how the Mline is capped at its ends, and whether joints are displayed.

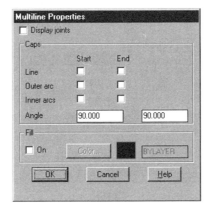

To turn a cap on, click on the check box next to the type of cap you want. If you prefer, you can give your Multiline style a solid fill, using the Fill check box.

Joining and Editing Multilines

Multilines are unique in their ability to combine several linetypes and colors into one entity. For this reason, you need special tools to edit them. On the Modify menu, the Edit Multiline option (and the Mledit command) has the sole purpose of allowing you to join Multilines in a variety of ways, as demonstrated in Figure 5.24.

FIGURE 5.24

The Mledit options and their meanings

Option	Description	
CLOSED CROSS	Trims one of two intersecting multilines so that they appear overlapping.	⊥⊤
OPEN CROSS	Trims the outer lines of two intersecting multilines.	⌐⌐
MERGED CROSS	Joins two Multilines into one multiline.	
CLOSED TEE	Trims the leg of a tee intersection to the first line.	
OPEN TEE	Joins the outer lines of a multiline tee intersection.	
MERGED TEE	Joins all the lines in a multiline tee intersection	
CORNER JOINT	Joins two multilines into a corner joint.	
ADD VERTEX	Adds a vertex to a multiline. The vertex can later be moved.	||| → ⟨⟨⟨
DELETE VERTEX	Deletes a vertex to straighten a multiline.	⟩⟩⟩ → |||
CUT SINGLE	Creates an opening in a single line of a multiline.	
CUT ALL	Creates a break across all lines in a multiline.	
WELD	Closes a break in a multiline.	

Here's how the Mledit command is used:

1. Type **mledit**↵ at the command prompt. The Multiline Edit Tools dialog box appears (see Figure 5.25), offering a variety of ways to edit your Multilines.

FIGURE 5.25

Multiline Edit Tools

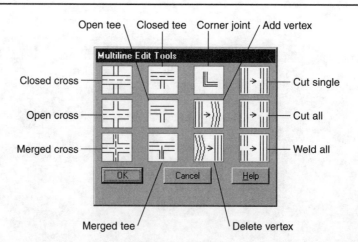

2. Click on the graphic that best matches the edit you want to perform.

3. Select the Multilines you want to join or edit.

Another option is to explode Multilines and edit them using the editing tools you've used in this and previous chapters. When a Multiline is exploded, it is reduced to its component lines. Linetype assignments and layers are maintained for each component.

If you are doing a lot of work with Multilines, you can open the Modify II toolbar. It contains a Multiline Edit tool that opens the Multiline Edit Tools dialog box. To open the Modify II toolbar, right-click on any toolbar and then click on the Modify II check box in the Toolbars dialog box.

Eliminating Blocks, Layers, Linetypes, Shapes, and Styles

A prototype may contain blocks and layers you don't need in your new file. When you erase a block, it remains in the drawing file's database. The erased block is considered "unused" because it doesn't appear as part of the drawing. Such extra blocks can slow you down by increasing the amount of time needed to open the file. They will also increase the size of your file unnecessarily. There are two commands for eliminating unused elements from a drawing: Purge and Wblock.

Selectively Removing Unused Elements

The *Purge* command is used to remove unused individual blocks, layers, linetypes, shapes, and text styles from a drawing file. To help keep the file size down and to make layer maintenance easier, you will want to purge your drawing of unused elements.

As you will see in the File ➢ Drawing Utilities ➢ Purge cascading menu and the Purge command prompt, you can purge other unused drawing elements, such as linetypes and layers, as well. Bear in mind, however, that Purge will not delete certain primary drawing elements—namely, layer 0, the Continuous linetype, and the Standard text style.

1. Click on File ➢ Open and open the DTO100 file.

2. Click on File ➢ Drawing Utilities ➢ Purge ➢ Blocks.

3. At the Names to purge <*>: prompt, you can enter the name of a specific block or press ↵ to purge all the blocks.

4. Go ahead and press ↵; you will see the prompt: Verify each name to be purged? <Y>:. This lets you selectively purge blocks by displaying each block name in succession.

5. Press ↵ again.

6. At the Purge block BKT-V2? <N>: prompt, enter **y**↵. The Purge block prompt will repeat for each unused block in the file. Continue to enter **y**↵ at each prompt until the Purge command is completed.

The DT0100 file is now purged of most, but not all, of the unused blocks. Now let's take a look at how to delete all the unused elements at once.

Opening a File as Read-Only

When you open an existing file, you might have noticed the Read Only Mode check box in the Open Drawing dialog box. If you open a file with this option checked, AutoCAD will not let you save the file under its original name. You can still edit the drawing any way you please, but if you attempt to use File ➢ Save, you will get the message "Drawing file is write-protected." You can, however, save your changed file under another name.

The read-only mode provides a way to protect important files from accidental corruption. It also offers another method for re-using settings and objects from existing files by letting you open a file as a prototype, and then save the file under another name.

Removing All Unused Elements

Purge does not remove nested blocks on its first pass. For example, say you had a bolt block inserted into a bracket block which is inserted into the bar drawing. To remove them using Purge, you would have to start the command again and remove the nested blocks. For this reason, Purge can be a time-consuming way to delete a large number of elements.

Cross-referenced (Xref) files do not have to be purged, because they never actually become part of the drawing's database. You must use the Xref command to detach these files.

In contrast, the File ➤ Export option (Wblock) enables you to remove *all* unused elements—including blocks, nested blocks, layers, linetypes, shapes, and styles—all at once. You cannot select specific elements or types of elements to remove.

Be careful: In a given file, there may be a block that is unused but that you want to keep, so you might want to keep a copy of the unpurged file.

If You Want to Experiment...

Try using the techniques you learned in this chapter to create a new file. Then draw the top view of the bracket, using Xlines, rays, Osnaps and point filters.

Chapter 6

Enhancing Your Drawing Skills

FEATURING

Reducing Wait Times by Controlling Regenerations

Improving Performance through Smart Use of the Display

Improving Performance Using Layers

Saving Views of Your Drawing

Enhancing Your Drawing Skills

As your drawing becomes more complex, you will find that you need to use the Zoom and Pan commands more often. Mechanical design requires accuracy whenever you are creating geometry, and it is often difficult to see the detail when you are standing back looking at the big picture. Larger drawings also require some special editing techniques. You will learn how to assemble and view drawings in ways that will save you time and effort as your design progresses.

You can use the Grid and Snap functions for simple layout and sketches. If you continue to use Grid for larger, more complex drawings, you will find that the grids take time to regenerate as you zoom in and out. You will have to set the grid to be more coarse and thus less helpful—or suffer the cost in productivity. Snap is difficult to use in larger drawings because the cursor resolution is relative to the screen, and it can get very difficult to move the cursor one snap interval on large drawings. For these reasons, we recommend that you dispense with using Snap and Grid for the rest of these tutorials—and for much of your production work as well.

Start by opening a new file using the Versions template.

1. Select File ➢ New. At the Create New Drawing dialog box, click on the Use a Template button.
2. Find and select the template titled Versions.dwt and click on OK.
3. Find the Grid toggle button on the status bar and double-click on it to toggle it off. You may toggle this feature on and off whenever you need the grid.
4. Do the same operation for Snap.

You may toggle either of these functions on and off by the double-click method or by typing **grid** or **snap** and following the prompts. Both Grid and Snap will continue to use the .125 grid functions that you had originally set until you change them. Next save the file as Versions.dwt.

5. Type **saveas**↵. In the dialog box, click on the Save as Type drop-down list and select the Drawing File Template (.dwt) option. The list of templates appears.
6. Select the template named Versions.dwt. Click on the Save button. You'll get a Warning: "<path>\versions.dwt This file already exists. Replace existing file?" and two buttons marked Yes and No.

You *do* want to replace the existing file, which will redefine it.

7. Click on the Yes button. The Template Description dialog box appears. You don't need to change the description. Click on OK to continue.

From now on, when you start a new drawing with this template, both Grid and Snap will be off. If you want to use them for any reason, just toggle them on. When you open the new version of the bar drawing named 6_bar.dwg, you will find that it was begun using the new template.

Assembling the Parts

Start by using a slightly modified version of the file that you have been working on. We have used the version of the bracket with two views, and we have changed the size of the tapped holes to ⅜" in the bracket and bar. You will also find three views of a ⅜" hex-head screw.

1. Open the 6_bar.dwg file from the companion CD-ROM. Use the Save As option from the Files menu to save the file as **6_bar** to your hard disk. The choice of directories is yours, but remember that you may have to find this file later, so make a note of the file's name and location.
2. Choose View ➢ Zoom ➢ All, or type **z**↵**a**↵ to get an overall view of the drawing area. Zoom in to the end of the bar. Include the bar end, the bracket, and the bolt details.

3. For these and all future exercises turn Grid and Snap off.

4. Create a group from the front view bracket objects. Type **g↵** to open the Object Grouping dialog box (see Figure 6.1). Name the group **bkt_front**. Create another group of the objects of the side view of the bracket. Name these **bkt_side**. Continue with the bolt and call the hex view **bolt_plan**, the lower view **bolt_front**, and the last view **bolt_side**. Remember that AutoCAD is not sensitive to typing upper- or lowercase names. You may use whichever style of data entry that you like or find convenient at the time.

FIGURE 6.1

The Object Grouping dialog box showing the group names

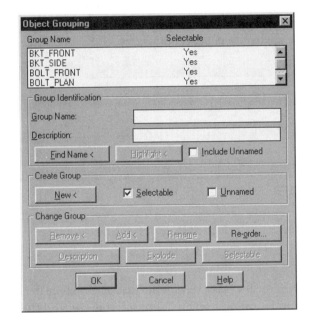

5. Move the `bkt_front` group and the `bkt_side` group together into position on the bar. Use the Center Osnap to help you align the top mounting hole in the bracket with the top leftmost mounting hole on the bar.

6. Zoom All to see the overall view. You may need to use the scroll bar to see all of the right side of the objects.

7. Be sure that Ortho mode is on and move the `bkt_side` group slightly beyond the right end of the bar.

This is an example of the great power of the Ortho mode for mechanical design. All of the views of a multiple-view drawing are supposed to align with one another horizontally and vertically. You will find that it is necessary to move entire views to better place them for dimensioning and positioning on title blocks. Notice how easy it was to move the side view group horizontally and maintain alignment with the front view.

8. Select all three groups of the bolt and use the Center Osnap to move them so that the bolt plan view is aligned with front view of the bracket and bar.
9. With Ortho still on, move the `bolt_side` group of the bolt until it is nearly in position next to the side view.
10. Move the front view of the bolt down a few units beyond the bottom of the bar.
11. If you have had to zoom in for any of these moves you should now Zoom All to see the overall effect of your work. Your drawing should be similar to Figure 6.2.

FIGURE 6.2

The bar, bracket, and bolt views

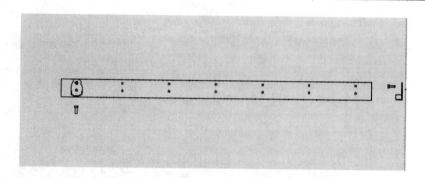

12. Array the front view of the bracket along the bar. It is a rectangular pattern with one row and seven columns. You can find the horizontal spacing with the Distance command and by checking the center-to-center distance of the tapped holes.
13. Array the bolt using a rectangular pattern of two rows and seven columns. Use the existing geometry to provide the horizontal and vertical distances. At the `Unit cell or distance between rows (--):` prompt, pick the center of the top left hole; at the `Other corner:` prompt, pick the center of the bottom hole in the next row. This will give AutoCAD the incremental distance between the sets of objects. See Figure 6.3 for the pick points.

FIGURE 6.3

The array pick points for the Unit cell distances between the bolt heads

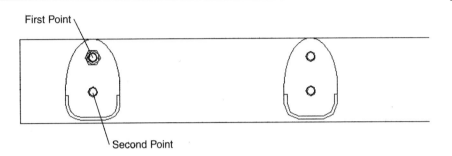

The bar is a custom aluminum extrusion. You can find the cross section of the extrusion on the CD-ROM as Extrusion.dwg.

14. Use the Insert Block tool on the Draw toolbar to import the extrusion cross section and place it near the right end of the bar.

Next make the extrusion cross section become the side view of the bar. You will use point filters to help you move the extrusion into place.

15. Click on the Move tool from the Modify toolbar. Pick the extrusion, then press ↵ to complete the selection set. Using the Running Osnaps (which have been on since you started this exercise), pick the lower-left endpoint of the extrusion. At the to point: prompt type .y↵, and at the .Y of: prompt, pick a point anywhere along the bottom of the bar. At the of (need XZ): prompt, pick a point anywhere to the right of the bracket.

Next insert (as exploded) a block from the CD-ROM called Washer.dwg.

NOTE

The hardware is being provided as blocks here because most likely you will either create, purchase, or subscribe to a library of these parts. Very extensive libraries are becoming available on CD (we have included a few samples) and via the Internet.

1. Select the Insert Block tool from the Draw toolbar.
2. Use the Files button to find the file named Washers.
3. Click on the Explode check box. Click on OK.
4. Pick a point near the right end of the bar and insert the washers at full scale. Zoom in to get a close-up view of the washers and bolt.
5. Next, create a stack of the washers and locate them on the bolt.

Highlight the grips of the edge view of the lock washer (the smaller washer) by picking any point on any object in the view. Note that the view is a block because only one grip appears. Pick the grip to make it hot.

1. Drag the cursor to the right, with the edge view in tow, until the midpoint triangle marker appears on the left side of the edge view of the flat washer, then left-click. The two side views are stacked.

2. Use either grips or the Move tool to move the stacked washers from the midpoint of the left edge of the lock washer to the midpoint of the vertical line of the right edge of the bolt. You may want to zoom in even closer to see that you have picked the right points. Your drawing should look like Figure 6.4.

FIGURE 6.4

The side view of the washers and bolt

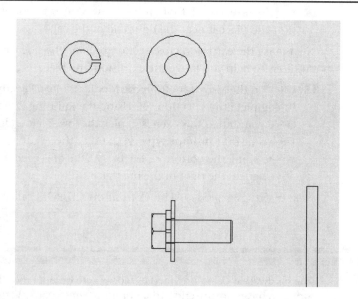

Locate the bolt and washers on the bracket. You will need two bolt and washer sets, so make a copy. You can use the distance between the tapped holes in the bar for the displacement of the copy.

3. Use Move and pick the bolt and both washers. At the `from point:` prompt, pick the midpoint of right side of the of the flat washer and at the `second point:` prompt, Shift+right-click, or type **per**↵, to select the Perpendicular object snap.

4. Move the mouse until the Perpendicular Osnap marker appears, then left-click the mouse.

5. Click on the Copy tool on the Modify toolbar and type **p**⏎ to use the previous selection set of objects, then press ⏎ to complete the selection set and begin the next part of the command. At the `first point or displacement:` prompt, use the Center Osnap to select any of the top row of circles that represent the tapped holes in the bar (look for the Center Osnap marker to appear as you hover the mouse over any of the circles in the bar, bracket, and bolts). Next, use the Center Osnap to pick the hole directly below the first one. AutoCAD measures this distance and moves the bolt and washers accordingly.

Next, move the extrusion into place.

6. Use Move tool again and the Perpendicular Osnap to move the extrusion into position. Your drawing should look like Figure 6.5.

FIGURE 6.5

The side view of the bracket, bolt, washer, and bar assembly

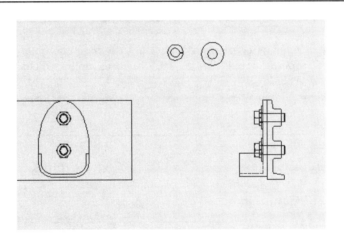

Next you will stack the front view of the washers, move them into place, and array them.

7. Select the front view of either washer and use the running Center Osnap to identify and pick the center of the view as the location of the `Base point or displacement:` of the washer that you want to move. At the `Second point of displacement:` prompt, use the running Center Osnap to identify and pick the new location of the front view of the washer.

8. Use the Move tool. Select both washers and use the center-to-center method again to position them at the lower-right bolt location on the bar.

9. Use the Array tool (see Chapter 5) to create a rectangular array of the previous selection set. The array will have two rows and seven columns with a distance of 1.375 units between the rows and –9 units between the columns. Use the center points of two of the holes to give the distances.
10. Zoom All to see the full effect of you handiwork. You should see something like Figure 6.6.
11. Save your work.

You could do much more to clean up these views. There are hidden lines and center lines to be added, and object lines to be removed before the views in this drawing could be considered finished. You have had to zoom in and out to see your work as you went along creating these views. Let's take a look at the tools AutoCAD has provided to make this task easier.

If you happen to insert a block in the wrong coordinate location, you can use the Properties tool in the Object Properties toolbar to change the insertion point for the block.

FIGURE 6.6

The overall view of the bar, bracket, bolt, and washers

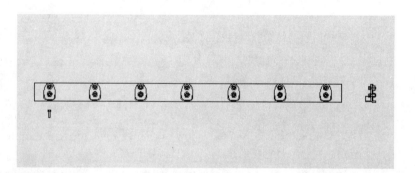

Taking Control of the AutoCAD Display

By now you should be familiar with the Pan and Zoom functions in AutoCAD. There are many other tools at your disposal that can help you get around in your drawing. In this section, you'll get a closer look at the different ways you can view your drawing.

Understanding Regeneration and Redraw

AutoCAD uses two methods for refreshing your drawing display: drawing regeneration (or *Regen*) and *Redraw*. Each serves a particular purpose, although these may not be clear to a new user.

AutoCAD stores drawing data in two ways; one is like a database of highly accurate coordinate information and object properties. This is the core information you supply as you draw and edit your drawing. For the purposes of this discussion, we'll call this simplified database the "virtual display," since it is like a computer model of the overall display of your drawing. This virtual display is in turn used as the basis for what is shown in the drawing area. When you issue a Redraw, you are telling AutoCAD to re-read this virtual display data and display that information in the drawing area. A Regen, on the other hand, causes AutoCAD to rebuild the virtual display based on information from the core drawing database.

As you edit drawings, you may find that some of the lines in the display disappear or otherwise appear corrupted. Redraw will usually restore such distortions in the display. In earlier versions of AutoCAD, the Blipmode system variable was turned on by default, causing markers called blips to appear wherever points were selected. Redraw was, and still is, useful in clearing the screen of these blips.

Regens are used less frequently, and are brought to bear when changes occur to settings and options that have a global effect on a drawing, such as a linetype scale change, layer color change, or text style changes (you'll learn more about text styles in Chapter 7). In fact, Regens are normally issued automatically when such changes occur. You usually don't have to issue the Regen command on your own, except under certain situations.

Regens can also occur when you select a view of a drawing that is not currently included as part of the virtual display. The virtual display contains display data for a limited area of a drawing. If you zoom or pan to a view that is outside that virtual display area, a regen occurs.

NOTE You may notice that Pan Realtime and Zoom Realtime will not work beyond a certain area in the display. When you've reached a point when these commands seem to stop working, you've come to the limits of the virtual display data. In order to go beyond these limits, AutoCAD must rebuild the virtual display data from the core data; in other words, it must do a drawing regeneration.

In past versions of AutoCAD, regens were to be avoided at all cost, especially in large files. A regen of a very large file could take several minutes to complete. Today, with faster processors, large amounts of RAM, and a retooled AutoCAD, regens are not the problem they once were. Still, they can be annoying in multi-megabyte files, and if you are using an older Pentium-based computer, regens can still be a major headache. For these reasons, it pays to understand the finer points of controlling regens.

In this section, you will discover how to manage regens, thereby reducing their impact on complex drawing. You can control how regens impact your work in three ways:

- By taking advantage of AutoCAD's many display-related tools
- By setting up AutoCAD so that regens do not occur automatically
- By freezing layers that do not need to be viewed or edited

We will explore these methods in the upcoming sections.

Exploring Other Ways of Controlling AutoCAD's Display

Perhaps one of the easiest ways of avoiding regens is by making sure you don't cross into an area of your drawing that falls outside of the virtual display's area. If you use Pan Realtime and Zoom Realtime, you are automatically kept safely within the bounds of the display list. In this section, you'll be introduced to other tools that will help keep you within those boundaries.

Controlling Display Smoothness

The virtual display can be turned on or off using the Viewres command. The Viewres system variable setting is on by default, and for the most part, should remain on. You can turn it off by typing **viewres**↵**no**↵ at the command prompt. This is not recommended. With Viewres off, a regen occurs every time you change your view using Pan or Zoom.

The Viewres command also controls the smoothness of linetypes, arcs, and circles when they appear in an enlarged view. With the display list active, linetypes sometimes appear as continuous even when they are supposed to be dotted or dashed. You may have noticed in previous chapters that on the screen, arcs appear to be segmented lines, though they are always plotted as smooth curves. You can adjust the Viewres value to control the number of segments an arc appears to have: the lower the value, the fewer the segments and the faster the redraw and regeneration. However, a low Viewres value will cause non-continuous linetypes, such as dashes or center lines, to appear continuous.

You can set the Viewres value in the Performance tab of the Preferences dialog box under the setting labeled Arc and Circle Smoothness.

Another way to accelerate screen redraw is to keep your drawing limits to a minimum area. If the limits are set unnecessarily high, AutoCAD may slow down noticeably. Also, make sure the drawing origin falls within the drawing limits.

A good value for the Viewres setting is 500. At this setting, linetypes display properly, and arcs and circles have a reasonably smooth appearance. At the same time, redraw speed is not noticeably degraded. However, you may want to keep Viewres lower still if you have a limited amount of RAM. High Viewres settings can adversely affect AutoCAD's overall use of memory.

Using the Aerial View

Let's take a tour of a tool that lets you navigate drawings that represent very large areas. It's called the Aerial View.

1. Click on Aerial View on the Standard Toolbar. The Aerial View window appears, as shown in Figure 6.7.
2. Move your cursor over the Aerial View window. Notice that you have a crosshair cursor in the window. This is the Aerial View zoom cursor.

FIGURE 6.7
The Aerial View window and its components

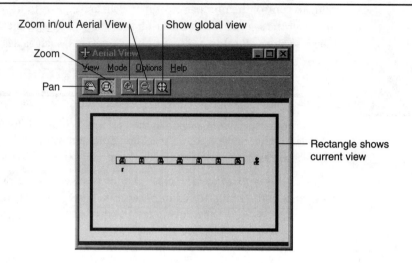

3. Click and drag on a point in the lower-left corner of the Aerial View. A view window appears.

4. Click on a point above and drag to the right of the first point you selected. Your view in the main AutoCAD Drawing area enlarges to display the area you just selected. You also see a bold rectangle in the Aerial View window showing the location of your Drawing Area view.

The Aerial View window doesn't zoom in. It continues to display the overall view of your drawing. The bold rectangle shows you exactly where you are in the overall drawing at any given time. This feature is especially useful in drawings of large areas that may take several pans to cross.

The View ➤ Zoom ➤ Dynamic option performs a similar function to the Aerial View window, but instead of opening a separate window, Dynamic temporarily displays the overall view in the drawing area.

Now let's look at the Pan feature on the Aerial View toolbar. The Pan feature can be helpful if you are moving or copying objects from one part of a drawing to another.

1. Choose Mode ➢ Pan from the Aerial View menu bar or click on the Pan tool on the Aerial View toolbar. Now as you move your cursor over the Aerial View window, you see a dotted rectangle. This lets you pan to any location in the view.

2. Move the dotted rectangle so it encloses an area in the upper-right corner and left-click the mouse. Your view in the AutoCAD Drawing area immediately pans to that area.

Once again, the view in the Aerial View window doesn't change during the pan, allowing you to see where you are in the overall drawing.

The Aerial View window is a great tool when you are working on a drawing that requires a lot of magnification in your zoomed-in views. You may not find it very helpful on drawings that don't require lots of magnification, like the bracket drawing you worked on in Chapters 3 and 4.

You were able to use the major features of the Aerial View in this exercise. Here are a few more features you can try on your own:

View ➢ Zoom In zooms in on the view in the Aerial View.

View ➢ Zoom Out zooms out on the view in the Aerial View.

View ➢ Global displays an overall view of your drawing in the Aerial View window. Global is like a View ➢ Zoom ➢ Extents option for the Aerial View.

Options ➢ Auto Viewport controls whether a selected viewport is automatically displayed in the Aerial View window. When checked, this option will cause the Aerial View window to automatically display the contents of a viewport when it becomes active. (See Chapters 9 and 12 for more on viewports.)

Options ➢ Dynamic Update controls how often changes in your drawing are updated in the Aerial View. When this option is checked, the Aerial View window is updated as you work. You may want to turn this feature off in very complex drawings as it can slow down redraw times.

Saving Views

Another way of controlling your views is by saving them. You might think of saving views as a way of creating a bookmark or placeholder in your drawing. You'll see how to save views in the following set of exercises.

A few details in the side view of the drawing are not complete. You'll need to zoom in on the areas that need work to add the lines, but these areas are spread out over the drawing. You could use the Aerial View window to view each area. There is, however, another way to edit widely separated areas: First, save views of the areas you want to work on, then jump from saved view to saved view. This technique is especially helpful when you know you will often want to return to a specific area of your drawing.

Stored views are especially useful for controlling what appears in your plot. The Plot command allows you to plot a saved view.

1. First close the Aerial View window by clicking on the Close button in the upper-left corner of the window.
2. Click on View ➣ Zoom ➣ All, or type **z**↵**a**↵, to get an overall view of the plan.
3. Click on View ➣ Named Views. The View Control dialog box appears.

From this dialog box, you can call up an existing view (Restore), create a new view (New), or get detailed information about a view (Description).

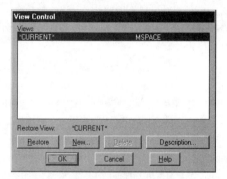

4. Click on the New button. The Define New View dialog box appears.

5. Click on the Define Window radio button. Notice that the grayed options become available.
6. Click on Window. The dialog box momentarily disappears.
7. At the `First corner:` prompt, click near one of the corners of the side view. (Don't pick a point where a marker has appeared; if you see one of the Osnap markers appear, move the cursor a little farther away.) You don't have to be exact, because you are selecting view windows.
8. At the `Other corner:` prompt, drag the mouse to create a window around the view and left-click to pick the point. Again, if you see an Osnap marker, move the mouse a little farther away from the object. The dialog boxes will reappear.
9. Type **first** for the name of the view you just defined. As you type, the name appears in the New Name input box.
10. Click on the Save View button. The Define New View dialog box closes, and you see FIRST listed in the View Control list.
11. Repeat steps 3–9 to define another view, named Second. Use Figure 6.8 as a guide for where to define the windows. Click on OK when you are done.

A quick way to restore saved views is to type **view.↵r.↵**, then enter the name of the view you wish to restore.

FIGURE 6.8

Save view windows in these locations for the Plan drawing.

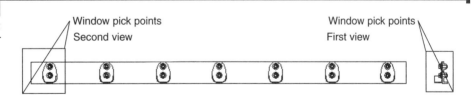

Now let's see how to recall the views that you've saved.

1. With the View Control dialog box open, click on FIRST in the list of views.
2. Click on the Restore button and then click on OK. Your screen displays the first view you selected.
3. Use the View Control dialog box again to restore the view named SECOND.

Next, you will zoom out to the extents of the your drawing and then create a new View to be named Overall.

1. Choose View ➢ Zoom ➢ Extents or type z↵e↵.
2. Enter **view**↵**s**↵ at the command prompt.
3. At the View name to save: prompt, enter **overall**↵.
4. Now save your work.

In prior versions of AutoCAD, Zoom ➢ Extents caused a regen to occur. Regens no longer occur with this command in Release 14.

As you can see, this is a quick way to save a view. With the name Overall assigned to this view, you can easily recall the overall view at any time. (The View ➢ Zoom ➢ All option gives you an overall view, too, but it may zoom out too far for some purposes, or it may not show what you consider to be an overall view.)

Another useful tool for getting around in your drawing is the Zoom toolbar. It contains tools for Zoom Window, Dynamic, Scale, Center, In, Out, All, and Extents. To open the Zoom toolbar, right-click on any toolbar, then click on the Zoom check box in the Toolbars dialog box that appears.

Opening a File to a Particular View

The Open Drawing dialog box contains a Select Initial View check box. If you open an existing drawing with this option checked, you are greeted with a Select Initial View dialog box just before the opened file appears on the screen. This dialog box lists any views saved in the file. You can then go directly to a view by double-clicking on the view name. If you have saved views and you know the name of the view you want, using Select Initial View saves time when you're opening very large files.

Freezing Layers to Control Regeneration Time

As mentioned earlier, you may wish to turn certain layers off altogether to plot a drawing containing only selected layers. But even when layers are turned off, AutoCAD still takes the time to redraw and regenerate them. The Layer & Linetype Properties dialog box offers the Freeze option. This acts like the Off option, except that Freeze causes AutoCAD to ignore frozen layers when redrawing or regenerating a drawing. By freezing layers that are not needed for reference or editing, you can reduce the time AutoCAD takes to perform regens.

You can freeze and thaw individual layers by clicking on the Sun icon in the layer drop-down list in the Object Properties toolbar.

Block Visibility with Freeze and Thaw

When the layer of a block is frozen, the entire block is made invisible, regardless of the layer assignments of the objects contained in the block.

Keep in mind that when blocks are on layers that are not frozen, the individual objects that are a part of a block are still affected by the status of the layer to which they are assigned.

You can take advantage of this feature by using layers to store parts of a drawing that you may want to plot separately. With respect to Freeze/Thaw visibility, external referenced files inserted using the external reference (Xref) command also act like blocks. For example, you can Xref several drawings on different layers. Then, when you want to view a particular Xref drawing, you can freeze all the layers except the one containing that drawing.

Using layers and blocks requires careful planning and record keeping. If used successfully, however, this technique of managing block visibility with layers can save substantial time when you're working with drawings that use repetitive objects or that require similar information that can be overlaid.

Taking Control of Regens

Another way to control regeneration time is by setting the Regenmode system variable to 0 (zero). You can also use the Regenauto command to accomplish the same thing, by typing **regenauto↵off↵**.

If you then issue a command that normally triggers a regen, AutoCAD will give the message Regen queued. For example, when you globally edit attributes, redefine blocks, thaw frozen layers, change the Ltscale setting, or in some cases, change a text's style, you will get the Regen queued message. You can "queue up" regens at a time you choose, or you can issue a Regen to update all the changes at once by choosing View ➢ Regen or by typing **re**↵. This way, only one regen occurs instead of several.

By taking control of when regens occur, you can reduce the overall time you spend editing large files.

Creating Multiple Views

So far, you've looked at ways to help you get around in your drawing while using a single view window. You also have the capability to set up multiple views of your drawing, called viewports. With viewports, you can display more than one view of your drawing at one time in the AutoCAD drawing area. For example, you can have one viewport to show a close-up of the extrusion, another viewport to display the overall Plan view, and yet another to display the a bolt detail.

When viewports are combined with AutoCAD's Paper Space feature, you can plot multiple views of your drawing. Paper Space is a display mode that lets you "paste up" multiple views of a drawing, much like a page layout program. To find out more about viewports and Paper Space, see Chapters 9 and 12.

If You Want to Experiment...

Open some of the sample files in the Sample subdirectory. Azimuth is a very good one. If you did not load them, they are available on your AutoCAD CD-ROM. Try Aerial View, the Regen command, and layer control with these files to better understand these options for improving productivity.

Chapter 7

Adding Text to Drawings

FEATURING

Adding Text

Setting Justification of Text

Setting Text Scale

Selecting Fonts

Creating a Text Style

Editing Existing Text

Importing Text from Outside AutoCAD

Using the Spelling Checker

Adding Text to Drawings

One of the more tedious drafting tasks is applying notes to your drawing. Anyone who has had to draft a large drawing containing a lot of notes knows the true meaning of writer's cramp. AutoCAD not only makes this job go faster by allowing you to type your notes right into the same document as the corresponding drawing, it also helps you to create more professional-looking notes by using a variety of fonts, type sizes, and type styles. And with Release 14, you have an improved text tool that simplifies access to all the text features.

In this chapter, you will add notes to your bracket drawing. In the process, you will explore some of AutoCAD's text-creation and editing features. You will learn how to control the size, slant, type style, and orientation of text, and how to import text files.

Adding Text to a Drawing

In this section you will type a set of notes that you might find on a working drawing.

1. Start AutoCAD and open the file named Bkt_v1. If you haven't been following the book so far, you can find the file, named ch6_bkt_v1.dwg, on the CD-ROM. Use File ➤ Save As to save as a file called notes.
2. Use the Layer drop-down list to make the Dim layer current. Some users create a layer called Notes and make it the current layer. Dim is the layer on which you will keep all your text information in this exercise.

NOTE

You can place text on any layer. The text will inherit the color and linetype of the layer that you choose. If you placed text on the Hidden layer, it would be red and composed of dashed lines.

3. Choose Multiline Text from the Draw toolbar or type **mt**↵.

4. Click on the first point indicated in the top image of Figure 7.1 to start the text boundary window. This boundary window indicates the area in which to place the text. Notice the arrow near the bottom of the window. It indicates the direction of the text flow.

NOTE

You don't have to be too precise about where you select the points for the boundary because you can make adjustments to the location and size later. The height of the window is unimportant. AutoCAD will use the width of the window to determine where the word wrap (the automatic width limit of the sentence or paragraph) will occur. AutoCAD will continue to write lines of text in the direction of the arrow as long as you do.

5. Click the second point indicated in the top image of Figure 7.1. The Multiline Text Editor appears.

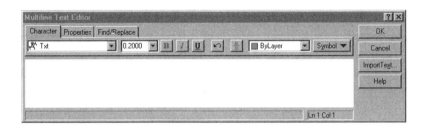

6. You could start typing the text, but first you need to select a size. Point to the Font Height drop-down list and click on it. The default size highlights.

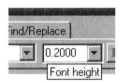

7. Enter **.12**↵ to make the default height .12 inches.

8. Click in the Main text window and type **NOTES: UNLESS OTHERWISE SPECIFIED**. As you type, the words appear in the text window, just as they will appear in your drawing. As you will see later, the text also appears in the same font as the final text.

The default font is a native AutoCAD font called Txt.It. As you will see, you can also use TrueType fonts and PostScript fonts.

9. Press ↵ to advance one line, then enter **1. PERMANENT MARK PART NUMBER AND LATEST REVISION LETTER USING .10 MINIMUM HIGH CONTRASTING CHARACTERS APPROXIMATELY WHERE SHOWN**.

10. Press ↵ again to advance another line and enter **2. THIS DRAWING SHALL BE INTERPRETED PER ANSI Y14.5 1994**.

11. Click on OK. The text appears in the drawing (see the second image in Figure 7.1).

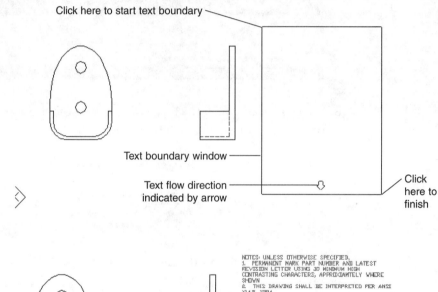

FIGURE 7.1
The top image shows the points to pick to place the text boundary window. The bottom image shows the completed text.

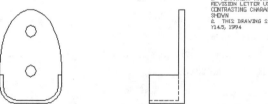

The Text window works like any text editor, so if you make a typing error, you can highlight the error and then retype the letter or word. You can also perform many other functions such as search and replace, import text, or make font changes.

The following sections discuss some of the many options available for formatting text.

TIP If text is included in an area where a hatch pattern is to be placed, AutoCAD will automatically avoid hatching over the text. If you add text over a hatched area, you must re-hatch the area to include the text in the hatch boundary.

Understanding Text Formatting in AutoCAD

AutoCAD offers a wide range of text formatting options. You can control fonts, text height, justification, and width. You can even include special characters such as degree symbols or stacked fractions. In a departure from the somewhat clumsy text implementation of earlier AutoCAD versions, you now have a much wider range of controls over your text.

Adjusting the Text Height and Font

Let's continue our look at AutoCAD text by making a proprietary note. You'll use the Multiline Text tool again, but this time you'll get to try out some of its other features. In this first exercise, you'll see how you can adjust the size of text in the editor.

1. Click the Multiline Text tool again; then select a text boundary window, as shown in the top image of Figure 7.2.
2. At the Multiline Text dialog box, start typing the following text:

   ALL OF THE INFORMATION ON THIS DOCUMENT IS PROPRIETARY. PERMISSION FROM AN OFFICIAL OF THIS COMPANY IS REQUIRED PRIOR TO THE PRODUCTION OF ANY COPIES OF THIS DOCUMENT. DISTRIBUTION OF THIS DOCUMENT REQUIRES PRIOR APPROVAL OF AN OFFICER OF THIS COMPANY.

   As you type, you will notice that the sentences break and appear on separate lines even though you did not press ↵ between them. AutoCAD uses word wrap to fit the text inside the text boundary area.

3. Highlight the text ALL OF THE INFORMATION as you would in any word processor. For example, you can click on the end of the line to place the cursor there; then Shift+click on the beginning of the line to highlight the whole line.
4. Click on the Font Height drop-down list and enter **.16**↵. The highlighted text changes to a smaller size.
5. Highlight the word PROPRIETARY.

6. Click on the Font drop-down list. A list of font options appears.

7. Scroll up the list until you find Complex. This font is available in all installations of AutoCAD. Notice that the text changes to reflect the new font.

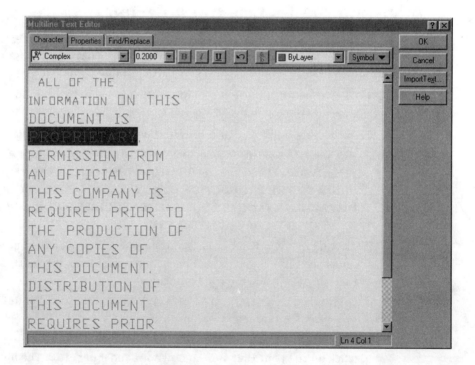

8. With PROPRIETARY still highlighted, click on the Underline tool.

9. Click on OK. The note appears in the area you indicated in step 2 (see the bottom image of Figure 7.2).

FIGURE 7.2
Placing the text boundary window for the proprietary note

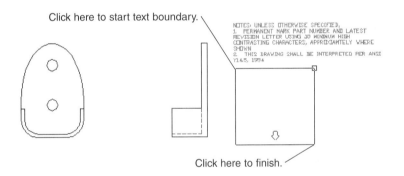

Click here to start text boundary.

Click here to finish.

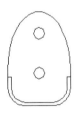
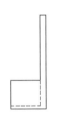

> ### Using PostScript Fonts
>
> If you have PostScript fonts that you would like to use in AutoCAD, you will need to compile them into AutoCAD's native font format. Here's how it's done:
>
> 1. Type **compile**↵. The Compile Shape or Font File dialog box appears.
> 2. Select PostScript Font (*.pfb) from the File of Type drop-down list.
> 3. Double-click on the PostScript font you want to convert into the AutoCAD format. AutoCAD will work for a moment; then you'll see this message:
>
>    ```
>    Compiling shape/font description file
>    Compilation successful. Output file E:\ACADR13\COMMMON\FONTS\
>    ```
>   *fontname*.shx contains 59578 bytes.
>
> When AutoCAD is done, you will have a file with the same name as the PostScript font file, but with the .shx file name extension. If you place your newly compiled font in Auto-CAD's Fonts folder, it will be available in the Style dialog box.
>
> When you work with AutoCAD's SHX font files, it is important to remember that
>
> - License restrictions still apply to the AutoCAD-compiled version of the PostScript font.
> - Like other fonts, compiled PostScript fonts can use up substantial disk space, so compile only the fonts you need.

While using the Multiline Text tool, you may have noticed the [Height/Justify/Rotation/Style/Width]: prompt immediately after you picked the first point of the text boundary. You can use any of these options to make on-the-fly modifications to the height, justification, rotation style, or width of the Multiline text.

For example, after clicking on the first point for the text boundary, you can type **r**↵ and then specify a rotation angle for the text windows, either graphically with a rubber-banding line or by entering an angle value. Once you've entered a rotation angle, you can resume selecting the text boundary.

Adding Color, Stacked Fractions, and Special Symbols

In the previous exercise, you were able to adjust the text height and font, just as you would in any word processor. You saw how you can easily underline portions of your text using the tool buttons in the editor. Other tools allow you to set the color for individual characters or words in the text, create stacked fractions, or insert special characters. Here's a brief description of how these tools work:

- To change the color of text, highlight the text and then select the color from the Text Color drop-down list.

- To turn a fraction into a stacked fraction, highlight the fraction and then click the Stacked Fraction tool.

- To add a special character, place the cursor at the location of the character and then click on the Symbol tool. A drop-down list appears, offering options for special characters.

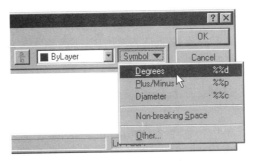

The Symbol tool offers three standard options that are typical for most technical drawings: the Degree, Plus/Minus, and Diameter signs. When you select these options, AutoCAD will insert the proper AutoCAD text code that corresponds to these symbols. They won't appear in the editor as symbols. Instead, they will appear as a special code.

However, once you return to the drawing, you will see the text with the proper symbol. You'll get a more detailed look at special symbols later in this chapter.

Adjusting the Width of the Text Boundary Window

Although your text font and height is formatted correctly, it appears stacked in a way that is too tall and narrow. The following exercise will show you how to change the boundary to fit the text.

1. Click on any part of the PROPRIETARY text you just entered to highlight it.
2. Click on the upper-right grip.
3. Drag the grip to the right to the location shown in Figure 7.3; then click that point.
4. Click on any grip and then right-click on the mouse and select Move.
5. Move the text down slightly.

FIGURE 7.3

Adjusting the text boundary window

Stretch grip to here.

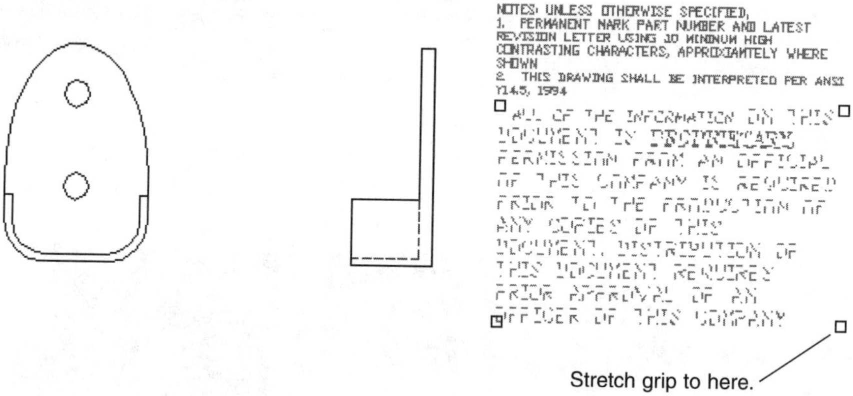

AutoCAD's word wrap feature automatically adjusts the text formatting to fit the text boundary window. This feature is especially useful to AutoCAD users because other drawing objects often affect the placement of text. As your drawing changes, you will need to make adjustments to the location and boundary of your notes and labels.

Adjusting the Text Alignment

The text is currently aligned on the left side of the text boundary. It would be more appropriate to align the text to the right. Here's how you can make changes to the text alignment:

1. If the text is not yet selected, click on it.
2. Click on the Properties tool on the Object Properties toolbar. The Modify Mtext dialog box appears.

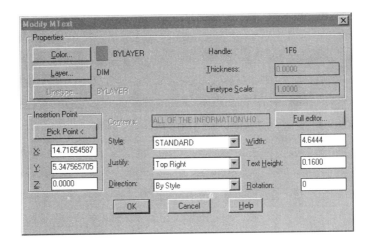

NOTE

If you click on multiple text objects while using the Properties tool, you will only get the abbreviated Change Properties dialog box.

3. Notice that the text appears in the Contents input box. Also included is some special code that helps format the text. If you only want to make changes to the text, this is one place you can do it.

WARNING

The code you see mixed in with the text in the Contents input box is normally hidden from you in the Multiline Text Editor, and you don't really need to concern yourself with it. If you edit the text in the Contents input box, make sure you don't change the coding unless you know what you are doing.

4. Click on the button labeled Full Editor. The Multiline Text Editor appears with the text.

You can go directly to the Multiline Text Editor by using the Ddedit command. Type **ddedit**↵ or **ed**↵ for the keyboard shortcut; then select the text you wish to edit.

5. Click on the Properties tab. The editor changes to display a different set of options.

6. Click on the Justification drop-down list. The alignment options appear.

7. Click on Top Right. All of the text in the window is now aligned to the right.

8. Click on OK and then at the Modify Mtext dialog box, click on OK again. The text changes to align at the top right of the text, as shown in Figure 7.4.

FIGURE 7.4

The text aligned using the Top Right alignment option

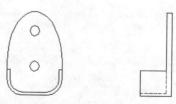

Text Alignment and Osnaps

While it's clear that the text is now aligned at the top right of the text, one important, but less obvious, change also occurred. You may have noticed that the Object Alignment list offered three centered options: Top Center, Middle Center, and Bottom Center. All three of these options will have the same effect on the text's appearance, but each has a different effect on how Osnaps act upon the text. Figure 7.5 shows where the Osnap point occurs on a text boundary, depending on which alignment option is selected. A Multiline text object will only have one insertion point on its boundary that you can access with the Insert Osnap.

FIGURE 7.5

The location of the Insert Osnap points on a text boundary based on its alignment setting

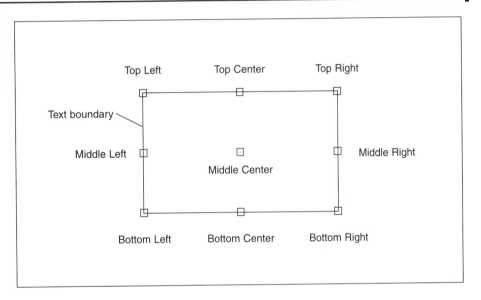

The Osnap point will also appear as an extra grip point on the text boundary when you click on the text. If you click on the text you just entered, you will see that a grip point now appears at the top center of the text boundary.

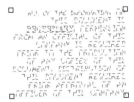

Knowing where the Osnap points occur can be helpful when you want to align the text with other objects in your drawing. In most cases, you can use the grips to align your text boundary, but the Center and Middle alignment options allow you to use the center and middle portions of your text to align the text with other objects.

Editing Existing Text

It is helpful to think of text in AutoCAD as a collection of text documents. Each text boundary window you place is like a separate document. To create and edit these documents, use the Multiline Text Editor.

You've already seen how you can access existing text using the Properties tool on the Object Properties toolbar when you modified the formatting of the NOTES label. Of course, you can use the same tool to change the content of the text. In the following example, you'll use a shortcut to the Multiline Text dialog box to add more text to the NOTES label.

1. Type **ed**↵ or choose Modify ➢ Object ➢ Text. This issues the Ddedit command.

NOTE

The Ddedit command does not allow Noun/Verb selection.

2. Click on the words NOTES: UNLESS. The Multiline Text dialog box appears.
3. Place and click the text cursor on the end of the line that reads 1994, then press ↵ and type **3. ALL RADII TO BE .020 MAXIMUM.**
4. Click on OK. The text appears in the drawing with the additional line.
5. The Ddedit command is still active, so press ↵ to exit Ddedit.

You can also add or delete text using the Properties tool in the Object Properties toolbar. When you select a single text object, the Modify Mtext dialog box appears, displaying the text in the Contents input box; you can change the text here.

In step 5, the Ddedit command remains active, so you can continue to edit other text objects. Besides pressing ↵ to exit the command, you can select another text object.

As with the prior exercise, you can change the formatting of the existing or new text while in the Multiline Text dialog box. Notice that the formatting of the new text is the same as the text that preceded it. Just like in Microsoft Word, the formatting of text is dependent on the paragraph or word to which it is added. If you had added the text after the last line, it would appear in the AutoCAD Txt font and in the same 6-inch height.

Understanding Text and Scale

In the first exercises of this chapter, you made the text height .12 inches. Just as in Chapter 3, where you applied a scale factor to a drawing's final sheet size to accommodate a full-scale drawing, you will sometimes need to make a scale conversion for your text size to make the text conform to the drawing's intended plot scale.

Text-scale conversion is a concept many people have difficulty grasping. As you discovered in previous chapters, AutoCAD allows you to draw at full scale—that is, to represent distances as values equivalent to the actual size of the object. When you plot the drawing later, you tell AutoCAD at what scale you wish to plot, and the program reduces or enlarges the drawing accordingly. This allows you the freedom to input measurements at full scale and not worry about converting them to various scales every time you enter a distance. Unfortunately, this feature can also create problems when you enter text and dimensions. Just as you had to convert the plotted sheet size to an enlarged size equivalent at full scale in the drawing editor, you must convert your text size to its equivalent at full scale.

To illustrate this point, imagine you are drawing a large piping plan at full size on a very large sheet of paper. When you are done with this drawing, it will be reduced to a scale that will allow it to fit on an 8½" × 11" sheet of paper. So you have to make your text quite large to keep it legible once it is reduced. This means that if you want text to appear ⅛" high when the drawing is plotted, you must convert it to a considerably larger size when you draw it. To do this, multiply the desired height of the final plotted text by a scale conversion factor, which is the inverse of (1 divided by) the plot scale. For example, if you are going to plot the drawing at ¼ the actual size of the object, use the formula text size=1 divided by the plot scale times the desired text size when plotted. To do the math, the text size=1 divided by ¼ times 0.125, giving you an actual size of 0.5.

If your drawing is at ⅛"=1" scale, you multiply the desired text height, ⅛", by the scale conversion factor of one divided by ⅛ to get a height of 1 (Table 3.3 showed scale factors as they relate to standard drawing scales). This is the height you must make your text to get ⅛"-high text in the final plot. Table 7.1 shows you some other examples of text height to scale.

TABLE 7.1: ⅛"-HIGH TEXT CONVERTED TO SIZE FOR VARIOUS DRAWING SCALES		
Drawing Scale	Scale Factor	AutoCAD Drawing Height for ⅛"-High Text
1/16"=1"	16	2"
⅛"=1"	8	1"
¼"=1"	4	.5"
½"=1"	2	.25"
¾"=1"	1.33	.166"
1"=1"	11	.125"
2"=1"	.5	.062"
4"=1"	.25	.031"

Organizing Text by Styles

If you understand the Multiline Text Editor and text scale, you know all you need to know to start labeling your drawings. As you expand your drawing skills and your drawings become larger, you will want to start organizing your text into *styles*. AutoCAD used the Standard text style to set text height when you created the proprietary text. You can think of text styles as a way to store your most common text formatting. Styles will store text height and font information, so you don't have to reset these options every time you enter text. But styles also include some settings not available in the Multiline Text editor. You can also make global changes to all of the text created with a specific style. You could change the height, obliquing angle, justification, and font for all text of a certain style.

Creating a Style

In the previous examples, you entered text using the AutoCAD default settings for text. Whether you knew it or not, you were also using a text style: AutoCAD's default style called Standard. The Standard style uses the AutoCAD Txt font and numerous other settings that you will learn about in this section. These other settings include width factor, obliquing, and default height.

If you don't like the way the AutoCAD default style is set up, open the Acad.dwt file and change the Standard text style settings to your liking. Or add other styles that you use frequently.

The previous exercises in this chapter demonstrate that you can modify the formatting of a style as you enter the text. But for the most part, once you've set up a few styles, you won't need to adjust settings like fonts and text height each time you enter text. You will be able to select from a list of styles you've previously created, and just start typing.

To create a style, use Format ➢ Text Style, and then select from the fonts available. The next exercise will show you how to create a style.

1. Click on Format ➢ Text Style or type **st**↵. The Text Style dialog box appears.

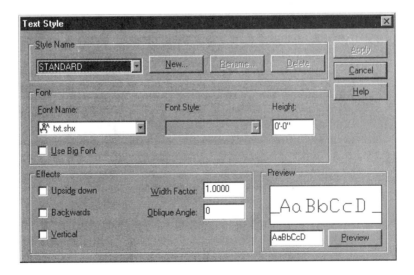

2. Click the New button in the Style Name group. The New Text Style dialog box appears.
3. Enter **Note2** for the name of your new style, then click on OK.
4. Now select a font for your style. Click on the Font Name in the Font group.
5. Locate the Courier New TrueType font and select it.
6. In the Height input box, enter **.12**.
7. Click on Apply and then click on Close.

Using a Type Style

You could have set up the style before creating the text or after the fact. All new text will be created with the Note2 style until you change it. Let's see what happens to text that we want to change to the new Note2 style.

1. Click on the notes text, then select the Change Properties button on the Standard toolbar. The Modify Mtext dialog box appears.
2. Click on the Style drop-down list. You will see the Note2 style.
3. Select the Note2 style.
4. Click on the OK button and look at the change in your notes text. Your drawing looks like Figure 7.6.

FIGURE 7.6

Modified notes text using the Note2 text style

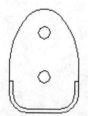
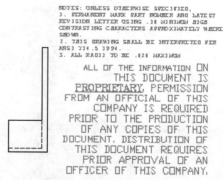

WARNING

When you change the style of a text object, it loses any custom formatting it may have, such as font or height changes that are different from those of the text's default style settings.

TIP

The Style input box in the Modify Mtext dialog box (found via the Properties tool) will allow you to select a new style for a text object. This option does not affect the style of text if the text has other custom format changes, such as a different font and size from its default settings.

Setting the Current Default Style

The last exercise showed you how you can change the style of existing text. But suppose you want all the new text you create to be of a different style than the current default style. You can change the current style by using the Style dialog box. Here's how it's done.

1. Click on Format ➢ Text Style or type **st**↵. The Text Style dialog box appears.
2. Select a style name from the Style drop-down list. For this exercise, choose Standard to return to the Standard style.
3. Click on Close.

Once you've done this, the selected style will be the default until you select a different style. AutoCAD will record the current default style with the drawing data when you issue a File ➢ Save command, so the next time you work on the file it will still have the same default style.

Understanding the Text Style Dialog Box Options

Now you know how to create a new style. As we mentioned before, there are other settings in the Text Style dialog box that you didn't apply in an exercise. Here is a list of those settings and their purposes. Some of them, like the Width factor, can be quite useful. Others, like the Backward and Vertical options, are rarely used.

Style Name

New lets you create a new text style.

Rename lets you rename an existing style. This option is not available for the Standard style.

Delete deletes a style. This option is not available for the Standard style.

Font

Font Name lets you select a font from a list of available fonts. The list is derived from the font resources available to the Windows NT or 95 System plus the standard AutoCAD Fonts.

Font Style offers variations of a font such as italic or bold, when they are available.

Height lets you enter a font size. A 0 height has special meaning when you are entering text using the Dtext command described later in this chapter.

Effects

Upside down prints the text upside down.

Backwards prints the text backwards.

Width Factor adjusts the width and spacing of the characters in the text. A value of 1 keeps the text at its normal width. Values greater than 1 will expand the text; values less than 1 will compress the text.

```
This is the Simplex font expanded by 1.4
This is the simplex font using a width factor of 1
This is the simplex font compressed by .6
```

Oblique Angle skews the text at an angle. When this option is set to a value greater than 0, the text appears to be italicized. A value of less than 0 (–12, for example) will cause the text to "lean" to the left.

```
This is the simplex font
using a 12-degree oblique angle
```

Renaming a Text Style

You can use the Rename option in the Text Style dialog box to rename a style. An alternate method is to use the Ddrename command. This is a command that allows you to rename a variety of AutoCAD settings. Here's how to use it.

NOTE This exercise is not part of the main tutorial. If you are working through the tutorial, make note of it and then try it out later.

1. Click on Format ➤ Rename or enter **ren**↵ at the command prompt. The Rename dialog box appears.

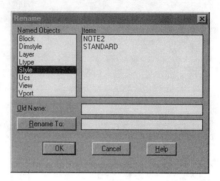

NOTE

The Ddrename command allows you to rename blocks, dimension styles, layers, linetypes, user coordinate systems, viewports, and views, as well as text styles.

2. In the Named Objects list box, click on Style.
3. Click on the name of the style you wish to change from the right-hand list; the name appears in the Old Name input box below the list.
4. In the input box next to the Rename To button, enter the new name, click on the Rename To button, and then click on OK.

NOTE

If you are an experienced AutoCAD user and accustomed to entering the Rename command at the command prompt, you can still. Then answer the prompts that appear.

What Do the Fonts Look Like?

You've already seen a few of the fonts available in AutoCAD. Chances are, you are familiar with the TrueType fonts available in Windows. You have some additional AutoCAD fonts from which to choose. In fact, you may want to stick with the AutoCAD fonts for all but your presentation drawings, as other fonts can consume more memory.

Figure 7.7 shows the basic AutoCAD text fonts. The Roman fonts are perhaps the most widely used, because they are easy to read without consuming much memory. Figure 7.8 shows the Symbols, Greek, and Cyrillic fonts.

FIGURE 7.7

The Standard AutoCAD text fonts

Font Sample	Description
This is Txt	
This is Monotxt	
This is Simplex	(Old version of Roman Simplex)
This is Complex	(Old version of Roman Complex)
This is Italic	(Old version of Italic Complex)
This is Romans	(Roman Simplex)
This is Romand	(Roman double stroke)
This is Romanc	(Roman Complex)
This is Romant	(Roman triple stroke)
This is Scripts	(Script Simplex)
This is Scriptc	(Script Complex)
This is Italicc	(Italic Complex)
This is Italict	(Italic triple stroke)
Τηισ ισ Γρεεκσ	(This is Greeks - Greek Simplex)
Τηισ ισ Γρεεκχ	(This is Greekc - Greek Complex)
Узит ит Вшсиллив	(This is Cyrillic - Alphabetical)
Тхис ис Чйрилтлч	(This is Cyriltlc - Transliteration)
This is Gothice	(Gothic English)
Thif if Gothicg	(Gothic German)
This is Gothici	(Gothic Italian)

FIGURE 7.8

The AutoCAD Symbols, Greek, and Cyrillic fonts

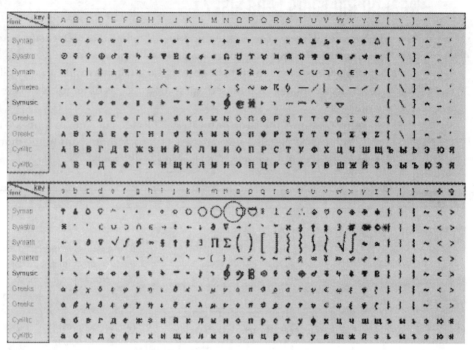

The Textfill System Variable

Unlike the standard stick-like AutoCAD fonts, TrueType and PostScript fonts have filled areas. These filled areas take more time to generate, so if you have a lot of text in these fonts, your redraw and regen times will increase. To help reduce redraw and regen times, you can set AutoCAD to display and plots these fonts as outline fonts, even though they are filled in their true appearance.

To change its setting, type **textfill**↵ and then type **0**↵. This turns off text fill for PostScript and TrueType fonts. For plots, you can remove the check mark on the option labeled Text Fill (this is the same as setting the Textfill system variable to 0).

We've shown you samples of the AutoCAD fonts in this section. You can see samples of all the fonts, including TrueType fonts, in the Preview window of the Text Style dialog box. If you use a word processor, you're probably familiar with at least some of the TrueType fonts available in Windows and AutoCAD.

Adding Special Characters

We mentioned earlier that you can add special characters using the Symbol button in the Multiline Text Editor. For example, the Degree symbol to designate angles, the Plus/Minus symbol for showing tolerance information, and the Diameter characters are already available as special characters. AutoCAD also offers a nonbreaking space. You can use the nonbreaking space when there is a space between two words but you do not want the two words to be separated by a line break.

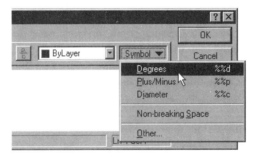

By clicking the Other option in the Symbols drop-down list, you can also add other special characters from the Windows Character Map dialog box.

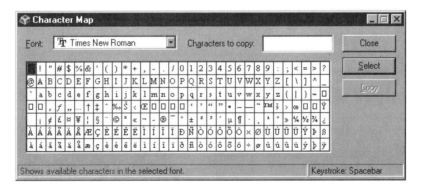

Characters such as the trademark (™) and copyright (©) symbols are often available. The contents of the list will vary depending on the current font. You can click and drag or just click your mouse over the character map to see an enlarged view of the character you are pointing to.

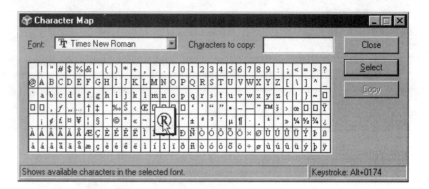

To use the characters from the Character Map dialog box, follow these steps.

NOTE

This is not part of the regular tutorial in this chapter, although you can experiment with these steps on your own.

1. Choose Other from the Symbols button of the Multiline Text Editor.
2. Highlight the character you want.
3. Either double-click on the character or click on the Select button. The character appears in the box at the upper-right corner of the dialog box.
4. Click Copy to copy the character to the Clipboard.
5. Close the dialog box.
6. In the Editor, place the cursor where you want the special character to appear.
7. Press Ctrl+v to paste the character into your text. You can also right-click on the mouse and choose Paste from the pop-up menu.

> ### Importing Text Files
>
> With Multiline text objects, AutoCAD allows you to import ASCII text or Rich Text Format files. Here's how you go about importing text files.
>
> **1.** From the Multiline Text Edit dialog box, click on Import Text.
>
> **2.** At the Open dialog box, locate a valid text file. It must be a file in a raw text (ASCII) format, such as a Notepad (.txt) file, or a Rich Text Format (.rtf) file. RTF files are capable of storing formatting information, such as boldface and varying point sizes.
>
> **3.** Once you've highlighted the file you want, double-click on it or click on OK. The text appears in the Edit Mtext window.
>
> **4.** You can then click on OK and the text will appear in your drawing.
>
> In addition, you can use the Windows Clipboard and Cut-and-Paste features to add text to a drawing. To do this, take the following steps:
>
> **1.** Use the Cut or Copy option in any other Windows program to place text onto the Windows Clipboard.
>
> **2.** Go to AutoCAD and choose Edit ➤ Paste. The text appears in the upper-left corner of the AutoCAD drawing window. You cannot, however, edit this text within AutoCAD.
>
> Because AutoCAD is an OLE client, you can also attach other types of documents to an AutoCAD drawing file. See Chapter 18 for more on AutoCAD's OLE support.

Adding Simple Text Objects

You might find that you are entering a lot of single words or simple labels that don't require all the bells and whistles of the Multiline Text Editor. AutoCAD offers the *single-line text object*, which is simpler to use and can speed text entry if you are adding only small pieces of text.

Continue the tutorial by trying the following exercise:

1. Enter **dt**↵ or choose Draw ➤ Text ➤ Single Line Text. This issues the Dtext command.

2. At the DTEXT Justify/Style/<Start point>: prompt, pick the starting point for the text you are about to enter, just below the side view.

3. At the Height: prompt, enter **.12**↵ to indicate the text height.

4. At the Insertion angle <0>: prompt, press ↵ to accept the default, 0°. You can specify any angle other than horizontal (for example, if you want your text to be aligned with a rotated object). You'll see a text I-beam cursor at the point you picked in step 2.

5. At the Text: prompt, enter **VERSION -01 SHOWN**↵**SEE VERSIONS TABLE**↵**FOR OTHER OPTIONS**. As you type, the words appear in the drawing as well as in the Command window.

NOTE
If you make a typing error, use the ← and → keys to move the text cursor in the Command window to the error; then use the Backspace key to correct the error. You can also paste text from the Clipboard into the cursor location by using Ctrl+v or by right-clicking in the Command window to access the pop-up menu.

6. Press ↵ to move the cursor above the front view.

7. This time, type **PAINT SURFACE**↵. Figure 7.9 shows how your drawing should look.

8. Press ↵ again to exit the Dtext command.

FIGURE 7.9

Adding simple labels using the Dtext command

ADDING SIMPLE TEXT OBJECTS   293

If for some reason you need to stop entering single-line text objects to do something else in AutoCAD, you can continue entering text where you left off by pressing ↵ at the Start point: prompt of the Dtext command. The text will continue immediately below the last line of text entered.

Here you were able to add two single lines of text in different parts of your drawing fairly quickly. Dtext will use the current default text style settings (remember that earlier you set the style to Standard), so the VERSION and PAINT labels used the Standard style.

Editing Single-Line Text Objects

Editing single-line text uses the same tools as those for Multiline text, although the dialog boxes that result are different. In this exercise, you'll change the PAINT label using the Ddedit command.

1. Type **ed**↵ or choose Modify ➤ Object ➤ Text.
2. Click on the PAINT label. A small Edit Text dialog box appears.

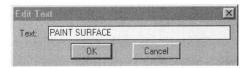

3. Using the cursor, click on the end of the line and type **4 PLACES**.
4. Click on OK and then press ↵ to exit the Ddedit command.

As you can see, even the editing is simplified. You are limited to editing the text only. This can be an advantage, however, when you need to edit several pieces of text. You don't have other options to get in the way of your editing.

You can change other properties of a single-line text object using the Properties dialog box. For example, suppose you want to change the PAINT label to a height of .18 inches.

1. Click on the Properties tool in the Object Properties toolbar.

2. Click on the PAINT text and then press ↵. The Modify Text dialog box appears.

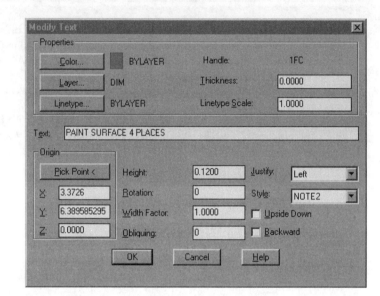

Notice that in this case, you do not see the Full Editor option.

3. Double-click on the Height input box and enter **.18**↵.

4. Click on OK. The text increases in size to .18" high.

5. Click on the Undo tool on the menu bar to undo the change in text height.

6. Choose File ➢ Save to save the changes you've made thus far.

The Modify Text dialog box lets you change the Height, Rotation, Width Factor, Obliquing, Justification, and style of a single-line text object. You can also modify the text content.

Unlike prior versions of AutoCAD, the height prompt appears in the Dtext command even if the current style has a non-zero height.

Justifying Single-Line Text Objects

Justifying single-line text objects works in a slightly different way from justifying Multiline text. For example, if you change the justification setting to Center, the text will move so that the center of the text is placed at the text insertion point. In other words, the insertion point stays in place while the text location adjusts to the new justification setting.

To set the justification of text as you enter it, you must enter **j**↵ at the `Justify/Style/<Start point>:` prompt after issuing the Dtext command.

NOTE
You can also change the current default style by entering **s**↵ and then the name of the style at the `Justify/Style/<Start point>:` prompt.

Once you've issued the Dtext's Justify option, you will then get the prompt:

`Align/Fit/Center/Middle/Right/TL/TC/TR/ML/MC/MR/BL/BC/BR:`

Here are descriptions of each of these options (we've left Fit and Align until last because these options require a bit more explanation).

Center

Center causes the text to be centered on the start point, with the baseline on the start point.

Middle

Middle causes the text to be centered on the start point, with the baseline slightly below the start point.

Right

Right causes the text to be justified to the right of the start point, with the baseline on the start point.

TL, TC, and TR

TL, TC, and TR stand for top left, top center, and top right. Text using these justification styles appears entirely below the start point and justified left, center, or right, depending on which of the three options you choose.

ML, MC, and MR

ML, MC, and MR stand for middle left, middle center, and middle right. These styles are similar to TL, TC, and TR, except that the start point will determine a location midway between the baseline and the top of the lowercase letters of the text.

BL, BC, and BR

BL, BC, and BR stand for bottom left, bottom center, and bottom right. These styles, too, are similar to TL, TC, and TR, but here the start point determines the bottom-most location of the letters of the text (the bottom of letters that have descenders, such as *p*, *q*, and *g*).

Figure 7.10 shows the relationship between the text start point and text justified with these options.

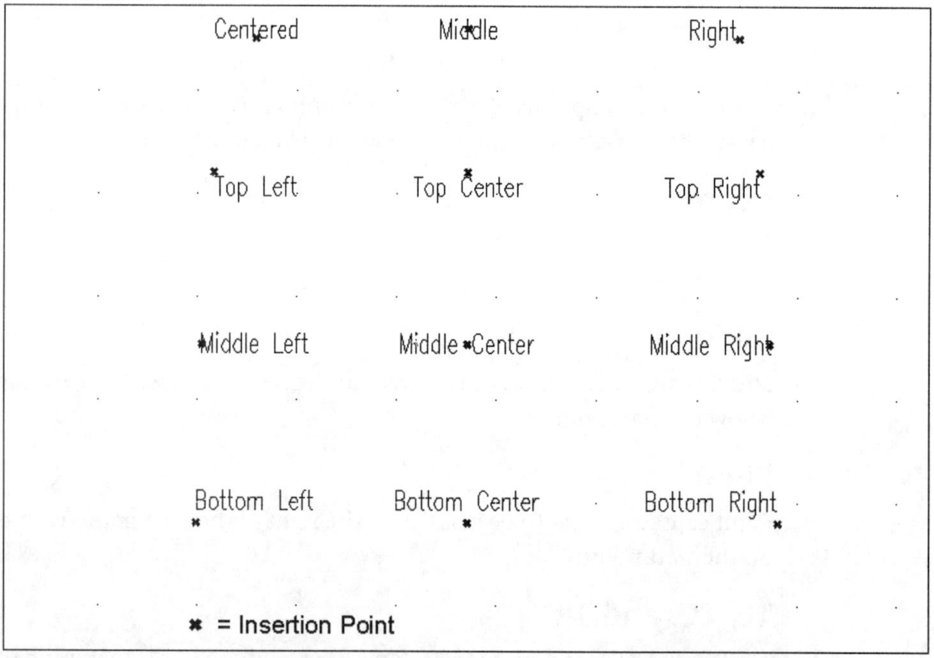

FIGURE 7.10

Text inserted using the various Justify options. The X indicates the location of the start point in relation to the text. If you use the Properties tool to change the text justification, the text will move while the text insertion point remains in place.

Align and Fit Options

With the Fit and Align justification options, you must specify a dimension within which the text is to fit. For example, suppose you want the words "SECTION A-A" to fit within the 1.25"-wide area at the bottom of a section view. You can use either the Fit or the Align option to accomplish this. With Fit, AutoCAD prompts you to select start and end points, and then stretches or compresses the letters to fit within the two points you specify. Use this option when the text must be a consistent height throughout the drawing and you don't care about distorting the font. Align works like Fit, but instead of maintaining the current text style height, Align adjusts the text height to keep it proportional to the text width, without distorting the font. Use this option when it is important to maintain the font's shape and proportion. Figure 7.11 demonstrates how Fit and Align work.

FIGURE 7.11
The phrase "SECTION A-A" as it appears normally and as it appears with the Fit and Align options selected

Using Special Characters with Single-Line Text Objects

Just as with Multiline text, you can add a limited set of special characters to single-line text objects. For example, you can place the degree symbol (°) after a number, or you can *underscore* (underline) text. To accomplish this, use double percent (%%) signs in conjunction with a special code. For example, to underscore text, enclose that text with the %% signs and follow it with the underscore code. So, to get this text:

you would enter this at the prompt:

This is %%uunderscored%%u text.

Overscoring (putting a line above the text) operates in the same manner. To insert codes for symbols, just place the codes in the correct positions for the symbols they represent. For example, to enter 100.5°, type **100.5%%d**.

Table 7.2 shows a list of the codes you can use.

TABLE 7.2: CODES FOR INSERTING SPECIAL CHARACTERS INTO SINGLE-LINE TEXT OBJECTS

Code	Special Character
%%o	Toggles overscore on and off.
%%u	Toggles underscore on and off.
%%d	Places a degree sign (°) where the code occurs.
%%p	Places a plus-minus sign where the code occurs.
%%%	Forces a single percent sign; useful when you want a double percent sign to appear, or when you want a percent sign in conjunction with another code.
%%nnn	Allows the use of extended Unicode characters when these characters are used in a text-definition file; nnn is the three-digit value representing the character.

Using the Character Map Dialog Box to Add Special Characters

You can add special characters to a single line of text in the same way you would with Multiline text. You may recall that to access special characters, you use the Character Map dialog box. This dialog box can be opened directly from the Windows Explorer.

Using the Explorer, locate the file Charmap.exe in the Windows folder. Double-click on it and the Character Map dialog box appears. You can then use the procedure spelled out in the *Adding Special Characters* section earlier in this chapter to cut and paste a character from the Character Map dialog box. If you find you use the Character Map dialog box often, create a shortcut for it and place the shortcut in your AutoCAD Program group.

Keeping Text from Mirroring

At times you will want to mirror a group of objects that contain some text. This operation will cause the mirrored text to appear backward. You can change a setting in AutoCAD to make the text read normally, even when it is mirrored.

1. Enter **mirrtext**↵.
2. At the `New value for MIRRTEXT <1>:` prompt, enter **0**↵.

Now any mirrored text that is not in a block will read normally. The text's *position*, however, will still be mirrored, as shown in the graphic. Mirrtext is set to 0 by default.

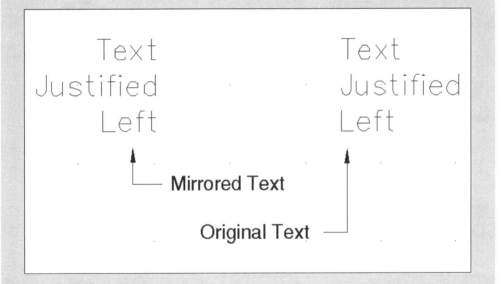

Checking Spelling

Although AutoCAD is primarily a drawing program, you will find that some of your drawings contain more text than graphics. At long last, Autodesk has recognized this and has included a spelling checker in AutoCAD Release 14. If you've ever used the spelling checker in a typical Windows word processor, such as Microsoft Word, the AutoCAD spelling checker's operation will be familiar to you. Here's how it works:

1. Click on the Spelling tool on the Standard toolbar or type **sp**↵.

2. At the `Select objects:` prompt, select any text object you want to check. You can select a mixture of Multiline and single-line text. When the spelling checker finds a word it does not recognize, the Check Spelling dialog box appears.

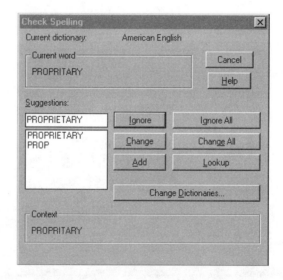

In the Check Spelling dialog box, you'll see the word in question, along with the spelling checker's suggested alternate word in the Suggestions input box. If the spelling checker finds more than one suggestion, a list of suggested replacement words appears below the input box. You can then highlight the desired replacement and click on the Change button to change the misspelled word, or click on Change All to change all occurrences of the word in the selected text. If the suggested word is inappropriate, choose another word from the replacement list (if any), or enter your own spelling in the Suggestions input box. Then choose Change or Change All.

Here is a list of the options available in the Check Spelling dialog box:

Ignore skips the word.

Ignore All skips all the occurrences of the word in the selected text.

Change changes the word in question to the word you have selected (or entered) from the Suggestions input box.

Change All changes all occurrences of the current word, when there are multiple instances of the misspelling.

Add adds the word in question to the current dictionary.

Lookup checks the spelling of the word in question. This option is for the times when you want to find another word that doesn't appear in the Suggestions input box.

Change Dictionaries lets you use a different dictionary to check spelling. This option opens the Change Dictionaries dialog box, described in the next section.

Choosing a Dictionary

The Change Dictionaries option opens the Change Dictionaries dialog box, where you can select a particular main dictionary for foreign languages, or create or choose a custom dictionary. Main dictionary files have the .dct extension. The Main dictionary for the U.S. version of AutoCAD is Enu.dct.

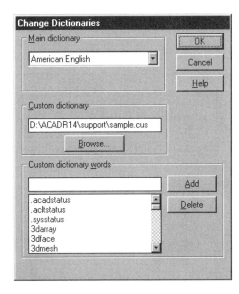

In this dialog box, you can also add or delete words from a custom dictionary. Custom dictionary files are ASCII files with the .cus extension. Because they are ASCII files, they can be edited outside of AutoCAD. The Browse button lets you view a list of existing custom dictionaries.

If you prefer, you can also select a main or custom dictionary using the Dctust and Dctmain system variables. See Appendix D for more on these system variables.

A third place where you can select a dictionary is in the Files tab of the Preferences dialog box (Tools ➤ Preferences). You can find the Dictionary listing under Text Editor, Dictionary, and Font File Names. Click on the plus sign next to this listing and then click on the plus sign next to the Main Dictionary listing to expose the dictionary options.

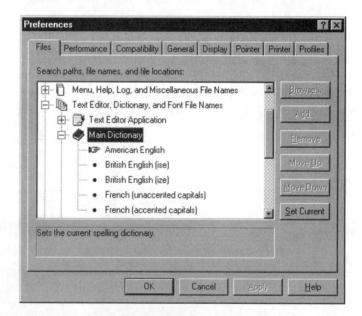

From here, you can double-click on the dictionary you prefer. The pointing hand icon will move to the selected dictionary.

Substituting Fonts

There will be times when you will want to change all the fonts in a drawing quickly. For instance, you may want to convert PostScript fonts into a simple Txt.shx font to help shorten redraw times while you are editing. Or you may need to convert the font of a drawing received from another office to a font that conforms to your own office standards. In AutoCAD Release 14, the Fontmap system variable works in conjunction with a font-mapping table, allowing you to substitute fonts easily in a drawing.

The font-mapping table is an ASCII file called `Acad.fmp`. You can also use a file you create yourself. You can give this file any name you choose, as long as it has the `.fmp` extension.

This font-mapping table contains one line for each font substitution you want AutoCAD to make. A typical line in this file would read as follows:

```
romant; C:\acadr14\common\font\Txt.shx
```

In this example, AutoCAD is directed to use the `Txt.shx` font in place of the Roman font. To execute this substitution, you would type **fontmap**↵ *Fontmap_filename*↵ where *Fontmap_filename* is the font-mapping table you've created. This tells AutoCAD where to look for the font-mapping information. Then you would issue the Regen command to view the font changes. To disable the font-mapping table, type **fontmap**↵, then a period (.), then press ↵.

You can also specify a font-mapping file in the Files tab of the Preferences dialog box. Look for the Text Editor, Dictionary, and Font File listing. Click on the plus sign next to this listing and then click on the plus sign next to the Font Mapping File listing to expose the current default font-mapping file name.

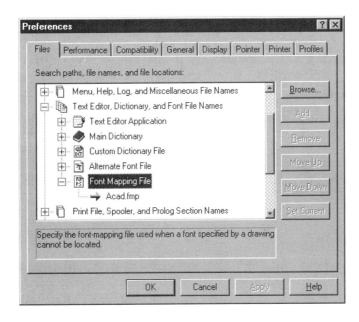

You can double-click on this file name to open a Select a File dialog box. From there you can select a different font-mapping file.

See Appendix D for more on Fontmap and other system variables.

Making Substitutions for Missing Fonts

When text styles are created, the associated fonts do not become part of the drawing file. Instead, AutoCAD loads the needed font file at the same time that the drawing is loaded. So if a text style in a drawing requires a particular font, AutoCAD looks for the font in the AutoCAD search path; if the font is there, it is loaded. Usually this isn't a problem if the drawing file uses the standard fonts that come with AutoCAD or Windows. But occasionally you will encounter a file that uses a custom font.

In earlier versions of AutoCAD, when you attempted to open such a file, you would see an error message. This missing-font message would often send the new AutoCAD user into a panic.

Fortunately, Release 14 offers a solution: AutoCAD automatically substitutes an existing font for the missing font in a drawing. By default, AutoCAD substitutes the Txt.shx font, but you can specify another one using the Fontalt system variable. Type **fontalt**↵ at the command prompt and then enter the name of the font you want to use as the substitute.

You can also select an alternate font through the Files tab of the Preferences dialog box. Locate the Text Editor, Dictionary, and Font File Names listing and then click on the plus sign at the left. Locate the Alternate Font File listing that appears and click on the plus sign at the left. The current alternate is listed. You can double-click on the font name to select a different font through a standard File dialog box.

Be aware that the text in your drawing will change in appearance, sometimes radically, when you use a substitute font. If the text in the drawing must retain its appearance, you will want to substitute a font that is as similar in appearance to the original font as possible.

Accelerating Zooms and Regens with Qtext

If you need to edit a drawing that contains a lot of text, but you don't need to edit the text, you can use the *Qtext* command to help accelerate redraws and regens when you are working on the drawing. Qtext turns lines of text into rectangular boxes, saving AutoCAD from having to form every letter. This allows you to see the note locations so you don't accidentally draw over them.

> **TIP**
>
> Selecting a large set of text objects for editing can be annoyingly slow. To improve the speed of text selection (and object selection in general), turn off the Highlight and Drag mode system variables. This will disable certain convenience features but may improve overall performance, especially on large drawings. See Appendix D for more information.

To turn on Qtext:

1. Select Tools ➤ Drawing Aids and turn on the Quick Text check box, or enter **qtext**↵ at the command prompt.
2. At the ON/OFF <OFF>: prompt, enter **on**↵.
3. To display the results of Qtext, issue the Regen command from the prompt.

When Qtext is off, text is generated normally. When Qtext is on, rectangles show the approximate size and length of text, as shown in Figure 7.12.

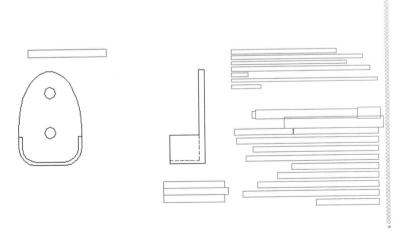

FIGURE 7.12
View of the bracket drawing with the Qtext system variable turned on

> ### Manipulating Text beyond Labels
>
> This chapter concentrates on methods for adding labels to your drawing, but you also use text in other ways with AutoCAD. Many of the inquiry tools in AutoCAD, such as Dist and List, produce text data. You can use the Windows Clipboard to manipulate such data to your benefit.
>
> For example, you can duplicate the exact length of a line by first using the List command to get a listing of its properties. Once you have the property list in the AutoCAD Text window, you can highlight its length listing and then press Ctrl+c to copy it to the Windows Clipboard. Next, you can start the line command and then pick the start point for the new line. Press Ctrl+v to paste the line-length data into the Command window; then add the angle data or use the direct distance method to draw the line.
>
> Any text data from dialog box input boxes or the AutoCAD Text window can be copied to the Clipboard using the Ctrl+c keyboard shortcut. That data can likewise be imported into any part of AutoCAD that accepts text.
>
> Consider using the Clipboard the next time you need to transfer data within AutoCAD, or even when you need to import text from some other application.

Bonus Text Editing Utilities

Finally, before finishing this chapter, you will want to know about a set of Bonus utilities that give you the following capabilities:

- Draw text along an arc. If the arc changes, the text follows.
- Globally change the Height, Justification, Location, Rotation, Style, Text, and Width factor of a set of text objects you select.
- Adjust the width of a single-line text object to fit within a specified area.
- Explode text into lines.
- Mask areas behind text so the text is readable when placed over hatch or solid filled patterns.
- Search and replace text for a set of single-line text objects.

These functions can save hours of your time when editing a complex drawing that is full of text. You can find out how access these tools in Chapter 19.

If You Want to Experiment...

Try adding some notes to drawings you have created in other *If You Want to Experiment* sections of this book. Most companies do not like to have mechanical drawings created with multiple fonts or text sizes. Change all of the text to one size and one font.

Chapter 8

Using Dimensions

FEATURING

- Setting Up a Dimension Style
- Drawing Linear Dimensions
- Continuing a Dimension String
- Drawing Dimensions from a Common Baseline
- Appending Data to a Dimension's Text
- Dimensioning Radii, Diameters, and Arcs
- Using Ordinate Dimensions
- Adding Tolerance Notation

Chapter 8

Using Dimensions

In the exercises of previous chapters, you have diligently drawn your design to scale. In this chapter, you will learn how easy it is to create dimensions in AutoCAD. The importance of dimensions cannot be overstated. The reason that you draw these models is to communicate concepts and specifications. Dimensioning can be crucial to how well a design works and how quickly it develops. The dimensions answer questions about tolerances, fit, and interference. When a design team is making decisions, communicating even tentative dimensions to others on the team can accelerate design development.

With AutoCAD, you can easily add dimensions to any drawing. AutoCAD gives you an accurate dimension without your having to take measurements. You simply pick two points of an object to be dimensioned and the location of the dimension line, and AutoCAD does the rest. AutoCAD's *associative dimensioning* capability automatically updates dimensions whenever the size or shape of the dimensioned object is changed. These dimensioning features can save you valuable time and reduce the number of dimensional errors in your drawings.

AutoCAD's dimensioning feature has a substantial number of settings. Though they give you an enormous amount of flexibility in formatting your dimensions, all these settings can be somewhat intimidating to the new user. We'll ease you into dimensioning by first showing you how to create a *dimension style*.

Creating a Dimension Style

Dimension styles are similar to text styles. They determine the look of your dimensions as well as the size of dimensioning features, such as the dimension text and arrows. You might set up a dimension style to have special types of arrows, for instance, or to position the dimension text above or in line with the dimension line. Dimension styles also make your work easier by allowing you to store and duplicate your most common dimension settings.

AutoCAD gives you a default dimension style called *Standard*, which is set up fairly close to a mechanical drafting standard. You will doubtless add many other styles to suit the style of drawings you are creating. You can also create variations of a general style for those situations that call for only minor changes in the dimension's appearance.

In this section you'll see how to set up a dimension style that is more appropriate than Standard for the latest revision of the American National Standards Institute (ANSI) drafting standard of 1994 (see Figure 8.1).

FIGURE 8.1
AutoCAD's standard dimension style compared with an ANSI-style dimension

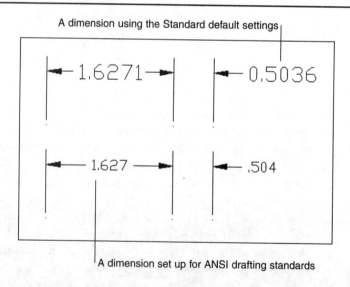

 1. Open the Plate.dwg file from the companion CD-ROM. The plate drawing is a better choice to demonstrate dimensioning than the parts you have drawn so far.

2. Issue a Zoom All to display the entire plate. Type **z↵a↵** or click on the Zoom tool from the Standard toolbar.
3. Choose Format ➤ Dimension Style, or type **d↵** at the command prompt. The Dimension Styles dialog box appears.
4. Click on the Current drop-down list at the top of the dialog box, and then click on Standard.
5. Double-click on the Name input box to highlight STANDARD, and then type in **designer**.
6. Click on Save. Notice the message at the bottom of the dialog box telling you that the Designer style was created based on the Standard style.

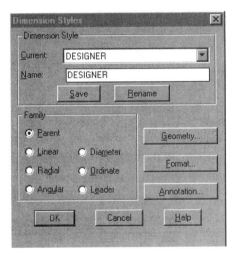

You've just created a dimension style called Designer; but at this point, it is identical to the Standard style on which it is based. Nothing has happened to the Standard style; it is still available if you need to use it. From here you will start to modify the Designer style.

Setting the Dimension Unit Style

Now you need to modify the new Designer dimension style so that it conforms to the mechanical style of dimensioning. Let's start by changing the unit style for the dimension text. Just as you changed the overall unit style of AutoCAD to three-place decimal style for the bracket in Chapter 3, you must do the same for your dimension styles. Setting the overall unit style does not automatically set the dimension unit style.

1. In the Dimension Styles dialog box, click on the Annotation button. The Annotation dialog box appears.

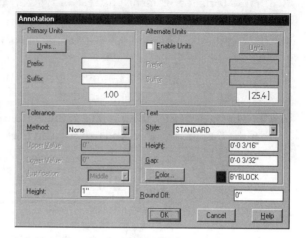

2. In the Primary Units group, click on the Units button. The Primary Units dialog box appears.

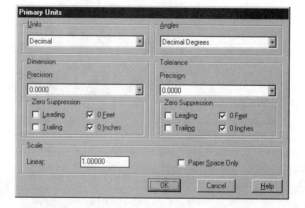

3. Open the Units drop-down list and choose Decimal. Notice that this drop-down list contains the same unit styles as the main Units dialog box (Format ➢ Units).

4. Click in the Precision window of the Dimension box, and a drop-down list appears that offers decimal precision of up to eight places to the right of the decimal point. The default is four places. For these exercises you will primarily use three-place decimal dimensions. Select the three-place option.

The Units drop-down list also offers the Fractional (stacked) option. This option produces stacked fractions in dimensions. Choose this option if you prefer stacked fractions. You can always make changes to individual dimensions later if you change your mind.

5. Every dimension style setting has an equivalent system variable. See Appendix D for more on system variables that are directly associated with dimensions. You will find four options in the Zero Suppression check box group, just below the Units list. If the Leading check box does not have a check mark in it, click in the box. (These check boxes operate like toggle switches: One click toggles the check on or off, and another click toggles the box to the previous setting.) If the Leading box has a check mark, this means that leading zero suppression is *on* and all dimensions with a value of less than 1 *will not* have a leading zero. The ANSI standard is not to have leading zeros.

6. Look to the right to find the Tolerance box. Follow steps 4 and 5 again to have your tolerance values default to the same precision and appearance as the dimension values.

7. Click on OK to close the Primary Units dialog box.

You have set up the Designer dimension unit style to show dimensions and tolerances to three decimal places, rather than four.

Setting the Height for Dimension Text

Along with the unit style, you will want to adjust the style that is used for the dimension text. The Text button group of the Annotation dialog box lets you select a text style from the Style drop-down list. You can also adjust the height in case you prefer a slightly smaller or larger height than the default style offers. The height now shows .188. Change it to .12" by following these steps:

1. Highlight the contents of the Height input box.
2. Type **.12**⏎ to make the text height .12" high.
3. Click on OK to return to the Dimension Styles dialog box.

To find out more about the Annotation dialog box and its options, see Appendix D.

Setting the Location of Dimension Text

AutoCAD's default setting for the placement of dimension text puts the text centered between the extension lines, as shown in the example at the top of Figure 8.1. The new Designer style should allow you to place the text anywhere between or even outside the extension lines. To do that, you will use the dimension style Format options.

1. In the Dimension Styles dialog box, click on the Format button. The Format dialog box appears.

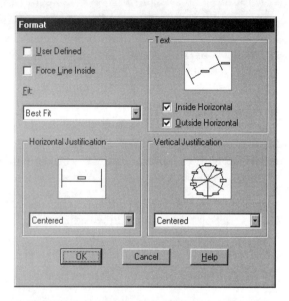

2. Click in the User Defined check box to allow the user to position the text. In the Vertical Justification box that occupies the lower-right area of the dialog box, open the drop-down list and choose Above. The graphic changes to show you what this format will look like in your dimensions.

3. In the Text box (upper-right corner of the dialog box), verify that the Inside Horizontal and Outside Horizontal check boxes are both turned on. This forces the dimension text to be Horizontal without regard to the orientation of the dimension. Most companies and ANSI require that all dimensions except ordinate dimensions read from the bottom of the page.

4. Now click on OK to return to the Dimension Styles dialog box.

Choosing an Arrow Style and Setting the Dimension Scale

Next, you will want to verify that you are using the correct type of arrow for your new dimension style. For linear dimension in mechanical drawings, a filled arrowhead is normally used; this is the AutoCAD default on startup. If you are not the only one to use the computer or if someone else has configured AutoCAD, make sure that you are using the filled arrowhead. For some layout and architectural purposes a diagonal line or "tick" mark is used, rather than an arrow. We will also take a look at this style of arrow.

If you know the scale to use when printing the drawing, set it here. You can also wait until more is known about the size of the project, (how many views are going to be required, how many dimensions are needed to describe the part, what size format) before making this decision. The preferred plot scale for mechanical drawings is full size (1"=1"), although a different plot scale may be used for clarity. If a part or assembly is too large or too small or if the detail is too fine or too coarse, it's probably a good idea to use a different scale. Recall from Chapter 7 that text may be scaled up or down in size in order to be the proper size in the final output of the drawing. Dimensions, too, may be scaled so they look right when the drawing is plotted. For both the arrow and the scale settings, you will use the Geometry settings.

1. In the Dimension Styles dialog box, click on the Geometry button. The Geometry dialog box appears.

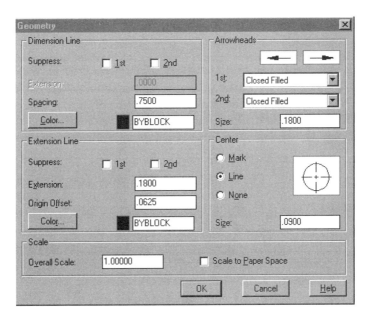

2. In the Arrowheads group, open the first drop-down list and choose Architectural Tick. The graphic shows you what the arrow looks like. Now select Closed Filled.

See Appendix D for details on how you can create your own arrowheads.

3. In the Scale group, locate the Overall Scale input box at the lower-left of the dialog box, and change this value to 1.000. This is the scale factor for a full size, or 1"=1", scale drawing.

Here's a simple way to figure out architectural scale factors: Divide 12 by the decimal equivalent of the inch scale. So for ¼", divide 12 by 0.25 to get 48. For ⅛", divide 12 by 0.125 to get 96. For 1½", divide 12 by 1.5 to get 8, and so on. For engineering scales, multiply the scale of the drawing by 12. For example, for a scale of 1"=10' scale, multiply the 10 by 12 to get 120.

4. In the Center group, we can tell AutoCAD to use two different types of center marks or none. The size box defines the size of the mark. Click on Line and watch the graphic change. Now the Size box also controls how far beyond the circle the lines extend.

5. Click on OK here, and click on Save in the Dimension Styles dialog box.

As with other dialog boxes in AutoCAD, you can cycle through options by clicking on the graphic that is displayed in a button group. For example, you can click on the arrowhead graphics to cycle through the different arrowheads that are available.

In a way similar to text styles, the most recently created dimension style becomes the default current style, so when you start dimensioning in the exercises in this chapter, you will be using the Designer style you just created. To switch to another style, open the Dimension Styles dialog box again, and then select the style from the Current drop-down list.

In this section, we've introduced you to the various dialog boxes that let you set the appearance of a dimension style. We haven't been able to discuss every option, so if you want to learn more about the other dimension style options, consult Appendix D.

There you'll find descriptions of all the items in the Dimension Styles dialog box, plus reference material covering the system variables associated with each option.

If your application is strictly mechanical, you may want to make these same dimension style changes to the Acad.dwt template file, or create a set of template files specifically for mechanical drawings of differing scales.

Drawing Linear Dimensions

The most common type of dimension you'll be using is the *linear dimension*, which is an orthogonal dimension measuring the width and length of an object. AutoCAD offers three dimensioning tools for this purpose: Linear (Dimlinear), Continue (Dimcont), and Baseline (Dimbase). These options are readily accessible from the Dimension tool palette.

Finding the Dimension Toolbar

Before you apply any dimension, you'll want to open the Dimension toolbar. This toolbar contains nearly all the commands necessary to draw and edit your dimensions.

Right-click on any toolbar; then at the Toolbars dialog box, click on Dimension from the list of toolbars. The Dimension toolbar appears. Click on OK to exit the Toolbars dialog box.

The Dimension commands are also available from the Dimension pull-down menu.

To help keep your you screen organized, you may want to dock the Dimension toolbar to the right side of the AutoCAD window. See Chapter 1 for more about docking toolbars.

Now you're ready to begin dimensioning.

Placing Horizontal and Vertical Dimensions

Let's start by looking at the basic dimensioning tool, Linear. The Linear Dimension button (the Dimlin command) on the Dimension toolbar accommodates both the horizontal and vertical dimensions.

In this exercise, you'll add a vertical dimension to the left side of the plate.

1. To start either a vertical or horizontal dimension, click on Linear Dimension from the Dimension toolbar, or enter **dli**↵ at the command prompt. You can also choose Dimension ➤ Linear from the menu bar.

2. The prompt First extension line origin or RETURN to select: is asking you for the first point of the distance to be dimensioned. An *extension line* is the line that connects the object being dimensioned to the dimension line. Use the Endpoint Osnap override and pick the lower-left corner of the plate at the bottom of the fillet. Care is required here because you do not want to select the other end of the fillet.

NOTE Notice that the prompt in step 2 gives you the option of pressing ↵ to select an object. If you do this, you are prompted to pick the object you wish to dimension, rather than the actual distance to be dimensioned. We'll look at this method later in this chapter.

3. At the Second extension line origin: prompt, use the Endpoint Osnap override to pick the top-left corner of the plate.

4. In the next prompt, (Mtext/Text/Angle/Horizontal/Vertical/Rotated):, the dimension line is the line indicating the direction of the dimension and containing the arrows or tick marks. Move your cursor from left to right, and you'll see a temporary dimension appear. This allows you to visually select a dimension line location.

DRAWING LINEAR DIMENSIONS | 321

NOTE

In step 4, you have the option to append information to the dimension's text or change the dimension text altogether. You'll see how in *Editing Dimensions*, later in this chapter.

5. Enter **@1.5<180**↵ to tell AutoCAD you want the dimension line to be 1.50 inches to the left of the last point you selected. (You could pick a point using your cursor, but this doesn't let you place the dimension line as accurately.) After you've done this, the dimension is placed in the drawing, as shown in Figure 8.2.

FIGURE 8.2

The dimension line added to the plate drawing

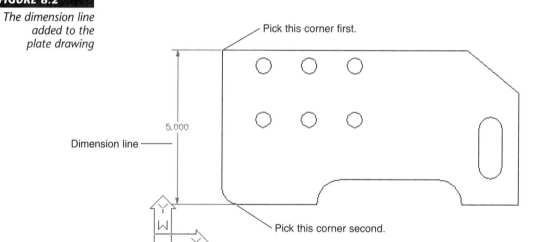

Continuing a Dimension

You will occasionally want to input a group of dimensions strung together in a line. For example, you may want to continue dimensioning a feature because a special dimension tolerance is required. To do this, use the Continue option found in both the Dimension toolbar and the Dimension pull-down menu. We will dimension the vertical location of the six-hole pattern.

Using Osnaps While Dimensioning

You may find that when you pick intersections and endpoints frequently, as during dimensioning, it is a bit inconvenient to use the Osnap pop-up menu. In situations where you know you will be using certain Osnaps frequently, you can use Running Osnaps. You can do so in either of the following two ways:

- Click on Tools ➢ Object Snap Settings. In the Osnap Settings dialog box, make sure the Running Osnap tab is selected and then select the Endpoint, Midpoint, and Center Osnap modes by clicking in the check box for each Osnap setting. With Running Osnaps set in this way, your cursor will automatically select the endpoint of a line or an arc, the midpoint of a line or an arc, or the center of a circle or an ellipse, depending on which object the cursor aperture is crossing.

If you have difficulty obtaining the Osnap that you want, you can use the Osnap Cycle mode. To do this, hover the cursor over the objects until the snap marker appears, then press the Tab key repeatedly until you see the marker representing the Osnap that you want.

You can pick as many of the Osnaps as you like to be Running Osnaps; however, a few Osnaps can preclude the use of others. For example, suppose you have Center, Quadrant, and Tangent set. When you hover the cursor over a circle, only one marker will show (according to Murphy's law, the marker that you see will rarely be the right one). If you find that you must frequently use the Tab key to select another Osnap, are frequently using the Osnap cursor menu (Shift+right click), or are frequently typing in Osnap overrides, you should take a look at changing the Running Osnap settings.

- Once you've designated your Running Osnaps, the next time you are prompted to select a point, the selected Osnap modes will be activated automatically. You can still override the default settings using the Osnap pop-up menu (Shift+right-click). However, you don't have to worry about accidentally using a Running Osnap during panning or zooming. Only explicitly issued Osnap overrides affect pans and zooms.

WARNING

There is a drawback to setting a Running Osnap mode: When your drawing gets crowded, you might end up picking the wrong point by accident. However, you can easily toggle the Running Osnap mode off by double-clicking on the OSNAP label in the status bar, or by pressing F3 to cycle through the available Osnaps.

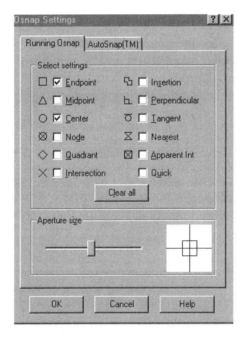

You may want to use Zoom to get a closer look when trying to select geometry. Dimensions are composed of lines, and you could as easily snap to one of these as to the intended geometry.

The Continue mode of linear dimensioning assists in making chain dimensions. When you use this mode, you will be asked only for the second extension line because the chain *continues* from the second extension line origin of the last dimension that you created before beginning the Continue mode.

1. Click on Linear Dimension from the Dimension toolbar.
2. At the `First extension line origin or press ENTER to select:` prompt, pick the upper-left corner of the plate.
3. At the `Second extension line origin:` prompt, pick the top left circle.
4. At the `Dimension line location (Mtext/Text/Angle/Horizontal/Vertical/Rotated):` prompt, type **from**↵ to use the From Osnap. At the `Base point:` prompt, pick the upper-left corner of the plate. At the `<offset>` prompt, enter **@.75<180**↵ to tell AutoCAD you want the dimension line to be .75 inches to the left of the From point.
5. Click on the Continue Dimension option from the Dimension toolbar, or enter **dco**↵. You can also choose Dimension ➤ Continue from the menu bar.

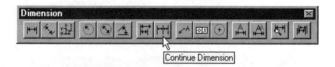

6. At the `Specify a second extension line origin or (Undo/<Select>):` prompt, pick the circle just below the last one you picked. See Figure 8.3 for the results.
7. Press ↵ twice to exit the command.

FIGURE 8.3

The dimension chain using the Continue Dimension option

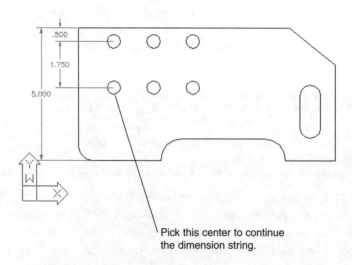

> If you find you've selected the wrong location for a continued dimension, you can click on the Undo tool or press **u**↵ to back up your dimension.

The Continue Dimension option adds a dimension from where you left off. The last drawn extension line is used as the first extension line for the continued dimension. AutoCAD will keep adding dimensions as you continue to pick points, until you press ↵.

Continuing a Dimension from a Previous Dimension

If you need to continue a string of dimensions from an older linear dimension instead of the most recently added one, press ↵ at the `Specify a second extension line origin or (Undo/<Select>):` prompt you saw in step 6 of the previous exercise. Then, at the `Select continued dimension:` prompt, click on the extension line from which you wish to continue.

Drawing Dimensions from a Common Base Extension Line

Another method for dimensioning objects is to have several dimensions originate from the same extension line. To accommodate this, AutoCAD provides the Baseline option. To see how this works, you will start another dimension—this time a horizontal one—across the top of the plate.

1. Click on Linear Dimension from the Dimension toolbar. Or, just as you did for the vertical dimension, you can type **dli**↵ to start the horizontal dimension. This option is also on the Dimension pull-down menu.

2. At the `First extension line...` prompt, use the Endpoint Osnap to pick the upper-left corner of the plate.

3. At the `Second extension line...` prompt, pick the upper-left circle again.

4. At the `Dimension line...` prompt, use the From Osnap override. At the `Base point:` prompt, pick the upper-left corner of the plate. At the `<offset>` prompt, enter **@.75<90**↵ to tell AutoCAD you want the dimension line to be .75 inches above the From point. After placing the dimensions, Zoom All to see the dimensioned plate so far. It should look like the one in Figure 8.4.

FIGURE 8.4

The plate with a horizontal dimension

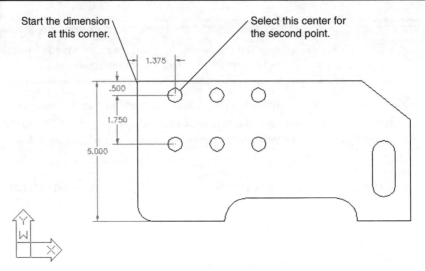

Now you're all set to draw another dimension continuing from the first extension line of the dimension you just drew.

5. Click on the Baseline Dimension option from the Dimension toolbar. Or you can type **dba**↵ at the command prompt to start a baseline dimension.

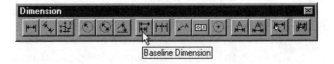

6. At the Second extension line... prompt, pick the upper middle circle in the set of six circles.

7. Continue to pick the upper-right circle, the upper-left of the bevel, and the upper-right end of the plate. Your drawing will look like Figure 8.5.

8. Press ↵ twice to exit the Baseline Dimension command.

9. Pan and Zoom your view so it looks similar to Figure 8.5.

FIGURE 8.5

The plate with the baseline dimensions

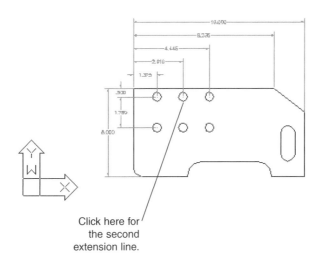

Click here for the second extension line.

In this example, you see that the Baseline Dimension option is similar to the Continue Dimension option, except that Baseline allows you to use the first extension line of the previous dimension as the base for a second dimension.

Continuing from an Older Dimension   You may have noticed in step 8 that you had to press ↵ twice to exit the command. As with Continue Dimension, you can draw a baseline dimension from an older dimension by pressing ↵ at the Specify a second extension line origin (<select>/Undo): prompt. You then get the Select base dimension: prompt, at which you can either select another dimension or press ↵ again to exit the command.

Editing Dimensions

As you begin to add more dimensions to your drawings, you will find that AutoCAD will occasionally place a dimension text or line in an inappropriate location, or you may need to make a modification to the dimension text. In this section, you'll take an in-depth look at how dimensions can be modified to suit those special circumstances that always crop up.

Appending Data to Dimension Text

So far in this chapter, you've been accepting the default dimension text. You can append information to the default dimension value, or change it entirely if you need to. When you see the temporary dimension dragging with your cursor, enter **t**↵. Then, by using the less than (<) and greater than (>) symbols, you can add text either before or after the default dimension or replace the symbols entirely to replace the default text. The Properties button on the Object Properties toolbar lets you modify existing dimension text in a similar way. Let's see how this works by changing an existing dimension's text in your drawing.

1. Click on the Properties tool in the Object Properties toolbar.
2. Next, click on the horizontal dimension to the bevel you added to the drawing in step 7 of the last exercise. The dimension is at the top of the screen.

With the Dimension Edit tool you can append text to several dimensions at once.

You can also use the Multiline Text Editor (**ed**↵ or Modify ➢ Object ➢ Text) to edit the dimension text.

3. Press ↵. The Modify Dimension dialog box appears.

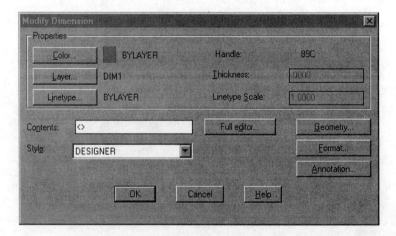

4. Click on the Contents input box, move the cursor behind the <> sign, and then type **PAINTED SURFACE**.
5. Click on OK. The dimension changes to read 8.235 PAINTED SURFACE. The text you entered is appended to the dimension text.
6. Because you don't really need the new appended text for the tutorial, click on the Undo button in the Standard toolbar to remove the appended text.

Place your appended text in front of the <> symbols if you want to add text to the beginning of the dimension text. You can also replace the dimension text entirely by replacing the <> sign in the Contents input box with new text. If you want to restore a dimension that has been modified, delete everything in the Contents input box, including space. Or include a space to leave the dimension text blank.

In this exercise, you were only able to edit a single dimension. To append text to several dimensions at once, you need to use the Dimension Edit tool. See the *Making Changes to Multiple Dimensions* sidebar in this chapter for more on this command.

You can also have AutoCAD automatically add a dimension suffix or prefix to all dimensions, instead of just a chosen few, by using the Annotation option in the Dimension Styles dialog box. See Appendix D for more on this feature.

Besides appending text to a dimension, the Modify Dimension dialog box lets you modify a dimension's other properties. Take a look at the Modify Dimension dialog box in step 3. Notice that it offers the Geometry, Format, and Annotation buttons. These buttons open the same dialog boxes you saw in the beginning of this chapter. Use these buttons to make changes to the formatting of individual dimensions.

Making Changes to Multiple Dimensions

The Dimension Edit tool offers a quick way to edit existing dimensions. It adds the ability to edit the text of more than one dimension at one time. One common use would be to change a string of dimensions to read Equal, instead of showing the actual dimensioned distance. The following example shows an alternative to the Properties tool for appending text to a dimension.

1. Click on the Dimension Edit tool in the Dimension toolbar or type **ded**↵.
2. At the Dimension Edit (Home/New/Rotate/Oblique)<Home>: prompt, type **n**↵ to use the New option. The Multiline Text Editor appears, showing the <> brackets in the text box.

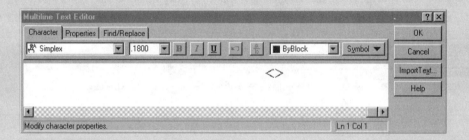

3. Click on the space behind or in front of the <> brackets, and then enter the text you want to append to the dimension. Or you can replace the brackets entirely to replace the dimension with your text.
4. Click on OK.
5. At the Select object: prompt, pick the dimensions you wish to edit. The Select object: prompt remains, allowing you to select several dimensions.
6. Press ↵ to finish your selection. The dimension changes to include your new text or to replace the existing dimension text.

Dimedit is useful for editing dimension text, but you can also use this command to make graphical changes to the text. Here is a listing of the other Dimedit options:

Home moves the dimension text to its standard default position and angle.

Rotate allows you to rotate the dimension text to a new angle.

Oblique skews the dimension extension lines to a new angle. See *Skewing Dimension Lines* later in this chapter.

Locating the Definition Points

AutoCAD provides the associative dimensioning capability to automatically update dimension text when a drawing is edited. Objects called *definition points* are used to determine how edited dimensions are updated.

The definition points are located at the same points you pick when you determine the dimension location. For example, the definition points for linear dimensions are the extension line origin and the intersection of the extension line/dimension line. The definition points for a circle diameter are the points used to pick the circle and the opposite side of the circle. The definition points for a radius are the points used to pick the circle, plus the center of the circle.

Definition points are actually point objects. They are very difficult to see because they are usually covered by the feature they define. You can, however, see them indirectly, using grips. The definition points of a dimension are the same as the dimension's grip points. You can see them by simply clicking on a dimension. Try the following:

1. Make sure the Grips feature is turned on (see Chapter 2 to refresh your memory on the Grips feature).

2. Click on the dimension you drew in the earlier exercise that defines the vertical distance between the top of the plate and the nearest of the two rows of circles. You will see the grips of the dimension, as shown in Figure 8.6.

FIGURE 8.6
The grip points are the same as the definition points on a dimension.

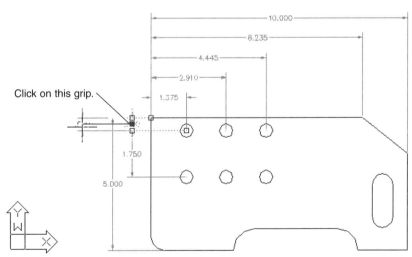

Making Minor Adjustments to Dimensions Using Grips

The definition points, whose location you can see through their grips, are located on their own unique layer, called Defpoints. Definition points are displayed regardless of whether the Defpoints layer is on or off. To give you an idea of how these definition points work, try the following exercises, which show you how to directly manipulate the definition points.

1. With the grips visible, click on the grip near the dimension text.

Since the Defpoints layer has the unique feature of being visible even when turned off, you can use it as a layer for laying out your drawing. While Defpoints is turned off, you can still see objects assigned to it, but the objects won't plot.

2. Move the cursor around. Notice that when you move the cursor vertically, the text moves along the dimension line. When you move the cursor horizontally, the dimension line and text move together, keeping their parallel orientation to the dimensioned plate.

Here the entire dimension line moves, including the text. In a later exercise, you'll see how you can move the dimension text independently of the dimension line.

3. Enter **@.38<180.↵**. The dimension line, text, and the dimension extensions move to the new location to the left of the previous location (see Figure 8.7).

If you need to move several dimension lines at once, select them all at the command prompt; then Shift+click on one set of dimension-line grips from each dimension. Once you've selected the grips, click on one of the hot grips again. You can then move all the dimension lines at once.

FIGURE 8.7
Moving the dimension line using its grip

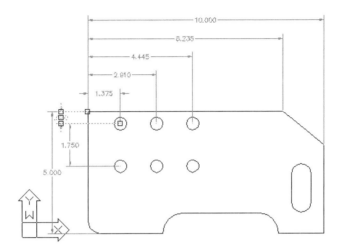

In step 3 of the last exercise, you saw that you can specify an exact distance for the dimension line's new location by entering a relative polar coordinate. Cartesian coordinates work just as well, and with Ortho on you can use dynamic distance entry (DDE). You can even use object snaps to relocate dimension lines. Next, try moving the dimension line back using the Perpendicular Osnap.

1. Click on the grip at the bottom of the dimension line.
2. Shift+right-click and choose Perpendicular from the Osnap pop-up menu.
3. Place the cursor on the vertical dimension line that dimensions the distance between the two rows of circles.
4. The selected dimension line moves back to its original location to align with the other vertical dimension.

Changing Style Settings of Individual Dimensions

In some cases, you will have to make changes to an individual dimension's style settings in order to edit that dimension. For example, if you try to move the text of a typical linear dimension, you'll find that the text and dimension lines are inseparable. You need to make a change to the dimension's style setting that controls how AutoCAD locates dimension text in relation to the dimension line. This section describes how you can make changes to the style settings of individual dimensions to facilitate changes in the dimension.

TIP If you need to change the dimension style of a dimension to match that of another, you can use the Match Properties tool.

Moving a Fixed Dimension Text

In some instances, you will want to manually move a dimension text away from the dimension line, but as you saw in an earlier exercise, this cannot be done with the current settings.

In the next exercise, you will make a change to a single dimension's style settings. Then you'll use grips to move the dimension text away from the dimension line.

1. Press Esc twice to cancel the grip selection from the previous exercise.

2. Click on Properties from the Object Properties toolbar and then click on the vertical dimension that measures the overall plate height—the 5.00 dimension—and press ↵ to finish your selection. The Modify Dimension dialog box appears. This dialog box contains the same three buttons—Geometry, Format, and Annotation—that are in the Dimension Styles dialog box.

3. Click on Format. The same Format dialog box appears that you saw in the early part of this chapter.

4. In the Format dialog box, choose Leader from the Fit drop-down list.

5. While still in the Format dialog box, verify that the Vertical Justification group setting is Centered. We'll explain why in the following paragraphs.

6. Click on OK here and then again in the Modify Dimension dialog box.

7. Now click on the 5.00" dimension to display its grips.

8. Make sure Ortho mode is off. Click on the grip at the center of the dimension text and move the dimension text above and to the left of its current location, as shown in Figure 8.8. The dimension text is now horizontal and shows a leader from the text to the dimension line.

9. Type **u**↵ to undo the last command or use the Undo tool in the Standard toolbar.

EDITING DIMENSIONS | **335**

FIGURE 8.8

Moving the dimension text using grips

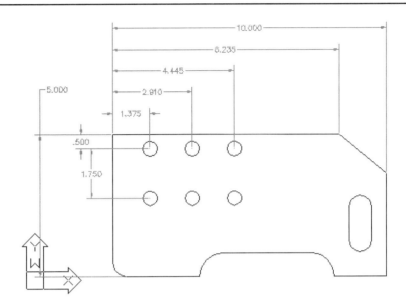

In the Format dialog box, the Leader option in the Fit drop-down list lets you move the dimension text independently of the dimension line. It also causes a leader to be drawn from the dimension line to the text. We asked you to change the Vertical Justification option to Centered because otherwise the leader line will be drawn as an underline beneath the dimension text.

Both the Fit and Vertical Justification settings can be made using system variables. See Appendix D for more on these settings.

To change an existing dimension to the current dimension style, use the Dimension Update tool. Click on Dimension Update from the Dimension toolbar, or choose Dimension ➢ Update from the menu bar. Then select the dimensions you want to change. Press ↵ when you are done selecting dimensions. The selected dimensions will be converted to the current style.

Rotating a Dimension Text

Once in a while, a dimension text works better if it is in a vertical orientation, even if the dimension itself is not vertical. If you find you need to rotate dimension text, here's how to do it:

1. Click on the Dimension Text Edit tool in the Dimension toolbar.

NOTE

You can also use the Dimension Edit tool (Dimedit command) in the Dimension toolbar to rotate the dimension text. Click on Dimension Edit or type **ded↵r↵**, enter the rotation angle, and then select the dimension text.

2. At the Select object: prompt, click on the 5.00" dimension text again.
3. Type **a↵**.
4. At the Enter text angle: prompt, type **45↵** to rotate the text to a 45° angle.

TIP

You can also choose Dimension ➢ Align Text ➢ Angle, select the dimension text, and then enter an angle. A 0 angle will cause the dimension text to return to its default angle.

The Dimension Text Edit tool (Dimtedit command) also allows you to align the dimension text to either the left or right side of the dimension line. This is similar to the Alignment option in the Multiline Text Editor that controls text justification.

NOTE

You can use the Home option of the Dimension Text Edit tool or Dimension ➢ Align Text to move dimension text back to its original location.

You may want to make other adjustments to the dimension text, such as its location along the dimension line or its rotation angle.

As you have seen in this section, the Grips feature is especially well suited to editing dimensions. With grips, you can stretch, move, copy, rotate, mirror, and scale dimensions.

Modifying the Dimension Style Settings for Groups of Dimensions

In *Moving a Fixed Dimension Text*, you used the Properties button on the Object Properties toolbar to facilitate the moving of the dimension text. You can also use the Dimension ➢ Override option (Dimoverride command) to accomplish the same thing. The Override option allows you to make changes to an individual dimension's style settings. The advantage to Override is that it allows you to effect changes to groups of dimensions, not just one dimension. Here's an example showing how Override can be used in place of the Properties button in the first exercise of the *Moving a Fixed Dimension Text* section.

1. Press Esc twice to make sure you are not in the middle of a command. Then choose Dimension ➢ Override from the menu bar.

2. At the next prompt:

   `Dimension variable to override (or Clear to remove overrides):`

   type **dimfit↵**.

3. At the `Current value <3>:` prompt, enter 4↵. This has the same effect as selecting Leader from the Fit pop-up list of the Format dialog box.

4. The `Dimension variable to override...:` prompt appears again, allowing you to enter another dimension variable. Press ↵ to move to the next step.

5. At the `Select object:` prompt, select the dimension you want to change. You can select a group of dimensions if you want to change several dimensions at once. Press ↵ when you are done with your selection. The dimension settings will be changed for the selected dimensions.

As you can see from this example, Dimoverride requires that you know exactly which dimension variable to edit in order to make the desired modification. In this case, setting the Dimfit variable to 4 will let you move the dimension text independently of the dimension line. If you find the Dimoverride command useful, consult Appendix D to find which system variable corresponds to the Dimension Styles dialog box settings. If you have already applied a Dimoverride to a dimension, you can also use the Match Properties icon or type **matchprop↵** to allow other dimensions to inherit the properties. Of course, they will also inherit the layer and linetype of the source object if those properties are set for matching as well.

Editing Dimensions and Other Objects Together

Certainly it's helpful to be able to edit a dimension directly using its grips. But the key feature of AutoCAD's associative dimensions is their ability to *automatically* adjust themselves to changes in the drawing. As long as you include the dimension's definition points when you select objects to edit, the dimensions themselves will automatically update to reflect the change in your drawing. To see how this works, try moving the six-hole pattern to the left edge of the plate. You can move a group of objects and vertices using the Stretch command and the Crossing option.

1. Click on the Stretch tool from the Modify toolbar, or type **s**↵.
2. At the Select objects to stretch by crossing-window or -polygon. Select objects: prompt, use your mouse to create a crossing window by picking one of the right corners first, then dragging the cursor diagonally to the left, as illustrated in Figure 8.9. Next press ↵ to confirm your selection. You must use either a crossing window or a crossing polygon to select objects (the default for the prompt is a crossing window). Remember that a crossing window is created from right to left and a standard window is created from left to right.

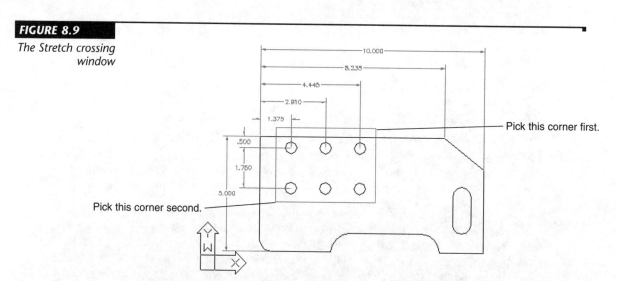

FIGURE 8.9
The Stretch crossing window

3. At the Base point: prompt, pick any point on the screen.

4. At the New point: prompt, enter **@.5,0<90↵** to move the circles –.5" in the positive X direction. The circles move, and the dimension text changes to reflect the new dimension, as shown in Figure 8.10.

5. When you are done reviewing the results of this exercise, close the file without saving it.

In some situations, you may find that a crossing window selects objects other than those you want to stretch. This frequently occurs when many objects are close together at the location of a vertex you want to stretch. To be more selective about the vertices you move and their corresponding objects, use a standard window instead of a crossing window to select the vertices. Then pick the individual objects whose vertices you wish to move.

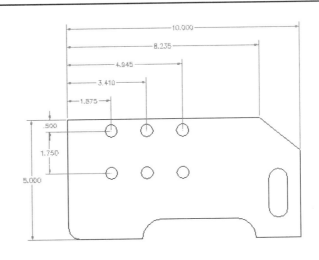

FIGURE 8.10

The moved circles, with the updated dimensions

When you selected the crossing window corners, you included the definition points of the horizontal dimensions. This allowed you to move the dimension extension lines along with the wall, thereby updating the dimensions automatically.

Understanding the Stretch Command

The tool you used for moving the line and the dimension line extensions is the Stretch command. This is one of the most useful, yet least understood commands offered by AutoCAD. Think of Stretch as a vertex mover: Its sole purpose is to move the vertices (or endpoints) of objects.

Stretch actually requires you to do two things: select the objects you want to edit, and then select the vertices you wish to move. The crossing window and the crossing polygon window offer a convenient way to do two things at once, because they select objects and vertices in one operation. But when you want to be more selective, you can click on objects and window vertices instead. For example, consider the exercise in this chapter where you moved the circles with the Stretch command. If you wanted to move the left edge and lower fillet but not the dimension-line extensions, you would do the following:

1. Click on Stretch from the Modify toolbar or from the Modify pull-down menu. You may also type **s**↵.
2. At the Select object: prompt, enter **w**↵ (Window) or **wp**↵ (Window Polygon) or pick an implied Automatic Window starting from the left.
3. Window the vertices you wish to move. Since the Window and Window Polygon selection options select objects completely enclosed within the window, enclose the vertical line and the fillet.
4. Click on the top and bottom adjacent horizontal lines to include them in the set of objects to be edited.
5. Press ↵ to finish your selection.
6. Indicate the base point and second point for the stretch.

You could also use the Remove selection option and click on the dimensions to deselect them in the previous exercise. Then, when you enter the base and second points, the line and fillet would move but the dimensions would stay in place.

Stretch will stretch only the vertices included in the last window, crossing window, crossing polygon, or window polygon (see Chapter 2 for more on these selection options). Thus, if you had attempted to window another part of your drawing in the circle-moving exercise, nothing would have moved. Before Stretch will do anything, objects need to be highlighted (selected) and their endpoints windowed.

Continued

> **CONTINUED**
>
> The Stretch command is especially well suited to editing dimensioned objects, and when you use it with the Crossing Polygon (CP) or Window Polygon (WP) selection options, you have substantial control over what gets edited.

You can also use the Mirror, Rotate, and Stretch commands with dimensions. The polar arrays will also work, and Extend and Trim can be used with linear dimensions.

When editing dimensioned objects, be sure you select the dimension associated with the object being edited. As you select objects, using the Crossing (C) or Crossing Polygon (CP) selection options will help you include the dimensions. For more on these selection options, see the *Other Selection Options* sidebar in Chapter 2.

You can turn off the associative feature, which is controlled by the system variable *dimaso* (enter **dimaso↵off↵**), in your drawing or in any template file (acad.dwt), and you can explode an existing dimension. If you insert a dimension with associativity turned off, AutoCAD draws the components of the dimension as individual objects (lines, arrowheads, and text). If you explode an associative dimension, AutoCAD turns the dimension into individual objects. You will undoubtedly find drawings in industry where this has happened. You can not restore associativity to dimensions by turning this feature back on in a drawing. Exploding dimensions was a common industry practice a few years ago, but today it is mostly unnecessary because of the wide range of control over dimensions now in place. You cannot restore associativity to dimensions by turning this feature back on in a drawing, but keep in mind that dimensions you create can be associative—you can replace or add new dimensions to a drawing that has exploded dimensions.

If you have some dimension text that overlaps a hatch pattern, and the hatch pattern obscures the text, you can use the Wipeout bonus tool on the Bonus Standard toolbar to mask out portions of the hatch. If a hatch pattern or solid fill completely covers a dimension, you can use the Draworder command to have AutoCAD draw the dimension over the hatch or solid fill. See Chapter 19 for more on the Bonus tools and Chapter 16 for more on the Draworder command.

Dimensioning Nonorthogonal Objects

So far, you've been reading about how to work with linear dimensions. You can also dimension nonorthogonal objects, such as circles, arcs, triangles, and trapezoids. Dimalign, Dimdia, Dimrad, and Dimang are powerful automatic tools that will assist you in dimensioning the myriad of shapes not describable orthogonally. AutoCAD can read the size of circles, arcs, and angles, allowing you to dimension these objects, just as it could read the distance between objects when you were placing linear dimensions.

Dimensioning Nonorthogonal Linear Distances

Now you will dimension the true length of the beveled corner. The unusual shape of the bevel prevents you from using the horizontal or vertical dimensions you've used already. However, the Dimension ➢ Aligned option will allow you to dimension at an angle.

1. Click on the Aligned Dimension tool from the Dimension toolbar or enter **dal**↵ to start Aligned Dimension. You can also select Dimension ➢ Aligned from the menu bar.

2. At the `First extension line origin or RETURN to select:` prompt, press ↵. You could have picked extension line origins as you did in earlier examples, but using the ↵ will show you firsthand how the Select option works.

3. At the `Select object to dimension:` prompt, pick the line that represents the bevel. As the prompt indicates, you can also pick an arc or circle for this type of dimension.

4. At the `Dimension line location (Text/Angle):` prompt, pick a point clear of the overall horizontal length dimension. The dimension appears in the drawing, as shown in Figure 8.11.

Just as with linear dimensions, you can enter **t↵** at step 4 to enter alternate text for the dimension.

FIGURE 8.11
The aligned dimension of a nonorthogonal line

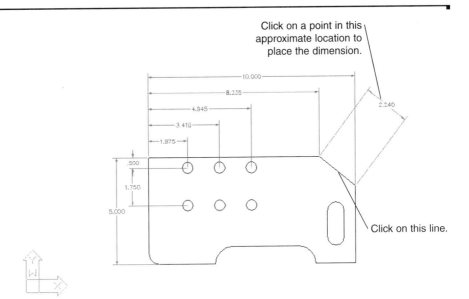

Next, you will dimension a face of a hexagon. Instead of its actual length, however, you will dimension a distance at a specified angle—the distance from the center of the face. First, we will draw a polygon in the space to the right of the plate.

1. Click on Polygon from the Draw toolbar or type **pol↵**.
2. At the Number of sides: prompt, enter **6↵**.
3. At the edge/<center of polygon>: prompt, pick a point for the center of the polygon approximately 3 AutoCAD units to the right of the plate. Use the coordinate display to locate this point. You may want to scroll to the right to see the area.

4. Enter **c↵** at the `Inscribe in circle/circumscribe:` prompt to select the Circumscribe option. This tells AutoCAD to place the Polygon outside the temporary circle used to define the polygon.

5. At the `Radius of circle:` prompt, you will see the hexagon drag along with the cursor. You could pick a point with your mouse to determine its size.

6. Enter **1↵** to get an exact-size hexagon. Your drawing will look like Figure 8.12.

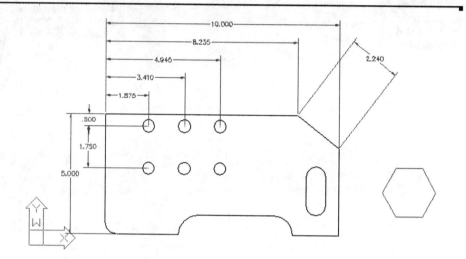

FIGURE 8.12

The completed polygon

Now let's locate the dimension.

1. Click on Linear Dimension from the Dimension toolbar.
2. At the `First extension line origin or RETURN to select:` prompt, press ↵.
3. At the `Select object to dimension…:` prompt, pick the lower-right face of the hexagon near the coordinate.
4. At the `Dimension line location (Text/Angle/Horizontal/Vertical/Rotated):` prompt, type **r↵** to select the rotated option.
5. At the `Dimension line angle <0>:` prompt, enter **30↵**.
6. At the dimension line location prompt, pick a point a little below and to the right of the hexagon. Your drawing should look like Figure 8.13.

DIMENSIONING NONORTHOGONAL OBJECTS

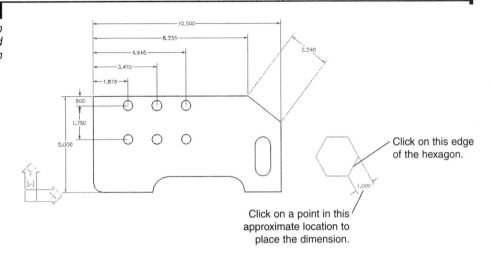

FIGURE 8.13
A linear dimension using the Rotated option

Dimensioning Angles

To dimension angles, you use another option from the Draw ➢ Dimension menu.

1. Click on Angular Dimension from the Dimension toolbar. Or you can enter **dan↵** or choose Dimension ➢ Angular from the menu bar to start the Angular Dimension tool.

2. At the Select arc, circle, line, or RETURN: prompt, pick the top line of the plate.
3. At the Second line: prompt, pick the bevel line.
4. At the Dimension line arc location (Mtext/Text/Angle): prompt, notice that as you move the cursor around the apex of the angle formed by the two edges, the dimension changes to measure either the acute angle or the obtuse angle in four quadrants.
5. Pick a point to the right of the apex, clear of the object and measuring the acute angle. The dimension is fixed in the drawing (see Figure 8.14).

If you need to make subtle adjustments to the dimension line or text location, you can do so using grips, after you have placed the angular dimension.

FIGURE 8.14

The angular dimension added to the plate

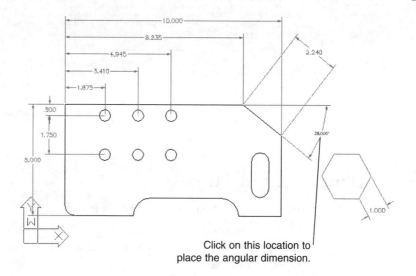

Click on this location to place the angular dimension.

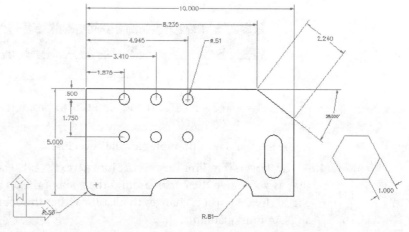

Dimensioning Radii, Diameters, and Arcs

To dimension circular objects, you use another set of tools from the Dimension toolbar. Now try the Diameter tool, which shows the diameter of a circle.

1. Click on Diameter Dimension from the Dimension toolbar. Or you can select Dimension ➤ Diameter from the menu bar or type **ddi**↵ to start the Diameter Dimension tool.

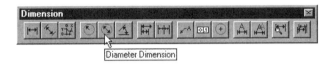

2. At the Select arc or circle: prompt, pick the top right circle in the set of six circles.
3. At the Dimension line location (Mtext/Text/Angle): prompt, you will see the diameter dimension drag along the circle as you move the cursor.

If the dimension text can't fit within the circle, AutoCAD gives you the option to place dimension text outside the circle as you drag the temporary dimension to a horizontal position.

4. Pick a point parallel with the third horizontal baseline dimension at about a 60° angle from the circle. AutoCAD has drawn a center mark at the center of the circle and a leader pointing at the circle towards the center of the circle; it has also included the diameter symbol in the text, as shown in Figure 8.15.

FIGURE 8.15

Dimension showing the diameter of a circle and the radius of two arcs

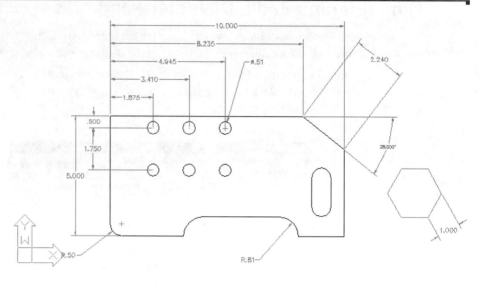

The Radius Dimension tool on the Dimension toolbar gives you a radius dimension just as Diameter provides a circle's diameter.

Figure 8.15 shows a radius dimension on the outside of the fillet at the bottom left of the plate. The leader points at the center of the arc, and AutoCAD has added the R prefix to the text. AutoCAD will also dimension the inside of an arc, as shown by the arc dimension to the inside of the recess at the bottom of the plate. The Center Mark tool on the Dimension toolbar just places a cross mark in the center of the selected arc or circle.

You can alter the format of diameter dimensions by changing the Dimtix and Dimtofl dimension variable settings. For example, to have two arrows appear across the diameter of the circle, turn both Dimtix and Dimtofl on. See Appendix D for more details on these settings.

Using Dimension Families to Fine-Tune Dimension Styles

In the first part of this chapter, you set up a mechanical dimension style called Designer. If you try to draw a radial or diameter dimension with this style, you will get a center mark in the circle instead of a center line, and a center mark in the center of the radius instead of nothing. The angular dimension is measured to three decimal places instead of none, or at most one place. This can be quite annoying because you would have to change the current dimension style every time you wanted to draw a different type of dimension.

AutoCAD offers the *dimension family* to help you set up each different type of dimension to have its own settings. When you create a new style, it is a "parent" style by default. You can set up a "child" style for radial, diameter, or other types of dimension. You can use different arrows, text formats, and scales for each child style you set up.

For example, you can have the Designer style as the parent style, and then create a child style for radius dimensions. This child would be set up to use a "none" for the center mark for radial dimensions. The next time you add a radial dimension, no center mark will be used.

1. Open the Dimension Styles dialog box.
2. Select the dimension style to which you want to add a child family member.
3. In the Family button group, click on the type of dimension you wish to change. The options are Linear, Radial, Angular, Diameter, Ordinate, and Leader. Each of these Family options controls the type of dimension of the same name.
4. Modify the child style to your needs (for example, select the Radial option, click on the Geometry button, and for Center, select None. Click on OK.).
5. Click Save to save the child.

The next time you draw the type of dimension associated with the child dimension style, AutoCAD will automatically use the child settings, instead of the parent settings. Note that the child settings will inherit the parent settings, but modifications to the parent settings will not affect the child settings. For example, if you change the scale factor for the parent, the child scale factor will not change.

Adding a Note with an Arrow

Finally, there is the Dimension ➢ Leader option, which allows you to add a note with an arrow pointing to the object the note describes.

1. Click on Leader from the Dimension toolbar, or enter **le↵**, or select Dimension ➢ Leader from the menu bar.
2. At the `Leader start:` prompt, pick a point near the middle of the right side of the slot.
3. At the `To point:` prompt, pick a point below the hexagon and away from the plate.
4. At the `To point (Format/Annotation/Undo)<Annotation>:` prompt, you can continue to pick points just as you would draw lines. For this exercise, however, press ↵ to finish drawing leader lines.

TIP You can also add Multiline text at the leader. See the next section.

5. At the `Annotation (or RETURN for options):` prompt, type **NO PAINT PERMISSIBLE IN THE SLOT**↵↵ as the label for this leader. Your drawing will look like Figure 8.16.

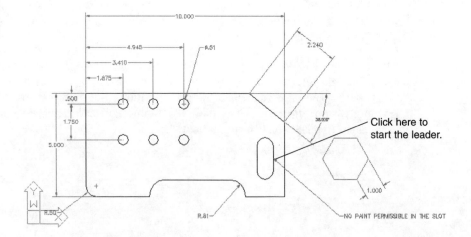

FIGURE 8.16

The leader with a note added

In this exercise, you used the default Annotation option in step 4 to add the text for the note. The Format option in this prompt lets you control the graphic elements of the leader. When you select the Format option by typing **f**↵, you get the following prompt: Spline/STraight/Arrow/None/<Exit>:. Choose Spline to change the leader from a series of line segments to a spline curve (see Figure 8.17). Often a curved leader shows up better than straight lines do. The STraight option changes the leader line to straight lines. None suppresses the arrowhead altogether, and Arrow restores it.

FIGURE 8.17
The Leader format options

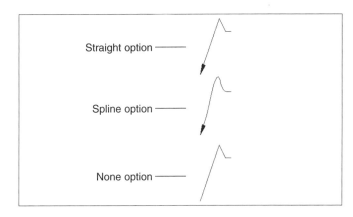

Using Multiline Text with Leaders

You also have the option to add Multiline text at the leader. Take another look at step 5 in the previous exercise. To add Multiline text, you would press ↵ at the Annotation (or RETURN for options): prompt. The next prompt to appear is

Tolerance/Copy/Block/None/<Mtext>:

Press ↵ again, and AutoCAD opens the Multiline Text Editor. From there, the Leader command acts just like the Mtext command by opening the Mtext dialog box (see Chapter 7 for more on Mtext).

The notes and leader act like a single object which you can easily edit using grips.

The other options in the Annotation prompt are as follows:

Tolerance lets you insert a tolerance symbol. A dialog box opens, where you specify the tolerance information (see *Adding Tolerance Notation* in this chapter and Appendix D for more on the Tolerance feature).

Copy lets you copy text from another part of the drawing. You are prompted to select a text object to copy to the leader text location.

Block lets you insert a block. You are asked for a block name, and then the command proceeds to insert the block—just like the Insert command (see Chapter 4).

None is used if you don't want to do anything beyond drawing the leader arrow and line.

Just as with other dimensions and objects in general, you can modify any property of a leader, including its text, using the Properties tool on the Object Properties toolbar. You can, for example, change a straight leader to a spline leader.

Skewing Dimension Lines

At times, you may find it necessary to force the extension lines to take on an angle other than 90° to the dimension line. This is a common requirement of isometric drawings, where most lines are at 30° or 60° angles instead of 90°. To facilitate non-orthogonal dimensions like these, AutoCAD offers the Oblique option.

1. Use the Linear Dimension icon in the Dimension tool bar to create a linear dimension between the lower-right two circles of the set of six circles. The dimension will look like the one in the top image of Figure 8.19.

2. Choose Dimension ➤ Oblique or type **ded**↵**o**↵. You can also click on the Dimension Edit tool from the Dimension toolbar, and then type **o**↵.

3. At the Select objects: prompt, pick the dimension that you just created and press ↵ to confirm your selection.
4. At the Enter obliquing angle (RETURN for none): prompt, enter **80**↵ for 80°. The dimension will skew so that the extension lines are at 80°, as shown in Figure 8.18.

FIGURE 8.18
A dimension using the Oblique option

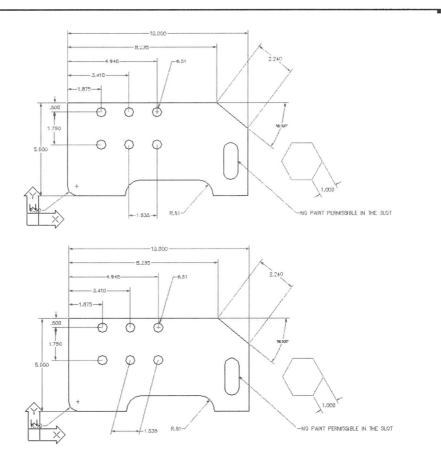

Applying Ordinate Dimensions

In mechanical drafting, *ordinate dimensions* are used to maintain the accuracy of machined parts by establishing an origin on the part. All major dimensions are described as X or Y coordinates of that origin. The origin is usually an easily locatable feature of the part, such as a machined bore or two machined surfaces. Figure 8.19 shows a typical application of ordinate dimensions. In the lower-left, notice the two dimensions whose leaders are jogged. Also notice the origin location in the upper-right.

FIGURE 8.19
A drawing using ordinate dimensions

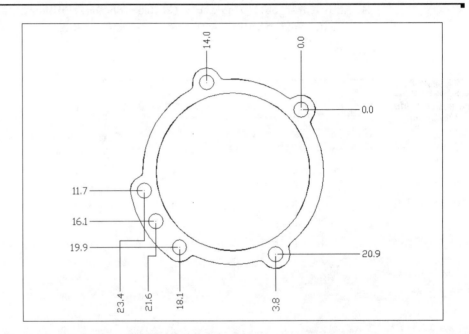

To use AutoCAD's Ordinate Dimension command, follow these steps:

1. Choose Tools ➤ UCS ➤ Origin or type **ucs↵or↵**.
2. At the Origin point <0,0,0>: prompt, click on the exact location of the origin of your part.
3. Toggle the Ortho mode on.

4. Click on Ordinate Dimension from the Dimension toolbar. You can also enter **dor**↵ to start the Ordinate Dimension tool.

5. At the Select feature: prompt, click on the item you want to dimension.

NOTE The direction of the leader will determine whether the dimension will be of the Xdatum or the Ydatum.

6. At the Leader endpoint (Xdatum/Ydatum/Mtext/text): prompt, indicate the length and direction of the leader. Do this by positioning the rubber-banding leader perpendicular to the coordinate direction you want to dimension, and then clicking on that point.

In steps 1 and 2, you used the UCS feature to establish a new origin in the drawing. The Ordinate Dimension tool then uses that origin to determine the ordinate dimensions. You will get a chance to work with the UCS feature in Chapter 11.

You may have noticed options in the Command window for the Ordinate Dimension tool. The Xdatum and Ydatum options force the dimension to be of the X or Y coordinate no matter what direction the leader takes. The Mtext option opens the Multiline Text Editor, allowing you to append or replace the ordinate dimension text. The Text option lets you enter replacement text directly through the Command window.

TIP As with all other dimensions, you can use grips to make adjustments to the location of ordinate dimensions.

With Ortho mode on, the dimension leader is drawn straight. If you turn Ortho mode off, the dimension leader will be drawn with a jog to whichever side of horizontal or vertical that you have picked for the text point. The first choice for an ordinate dimension should be straight from the object, and you should only use the jog to avoid writing over or through another dimension (see Figure 8.20).

FIGURE 8.20

A drawing using ordinate dimensions

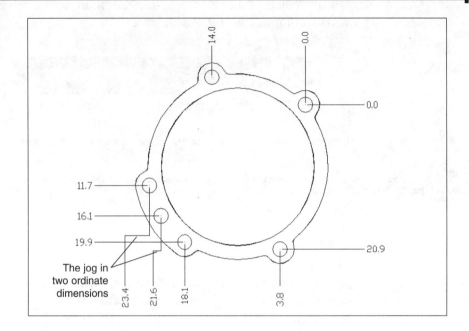

Adding Tolerance Notation

In mechanical drafting, tolerances are a key part of a drawing's notation. They specify the allowable variation in size and shape that a mechanical part can have. To help facilitate tolerance notation, AutoCAD provides the Tolerance command, which offers common ISO tolerance symbols together with a quick way to build a standard *feature control symbol*. Feature control symbols are industry standard symbols used to specify tolerances. If you are a mechanical engineer or drafter, AutoCAD's tolerance notation options will be a valuable tool. However, a full discussion of tolerances requires a basic understanding of mechanical design and drafting and is beyond the scope of this book.

To use the Tolerance command, choose Tolerance from the Dimension toolbar, type **tol**↵ at the command prompt, or select Dimension ➢ Tolerance from the menu bar.

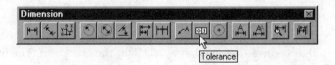

ADDING TOLERANCE NOTATION | **357**

The Symbol dialog box appears.

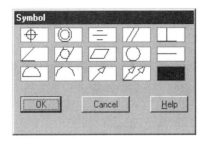

The top image of Figure 8.21 shows what each symbol in the Symbol dialog box represents. The bottom image shows a sample drawing with a feature symbol used on a cylindrical object.

FIGURE 8.21

The tolerance symbols

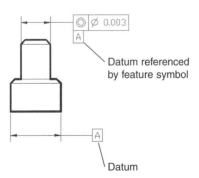

Once you select a symbol, you see the Geometric Tolerance dialog box. This is where you enter tolerance and datum values for the feature control symbol. You can enter two tolerance values and three datum values. In addition, you can stack values in a two-tiered fashion.

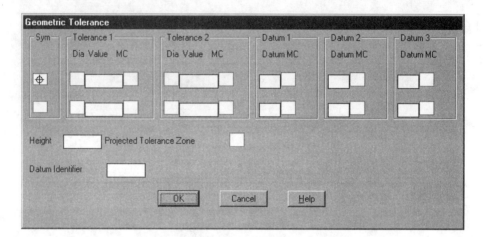

Understanding the Power of the Properties Tool

In this and the previous chapter, you have made frequent use of the Properties tool in the Object Properties toolbar. By now, you may have recognized that the Properties tool is like a gateway to editing virtually any object. It allows you to edit the general properties of layer, color, and linetype assignments. And when used with individual objects, it allows you to edit properties that are unique to the selected object. For example, using this tool, you can to change a spline leader with an arrow into one with straight line segments and no arrow.

If the Modify Properties dialog box does not offer specific options to edit the object, then it will offer a button to open a dialog box that will. If you edit a Multiline text object with the Properties tool, for example, you will have the option to open the Multiline Text Editor. The same is true for dimension text.

Continued

> **CONTINUED**
>
> With R14, Autodesk has made a clear effort to make AutoCAD's interface more consistent. The Text editing tools now edit text of all types—single-line, multiline, and dimension text—so you don't have to remember which command or tool you need for a particular object. Likewise, the Properties tool offers a powerful means to editing all types of objects in your drawing.
>
> You may want to experiment with the Properties tool on new objects you learn about. In addition to allowing you to edit properties, the Properties tool can show you the status of an object, much like the List tool.

If You Want to Experiment...

At this point, you might want to experiment with the settings described in this chapter to identify the ones that are most useful for your work. You can then establish these settings as defaults in a template file or the Acad.dwt file.

It's a good idea to experiment even with the settings you don't think you will need often—chances are you will have to alter them from time to time.

As an added exercise, try using Ordinate Dimension to define the plate drawing. Start a new file using the plate drawing from the companion CD-ROM.

Chapter 9

Advanced Productivity Tools

FEATURING

Editing More Efficiently

Using Grips to Simplify Editing

Using External References (Xrefs)

Switching to Paper Space

Advanced Tools: Selection Filter and Calculator

Advanced Productivity Tools

As you have seen so far, drawing in AutoCAD is more than just drawing a line, then drawing another, then drawing a circle. You can save a great deal of time by re-using existing geometry already created in this or another drawing. If there is symmetry in your design, you need only draw half of an object and use the mirror tool to make the other half. Editing is not used just for making changes—it is also used to create. In this chapter, you will learn some refinements of the editing tools you have already encountered.

You have worked with blocks in some of the previous chapters. Drawings imported as blocks become part of your drawing and can swell the size of a drawing file enormously. If you don't need to use the geometry in a block, you can use an external reference. You will have the opportunity to experiment with external references later on in this chapter.

In preparation for 3D modeling and as an alternative to using views to move around your work, you will experiment with Paper Space, Model Space, and Tilemode. You will see that you can set up your work in viewports, which allow you to see multiple details of your work simultaneously.

And you'll take a tour of the geometry calculator and selection filters. The Calculator gives the AutoCAD user access to higher math functions while returning the answer to the command line. You can also use hooks into AutoCAD commands that enable you to use the position and size of existing geometry in your calculations. The selection filter allows you to choose specific objects by name from a selection set for editing.

Editing More Efficiently

The bar and bracket assembly that you have been working on is incomplete. The side view contains a number of hidden lines that are shown as solid. In the following exercise you'll change the layer so that the lines will be on the Hidden layer and will inherit the linetype of the Hidden layer. You will be working in some tight areas full of detail, so you will also take a look at how to be more precise when making your selections.

> ### Quick Access to Your Favorite Commands
>
> As you continue to work with AutoCAD, you'll find that you use a handful of commands 90 percent of the time. You can collect your favorite commands into a single toolbar using AutoCAD's toolbar customization feature. This way, you can have ready access to your most frequently used commands. Chapter 21 gives you all the information you need to create your own custom toolbars.

Editing an Existing Drawing

First, let's look at how you can modify the bolts in the bar plan. You'll begin by trimming the existing bolt lines.

1. Open the Bar file.
2. Make layer 0 the current layer by trying the following: Click on the Make Object Layer Current button on the Object Properties toolbar, and then click on an object line.

TIP If you didn't create a bar, you can use bar_9.dwg from the companion CD-ROM.

3. If they are not already on, turn on Noun/Verb Selection (Ddselect) and the Grips feature (Ddgrips).
4. Zoom in to the right side view. It might take you a couple of tries to get the view centered in the viewport.

The bolts and interior of the washer should not show through the extrusion or each other. You will need to trim away the offending geometry and recreate it on the Hidden layer. You could use the Break tool, but this method is usually much faster.

Using the Fence Object Selection in the Trim Command

You are about to form a line that can be composed of one or many segments. Each point that you pick becomes a vertex, and the line continues until you press ↵. At that time any object crossed by any of the line segments will be selected. You do not "fence in" or surround the selection set.

1. Enter **tr**↵ to start the Trim command or click on the Trim tool from the Modify toolbar.

2. Pick the right side of each boss on the extrusion shown in Figure 9.1. Complete the selection set.
3. Type **f**↵ to invoke the Fence option for selecting objects. Press ↵ to complete the selection set.
4. Pick the points for the fence vertices in the order shown in Figure 9.2.
5. Press ↵ to complete the selection set. All four lines representing the bolt have been trimmed.

NOTE The Fence is the only tool that you can use to trim multiple objects simultaneously. This handy little tool works with any of the other commands that ask you to select objects. Remember that Trim and Extend are very similar commands. The Fence tool will also select multiple objects for the Extend command.

FIGURE 9.1
Lines on the extrusion to be used as cutting edges during the trim

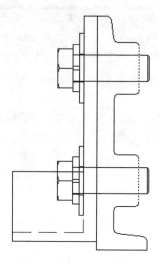

FIGURE 9.2
The points to pick to create the fence

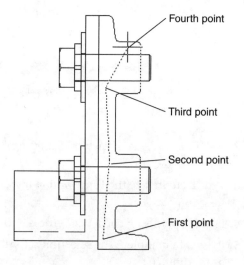

Changing the Layer of an Object

When you created the Versions template file, you created a layer named Hidden, giving it a hidden linetype property and the color red. You used the Versions template file when you started the Bar drawing file. The AutoCAD rule is, any object (that is not a block or Xref) whose linetype property and/or layer property is Bylayer (the default linetype and layer property) will inherit the linetype and/or color of the layer.

The following exercise demonstrates how you can use this rule to change lines to be hidden lines and to change their color to make them easier to see on the screen. Let's change a few lines to the Hidden layer, then draw the lines for the top bolt back in and change them to the Hidden layer.

1. Zoom into the top bolt head and washers. You will see six horizontal lines: Four of these represent the holes through the washers and the other two (close together near the centerline of the washer) represent the split in the lock washer. Pick the four lines that represent the hole in the washer.

2. Open the layer drop-down list from the Object Properties toolbar and select the Hidden layer. The lines are changed to the Hidden layer.

3. Pick one of the new hidden lines and select the Make Object's Layer Current tool from the Object Properties toolbar. Press Esc twice to exit Grips. Notice the Object Properties toolbar layer listing now shows the Hidden layer as current.

NOTE

Just a reminder: When using grips, as you have been in the previous steps, you must press the Esc key twice to complete the command. If an object's grips are still highlighted when you enter the next command, AutoCAD will assume that you want to apply the next command to the highlighted objects.

You would normally erase the lines in the washers because they would only confuse when the drawing is printed. For this exercise, we will change them to the Hidden layer.

4. Use the Line tool to draw a line to replace the portion of trimmed line that was the top of the bolt through the extrusion.

5. Pick the line and highlight the right-end grip. Right-click to pop the Grips menu and select Move, right-click again to select Copy, then pick the endpoint of the remaining bottom of the screw. Press ↵ once to complete the command, then press Esc twice to exit Grips. Your drawing will look like Figure 9.3.

FIGURE 9.3
The hidden lines for the top bolt

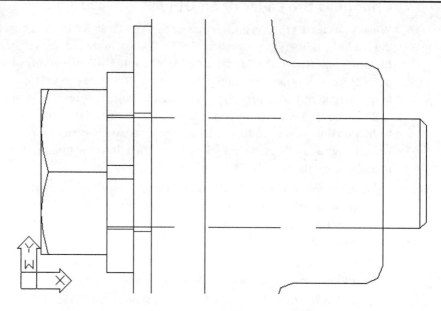

Using the Copy Command to Copy Objects on any Layer

You have created the objects for the top bolt that can now be used for the bottom bolt. Now you'll copy the top bolt's hidden lines to the bottom bolt.

1. Zoom out until you see all of the side view again.
2. Use the Copy command, grips, or the Copy tool from the Modify toolbar and select the two new hidden lines for the bolt.
3. At the `Base point or displacement:` prompt, pick the endpoint of the line of the top bolt that was part of the bolt after is was trimmed. At the `Second point of displacement:` prompt, choose the corresponding point on the lower bolt. A copy appears like the one in Figure 9.4.

You can use the Match Properties tool to make a set of objects match the layer of another object. Click on Match Properties on the Standard toolbar, select the objects whose layer you want to match, and then select the objects you want to assign to the objects layer.

FIGURE 9.4
The side view of both bolts, with the hidden lines

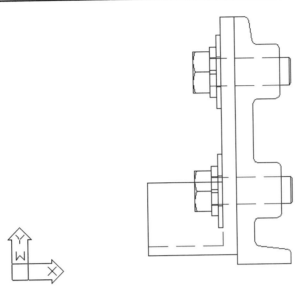

In this exercise, you used the Fence option to select the objects you wanted to trim. You could have selected each line individually by clicking on it, but the Fence option offered you a quick way to select a set of objects without having to be too precise about where they are selected.

This exercise also shows that it's easier to trim lines back and then draw them back in than to try to break them precisely at each location. At first this may seem counterproductive, but trimming and then redrawing can be much faster because doing so requires fewer steps.

You did not complete all of the changes required to generate the side view. A portion of the lower bolt head is obscured by the ledge and should be drawn with hidden lines. You are welcome to finish the view on your own. It would be good practice, but it is not essential for continuing with the tutorial.

Singling Out Proximate Objects

In Chapter 3 we mentioned that you will encounter situations where you need to select an object that is overlapping or very close to another object. Often in such a situation, you end up selecting the wrong object. To help you select the exact object you want, AutoCAD offers the Selection Cycling tool and the Object Selection Settings dialog box.

Selection Cycling

Selection cycling lets you cycle through objects that overlap until you select the one you want. To use this feature, hold down the Ctrl key and click on the object you want to select. If the first object highlighted is not the one you want, click again, but this time don't hold down the Ctrl key. When several objects are overlapping, just keep clicking until the right object is highlighted and selected. When the object you want is highlighted, press ↵, and then go on to select other objects or press ↵ to finish the selection process. You may want to practice using selection cycling a few times to get the hang of it. It can be a bit confusing at first, but once you've gotten accustomed to how it works, selection cycling can be an invaluable tool.

Object Sorting

If you are a veteran AutoCAD user, you may have grown accustomed to clicking on the most recently created object of two overlapping objects. With Release 12, AutoCAD introduced user-definable controls that set the method of selecting overlapping objects. These controls changed the way AutoCAD selected overlapping objects. You didn't always get to the most recently drawn object when you clicked on overlapping objects.

If you prefer to use pre-Release 12 versions, in which AutoCAD selected the most recently drawn object, you can use the Object Sort Method dialog box to revert to the old selection method. This dialog box is buried in the Object Selection Settings dialog box. To get there, click on Tools ➤ Selection, and then click on the Object Sort Method button.

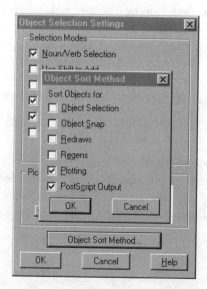

The Object Sort Method dialog box lets you set the sort method for a variety of operations. If you enable any of the operations listed, AutoCAD will use the pre-Release 12 sort method for that operation. You will probably not want to change the sort method for object snaps or regens. But by checking Object Selection, you can control which of two overlapping lines is selected when you click on them. For plotting and for PostScript output, you can control the overlay of screened or hatched areas.

These settings can also be controlled through system variables. See Appendix D for details.

NOTE When you use the Draworder command (Tools ➢ Display Order), all the options in the Object Sort Method dialog box are turned on.

Using External References (Xrefs)

Concurrent engineering has been the goal of a lot of companies over the last decade. AutoCAD devised external references to facilitate design teams working in this manner. Think of a file with Xrefs as a room with many windows into other rooms where work is being done. The Xref is the window. As the work in another room progresses, each team can look through the window to see how the changes are affecting their work. Each time that you reload the Xref, it is refreshed and represents the current version of the file. For example, imagine you are working on a hub and someone else is working on the shaft. The horsepower requirements of the shaft have been reduced, and the other designer has redrawn it with the new diameters and the new-sized bearings. You have the shaft in your layout as an Xref and you get an e-mail informing you of the change. If you are working in the file, you must reload the Xref. If you are not, the latest version (the last saved version) of the Xref will appear when you load your drawing.

Any file can be referenced into any other file. We will call the file that is being referenced the original and the file that is using the Xref the working drawing. You can only change the Xref by opening the original file. If you have an Xref inserted into a working drawing and you want to change some aspect of the Xref, you must open the original and make the change, then to see the change you must reload the Xref in the file that is using it.

There are some problems with trying to use Xrefs in production working drawings such as part and assembly drawings. ISO 9000 and nearly all company standards require that drawing changes be controlled so that when a drawing is distributed to the enterprise, the drawing can not change unless the distribution channel is notified. If a drawing contains an Xref, and even if the Xref is changed correctly (with a change order and distribution), there is no process within AutoCAD to warn of the change's effect on other files that use it as an Xref. Each drawing contains a list of the Xrefs that it is using, but there is no AutoCAD command that can tell if other drawings are using a file as an Xref.

In this section you will see firsthand how to create and use external references (Xrefs) to help reduce design errors, share information in workgroups, and reduce the size of assembly and layout drawings.

Attaching a Drawing as an External Reference

The next exercise shows how you can use an external reference in place of an inserted block to construct a gearbox and motor assembly. You'll start by opening the gearbox file, then import a motor.

1. Open the Gearbox.dwg file from the companion CD-ROM.

2. Choose Save As and save the file to your hard disk under the name Gearbox in your favorite directory (My Documents will serve if you have trouble deciding which directory to use). The current file is now Gearbox. Use Microsoft Explorer to locate the Motor.dwg file on the companion CD-ROM, then copy the file to your hard disk in Acadr14/Sample or any other directory that you like.

3. Click on Insert ➤ External Reference or type **xr**↵ to open the External Reference dialog box.

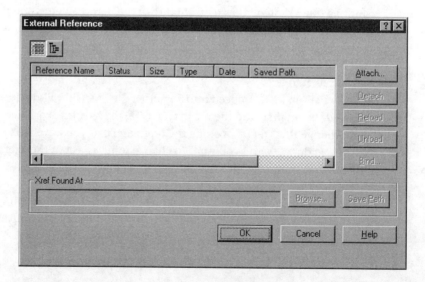

4. Click on the Attach button in the dialog box. The Select File to Attach dialog box appears. This is a typical AutoCAD File dialog box, complete with a preview window.

5. Locate and select the Motor file from your hard disk, then click on Open. The Attach Xref dialog box appears.

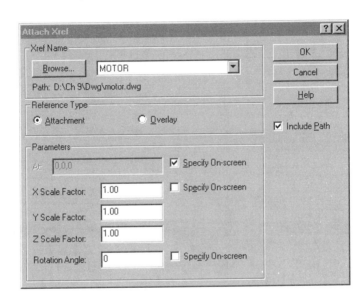

6. Later, you'll get a description of the options presented in this dialog box. For now, click on OK.

7. Pick a point near the gearbox for the insertion point.

8. Align the motor with the gearbox. Use grips to rotate and move the motor Xref and Osnaps to align the top view of the motor and gearbox. Your drawing should look like Figure 9.5.

9. Save the Gearbox file. Then open `motor.dwg`.

10. Move the label onto the motor slightly above the electrical box.

11. Click on File in the menu bar. Read down the choices until you see a list of recently opened files. Pick the Gearbox file, then at the Save this File dialog box, click on Yes. After the `Gearbox.dwg` file opens, notice that the Motor Xref has been updated to reflect the change that you made to the `Motor.dwg` file.

FIGURE 9.5

The motor aligned with the gearbox

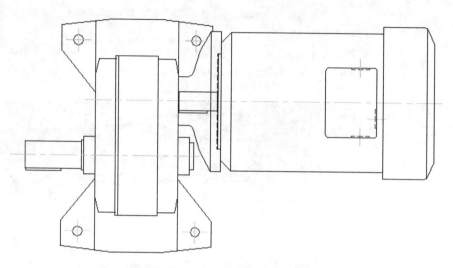

Importing Blocks, Layers, and Other Named Elements from External Files

You can use the tool labeled External Reference Bind in the Reference toolbar to import blocks and other drawing components from another file. First use External Reference Attach to cross-reference a file; then click on External Reference Bind. At the Xbind dialog box, click on the plus sign next to the external reference file name, then select Block. Locate the name of the block you want to import, click on the Add button and click on OK. Finally, use Detach from the Reference dialog box to remove the Xref file. The imported block will remain as part of the current file.

Here you saw how an Xref file doesn't have to be updated the way blocks do. Also, you avoid the task of having to update nested blocks, since AutoCAD updates nested Xrefs, as well as non-nested Xrefs.

TIP If you find that you use Xrefs frequently, you may want to place the Insert toolbar permanently in your AutoCAD window. It contains tools for inserting both blocks and Xrefs. To open the Insert toolbar, right-click on any toolbar, then click on the Insert check box in the Toolbars dialog box.

Other Differences between External References and Blocks

Here are a few other differences between Xrefs and inserted blocks that you will want to keep in mind:

- Any new layers, text styles, or linetypes brought in with cross-referenced files do not become part of the current file. You must use the Xbind command.

TIP Another way to insure that layer settings for Xrefs are retained is to type **visretain**↵ at the command prompt. At the New value for VISRETAIN <0> prompt, enter **1**.

- If you make changes to the layers of a cross-referenced file, those changes will not be retained when the file is saved, unless the Retain Changes to Xref-dependent Layers option is checked. This is a check box on the Layers tab of the Layer & Linetype Properties dialog box. Click on the Details button to find it. This option will then instruct AutoCAD to remember any layer color or visibility settings from one editing session to the next. In the standard AutoCAD settings, this option is on by default.
- To segregate layers on a cross-referenced file from the ones in the current drawing, the cross-referenced file's layers are prefixed with their file's name. A vertical bar separates the file-name prefix and the layer name, as in motor | 0.

You can convert an Xref into a block by using the Bind option of the External Reference dialog box.

- Xrefs cannot be exploded. You can, however, convert an Xref into a block, then explode it. To do this, you must use the Bind button in the External Reference dialog box. This opens the Bind Xrefs dialog box that offers two ways of converting an Xref into a block. See the next section, *Options in the External Reference Dialog Box,* for more information.
- If an Xref is renamed or moved to another location on your hard disk, AutoCAD won't be able to find the moved or renamed file when it opens other files to which the Xref is attached. If this happens, you must use the Browse option of the External Reference dialog box to tell AutoCAD where to find the cross-referenced file.

Take care when relocating an Xref file with the Browse button. The Browse button can assign a file of a different name to an existing Xref as a substitution.

Other External Reference Options

There are many other features unique to cross-referenced files. Let's look briefly at some of the options in the External Reference dialog box we haven't discussed yet.

Options in the External Reference Dialog Box

You'll find the following options in the External Reference dialog box. All but the Attach option are available only when an Xref is present in the current drawing and its name is selected from the list of Xrefs shown in the main part of the dialog box.

Attach opens the Attach Xref dialog box, allowing you to select a file to attach and set the parameters for the attachment.

Detach detaches an Xref from the current file. The file will then be completely disassociated from the current file.

Unload is similar to Detach, but Unload maintains a link to the Xref file so that it can be quickly reloaded. This has an effect similar to freezing a layer and can reduce redraw, regen, and file-loading times.

Reload restores an unloaded Xref.

Bind converts an Xref into a block. Bind offers two options: Bind (again) and Insert. The Bind option will maintain the Xref's named elements (layers, linetypes, and text and dimension styles) by creating new layers in the current file with the Xref's filename prefix. The Insert option will not attempt to maintain the Xref's named elements but will merge them with named elements of the same name in the current file. For example, if both the Xref and the current file have layers of the same name, the objects in the Xref will be placed in the layers of the same name in the current file.

Browse opens the Select New Path dialog box, from which you can select a new file or location for a selected Xref.

Save Path saves the file path displayed in the Xref Found At input box. See Include Path in the following section.

List View/Tree View are two buttons in the upper-left corner of the dialog box. They let you switch between a List view of your Xrefs and a hierarchical Tree view. The Tree view can be helpful in determining how Xrefs are nested.

The Xref list works like other Windows lists, offering the ability to sort by name, status, size, type, date, or path. To sort by name, for example, click on the Reference Name button at the top of the list.

The Attach Xref Dialog Box

The Attach Xref dialog box shown in the previous exercise offers these options:

Attachment causes AutoCAD to include other Xref attachments that are nested in the selected file.

Overlay causes AutoCAD to ignore other Xref attachments that are nested in the selected file. This avoids multiple attachments of other files and eliminates the possibility of circular references (referencing the current file into itself through another file).

Include Path lets you determine whether AutoCAD stores or discards the path information to the Xref in the current drawings database. If you choose not to use this option, AutoCAD will use the default file search path to locate the Xref the next time the current file is open. If you plan to send your files to someone else, you may want to turn this option off; otherwise the recipient of the file will have to duplicate the exact file path structure of your computer before the Xrefs will load properly.

Specify On-screen gives you the option to enter insertion points, scale factors, and rotation angles within the dialog box or in the command window, in a way similar to inserting blocks. If you check this option for any of the corresponding parameters, the parameters change to allow input. If they are not checked, you are prompted for those parameters after you click OK to close the dialog box. With all three Specify On-screen check boxes unchecked, you receive the same set of prompts as those for the Insert dialog box.

Clipping Xref Views and Improving Performance

Xrefs are frequently used to import large drawings for reference or backgrounds. Multiple Xrefs, such as weldments, purchased detail parts, and fabricated parts, might be combined into one file. One drawback to multiple Xrefs in earlier versions of AutoCAD was that the entire Xref was loaded into memory, even if only a small portion of the Xref was used for the final plotted output. In computers with limited resources, multiple Xrefs could slow the system to a crawl.

Release 14 offers two tools that will help make display and memory use more efficient when using Xrefs: the Xclip command and the Demand load option in the Preferences dialog box.

Clipping Views

Xclip is the name of a command accessed by choosing Modify ➢ Object ➢ Clip. This command allows you to clip the display of an Xref or block to any shape you desire, as shown in Figure 9.6. For example, you may want to display only an L-shaped portion of a layout to be part of your current drawing. Xclip lets you define such a view.

FIGURE 9.6

The top panel shows a polyline outline of the area to be isolated with Xclip. The middle panel shows how the Xref appears after Xclip is applied. The bottom panel shows a view of the plan with the polyline's layer turned off.

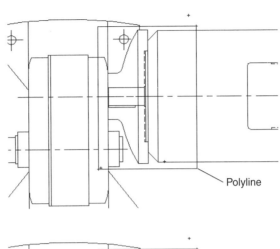

Polyline

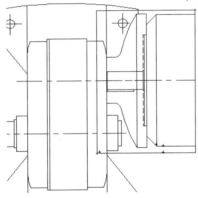

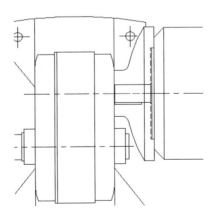

Blocks and Multiple Xrefs can be clipped as well. You can also specify a front and back clipping distance so that visibility of objects in 3D space can be controlled. You can define a clip area using polylines or spline curves (see Chapter 10 for more on polylines and spline curves).

Controlling Xref Memory Use

The Demand load option in the Performance tab of the Preferences dialog box limits how much of an Xref is loaded into memory. Only the portion that is displayed gets loaded into memory.

Demand load has a drop-down list with three settings: Disabled, Enabled, and Enabled with Copy. Demand load is enabled by default in the standard AutoCAD drawing setup. Besides reducing the amount of memory an Xref consumes, Demand load also prevents other users from editing the Xref while it is being viewed as part of your current drawing. This is done to aid drawing version control and drawing management. The third Demand load option, Enabled with Copy, creates a copy of the source Xref file, then uses the copy, thereby allowing other AutoCAD users to edit the source Xref file.

Demand loading improves performance by loading only the parts of the referenced drawing that are needed to regenerate the current drawing. You can set the location for the Xref copy in the Files tab of the Preferences dialog box under Temporary External Reference Location.

Special Save As Options That Affect Demand Loading

AutoCAD offers a few additional settings that will boost the performance of the Demand load feature mentioned in the previous section. When you choose File ➢ Save As to save a file in the standard Dwg format, you will see a button labeled Options. If you click on the Options button, the Export Options dialog box appears. This dialog box offers the Index Type drop-down list. The index referred to can help improve the speed of demand loading at the cost of a 6 percent or so increase in file size per index. The index options are

None creates no index.

Layer loads only layers that are both turned on and thawed.

Spatial loads only portions of an Xref or Raster image within a clipped boundary.

Layer & Spatial turns on both the Layer and Spatial options.

WARNING

Remember, if you move an external reference file to another directory after you have inserted it into a drawing, AutoCAD may not be able to find it later when you attempt to open the drawing. If this happens, you can use the Browse option in the External Reference dialog box to tell AutoCAD the new location of the external referenced file.

External references do not have to be permanent. You can attach and detach them easily at any time. This means that if you need to get information from another file—to see how well an object aligns, for example—you can temporarily external reference the other file to quickly check alignments, and then detach it when you are done.

Think of these composite files as final plot files that are only used for plotting and reviewing. Editing can then be performed on the smaller, more manageable external referenced files. Figure 9.7 diagrams the relationship of these files.

FIGURE 9.7

Diagram of external referenced file relationships

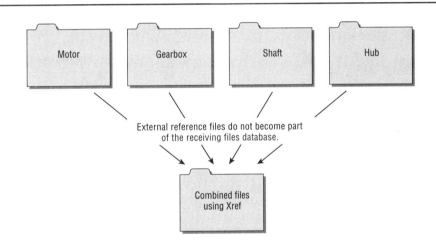

The combinations of external references are limited only by your imagination, but avoid multiple external references of the same file.

TIP

Because Xref files do not become part of the file they are referenced into, you must take care to keep Xref files in a location where AutoCAD can find them when the referencing file is opened. This can be a minor annoyance when you need to send files to others outside your office. To help you keep track of external references, AutoCAD offers the Pack 'n Go tool in the Bonus Standard toolbar. See Chapter 19 for details.

Importing Named Elements from External References

In Chapter 5, we discussed how layers, blocks, linetypes, and text styles—called *named elements*—are imported along with a file that is inserted into another. External reference files, on the other hand, do not import named elements. You can, however, review their names and use a special command to import the ones you want to use in the current file.

You can set the Visretain system variable to 1 to force AutoCAD to remember layer settings of external referenced files (see Appendix D for details). You can also use the Layer Manager bonus utility to save layer settings for later recall. The Layer Manager is described in detail in Chapter 19.

When the Xref is attached to the drawing, AutoCAD renames named elements to eliminate conflict between the Xref's named elements and the named elements of the current drawing. AutoCAD renames the elements by prefixing them with the name of the file they've come from. In our example, the Xref file name is Motor and there is a layer in the Motor file named Hidden. When the Xref is attached, the layer is added and renamed Motor|Hidden. You cannot draw on any layer created for an Xref, but you can view and manage external referenced layers in the Layer & Linetype Properties dialog box and in the Layer Control dialog box. You can also view external referenced blocks using the Insert dialog box.

Next you'll look at how AutoCAD identifies layers and blocks in external referenced files, and you'll get a chance to import a layer from an Xref.

1. While in the Gearbox file, open the Layer & Linetype Properties dialog box. Notice that the names of the layers from the external referenced files are all prefixed with the file name and the vertical bar (|) character. Exit the Layer & Linetype Properties dialog box.

You can also open the Layer Control drop-down list in the Object Properties toolbar to view the layer names.

2. Click on External Reference Bind from the Reference toolbar or enter **xb**↵.

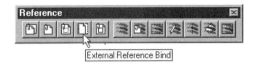

The Xbind dialog box appears.

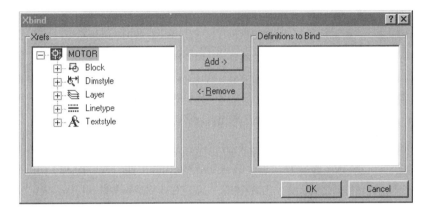

You see a listing of the current Xrefs. Each listing shows a plus sign to the left. This list box follows the Microsoft Windows 95 format for expandable lists, much like the directory listing in the Windows Explorer.

3. Click on the plus sign next to the MOTOR|Xref listing. The list expands to show the types of element available to bind.

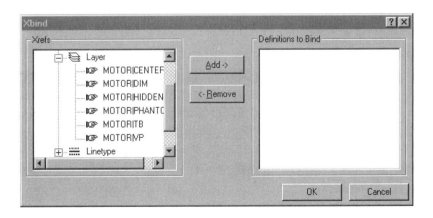

4. Now click on the plus sign next to the Layer listing. The list expands further to show the layers available for binding.

5. Locate MOTOR|BL_L_NAME in the listing, click on it, and then click on the Add button. MOTOR|BL_L_NAME is added to the list to the right: Definitions to Bind.
6. Click OK to bind the MOTOR|BL_L_NAME layer.
7. Now open the Layer & Linetype Properties dialog box.
8. Scroll down the list and look for the MOTOR|BL_L_NAME layer. You will not find it. In its place is a layer called MOTOR0BM_L_NAME.

As you can see, when you use Xbind to import a named item, such as the MOTOR|BL_L_NAME layer, the vertical bar (|) is replaced by two dollar signs surrounding a number, which is usually zero. (If for some reason the imported layer name MOTOR0BM_L_NAME already exists, then the 0 in that name is changed to 1, as in MOTOR1BM_L_NAME.) Other named items are also renamed in the same way, using the 0 replacement for the vertical bar.

Although you used the Xbind dialog box to bind a single layer, you can also use it to bind multiple layers, as well as other items from Xrefs attached to the current drawing.

You can bind an entire Xref to a drawing, converting it into a simple block. By doing so, you have the opportunity to maintain unique layer names of the Xref being bound, or merge the Xref's similarly named layers with those of the current file.

Nesting External References and Using Overlays

External references can be nested. For example, the gearbox.dwg file created in this chapter uses the motor.dwg file as an external reference. In this situation, motor.dwg is nested in the gearbox.dwg file, which could in turn be an external reference in another drawing file.

Continued

> **CONTINUED**
>
> This combination of nested external references could be used numerous times in another file being drawn to lay out the entire machine. AutoCAD would dutifully load each instance of the motor gearbox and thus use substantial memory, slowing your computer down.
>
> If you don't need to see the motor information in your layout, you can use the Overlay option when you attach the Xref. The Attach Xref dialog box contains a Reference box, which in turn contains two check boxes: Attachment and Overlay. If you select the Overlay option when you attach the gearbox, the nested motor Xref will be ignored.

Using Tiled Viewports

To gain a clear understanding of Model Space, Paper Space, and Tiled Model Space, imagine that the object that you have been drawing is actually the full-sized object (since you have been drawing the object to full scale). Your computer screen is your window into a "room" where this object is being constructed, and the keyboard and mouse are your means of access to this room. You can control your window's position in relation to the object through the use of Pan, Zoom, View and other display-related commands. You can also construct or modify the model by using drawing and editing commands. Think of this room as your *Model Space*.

Now suppose you have the ability to step back and add windows with different views looking into your Model Space. The effect is the same as having several video cameras in your Model Space "room," each connected to a different monitor. You can view all your windows at once on your computer screen, or enlarge a single window to fill the entire screen. These are *tiled* viewports. The overall environment in which you see tiled viewports is called *Tiled Model Space*.

In Chapter 6 you learned how to save views so that you could more easily move around a large drawing. Tiled viewports offer a similar advantage. In addition, each tiled viewport can have different Zoom and Pan settings, which means that each tiled viewport can offer a different view of you drawing. We will take a quick look at viewports now, and we'll return to them in Chapter 12.

1. Open the AutoCAD sample file named `Azimuth.dwg`. Notice that the UCS icon looks quite different from what you've seen before. Instead of the familiar icon with two arrows pointing in X and Y, the icon is now a triangle with a W, a square, and an X inside.

2. Double-click on the word Tile in the status bar. You were momentarily in Paper Space; double-clicking here changes you to Model Space. The next section of this chapter will explain the difference between Model Space and Paper Space.

3. Go to View ➤ Tiled Viewports ➤ Layout. The Tiled Viewport Layout dialog box appears.

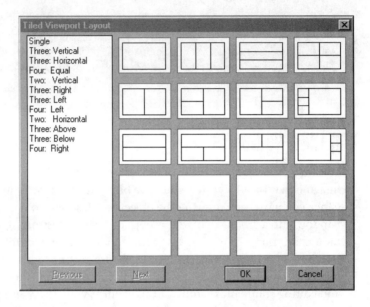

4. Click on the Three: Right option. Notice the three-box version with the large box on the right is highlighted. Click on OK. The screen takes some time to regen. When the computer is finished, your drawing will look like the top image of Figure 9.8.

5. As an experiment, click in each of the three boxes, called *tiles*, and notice the outline change. As each tile is highlighted, it becomes the current tile.

6. Click in the right tile then type **z↵2x↵** to zoom the viewport 2x.

7. Click in the lower-left tile and type **zoom↵c↵ 52,38↵** to center the zoom on the bearing, then at the Magnification or height: prompt, type **8x↵**. Your drawing looks like the bottom image of Figure 9.8.

FIGURE 9.8

The Azimuth before and after changing the zoom factor

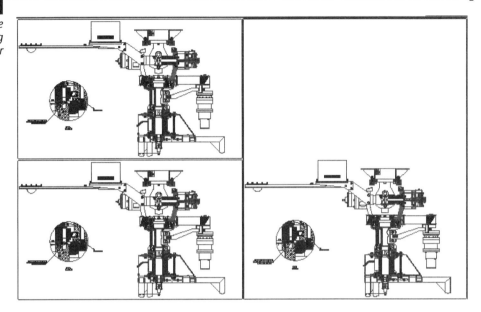
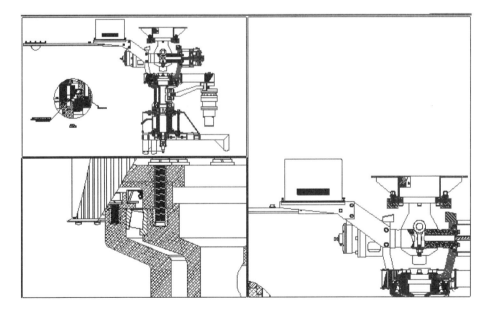

Use the Line tool from the Draw toolbar to construct a line from the middle of the roller in the bearing to the midpoint of the top flange. See the top of Figure 9.9 for these points. You will need to use Osnaps for this part of the exercise. Set the Endpoint and Midpoint Running Osnaps.

8. Click in the lower-left tile to set the focus. Click on the Line tool in the Draw toolbar and find the midpoint of the lower side of the roller in the bearing. At the `to point:` prompt, click in the large window to change the focus, then click in the midpoint of the top flange. Your drawing should look like the bottom of Figure 9.9.

You have just seen that the tiles can be used in real time. Notice that the new line appeared in all three tiled viewports.

You can change the number of viewports and the viewpoint in any of the tiled viewports whenever you wish. The restrictions are

- The viewports must fill the entire screen.
- If you are to print or plot your drawing only the current tiled viewport will plot. You cannot plot all that you see.

You have just experienced Tiled viewports in Model Space. There is another viewport option: You can create multiple viewports in Paper Space mode.

Understanding Model Space and Paper Space

You have been viewing your work in Tiled Model Space since the beginning of this book. Most of the time, you have been working on your drawings by looking through a single "window," called a viewport, into Model Space from Tiled Model Space. As you saw in the previous exercise, you can create multiple viewports in Tiled Model Space. The greatest drawback to tiled viewports is that you may only print one of these viewports at a time. What you see is not what you can print.

Paper Space is another mode similar to Tiled Model Space. In Paper Space, you can position, size, and print as many viewports as you like. They need not be adjacent to one another and they do not have to fill the screen.

Paper Space is most valuable to mechanical engineers when they must plot multiple views in different scales. If you draw a detail at a different scale than the rest of your objects, you will not be able to dimension it using the AutoCAD automatic dimensions. (You could set a special dimension style with an appropriate linear scale. The automatic dimension that appears in the dimension line is the product of the linear scale factor and the measured dimension.) See *Dimension Style* in Appendix B and in Chapter 8.

FIGURE 9.9

The Osnap points and the line in the Azimuth drawing

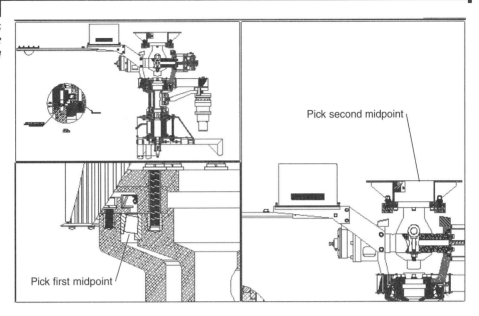
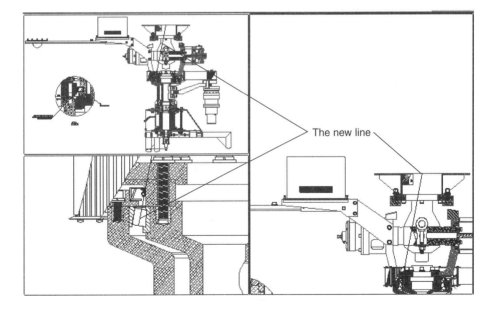

Paper Space allows you to create a viewport for detail and then change the scale in the viewport so that the objects shown in the viewport *appear* to be at a different scale than the rest of the views. You will still have to create a special dimension style for this view, but this time you are setting the overall scale for the text and arrowheads, not a linear scale factor which is applied to the dimension value. The advantage of overall scale versus linear scale is that overall scale, if used incorrectly, will only return a dimension that looks too big or too small. If the linear scale is used incorrectly, the dimension value is wrong and could result in manufacturing errors.

If you usually don't print your work at full scale, or if you are converting from board drafting and you usually don't draw your work to full scale, then you must determine a plot scale factor and use it whenever you print your work. Paper Space avoids these problems by allowing you to always plot to one scale. You do this by controlling the scale of the objects in the viewport. In this way, different viewports can have different scales, and it is the viewports that are printed at full scale with the objects within them printed at the scale of each individual viewport.

Perhaps one of the more powerful features of Paper Space is that you can plot several views of the same drawing on one sheet of paper. You can also include graphic objects, such as borders and notes, that appear only in Paper Space. In this function, Paper Space acts very much like a page layout program such as Quark or Adobe PageMaker. You can "paste up" different views of your drawing and then add borders, title blocks, general notes, and other types of graphic and textual data. Figure 9.10 shows the Plan drawing set up in Paper Space mode to display different views.

FIGURE 9.10

Different views of the same drawing in Paper Space

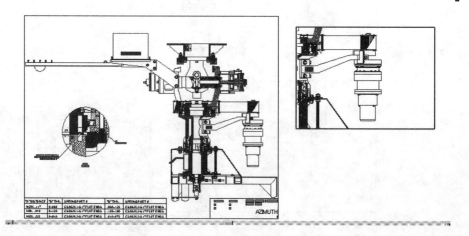

Taking a Tour of Paper Space

Your gateway to Paper Space is the Tilemode system variable. When Tilemode is set to 1 (On), the default setting, you are in Model Space. When it is set to 0 (Off), you are in Paper Space. This system variable is neatly packaged in pull-down menus as View ➢ Model Space (tiled) and View ➢ Paper Space. Let's start with the basics of entering Paper Space.

1. If it isn't already open, open the Bar file, making sure your display shows all of the drawing.

2. Click on View ➢ Paper Space, or enter **tilemode↵0↵**. You can also double-click on the word TILE in the status bar. Your screen goes blank and your UCS icon changes to a triangular shape. Also note the word PAPER replacing TILE in the status bar; this tells you at a glance that you are in Paper Space.

A third option, View ➢ Model Space (floating) switches you to Paper Space, and then places you in a Model Space window. You'll learn about floating Model Space later in this chapter.

The new UCS icon tells you that you are in Paper Space. But where did your drawing go? Before you can view it, you must create the windows, or *viewports*, that let you see into Model Space. But before you do that, let's explore Paper Space.

1. Click on Format ➢ Drawing Limits, or enter **limits↵** at the command prompt. Then note the default value for the lower-left corner of the Paper Space limits. It is 0.000,0.000".

2. Press ↵ and note the current default value for the upper-right corner of the limits. It is 12.000,9.000, which is the standard default for a new drawing. This tells us that the new Paper Space area is 12 units wide by 9 units high—an area quite different from the one you set up originally in this drawing. Press ↵ again to accept the default and exit the command.

3. Click on Format ➢ Drawing Limits again.

4. At the ON/OFF: prompt, press ↵.

5. At the upper-right corner: prompt, enter **34,22↵** to designate an area that is 34" × 22".

6. Click on View ➢ Zoom ➢ All.

Now you have your Paper Space set up. The next step is to create viewports so you can begin to paste up your views.

7. Click on View ➢ Floating Viewports ➢ 3 Viewports, or type **mv↵3↵**.
8. At the `Horizontal/Vertical/Above/Below/Left/<Right>:` prompt, enter **a↵** for Above. This option creates one large viewport along the top half of the screen, with two smaller viewports along the bottom.
9. At the `Fit/<first point>:` prompt, enter **f↵** for the Fit option. Three rectangles appear in the formation, shown in Figure 9.11. Each of these is a viewport to your Model Space. The viewport at the top fills the whole width of the drawing area; the bottom half of the screen is divided into two viewports.

FIGURE 9.11

The newly created viewports

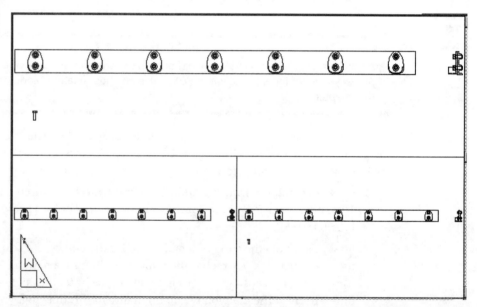

Now suppose you need to have access to the objects within the viewports in order to adjust their display and edit your drawing.

10. Click on View ➢ Model Space (floating). This gives you control over Model Space even though you are in Paper Space. (You can also enter **ms↵** as a keyboard shortcut.)

The first thing you notice is that the UCS icon changes back to its L-shaped arrow form. It also appears as if you had three AutoCAD windows in each viewport, instead of just one.

11. Move your cursor over each viewport. Notice that in one of the viewports, the cursor appears as the AutoCAD crosshair cursor, while in the other viewports, it appears as an arrow pointer. The viewport that shows the AutoCAD cursor is the active one; you can pan and zoom, as well as edit objects, in the active viewport.

If your drawing disappears from a viewport, you can generally retrieve it by using View ➤ Zoom ➤ Extents (**zoom**↵**e**↵).

12. Click on the lower-left viewport to activate it.
13. Click on View ➤ Zoom ➤ Window and window the side view of the bar.
14. Click on the lower-right viewport and use View ➤ Zoom ➤ Window to enlarge your view of the front view of the bolt. You can also use the Pan Realtime and Zoom Realtime tools.

When you use View ➤ Floating Model, the UCS icon again changes shape—instead of one triangular-shaped icon, you have three arrow-shaped ones, one for each viewport on the screen. Also, as you move your cursor into the currently active viewport, the cursor changes from an arrow into the usual crosshair. Another way to tell which viewport is the active one is by its double border.

You can move from viewport to viewport even while you are in the middle of most commands. For example, you can issue the Line command, then pick the start point in one viewport, then go to a different viewport to pick the next point, and so on. To activate a different viewport, simply click on it (see Figure 9.12). You can also switch viewports by typing Ctrl+r.

You cannot move between viewports while in the middle of the Snap, Zoom, Vpoint, Grid, Pan, Dview, or Vplayer commands.

FIGURE 9.12

The three viewports, each with a different view of the bar assembly

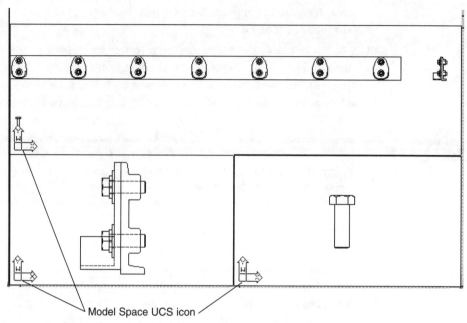

Your current view of the viewports is a bit constricted. Take the following steps to give yourself some room and to view a bit more of the Paper Space work area.

1. Choose View ➤ Paper Space or type **ps**↵. You can also double-click on the word MODEL in the status bar.

2. Use the Zoom Realtime tool to zoom the Paper Space view back just a bit so that it looks like Figure 9.13.

This brief exercise shows that you can use the Zoom tool in Paper Space just as you would in Model Space. All the display-related commands are available, including Pan Realtime.

Unlike prior versions of AutoCAD, zooms and pans in Paper Space do not trigger regens. R14 uses display list programming for Paper Space. This means that users can now use all of the transparent view commands they are accustomed to using in R13.

FIGURE 9.13
The view of Paper Space after zooming out

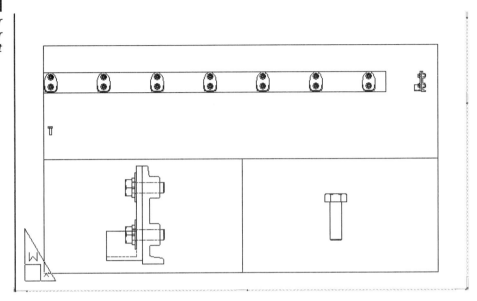

Getting Back to Full-Screen Model Space

Once you've created viewports, you can then re-enter Model Space through the viewport using View ➢ Floating Model Space. But what if you want to quickly get back into the old, familiar, full-screen Model Space you were in before you entered Paper Space? The following exercise demonstrates how this is done.

1. Click on View ➢ Tiled Model Space, or enter **tm↵1↵**. Your drawing returns to the original full-screen Model Space view—everything is back to normal.

2. Click on View ➢ Paper Space, or enter **tm↵0↵**. You are back in Paper Space. Notice that all the viewports are still there when you return to Paper Space. Once you've set up Paper Space, it remains part of the drawing until you delete all the viewports.

You may prefer doing most of your drawing in Model Space and using Paper Space for setting up views for plotting. Since viewports are retained, you won't lose anything when you go back to Model Space to edit your drawing.

Working with Paper Space Viewports

Paper Space is intended as a page-layout or composition tool. You can manipulate viewports' sizes, scale their view independently of one another, and even set layering and linetype scale independently. Let's try manipulating the shape and location of viewports, using the Modify command options.

1. Click on bottom edge of the lower-left viewport to expose its grips (see the top image of Figure 9.14).
2. Click on the upper-right grip, and then drag it down and to the left @–1,–1.
3. Press the Esc key twice and then Erase the lower-right viewport by clicking on Erase in the Modify toolbar, then clicking on the bottom edge of the viewport.
4. Move the lower-left viewport so it is centered in the bottom half of the window, as shown in the bottom image of Figure 9.14.

In this exercise, you clicked on the viewport edge to select it for editing. If, while in Paper Space, you attempt to click on the image within the viewport, you will not select anything. Later you will see, however, that you can use the Osnap modes to snap to parts of the drawing image within a viewport.

Viewports are recognized as AutoCAD objects, so they can be manipulated by all the editing commands just like any other object. In the previous exercise you moved, stretched, and erased viewports. Next, you'll see how layers affect viewports.

1. Remember the Vp layer you created in Chapter 4. Vp stands for viewport. Use the layers drop-down list to make Vp current.
2. Use the Properties button on the Object Properties toolbar to change the viewport borders to the Vp layer. The Vp color is magenta; notice that the outlines of the viewports become magenta.
3. Turn off the Vp layer. The viewport borders will disappear.

A viewport's border can be assigned a layer, color, or linetype. If you put the viewport's border on a layer that has been turned off or frozen, that border will become invisible, just like any other object on such a layer. Making the borders invisible is helpful when you want to compose a final sheet for plotting. Even when turned off, the active viewport will show a heavy border around it when you switch to floating Model Space, and all the viewports will still display their views.

FIGURE 9.14

Stretching, erasing, and moving viewports

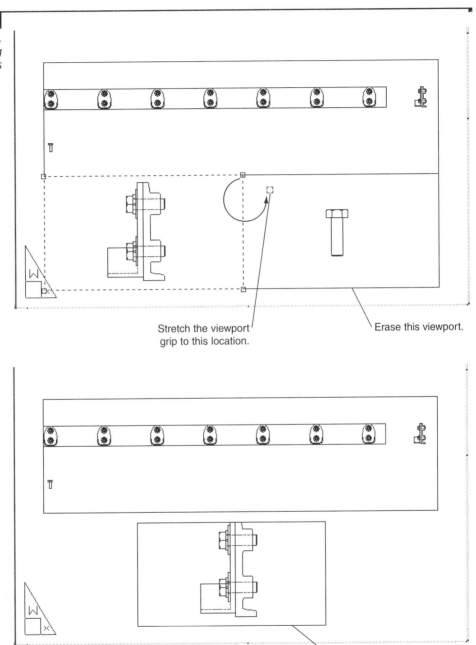

Disappearing Viewports

As you add more viewports to a drawing, you may discover that some of them blank out, even though you know you haven't turned them off. Don't panic. AutoCAD limits the number of active viewports at any given time to 48. (An *active* viewport is one that displays.) This limit is provided because too many active viewports can bog a system down.

If you are using a slow computer with limited resources, you can lower this limit to 2 or 3 to gain some performance. Then only two or three viewports will display their contents. (All viewports that are turned on will still plot, however, regardless of whether or not their contents are visible.) Zooming into a blank viewport will restore its visibility, thereby allowing you to continue to work with enlarged Paper Space views containing only a few viewports.

The Maxactvp system variable controls this value. Type **maxactvp**↵ and then enter the number of viewports you want to have active at any given time.

Scaling Views in Paper Space

Paper Space has its own unit of measure. You have already seen how you can set the limits of Paper Space independently of Model Space. When you first enter Paper Space, regardless of the area your drawing occupies in Model Space, you are given limits that are 12 units wide by 9 units high. This may seem incongruous at first, but if you keep in mind that Paper Space is like a paste-up area, then this difference of scale becomes easier to comprehend. Just as you might paste up photographs and maps representing several square feet onto an 11"×17" board, so can you use Paper Space to paste up views of scale drawings representing huge, complete assembly lines or micromechanisms measuring only a few thousandths of an inch. In AutoCAD, however, you have the freedom to change the scale and size of the objects you are pasting up.

NOTE

While in Paper Space, you can edit objects in a Model Space viewport, but to do so, you must use View ➤ Floating Model Space. You can then click on a viewport and edit within that viewport. While in this mode, objects that were created in Paper Space cannot be edited. View ➤ Paper Space brings you back to the Paper Space environment.

If you want to be able to print your drawing at a specific scale, then you must carefully consider scale factors when composing your Paper Space paste-up. Let's see how to put together a sheet in Paper Space and still maintain accuracy of scale.

1. Click on View ➤ Floating Model Space or enter **ms**↵ to return to Model Space. This will allow you to manipulate the views of each viewport.
2. Click on the top view to activate it.
3. Click on View ➤ Zoom ➤ Scale or enter **z**↵**s**↵.
4. At the All/Center/Dynamic/Extents... prompt, enter **1/2xp**↵. The xp suffix appended to the 1/2 tells AutoCAD that the current view should be scaled to 1/2 of the Paper Space scale. You get the 1/2 by taking the inverse of the drawing's scale factor, 2.
5. Click on the lower viewport.
6. Choose View ➤ Zoom ➤ Scale again and enter **2xp**↵ at the All/Center/Dynamic: prompt. Your view of the unit will be scaled to 2"=1' in relation to Paper Space (see Figure 9.15).

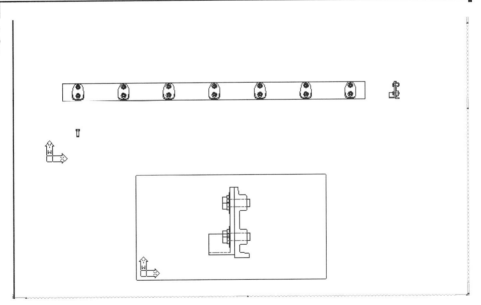

FIGURE 9.15

Paper space viewport views scaled to 1/2=1 and 2=1

It's easy to adjust the width, height, and location of the viewports so that they display only the parts of the unit you want to see; just go back to the Paper Space mode and use the Stretch, Move, or Scale commands to edit any viewport border. The view within the viewport itself will remain at the same scale and location, while the viewport changes in size. You can move and stretch viewports with no effect on the size and location of the objects within the view.

If you have a situation where you need to overlay one drawing on another, you can overlap viewports. Use the Osnap overrides to select geometry within each viewport, even while in Paper Space. This allows you to align one viewport on top of another at exact locations.

You can also add a title block in Paper Space at a 1:1 scale to frame your viewports, and then plot this drawing from Paper Space at a scale of 1:1. Your plot will appear just as it does in Paper Space, at the appropriate scale.

WARNING

While working in Paper Space, pay close attention to whether you are in Paper Space or floating Model Space mode. It is easy to accidentally perform a pan or zoom within a floating Model Space viewport when you intend to pan or zoom your Paper Space view. This can cause you to lose your viewport scaling or alignment with other parts of the drawing. It's a good idea to save viewport views using View ➤ Named Views, in case you happen to change a viewport view accidentally.

Setting Layers in Individual Viewports

Another unique feature of Paper Space viewports is their ability to freeze layers independently. You could, for example, display the usual drawing and dimensions in the overall view and show only a large scale detail in the other viewport.

You control viewport layer visibility through the Layer & Linetype Properties dialog box. You may have noticed that there are three sun icons for each layer listing.

You're already familiar with the sun icon farthest to the left. This is the Freeze/Thaw icon that controls the freezing and thawing of layers globally. Just to the right of that icon is a sun icon with a transparent rectangle. This icon controls the freezing and thawing of layers in individual viewports. The next exercise shows you firsthand how it works.

1. Activate the lower viewport.
2. Open the Layer & Linetype Properties dialog box.

3. In the Hidden layer listing, move your cursor over the icon that shows a transparent rectangle over a sun. You will see a tool tip describing its purpose: Freeze/Thaw in current viewport.

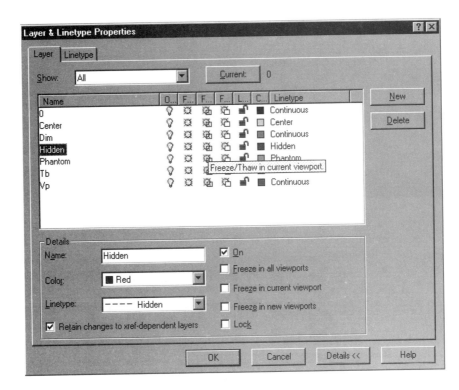

The Cur VP and New VP options in the Layer Control dialog box cannot be used while you are in tiled Model Space.

4. Click on the sun with the transparent rectangle icon. The sun changes into a snowflake, telling you that the layer is now frozen for the current viewport.
5. Click on OK. The active viewport will regenerate, with the Hidden layer of the Common Xref made invisible in the current viewport. However, the hidden lines remain visible in the other viewport (see Figure 9.16).

FIGURE 9.16
The drawing editor with the Hidden layer frozen in the active viewport

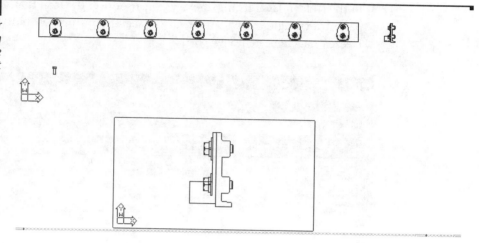

You might have noticed the other, similar-looking sun icon next to the one you used in the previous exercise. This icon shows an opaque rectangle over the sun.

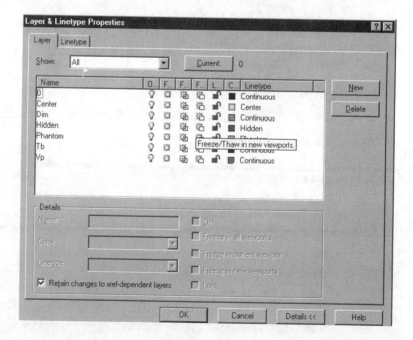

This other icon controls layer visibility in any new viewports you might create next, rather than controlling existing viewports.

6. If you prefer, you can also use the Layer drop-down list in the Object Properties toolbar to freeze layers in individual viewports. Select the layer from the list, and then click on the same Sun icon with the small rectangle below it.

7. Now save and exit the Xref-1 file.

Linetype Scales and Paper Space

As you have seen from previous exercises, drawing scales have to be carefully controlled when creating viewports. Fortunately, this is easily done by choosing View ➢ Zoom ➢ Scale and then entering the scale factor followed by **.xp**. While Paper Space offers the flexibility of combining different scale images in one display, it also adds to the complexity of your task in controlling that display. Your drawing's linetype scale, in particular, needs careful attention.

In Chapter 4, you saw how you had to set the linetype scale to the scale factor of the drawing in order to make the linetype visible. If you intend to plot that same drawing from Paper Space, you will have to set the linetype scale back to 1 to get the linetypes to appear correctly. This is because AutoCAD faithfully scales linetypes to the current unit system. Remember that Paper Space units differ from Model Space units. So when you scale a Model Space image down to fit within the smaller Paper Space area, the linetypes remain scaled to the increased linetype scale settings.

The *Psltscale* system variable allows you to determine how linetype scales are applied to Paper Space views. You can set Psltscale so that the linetypes will appear the same regardless of whether you view them directly in tiled Model Space, or through a viewport in Paper Space. By default, this system variable is set to 1. This causes AutoCAD to scale all the linetypes uniformly across all the viewports in Paper Space. You can set Psltscale to 0 to force the viewports to display linetypes exactly as they appear in Model Space.

This setting can also be controlled in the Linetype tab of the Layer & Linetype Properties dialog box. There, in the lower-right corner, you will see a setting called Use Paper Space Units for Scaling. When this is checked, Psltscale is set to 1. When it is unchecked, Psltscale is set to 0.

Dimensioning in Paper Space

At times, you may find it more convenient to add dimensions to your drawing in Paper Space rather than directly on your objects in Model Space. There are several dimension settings you will want to know about that will enable you to do this.

In order to have your dimensions produce the appropriate values in Paper Space, you need to have AutoCAD adjust the dimension text to the scale of the viewport from which you are dimensioning. You can have AutoCAD scale dimension values so that they correspond to a viewport zoom-scale factor. Here's how this setting is made:

1. Open the Dimension Styles dialog box.
2. Click on the Annotation button.
3. Click on the Units button in the Annotation dialog box.
4. In the Linear input box of the Scale group, enter the inverse of the scale factor of the viewport you intend to dimension. For example, if the viewport is scaled to a 1/4xp scale, enter 4. (If the Linear box is grayed out, remove the check from the Paper Space Only check box.)
5. Click on the Paper Space Only check box so that a check appears there.
6. Click on OK, and then click on OK again at the Annotation dialog box.
7. Click on the Geometry button in the Dimension Styles dialog box.
8. Click on the Scale to Paper Space check box to place a check there. This forces the dimension objects to be scaled to Paper Space units.
9. Click on OK, and then click on OK to exit the Dimension Styles dialog box.

You are ready to dimension in Paper Space. Remember that you can snap to objects in a floating viewport, so you can add dimensions as you normally would in Model Space.

WARNING

While AutoCAD offers the capability of adding dimensions in Paper Space, you may want to refrain from doing so until you have truly mastered AutoCAD. Because Paper Space dimensions won't be visible in Model Space, it is easy to forget to update your dimension when your drawing changes. Dimensioning in Paper Space can also create confusion for others editing your drawing at a later date.

Other Uses for Paper Space

The exercises presented in this section should give you a sense of how to work in Paper Space. We've given examples that reflect the more common uses of Paper Space. Remember that it is like a page-layout portion of AutoCAD—separate from Model Space, yet connected to Model Space through viewports.

You needn't limit your applications of Paper Space to drawing format. When used in conjunction with AutoCAD raster import capabilities, Paper Space can be a powerful tool for creating presentations.

Advanced Tools: Selection Filter and Calculator

Before finishing this chapter, you will want to know about two other tools that are extremely useful in your day-to-day work with AutoCAD: Selection Filters and the Calculator. We've saved the discussion of these tools for this chapter because you won't really need them until you've become accustomed to the way AutoCAD works. Chances are you've already experimented with some of AutoCAD's menu options not yet discussed in the tutorial. Many of the pull-down menu options and their functions are self-explanatory. Selection Filters and the Calculator, however, do not appear in any of the menus and require some further explanation.

We'll start with Selection Filters.

Filtering Selections

Suppose you need to take just the circles in your drawing and isolate them, in order to use the pattern in another version of the bar. This is a simple problem in our bar model, but suppose you have a plate like the one in Plate.dwg on the companion CD-ROM. One way to select only the circles would be to turn off all the layers except the layer that the circles are on; then you could use the Wblock command and select the remaining circles, using a window to write the circle information to a file. Filters can simplify this operation by allowing you to select groups of objects based on their properties.

1. Open the Plate.dwg file from the companion CD-ROM.
2. Start the Wblock command by clicking on File ➤ Export. In the Create Drawing File dialog box, enter **plate2.dwg** in the File input box, and click on OK.
3. Press ↵ at the prompt for a block name, and then enter **0,0** at the Insertion base: prompt.

4. At the object-selection prompt, type **'filter**↵. The Object Selection Filters dialog box appears.

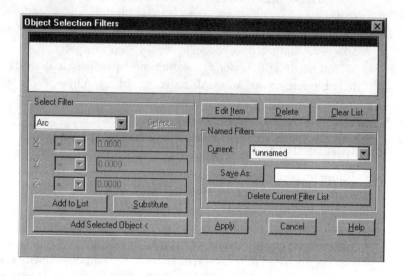

5. Open the drop-down list in the Select Filter button group.
6. Scroll down the list; find and highlight the Circle option.
7. At the Object Selection Filters dialog box, click on the Add to List button toward the bottom of the Select Filter button group. Object = Circle is added to the list box.
8. Click on Apply. The dialog box closes.
9. Type **all**↵ to select everything in the drawing. Only circle objects are selected. You'll see a message in the Command window indicating how many objects were found and how many were filtered out.
10. Press ↵ and you'll see the message Exiting Filtered selection. 20 found.
11. Press ↵ again to complete the Wblock command. All the circles are written out to a file called Plate2. You can type **oops**↵ to get the circles back into this drawing.

In this exercise, you filtered out objects using the command. Once a filter is designated, you then select the group of objects you want AutoCAD to filter through. AutoCAD finds the objects that match the filter requirements and passes those objects to the current command.

As you've seen from the previous exercise, there are many options to choose from in this utility. Let's take a closer look.

Working with the Object Selection Filters Dialog Box

To use the Object Selection Filters dialog box, first select the criteria for filtering from the drop-down list. If the criterion you select is a named item (layers, linetypes, colors, or blocks), you can then click on the Select button to choose specific items from a list. If there is only one choice, the Select button is grayed out.

Once you've determined what to filter, you must add it to the list by clicking on the Add to List button. The filter criteria then appear in the list box at the top of the dialog box. Once you have something in the list box, you can then apply it to your current command or to a later command. AutoCAD will remember your filter settings, so if you need to reselect a filtered selection set, you don't have to redefine your filter criteria.

Saving Filter Criteria

If you prefer, you can preselect a filter criterion. Then, at any `Select objects:` prompt, you can type '**filter**↵, highlight the appropriate filter criteria in the list box, and click on Apply. The specifications in the Object Selection Filters dialog box remain in place for the duration of the current editing session.

You can also save a set of criteria by entering a name in the input box next to the Save As button and then clicking on the button. The criteria list data is saved in a file called `Filter.nfl`. You can then access the criteria list at any time by opening the Current drop-down list and choosing the name of the saved criteria list.

Filtering Objects by Location

Notice the X, Y, and Z drop-down lists just below the main Select Filters drop-down list. These lists become accessible when you select a criterion that describes a geometry or a coordinate (such as an arc's radius or center point). You can use these lists to define filter selections even more specifically, using *relational operators*: greater than (>), less than (<), equal to (=), or not equal to (!=) comparisons.

For example, suppose you want to grab all the circles whose radii are greater than 4.0 units. To do this, choose Circle Radius from the Select Filters drop-down list; then, in the X list, select >. Enter **4.0** in the input box to the right of the X list, and click on Add to List. You see the item

    Circle Radius > 4.0000

added to the list box at the top of the dialog box. You have used > to indicate a circle radius greater than 4.0 units.

Creating Complex Selection Sets

There will be times when you will want to create a very specific filter list. For instance, say you need to filter out all the circles representing holes equal to .25 dia *and* all arcs with a radius equal to 1. To do this, use the *grouping operators* found at the bottom of the Select Filter drop-down. You'll need to build a list as follows:

```
** Begin OR
** Begin AND
Entity = circle
circle dia = .250
** End AND
** Begin AND
Entity = Arc
Arc Radius = 1.0000
** End AND
** End OR
```

Notice that the Begin and End operators are balanced; that is, for every `Begin OR` or `Begin AND`, there is an `End OR` or an `End AND`.

This list may look rather simple, but it can get confusing—mostly because of the way we normally think of the terms *and* and *or*. If criteria are bounded by the `AND` grouping operators, then the objects must fulfill *all* criteria before the objects are selected. If criteria are bounded by the `OR` grouping operators, then the objects fulfilling *any one* criterion will be selected.

Here are the steps to build the list shown just above:

1. In the Select Filter drop-down list, choose **Begin OR, and click on Add to List. Then do the same for **Begin AND.

2. Click on Block from the Select Filters list, and then click on Add to List.

3. For the layer, click on Layer from the Select Filter list. Then click on Select, choose the layer name, and click on Add to List.

4. In the Select Filter list, choose **End AND and click on Add to List. Then do the same for **Begin AND.

5. Select Arc from the Select Filter drop-down list and click on Add to List.

6. Select Arc Radius from the Select Filter list, and enter **1.0** in the input box next to the X drop-down. Be sure the = shows in the X drop-down, and then click on Add to List.

7. Choose **End AND and click on Add to List. Then do the same for **End OR.

If you make an error in any of these steps, just highlight the item, select an item to replace it, and click on the Substitute button instead of the Add to List button. If you only need to change a value, click on Edit Item near the center of the dialog box.

Finding Geometry with the Calculator

Another useful AutoCAD tool is the geometry Calculator. Like most calculators, it adds, subtracts, divides, and multiplies. If you enter an arithmetic expression such as 1+2, the calculator returns 3. This is useful for doing math on the fly, but the Calculator does much more than arithmetic, as you will see in the next examples.

Finding the Midpoint between Two Points

One of the most common questions heard from AutoCAD users is, "How can I locate a point midway between two objects?" You can draw a construction line between the two objects, and then use the Midpoint Osnap override to select the midpoint of the construction line. The Calculator offers another method that doesn't require drawing additional objects.

In the following exercise, you will start a line midway between the center of an arc and the endpoint of a line. Draw a line and an arc and try this out.

1. Start the Line command, and at the From point: prompt, type '**cal**↵.

2. At the >> Expression: prompt, enter **(end + cen)/2**↵.

3. At the >> Select entity for END snap: prompt, the cursor turns into a square. Place the square on the endpoint of a line and click on it.

4. At the >> Select entity for CEN snap: prompt, click on an arc. The line will start midway between the arc's center and the endpoint of the line.

Typing the Calculator expressions may seem a bit too cumbersome to use on a regular basis, but if you find you could use some of its features, you can create a toolbar macro to simplify the Calculator's use. See Chapter 21 for more on customizing toolbars.

Using Osnap Modes in Calculator Expressions

In the foregoing exercise, you used Osnap modes as part of arithmetic expressions. The Calculator treats them as temporary placeholders for point coordinates until you actually pick the points (at the prompts shown in steps 3 and 4 above).

The expression

```
(end + cen)/2
```

finds the average of two values. In this case, the values are coordinates, so the average is the midpoint between the two coordinates. You can take this one step further and find the centroid of a triangle using this expression:

```
(end + end + end)/3
```

Notice than only the first three letters of the Osnap mode are entered in Calculator expressions. Table 9.1 shows what to enter in an expression for Osnap modes.

TABLE 9.1: THE GEOMETRY CALCULATOR'S OSNAP MODES

Calculator Osnap	Meaning
End	Endpoint
Ins	Insert
Int	Intersection
Mid	Midpoint
Cen	Center
Nea	Nearest
Nod	Node
Qua	Quadrant
Per	Perpendicular
Tan	Tangent
Rad	Radius of object
Cur	Cursor pick

We've included two items in the table that are not really Osnap modes, although they work similarly when they are used in an expression. The first is Rad. When you include Rad in an expression, you get the prompt

```
Select circle, arc or polyline segment for RAD function:
```

You can then select an arc, polyline arc segment, or circle, and its radius is used in place of Rad in the expression.

The other item, Cur, prompts you for a point. Instead of looking for specific geometry on an object, it just locates a point. You could have used Cur in the previous exercise in place of the End and Cen modes, to create a more general purpose midpoint locator, as in the following form:

(cur + cur)/2

Since AutoCAD does not provide a specific tool to select a point midway between two other points, the form shown here would be useful as a custom toolbar macro. You'll learn how to create macros in Chapter 21.

Finding a Point Relative to Another Point

Another common task in AutoCAD is starting a line at a relative distance from another line. The following steps describe how to use the Calculator to start a line from a point that is 2.5" in the X-axis and 5.0" in the Y-axis from the endpoint of another line.

1. Start the Line command. At the First point: prompt, enter **'cal**↵.
2. At the >> Expression: prompt, enter **end + [2.5,5.0]**↵.
3. At the >> Select entity for END snap: prompt, pick the endpoint. The line starts from the desired location.

In this example, you used the Endpoint Osnap mode to indicate a point of reference. This is added to the Cartesian coordinates in square brackets, describing the distance and direction from the reference point. You could have entered any coordinate value within the square brackets. Or you could have entered a polar coordinate in place of the Cartesian coordinates, as in the following:

end + [5.59<63]

You don't have to include the @, because the Calculator assumes you want to add the coordinate to the one indicated by the Endpoint Osnap mode. Also, it's not necessary to include every coordinate in the square brackets. For example, to indicate a displacement in only one axis, you can leave out a value for the other two coordinates, as in the following examples:

[4,5] = [4,5,0] [,1] = [0,1,0]
[,,2] = [0,0,2]
[] = [0,0,0]

Adding Feet and Inch Distances on the Fly

One of the more frustrating situations you may have run across is having to stop in the middle of a command to find the sum of two or more distances. Say you start the Move command, select your objects, and pick a base point. Then you realize you don't know the distance for the move, but you do know that the distance is the sum of two values—unfortunately, one value is in millimeters and the other is in inches. Usually in this situation you would have to reach for pen and paper or use the Windows accessory called the Calculator, figure out the distance, then return to your computer to finish the task. AutoCAD's Geometry Calculator puts an end to this runaround.

The following steps show you what to do if you want to move a set of objects a distance that is the sum of 12 ' 6 " -⅝" and 115- ¾".

1. Issue the Move command, select objects, and pick a base point.
2. At the Second point: prompt, start the Calculator.
3. At the >> Expression: prompt, enter **[@12'6" + 115-3/4" < 45]**. Press ↵, and the objects move into place at the proper distance.

You must always enter an inch symbol (") when indicating inches in the Calculator.

In this example, you are mixing inches and feet, which under normal circumstances is a time-consuming calculation. Notice that the feet-and-inch format follows the standard AutoCAD syntax (no space between the feet and inch values). The coordinate value in square brackets can have any number of operators and values, as in the following:

[@4 * (22 + 15) - (23.3 / 12) + 1 < 13 + 17]

This expression demonstrates that you can also apply operators to angle values.

Guidelines for Working with the Calculator

You may be noticing some patterns in the way expressions are formatted for the Calculator. Here are some guidelines to remember:

- Coordinates are enclosed in square brackets.
- Nested or grouped expressions are enclosed in parentheses.
- Operators are placed between values, as in simple math equations.
- Object snaps can be used in place of coordinate values.

Table 9.2 lists all the operators and functions available in the Calculator. You may want to experiment with these other functions on your own.

TABLE 9.2: THE GEOMETRY CALCULATOR'S FUNCTIONS

Operator/Function	What It Does	Example
+ or -	Adds or subtracts numbers or vectors	2 - 1 = 1 [a,b,c] + [x,y,z] = [a+x, b+y, c+z]
* or /	Multiplies or divides numbers or vectors	2 * 4.2 = 8.4 a*[x,y,z] = [a*x, a*y, a*z]
^	Exponentiation of a number	3^2 = 9
sin	Sine of angle	sin (45) = 0.707107
cos	Cosine of angle	cos (30) = 0.866025
tang	Tangent of angle	tang (30) = 0.57735
asin	Arcsine of a real number	asin (0.707107) = 45.0
acos	Arccosine of a real number	acos (0.866025) = 30.0
atan	Arctangent of a real number	atan (0.57735) = 30.0
ln	Natural log	ln (2) = 0.693147
log	Base-10 log	log (2) = 0.30103
exp	Natural exponent	exp (2) = 7.38906
exp10	Base-10 exponent	exp10 (2) = 100
sqr	Square of number	sqr (9) = 81.0
abs	Absolute value	abs (~-3.4) = 3.4
round	Rounds to nearest integer	round (3.6) = 4
trunc	Drops decimal portion of real number	trunc (3.6) = 3
r2d	Converts radians to degrees	r2d (1.5708) = 90.0002
d2r	Converts degrees to radians	d2r (90) = 1.5708
pi	The constant pi	3.14159

The Geometry Calculator is capable of much more than the typical uses you've seen here and extends beyond the scope of this text. Still, the processes described in this section will be helpful as you use AutoCAD. If you want to know more about the Calculator, consult the *AutoCAD Command Reference* and the *User's Guide* that come with Release 14.

If You Want to Experiment...

Finish changing the objects in the end view of the bar to hidden lines. You will find that the Fence and Inherit Properties commands continue to be very useful as you do this.

You might want to experiment further with Paper Space in order to become more familiar with it. You will be using it again in the chapter on solid modeling.

Chapter 10

Drawing Curves and Solid Fills

FEATURING

Introducing Polylines

Editing Polylines

Creating a Polyline Spline Curve

Using True Spline Curves

Marking Divisions on Curves

Sketching with AutoCAD

Filling In Solid Areas

Drawing Curves and Solid Fills

So far in this book, you've been using basic lines, arcs, and circles to create your drawings. Now it's time to add polylines and spline curves to your repertoire. Polylines offer many options for creating forms, including solid fills. Spline curves are perfect for drawing smooth, nonlinear objects. The splines are true *NURBS* curves. NURBS stands for Non-Uniform Rational B-Splines.

Introducing Polylines

Polylines are like composite line segments and arcs. A polyline may look like a series of line segments, but it acts like a single object. This characteristic makes polylines useful for a variety of applications, as you'll see in the upcoming exercises. You have already created a few polylines in the form of rectangles.

Drawing a Polyline

First, to become acquainted with the polyline, you will begin a drawing of the top view of the clevis in Figure 10.1.

FIGURE 10.1

A sketch of a clevis

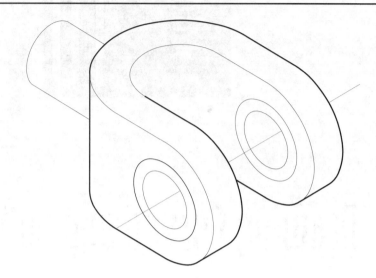

1. Open a new file and save it as **clevis2d**. Don't bother to make special setting changes, as you will do this drawing with the default settings.

2. Click on the Polyline tool on the Draw toolbar, or type **pl**↵.

3. At the From point: prompt, enter a point at coordinate 3,3 to start your polyline.

4. At the prompt Arc/Close/Halfwidth/Length/Undo/Width/<Endpoint of line>: enter **@3<0**↵ to draw a horizontal line of the clevis.

You can draw polylines just as you would with the Line command. Or you can use the other Pline options to enter a polyline arc, specify polyline thickness, or add a polyline segment in the same direction as the previously drawn line.

5. At the Arc/Close/Halfwidth... prompt, enter **a**↵ to continue your polyline with an arc.

NOTE

The Arc option allows you to draw an arc that starts from the last point you selected. Once selected, the Arc option offers additional options. The default Save option is the endpoint of the arc. As you move your cursor, an arc follows it, in a tangent direction from the first line segment you drew.

6. At the prompt:

Angle\CEnter\CLose\Direction\Halfwidth\Line\Radius\Second pt\Undo\Width\<Endpoint of arc>:

enter **@4<90**↵ to draw a 180° arc from the last point you entered. Your drawing should look like Figure 10.2.

FIGURE 10.2

A polyline line and arc

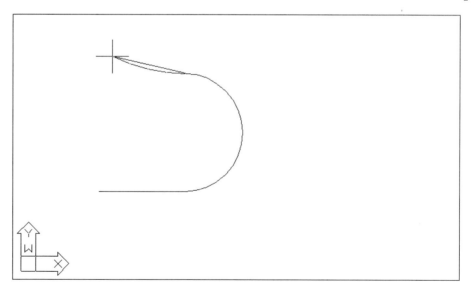

7. Continue the polyline with another line segment. To do this, enter **L**↵.

8. At the Arc\Close\Halfwidth... prompt, enter **@3<180**↵. Another line segment continues from the end of the arc.

9. Press ↵ to exit Pline.

You now have a sideways, U-shaped polyline that you will use in the next exercise to complete the top view of your joint.

Polyline Options

Let's pause from the tutorial to look at some of the Polyline options you didn't use.

Close draws a line segment from the last endpoint of a sequence of lines to the first point picked in that sequence. This works exactly like the Close option of the Line command.

Length enables you to specify the length of a line that will be drawn at the same angle as the last line entered.

Halfwidth creates a tapered line segment or arc by specifying half its beginning and ending widths (see Figure 10.3).

Width creates a tapered line segment or arc by specifying the full width of the segment's beginning and ending points.

Undo deletes the last line segment drawn.

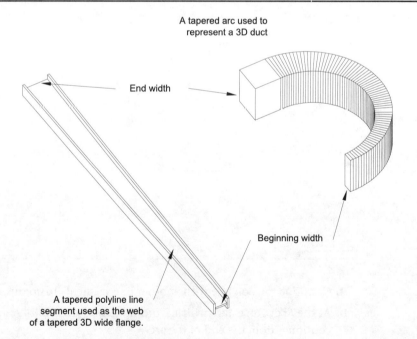

FIGURE 10.3

Tapered line segment and arc created with Halfwidth

If you want to break down a polyline into simple lines and arcs, you can use the Explode option on the Modify toolbar, just as you would with blocks. Once a polyline is exploded, it becomes a set of individual line segments or arcs.

To turn off the filling of solid polylines, click on Tools ➢ Drawing Aids, then at the Drawing Aids dialog box, remove the check from the Solid Fill option. (The Display options are explained in detail later in this chapter, in the section on solid fills.)

The Fillet tool on the Modify toolbar can be used to fillet all the vertices of a polyline composed of straight line segments. To do this, click on Fillet, and then set your fillet radius. Click on Fillet again, type **p**↵ to select Polyline option, and then pick the polyline you want to fillet.

Editing Polylines

You can edit polylines with many of the standard editing commands. To change the properties of a polyline, click on the Properties tool on the Object Properties toolbar (Ddmodify). The Stretch command on the Modify toolbar can be used to displace vertices of a polyline; the Trim, Extend, and Break commands on the Modify toolbar also work with polylines.

In addition, there are many editing capabilities offered only for polylines. For instance, later in this chapter you will see how you can smooth out a polyline using the Curve Fit option in the Pedit command and the Modify Polyline dialog box. In the following exercise, you'll use the Offset command on the Modify toolbar to add the inside portion of the joint.

1. Click on the Offset tool in the Modify toolbar, or type **o**↵.
2. At the Offset distance prompt, enter **1**↵.
3. At the Select object: prompt, pick the U-shaped polyline you just drew.
4. At the Side to offset: prompt, pick a point toward the inside of the U. A concentric copy of the polyline appears (see Figure 10.4).
5. Press ↵ to exit the Offset command.

FIGURE 10.4

The offset polyline

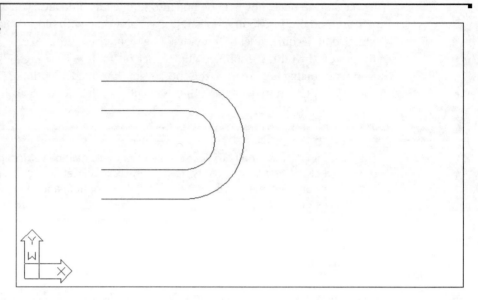

The concentric copy of a polyline made with Construct ➢ Offset can be very useful when you need to draw complex parallel curves like the ones in Figure 10.5.

FIGURE 10.5

Sample complex curves drawn using offset polylines

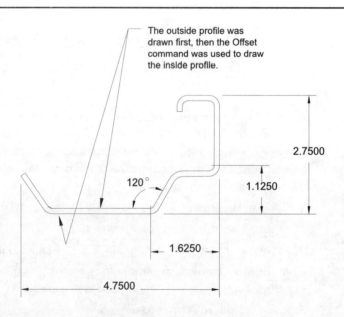

Next, complete the top view of the clevis.

1. Connect the two top ends of the polylines with a short line segment (see Figure 10.6).

The objects to be joined must touch the existing polyline exactly endpoint to endpoint, or they will not join. To insure that you place the endpoints of the lines exactly on the endpoints of the polylines, use the Endpoint Osnap override to select each polyline endpoint.

FIGURE 10.6

The joined polyline

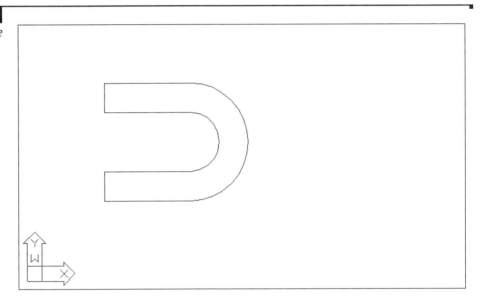

2. Choose Modify ➢ Object ➢ Polyline, or type **pe**↵. You can also choose Edit Polyline from the Modify II toolbar.

3. At the Select polyline: prompt, pick the outermost polyline.

4. At the Close/Join/Width... prompt, enter **j**↵ for the Join option.

5. At the Select objects: prompt, select all the objects you have drawn so far.

6. Once all the objects are selected, press ↵ to join them into one polyline. It appears that nothing has happened, though you will see the message `Segments added to polyline` in the Command window. The segments referred to by the message are the short line and the offset polyline composed of three objects in your drawing.

7. Press **c**↵ to use the Close option and a new line appears closing the shape. Notice the command line again. The Close option has been replaced with the Open. This means that the polyline is currently a closed figure and you only have the option to open it. Press ↵ again to exit the Pedit command.

8. Click on the drawing to expose its grips. The entire object is highlighted, telling you that all the lines have been joined into a single polyline.

By using the Width option under Edit Polyline, you can change the thickness of a polyline. Let's change the width of your polyline, to give some thickness to the outline of the joint.

1. Click on the Edit Polyline tool from Modify toolbar.
2. Click on the polyline.
3. At the `Close/Join/Width...` prompt, enter **w**↵ for the Width option.
4. At the `Enter new width for all segments:` prompt, enter **.03**↵ for the new width of the polyline. The line changes to the new width (see Figure 10.7), and you now have a top view of your joint.
5. Press ↵ to exit the Pedit command.
6. Save this file.

FIGURE 10.7

The polyline with a new thickness

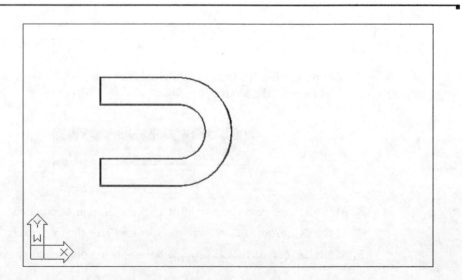

Now here's a brief look at a few of the Pedit options you didn't try firsthand.

Close connects the two endpoints of a polyline with a line segment. If the polyline you selected to be edited is already closed, this option changes to Open.

Open removes the last segment added to a closed polyline.

Spline/Decurve smooths a polyline into a spline curve (discussed in detail later in this chapter).

Edit Vertex lets you edit each vertex of a polyline individually (discussed in detail in the next section).

Fit turns polyline segments into a series of arcs.

Ltype Gen controls the way non-continuous linetypes pass through the vertices of a polyline. If you have a fitted or spline curve with a non-continuous linetype, you will want to turn this option on.

You can change the thickness of regular lines and arcs by using Pedit to change them into polylines, and then using the Width option to change their width.

Smoothing Polylines

There are many ways to create a curve in AutoCAD. If you don't need the representation of a curve to be exactly accurate, you can use a polyline curve. In the following exercise, you will draw a polyline curve to represent a contour on a topographical map.

1. Open the Topo.dwg drawing that is included on the CD-ROM that comes with this book. You will see the drawing of survey data shown in the top image of Figure 10.8. Some of the contours have already been drawn in between the data points.

2. Zoom in to the upper-right corner of the drawing, so that your screen looks like the middle image of Figure 10.8.

3. Click on the Polyline tool in the Draw toolbar. Using the Center Osnap, draw a polyline that connects the points labeled "254.00." Your drawing should look like the bottom image of Figure 10.8.

4. When you have drawn the polyline, press ↵.

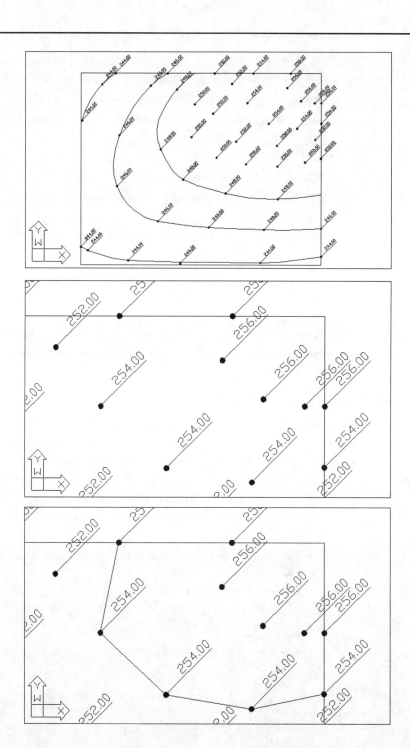

FIGURE 10.8
The Topo.dwg drawing shows survey data portrayed in an AutoCAD drawing. Notice the dots indicating where elevations were taken. The actual elevation value is shown with a diagonal line from the point.

TIP

If Running Osnaps are not set, you can double-click on the OSNAP label in the status bar to open the Osnap Settings dialog box. From there, you can select Center to open the Center Running Osnaps dialog box.

Next you will convert the polyline you just drew into a smooth contour line.

1. Choose Modify ➢ Object ➢ Polyline, or type **pe**↵.
2. At the PEDIT Select objects: prompt, pick the contour line you just drew.
3. At the prompt:

   Close/Join/Width/Edit vertex/Fit/Spline/Decurve/Ltype gen/Undo/Exit:

   press **f**↵ to select the Fit option. This causes the polyline to smooth out into a series of connected arcs that pass through the data points.
4. Press ↵ to end the Pedit command.

Your contour is now complete. The Fit curve option under the Pedit command causes AutoCAD to convert the straight-line segments of the polyline into arcs. The endpoints of the arcs pass through the endpoints of the line segments, and the curve of each arc depends on the direction of the adjacent arc. This gives the effect of a smooth curve. Next, you'll use this polyline curve to experiment with some of the editing options unique to the Pedit command.

Turning Objects into Polylines and Polylines into Splines

There may be times when you will want to convert regular lines, arcs, or even circles into polylines. You may want to change the width of lines, or join lines together to form a single object such as a boundary. Here are the steps to take to convert lines, arcs, and circles into polylines.

1. Choose Modify ➢ Object ➢ Polyline. You can also type **pe**↵ at the command prompt.
2. At the Select polyline: prompt, pick the object you wish to convert. If you want to convert a circle to a polyline, you must first break the circle (use the Break tool on the Modify toolbar) so that it becomes an arc of approximately 359°.

Continued

CONTINUED

3. At the prompt:

    Object selected is not a polyline. Do you want to turn it into one? <Y>

    press ↵ twice. The object is converted into a polyline.

If you have a polyline you would like to turn into a true spline curve, do the following:

1. Choose Modify ➢ Object ➢ Polyline, or type **pe**↵. Select the polyline you want to convert.
2. Type **s**↵ to turn it into a polyline spline; then type **x**↵ to exit the Pedit command.
3. Click the Spline tool in the Draw toolbar or type **spl**↵. You can also select Draw ➢ Spline from the menu bar.
4. At the Object/<Enter first point>: prompt, type **o**↵ for the Object option.
5. At the Select object: prompt, click the polyline spline. Though it may not be apparent at first, the polyline is converted into a true spline.

You can also use the Spline Edit tool (Modify ➢ Object ➢ Spline or **spe**↵) on a polyline spline. If you do, the polyline spline is automatically converted into a true spline.

Editing Vertices

One of the Pedit options we haven't yet discussed, Edit Vertex, is almost like a command within a command. Edit Vertex has numerous suboptions that allow you to fine-tune your polyline by giving you control over its individual vertices. We'll discuss this command in depth in this section.

To access the Edit Vertex options, follow these steps.

1. First, turn off the Data and Border layers to hide the data points and border.
2. Issue the Pedit command again. Then select the polyline you just drew.
3. Type **e**↵ to enter the Edit Vertex mode. An X appears at the beginning of the polyline, indicating the vertex that will be affected by the Edit Vertex options.

When using Edit Vertex, you must be careful about selecting the vertex to be edited. Edit Vertex has six options, and you often have to exit the Edit Vertex operation and use Pedit's Fit option to see the effect of Edit Vertex's options on a curved polyline.

Edit Vertex Suboptions

Once you've entered the Edit Vertex mode of the Pedit command, you have the option to perform the following functions:

- Break the polyline between two vertices
- Insert a new vertex
- Move an existing vertex
- Straighten a polyline between two vertices
- Change the Tangent direction of a vertex
- Change the width of the polyline at a vertex

These functions are presented in the form of the prompt:

    Next/Previous/Break/Insert/Move/Regen/Straighten/Tangent/Width/eXit <N>:

We'll examine each of the options presented in this prompt in the rest of this section, starting with the Next and Previous options.

Next and Previous The Next and Previous options enable you to select a vertex for editing. When you started the Edit Vertex option, an X appeared on the selected polyline to designate its beginning. As you select Next or Previous, the X moves from vertex to vertex to show which one is being edited. Let's try this out.

1. Press ↵ a couple of times to move the X along the polyline. (Because Next is the default option, you only need to press ↵ to move the X.)
2. Type **p**↵ for Previous. The X moves in the opposite direction. Notice that now the default option becomes P.

To determine the direction of a polyline, note the direction the X moves in when you use the Next option. Knowing the direction of a polyline is important for some of the other Edit Vertex options discussed below.

Break The Break option breaks the polyline between two vertices.

1. Position the X on one end of the segment you want to break.
2. Enter **b**↵ at the command prompt.

3. At the Next/Previous/Go/Exit <N>: prompt, use Next or Previous to move the X to the other end of the segment to be broken.

4. When the X is in the right position, select Go from the Edit Vertex menu or enter **g**↵, and the polyline will be broken (see Figure 10.9).

NOTE

You can also use the Break and Trim options on the Modify toolbar to break a polyline anywhere.

FIGURE 10.9

How the Break option works

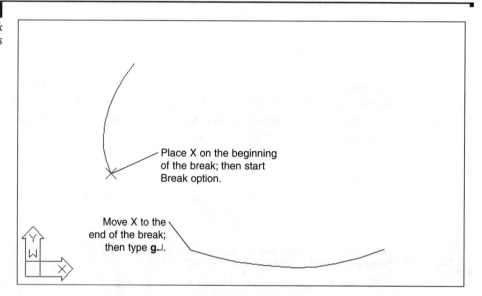

Insert Next, try the Insert option, which inserts a new vertex.

1. Type **x**↵ to temporarily exit the Edit Vertex option. Then type **u**↵ to undo the break.

2. Type **e**↵ to return to the Edit Vertex option, and position the X before the new vertex.

3. Press ↵ to advance the X marker to the next point.
4. Enter **i**↵ to select the Insert option.
5. When the prompt Enter location of new vertex: appears, along with a rubber-banding line originating from the current X position (see Figure 10.10), pick a point indicating the new vertex location. The polyline is redrawn with the new vertex.

FIGURE 10.10

The new vertex location

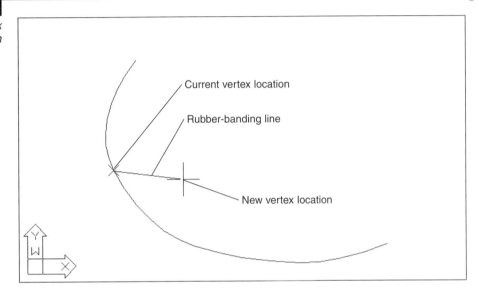

Notice that the inserted vertex appears between the currently marked vertex and the *next* vertex, so this Insert option is sensitive to the direction of the polyline. If the polyline is curved, the new vertex will not immediately appear as curved (see the top image of Figure 10.11). You must smooth it out by exiting the Edit Vertex option and then using the Fit option, as you did to edit the site plan (see the bottom image of Figure 10.11). You can also use the Stretch command (on the Modify toolbar) to move a polyline vertex.

FIGURE 10.11
The polyline before and after the curve is fitted

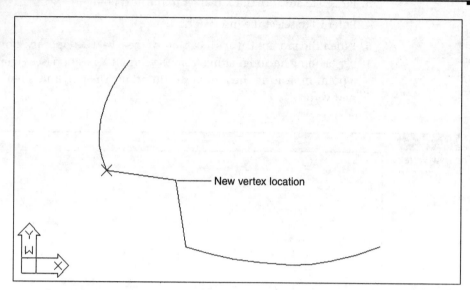
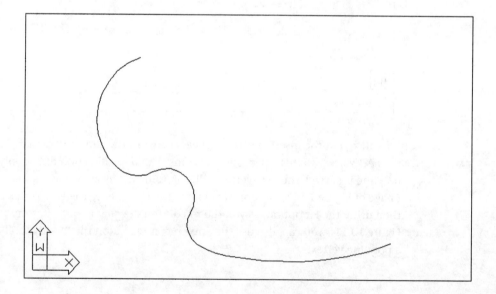

Move In this brief exercise, you'll use the Move option to move a vertex.

1. Undo the inserted vertex by exiting the Edit Vertex option (**x**↵) and typing **u**↵.
2. Restart the Edit Vertex option, and use the Next or Previous option to place the X on the vertex you wish to move.
3. Enter **m**↵ for the Move option.
4. When the `Enter new location:` prompt appears, along with a rubber-banding line originating from the X (see the top image of Figure 10.12), pick the new vertex. The polyline is redrawn (see the middle image of Figure 10.12). Again, if the line is curved, the new vertex appears as a sharp angle until you use the Fit option (see the bottom image of Figure 10.12).

You can also move a polyline vertex using its grip. This is usually much easier than using Pedit.

Straighten The Straighten option straightens all the vertices between two selected vertices, as shown in the following exercise.

1. Undo the moved vertex from the previous exercise.
2. Start the `Edit Vertex` option again, and select the starting vertex for the straight line.
3. Enter **s**↵ for the Straighten option.
4. At the `Next/Previous/Go/Exit:` prompt, move the X to the location for the other end of the straight-line segment.
5. Once the X is in the proper position, enter **g**↵ for the Go option. The polyline straightens between the two selected vertices (see Figure 10.13).

The Straighten option offers a quick way to delete vertices from a polyline.

FIGURE 10.12

Picking a new location for a vertex, with the polyline before and after the curve is fitted

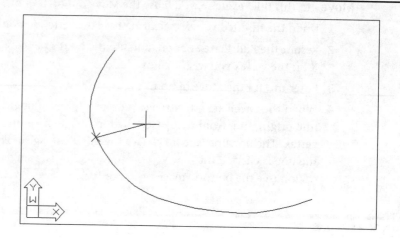

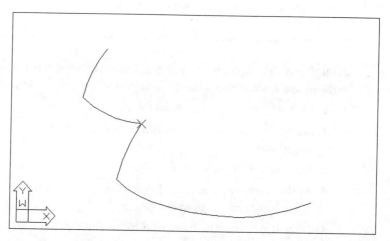

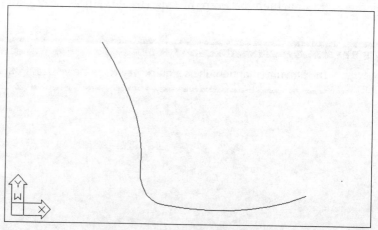

FIGURE 10.13

A polyline before and after straightening

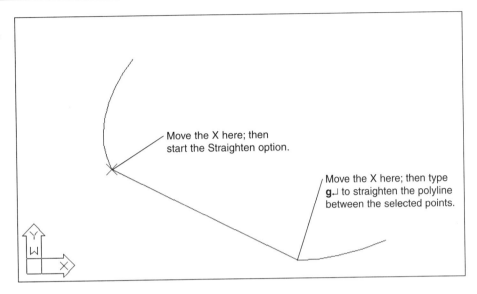

Tangent   Next is the Tangent option, which alters the direction of a curve on a curve-fitted polyline.

1. Undo the straightened segment from the previous exercise.
2. Restart the Edit Vertex option, and position the X on the vertex you wish to alter.
3. Enter **t**↵ for the Tangent option. A rubber-banding line appears (see the top image of Figure 10.14).
4. Point the rubber-banding line in the direction for the new tangent, and click the mouse. An arrow appears, indicating the new tangent direction (see the middle image of Figure 10.14).

Don't worry if the polyline shape does not change. You must use Fit to see the effect of Tangent (see the bottom image of Figure 10.14).

FIGURE 10.14

Picking a new tangent direction

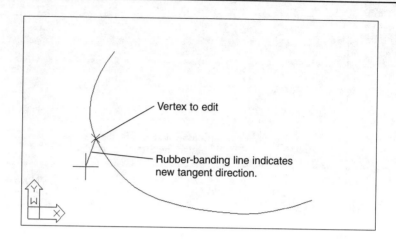

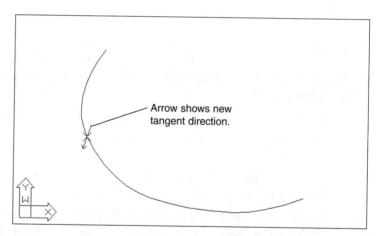

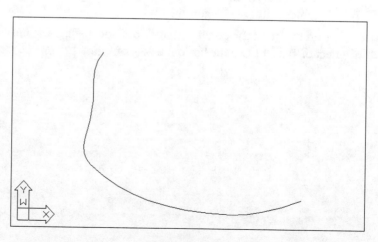

Width Finally, try out the Width option. Unlike the Pedit command's Width option, Edit Vertex/Width enables you to alter the width of the polyline at any vertex. Thus you can taper or otherwise vary polyline thickness.

1. Undo the tangent arc from the previous exercise.
2. Return to the Edit Vertex option, and place the X at the beginning vertex of a polyline segment you want to change.
3. Type **w**↵ to issue the Width option.
4. At the Enter starting width: prompt, enter a value, **1'** for example, indicating the polyline width desired at this vertex.
5. At the Enter ending width: prompt, enter the width, **2'** for example, for the next vertex.

Again (as with Tangent), don't be alarmed if nothing happens after you enter this width value. To see the result, you must exit the Edit Vertex command (see Figure 10.15).

NOTE

The Width option is useful when you want to create an irregular or curved area in your drawing that is to be filled in solid. This is another option that is sensitive to the polyline direction.

FIGURE 10.15

A polyline with the width of one segment increased

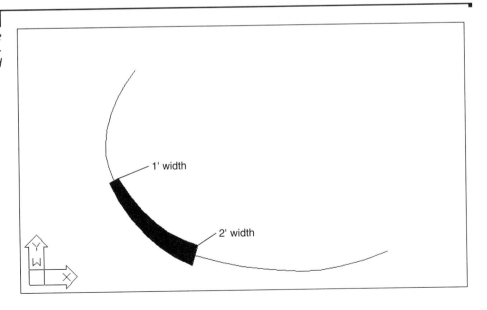

As you have seen throughout these exercises, you can use the Undo option to reverse the last Edit Vertex option used. And you can use the Edit option to leave Edit Vertex at any time. Just enter **x↵**, and this brings you back to the Pedit `Close/Join/Width...` prompt.

Creating a Polyline Spline Curve

The Pedit command's Spline option (named after the spline tool used in manual drafting) offers you a way to draw smoother and more controllable curves than those produced by the Fit option. A polyline spline does not pass through the vertex points as a fitted curve does. Instead, the vertex points act as weights pulling the curve in their direction. The polyline spline only touches its beginning and end vertices. Figure 10.16 illustrates this concept.

NOTE

A polyline spline curve does not represent a mathematically true curve. See *Using True Spline Curves* later in this chapter to learn how to draw a more accurate spline curve.

FIGURE 10.16

The polyline spline curve pulled toward its vertices

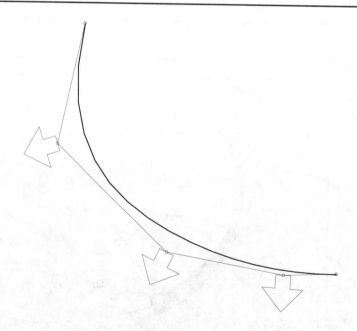

Let's see how using a polyline spline curve might influence the way you edit a curve.

1. Undo the width changes you made in the previous exercise.
2. To change the contour into a polyline spline curve, choose Modify ➢ Object ➢ Polyline.
3. Pick the polyline to be curved.
4. At the `Close\Join\Width...` prompt, enter **s**↵. Your curve will change to look like Figure 10.17.
5. Press ↵ to exit Edit Polyline.

FIGURE 10.17

A spline curve

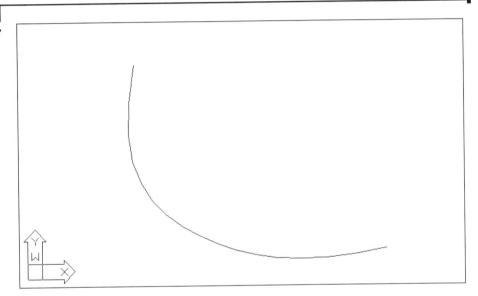

The curve takes on a smoother, more graceful appearance. It no longer passes through the points you used to define it. To see where the points went and to find out how spline curves act, do the following:

1. Make sure the Noun/Verb Selection mode and the Grips feature are turned on.
2. Click on the curve. You'll see the original vertices appear as grips (the top image of Figure 10.18).
3. Click on the grip that is second from the top of the curve, as shown in the middle image of Figure 10.18, and move the grip around. Notice how the curve follows, giving you immediate feedback on how the curve will look.
4. Pick a point. The curve is fixed in its new position, as shown in the bottom image of Figure 10.18.

FIGURE 10.18

The fitted curve changed to a spline curve, with the location of the second vertex and the new curve

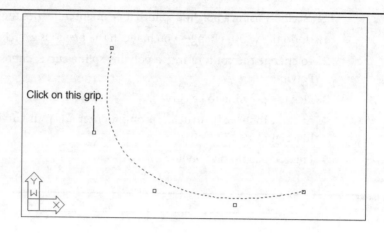

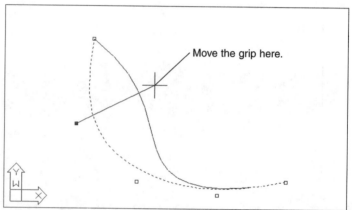

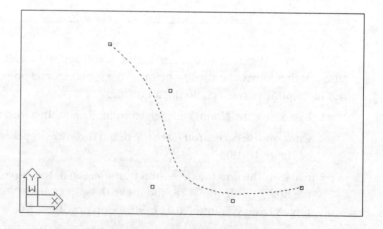

Using True Spline Curves

So far, you've been working with polylines to generate spline curves. The advantage to using polylines for curves is that they can be enhanced in other ways. You can modify their width, for instance, or join several curves together. But at times you will need a more exact representation of a curve. The Spline object, created with Draw ➢ Spline, offers a more accurate model of a spline curve, as well as more control over its shape.

Drawing a Spline

The next exercise demonstrates the creation of a spline curve.

1. Undo the changes made in the last two exercises.
2. Turn on the Data layer so you can view the data points.
3. Adjust your view so you can see all the data points with the elevation of 250.00 (see Figure 10.19).
4. Click on the Spline tool on the Draw toolbar or type **spl**↵.

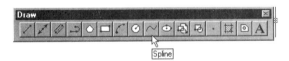

5. At the `Object/<Enter first point>:` prompt, use the Center Osnap to start the curve on the lowest right data point (see Figure 10.19). The prompt changes to `Close/Fit Tolerance/<Enter point>:`.

FIGURE 10.19

Starting the Spline curve at the first data point

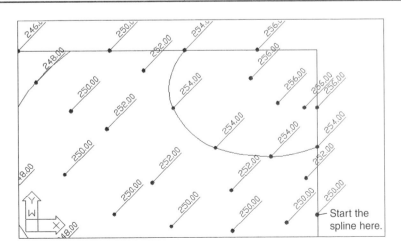

6. Continue to select the 250.00 data points until you reach the last one. Notice that as you pick points, a curve appears, and bends and flows as you move your cursor.

7. Once you've selected the last point, press ↵. Notice that the prompt changes to Enter start tangent:. Also, a rubber-banding line appears from the first point of the curve to the cursor. As you move the cursor, the curve adjusts to the direction of the rubber-banding line. Here, you can set the tangency of the first point of the curve (see the top image of Figure 10.20).

8. Press ↵. This causes AutoCAD to determine the first point's tangency based on the current shape of the curve. A rubber-banding line appears from the last point of the curve. As with the first point, you can indicate a tangent direction for the last point of the curve (see bottom image of Figure 10.20).

9. Press ↵ to exit the Spline command without changing the endpoint tangent direction.

See Chapter 2 for more detailed information on grip editing.

You now have a smooth curve that passes through the points you selected. These points are called the *control points*. If you click on the curve, you'll see the grips appear at the location of these control points, and you can adjust the curve simply by clicking the grip points and moving them. (You may need to turn off the Data layer to see the grips clearly.)

You may have noticed two other options—Fit Tolerance and Close—as you were selecting points for the spline in the last exercise. Here is a description of these options.

Fit Tolerance lets you change the curve so that it doesn't actually pass through the points you pick. When you select this option, you get the prompt Enter Fit Tolerance <0.0000>:. Any value greater than 0 will cause the curve to pass close to, but not through the points. A value of 0 causes the curve to pass through the points. (You'll see how this works in a later exercise.)

Close lets you close the curve into a loop. If you choose this option, you are prompted to indicate a tangent direction for the closing point.

FIGURE 10.20

The last two prompts of the Spline command let you determine the tangent direction of the spline.

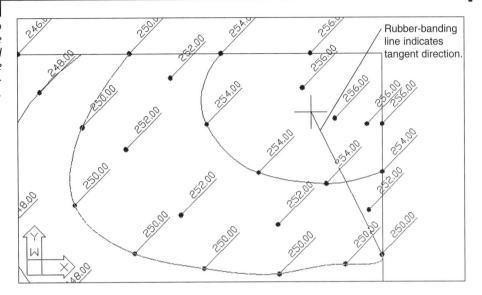

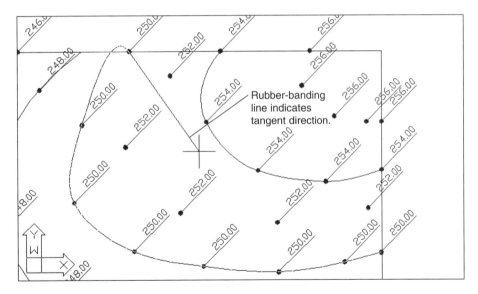

Fine-Tuning Spline Curves

Spline curves are different from other types of objects, and many of the standard editing commands won't work on splines. AutoCAD offers the Modify ➢ Object ➢ Spline option (Splinedit command) for making changes to splines. The following exercise will give you some practice with this command. You'll start by focusing on Splinedit's Fit Data option, which lets you fine-tune the spline curve.

Controlling the Fit Data of a Spline

The following exercise will demonstrate how the Fit Data option lets you control some of the general characteristics of the curve.

1. Choose Modify ➢ Object ➢ Spline, or type **spe**↵ at the command prompt.

2. At the Select Spline: prompt, select the spline you drew in the previous exercise.

3. At the prompt:

    Fit Data/Close/Move Vertex/Refine/rEverse/Undo/eXit <X>:

    type **f**↵ to select the Fit Data option.

> The Fit Data option is similar to the Edit Vertex option of the Pedit command, in that Fit Data offers a subset of options that let you edit certain properties of the spline.

Controlling Tangency at the Beginning and End Points

4. At the next prompt:

    Add/Close/Delete/Move/Purge/Tangents/toLerance/Exit <X>

    type **t**↵ to select the Tangents option. Move the cursor, and notice that the curve changes tangency through the first point, just as it did when you first created the spline (see Figure 10.20).

5. Press ↵. Now you can edit the other endpoint tangency.

6. Press ↵ again. You return to the Add/Close/Delete... prompt.

Adding New Control Points Now add another control point to the spline curve.

7. At the Add/Close/Delete... prompt, type **a**↵ to access the Add option.

8. At the Select point: prompt, click the second grip point from the bottom end of the spline (see the top image of Figure 10.21). A rubber-banding line appears from the point you selected. That point and the next point are highlighted. The

two highlighted points tell you that the next point you select will fall between these two points. You also see the `Enter new point:` prompt.

9. Click on a new point. The curve changes to include that point. In addition, the new point becomes the highlighted point, indicating that you can continue to add more points between it and the other highlighted point (see the bottom image of Figure 10.21).

10. Press ↵. The `Select point:` prompt appears, allowing you to select another point if you so desire.

11. Press ↵ again to return to the `Add/Close/Delete…` prompt.

FIGURE 10.21

Adding a new control point to a spline

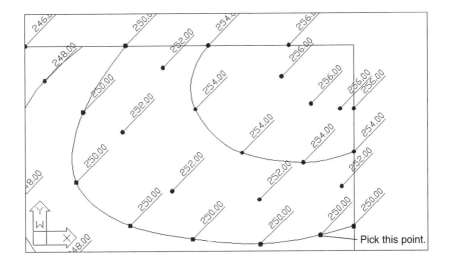

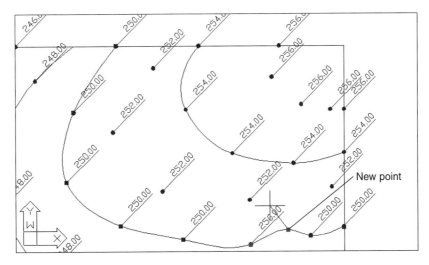

Adjusting the Spline Tolerance Setting   Before we end our examination of the Fit Data options, let's see how Tolerance works.

1. At the Add/Close/Delete… prompt, type **L↵** to select the Tolerance option. This option sets the tolerance between the control point and the curve.
2. At the Enter fit tolerance <0.0000>: prompt, type **30↵**. Notice how the curve no longer passes through the control points, except for the beginning and end points (see Figure 10.22). The fit tolerance value you enter determines the maximum distance from any control point the spline can be.
3. Type **x↵** to exit the Fit Data option.

FIGURE 10.22

The spline after setting the control point tolerance to 30

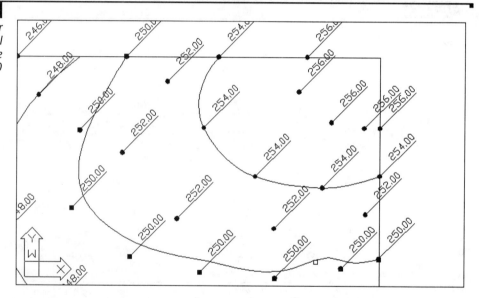

You've seen how you can control many of the shape properties of a spline through the Fit Data option. Here are descriptions of the other Fit Data options you didn't try in these exercises.

Delete   removes a control point in the spline.

Close   lets you close the spline into a loop.

Move   lets you move a control point.

Purge   deletes the fit data of the spline, thereby eliminating the Fit Data option for the purged spline. (See *When Can't You Use Fit Data?*)

When *Can't* You Use Fit Data? The Fit Data option of the Splinedit command offers many ways to edit a spline; however, this option is not available to all spline curves. When you invoke certain of the other Splinedit options, a spline curve will lose its fit data, thereby disabling the Fit Data option. These operations are as follows:

- Fitting a spline to a tolerance and moving its control vertices
- Fitting a spline to a tolerance and opening or closing it
- Refining the spline
- Purging the spline of its fit data using the Purge option (Splinedit ➢ Fit Data ➢ Purge)

Also, note that the Fit Data option is not available when you edit spline curves that have been created from polyline splines. See the *Turning Objects into Polylines and Polylines into Splines* sidebar earlier in this chapter.

Adjusting the Control Points with the Refine Option

While you are still in the Splinedit command, let's look at another of its options, Refine, with which you can fine-tune the curve.

1. Type **u**↵ to undo the changes you made in the previous exercise.
2. At the Fit Data/Close/Move Vertex/Refine… prompt, type **r**↵. The Refine option lets you control the "pull" exerted on a spline by an individual control point. This isn't quite the same effect as the Fit Tolerance option you used in the previous exercise.
3. At the prompt:

    Add Control Point/Elevate Order/Weight/Exit <X>:

    type **w**↵. The first control point is highlighted.
4. At the next prompt:

    Next/Previous/Select Point/Exit/<Enter new weight><1.0000> <N>:

    press ↵ three times to move the highlight to the fourth control point.
5. Type **25**↵. The curve not only moves closer to the control point, it also bends around the control point in a tighter arc (see Figure 10.23).

FIGURE 10.23
The spline after increasing the Weight value of a control point

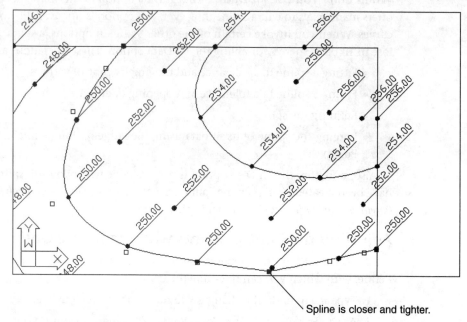

Spline is closer and tighter.

You can use the Weight value of Splinedit's Refine option to pull the spline in tighter. Think of it as a way to increase the "gravity" of the control point, causing the curve to be pulled closer and tighter to the control point.

Continue your look at the Splinedit command by adding more control points—without actually changing the shape of the curve. You do this using Refine's Add Control Point and Elevate Order options.

1. Type **1**↵ to return the spline to its former shape.
2. Type **x**↵ to exit the Weight option; then type **a**↵ to select the Add Control Point option.
3. At the `Select a point on the spline:` prompt, click the second-to-last control point toward the top end of the spline (see the top image of Figure 10.24). The point you select disappears and is replaced by two control points roughly equidistant from the one you selected (see the bottom image of Figure 10.24). The curve remains unchanged. Two new control points now replace the one control point you selected.
4. Press ↵ to exit the Add control point option.
5. Now type **e**↵ to select the Elevate Order option.

6. At the Enter new order <4>: prompt, type **6**↵. The number of control points increases, leaving the curve itself untouched.

7. Type **x**↵ twice to exit the Refine option and then the Splinedit command.

FIGURE 10.24

Adding a single control point using the Refine option

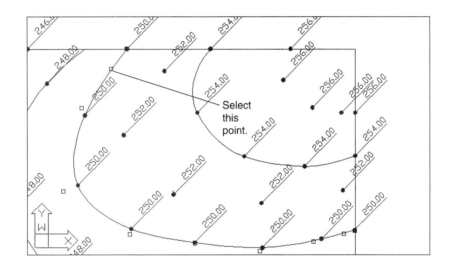

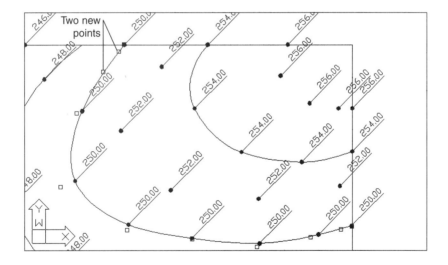

You would probably never edit contour lines of a topographical map in quite the way these exercises have shown. But by following this tutorial you have explored all the potential of AutoCAD's spline object. Aside from its usefulness for drawing contours, it can be a great tool for drawing free-form illustrations. Overall, it is an excellent tool for mechanical applications, where precise, non-uniform curves are required, such as drawings of cams or sheet metal work.

Marking Divisions on a Curve

One of the most difficult things to do in manual drafting is to mark regular intervals on a curve. AutoCAD offers the Divide and Measure commands to help you perform this task with speed and accuracy.

NOTE

Divide and Measure are discussed here in conjunction with polylines, but you can use these commands on any object except blocks and text.

Dividing Objects into Segments of Equal Length

Divide can be used to divide an object into a specific number of equal segments. For example, suppose you needed to mark off the contour you've been working on in this chapter into nine equal segments. One way to do this is to first find the length of the contour by using the List command, and then sit down with a pencil and paper to figure out the exact distances between the marks. But there is another, easier way.

Divide will place a set of point objects on a line, arc, circle, or polyline, marking off exact divisions. The next exercise shows how it works.

1. Open the file called 10a-divd.dwg from the companion CD-ROM. This is a file similar to the one we have been discussing in the prior exercises.
2. Choose Draw ➢ Point ➢ Divide, or type **div**↵.
3. At the Select objects to divide: prompt, pick the spline contour line.
4. The <Number of segments>/Block: prompt that appears next is asking for the number of divisions you want on the selected object. Enter **9**↵.

The command prompt now returns, and it appears that nothing has happened. But AutoCAD has placed several points on the contour which indicate the locations of the nine divisions you have requested. To see these points more clearly, do the following:

5. Click Format ➤ Point Style or type **ddptype**↵. The Point Style dialog box appears.

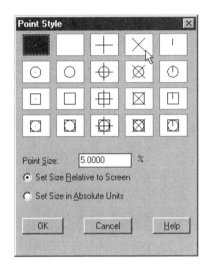

6. Click on the X point style in the upper-right side of the dialog box, click on the Set Size Relative to Screen radio button, and then click on OK.

7. Enter **re**↵. A set of Xs appears, showing the nine divisions (see Figure 10.25).

You can also change the point style by changing the Pdmode system variable. When Pdmode is set to 3, the point appears as an X. See Appendix D for more on Pdmode.

The Divide command uses point objects to indicate the division points. Point objects are created by using the Point command; they usually appear as dots. Unfortunately, such points are nearly invisible when placed on top of other objects. But, as you have seen, you can alter their shape using the Point Style dialog box. You can use these X points to place objects or as references to break the object being divided. (Divide does not actually cut the object into smaller divisions.)

FIGURE 10.25
Using the Divide and Measure commands on a polyline

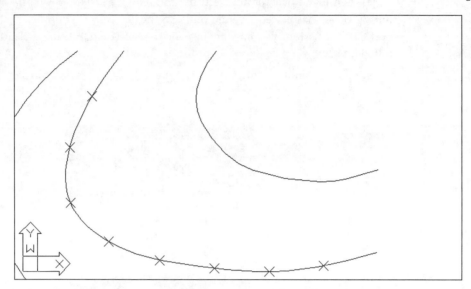

If you are in a hurry, and you don't want to bother changing the shape of the point objects, you can do the following: Set the Running Osnaps to Node. Then when you are in point selection mode, move the cursor over the divided curve. When the cursor gets close to a point object, the Node Osnap marker will appear.

Dividing Objects into Specified Lengths

The *Measure* command acts just like Divide; however, instead of dividing an object into segments of equal length, Measure marks intervals of a specified distance along an object. For example, suppose you need to mark some segments exactly 5' apart along the contour. Try the following exercise to see how Measure is used to accomplish this task.

1. Erase the X-shaped point objects.
2. Choose Draw ➢ Point ➢ Measure or type **me**↵.
3. At the Select object to measure: prompt, pick the contour at a point closest to its lower endpoint. We'll explain shortly why this is important.

4. At the <Segment length>/Block: prompt, enter **5'**↵. The X points appear at the specified distance.

5. Now exit this file.

NOTE Measure is AutoCAD's equivalent of the divider tool in manual drafting. A divider is a V-shaped instrument, similar to a compass, used to mark off regular intervals along a curve or line.

Bear in mind that the point you pick on the object to be measured will determine where Measure begins measuring. In the last exercise, for example, you picked the contour near its bottom endpoint. If you had picked the top of the contour, the results would have been different because the measurement would have started at the top, not at the bottom.

Marking Off Intervals Using Blocks Instead of Points

You can also use the Block option under the Divide and Measure commands to place blocks at regular intervals along a line, polyline, or arc. Here's how to use blocks as markers:

1. First be sure the block you want to use is part of the current drawing file.
2. Start either the Divide or Measure command.
3. At the Number of segments: prompt, enter **b**↵.
4. At the Block Name to Insert: prompt, enter the name of a block and press ↵.
5. At the Align Block with Object? <Y> : prompt, press ↵ if you wish the blocks to follow the alignment of the selected object. (Entering **n**↵ causes each block to be inserted at a 0 angle.)
6. At the Number of Segments: prompt, enter the number of segments and press ↵. The blocks appear at regular intervals on the selected object.

One example of using Divide's or Measure's Block option is to place a row of brackets equally spaced along a bar. Or you might use this technique to make multiple copies of an object along an irregular path defined by a polyline. In civil projects, a fence line can be indicated by using Divide or Measure to place Xs along a polyline.

Sketching with AutoCAD

No discussion of polylines would be complete without mentioning the Sketch command. Although AutoCAD isn't a sketch program, you *can* draw "freehand" using the Sketch command. With Sketch, you can rough in ideas in a free-form way, and later overlay a more formal drawing using the usual lines, arcs, and circles. You can use Sketch with a mouse, but it makes more sense to use this command with a digitizing tablet that has a stylus. The stylus affords a more natural way of sketching.

Freehand Sketching with AutoCAD

Here's a step-by-step description of how to use Sketch.

1. Make sure the Ortho and Snap modes are turned off. Then type **skpoly↵1↵**. This sets the Sketch command to draw using polylines.

2. Type **sketch↵** at the command prompt.

3. At the Record increment: prompt, enter a value that represents the smallest line segment you will want Sketch to draw. This command approximates a sketch line by drawing a series of short line segments; the value you enter here determines the length of those line segments.

4. At the Sketch. Pen eXit Quit Record Erase Connect: prompt, press the pick button and then start your sketch line. Notice that the message <Pen down> appears, telling you that AutoCAD is recording your cursor's motion.

You can also start and stop the sketch line by pressing P on the keyboard.

5. Press the pick button to stop drawing. The message <Pen up> tells you AutoCAD has stopped recording your cursor motion. As you draw, notice that the line is green. This indicates that you have drawn a temporary sketch line and have not committed the line to the drawing.

6. A line drawn with Sketch is temporary until you use Record to save it, so turn the sketch line into a polyline now by typing **r↵**.

7. Type **x↵** to exit the Sketch command.

Here are some of the other Sketch options we weren't able to cover in this brief exercise:

Connect allows you to continue a line from the end of the last temporary line drawn. Type **c** and then move the cursor to the endpoint of the temporary line. AutoCAD automatically starts the line, and you just continue to draw. This only works in the <Pen up> mode.

Period (.) allows you to draw a single straight-line segment by moving the cursor to the desired position and then pressing the period key. This only works in the <Pen up> mode.

Record, **Erase**, **Quit**, and **Exit** control the recording of lines and exiting from the Sketch command. Record is used to save a temporary sketched line; once a line has been recorded, you must edit it as you would any other line. With Erase you can erase temporary lines before you record them. Quit ends the Sketch command without saving unrecorded lines. On the other hand, the Exit option on the Sketch menu automatically saves all lines you have drawn and then exits the Sketch command.

Filling In Solid Areas

You have learned how to create a solid area by increasing the width of a polyline segment. But suppose you want to create a simple solid shape or a very thick line. AutoCAD provides the Solid, Trace, and Donut commands to help you draw simple filled areas. The Trace command acts just like the Line command (with the added feature of drawing wide line segments), so only Solid and Donut are discussed here.

You can create free-form solid filled areas using the new Solid hatch pattern. Create an enclosed area using any set of objects, and then use the Hatch tool to apply a solid hatch pattern to the area.

Drawing Solid Filled Areas

In the past, AutoCAD users used hatch patterns to fill in solid areas—and that was a great way to fill an irregular shape. However, hatches tended to increase the size of a file dramatically, thereby increasing loading and regeneration time. Autodesk has completely changed the way Release 14 handles hatches so they don't use nearly the amount of memory they once did. You can even use a new, predefined hatch pattern called *Solid* to create a memory-efficient solid fill of an irregular-shaped area.

Here is a short exercise to demonstrate how to use it.

1. Open file 13a-htch.dwg from the companion CD-ROM.
2. Click on Hatch from the Draw toolbar.
3. At the Boundary Hatch dialog box, click on the Pick Points button.
4. Click in the area bounded by the border and contour line in the upper-right corner of the drawing, as shown in the top image of Figure 10.26. Then press ↵. You have the option here to click in more areas if you so desire.
5. Back at the Boundary Hatch dialog box, click on the Pattern button.
6. At the Hatch Pattern Palette dialog box, click on SOLID from the list at the left.
7. Click on OK, and then click on Apply. A solid fill is applied to the selected area, as shown in the bottom image of Figure 10.26.

Drawing Filled Circles

If you need to draw a thick circle for a pad, or a solid filled circle, take the following steps.

1. Choose Draw ≻ Donut or type **do**↵ at the command prompt.
2. At the Inside diameter: prompt, enter the desired diameter of the donut "hole." This value determines the opening at the center of your circle.
3. At the Outside diameter: prompt, enter the overall diameter of the circle.
4. At the Center of doughnut: prompt, click on the desired location for the filled circle. You can continue to select points to place multiple donuts (see Figure 10.27).
5. Press ↵ to exit this process.

If you need to fill only a part of a circle, such as a pie slice, you can use the Donut command to draw a full, filled circle, and then use the Trim or Break options on the Modify toolbar to cut out the portion of the donut you don't need.

FIGURE 10.26

Locating the area to fill and the final result of the solid hatch

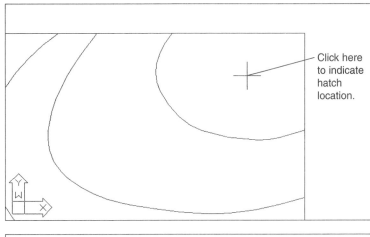
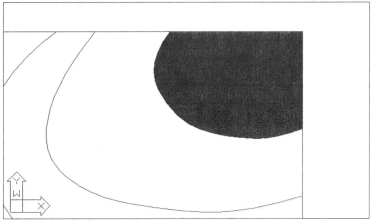

FIGURE 10.27

Drawing wide circles using the Donut command

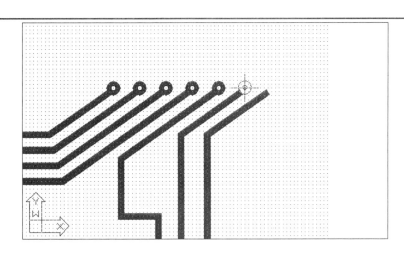

Toggling Solid Fills On and Off

Once you have drawn a solid area with the Pline, Solid, Trace, or Donut commands, you can control whether the solid area is actually displayed as filled in. Open the Drawing Aids dialog box by choosing Tools ➢ Drawing Aids or by typing **rm**↵. If the Solid Fill check box does not show a check mark, thick polylines, solids, traces, and donuts appear as outlines of the solid areas (see Figure 10.28).

You can shorten regeneration and plotting time if solids are not filled in.

FIGURE 10.28

Two polylines with the Fill option turned on (top) and turned off (bottom)

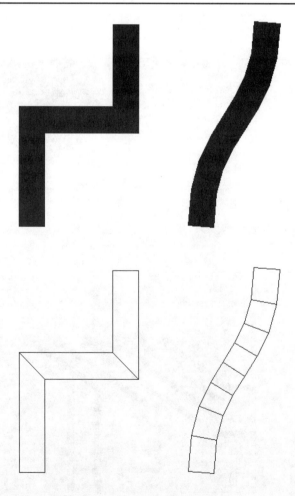

WARNING

If Regenauto is turned off, you will have to issue the Regen command to display the effects of the Fill command.

The Drawing Aids Solid Fill option is an easy-to-remember way to control the display of solid fills. Or you can enter **fill**↵ at the command prompt; then, at the ON/OFF <ON>: prompt, enter your choice of **on** or **off**.

If You Want to Experiment...

There are many valuable uses for polylines beyond those covered in this chapter. We encourage you to become familiar with the polyline so you can take full advantage of AutoCAD.

To further explore the use of polylines, try the following exercise illustrated in Figure 10.29. It will give you an opportunity to try out some of the options discussed in this chapter that weren't included in exercises.

1. Open a new file called **part13**. Set the Snap mode to .25 and be sure the Snap mode is on. Use the Pline command to draw the object shown at step 1 of Figure 10.29. Draw it in the direction indicated by the arrows, and start at the upper-left corner. Use the Close option to add the last line segment.

2. Start the Pedit command, select the polyline, and then type **e**↵ to issue the Vertex option. At the Next/Previous/Break: prompt, press ↵ until the X mark moves to the first corner shown in the figure to the right. Enter **s**↵ for the Straighten option. At the Next/Previous/Go: prompt, press ↵ twice to move the X to the other corner shown in the figure. Press **g**↵ for Go to straighten the polyline between the two selected corners.

3. Press ↵ twice to move the X to the upper-right corner, then enter **i**↵ for Insert. Pick a point as shown in the figure. The polyline changes to reflect the new vertex. Enter an **x**↵ to exit the Edit Vertex option, and then press ↵ to exit the Pedit command.

4. Start the Fillet command and use the Radius option to set the fillet radius to .30. Press ↵ to start the Fillet command again, but this time use the Polyline option and pick the polyline you just edited. All the corners fillet to the .30 radius. Add the .15 radius circles as shown in the figure and exit the file with the End command.

FIGURE 10.29

Drawing a simple plate with curved edges

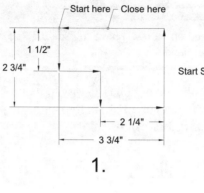
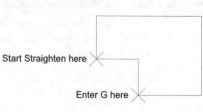
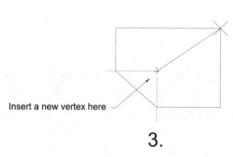
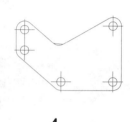

PART III

Modeling and Imaging in 3D

LEARN TO:

- Create Solid Models
- Manage and Save a UCS
- Work with Primitive Solids
- Find Mass Properties of Solid Objects
- Move Objects in 3D Space
- Add Special Effects

Chapter 11

Introducing 3D

FEATURING

Creating a Solid Model

Boolean Operations

Understanding the UCS

Viewing a 3D Drawing

Visualizing Your Model

Creating a Solid Model from a 2D Drawing

Building an Assembly of Parts

Chapter 11

Introducing 3D

Viewing an object in three dimensions gives you a sense of its true shape and form. It also helps you conceptualize the design, which will ultimately result in better design decisions. And using three-dimensional objects helps you visually communicate your ideas to the others on your design team.

Another advantage to creating parts and assemblies in three dimensions is that you can derive 2D drawings from your 3D model. For example, you could model a mechanical part in 3D and then quickly derive top, front, and right-side views using the techniques discussed in this chapter.

AutoCAD offers three methods for creating 3D models: wire-frame modeling, surface modeling, and solid modeling. All three methods will look pretty much the same as you view them in AutoCAD.

A *wire-frame model* represents a solid object using lines and arcs drawn along the boundaries of the surfaces of objects. The wire-frame model can not be rendered or shaded, because only the hollow outline of the model is displayed. You will not learn about wire-frame model making in this book. It is an old technique, and you will only use it to revise old drawings. The methods used to draw and edit new solids and surfaces will suffice to provide examples for editing old wire-frame files.

A *surface model* is drawn by creating the model's surfaces. So a box that has six flat sides requires six flat surfaces. If the box is hollow and the sides have thickness, six more surfaces are required to represent the interior. A surface model can be shaded and rendered by bouncing light off the surfaces, but surfaces only have area—neither a surface nor a collection of surfaces has volume in AutoCAD. You will get a taste of surface modeling in Chapter 13.

A *solid model* has both surfaces and volume. It can be rendered like the surface model, but unlike the surface model it can be analyzed for mass properties. Let's explore solids and learn about 3D.

NOTE Many 3D designs were created in AutoCAD Releases 10–12 and are still used in production today. The only method available to create 3D models in the earlier releases was wire frame, and you may be called on to edit or update some of these wire-frame models. Wire frame isn't a good way to learn 3D modeling, because a wire-frame model is just what it looks like: line, arc, and circle objects assembled in 3D space to represent the boundary of a 3D model. To edit wire frame, you use many of the same tools that you use to edit the properties of 2D objects (with a few additional options). As you proceed through the following 3D chapters, note the methods used to manipulate the UCS, the way in which objects are created in each UCS, and the visualization tools. Once mastered, these techniques will allow you to revise existing 3D models.

Creating a 3D Drawing

Let's start by drawing a 3D ⅜" socket-head machine screw. You will not draw every nuance of the screw because it's often a purchased part. (Of course, the manufacturer of the screw would want it drawn complete with every detail.)

First, you will use the Cylinder command to start the head of the screw. Then you'll use the Polygon command to draw the hexagon shape of the screw head drive, which you'll make into a solid with the Extrude tool. Next you'll edit the cylinder using the Boolean operation of subtracting one solid from another, and you'll learn that you can use the Chamfer command with solids by taking the sharp edge off the top of the screw head.

It sounds like a lot to do, but you'll see that these are pretty simple operations, following a pretty simple logic.

Preparing to Draw Solid Models

First, you'll need to tell AutoCAD which drawing tools you'll need.

1. Right-click on any toolbar, and the Toolbars dialog box will appear.
2. Scroll down the toolbar choices until you find the Modify II toolbar; click on the check box to the left to activate it.
3. Continue scrolling down to select the Solids, UCS, and Viewpoint toolbars.
4. Click on the Close button when you are finished.

NOTE

If you find that the toolbars are in your way, remember that you can turn them on and off by right-clicking on any toolbar. You can also dock a toolbar by dragging it to the side of your screen.

Drawing a Solid Cylinder

Let's start with the head of the screw.

1. Click on the Cylinder tool from the Solids toolbar or choose Draw ≻ Solids ≻ Cylinder. Or type **cylinder**↵.

2. At the `Elliptical/<center point> <0,0,0>:` prompt, pick a point in the lower-left corner of the screen.
3. At the `Diameter/<Radius>:` prompt, enter **d**↵. At the `Diameter:` prompt, enter **.562**↵. This is the head diameter of a standard ⅜ socket-head machine screw. So far, this is has been just like drawing a circle.
4. At the `Center of other end/<Height>:` prompt, type **.375**↵. A circle appears.

You need to change your point of view if you want to see the cylinder. Right now, you are looking straight down on the cylinder.

5. Drag the mouse across the Viewpoint toolbar while looking at the tool tips. You'll find the tool tips to be very clear and descriptive. Click on the S(outh) E(ast) Isometric View option.

There are a couple of important things to note about what happened when you chose Isometric View. The cylinder looks transparent, or as though it is composed of lines and ellipses. This is the wire-frame representation of the solid form. The X-Y icon, called the UCS icon, has changed in appearance as though it too is drawn in an Isometric view. But it has not actually changed—only your view of it has: A Zoom All command was issued at the end of the Viewpoint command.

Next, draw the hexagon-shaped drive socket.

1. Click on the Polygon tool from the Draw toolbar, choose Draw ➢ Polygon from the menu bar, or type **polygon**⏎ in the command line. At the Number of sides <4>: prompt, type **6**⏎, and at the Edge/<Center of polygon>: prompt, pick the top center of the cylinder.

2. At the Inscribed in circle/Circumscribed about circle (I/C) <I>: prompt, type **c**⏎, and at the Radius of circle: prompt, type **5/32**⏎.

Your drawing is starting to take on the shape of the screw head, even though it is still basically a hexagon and a cylinder.

Turning a Polyline into a Solid

Now you've got the basic shape outlined as a polyline. In order to make it a solid, you will assign mass to the shape with the Extrude tool.

1. To turn the polygon into a solid, click on the Extrude tool from the Solids toolbar, choose Draw ➢ Solids ➢ Extrude from the menu bar, or type **ext**↵. (Remember that **ex**↵ is what you type in for the Extend command.)

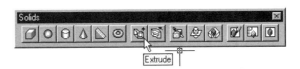

2. At the Select objects: prompt, type **L**↵ to select the last object drawn (the polygon), and press ↵ again to complete the selection set.
3. At the Path/<Height of Extrusion>: prompt, type **–.182**↵. The normal extrusion direction is in the positive Z direction, unless you tell AutoCAD otherwise. The minus sign tells AutoCAD to extrude in the negative Z direction (away from you).
4. At the Extrusion taper angle <0>: prompt, hit ↵ to accept the default of no taper angle.

Now you've taken another step in creating 3D objects—you've given a polyline a new dimension by extruding it down .182 units.

Removing the Volume of One Solid from Another

Next, you will remove the hexagon solid from the head of the screw to create the socket drive feature.

1. Click on the Subtract tool from the Modify II toolbar, choose Modify ➢ Boolean ➢ Subtract from the menu bar, or type **su**↵ in the command line.

2. At the Select solids and regions to subtract from: prompt, choose the cylinder. When the cylinder is highlighted, press ↵ to complete the selection set.

3. At the Select solids and regions to subtract: prompt, pick the extruded polygon, or type **L**↵ to use the last object created. Press ↵ again to complete the selection set. Your drawing should look like Figure 11.1.

FIGURE 11.1

The head of the hex-drive socket screw

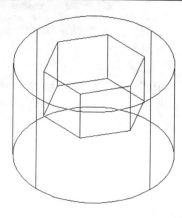

The hexagon shape was subtracted from the cylinder that represents the head of the screw.

Using the Hide Tool to Verify Your Design

Your model looks the same after the subtraction of the hexagon solid as it looked before. You'll use the Hide tool to see the difference. The Hide tool evaluates any solids or surfaces and displays each one relative to their positions and your viewpoint. If one shape is closer to you than another, the Hide command obscures the lines of the shape that is farther away, like a house might block your view of a shrub. So external surfaces hide internal surfaces, and near surfaces hide far surfaces.

The Hide tool can be very useful when your drawing gets more complex. This tool can clarify the model by eliminating extra lines that are hidden.

1. Click on the Hide tool from the Render toolbar, choose View ➢ Hide, or type **hi**↵ in the command line. The surfaces of the cylinder (and all non-planar surfaces in any model) are shown as triangles. This solid-looking object, with triangles for

CREATING A 3D DRAWING | 473

non-planar surfaces, is the AutoCAD default view. Your drawing should look like the top image of Figure 11.2.

2. Change the Hide view default to show the silhouette of the surfaces with the system variable Dispsilh (for display silhouette). Type **dispsilh**↵ in the command line. At the New value for DISPSILH <0>: prompt, type **1**↵.

3. Type **hi**↵ to see the results of changing the system variable. Your drawing should look like the bottom image of Figure 11.2.

FIGURE 11.2

The head of the hex-drive socket screw with Dispsilh set to 0 in the top view and set to 1 in the bottom view

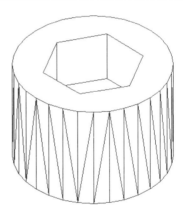

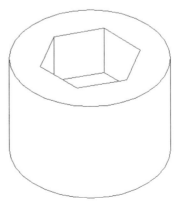

4. When you have finished looking at the model in Hidden view type **re↵**. This will regenerate the view and return the view to wire frame.

As matter of personal preference, you can continue to work with the Hidden view. This view will only allow you to select objects that you can see; this can be limiting. A number of normal view regenerations occur during solid modeling. For example, a regen occurs when you edit a model and when you change viewpoint. At each regen the hidden line view is redrawn as a wire-frame view and you will have to use the Hide tool to restore your hidden line view.

Editing a Solid Model with the Chamfer Tool

The Chamfer tool offers a way of breaking sharp edges. In a hand tool where sharp edges are a problem, or in some other part where a certain elegance of design is desired, chamfering will allow you to remove material using a few simple steps.

1. Click on the Chamfer tool from the Draw toolbar.

2. At the (TRIM mode) Current chamfer: prompt, type **Dist1 = 0.5000, Dist2 = 0.5000↵**.

3. At the Polyline/Distance/Angle/Trim/Method/<Select first line>: prompt, pick the circle that is the top of the screw head's cylinder.

4. At the Select base surface: Next/<OK>: prompt, look at the cylinder. If the top appears to be highlighted, press ↵; if not, press **n↵** until it is. If you pick the edge of a solid, there will be more than one surface that you could have intended for the base surface, so the Next option cycles through the available surfaces, allowing you to select the one that you want.

5. At the Base surface distance <0.5000>: prompt, type **.035↵**. You are telling AutoCAD the dimension of the top edge of the chamfer for the selected solid object (the highlighted surface) of the screw head.

6. At the Other surface distance <0.5000>: prompt, type **.035↵** to form the second edge of the chamfer a little distance down the side of the screw head.

7. At the Loop/<Select edge>: prompt, pick the top edge of the cylinder again.

CREATING A 3D DRAWING | 475

8. To tell AutoCAD that you are finished picking edges to be chamfered, at the Loop/<Select edge>: prompt, press ↵. The screw head is redrawn with the chamfer.

9. Use the Hide command to see your work. Your drawing should look like Figure 11.3.

FIGURE 11.3
The chamfered head of the hex-drive socket screw

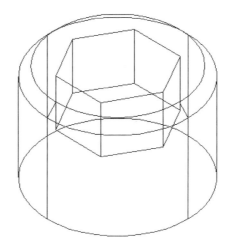

Next, you will add the body of the screw. AutoCAD can't draw the helical threads of the screw, so you will represent the threads of the screw with a simple cylinder.

Adding the Volumes of Two Solids

You've drawn the cylindrical head of the screw. Next you will need to draw the threaded portion of the screw. New solids are always drawn as separate parts, so you will then have and join the head and threads together.

1. Click on the Cylinder tool from the Solids toolbar.

2. At the Elliptical/<center point> <0,0,0>: prompt, use the Center Osnap tool to pick the center of the bottom of the head of the screw.

3. At the Diameter/<Radius>: prompt, type **d**↵, and at the Diameter: prompt, enter **.375**↵ (because you have been drawing a ⅜" socket-head screw).

4. At the Center of other end/<Height>: prompt, type **–1**↵ to draw a 1-unit-long cylinder down the negative Z axis to make the threads 1 unit long.

5. Click on the Union tool from the Modify II toolbar, choose Modify ➤ Boolean ➤ Union from the menu bar, or type **union**↵ at the command line.

6. At the `Select objects:` prompt, pick the head and the cylinder, and press ↵.
7. Choose Zoom All to see your handiwork. Use the Hide tool, and your drawing should look like Figure 11.4.
8. Save your work.

FIGURE 11.4

The completed socket-head screw

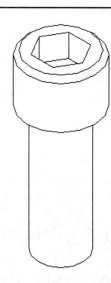

The prompts generated by the Cylinder command were straightforward, but let's take a closer look to understand them better. The first prompt was `Elliptical/<center point> <0,0,0>`. The default asks you to pick a center point on the screen. (You could also enter an absolute position relative to the world origin.) The Elliptical option allows you to draw an ellipse, instead of a circle, as the basis of your cylinder.

The `Diameter/<Radius>:` prompt comes directly from the Circle command. The `Center of other end/ <Height>:` prompt gives you the option to enter a value; a positive number generates the object along the positive Z axis and a negative number generates the object along the negative Z axis.

The Center option allows you to pick any point on the screen. AutoCAD will draw the cylinder from the first point you selected to this center point, regardless of the coordinate system.

The Extrude tool offered a few interesting prompts. At the `Select objects:` prompt, you selected a closed polyline. In another drawing, you might use this tool with other shapes, like a closed spline, a region, a circle, or an ellipse.

You cannot extrude text or dimensions using the Extrude tool because you have to use an object that can become a solid.

The next prompt was `Path/<Height of Extrusion>:`. You established the height of an extrusion using the same sort of prompt that you used in the Cylinder tool (defining positive or negative measurements along the Z axis). The Path option is something different. The path of an extrusion could follow a line, an arc, a 2D polyline, or a spline. If you are careful and don't draw a polyline that causes the extrusion to curl inside itself, you can sweep your extrusion along interesting shapes like an ellipse or a spiral. This can be very handy for representing, for example, piping or the plumbing on a pneumatic layout.

Next, let's explore the Box primitive and draw the post of an arbor press.

Using the Box Primitive

We've looked at some basic shapes: the line, the polyline, spline curves, and circles. Let's look at how we can manipulate another basic shape. Since we're now talking about more than two dimensions, we're not going to talk about a simple rectangle. We'll use the box shape to build the base and the post of an arbor press.

1. Click on the Box tool from the Solids toolbar, choose Draw ➢ Solids ➢ Box, or type **box**↵.

2. At the `Center/<Corner of box> <0,0,0>:` prompt, type **1,1↵** as your coordinates.

3. At the `Cube/Length/<other corner>:` prompt, type **1↵**. At the `Length:` prompt type **.98↵**; at the `Width:` prompt type **3.5↵**; and at the `Height:` prompt type **6↵**.

4. The post requires no extrusion taper angle, so enter **<0>:↵** at the `Extrusion taper angle:` prompt.

The post is going to be L-shaped, and you need to create more objects to use as tools to edit the post. These new objects will need to be created in other 3D planes. All new objects are created relative to the current User Coordinate System.

Let's take a closer look at the World Coordinate System and the User Coordinate System so you will be able to easily adjust the UCS to your needs.

The UCS and the WCS

The *World Coordinate System* (WCS) is the default when you start any new drawing (unless you have created a special template file with other settings).

The *User Coordinate System* (UCS) icon is the figure in the lower left corner of every drawing that you have made since the beginning of this book. It is represented by connected perpendicular X and Y arrows with a W at the apex. The purpose of the icon is to apprise you of the current UCS settings. The W means that the icon is showing the World Coordinate System X and Y axes. The arrows point in the positive X and Y directions. Take a look at Figure 11.5 to see the UCS icon showing the WCS from a Southeast Isometric viewpoint.

FIGURE 11.5

The World UCS icon

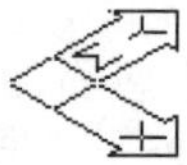

The UCS icon does not plot or print, and it remains the same size regardless of the zoom scale. In Chapter 8, you created a new UCS. The UCS icon stayed in place, but it changed to a simpler X- and Y-arrow unit, demonstrating that you were using a system other than the default World Coordinate System. If you'd wanted to, you could have set the UCS icon to follow the origin. Let's look at the UCS Icon controls.

Managing the UCS Icon's Appearance

The View ➢ Display ➢ UCS Icon cascading menu displays two options: On and Origin. These two options actually represent the On, Off, Origin, and No Origin options of the UCS icon command. Let's take a moment to examine the options offered by the UCS icon command.

Origin/No Origin   This option makes the UCS icon appear at the location of the current UCS's 0,0,0 origin point (Origin), or places it in the lower-left corner of the AutoCAD window (No Origin). If the UCS's origin is off the screen while the Origin option is active (shown by a check mark in the menu), the UCS icon appears in the screen's lower-left corner.

On/Off   This option controls whether or not the UCS icon is displayed. To use On/Off, select View ➢ Display ➢ UCS Icon ➢ On from the menu bar. When this option is checked, UCS is on, and the icon is displayed; when it is not checked, the icon is turned off. Notice that you need to deselect the On option to turn off this feature.

In addition, there are two other settings you will want to know about that do not appear on this pull-down menu.

All   This option lets you set the UCS icon's appearance in all the viewports on your screen at once. This option has no significance if you only have one viewport on the screen. To use All, type **ucsicon↵a↵**.

UCSFollow   This option is a system variable that, when set to 1, will cause the display in a given viewport to always show a Plan view of the current UCS. To use this option, highlight a viewport, type **ucsfollow↵**, and at the New value for UCSFOLLOW <0>: prompt type **L↵**.

The User Coordinate System

You have been working with the three-axis coordinate system since the beginning of this book. You know how to specify the two main X and Y coordinates for 2D drawings. And you might have noticed a number of commands where you had the option to specify a third coordinate, which AutoCAD has assumed was 0. The third point, had you changed the 0 to anything else, would have been along the Z axis, making your drawing three-dimensional.

> If you will never use 3D, all you need to know about the Z axis is that when you rotate anything, the point you pick about which those objects are to be rotated is a point on the Z axis.

The Z Axis

You may be familiar with the "right-hand rule," which is useful when you're looking at Cartesian coordinates and trying to recall which directions are positive. Hold your right-hand palm up and stretch out your thumb. Then align your thumb so that it points in the positive X axis direction; your index finger automatically points 90° from your thumb in the positive Y axis. You have been using these axes to define locations in 2D space since your first drawing in Chapter 2. With your right hand still stretched out, curl your middle finger until it is perpendicular to your thumb and index finger. Your middle finger is pointing in the positive Z axis. With the X, Y, and Z axes you can define any point in 3D space, just as the X and Y axes alone allowed you to define any point in 2D space. Both the World Coordinate System and the User Coordinate system allow you to enter values (both the positive and negative) for the Z axis location of a point. You can also enter cylindrical and spherical coordinates.

Coordinate Systems for UCS and WCS

You're pretty used to using the WCS because you have been using it since you started this book. Each object that you have created has been located on, or relative to, the X,Y plane of the WCS. When you created a circle, the plane of the circle was parallel to the X,Y plane of the WCS. You might think of WCS as a sort of cube of 3D space with a plane located in the middle. One point on the plane is the origin: 0,0,0. So far, you have been viewing the plane from a position normal to the origin.

But you are not limited to using the plane of the WCS. You can create a coordinate system of your own, relative to the WCS, called a User Coordinate System (UCS). A 2D UCS might be rotated to ease drawing and dimensioning of a view that is projected from an angled surface. If you have used a drafting machine, creating a rotated UCS is akin to rotating the head and scales to some angle, as if you were rotating the scales on the X-Y plane about the Z axis.

All objects are created on, or relative to, the X,Y plane of the current UCS. You could draw every solid object on, or relative to, the WCS and move and rotate the object into position using the Move, 3D Rotate, and Align commands. The UCS allows you to create, save, and restore any number of planes to draw on.

It's a good idea to save two UCSs in addition to the WCS, each one orthogonal to the others, whenever starting a new solid model. When you're creating 3D objects, having a number of UCSs defined can save you a good deal of time. You might let the WCS equal the Plan view and then set up one UCS parallel to X-Z called *f* for *front*, and another UCS parallel to Y-Z called *s* for *side*.

Controlling the UCS

Have a look at the UCS controls by clicking on the Tools menu so that the UCS menu cascades. Pull up the UCS toolbar. You've got lots of options.

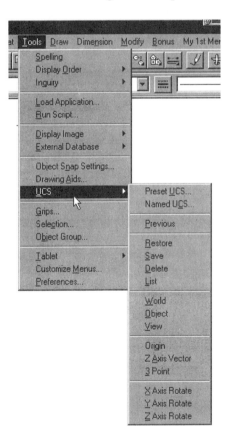

The UCS Toolbar Options

These are the settings for UCS options. You'll use many of them in the exercises to come.

Preset UCS allows you to choose a UCS supplied by AutoCAD. There are some standard views, so Autodesk provided you with them.

Named UCS allows you to use the command line to select a UCS that you'd previously saved.

Previous returns to the previous UCS.

Restore restores a saved UCS.

Save allows you to name and save a UCS.

Delete allows you to remove a saved UCS from the current drawing.

List lists previously saved UCSs.

World is the default option used to return to the World Coordinate System.

Object allows you to select an object to define the X-Y plane of a new UCS. See Table 11.1 for the description of how objects determine the UCS orientation.

View uses the current viewpoint to define a plane parallel to the screen. Use this option to define an isometric plane.

Origin sets a new origin (0,0,0) point of the UCS relative to the current UCS.

Z Axis Vector allows you to define a UCS by picking the point through which the new Z axis will extend.

3 point allows you to define three points which will become the new UCS X-Y plane. The first point is the origin, the second point is a positive point on the X axis and the last point is a positive point on the Y axis.

X Axis allows you to define a plane by rotating the current UCS plane about the X axis.

Y Axis allows you to define a plane by rotating the current UCS plane about the Y axis.

Z Axis allows you to define a plane by rotating the current UCS plane about the Z axis.

Setting the UCS Origin

When you create a UCS, you begin by defining the Origin. You can change the origin location of the current UCS with the UCS origin tool. First, let's set the UCS icon to follow the UCS origin so we can view a change in the origin location as it happens.

The origin will change to the new setting whether or not the UCS icon is set to follow. It's usually bad practice to set the UCS icon to follow the UCS origin because the UCS icon can be confused with object lines around the origin. Let's set the UCS icon to follow the origin here just so you can see what happens. Go back to the box you were drawing to be the base and the post of the press.

1. From the View ➢ Display ➢ UCS Icon cascading menu, select Origin. The UCS icon changes to include a + near the apex and moves to the current WCS origin. The UCS will look like the left image of Figure 11.6.

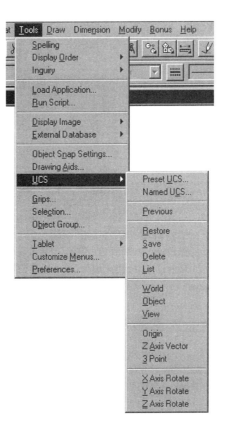

2. To move the UCS origin to the corner of the box, select the Origin tool from the UCS toolbar. Use the Endpoint Osnap to select the lower-left corner of the box. Now the icon should have moved to look like the right image of Figure 11.6.

Note that the X and Y axes remain pointing in the same direction.

FIGURE 11.6
Left: The UCS icon changed to follow the UCS origin and moved to the WCS point 0,0. Right: The UCS icon moved again to the new origin location.

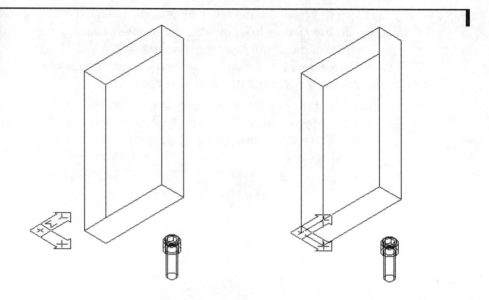

A UCS Defined with Three Points

You can easily define a UCS by locating three points in 3D space. The first point will be the origin, the second point will be a point on the positive X axis, and the third point will a point somewhere in the positive Y direction. Use this method when you know three points on a plane and you need the plane to define some objects. Next you'll create a UCS using three points.

1. Use the right side of the box to define a new UCS plane. From the UCS toolbar, click on the 3-Point UCS tool. At the Origin point <0,0,0>: prompt, pick the bottom back corner of the right side of the box (as shown in the left image of Figure 11.7).

2. At the Point on positive portion of the X-axis <1.9800,0.0000,0.0000>: prompt, pick the point (shown in the left image of Figure 11.7) at the bottom front corner of the right side of the box to set the X axis the same as the former Y axis.

3. At the Point on positive Y portion of the UCS XY plane <-0.0200,0.0000, 0.0000>: prompt, pick the corner of the box above the first corner that you selected (as shown in the right image of Figure 11.7). This sets the Y axis to the direction of the former Z axis.

As mentioned earlier, each object is drawn on the current UCS plane. You have now created your first UCS plane that is not parallel to the WCS. Draw a circle and watch how it is positioned relative to the current box. You will use this circle later to make a hole in the part for the rack gear.

Locate a circle 1.062 units horizontal and 5 units vertical.

1. From the Draw toolbar, select the Circle tool and at the CIRCLE 3P/2P/TTR/ <Center point>: prompt, Shift+right-click to use the From Osnap.

2. At the Base point: <Offset>: prompt, type **@1.062,5**↵ to locate the circle in the UCS, and at the Diameter/<Radius>: prompt, type **d**↵. Finish the command by typing **1.17**↵ at the Diameter: prompt.

A 1.17-unit diameter circle is drawn in on the side of the block. It looks like an ellipse, but this is an illusion due to your Isometric view of the object. It should look like Figure 11.7.

FIGURE 11.7
The circle drawn on the vertical plane of the block

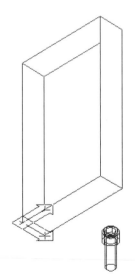
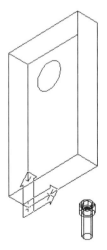

If your circle does not look like the figure, you have probably not set the UCS correctly, and you might need to take another look at *A UCS Defined with Three Points*.

NOTE If you choose to select points without the benefit of Osnaps while in a non-orthogonal viewpoint, you may not get the results that you expect. Any point you pick which is not attached to an object offers AutoCAD the choice of picking a point anywhere along an axis formed by the point receding directly into the screen. When you look at your object from another perspective, you might find that the object is far from where you expected it to be.

Using an Object to Define a UCS Plane

The circle in the previous exercise can be used to define a UCS. Use the Object UCS tool on the UCS toolbar to set a definition.

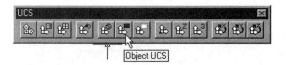

At the Select object to align UCS: prompt, pick the circle that you just created. The UCS origin has moved to the center of the circle and the UCS has rotated about –45°. See Table 11.1 to identify objects that can be used to establish a UCS origin and plane. See Figure 11.8 to see what the current UCS icon looks like.

FIGURE 11.8

Setting the UCS using an object

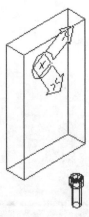

TABLE 11.1: EFFECTS OF OBJECTS ON THE ORIENTATION OF A UCS

Object Type	UCS Orientation
Arc	The center of the arc establishes the UCS origin. The X axis of the UCS passes through the pick point on the arc.
Circle	The center of the circle establishes the UCS origin. The X axis of the UCS passes through the pick point on the circle.
Dimension	The midpoint of the dimension text establishes the origin of the UCS origin. The X axis of the UCS is parallel to the X axis that was active when the dimension was drawn.
Line	The endpoint nearest the pick point establishes the origin of the UCS, and the X-Z plane of the UCS contains the line.
Point	The point location establishes the UCS origin. The UCS orientation is arbitrary.
2D Polyline	The starting point of the polyline establishes the UCS origin. The X axis is determined by the direction from the first point to the next vertex.
Solid	The first point of the solid establishes the origin of the UCS. The second point of the solid establishes the X axis.
Trace	The direction of the trace segment establishes the X axis of the UCS with the beginning point setting the origin.
3D Face	The first point of the 3D Face segment establishes the origin. The first and second points establish the X axis. The plane defined by the face determines the orientation of the UCS.
Shapes, Text, Blocks, Attributes, and Attribute Definitions	The insertion point establishes the origin of the UCS. The object's rotation angle establishes the X axis.

Rotating the UCS Plane about the Z Axis

Suppose you want to rotate the current X-Y plane about the Z axis. You can accomplish this by using the Z Axis Rotate option of the UCS command. Let's try rotating the UCS about the Z axis to see how this works.

1. Click on the Z Axis Rotate UCS tool on the UCS toolbar.

2. At the `Rotation angle about Z axis <0>:` prompt, enter **35** for 35°. The UCS icon rotates to reflect the new orientation of the current UCS (see Figure 11.9).

FIGURE 11.9

Rotating the UCS about the z-axis

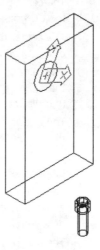

Returning to the WCS

Regardless of the location or plane of the current UCS, you can always restore the WCS. This can be a boon—especially for those times when you would rather start over to define a new UCS plane.

- From the UCS toolbar, select the World UCS tool. You can see that the UCS icon has returned to the original position and the W once more appears in the Y leg (see the left image of Figure 11.6).

Rotating the UCS Plane about the X, Y, or Z Axis

Next you will move the origin back to the bottom of the box again and rotate the UCS about the X axis.

1. On the UCS toolbar, click on the Origin UCS tool. Select the corner of the box, as shown in the bottom of Figure 11.6.

2. Rotate the UCS about the X axis using the X Axis Rotate UCS tool on the UCS toolbar. At the `Rotation angle about X axis <0>:` prompt, enter **90**↵. Your drawing should look like Figure 11.10. The UCS is now rotated 90° and appears aligned with the southwest plane of the box.

FIGURE 11.10

The UCS rotated about the X axis

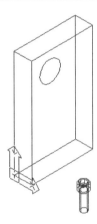

Similarly, the Y Axis Rotate option also allows you to rotate the UCS about the current Y axis.

Defining a UCS Origin and Plane with the Z Axis

You can align a UCS perpendicular to two points and set the origin in one simple command. Use this option when you can select two points on the Z axis of the plane that you want to define.

1. From the UCS toolbar, select the Z Axis Vector UCS tool.

2. At the Origin point <0,0,0>: prompt, use the Endpoint Osnap marker to find the upper-left front corner of the box, and pick this point.
3. At the next prompt, Point on positive portion of Z-axis <0'-0",0'- 0", 0'-1">:, use the Endpoint Osnap to pick the upper-right corner of the front of the box, as shown in Figure 11.11. The UCS twists to reflect the new Z axis of the UCS. See Figure 11.12 for the final result.

FIGURE 11.11

Picking points for the Z Axis Vector option

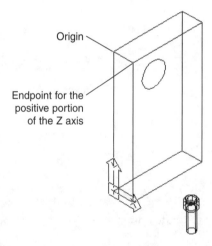

FIGURE 11.12

The new UCS after assigning the Z axis points

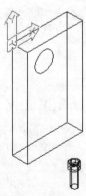

Orienting a UCS to the View Plane

You can define a UCS in the current view plane. This is useful if you want to quickly switch to the current view plane for editing or for adding text to a 3D view. To try this, let's create text in the current UCS, switch to the WCS, and then change the UCS to be parallel to the screen.

1. Click on the Text tool from the Draw toolbar.
2. At the Justify/Style/<Start point>: prompt, use the current origin by typing **0,0**↵.
3. At the Height <0.2000>: prompt, type **.2**↵, and at the Rotation angle <0>: prompt, press ↵ to accept the default.
4. At the Text: prompt type **MASTERING AUTOCAD**↵. It is not necessary to use capital letters. For this exercise, capital letters are a little bit easier to see. The left image of Figure 11.13 demonstrates how the text should look.

FIGURE 11.13

Text written on three different UCS planes

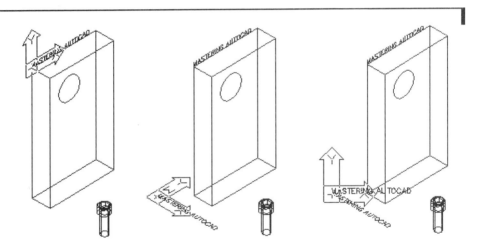

5. Use the World UCS tool from the UCS toolbar to return the UCS to the original position.
6. Use the Text tool again to write **MASTERING AUTOCAD** at the origin of the WCS. The middle pane of Figure 11.13 shows the text at the WCS center.

7. Next, change the UCS to be parallel to the screen. Click on the View UCS tool on the UCS toolbar, or choose View ➣ Set UCS ➣ View. You can also type **ucs↵v↵**. The UCS icon changes to show that the UCS is aligned with the current view.

8. Use the Text tool again to write **MASTERING AUTOCAD**. You can also cut and paste the words from one of the previous lines. Your work will look like the right pane of Figure 11.13.

AutoCAD uses the current UCS origin point for the origin of the new UCS. By defining a UCS as parallel to your monitor's screen (the view), you can enter text to label your drawing, as you would in a technical illustration. Text entered in a plane created in this way will appear normal (that is, it will be legible and organized relative to the plane on which it is typed).

Saving a Named UCS

The UCS Save option is designed for speed. You could type **ucs↵r↵i↵** very quickly if you wanted to restore a saved UCS named *i* which could stand for "a UCS parallel to the screen when the object is viewed from the southeast corner." But first, you must save a UCS named *i*.

1. Click on the UCS tool from the UCS toolbar. The prompt is the same as if you had typed **ucs↵**. At the `Origin/ZAxis/3point/Object/View/X/Y/Z/Prev/Restore/Save/Del/?/<World>:` prompt, type **s↵** to use the UCS Save option.

2. If you had selected UCS Save from the Tools menu, you would also see the prompt `?/Desired UCS name`. The ? allows you to list the current saved UCS, if any. At the prompt, type **i↵**. You have just saved your first UCS.

Restoring and Deleting a Saved UCS

You could use the UCS toolbar or the menu bar to issue the Restore command, but these are very slow compared to using the command line. If you find any of the commands hard to remember, the toolbars and pull-down menus are there to help (AutoCAD's Help system can also be very useful).

In the following exercise you will change the UCS back to the WCS again, recall the *i* UCS and, because you need to always be in control of the program, you will delete the *i* UCS.

1. Use the World UCS tool from the UCS toolbar to return to the WCS. You will see the icon return to the origin and orientation of the World UCS.
2. Type **ucs**↵, and at the Origin/ZAxis/3point/OBject/View/X/Y/Z/Prev/Restore/Save/Del/?/<World>: prompt, type **r**↵ to use the Restore option. At the ?/Name of UCS to restore: prompt, type **i**↵. You can see that the UCS is restored to normal on the screen. You could also have used the Named UCS tool from the UCS toolbar or chosen Tools ➤ UCS ➤ Named from the menu bar. Either of these options brings up the UCS Control dialog box, so you have complete control over naming, saving, and restoring (making current) a UCS.

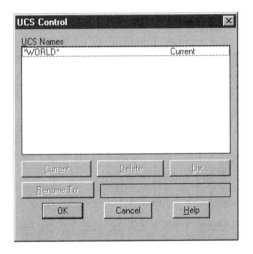

The UCS Presets Menu

Another way to define a UCS is with the help of the UCS Orientation dialog box. This tool gives you names and visual references for a set of orthogonal planes that you can select when you need them, or you can use the UCS Orientation dialog box to save a set of planes to use with the UCS Restore command. Let's try it.

1. Select the World UCS option from the UCS toolbar. This is a good starting point for the new or occasional user to define a new UCS.

NOTE The preset uses the current UCS plane to generate the new UCS. This can be quite disconcerting. You expect the preset to look like the one in your drawing, but it doesn't—it looks as if it were viewed from the previous UCS perspective.

2. Click on the Preset UCS tool on the UCS toolbar or, choose Tools ➢ UCS ➢ Preset UCS. The UCS Orientation dialog box appears.
3. Double-click in the box labeled Right. The dialog box closes and the UCS icon is set parallel to the southeast face of the block, as in Figure 11.14.

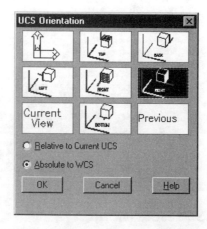

FIGURE 11.14
The UCS icon after the Right preset was selected

4. Choose Tools ➤ UCS ➤ Save. You will want to use this UCS to draw other geometry.
5. At the ?/Desired UCS name: prompt, type **s**↵ to save a UCS named *s*. The *s* stands for *side*—it's faster to type and less prone to error.

You could also try clicking on the UCS tool from the UCS toolbar or typing **ucs**↵. At the Origin/ZAxis/3point/OBject/View/X/Y/Z/Prev/Restore/Save/Del/?/<World>: prompt, type **s**↵. Repeat Step 4 and type **s**↵ at the prompt ?/Desired UCS name.

Viewing Your Model

Earlier in this chapter, you used the View toolbar to select the SE Isometric view. It's time to take a closer look at the View toolbar and the options it provides. You are about to take a virtual walk around your model.

As you go through the tutorial, be sure to notice the UCS icon. The UCS icon will remain attached and pointing in the direction of the positive X and Y axes. The only thing that changes is your position as you view the model.

You will not see the progress of the change in viewport, only the end result as Auto-CAD regenerates the screen. There is a practical reason for this. The current graphics are not fast enough to show it "live"—the effect would be to slow you down.

The Viewpoint command allows you to view your model from any point in 3D space. You may select a point that is relative to the World UCS.

Viewing a 3D Drawing

Your first 3D view of a model is a wire-frame view. This means that your model appears to be made of wire and none of the sides appears solid. This section describes how to manipulate this wire-frame view so you can see your drawing from any angle. You'll also learn how, once you have selected your view, you can view the 3D drawing as a solid object with the hidden lines removed. And you'll learn methods for saving views.

Finding Isometric and Orthogonal Views

First, let's start by looking at some of the viewing options available. You used one option already to get the current 3D view. That option, View ➤ 3D Viewpoint ➤ SE Isometric, brings up an Isometric view from a southeast direction, where north is the same direction as the Y axis. Figure 11.15 illustrates the three other Isometric View options: SE Isometric, NE Isometric, and NW Isometric. In Figure 11.15, the cameras represent the different viewpoint locations. You can get an idea of their location by looking at the grid and UCS icon shown in the figure.

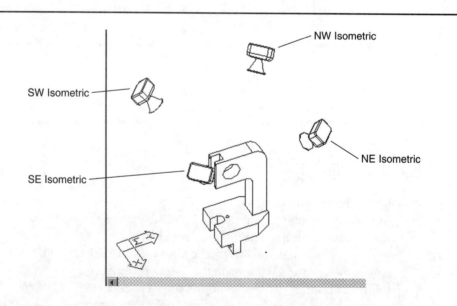

FIGURE 11.15
This diagram shows the viewpoints for the four Isometric views available from the View pull-down menu.

Another set of options available on the View ➤ 3D Viewpoint cascading menu contains Top, Bottom, Left, Right, Front, and Back. These are Orthogonal views that show the top, bottom, and sides of the model, as shown in Figure 11.16. In this figure, the cameras once again show the points of view.

FIGURE 11.16

The six viewpoints of the Orthogonal view options on the View ➢ 3D Viewpoint cascading menu

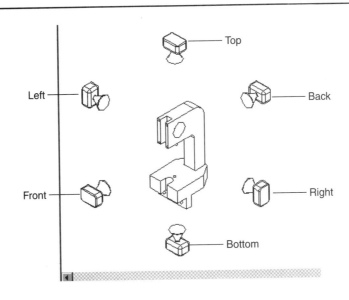

To give you a better idea of what an Orthogonal view looks like, Figure 11.17 shows the view that you see when you choose View ➢ 3D Viewpoint ➢ Right. This is a side view of the unit.

FIGURE 11.17

The view of the model you see when you choose View ➢ 3D Viewpoint ➢ Right

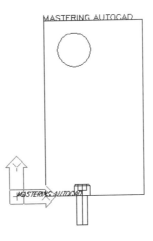

From time to time you will see the UCS icon change to a broken pencil. This will only happen if the viewpoint is orthogonal to the UCS but not aligned to the current X-Y plane. The icon will look like the one in Figure 11.18.

There are two remedies for this problem. Unless you intended the UCS to be in another plane, you can use the UCS View option from the UCS toolbar to align a UCS to your current viewpoint. Just as with Isometric view, you can set the UCS to be parallel to the current screen viewpoint. The second solution requires that you have saved and can restore a UCS parallel the current view. You could use the UCS command at the command line, choose the UCS option from the Tools menu, or click on the Named UCS tool from the UCS toolbar.

FIGURE 11.18

The UCS icon showing a broken pencil

When you use any of the View options described here, AutoCAD will display the extents of the drawing. You can then use the Pan and Zoom tools to adjust your view.

TIP

If you think you will change view options frequently, you might want to open the Viewpoint toolbar.

The Viewpoint toolbar offers quick, single-click access to all the options discussed in this section. To open it, right-click on any toolbar, and then click on Viewpoint in the Toolbars dialog box.

Using a Dialog Box to Select 3D Views

You now know that you can select from a variety of "canned" viewpoints to view your 3D model. You can also fine-tune your view by indicating an angle from the drawing's X axis and from the floor plane using the Viewpoint Presets dialog box. Let's try it.

1. Choose View ➤ 3D Viewpoint ➤ Select, or type **vp**↵ in the command line. The Viewpoint Presets dialog box appears (see Figure 11.19). The square dial to the left lets you select a viewpoint location in degrees relative to the X axis. The semicircle to the right lets you select an elevation for your viewpoint.

FIGURE 11.19

The Viewpoint Presets dialog box

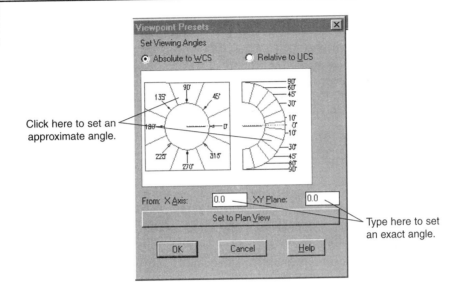

2. Click on the angle labeled 135° in the upper-left of the square dial. Then click on the area labeled 60° in the right-hand semicircle. Notice that the pointer moves to the angle you've selected and the input boxes below the graphic change to reflect the new settings.

3. Click on OK. Your view changes according to the new settings you just made.

Other settings in the Viewpoint Presets dialog box let you determine whether the selected view angles are relative to the WCS or to the current UCS. You can also go directly to a Plan view by clicking on the Set to Plan View button.

There are a couple of features of the Viewpoint Presets dialog box that are not readily apparent. First of all, you can select the exact angle indicated by the label of either graphic by clicking anywhere inside the outlined regions around the pointer (see Figure 11.19). For example, in the graphic to the left, click anywhere in the region labeled 90° to set the pointer to 90° exactly.

You can set the pointers to smaller degree increments by clicking within the pointer area. For example, if you click in the area just below the 90° region in the left graphic, the pointer will move to that location. The angle will be slightly greater than 90°.

If you want to enter an exact value from the X axis or X-Y plane, you can do so by entering an angle value in the input boxes provided. You can obtain virtually any view you want using the options offered in this dialog box.

NOTE Three other options—Rotate, Tripod, and Vector—are also available from the View pull-down menu. These options are somewhat difficult to use and duplicate the functions described here. If you'd like to learn more about these options, consult the AutoCAD Help system and look up the Vpoint command.

More Sculpturing of 3D Solids

The socket-head screw model required the use of two interesting commands to construct a solid. You used the Subtract Boolean operation to remove the hexagon shaped drive socket from the head of the screw, and you used the Union Boolean operation to add the cylinder for the threads. This is the normal method used in AutoCAD to construct solids. This process would be similar to one you would use if you were sculpting in clay. You would gather a lump of clay that is about the right size, and you might find that you need to add some more here or to cut away some over there. You might need to fashion a special tool to remove material or carefully create a piece to be added.

In the following exercise, you will continue to work with the Subtract and Union commands. You'll create the base and post of the press by the subtraction and union of material. You don't need the text any more, so erase it by clicking on the Erase tool from the Modify toolbar and selecting each line of text. Right-click to complete the command.

1. You should still have the UCS set to the right side of the block. If not, type **ucs↵r↵s↵** to restore the saved UCS called *s*, click on the Named UCS tool to bring up the UCS Control dialog box and restore the *s* UCS, or choose Tools ➢ UCS ➢ Restore and at the ?/Name of UCS to restore: prompt, type **s↵**. The UCS menu also offers the Named UCS option to open the dialog box.

2. Be sure that the UCS origin is located at the southwest corner of the right side of the block. Your drawing should look like Figure 11.20.

3. Draw a new box from the corner of the existing box. You will use the new box to remove material from the existing box. Type **box↵** or click on the Box tool from the Solids toolbar. At the Center/<Corner of box> <0,0,0>: prompt press ↵ to accept the default. At the Cube/Length/<other corner>: prompt, type **L↵** to enter the lengths of the sides.

FIGURE 11.20

The box and screw ready to use

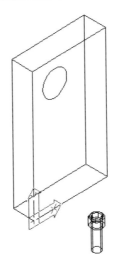

4. At the Length: prompt, enter **2.5**↵ to enter the X-axis value. At the Width: prompt, enter **4**↵ to enter the Y-axis value, and at the Height: prompt, enter **–2**↵ to enter the Z-axis value.

   - The negative value is used to make a box that has thickness in the –Z direction. These prompt names of Length, Height, and Width are a holdover from the original solid modeling add-on.

   - The thickness of the new box is not the same as the large box. Since the space occupied by the small box is going to be subtracted from the large box, you need only insure that the small box is large enough to remove all of the material that you want to remove.

WARNING

If surfaces are drawn collinear, AutoCAD can actually generate an error when processing Boolean operations.

5. To remove the small box from the larger and create the L shape of the post, type **su**↵. At the Select solids and regions to subtract from: prompt, select the large box. Press ↵ to complete the selection. This should be the box area that you want to keep. At the Select solids and regions to subtract: prompt, select the small box and press ↵ to complete the selection. You should have a drawing that looks like Figure 11.21.

FIGURE 11.21
The box after the subtraction

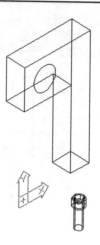

Using the Fillet Tool to Edit a Solid Model

You used the Chamfer tool earlier to modify the screw. The press' post needs an inside and an outside fillet to reduce stress. You can use the same Fillet tool that you first learned about in Chapter 3, with a small change in prompts, because you have selected a solid to work on instead of line or arc objects.

1. Type **f**↵, click on the Fillet tool from the Modify toolbar, or choose Modify ➢ Fillet from the menu bar.

2. At the `(TRIM mode) Current fillet radius = 0.3800/Polyline/Radius/Trim/<Select first object>:` prompt, pick the line that represents the corner between the top and back side. Your default may be different from the .03800 listed here.

3. This edge has a 1-unit radius. At the `Enter radius <0.3800>:` prompt, enter **1**↵. The solid object fillet prompt fillets edges and therefore looks a lot different from the prompts you are given to fillet a line or polyline.

4. At the `Chain/Radius/<Select edge>:` prompt, press ↵ to tell AutoCAD that this is the only edge that you want to fillet. This prompt is asking whether you want to change the radius default, define a chain of consecutive edges, or select more edges to fillet. Your drawing should look like the top of Figure 11.22.

MORE SCULPTURING OF 3D SOLIDS | 503

5. Fillet the inside corner the same way. Type **f**↵ for the Fillet command. Pick the inside corner, type **0.38**↵ for the radius, and press ↵ again to complete the command. Now your drawing should look like the bottom image of Figure 11.22.

FIGURE 11.22

The post before and after filleting

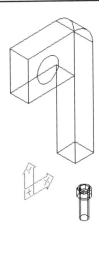

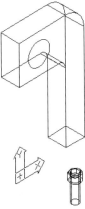

Next, let's see how fast we can complete the post. We are going to use very abbreviated commands, much the way that you will actually use AutoCAD to build and edit models. If you don't understand a command or sequence, be sure to pay attention to the prompts on the AutoCAD command line as you execute them.

1. Type **ucs**↵↵ to return to the World UCS.
2. Type **ucs**↵**x**↵**90**↵ to rotate the UCS plane 90° about the X axis.
3. Type **ucs**↵**s**↵**f**↵ to save a UCS named *f* for *front*.

4. See Panel 1 of Figure 11.23 to locate the lower-left corner of the front of the post. Type **box↵from↵**, then use the Endpoint Osnap to pick the lower-left corner of the front of the post. Type **@.24<0↵** for the offset. Type **@.5,3↵** for the other corner, and then type **–.625↵** for the height.

5. Type **su↵** to subtract the new solid from the post. Pick the post, press ↵, pick the new box, and then press ↵. To see what this should look like, see Panel 1 in Figure 11.23.

6. Type **cylinder↵** to draw the tapped holes in the front of the post. Use the From Osnap and pick the lower-left corner of the front. Type **@.125,.125↵**, and then type **d↵.138↵** to give the diameter value. Type **–.375↵** to give the depth of the threads.

If you need to show the tap drill depth, you could draw the tap drill diameter for the thread. You could even draw a cone to represent the point of the drill and subtract this from your solid. If you need to show the actual thread depth, you could draw another cylinder the diameter and length of the threads.

7. Type **ar↵** to use the Array command. There are four threaded holes in the front of the post spaced 0.75 in X and 1.75 in Y. Type **L↵↵** to use the last object. Type **r↵** for a rectangular array, **2↵** for the rows, and **2↵** for the columns. Type **1.75↵** for the distance between the rows and **.75↵** for the distance between the columns.

8. Type **su↵** to subtract the array from the post. Pick the post, press ↵, and then use a window to select the four cylinders. Press ↵ to complete the selection set. Your drawing should look like Panel 2 in Figure 11.23.

9. Type **ucs↵r↵s↵** to restore the UCS named s for *side*.

10. Type **ext↵** to begin the extrude command and pick the 1.17 diameter circle that you drew earlier. Type **–.778↵** for the depth of the extrusion and press ↵ for no taper angle.

11. Type **cylinder↵** to draw another cylinder for the drill through. Use the Center Osnap to use the center of the original circle. Type **d↵.5↵** for the diameter and **–2↵** for the depth.

12. Type **su↵** to subtract the last two cylinders from the post. Pick the post and ↵, and then the two cylinders and ↵. Your drawing should look like Panel 3 in Figure 11.23.

13. Type **ucs↵r↵f↵** to restore the UCS named f for *front*.

14. Type **ucs.⏎or.⏎** to pick a new origin for the UCS. Use the Endpoint Osnap to pick the bottom back-left corner.
15. To draw the counterbored hole, type **cylinder.⏎**. Type **from.⏎** to use the From Osnap, and type **0,0.⏎**, then **@.5,.5.⏎** for the distance. Type **d.⏎.38.⏎** for the diameter, and **2.⏎** for the height. To draw the counterbore, type **cylinder.⏎** and select the center of the cylinder that you just drew. Type **d.⏎.562.⏎** to give the diameter value, and then type **.375.⏎** for the height.
16. Type **su.⏎** to subtract the last two cylinders from the post. Pick the post and ⏎ and then the two cylinders and ⏎. Your drawing should look like the Panel 4 in Figure 11.23.
17. Save your work.

FIGURE 11.23

Panel 1: The box subtracted from the post. Panel 2: The array of tapped holes subtracted from the post. Panel 3: The large counterbore subtracted from the side of the post. Panel 4: The small counterbore subtracted from the back of the post.

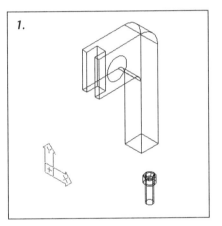
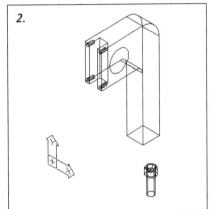
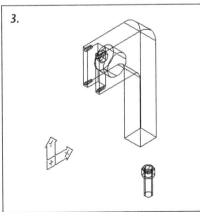
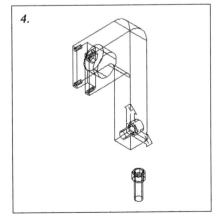

Making a Solid Model from a 2D Drawing

As you will see, making a solid model from a 2D drawing is not an automatic process. However, it is not a very difficult problem if you think your way through it. You learned about polylines in Chapter 10; now you can put that knowledge to work to create closed polylines from the 2D objects. Next, you will extrude the polylines into solid objects, align them, and use a new tool which will remove any material that is not common to both solids. A 2D drawing has been created of the base of the press, called Press Base on the companion CD-ROM.

1. Type **ucs.↵↵** to return to the World UCS.

2. Click on the Insert option of the Draw toolbar to open the Insert dialog box. Click on the File button to find and select the file named `Press Base.dwg` on the CD-ROM. A Warning appears at the bottom of the box: `Block name must be less than 32 characters, no spaces`. Click in the Block input box and rename the block **base**. The Options box contains the Specify Parameters on Screen check box, which is normally checked by default. Click on this check box to remove the check mark. The Explode check box below the Options box normally does *not* have a check mark. Click on the Explode check box to add a check mark. Click on OK. See Chapter 4 if you want to review the block insertion process.

3. Zoom All to see all of the new block and your solids. Your drawing should look like Figure 11.24.

FIGURE 11.24

The base block inserted

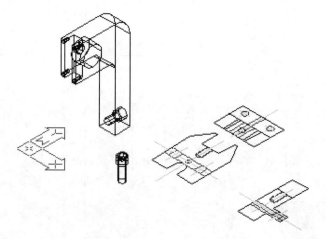

4. Type **pe**↵ to begin the Polyline Edit command. At the `PEDIT Select polyline:` prompt, pick any outside object line in the plan view of the base. At the prompt `Object selected is not a polyline Do you want to turn it into one? <Y>`, press ↵ to accept the default and turn your selection into a polyline.

5. Join the other object lines to the first to create a closed polyline. At the `Close/Join/Width/Edit vertex/Fit/Spline/Decurve/Ltype gen/Undo/eXit <X>:` prompt, enter **j**↵ to use the Join option.

6. At the `Select objects:` prompt, use a window to select all of the objects in the Plan view. You do not need to be careful when you make this window because AutoCAD is only looking for lines that join the first line that you selected at the endpoint. Any others will be ignored, and AutoCAD will continue to test each line that it adds to the polyline to see if another line in your selection set has the same endpoint. The program will continue this way until all objects have been tested. If a closed polyline is formed, the first option in the prompt line changes from `Close` to `Open`, to see if you now want to open a closed polyline.

7. You should see the statement `13 segments added to polyline` and the prompt `Open/Join/Width/Edit vertex/Fit/Spline/Decurve/Ltype gen/Undo/eXit <X>`. Press ↵ to exit the command.

8. To do the same thing to the side view, press ↵ again to reenter the Pedit command and at the `Select polyline:` prompt, pick any outside object line in the side view. Repeat steps 5–7 to complete the process. You will see the polyline close with seven segments added.

9. Save your work.

Next, you will extrude to create solids and align the solids to take the interference.

1. Type **ext**↵ to use the Extrude command. You are going to extrude both views and the hole that you see as a circle in the plan view. At the `Select objects:` prompt, pick the two polylines and window the circle.

2. At the `Path/<Height of Extrusion>:` prompt, enter **4**↵. Once again, there is no need for precision. The extrusion height only needs to be greater than the width and thickness of the actual part. You drawing should look like Figure 11.25.

FIGURE 11.25

The two views of the base after extrusion

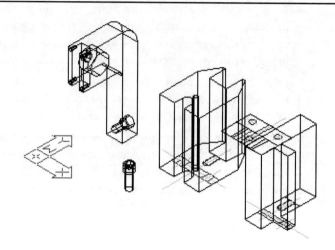

Aligning Two 3D Objects

The Align tool allows you to define three points on two objects. The points will be used to define one plane on each object and the orientation of the objects to each other. In our example, the Plan view extrusion will remain where it is and the Side view extrusion will be aligned to the Plan view. See the top panel of Figure 11.26 to preview the points that you will be selecting.

You could also use a series of discrete Move and Rotate commands to align the two extrusions.

The purpose of this alignment will become clear in just a few more steps.

Select Modify ➢ 3D Operations ➢ Align or type **align**↵. At the Select objects: prompt, pick the side view extrusion. The prompt is asking for the object that you want to move.

1. Look at the top image of Figure 11.26 to find the points in the correct order to align the two views. To select the objects, at the Specify 1st source point: prompt, pick the corresponding point; at the Specify 1st destination point: prompt, pick the corresponding point; and so on, through all three sets of points. You can use the Undo command or click on the Undo tool on the Standard toolbar if you pick points out of order, or if the result is not as shown in the bottom image of Figure 11.26.

FIGURE 11.26

Aligning the two extrusions

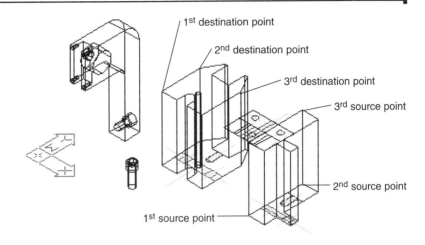

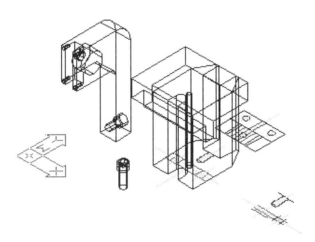

This would be a good time to look at your work from a few different views. Use the Front View option to look at your models from the front orthogonal view. Your drawing will look like the top image of Figure 11.27.

1. Click on the Right View tool on the View toolbar to look at the right side of your models. Your drawing should look like the bottom image of Figure 11.27.
2. Click on SE Isometric View to return to your previous viewpoint.

FIGURE 11.27
The front and side views of the models

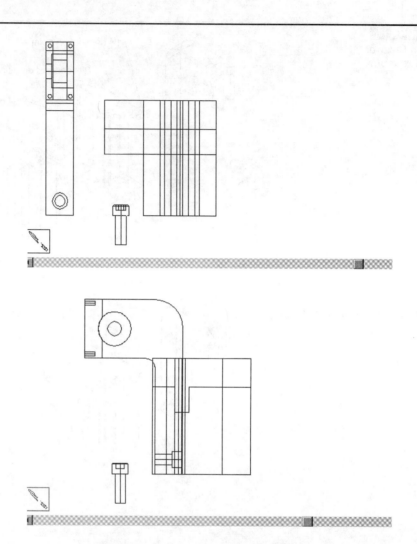

3. Click on the Intersect tool on the Modify II toolbar, choose Modify ➢ Boolean ➢ Intersect from the menu bar, or type **intersect**↵ in the command line.

MAKING A SOLID MODEL FROM A 2D DRAWING | **511**

4. At the Select objects: prompt, pick both extrusions and press ↵ to complete the selection set. Your drawing should look like Figure 11.28. This Boolean operation evaluated the two models for all common volumes and discarded any portion of either model not found to be common to both.

FIGURE 11.28

The two extrusions after the Intersection Boolean operation

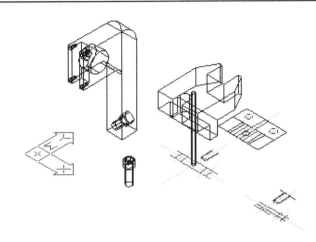

To complete the base, you still need to add the mounting holes to the model. If drawings are created to scale, you can use the objects to make solids. The cylinder that you need is the extruded circle that is a hole in the 2D model. First you will subtract the cylinder from the solid. Three holes are shown in the remaining view and you will use them to finish editing your model. As you will see, you only need the objects in the remaining view to finish editing the solid, so the second operation will be to change the viewpoint and make it easier to erase these objects.

1. Use the Subtract command to make a hole by subtracting the cylinder from the base.
2. Click on Top View on the View toolbar to change your viewpoint to the Top view.
3. Type **e**↵**w**↵ or use your favorite method to issue the Erase command and the Window option to select objects. Figure 11.29 illustrates what to window in order to erase all unnecessary objects.
4. Click on the SE Isometric View on the View toolbar to change your viewpoint to the previous view.

5. The remaining 2D view would be much easier to use if it were parallel to the correct plane of the solid. Type **rotate3d↵**, or select Modify ➢ 3D Operation ➢ Rotate 3D, to rotate the view without changing the UCS.

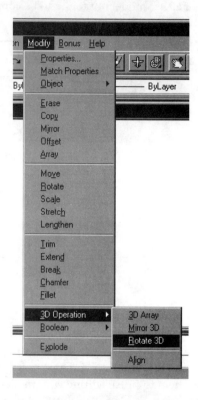

6. At the Select objects: prompt, type **w↵** to use a window to capture all of the objects in the remaining view. Remember, if you accidentally get any objects that you don't want, type **r↵** for Remove and select the objects that you want to remove from the selection set. When you are satisfied with your choices, press ↵ to go to the next step.

7. You are going to pick a line to be the axis and rotate a set of objects about the line. At the Axis by Object/Last/View/Xaxis/Yaxis/Zaxis/<2points>: prompt, type **o↵** to pick an object. At the Pick a line, circle, arc or 2D-polyline segment: prompt, pick the top line in the view. The line is shown in the top image of Figure 11.29.

FIGURE 11.29

Top: The line to select to be the Z-axis of rotation. Bottom: The rotated view.

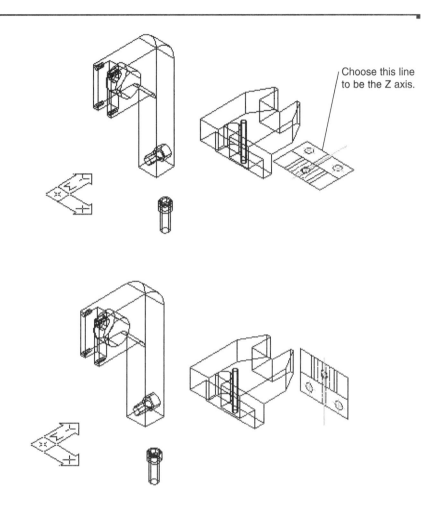

8. At the <Rotation angle>/Reference: prompt, type **–90**↵ to rotate the object clockwise 90°. See the bottom image of Figure 11.29.

> ### The Rotate 3D Options
>
> The following list shows several ways to rotate an object.
>
> **Axis by Object** allows you to align the axis of rotation with an existing object (line, circle, arc, or 2D polylines).
>
> **Last** uses the last axis of rotation.
>
> **View** aligns the axis of rotation with the viewing direction of the current viewport that passes through a selected point. At the `Point on view direction axis <0,0,0>:` prompt, select a point.
>
> **X/Y/Z Axis** aligns the axis of rotation on the axes (X, Y, Z) that pass through the selected point. At the `Point on (X, Y, Z) axis <0,0,0>:` prompt, select a point.
>
> **2 Points <the default>** allows you to specify two points to define the axis of rotation. You will be prompted for `1st point on axis` and `2nd point on axis`. Select two points.

Tidying Up

You must align the view with the solid before you move the view into place.

1. Type **m**↵ to use the Move command. Type **p**↵ to use the previous selection set and use the Endpoint Osnaps, as shown in the top panel of Figure 11.30, to move the view. The middle panel of Figure 11.30 shows the view after the move.
2. Be sure that Ortho is on and move the view by dragging it close to the final position, as shown in the bottom panel of Figure 11.30.
3. Save your work.

The ⅜ tapped hole is .625 deep, and the other two base mounting holes go all the way through. You still need to extrude the circles and use them to subtract the holes. We won't draw the tap drill and point, but you can try this if you like. The tap drill is nearly always drilled .06 to .10 deeper than the threads (if you can't drill through) and has a 120° conical point. You could extrude the tap drill circle, create a 120° cone at the end of the extrusion, and subtract both the cylinder and the cone from the base, then begin step 1.

FIGURE 11.30

Top panel: Osnap points used to move the view. Middle panel: The moved view. Bottom panel: The view moved again.

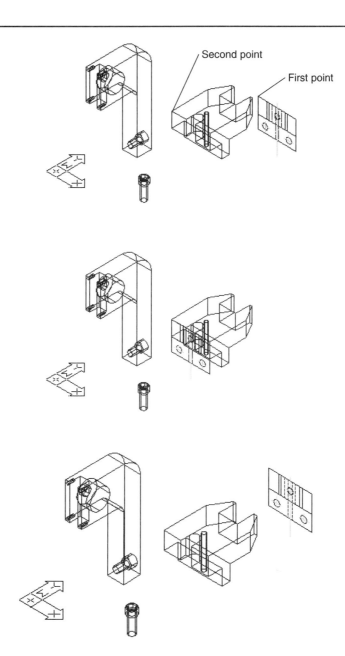

1. Type **ext**↵ and pick the .375 diameter circle in the view that represents the outside diameter of the threads. You will probably want to zoom in close enough to clearly see these objects. You can use the Transparent Zoom by typing **'z**↵. Window is the default, so just drag a small window around the view and pick the circle. (The circle is drawn with a hidden line.) Type **–.625**↵ at the Height of Extrusion: prompt. There is no taper, so press ↵ again.

   - The cylinder was created on the current layer. You can change it to the 0 layer if you like.

2. Press ↵ again to reenter the Extrude command. This time you want to select the two mounting holes at the bottom of the view. At the Height of Extrusion: prompt, type **–8**↵↵ to create two cylinders –8 units long without taper. Your drawing should look like Figure 11.31.

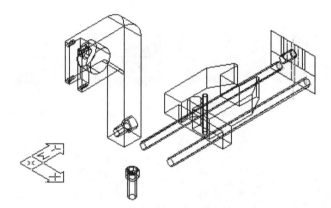

FIGURE 11.31

The screw with extruded thread hole and mounting holes

3. You made the mounting hole extrusions long enough to simply subtract them from the solid. Type **su**↵ to start the Subtract command, select the base, and press ↵. At the Select object to subtract: prompt, select the cylinders and press ↵ to complete the selection and the command. Your drawing will look like Figure 11.32.

FIGURE 11.32

The mounting holes subtracted from the base

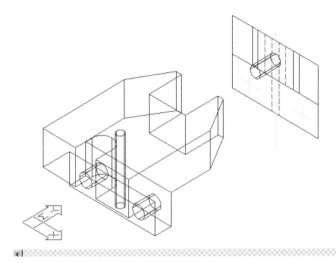

4. You need to move the cylinder that represents the tapped hole into place in the cutout on the base. Use two Osnap points to move the cylinder the exact distance. See Figure 11.33 for these points. Type **m**↵, select the cylinder, press ↵, and use your Endpoint Osnap markers to find the points.

FIGURE 11.33

The points needed to move the cylinder

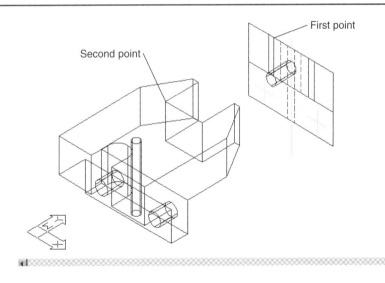

5. Type **su↵**, select the base and press ↵, and select the cylinder and press ↵. You have issued the Subtract command, selected an object to subtract from, and then selected the object to be subtracted. Your drawing should look like Figure 11.34.

FIGURE 11.34

The threaded hole subtracted from the base

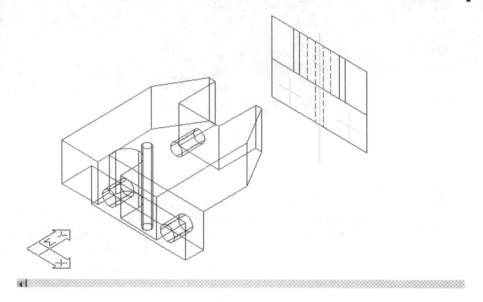

6. You no longer need the lines of the view, so you can erase them. Draw a window around the remaining view to turn on the grips, and type **e↵**.

7. Save your work.

You can begin to see what the assembly will look like. The base and post are oriented correctly, but the screw will have to be rotated and moved into position. The easiest way to align the base and post is to find a common point, like the hole for the screw in the post and base. Use Osnaps to move the base to the post.

1. Type **m↵** to start the Move command, select the base, and press ↵. At the `base point` prompt, select the center of the threaded hole, and at the `Base point or Displacement:` prompt, select the mating hole in the post. See the top of Figure 11.35 for help in finding these points. See the bottom image of Figure 11.35 for the results of the move.

FIGURE 11.35

The points required to move the base, and the base after it has been moved

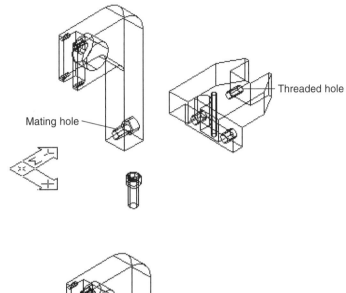
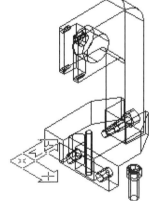

2. Use the 3DRotate command to rotate the screw. Type **rotate3d**↵, select the screw, and press ↵. At the Axis by Object/Last/View/Xaxis/Yaxis/Zaxis/<2points>: prompt, pick the bottom of the threads using the Center Osnap, again at the 2nd point on axis: prompt, and with Ortho on, drag the mouse to the right and pick any point. You have now defined an axis through the end of the screw.

3. At the <Rotation angle>/Reference: prompt, type **–90**↵. The screw should look like Figure 11.36.

FIGURE 11.36

The rotated screw

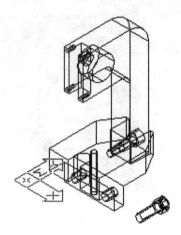

4. Pick the screw to highlight the grips. Zoom in so that you can easily see, and pick the grip that is at the center of the base of the head of screw—this grip is at the center of the cylinder shape. Use the Transparent Zoom again to Zoom Previous, and zoom into the counterbore on the post. Drag the cursor over the counterbore until you find the center of its the depth; pick this point. Your model is assembled.

You could take a look at the results of your work with the Hide tool. Let's do that now, and then use the Shade tool to do the same thing. The Shade tool is not often used, but you should know it is there if you need it.

1. Type **hi**↵, choose View ➣ Hide from the menu bar, or click on the Hide button on the Render toolbar.

2. Type **shade**↵, choose View ➣ Shade, or click on the Shade button on the Render toolbar.

FIGURE 11.37

The assembly using Hide in the top panel and Shade in the bottom panel

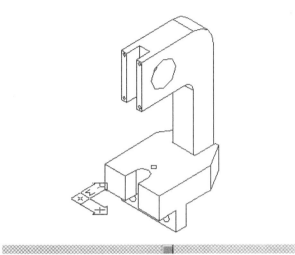

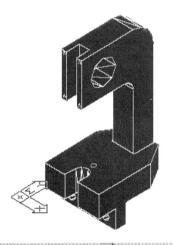

If You Want to Experiment...

- Dig out some of your old drafting books and try using Boolean addition and subtraction to construct some interesting parts.
- Use the View toolbar to walk around your model. View it from each of the orthogonal viewpoints. Then view it from each of the isometric angles.
- At each isometric viewpoint, use both Hide and Shade to see your model more clearly.

Chapter 12

12

Mastering 3D Solids

FEATURING

The Union and Subtract Commands for Complex Models

Primitive Solids: Box, Cylinder, Torus, Sphere, Cone, and Wedge

Creating a Solid with Revolve

Creating and Viewing Slides

Using Slice to Edit a Solid and Section to Copy a Cross Section

Regions and Boolean Operations

Using the Clipboard with Solid Models

3D to 2D Solid Model Conversion Tools: Solview, Soldraw, and Solprof

Mastering 3D Solids

You should now have a basic understanding of the power of 3D solids. The parts that you constructed in Chapter 11 can be used to create an arbor press; in this chapter you will generate more parts for the press. Chapter 11 included the box and the cylinder primitives, and this chapter will illustrate the rest of the primitive solids. In Chapter 11, you learned about Shade and Hide as techniques for making your two-dimensional drawing appear to be three-dimensional. Next, you'll learn some more subtle 3D effects.

First you'll put two solids together. Then you'll try revolving and sweeping closed polylines to create solids. You'll learn to make two solids from one using the Slice command, you'll see how to find out if two solids are occupying the same space by doing some interference checking, and you'll try rendering to see the work that you have done. Along the way, you will learn about making slides, saving and viewing slides, and how to make and use other important snapshots of your work. You'll make a useful 2D drawing from a 3D solid image, and you'll learn how to manage these drawings. Finally, we'll look at exporting these files to in-house and out-of-house fabrication shops.

Putting Two Solids Together

You've already used the Union command to add material as you built a model. Now you'll use the Union command to edit two existing models. The base and the post from Chapter 11 need to be joined, because the stress on the bolt would be enormous if the press were used for heavy duty work. The hole for the bolt needs to be removed, the clearance between the base and the post needs to be filled, and the screw should be moved away from the assembly.

1. Open the file named press.dwg that you saved in Chapter 11. If you don't have this file or if you aren't sure that it is correct, you can start with the file named ch12press.dwg from the companion CD-ROM.

2. Type **m**↵ to begin the Move command. At the Select objects: prompt, pick the screw. You can use selection cycling, or zoom in, to make it easier to pick the screw.

3. At the Select objects: prompt, press ↵ to complete the selection. Then at the Base point or displacement: Second point of displacement: prompt, type **4<90**↵↵ to move the bolt 4 units on the Y axis. You have moved the bolt away from the base.

4. You will join the post and base next. Type **uni**↵ to begin the Union command. At the Select objects: prompt, pick the post. At the new Select objects: prompt, pick the base, and at the next Select objects: prompt, press ↵ to complete the selection set. The two parts are now one. Notice that there are some gaps on either side of the post, and that the holes for the screw are still there. Your drawing should look like Figure 12.1.

FIGURE 12.1

The union of the post and base

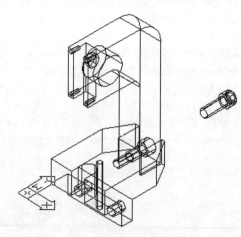

5. Use the Zoom Window tool to get a better look at the area between the base and the post.
6. The spaces to either side of the post need to be filled in. You can use the Box tool to create a solid that will fill in the space, and then union it to the new base/post. Type **box**.↵ and at the Center/<Corner of box> <0,0,0>: prompt, use the Endpoint Osnap to find any one of the corners of the cutout in the base. At the Cube/Length/<other corner>: prompt, pick the diagonal corner. See Figure 12.2 for these points. You can barely see any change in your drawing, because you just drew a box the exact same size as an existing box in your drawing.
7. Type **uni**.↵ to begin the Union command. At the Select objects: prompt, pick the post/base, and at the next Select objects: prompt, type **L**.↵ to use the last object created. At the next Select objects: prompt, press ↵ to complete the selection set.

FIGURE 12.2

Picking the point for the box on the base/post, and the base/post after the union

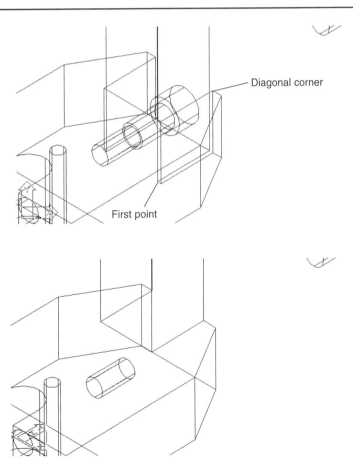

Your drawing should look like the bottom of Figure 12.2. The spaces, counterbore, and the through hole that were in the post are gone. The cylinder that was the tapped hole in the base still remains, and you will need to fill it. AutoCAD does not check to see if it is possible to make a part, so you must be aware of the feasibility of what you draw—the hole you are looking at is actually a void in the middle of the part. You must fill the hole so that AutoCAD can return the accurate properties of the part, and so that the 2D drawing won't include hidden lines that represent the void.

1. Type **cylinder**↵, and at the `Elliptical/<center point> <0,0,0>:` prompt, use the Osnap marker to find and pick the center of either end of the cylinder that you want to fill. At the `Diameter/<Radius>:` prompt, use the Quadrant Osnap to pick any of the four quadrants of the existing cylinder. If you remember the size of the existing cylinder, you can type in a diameter the same as, or slightly larger than, the diameter of the existing cylinder. The size of the cylinder must be larger but does not have to be exact to fill in the void.

2. At the `Center of the Other end/<Height>:` prompt, type **c**↵ to specify the location of the other end, rather than the height of the cylinder. You could give the location by coordinates, or use the geometry on the screen. At the `Center of other end:` prompt, use the Center Osnap to locate and pick the other end of the existing cylinder.

3. Union the base/post and the cylinder. Type **uni**↵ and at the `Select objects:` prompt, pick the base/post, and at the next `Select objects:` prompt, type **L**↵ to select the cylinder (which was the last object created). At the `Select objects:` prompt, press ↵ to complete the selection set and the command.

Using the Fillet Tool with Solids

A fillet would greatly improve the strength of the base/post.

1. Type **f**↵ and at the `(TRIM mode) Current fillet radius = 0.3800 Polyline/Radius/Trim/<Select first object>:` prompt, pick the line/edge between the front of the post and base.

2. At the `Enter radius <0.3800>:` prompt, type **.38**↵ and at the `Chain/Radius/<Select edge>:` prompt, press ↵ to complete the selections and complete the command. Your drawing will look like Figure 12.3.

3. Save your work. If you opened this drawing from the CD-ROM, save your work as **press.dwg**.

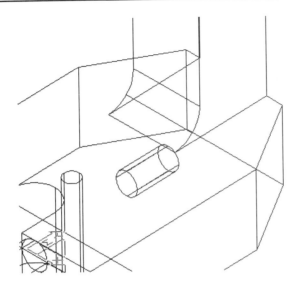

FIGURE 12.3

The fillet between the front of the post and the base

You can use the same Boolean operations to add and subtract complex models as well as primitives. Once the model is whole, the same editing commands that you used in Chapter 11 can be used on more complex modes.

Drawing a Few More Parts

Next, you'll be drawing a few new parts very quickly, using the techniques you learned in the last chapter. If you've forgotten how to use a command or tool, you can review the appropriate chapter, or use the AutoCAD Help feature by choosing the question mark icon after you have begun a command. We don't have space here to draw all of the parts of the press. The CD-ROM contains the missing parts' drawings that you will need to complete the press assembly. The next part to construct is called a table. This part will sit on the base and will have four cutouts. See Figure 12.4 for the table's dimensions.

FIGURE 12.4

The table with dimensions for the press

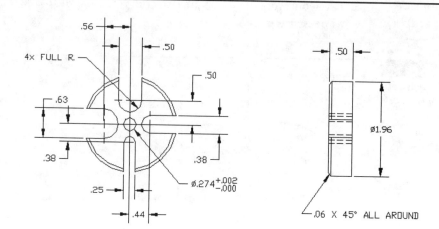

1. Start a new drawing using the Versions template. If AutoCAD is already started, type **new**↵. At the Save changes to press.dwg: prompt, type **yes**↵ if you have not saved your changes; type **no**↵ if you have already saved them.

2. Click on the Save tool in the Standard toolbar to give your new drawing a name. When the Select File dialog box opens, type **table** in the File name input box and click on the Open button. A new file named "Table" has been generated.

3. Change the viewpoint to the Isometric view so that you can more easily see and edit the table. Click on the SE Isometric tool on the View toolbar.

4. See Figure 12.4 for the dimensions of the table. Start by drawing a cylinder at coordinate 2.5,2.5. Make the cylinder 2 units in diameter and .5 units in height by typing **cylinder**↵, then **2.5,2**↵, then **d**↵, then **2**↵, and finally **.5**↵.

5. To create the hole at the center of the table, draw another cylinder .274 units in diameter and 1 unit high. Type **cylinder**↵ and use the Center Osnap to find the bottom center of the existing cylinder. Type **d**↵, then **.274**↵, then **1**↵.

6. To subtract the second cylinder from the first cylinder, click on the Subtract tool on the Modify II toolbar or type **su**↵. At the Select solids and regions to subtract from... Select objects: prompt, pick the first cylinder. At the

Select solids and regions to subtract...Select objects: prompt, pick the large cylinder and then type **L**↵ to select the last object created. Finally, hit ↵ to finish selecting objects and execute the command.

You have just created the basic shape of the circular table for the press. If you want to see how this part will be used in the assembly of the press you can look ahead in this chapter.

Using the Chamfer Tool on a Solid

Because this is a hand tool and sharp edges tend to cut hands, it's a good idea to add a chamfer around the top of the table.

1. Type **cha**↵, click on the Chamfer tool from the Modify toolbar, or choose Modify ➤ Chamfer to begin the Chamfer command.

2. You need to reenter the command, so press ↵ to reissue the last command, and at the Polyline/Distance/Angle/Trim/Method/<Select first line>: prompt, pick the circle that is the top of the table. The surface—not just the line—is highlighted.

3. At the Enter base surface distance <0.0600>: prompt, unless the prompt is <0.0600>, type **.06**↵ to chamfer the surface .06 units along the top surface.

4. At the Enter other surface distance <0.0600>: prompt, the default should be <0.0600>; if it is press ↵, if it is not type **.06**↵ to create a .06 × .06 chamfer.

5. At the Loop/<Select edge>: prompt, pick the top circle again to define the edge that you want to chamfer. At the Loop/<Select edge>: prompt, press ↵ to finish selecting edges and execute the command.

Now you need to make four cutouts. Use the Polyline tool to make the first shape.

1. To draw the small cutout, draw a closed polyline that is a little longer than it needs to be. Type **pl**↵, and at the From point: prompt, Shift+click, select the Quadrant Osnap, and pick the front quadrant of the either circle. See Figure 12.5 for this point. Next enter the distance from the quadrant to begin the polyline. At the from Base point: <Offset>: prompt, type **@-.125,-.375**↵.

2. Draw the polyline counterclockwise from this point. (However, you could draw it in either direction.) Type **@2<270↵** at the Arc/Close/Halfwidth/Length/Undo/Width/<Endpoint of line>: prompt.

3. At the Arc/Close/Halfwidth/Length/Undo/Width/<Endpoint of line>: prompt, type **@.25<0↵**. At the Arc/Close/Halfwidth/Length/Undo/Width/<Endpoint of line>: prompt, type **@2<90↵**.

4. To close the polyline and draw the arc, at the fourth Arc/Close/Halfwidth/Length/Undo/Width/<Endpoint of line>: prompt, type **a↵**. To draw an arc rather than another line, at the <Endpoint of arc>: prompt, use the Endpoint Osnap marker to find and pick the beginning of the polyline. Your drawing will look like Figure 12.5.

FIGURE 12.5

The table with the polyline drawn under it

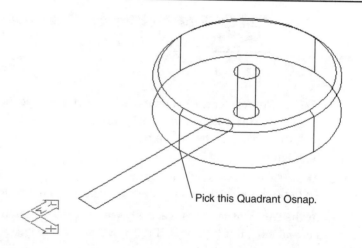

Next, make the polyline into a solid using the Extrude tool.

1. Type **ext↵**, and at the Select objects: prompt, pick the polyline.

2. To make the cut-out three dimensional, give a 1-unit height at the Path/<Height of Extrusion>: prompt by typing **1↵**, and at the Extrusion taper angle <0>: prompt, press ↵ to make the sides of the cut-out perpendicular to the top and bottom of the table.

Using the Polar Array Tool with Solid Objects

The Table drawing (Figure 12.4) shows four evenly spaced cut-outs around the edge of the table. You used a rectangular array in Chapter 5. Here is an opportunity to use the same technique to draw a polar array; this time you will use solid objects.

1. Type **ar**↵ and at the `Select objects:` prompt, pick the extruded polyline and press ↵.

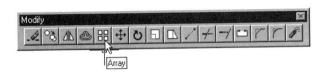

2. At the `Rectangular or Polar array (<R>/P):` prompt, type **p**↵ to begin a polar array, and at the `Base/<Specify center point of array>:` prompt, use the Center Osnap to find and pick the center of the table.

3. You will need four of the extruded polylines, so at the `Number of items:` prompt, type **4**↵.

4. At the `Angle to fill (+=ccw, -=cw) <360>:` prompt, your choice is to accept the default (360°) or enter an angle less than 360°. Choosing the default results in four extruded polylines equally spaced along the arc equal to the angle. Let's use the default: Press ↵.

5. At the `Rotate objects as they are copied? <Y>:` prompt, you can accept the default or enter **n**. Type **n**↵. Three copies of the original solid are created and rotated so that each is in a good orientation.

    - Entering No (n) results in four copies equally spaced on the circumference of the circle. AutoCAD measures the distance and angle from the center of the circle to the beginning point of the polyline. The distance and angle are used to create and locate three more solids. Each copy will be oriented the same as the original. The result of such an array of objects is not intuitive. You could try this to better understand it and then undo it to continue the tutorial. Press ↵ to accept the default.

6. Save your work.

7. Your drawing should look like Figure 12.6.

FIGURE 12.6
The Polar array of the extruded polyline

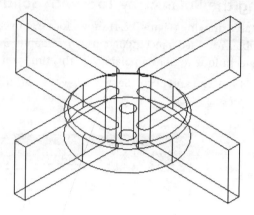

You now have four equally-sized solids which you'll use to cut out the four solids on the table. Next, you need to change the size of the solids to agree with the four different-sized solids shown in the dimensioned drawing (Figure 12.4). The original solid is the correct size, so only the other three need adjustment. The first solid located counterclockwise from the original should be .375 across, or 1.5 times larger, and .062 farther from the center of the table. The next solid should be two times larger and .125 farther from center. The third solid should be 2.5 times larger and .188 farther from center. Let's make the changes.

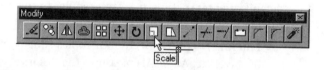

Using the Scale Tool to Work on Solids

Let's use the tools in AutoCAD that allow us to make proportional changes.

1. Type **sc**↵ to start the Scale command, click on the Scale tool from the Modify toolbar, or select Modify ➤ Scale from the menu bar.

2. At the `Select objects:` prompt, select the first solid going counterclockwise from the original. Press ↵.

3. At the `Base point:` prompt, use the Center Osnap to locate and pick the center of the arc of the polyline. The scale value will be applied from this point in all directions, which will leave the center located where it is now. If you were to

pick any other point, the solid would scale from that point and move the center. (You might try choosing different points to see what happens, and then undo it when you are finished.)

4. At the `<Scale factor>/Reference:` prompt, type **1.5**↵ to get a shape 1.5 times larger than the original.

5. Type **m**↵ to begin the Move command and at the `Select objects:` prompt, pick the larger solid and press ↵.

6. At the `Base point or displacement:` prompt, pick any point on the screen to choose a displacement. At the `Second point of displacement:` prompt, type **.0625<0**↵↵ to move the solid .0625 units to the right to agree with the dimensioned drawing.

Make the same changes to the next solid (the one pointing in the Y axis away from the original solid), except this solid will be twice the width and .125 farther from the center. Type **sc**↵, and select the third solid.

1. At the `Base point:` prompt, use the Center Osnap to find and pick the center of the arc on the solid. At the `<Scale factor>/Reference:` prompt, type **2**↵.

2. Type **m**↵ to move this solid into position.

3. At the `Select objects:` prompt, pick the new, larger solid and press ↵. At the `Base point or displacement:` prompt, pick any point on the screen, and at the `Second point of displacement:` prompt, type **.125<90**↵.

For the last solid, the one that is 2.5 times the size of the original and .188 farther from the center, follow these steps:

1. Type **sc**↵ to begin the Scale command. At the `Select objects:` prompt, pick the left-facing solid and press ↵.

2. At the `Base point:` prompt, pick the center of the arc of the solid. At the `<Scale factor>/Reference:` prompt, type **2.5**↵.

3. Type **m**↵ to start the Move command, and at the `Select objects:` prompt, pick the last solid that you scaled and press ↵.

4. At the `Base point or displacement:` prompt, pick any point on the screen. At the `Second point of displacement:` prompt, type **@.188<180**↵.

Now you need to subtract the solid shapes from the table.

1. Type **su**↵ to begin the Subtract command again. At the `Select solids and regions to subtract from:` prompt, pick the table cylinder. At the `Select solids and regions to subtract:` prompt, pick the four solids; press ↵ to finish. Your drawing should look like Figure 12.7.

2. Save your drawing. You can take a breather now if you like.

FIGURE 12.7
The completed table

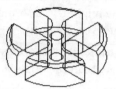

Using The Revolve Tool to Make New Parts

Start a new drawing using the Versions template and name it **revolutions** when you save it, just as you did with the Table drawing. You'll be drawing more of the press' parts: a sleeve, a button, an end cap, and a special screw. You will draw the sleeve using a rectangle (a closed polyline). The button, end cap, and special screw will be constructed of lines, arcs, and circles gathered into closed polylines and revolved to make them solid.

Let's start by making the rectangle, which you will revolve to become the sleeve.

1. Type **rec**↵ to begin the Rectangle command. You could also click on the Rectangle tool from the Draw toolbar or choose Draw ➢ Rectangle from the menu bar.

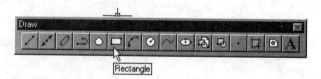

2. At the `Chamfer/Elevation/Fillet/Thickness/Width/<First corner>:` prompt, type coordinates **3,1**↵ to begin the rectangle in the lower-left corner of the drawing space. At the `Other corner:` prompt, type **@.208,1.68**↵ to draw the profile of the cross section of a cylinder that is 0.75 ID × 1.166 OD × 1.68 L. The arithmetic for the size of the rectangle is (OD–ID)/2.

3. Type **cha.↵** to chamfer the bottom right corner by 0.06 × 45 degrees. At the (TRIM mode) Current chamfer Dist1 = 0.5000, Dist2 = 0.5000 Polyline/Distance/ Angle/Trim/Method/<Select first line>: prompt, type **d.↵** because the distances are the same for 45°, and accepting the defaults in the distance version of the prompts will require less typing. At the Enter first chamfer distance <0.5000>: prompt, type **.06.↵**, and at the Enter second chamfer distance <0.0600>: prompt, press ↵ to accept the default. You have now set the distance that will be used at the next Chamfer command.

4. Press ↵ again to begin the Chamfer command again. After the CHAMFER (TRIM mode) Current chamfer Dist1 = 0.0600, Dist2 = 0.0600 Polyline/Distance/Angle/Trim/Method/<Select first line>: prompt, pick the bottom line and then at the Select second line: prompt, pick the right side line.

Next you will use the sphere primitive to construct the end cap.

Using the Revolve Tool to Make Complex Round and Cylindrical Solid Objects

The *Revolve* command is similar to the Extrude command, but instead of extruding the shape in a straight line, you define an axis which is the center of the revolution; Auto-CAD revolves the shape about the axis to form a solid. In this case, the axis is located .375 units from the left side of the polyline, which is the ID of the sleeve. The chamfered rectangle revolved around the axis defines a hollow cylinder.

1. Type **rev.↵**, click on the Revolve tool from the Solids toolbar, or choose Draw ➢ Solids ➢ Revolve to begin the Revolve command.

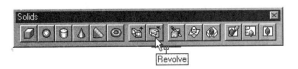

2. At the Select objects: prompt, pick the polyline and press ↵.

3. At the Axis of revolution-Object/X/Y/<Start point of axis>: prompt, use the From Osnap. At the Base point: prompt, use the Endpoint Osnap to pick the endpoint of the polyline at the lower-left corner. At the <Offset>: prompt, type **@.375<180.↵** to begin the axis definition .375 units to the left of the lower-left corner. At the <End point of axis>: prompt, drag the mouse up (insure that Ortho mode is on or type **@1<90.↵**) and pick a point. This will define an axis parallel to the rectangle and .375 units to the left.

4. At the `Angle of revolution <full circle>:` prompt, press ↵ to accept the default. You could draw an arc less than a full circle by entering an angle less than 360°.

5. Click on the SE Isometric option of the View toolbar or type **vpoint**↵, and at the `Rotate/<View point> <0.0000,0.0000,1.0000>:` prompt, type **1,–1,1**↵.

The sleeve has a radial hole through it that is 1.178 units from the far end. See Figure 12.8 to better understand this verbal picture.

1. Type **cylinder**↵ to begin the Cylinder command, and at the `Elliptical/<center point> <0,0,0>:` prompt, use the Quadrant Osnap by typing **qua**↵ or Shift+right-click to use the pop-up menu. At the `qua of:` prompt, pick the lower quadrant of the circle defining the end of the cylinder farthest from you (opposite the chamfered end).

If you aren't sure which end is which, use the Hide command (**hi**↵). Hide is not a transparent command, so you will have to escape the Cylinder command to use it.

2. At the `Diameter/<Radius>:` prompt, type **d**↵ and at the `Diameter:` prompt, type **3/8**↵ to set the interior dimensions of the hole.

3. At the `Center of other end/<Height>:` prompt, type an arbitrary number, such as **2**↵.

4. Type **m**↵ to move the cylinder (hole) 1.178 units from the end of the sleeve. At the `Select objects:` prompt, type **L**↵ to use the last object created. At the `Select objects:` prompt, press ↵ to complete the selection set. At the `Base point or displacement:` prompt, pick any point on the screen, and at the `Second point of displacement:` prompt, type **–1.178,0**↵.

5. Type **su**↵ to begin the Subtract command. At the `Select solids and regions to subtract from:` prompt, pick the sleeve. At the `Select objects:` prompt, press ↵ to complete the command.

6. At the `Select solids and regions to subtract:` prompt, pick the cylinder. At the `Select objects:` prompt, press ↵ to complete the selection set.

7. Change the Display Silhouette system variable to display the silhouette only. Type **dispsilh**↵, and at the `New value for DISPSILH <0>:` prompt, type **1**↵.

8. Issue the Hide command by typing **hi**↵ to see the finished sleeve. Your drawing should look like Figure 12.8.

9. Save your work.

PUTTING TWO SOLIDS TOGETHER

FIGURE 12.8
The completed sleeve model

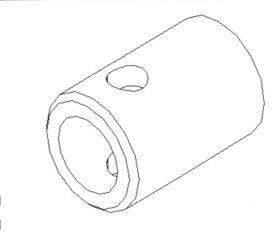

Using the Torus Tool

The torus (donut) shape may be unfamiliar to you. This shape can be very useful in the creation of complex geometry. In this exercise, you are going to draw a button with the inverse of a fillet cut out of it. The dimensioned drawing in Figure 12.9 describes the shape.

FIGURE 12.9
The dimensions for the button

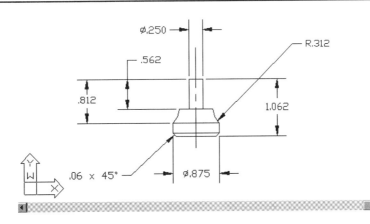

NOTE

The following exercise is only one of many ways to create this shape. Another way would be to draw half of the outline as a closed polyline, as you did with the sleeve, and then revolve the shape.

1. Type **cylinder**↵ to begin the Cylinder command, and at the `Elliptical/<center point> <0,0,0>:` prompt, type **1,1**↵ to place the center of the cylinder.
2. At the `Diameter/<Radius>:` prompt, type **7/16**↵.
3. At the `Center of other end/<Height>:` prompt, type **.5**↵.
4. Click on Torus on the Solids toolbar, select Draw ➢ Solids ➢ Torus, or type **tor**↵ to begin the Torus tool. Look at the icon on the toolbar button if this donut shape is unfamiliar to you.

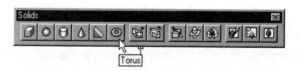

5. At the `Center of torus <0,0,0>:` prompt, use the Center Osnap to find and pick the top center of the cylinder.
6. At the `Diameter/<Radius> of torus:` prompt, type **.625**↵. This is the dimension to the center of the tube, *not* the outside diameter.
7. At the `Diameter/<Radius> of tube:` prompt, type **5/16**↵ (as on the dimensioned drawing).

You might have some trouble seeing this object. Take a look at it from some different angles using the View toolbar, and selecting the Front, Side, and Top tools. Return to the SE Isometric view at the end of your exploration.

8. Next you will subtract the torus from the cylinder. Type **su**↵, press ↵, and at the `SUBTRACT Select solids and regions to subtract from:` prompt, pick the cylinder. At the `Select objects:` prompt, press ↵ to complete the selection set.

PUTTING TWO SOLIDS TOGETHER | 541

9. At the Select solids and regions to subtract: prompt, type **L**⏎ to use the last object created, and at the Select objects: prompt, press ⏎ to complete the selection set and the command. Your drawing should look like Figure 12.10.

FIGURE 12.10

The cylinder with the torus subtracted

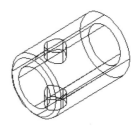

Next you will add the rest of the detail to the button. The bottom of the button gets a chamfer and the top has a pin projection.

1. Type **ch**⏎ to begin the Chamfer command.
2. At the (TRIM mode) Current chamfer Dist1 = 0.0600, Dist2 = 0.0600 Polyline/Distance/Angle/Trim/Method/ <Select first line>: prompt, pick the circle at the bottom of the cylinder. At the Select base surface: Next/<OK>: prompt, remember that the highlighted surface is the base surface and that this is how the distance dimensions are applied. In this case the two distances are the same, so either surface is acceptable.
3. At the Enter base surface distance <0.0600>: prompt, press ⏎ only if your default is 0.06. If it's not, type **.06**⏎ to match Figure 12.9. At the next prompt, Enter other surface distance <0.0600>:, do the same thing: Press ⏎ to accept a default of 0.06 or enter **.06**⏎ if you have a different default distance.
4. At the Loop/<Select edge>: prompt, pick the bottom edge again to tell AutoCAD which edge you want to chamfer. There is only one edge on the surface you selected, but there could have been many, and AutoCAD needs to know which one you want to use.
5. At the Loop/<Select edge>: prompt, press ⏎ to tell AutoCAD that you have finished selecting edges.

This also completes the command, and you'll see the chamfered edge. Next you need to draw the pin projection.

1. Type **cylinder.↵** to begin the Cylinder command and draw the pin projection.
2. At the `Elliptical/<center point> <0,0,0>:` prompt, use the Center Osnap marker to find and pick the center of the top of the button. At the `Diameter/<Radius>:` prompt, type **d.↵**.
3. At the `Diameter:` prompt, type **.25↵** for the dimension (from Figure 12.9) and at the `Center of other end/<Height>:` prompt, type **9/16↵** (also from Figure 12.9).

Let's put the chamfered button and the pin together.

1. Type **uni.↵** to start the Union tool, and at the `Select objects:` prompt, pick the button and the cylinder. Then, at the `Select objects:` prompt, press ↵ to complete the selection and complete the command. Your drawing should look like Figure 12.11.
2. Save your work now.

FIGURE 12.11

The completed button

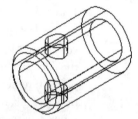

The Sphere Tool

Let's try a new tool to make a 9/16 diameter spherical end cap to be pressed onto the press' handle bar.

1. Click on the Sphere tool from the Solids toolbar, or select Draw ➤ Solids ➤ Sphere from the menu bar, or type **sphere**↵ to begin the Sphere tool. Look at the icon on the toolbar button if this ball shape is unfamiliar to you.

2. At the Center of sphere <0,0,0>: prompt, type **1,3**↵ to locate the sphere.
3. At the Diameter/<Radius> of sphere: prompt, type **d**↵ to enter the diameter of the sphere.
4. At the Diameter: prompt, type **9/16**↵ to draw a 9/16-unit diameter sphere.

The sphere is truncated .21 units from the center and a .4-unit deep hole is drilled from the flat into the sphere. One way to create the flat is to draw a box using the Center Placement option of the Box command and then move the box into position to subtract it from the sphere.

1. Type **box**↵ to begin the Box tool. At the Center/<Corner of box> <0,0,0>: prompt, type **c**↵ to use the Center option to locate the box. At the Center of box <0,0,0>: prompt, use the Center Osnap to pick the center of the sphere. You want the box located in a known position relative to the sphere so that it is easy to move the box in relationship to the sphere. If you had not placed the box using the sphere, you would have had to move it into a known, identifiable position—which would require a number of additional steps.
2. To use the Cube option of the Box tool, at the Cube/Length/<corner of box>: prompt, type **c**↵ and at the Length: prompt, type **.75**↵ to create a cube .75 units on a side. You could use any value to define the cube (but the math in step 4 would have to change to suit a different value).

When you created the cube in step 2, you created a tool to slice off one side of the sphere and make one surface flat, because it is very hard to drill into a small spherical object without its rolling.

3. The box completely covers the sphere. Type **m↵** to start the Move tool, and at the Select objects: prompt, type **L↵** to move the last object created. At the Select objects: prompt, press ↵ to complete the selection set.

4. At the Base point or displacement: prompt, pick any point on the screen, and at the Second point of displacement: prompt, type **@.585<0↵** to move the box .375 (half the length of a side of the box) + .21 (the distance from the center of the sphere to the flat). The box is now positioned to slice off a portion of the sphere.

You could take a quick look from the Top view (Top tool on the View toolbar). If you do, be sure to return to the SE Isometric view so that your drawing will look the same as the figures.

5. Type **su↵** to subtract the box from the sphere and create the flat.

6. At the SUBTRACT Select solids and regions to subtract from...Select objects: prompt, pick the sphere, and at the Select objects: prompt, press ↵.

7. At the Select solids and regions to subtract...Select objects: prompt, pick the box, and at the Select objects: prompt, press ↵. The box disappears and you have only the remaining flat on the sphere.

8. Use the Cylinder command to create the solid to subtract from the sphere that creates the hole. Type **cylinder↵** and at the Elliptical/<center point> <0,0,0>: prompt, type **c↵**. At the cen of: prompt, use the Center Osnap to pick the center of the flat. At the Diameter/<Radius>: prompt, type **d↵** and at the Diameter: prompt, type **.25↵**.

9. At the Center of other end/<Height>: prompt, type **c↵** to enter the location of the center of the other end of the cylinder. At the Center of other end: prompt, type **@.4<180↵** to define the center of the other end as a point .4 units in the –X axis.

If the UCS is still set to World UCS (look at the icon if you aren't sure), the cylinder will be drawn with the axis in the positive Z direction. However the Center option of the Cylinder command allows you to define the center as any coordinate in any direction.

PUTTING TWO SOLIDS TOGETHER | 545

10. Subtract the cylinder from the sphere. Type **su**↵ and at the SUBTRACT Select solids and regions to subtract from...Select objects: prompt, pick the sphere. At the Select objects: prompt, press ↵ and at the Select solids and regions to subtract...Select objects: prompt, pick the cylinder. At the Select objects: prompt, press ↵.

11. Save your work and close this drawing.

If you like, take a look at a Hidden Line view of your work. Type **hi**↵ to use the Hide command.

That's it! You have finished the sleeve, button, and handle end cap. Your drawing should look like Figure 12.12.

FIGURE 12.12

The complete handle end cap

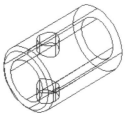

If you want to practice a little more, we used the Cylinder, Array, Subtract, and Osnap commands that you have already learned to make the screw holes and the cover. If you wish to see how this was done, the images are included on the CD-ROM. Try it yourself and compare your results to the drawings called pressparts.dwg on the CD-ROM.

Making Slides to Improve Communication

A *slide* is a presentation tool that allows you to share the work you've done. It can be viewed by anyone with a copy of AutoCAD Release 12 through 14. A slide cannot be edited, measured, zoomed, or panned, but even with these limitations it is a very good tool for informing others of the current status of your work, or for providing a historical record of the status of a specific drawing at a specific time.

Mslide and Vslide

Mslide and Vslide are tools that make and view an externally saved file with the extension .sld, which is a snapshot (raster image) of your work. The slide file is independent of the current drawing file, and it can be viewed at any time from within any AutoCAD editing session with Vslide command. In this exercise, you will zoom the extents of the press, change the UCS to View so that you can write an easy-to-read note, hide the view to get rid of any hidden lines, and make a slide. Then you will zoom into the face of the press so that you are looking at a different zoom scale. Finally, you will view the slide.

First, let's zoom the extents of the press and place a note.

1. Open ch12press2.dwg from the companion CD-ROM.
2. Type z↵e↵, to zoom the extents of the drawing.
3. Type **ucs**↵ to change the UCS to a view in which the text will be easy to read. At the Origin/ZAxis/3point/OBject/View/X/Y/Z/Prev/Restore/Save/Del/?/<World>: prompt, type **v**↵.
4. Type **t**↵ to begin the Mtext tool.
5. At the Justify/Style/<Start point>: prompt, pick a point above and to the left of the press. At the Height <0.2500>: prompt, type **.25**↵ (your prompt may have a different default value). At the Rotation angle <0>: prompt, press ↵ unless your default is something other than 0.
6. At the Text: prompt, type **This is the press before adding the rack, pinion, and the press parts.**
7. The UCS icon is in the way of the drawing, so let's turn it off by changing its setting to No Origin. Type **ucsicon**↵ and at the ON/OFF/All/Noorigin/ORigin <ON>: prompt, type **n**↵.
8. Type **hi**↵. The hidden lines are gone, but the text remains. Your drawing should look like Figure 12.13.

MAKING SLIDES TO IMPROVE COMMUNICATION | 547

FIGURE 12.13

The press with a note

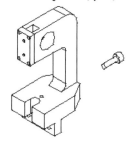

Making a Slide

Now let's make a slide.

NOTE There aren't any toolbar or menu shortcuts to making a slide. You have to use the command line, and type **mslide**↵.

1. To make a slide, type **mslide**↵. The Create Slide File dialog box appears. Check the file name; the dialog box automatically brings up the same file name as the drawing name. You can either accept the default or rename the file. In either case, the file extension must be .sld. The location of the slide file will be the same as the drawing file unless you change the location in the dialog box. You can always move the file later. Accept the defaults by clicking on the Save button in the dialog box.

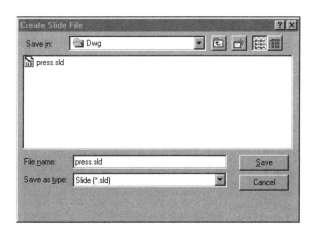

Modeling and Imaging in 3D

2. As with the Hide and Shade tools, as soon as you do anything that forces a regeneration of the screen, the image becomes the normal, editable version. Type **z↵** and window the area around the cover.

NOTE As with Mslide, there aren't any toolbar or menu shortcuts for the View Slide command. You have to type **vslide↵**.

3. To view the slide, type **vslide↵**. The Select Slide File dialog box comes up. The window shows you the files available in the current directory, and you have the option of searching other directories. The default slide is the one that matches the name of the file that you are working in. Accept the default by clicking on Open, and up pops the slide that you made. To get back to your drawing, type **r↵** or run any command (such as Redraw) that causes a regeneration. The slide should look like Figure 12.13.

4. Save your work.

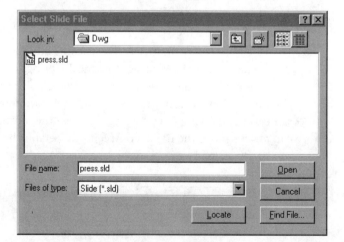

More Primitives

You've probably noticed that there are two more solid primitives that we have not yet used or talked about. The Wedge and the Cone are tools that are easy to define now that you have used the other primitives. You will find a use for them from time to time.

The cone could be the point on a set screw or the nose cone on a rocket. The wedge can be used to add inclined surfaces to solids, and as the name implies, to create a wedge that takes up tolerance in a design. We are only going to discuss these two tools, rather than use them to create press parts.

Wedge

The *Wedge* tool can be found on the Solids toolbar, by selecting the Draw ➤ Solids ➤ Wedge from the menu bar, or by typing **we**↵. The prompts are Center/<Corner of wedge> <0,0,0>:, giving you the same options as the Box tool to either locate a corner or the center of the wedge. The Cube/Length/<other corner>: options allow you to define the wedge as a cube, either by defining lengths or by selecting points for the opposite corners and then giving the height. The length is defined by the size in the current X axis, the width is defined by the size in the current Y axis, and the height is defined by the size in the current Z axis. Figure 12.14 shows the wedge drawn squarely in a cube, and a variation with a length of 2, a width of 1, and a height of .5.

FIGURE 12.14

A wedge drawn as a cube in a 2-unit cube where Length=2, Width=1, and Height=.5

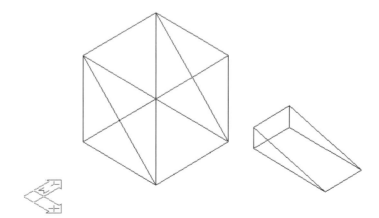

Cone

The *Cone* tool resembles the Cylinder tool in the same way that the Wedge tool resembles the Box tool. Many of the same options appear when you use this tool. The Cone tool is found on the Solids toolbar, by choosing Draw ➢ Solids ➢ Cone from the menu bar, or by typing **cone**↵. The prompts are Elliptical/<center point> <0,0,0>:, allowing you to create a cone based on an ellipse or a circle. Type **c**↵ to use the Circle option. At the prompts for Diameter/<Radius>: prompt, type your choice. At the Apex/<Height>: prompt, type your choice of height (where the height is a point along the Z axis in the current UCS) or enter a point that will be the apex (it can be any point in space). If you use the Apex option, the end of the cone will be perpendicular to the axis. This system works in the same way as the Cylinder tool when you select the Center option.

If you select the elliptical option from the Elliptical/<center point> <0,0,0>: prompt, the rest of the prompts look a little different from the circular option, because first you must define the ellipse. You will see the prompts Center/<Axis endpoint>: (to pick a center point or axis endpoint and the default), Other axis distance: (to complete the definition of the ellipse), and Apex/<Height>: (to determine whether the height is along the Z axis or perpendicular to the axis). Figure 12.15 shows a circular cone and elliptical cone.

FIGURE 12.15

A circular cone drawn with a 1-unit radius and a 2-unit height, and an elliptical cone drawn from the center option with a 1-unit first axis endpoint, a .5-unit other axis endpoint and a 1-unit height

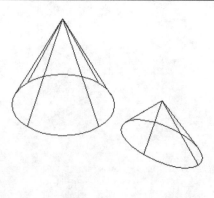

Editing Solids with Slice

Although you have been editing solids all along, the *Slice* tool may be the most valuable for editing. The Slice tool allows you to define a plane and cut a shape in two along the plane. You can choose whether to save the two parts or to discard one of them.

The post on the press needs to be lengthened .5 units. This would be very difficult without the Slice tool. You would have to delete the top or the bottom section and redraw it.

So let's try lengthening the post using the Slice tool.

1. If you still have the press file open, you can continue with this exercise. If not, or if you aren't sure that it is right, open the one on the CD-ROM called c12 slice.dwg.

2. Click on the Slice tool on the Solids toolbar, choose Draw ➤ Solids ➤ Slice from the menu bar, or type **slice**↵.

3. At the Select objects: prompt, pick the post/base and at the next Select objects: prompt, press ↵.

4. At the Slicing plane by Object/Zaxis/View/XY/YZ/ZX/<3points>: prompt, you have several choices:

    - The Object option allows you to define a plane by the object that you select. Any object that describes a plane is eligible. See Chapter 11 for some examples.

    - The Zaxis option defines the cutting plane by a specified origin point on the Z axis (normal) of the X-Y plane.

    - The View option defines a plane parallel to the current view, and the point that you pick is a point on that plane.

    - The XY/YZ/ZX options define the cutting plane parallel to the plane of the axes indicated and the point that you pick is a point on that plane.

    - The 3-point option allows you to identify three specific points to form your own plane. The first point is the origin, the second point is a point on the X axis, and the third point is a point on the Y axis.

5. For this exercise, press ↵ to use the 3-point (default) option. At the `1st point on plane:` prompt, pick the midpoint of one of the lines from the vertical section of the post. At the `2nd point on plane:` prompt, with Ortho mode on, drag the mouse to the right and pick a point below and to the right of the original point. At the `3rd point on plane:` prompt, drag the mouse and pick a point to the right and above the original point.

6. At the `Both sides/<Point on desired side of the plane>:` prompt, you have a choice of saving either side or both. Type **b**↵ to save both sides. (You would point to the side you wanted to preserve if you only needed half of your drawing.) A square appears at the point of the slice. Although it is not obvious, you now have two separate solids.

7. Type **m**↵ to start the Move tool and at the `Select objects:` prompt, pick the top half of the post. At the `Select objects:` prompt, press ↵. At the `Base point or displacement:` prompt, pick any point on the screen, and at the `Second point of displacement:` prompt, type **@0,0,.5**↵. Now your two solids are easy to identify. Your drawing should look like Figure 12.16.

FIGURE 12.16

The post sliced in two and separated

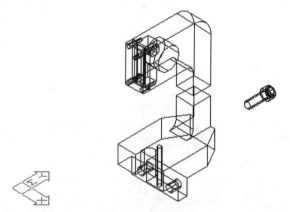

Now, let's add a section to extend the length of the post.

1. Use the Box command to create a solid to fill in the gap between the two parts. Type **box**↵.

2. At the Center/<Corner of box> <0,0,0>: prompt, pick one corner of the slice on the base, and at the Cube/Length/<other corner>: prompt, pick the opposite corner—the one the first corner used to be attached to—on the post. A box appears to fill in the gap.

3. Union the three parts. Type **uni**↵.

4. At the Select objects: Other corner: prompt, pick all three parts. You could use a crossing window (as described in Chapter 2). At the Select objects: prompt, press ↵ and the post/base is again one piece, but longer. Your drawing should look like Figure 12.17.

5. Save your work. If you opened this file from the CD-ROM, save your work as **press.dwg** to your hard disk.

FIGURE 12.17

The post/base united but longer

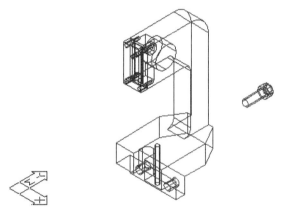

The Section Tool

Section is not a drafting tool in the classic sense of drawing sections. Rather, it creates a region on a cutting plane that you define. The region can be hatched as a separate command.

You will find that you can draw sections very well in the 3D to 2D drawing section using other tools. But the Section tool allows you to capture geometry that you can use to create additional solids. You also can isolate areas that may not be clear after numerous edits and Boolean operations.

To explore the Section tool, let's take a section of the press assembly. We will not go over the methods for selecting the plane, because they are the same as those for slice. The result will be an object called a region. We will discuss regions in the next section.

Do not save any of the work that you do in this exercise.

1. Click on the Section tool on the Solids toolbar, choose Draw ➤ Solids ➤ Section from the menu bar, or type **sec**↵.

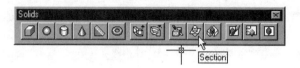

2. At the Select objects: prompt, use a crossing window to select the post/base, the cover, and the gib.

3. At the Select objects: prompt, press ↵.

4. At the Section plane by Object/Zaxis/View/XY/YZ/ZX/<3points>: prompt, type **yz**↵ so you can pick a point and generate a plane on the Y-Z axes of the current UCS.

5. At the Point on YZ plane <0,0,0>: prompt, use the Midpoint Osnap to pick the midpoint of one of the top lines that you can see on the cover. This is a point through which the Y-Z plane will pass and at which the section will be created. See Figure 12.18 to help you find the midpoint.

6. Type **z**↵ and adjust Zoom so that you can see the three sections.

7. Move these new sections so you can see them. Type **m**↵ to start the Move tool, and at the Select objects: prompt, pick the three new sections. They will be slightly difficult to pick. You can use object cycling or the Add and Remove options to pick the sections. At the Select objects: prompt, press ↵.

8. At the Base point or displacement: prompt, pick any point on the screen, and at the Second point of displacement: prompt, type **@10<270**↵.

9. Zoom all to see the move. Your drawing will look like Figure 12.18.

FIGURE 12.18

The Section of the post/base, cover, and gib

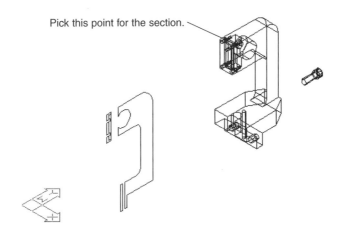

Don't save these changes.

The Region: A 2D Object for Boolean Operations

The sections you created (and didn't save) are objects called *regions*. A region acts like a surface in that it is a closed area that will obscure objects behind it, but unlike a surface, a region can be trimmed or edited with other regions. A region can be extruded to become a 3D solid. When the region is exploded it becomes a collection of line and arc objects. All of the Boolean operations will work with regions.

Let's look at what would happen if we were to add a rectangle 1 unit wide and 2 units high to the post section of our press drawing.

NOTE

Regions can be used to draw in a 2D mode. Although we will not demonstrate this, except to show how the Boolean operation would affect the section, you might want to experiment with this on your own.

We're just going to take a quick look at Boolean operations and regions, so don't save any of the work from this exercise.

> **If Your UCS Doesn't Look the Way You Expected**
>
> If your UCS icon does not look like the one in Figure 12.19, take the following steps:
>
> 1. Type **ucs**↵↵ to return to the World UCS from wherever you are.
> 2. Press ↵ again to reissue the last command, then type **x**↵ to rotate the UCS about the X axis.
> 3. Type **90**↵ to rotate the UCS 90°.
> 4. Press ↵ again to reissue the last command, and then type **s**↵ to save the UCS.
> 5. Type **s**↵ to save a UCS named *s*.

1. Type **ucs**↵**r**↵**s**↵ to restore the UCS you saved in the press drawing in Chapter 11 called *s* for *side*. If you didn't save it, be sure that the UCS is set to World and use the UCS presets to change the UCS to *s*.

2. Draw a rectangle near the section of the post. Type **rectang**↵.

3. At the Chamfer/Elevation/Fillet/Thickness/Width/<First corner>: prompt, pick any point on the screen.

4. At the Other corner: prompt, type **@1,2**↵.

5. Type **m**↵ to move and align the rectangle, and at the Select objects: prompt, type **L**↵. At the next Select objects: prompt, press ↵ to complete the selection set.

6. At the Base point or displacement: prompt, pick the midpoint of the left side of the rectangle, and at the of Second point of displacement: prompt, pick the midpoint of the right side of the post.

7. You need to pick one more point to move the rectangle .125 past the edge of the post and create a notch. Press ↵ to reissue the Move tool, and at the Select objects: prompt, type **L**↵ to use the last object. At the Select objects: prompt, press ↵.

8. At the Base point or displacement: Second point of displacement: prompt, type **@.125<180**↵. Your drawing should look like Figure 12.19.

THE SECTION TOOL | 557

FIGURE 12.19

The section and rectangle

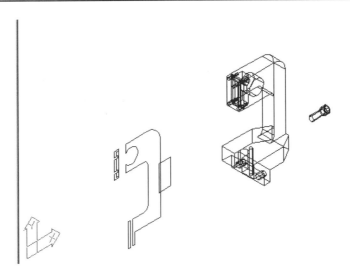

You will need to make the polyline rectangle into a region before you can try out some of the Boolean operations.

1. Click on the Region tool from the Draw toolbar, select Draw ➤ Region, or type **reg**↵.

2. At the Select objects: prompt, pick the rectangle, and at the Select objects: prompt, press ↵. You will see the validation 1 loop extracted. 1 Region created.
3. Type **uni**↵ to union the section region and the rectangle, and at the Select objects: prompt, pick both regions. At the next Select objects: prompt, press ↵ and you have added the two shapes together. Your drawing will look like the top half of Figure 12.20.
4. Type **u**↵ to undo the last command, so you can try a subtraction instead of a union.
5. Type **su**↵ to subtract the rectangle from the section.

6. At the Select solids and regions to subtract from...Select objects: prompt, pick the section and at the Select objects: prompt, press ↵.

7. At the Select solids and regions to subtract...Select objects: prompt, pick the rectangle. At the Select objects: prompt, press ↵. Now your drawing should look like the bottom half of Figure 12.20.

Don't save the changes you have made. You could also have tried the Intersection tool to save only the part of the two regions common to both regions.

FIGURE 12.20

The section and the rectangle. The top is after a Boolean union and the bottom is after a Boolean subtract

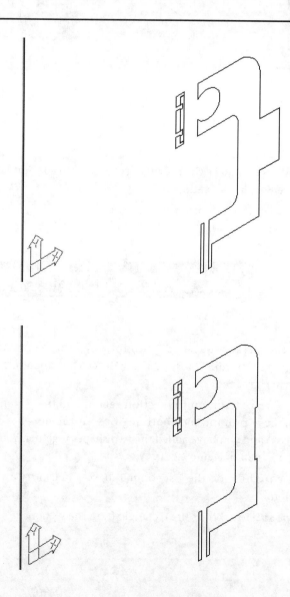

Interference Checking: Looking for the Not-So-Obvious

Interference checking is a buzz phrase in the industry today. In AutoCAD, the interference you can check for uses the nominal dimensions used to create your solids. You may have noticed that you haven't applied any tolerances to the size of your solids. This is because you can't. AutoCAD has no capacity to do this. You can, however, check the nominal fit of your parts.

So far you have drawn parts that fit. Now, you'll move the cover .125 units on the Y axis to create an obvious interference between the cover and the post of the press, and let AutoCAD check for it. Restore the last saved version of the press, and let's see how interference checking works.

1. Type **open**↵, and in the Select File dialog box, be sure that the file name is press.dwg. If you can't use yours, you'll find one on the CD-ROM called c12 Interfere.dwg.

2. Type **m**↵ to start the Move command. At the Select objects: prompt, pick the cover, and at the next Select objects: prompt, press ↵.

3. At the Base point or displacement: prompt, pick any point on the screen, and at the Second point of displacement: prompt, type **@.125<90**↵.

This move has created a definite interference between the post and the cover.

NOTE

If you had not moved the post, there would be no interference and AutoCAD would have returned a null finding. This would not have made a very interesting exercise. If you had made some error earlier in the tutorial, the Interfere tool would find it for you.

1. Start the Interfere tool by clicking on Interfere on the Solids toolbar, by selecting Draw ➢ Solids ➢ Interfere, or by typing **interfere**↵.

2. At the Select the first set of solids: Select objects: prompt, pick the post, and at the Select objects: prompt, press ↵.

3. At the Select the second set of solids: Select objects: prompt, pick the cover, and at the Select objects: prompt, press ↵.

4. You'll see the following validation:

   Comparing 1 solid against 1 solid. Interfering solids (first set): 1 (second set): 1 Interfering pairs: 1.

5. At the Create interference solids ? <N>: prompt, type **y**↵. You do not want to create the solid but simply to check for interference.

NOTE If you do not let AutoCAD make the solid, all that you will know is that the two objects interfere. You will not know *how* they interfere.

6. Type **m**↵ to move the new solid so that you can see it. At the Select objects: prompt, type **L**↵, and at the next Select objects: prompt, press ↵.

7. At the Base point or displacement: prompt, pick any point on the screen, and at the Second point of displacement: prompt, type **@3<180**↵. You can now see the solid, which should look like Figure 12.21.

FIGURE 12.21
The interference solid generated by the Interfere tool

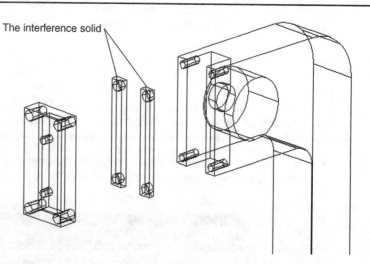

The block created by the Interfere tool could be used to remove material from one of the parts in order to make them fit, or it could just be used to create new solids. Move the cover back to the original position, erase the interference solid, and check for interference again.

1. Type **m↵** to move the cover .125 units in the Y axis. At the Select objects: prompt, pick the cover, and at the Select objects: prompt, press ↵. At the Base point or displacement: prompt, pick any point on the screen, and at the Second point of displacement: prompt, type **@.125<270↵**.

2. Type **e↵**, and at the Select objects: prompt, pick the interference solid. At the Select objects: prompt, press ↵.

3. Type **interfere↵** or click on the Interfere tool on the Solids toolbar.

4. At the Select the first set of solids: Select objects: prompt, pick the post and at the Select objects: prompt, press ↵.

5. At the Select the second set of solids: Select objects: prompt, pick the cover, and at the Select objects: prompt, press ↵.

6. You should see the following validation:

    Comparing 1 solid against 1 solid. Solids do not interfere.

7. Save your work. If you opened this file from the CD-ROM, save your work as **press.dwg**.

If you want to see what your work looks like now, skip ahead to the rendering section in Chapter 15. Rendering consumes huge amounts of memory, so if you don't have a lot of time, save this slow task for later.

Enhancing the 2D Drawing Process

You have already created a solid from a 2D drawing in Chapter 11. The technique in the next section is a good way to re-use any existing 2D drawing work you might have already drawn, rather than trying to redraw it using 3D techniques.

AutoCAD has provided a way to make 2D multiple-view working drawings from the solid models that you have created. These drawings are often required by fabrication shops and inspection departments. Two-dimensional drawings are often used to estimate the cost of making a part, even if the part will be made from the data in the model.

Two-dimensional drawings of parts are often required to meet the legal requirements of contracts between companies and their vendors. US law is decidedly unclear on the responsibility of the parties to a contract if the documents that define the part are not drawn in a format that is easily interpreted using a standard such as the one available from the American National Standards Institute (ANSI) or other international standard format. As of this writing, none of the international standards yet supports solid modeling as legal documentation.

The next few exercises will show you one way to let your computer convert your 3D work into more widely and easily understood 2D drawings. AutoCAD will take care of all of the projections, view creation, layer creation, and layer management. The only bad news is that these drawings appear in Paper Space, which you might find difficult to manage at first.

You will probably want to take a quick review of the Chapter 9 discussion of Paper Space before going on to make the 2D drawing view. Be sure to pay particular attention to viewport scale.

The status bar is the best way to know whether you are in Paper Space or Model Space. If you are in Model Space, the cursor and UCS icon are visible in the viewport.

Copying Solid Models Using the Clipboard

You could create the 2D drawings of each part without the next step, but you will find this process difficult to manage as the number of Paper Space viewports and layers increases. We recommend that you create a separate file for each part, which will keep the number of layers and viewports to a minimum. You could use the Wblock tool that you learned about in Chapter 4 to export the part to a new file and then open the new file to begin the 2D conversion, but if you do this you will have to reset the system variables in the new file to be the way that you want them. Remember that the new file used the default system variables. Or you could begin a new file using the template of your choice, and then insert the Wblock-created file.

There is another alternative that requires fewer steps and won't clutter up the hard disk with unnecessary files. You can use the Copy to Clipboard and Paste from Clipboard tools. The *Clipboard* is a feature of the Windows 95 operating system and is available in most programs designed to run in Windows 95. The Clipboard is temporary memory space that loses any information stored there when the computer is restarted.

There are several ways to use the Copy to Clipboard command: Click on the Copy tool on the Standard toolbar, choose Edit ➤ Copy from the menu bar, press Ctrl+c, or type **copyclip**↵. Select the item you wish to duplicate, and the copy is stored temporarily as a block on the Clipboard.

ENHANCING THE 2D DRAWING PROCESS | 563

WARNING

The Clipboard is a temporary space and can only hold one item or selection set at a time. New items copied to the Clipboard will overwrite the existing items.

The next step is to open a file that is set up the way that you like—in our case this is the Versions template—and paste the contents from the Clipboard. Click on the Paste tool on the Standard toolbar, choose Edit ➢ Paste from the menu bar, press Ctrl+v, or type **pasteclip**↵.

Let's try it.

1. Open the assembly of the press. If you don't think your version is correct or if you are just starting here, you can open the file named ch12post.dwg on the companion CD-ROM.

2. Click on the Copy tool on the Standard toolbar, choose Edit ➢ Copy, press Ctrl+c, or type **copyclip**↵.

3. Pick the post, then press ↵ to complete the selection set and execute the command.

NOTE

If you are a DOS holdout, you may be wondering why you can't use Ctrl+c to exit or escape from a command. The Windows operating system uses that particular combination to copy text and graphics to the Clipboard. AutoCAD is Windows-compliant, so if you try Ctrl+c, the Copyclip command will be issued.

4. Start a new drawing using the Versions template. Save the new drawing as **base.dwg**.

5. To insert or paste the contents of the Clipboard into the base drawing, click on the Paste tool on the Standard toolbar, choose Edit ➤ Paste from the menu bar, press Ctrl+v, or type **pasteclip**↵.

6. At the `insertion point:` prompt, type **4,4**↵ to place the base point of the base at the coordinate 4,4. This is an arbitrary location that is some distance from 0,0. At the `X scale factor <1>/Corner/XYZ:` prompt, press ↵. At the `Y scale factor (default=X):` prompt, press ↵, and at the `Rotation angle <0>:` prompt, press ↵.

7. You have inserted a block which will not act like a solid until it is exploded. Use the Explode tool from the Modify toolbar and pick the base.

NOTE

A block is a special group of AutoCAD objects that cannot be edited (see Chapter 4). The Solprof, Solview, and Soldraw commands that you are about to use will not execute using solids grouped in a block. For these commands to function, a block must be exploded so that the solids can be edited and the block identity of the solids removed. Warning: The solid or mass identity will be lost if you explode the objects again.

It's time to look at the post from an Isometric viewpoint.

1. Click on the SE Isometric View tool to enter this command. Your drawing should look like Figure 12.22.

2. Save this drawing as **base.dwg**.

FIGURE 12.22

The post/base in the new drawing

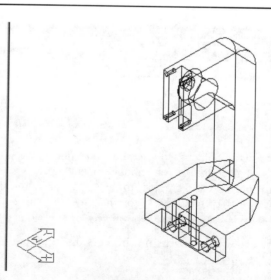

The Setup View Tools

The next step is to create and arrange the drawing views that you will need for your multiple-view 2D drawing. You will use the Solview tool for this purpose. Then you will use the Soldraw tool to project the view onto a single plane, determine which lines are hidden lines and place them on a hidden lines layer, and manage the viewport layer visibility. Finally, you will create a few dimensions for the part.

The Setup Views Tool: Solview

Solview is used to create the views on your drawing. Next, you will change to Paper Space and create floating viewports easily with the aid of Solview.

1. Click on the Setup View tool from the Solids toolbar, choose Draw ➤ Solids ➤ Setup ➤ View, or type **solview**↵.

2. At the next prompt—Entering Paper space. Use MVIEW to insert Model space viewports. Regenerating paperspace. Ucs/Ortho/Auxiliary/Section <eXit>:—type **u**↵ to use the UCS option.

Use a UCS to define the plane from which your first view will be generated. Throughout this book, we have used a convention of drawing the base with the bottom parallel to the World UCS. If you make this first view the Top or Plan view, it's a good idea to use the World UCS plane option.

3. At the Named/World/?/<Current>: prompt, type **w**↵ to use the WCS. You could choose either a named UCS or the current UCS (if the current UCS is different from the WCS). All of these options are designed to allow you to place your 3D object in this first view with the X and Y axes in a good relationship to the part. You will be placing dimensions on the 2D drawing which will be oriented on the drawing relative to the UCS.

4. At the `Enter view scale<1.0000>:` prompt, press ↵ to accept the default. You could enter any scale here—your drawing will work at any size, including full size.

5. At the `View center:` prompt, pick a point approximately in the middle of the screen.

6. At the `View center:` prompt, press ↵ to complete the command.

7. At the `Clip first corner:` prompt, pick a point near the lower-left corner.

8. At the `Clip other corner:` prompt, pick a point near the upper-right corner.

9. At the `View name:` prompt, type **top**↵.

10. At the `Ucs/Ortho/Auxiliary/Section/<eXit>:` prompt, press ↵ to accept the default and exit. Your drawing should look like Figure 12.23.

FIGURE 12.23

The first view created by Setup View

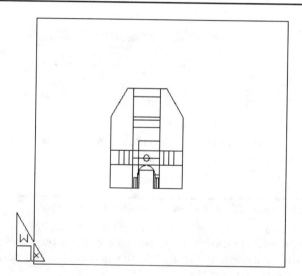

Paper Space

Take a look at the state of your work. You are in Paper Space and you have one view showing. You will need more views to define your part properly. The Paper Space limits are different from Model Space limits and can be set separately. Set the Paper Space limits to D size (34 units × 22 units) and zoom all. Let's have a look.

1. Change to Paper Space by typing **ps**↵.
2. Reset the Paper Space limits to a D-size drawing. Type **limits**↵, and at the `Reset Paper space limits: ON/OFF/<Lower left corner><0.0000,0.0000>:` prompt, press ↵ to accept the default. At the `Upper right corner<12.0000,9.0000>:` prompt, type **34,22**↵.
3. Use Zoom All to see the change.

The first view that you have defined is a Plan or Top view and was placed on the screen arbitrarily. Move the view to a more appropriate place towards the upper left of the screen: up and to the right. Type **m**↵ to move the viewport, pick the viewport—not the objects in the viewport—and drag the viewport roughly into position.

You need to start Solview again to create a Front view, a Side view, and what will be an Isometric view.

1. Type **solview**↵ and at the `Ucs/Ortho/Auxiliary/Section/<eXit>:` prompt, type **o**↵ to create a new orthogonal Front view.
2. At the `Pick side of viewport to project:` prompt, pick the bottom of the Top view.
3. At the `View center:` prompt, drag the mouse down about 5.5 units and pick this point.
4. At the `View center:` prompt, press ↵ to set the Center view.
5. At the `Clip first corner:` prompt, pick a point approximately in line with the left side of the top viewport and very near the bottom of the screen.
6. At the `Clip other corner:` prompt, pick a point approximately in line with the right side of the top viewport and very near the bottom of the top viewport.
7. At the `View name:` prompt, type **side**↵ to name this view.

Creating an Isometric view is a similar process.

1. At the `Ucs/Ortho/Auxiliary/Section/<eXit>:` prompt, type **o**↵ to create a new Isometric view.
2. At the `Pick side of viewport to project:` prompt, pick the top of the Side view.
3. At the `View center:` prompt, drag the mouse up about 4.5 units and pick this point.
4. At the `View center:` prompt, press ↵ to set the center of the view.

5. At the `Clip first corner:` prompt, pick a point very near the top right corner of the side viewport.

6. At the `Clip other corner:` prompt, pick a point very near the upper-right corner of the top viewport.

7. At the `View name:` prompt, type **iso**↵ to name this view.

8. At the `Ucs/Ortho/Auxiliary/Section/<eXit>:` prompt, press ↵ to accept the default. You should have four viewports, and your drawing should look like Figure 12.24.

9. Save your work.

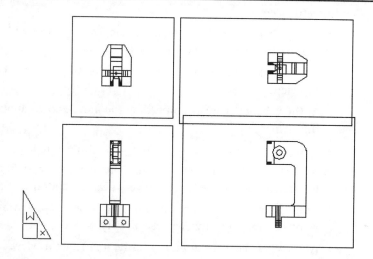

FIGURE 12.24

The drawing layout with all four views

The Setup Drawing Tool: Soldraw

The arrangement of the views was a neat trick. Did you notice that all of the objects are aligned vertically and horizontally? The floating viewports are still showing the model. Next, you will want AutoCAD to create the 2D drawings. This is the job of the *Setup Drawing* tool. Let's see what it does.

1. Click on the Setup Drawing tool from the Solids toolbar, choose Draw ➢ Solids ➢ Setup ➢ Drawing from the menu bar, or type **soldraw**↵.

2. At the `Select viewports to draw:` prompt, pick the Top, Front, and Side views. Do not pick the Iso view because you still have a little work to do with it.
3. At the `Select objects:` prompt, press ↵. The Hidden Line views have been created.

You do not want your scaled and aligned views to change in scale and alignment unless you are controlling such changes. Remember not to move the viewports or zoom while working in Model Space on any of the viewports, because these machinations will change the alignment of the view and the scale. You can occasionally restore the zoom scale by using Zoom and the previous option. If zoom is not the problem, use the Undo tool until you are back to the original arrangement, or realign the viewports by using the Move command.

The Setup Drawing tool only works in viewports created with the Setup View tool.

The Setup Profile Tool: Solprof

The *Setup Profile* tool is a subset of the Setup View tool, and with it you can create Hidden Line views. The Setup Profile tool works in any floating viewport, not just those created by the Setup View tool.

In the Setup Drawing section, we didn't finish the isometric viewport. First, change the UCS to view to be parallel to the viewport, and then use Setup Profile.

Look at the status bar. If the box next to tile has the word `model`, double-click on the word model to toggle to Paper Space. This is the easiest way to change to and from Model Space. You can also type **ms**↵ or **ps**↵ to toggle between the two.

1. Type **ucs↵** to begin the UCS command. Pick the Iso window, and at the `Origin/ZAxis/3point/OBject/View/X/Y/Z/Prev/Restore/Save/Del/?/<World>:` prompt, type **v↵** to align the UCS with the view. The Setup Profile tool generates the Hidden Line view parallel to the current UCS.

2. Click on the SE Isometric tool from View toolbar to set the view to Isometric.

3. To make the Isometric view 1/2 the scale of the other views, use the Zoom tool to set the scale of the viewport to .5 Paper Space. Type **z↵**, and at the `All/Center/Dynamic/Extents/Previous/Scale(X/XP)/Window/<Realtime>:` prompt, type **.5xp↵**.

4. Click on the Setup Profile tool from the Solids toolbar, choose Draw ➤ Solids ➤ Setup ➤ Profile, or type **solprof↵**.

5. At the `Select objects:` prompt, pick the solid (not the viewport).

6. At the `Select objects:` prompt, press ↵.

7. At the `Display hidden profile lines on separate layer? <Y>:` prompt, press ↵ to accept the default. If you answer No, you won't be able to manage the hidden lines with the Layer tool.

8. At the `Project profile lines onto a plane? <Y>:` prompt, press ↵ to accept the default. If you answered No, AutoCAD could place the lines randomly in the space. Projecting them onto a plane makes them easier to manage, and they arrive predictably.

9. At the `Delete tangential edges? <Y>:` prompt, press ↵ to accept the default, or AutoCAD can project each edge as many times as it finds one. Choosing Yes here simplifies the drawing should it be necessary to edit it.

10. You'll see the validation `One solid selected`, and the Setup Profile is finished. Your drawing will look like Figure 12.25.

11. Save your work.

FIGURE 12.25

The four views with an Isometric view

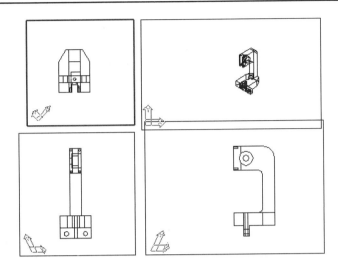

NOTE

You could have made the Iso view with the Setup Drawing tool. Using the Setup Drawing tool would have changed the viewpoint and set the UCS to the Iso view, just like the Setup Profile tool, but you would have used the Setup Drawing tool and picked the viewport instead of the solid. The results are actually a little better using the Setup Drawing tool, because the Hidden Line layer uses the hidden linetype.

Managing the New Layers

These Setup tools have made a few not-so-obvious changes to your drawing. A number of new layers have been created and special properties have been assigned to them.

Click on the layer drop-down list in the Layer & Linetype Properties dialog box (see Figure 12.26). Each of the new views that you created using Setup View has three layers assigned to it, beginning with the view name. The layer called **hid** is where the hidden lines are, the **vis** layer is where the object lines are, and the **dim** layer is were you place the dimensions for that view. The **ph** (Profile Hidden) and **pv** (Profile Visible) layers were created by the Setup Profile tool. You may need to use the scroll bar to see all of the available layers.

To see the Iso view correctly, you have to freeze the ph layer and also the 0 layer, which is now the current layer. You can't freeze the current layer, so you need to set another layer to be current. You also need to create some dimensions, so make the front-dim layer current by highlighting it and clicking on OK.

FIGURE 12.26

The Layer & Linetype Properties dialog box with some of the new layers shown

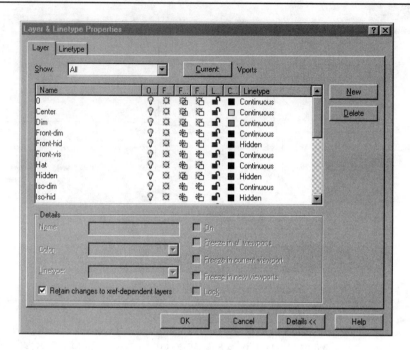

Adding Dimensions

Before placing a dimension in a viewport, see that you are in Model Space. If not, type **ms.↲** or double-click in the MODEL/PAPER toggle box in the status bar.

Next, prepare and place a few dimensions.

1. Decide which view you are going to dimension first and then use the layer drop-down list in the Layer & Linetype Properties dialog box to set the <View name>_Dim layer current. Pick the layer, and then pick anywhere on the screen. The drop-down list recedes, and you will see that the layer that you chose is now current.

2. Set the UCS to the view of the viewport.

Do not zoom or pan while you are in Model Space. You may zoom to your heart's content while in Paper Space—zooming in Paper Space does no damage.

FINDING THE MASS PROPERTIES | **573**

TIP

You might find it easier to dimension the various views if, while in Paper Space, you use the Zoom tool to fill the screen with the view and then change to Model Space to pick and place the dimensions.

3. Look at Figure 12.27 to place the dimensions as shown. If you aren't sure how to pick and place these dimensions or which commands to use, refer to Chapter 8.
4. Save your work.

FIGURE 12.27

The part with some dimensions

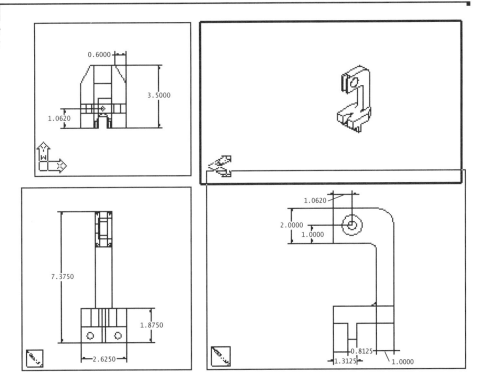

Finding the Mass Properties

You will need to know some of the *mass properties* of your solid objects. Calculating mass can be done for 2D regions as well as 3D objects. The regions that you created and edited earlier in this chapter are candidates for mass properties.

To find the mass properties of a solid or region, use the Mass Properties tool. Click on Mass Properties on the Standard toolbar (a flyout usually located under the Distance tool), choose Tools ➢ Inquiry ➢ Setup ➢ Mass Properties, or type **massprop**↵. Pick a solid (in this case we chose the base). AutoCAD will automatically bring up the text screen and list all the information shown in Table 12.1. At the end of the list, you will see the prompt Write to a file ? <N>:. If you answer Yes (y) you will be able to save this information to a file for use in another document or spreadsheet.

TABLE 12.1: MASS PROPERTY DATA RETURNED BY THE MASS PROPERTIES TOOL

Property	Data
Mass	17.7384
Volume	17.7384
Bounding box	X: −0.3125–2.3125 Y: 1.5000–5.0000 Z: −1.3750–6.0000
Centroid	X: 0.9909 Y: 3.5108 Z: 1.8372
Moments of inertia	X: 387.8420 Y: 175.5605 Z: 258.7756
Products of inertia	XY: 61.7523 YZ: 121.8875 ZX: 31.8773
Radii of gyration	X: 4.6760 Y: 3.1460 Z: 3.8195
Principal moments and X-Y-Z directions about centroid	I: 109.3338 along [1.0000–0.0076 0.0054] J: 99.0065 along [0.0081 0.9952–0.0975] K: 21.9833 along [−0.0047 0.0975 0.9952]

Taking Advantage of Rapid Prototyping

The corporate rush to get products to market as fast as possible has generated a demand to create prototypes from solid models as fast as possible. *Rapid prototyping* has been used to describe a process of creating parts as solid models which are directly interpreted and fabricated by methods exotic (like Stereolithography) and humble (like Computer Numerically Controlled, or CNC, mill). As an AutoCAD user, you have an excellent set of industry-standard tools to transfer your data directly from the model to the fabrication shop. This is euphemistically known as *art-to-part*. Many corporations use various forms of art-to-part to make prototypes—and even production parts—that have no conventional drawings to support them. Art-to-part is not yet considered a standard method of operation.

Here are some suggestions for solid modelers who wish to use rapid prototyping, or even use their models for production the same day:

- Draw all of your models to perfect nominal scale. Your model will not be evaluated by a human eye, and whatever discrepancies exist in the model will also exist in the product. If you are unable to draw your design in complete, accurate detail because of a limitation in the software, you will have to note the problem area so that it can get special attention in the shop.

- Add as much information to the file as you think is necessary to build the part—and then add a little more. You can use leaders (as you will use with an attribute balloon in Chapter 17) to point to surfaces requiring special finishes, such as with paint, without paint, ground, or ground to a specific finish. And you can point to welds or even highlight surfaces and use GD&T symbology to define contour, texture, and tolerance.

- Remember that the solid model generates much less ambiguity in definition than conventional Orthogonal view drawings.

You will want to export your solid and surface models to Computer Aided Manufacturing (CAM) software that is the front end for rapid prototyping machines. A few of these systems can accept AutoCAD drawing files directly. These are rare, because the drawing file format of the AutoCAD drawing file is binary and very difficult to read. If the receiving software can't read the AutoCAD binary file, the file must be translated into a readable format.

Data Translation and AutoCAD

AutoCAD has three primary data translation tools: DXF, SAT, and stereolithography (STL). Each of these translators has a significantly different purpose.

NOTE You may have heard of the International Graphics Exchange Standard (IGES) translation, or a vendor may have requested an IGES translation. Autodesk offers an add-on package for IGES translation.

This section is not intended to be an exercise in file export options; rather, it is an exploration of the concept of importing files as it relates to rapid prototyping. All of the File Export options are combined in a discussion of tasks in Chapter 18. Go there for specific information on the mechanics of importing and exporting models, text, and graphics.

Read on here to discover how file translation is applied to rapid prototyping.

DXF

By far the most commonly used of the translators included with AutoCAD is the *DXF translator*. Such a file carries the file extension .dxf. This translator is capable of exporting 2D data, 3D surface data, and solid-model data in a simple-to-read ASCII format. The ASCII format can be viewed and edited in your Write and Notepad programs, as well as in all desktop publishing programs. The DXF standard is published in the documentation that came with your copy of AutoCAD, so we won't go into it here. You need to know quite a lot about AutoCAD to edit a DXF file.

SAT and ACIS

At the core of AutoCAD solid modeling is the *ACIS solid-modeling engine*. This tool makes all of the wonderful solid models possible. A translation of a drawing file into an SAT file is not strictly a translation; rather, it is an export of the ACIS data. Some users consider this lack of translation to be a great advantage, because no errors induced by translation can affect the model. You could say that no translation occurred at all, but that the drawing data was stripped from the ACIS model. An SAT-translated file carries the file extension .sat, and it is not an editable or user-readable file. Like the AutoCAD drawing file, the SAT file is binary.

Unlike the DXF file, no other drawing data is exported. If you have placed annotation in this file, you will have to export that data using another tool, and then have users merge the SAT file and the data export files when they import these files into their systems.

One of the advantages of ACIS is that it is a standard engine used by a good number of CAD and CAM system manufactures.

Stereolithography

The earliest version of the *stereolithography* process used a vat of photosensitive plastic that hardened in the presence of a laser. When the machine read the file, it played a laser across the surface, creating a thin layer of solid plastic. The machine then lowered the solid by one-slice increments, and the laser hardened another layer. The process continued until the entire model was described, and the part was lifted from the vat and cleaned to provide a reasonable simulation of the finished part. Today, many machines use similar processes to create parts from materials ranging from paper to sintered metals.

Complex models can be created from a number of materials using processes similar to stereolithography, and many tools can accept the STL file. The translation of a solid model into the STL format creates a number of slices through your model at a pitch defined by you. STL files do not carry any annotation, so you will have to provide supporting documentation.

Talking to the Shop

A lot of 2D data goes to fabrication shops in the form of 2D multiple-view drawings saved to disk and sent by email, bulletin board, or floppy disk. If the shop can use your accurate drawing's geometry in their process of creating the program for the CNC, you'll save them time *and* reduce the possibility of making an error in transposing the information.

If you have drafting experience, you know that there can be a lot of ambiguity in an orthographically projected view using hidden lines and the silhouettes of parts to describe the shape of the part. These problems do not exist in 3D modeling, because the CAM software creates the evaluated shape. You will still need a creative approach, however, to indicate thread size, pitch, and type. Tolerances and surface contour specifications are not covered in any ANSI specification, so it's a good idea to keep a running dialog going with the fabrication shop, whether they are in-house or a vendor. If you and the shop are both proactive, you will soon arrive at the appropriate methods for defining this information. For instance, you might use an inspection document, which can easily be created using the Solview and Soldraw commands, to help you communicate with the fabrication shop. The inspection document would carry special instructions for the shop and assist your inspection department to interpret the part.

3D and Solid Modeling Design Tips

The joy and the curse of solid modeling is the accuracy of the model. When you design a part, keep in mind that some kind of machine will have to make it. For instance, the cover that you drew on the face of the post earlier in this chapter has a pocket for the gib. The pocket has square inside corners, and the outside has square corners all around. If the part were to be machined, the pocket could not be made as drawn. If the cover were fabricated by a mill, it would have either rounded corners or corner reliefs. And the shop would certainly call, asking you which way to proceed.

As you'll recall from chamfering the edges of the table, you also have to keep in mind how the piece will be used once it is manufactured. Our press' cover on the face of the post might need chamfered outside corners, because this is a hand tool, and you must consider how hands will be placed when someone is using the tool.

A machine shop is only one of many places where AutoCAD solid models can be found. The sheet metal shop, the tool shop, and the welding shop often see the work of AutoCAD solid modelers. Each of these specialties has certain standards and process limitations. A precision sheet metal shop, for example, has a certain minimum-bend radius for the thickness and type of material, a need for corner relief to keep the material from tearing, and special features that can be made, such as half shears and extruded tapped holes. Each of these must be drawn and defined with the help of the shop.

If You Want to Experiment...

You've covered a lot of territory in this chapter, so it would be a good idea to play with these commands to help you remember what you've learned. Try finishing the dimensions in the 2D base drawing that you made.

Other options for creating views in Setup View are a Section view and an Auxiliary view. Open the Base drawing and use Setup View and Setup Drawing to create two additional views. You will have to change the limits to 44,34 (E size) to accommodate the new views.

The rack, pinion, shaft collar for the pinion, handle and table pin, jack screws, nuts, and cover mounting screws are included in a file called `ch12 parts for the press`. Open the Press drawing, insert the block, explode it, and place all of the parts in the assembly to make your drawing look like Figure 12.28.

FIGURE 12.28

The complete press assembly

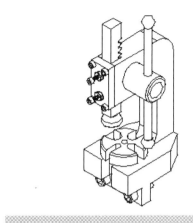

Chapter 13

Using 3D Surfaces

FEATURING

Creating a Surface Model

Getting the 3D Results You Want

Drawing 3D Surfaces

Creating Complex 3D Surfaces

Other Surface Drawing Tools

Editing a Mesh

Moving Objects in 3D Space

Viewing Your Model in Perspective

Using 3D Surfaces

A *3D surface* is an object that can take any shape—from a simple plane to the convoluted undulations of a contour map. A surface has no thickness, only area, thus a surface has no volume. You can create surfaces that cannot be described by AutoCAD solid modeling because they are not solids—they're several combined surfaces.

For example, the elliptical shape of an airplane fuselage and the blend of the wing to the body require complex shape descriptions. The complex contours of many consumer products that have molded components, like a computer keyboard, are good candidates for surface models. The hull of a boat or the body of a car are very difficult solid models because they are not confined to a single plane, but these represent much less of a challenge to surface modeling than something with many parts. AutoCAD surface models, like their solid-model counterparts, are basic tools; Mechanical Desktop, an add-on product from Autodesk, provides the sophisticated tools necessary to manage very complex surface models.

Creating a Surface Model

With surface modeling, you use two types of objects. One is called a 3D Face, which you will learn about later in this chapter. The other is the standard AutoCAD set of objects you've been using all along, but with a slight twist. By changing the thickness property of objects, you can create 3D surfaces. These surfaces, along with some 3D editing tools, let you create some very interesting forms (see Figure 13.1).

FIGURE 13.1
A cube created using thickness

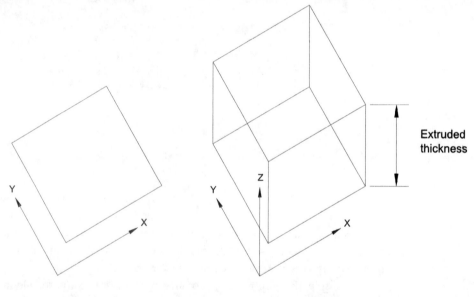

Square drawn with lines   Lines extruded to form cube

Elevation is the drafting term for the height of an object.

Another object property related to 3D is elevation. You can set AutoCAD so that everything you draw has an elevation. By default, objects have a zero elevation. This means that objects are drawn on the imagined 2D plane of Model Space, but you can set the Z coordinate for your objects so that whatever you draw is above or below that surface. An object with an elevation value other than 0 rests not *on* the imagined drawing surface but *above* it (or *below* it if the Z coordinate is a negative value). Figure 13.2 illustrates this concept.

FIGURE 13.2
Two identical objects at different Z coordinates

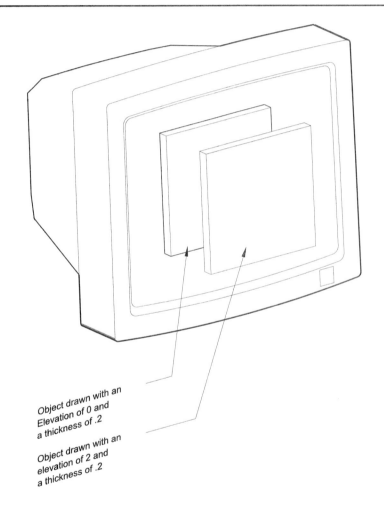

Object drawn with an Elevation of 0 and a thickness of .2

Object drawn with an elevation of 2 and a thickness of .2

Changing a 2D Polyline into a 3D Model

In this exercise, you will change the thickness of a 2D polyline to create 3D surfaces by changing the polyline properties.

1. Start AutoCAD and open a new file using the Versions template and save it as **surfaces**. (See Chapter 5 if you need a refresher on templates.)
2. Turn off the grid if it's on.
3. Click on View ➢ 3D Viewpoint ➢ SW Isometric.
4. Draw a rectangle 4 × 7 from point 2,2.
5. Click on the Fillet button of the Modify toolbar.
6. At the `Polyline/Radius/Trim/<Select first object>`: prompt, type **r**↵.
7. At the `fillet radius <0.000>`: prompt, enter **.56**↵↵.
8. At the `Polyline/Radius/Trim/<Select first object>`: prompt, type **p**↵ to use the Polyline option and pick the polyline. You have now filleted all four corners.
9. Click on the Properties button on the Object Properties toolbar.
10. At the `Select object`: prompt, pick the polyline and press ↵.
11. At the Change Properties dialog box, double-click on the Thickness input box, enter **5.5**, and click on OK.
12. Zoom All to see that the polyline has tall sides.

Figure 13.3 shows the extruded polyline. You can see through the "surfaces" because this is a *Wire-frame view*. Just as with solids, a Wire-frame view shows the volume of a 3D object by showing the lines representing the intersections of surfaces.

13. Type **hi**↵ to use the Hide command. You will see that the polyline is a surface. It can obscure all that is behind it from your point of view.

Next you will offset the polyline and change the elevation of the original polyline. The Change command is only available from the command line.

1. Offset the polyline. Click on the Offset tool from the Modify toolbar.
2. At the `Offset distance or Through <0.000>`: prompt, type **.75**↵.
3. At the `Select object to offset`: prompt, pick the polyline.
4. At the `Side to offset?` prompt, pick anywhere outside of the polyline.
5. Finally, at the `Select object to offset`: prompt, press ↵ to finish the command.

You have created a new polyline with the same thickness as the first one, and offset from it. Now you need to change the elevation of the inside line.

1. Type **hi**↵ to see the two polylines with the hidden lines removed.
2. Type **change**.↵ to begin the change command.
3. At the Select objects: prompt, pick the inside polyline and press ↵ to complete the selection set.
4. At the Properties/<Change point>: prompt, type **p**↵ to use the Properties option. This prompt provides the opportunity to change the endpoint of a line or arc, or to change the properties of the object that you have selected.
5. At the Change what property (Color/Elev/LAyer/LType/ltScale/Thickness)?: prompt, type **e**↵ to use the Elevation option. You can use this prompt to change the properties of an object by choosing from a selection of object properties. All the command line properties are available in the Object Properties dialog box except one: the option that allows you to change the elevation of the base point of an object.

You can use the Move tool to move the object or the grips stretch to change properties like elevation and thickness.

6. You have only to enter the new value. At the New elevation <0.0000>: prompt, type **2**↵ and at the Change what property (Color/Elev/LAyer/LType/ltScale/Thickness)?: prompt, press ↵.

At one time the only way that you could change the list of properties in the command was through the prompt, Change what property (Color/Elev/LAyer/LType/ltScale/Thickness)?.

The drawing in Figure 13.3 shows the effects of the change of thickness, the offset, and the change in elevation of the polylines after the Hide command has obscured the hidden surfaces.

FIGURE 13.3
The two polylines after they have been given thickness, and one has had the elevation changed

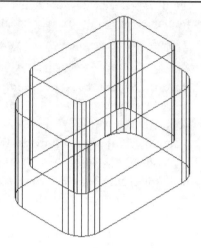

Creating a 3D Object

Although you may visualize a design in 3D, you will often start sketching it in 2D and later generate the 3D views. When you know from the start what the thickness and elevation of an object will be, you can set these values so that you don't have to extrude the object later. The following exercise shows you how to set elevation and thickness before you start drawing.

1. Choose Format ➢ Thickness.
2. Enter **1.2**↵ at the New Current Thickness: prompt. Now as you draw objects, they will appear 1.2 units thick.
3. Draw a circle next to the polylines at approximately coordinate 9,2. Make your circle 3 units in diameter. The circle appears as a 3D object with the current thickness and elevation settings, as in Figure 13.4.

If you use the same thickness and elevation often, you can create a template file with these settings so they are readily available when you start your drawings. The command for setting thickness and elevation is Elev. You can also use the Elevation and Thickness system variables to set the default elevation and thickness of new objects.

FIGURE 13.4

The circle and polylines

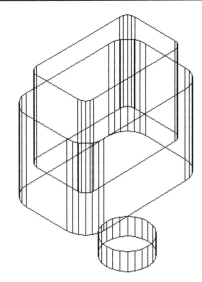

After you have set the thickness setting to 1.2, everything you draw will have a thickness of 1.2 units until you change it back to zero or some other setting.

NOTE

You can change the default elevation from 0 to some other (positive or negative) value. To do this, type **elev**↵ and then enter the elevation you want. You are then prompted for a thickness: Enter the value you want or hit ↵ to accept the default setting.

Giving objects thickness and modifying their elevation is a very simple process, as you have seen. With these two properties, you can create quite a few three-dimensional forms.

Getting the 3D Results You Want

Working in 3D is tricky because you can't see exactly what you are drawing. You must alternately draw and then hide or shade your drawing from time to time to see exactly what is going on. Here are a few tips on how to keep control of your 3D drawings.

Making Horizontal Surfaces Opaque

To make a horizontal surface appear opaque, you must draw it with a wide polyline, a solid hatch, a 3D Face, or a region. For example, consider the polylines you drew for the changing elevation process; the shape formed by the two lines is open at both the top and the bottom. Only the sides of the polyline have become opaque. To make the entire top "surface" opaque, you can use a Solid Hatch or an object called a region. (If you need a refresher on regions, see Chapter 12.) When the lines are hidden, the top appears to be opaque (see Figure 13.5).

FIGURE 13.5

The polyline before and after adding a region the to the top

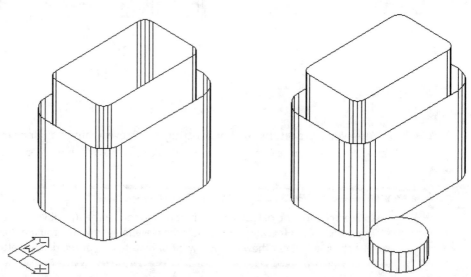

When a circle is extruded, the top surface appears opaque when you use the Hide command. Where you want to show an opening at the top of a circular volume, as in a circular chimney, you can use two 180° arcs (see Figure 13.6).

To create more complex horizontal surfaces, you can use a combination of wide polylines, solids, and 3D Faces.

FIGURE 13.6

A circle and two joined arcs

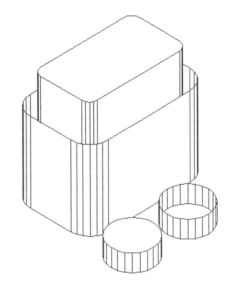

Setting Layers Carefully

Bear in mind that the Hide command hides objects that are obscured by other objects on layers that are turned off. For example, if the inside polyline is on a layer that is off when you use Hide, the lines behind the polyline are hidden, even though the polyline does not appear in the view (see Figure 13.7). You can, however, freeze any layer containing objects that you do not want to affect the hidden-line removal process. You can also use solid modeling (described in Chapters 11 and 12) to draw complex 3D objects with holes.

FIGURE 13.7

A polygon hiding a line, when the polygon's layer is turned off

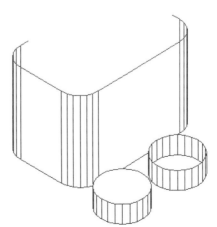

Drawing 3D Surfaces

In your work with 3D so far in this chapter, you have simply extruded existing forms, or you have set AutoCAD to draw extruded objects. But extruded forms have their limitations. Using just extruded forms, it's hard to draw diagonal surfaces in the Z axis. AutoCAD provides the 3D Face object to give you more flexibility in drawing surfaces in three-dimensional space. The 3D Face produces a 3D surface where each corner can be given a different X, Y, and Z value. By using 3D faces in conjunction with extruded objects, you can create a 3D model of just about anything. When you view these 3D objects in a 2D Plan view, you will see them as 2D objects showing only the X and Y positions of their corners or endpoints.

If you plan to view a 3D drawing in 2D at some later stage, you can use Dview or 3D Viewpoint presets.

Using Point Filters

Before you start working with 3D surfaces, you should have a good idea of what the Z coordinate values are for your model. The simplest way to construct surfaces in 3D space is to first create some layout lines to help you place the endpoints of 3D faces.

AutoCAD offers a method for 3D point selection, called *point filtering*, that simplifies the selection of Z coordinates. Point filtering allows you to enter an X, Y, or Z value by picking a point on the screen and telling AutoCAD to use only the X, Y, or Z value of that point, or any combination of those values. If you don't specify a Z coordinate, the current elevation setting is assumed.

If you use filters, you will be asked to supply any value that is not already set.

In the following exercises, let's imagine you have been asked to design a decorative addition to a new product. The Marketing head wants a five-pointed star projected to a point as a logo on the injection-molded front panel of the Mark Five Widget. In creating the star, you will practice using 3D faces and filters. You'll start by doing some setup, so you can work on a like-new drawing.

Laying Out a 3D Form Object

Spherical and Cylindrical Coordinate Formats

In many of these exercises, you used relative Cartesian coordinates to locate the second point for the Move command. For commands that accept 3D input, you can also specify displacements by using the *spherical* and *cylindrical coordinate* formats.

The spherical coordinate format lets you specify a distance in 3D space while specifying the angle in terms of degrees from the X axis of the current UCS, and degrees from the X-Y plane of the current UCS (see the top image of Figure 13.8). In other words, you create an object formed by defined distances from the X and Y axes, rather than spherically exploding a two-dimensional object along three planes. You don't need to be creating a sphere to use this coordinate formatting; you just need to be able to think of the shape as a piece of a sphere.

The cylindrical coordinate format, on the other hand, lets you specify a location in terms of a distance on the plane of the current UCS and a distance on the Z axis. You also specify an angle from the X axis of the current UCS (see the bottom image of Figure 13.8). In other words, you create a cylindrical object at a defined distance from the Z axis along the planes formed by the UCS and the X axis, rather than expanding a two-dimensional X-Y oriented rectangle along the Z axis. Again, you don't need to be drawing a cylinder to use this tool; you just need to be able to think of your object as a segment from a cylinder.

Now you are ready to lay out your star.

1. Choose Format ➢ Thickness and set the thickness to 0.

2. Draw a circle 4 units in diameter at coordinates 2,2. Click on the Circle tool in the Draw toolbar. At the `3P/2P/TTR/<Center point>:` prompt, type **2,2**↵ and at the `Diameter/<Radius> <0.000>:` prompt, type **d**↵. Then, at the `Diameter <0.000>:` prompt, type **4**↵.

3. Place a 5-sided polygon about the circle. Click on the Polygon tool on the Draw toolbar and at the `Number of sides <4>:` prompt, type **5**↵. At the `Edge/<Center of polygon>:` prompt, use the Center Osnap marker to pick the center of the circle, and at the `Inscribed in circle/Circumscribed about circle (I/C) <C>:` prompt, press ↵ to accept the default. At the `Radius of circle:` prompt, use the Quadrant Osnap marker to pick the bottom quadrant.

FIGURE 13.8

The spherical and cylindrical coordinate formats

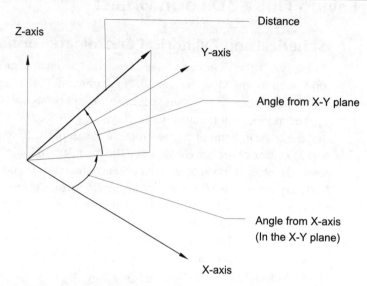

[Distance] < [Angle from X-axis] < [Angle from X-Y plane]

The Spherical Coordinate Format

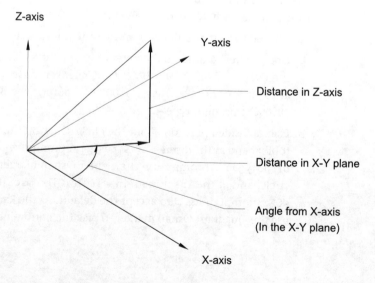

[Distance in X-Y plane] < [Angle from X-axis] , [Distance in Z-axis]

The Cylindrical Coordinate Format

Next you will draw the lines that make up a five-pointed star. If you aren't sure what this shape looks like, see Figure 13.9.

FIGURE 13.9

The five-pointed star drawn from the points of the pentagon circumscribed about the circle

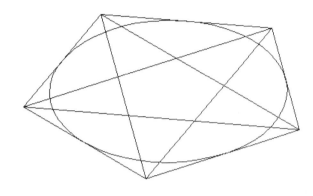

4. Use the Line tool from the Draw toolbar and the Endpoint Osnap marker to pick the endpoints of the lines of the pentagon.
5. Draw another line from the center of the circle 2 units in the positive Z axis. Press ↵ to start the Line command and at the `From point:` prompt, use the Center Osnap marker to locate and pick the center of the circle. At the `To point:` prompt, type **@0,0,2**↵ to draw the vertical line. Press ↵ to finish the command.

Now we need to use the Surfaces tools to begin giving the star three dimensions.

Loading the Surfaces Toolbar

The 3D Face command and AutoCAD's 3D shapes are located on the Surfaces toolbar. Right-click on any toolbar to open the Toolbars dialog box, and then click on the Surfaces check box. Click on Close to close the Toolbars dialog box.

Adding a 3D Face

Now that you've opened the Surfaces toolbar, you can begin to draw 3D Faces on the star.

1. Zoom in to the lines you just created, so you have a view similar to Figure 13.10.

FIGURE 13.10

Zooming in on the star lines so you can pick the points to make the 3D Face

2. Click on the 3D Face button on the Surfaces toolbar, or type **3f↵**. You can also choose Draw ➢ Surfaces ➢ 3D Face from the menu bar.

3. At the First Point: prompt, use the Osnap overrides to pick the first of the five endpoints of the 3D lines you drew. (You could pick any endpoint for a starting point.) Be sure the Ortho mode is off.

DRAWING 3D SURFACES | 597

The Running Osnap mode can help you select endpoints quickly in this exercise. See Chapter 4 if you need to review how to set up the Running Osnap mode.

   4. As you continue to pick the endpoints, you will be prompted for the second, third, and fourth points.

With the 3D Face tool, you pick three points per surface of the star in a circular fashion, as shown in Figure 13.10. When prompted for the fourth point, press ↵. Once you've drawn one 3D Face, you can continue to add more by selecting more points.

   5. When the Third point: prompt appears again, press ↵ to end the 3D Face command. A 3D face appears between the two 3D lines. It is difficult to tell if the 3D points are actually there until you use the Hide command, but you should see vertical lines connecting the endpoints of the 3D lines. These vertical lines are the edges of the 3D face (see Figure 13.11).

FIGURE 13.11
The 3D face

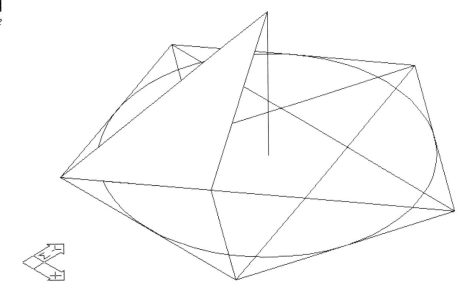

PART
III

Modeling and Imaging in 3D

NOTE When the `Third point:` prompt reappears, you can draw more 3D faces if you like. The next 3D face will use the last two points selected as the first two of its four corners—hence the prompt for a third point.

6. Use the 3D Face command to put another surface on the star, as demonstrated in the top image of Figure 13.12.

7. Use the Intersection Endpoint Osnap override to snap to the corners of the 3D faces.

8. Use the Hide command on the Render toolbar to get a view that looks like the bottom image of Figure 13.12.

Use the Array tool on the Modify toolbar to complete the star. The Polar Array option will create five sets of the surfaces around the center of the circle.

1. Select the Array tool and at the `Select objects:` prompt, pick the two surfaces that you just created. At the next `Select objects:` prompt, press ↵. At the `Rectangular or Polar array (<R>/P):` prompt, type **p**↵ to create a polar array, and at the `Base/<Specify center point of array>:` prompt, pick the center of the circle using the Center Osnap marker. At the `Number of items:` prompt, type **5**↵ to make five sets.

2. At the `Angle to fill (+=ccw, -=cw) <360>:` prompt, press ↵ to accept the default of 360°.

3. At the `Rotate objects as they are copied? <Y>:` prompt, press ↵ to accept the default of Yes.

4. Use the Hide tool again and your star will look like the one in Figure 13.13.

5. Save the `faces.dwg` file.

You might be pretty happy about the way the star looks, but if you want to remove some of the lines, read the next section. Otherwise, skip to the *Creating Complex 3D Surfaces* section, and we'll play with our star shape a little more.

FIGURE 13.12

The two faces of the star and the star with hidden lines removed

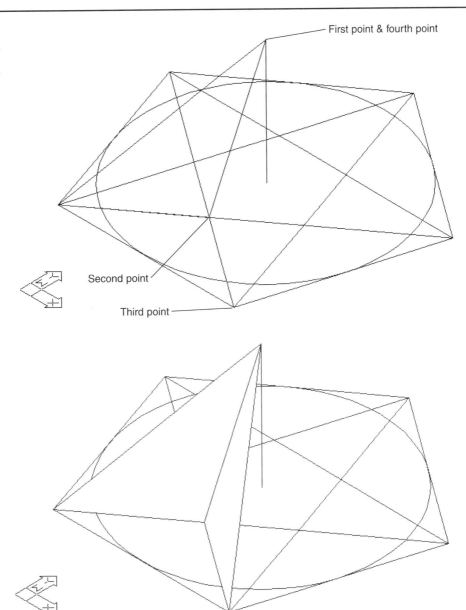

FIGURE 13.13
The complete star with hidden lines removed

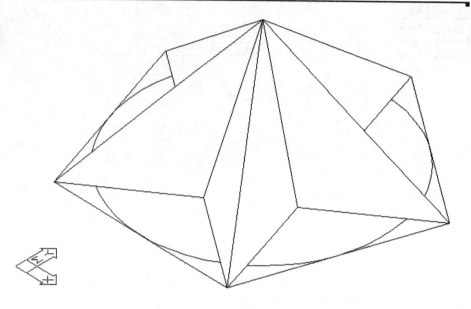

Hiding Unwanted Surface Edges

When using the 3D Face command, you are limited to drawing surfaces with three or four straight sides. You can, however, create more complex shapes by simply joining several 3D faces. Figure 13.14 shows an odd shape constructed of three joined 3D faces. Unfortunately, you are left with extra lines that cross the surface, as shown in the top image of Figure 13.14. You can hide those lines by using the Invisible option under the 3D Face command, in conjunction with the Splframe variable.

To make an edge of a 3D face invisible, start the 3D Face command as usual. While selecting points, just before you pick the first point of the edge to be hidden, enter **i↵**, as shown in the bottom image of Figure 13.14. When you are drawing two 3D faces sequentially, only one edge needs to be invisible to hide their joining edge.

You can make invisible edges visible for editing by setting the Splframe system variable to 1. Setting Splframe to 0 will cause AutoCAD to hide the invisible edges. Bear in mind that the Splframe system variable can be useful in both 3D and 2D drawings.

DRAWING 3D SURFACES   601

FIGURE 13.14

Hiding the joined edge of multiple 3D faces

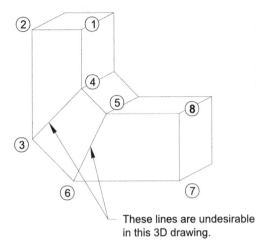

Drawing an odd-shaped surface using 3D Face generates extra lines. The numbers in the drawing to the left indicate the sequence of points selected to create the surface.

These lines are undesirable in this 3D drawing.

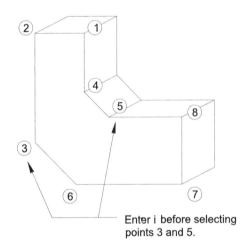

If you draw the same surface using the i option before selecting the appropriate points, the unwanted lines will be hidden. This drawing indicates where the i option is issued in the point-selection process.

Enter i before selecting points 3 and 5.

The Edge option on the Surfaces toolbar lets you change an existing, visible 3D face edge to an invisible one. Click on the Edge button and then select the 3D face edge to be hidden.

Using Pre-Defined 3D Surface Shapes

You may have noticed that the Surfaces toolbar offers several 3D surface objects, such as cones, spheres, and tori (which are donut-shaped). All are made up of 3D faces, which may not be apparent until after you have tried exploding one of these primitive surfaces. To use them, click on the appropriate button on the Surfaces toolbar. When you select an object, AutoCAD will prompt you for the points and dimensions that define that 3D object; then AutoCAD will draw the object. This provides quick access to shapes that would otherwise take substantial time to create.

Things to Watch Out for When Editing 3D Objects

You can use the Move Copy and Stretch commands on 3D lines, 3D faces, and 3D shapes to modify their Z coordinate values—but you have to be careful with these commands when editing in 3D. Here are a few tips to keep in mind:

- The Scale command will scale an object's Z coordinate value, as well as the standard X and Y coordinates. (Click and drag the Resize button on the Modify toolbar and select Scale.) Suppose you have an object with an elevation of two units. If you use the Scale command to enlarge that object by a factor of 4, the object will have a new elevation of 2 units times 4, or 8 units. If, on the other hand, that object has an elevation of 0, its elevation will not change, because 0 times 4 is still 0. But you could still force the change by selecting a base point on the object in question using an Osnap.

- Array, Mirror, and Rotate (on the Modify toolbar) can also be used on 3D lines, 3D faces, and 3D objects, but these commands won't affect their Z coordinate values. Z coordinates can be specified for base and insertion points, so you may be surprised by the results when you use these commands with 3D models.

- Using the Move, Stretch, and Copy commands (on the Modify toolbar) with object snaps can produce some unpredictable and unwanted results. As a rule, it is best to use point filters when selecting points with Osnap overrides. For example, to move an object from the endpoint of one object to the endpoint of another on the same Z coordinate, invoke the .XY point filter at the Base Point: and Second Point: prompts before issuing the endpoint override. Pick the endpoint of the object you want; then enter the Z coordinate, or just pick any point to use the current default Z coordinate.

- When you create a block, the block will use the UCS that is active at the time the block is created to determine its own Local Coordinate System. When that block is later inserted, it will orient its own coordinate system with the current UCS. (UCS is discussed in more detail in Chapter 11.)

Creating Complex 3D Surfaces

In the previous examples, you drew objects that were mostly straight lines or curves with a thickness. All the forms in your star were defined in planes perpendicular to each other. At times, however, you will want to draw objects that do not fit so easily into perpendicular or parallel planes. The following exercise demonstrates how you can create more complex forms using some of AutoCAD's other 3D commands.

Creating Curved 3D Surfaces

Next, imagine that the people in Marketing have changed their minds about the shape of the star. They would like to soften and contour some of the edges of the star while keeping its basic shape. They'd like to clip the point of the star, but they're not sure whether they'd like it clipped a lot or if it should just get a trim.

To soften the contour, first you need to define the perimeter of one of the sides of the star using arcs, and then use the Edge Surface tool on the Surfaces toolbar to form the shape of the draped surface. Edge Surface creates a surface based on four objects defining the edges. In this example, you will use arcs to define the edges of the star.

Before you draw the arcs defining the star edge, you must erase the ten surfaces that you drew in the last exercise, then create the top and bottom arcs and create a UCS in the plane of one of the edges.

1. Click on the Erase tool on the Modify toolbar and pick the surfaces one at a time until you have them all, then press ↵. Or you can pick and erase each surface one at a time to be sure that you are not erasing geometry that you want to keep. After you have erased all the surfaces that need to be erased, erase the lines that you drew inside the polygon. Do not erase the line that you drew in the Z axis.

You might consider copying the objects that are required for the new version to another drawing, maintaining old objects on a frozen layer for historical reasons.

2. Click on the Arc tool in the Draw toolbar. You are going to create an arc at the bottom of the star that begins at one point, ends at the adjacent point, and has a radius of 2.00 units. Click in the upper-left viewport to make it active. You can do this either before you start a new command or after you have started the command.

WARNING

You cannot change the viewpoint focus after you have entered the Pan or Zoom commands.

3. At the ARC Center/<Start point>: prompt, pick the lower-right corner of the polygon. At the Center/End/<Second point>: prompt, type **e↵** to use the End option to specify the end of the arc instead of the midpoint (Midpoint is the Second point default option). At the End point: prompt, pick the lower-left corner of the polygon, and at the Angle/Direction/Radius/ <Center point>: prompt, type **r↵** to specify the radius value. At the Radius: prompt, type **2↵**. Your drawing should look like Figure 13.15.

FIGURE 13.15

The polygon with an arc across the corners

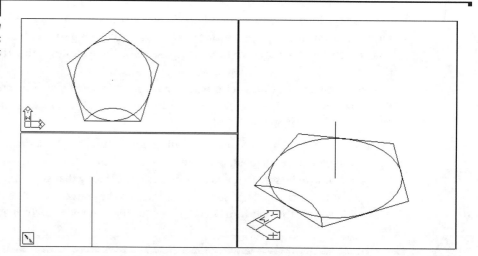

The next thing that you will do is draw an arc from the corner of the polygon to the top endpoint of the Z-axis line. Before you can draw the arc, you must align a UCS with this plane, which the arc will also align with. Remember that all 2D geometry is drawn on the current UCS plane.

1. To create a UCS that is aligned with the center and one point of the polygon, click on the 3 Point UCS tool on the UCS toolbar, or choose Tools ➢ UCS ➢ 3 Point, then create a UCS using the three points shown in the upper-left image of Figure 13.16.

FIGURE 13.16

The points to pick to create the UCS

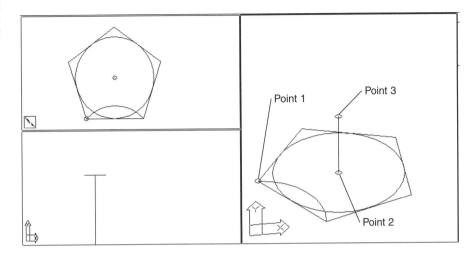

2. Choose Draw ➢ Arc ➢ Start, End, Radius.
3. Draw the arc defining another edge of the star (see Figure 13.17). Use the Endpoint Osnap marker to pick the endpoints of the polygon and the Z-axis line. Pick the corner of the polygon first, then pick the top end of the Z-axis line, and then enter the radius (2 units will do it) in response to the prompts. (If you need help with the Arc command, refer to Chapter 3.)

FIGURE 13.17

Drawing the arc in the new UCS

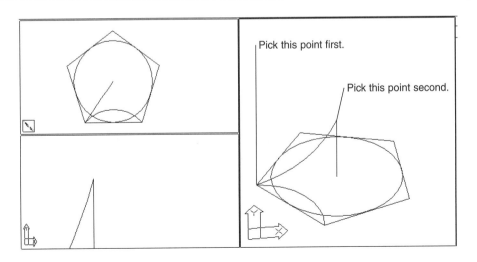

You still need to create the arc on the other side of the surface. Click on the World UCS tool on the UCS toolbar to return the UCS to the World plane.

Next, you will mirror the side-edge arc to the opposite side. This will save you from having to define a UCS for that side.

1. Click on the Mirror tool in the Modify toolbar. Click in the upper-left viewport to change the focus to that viewport. The arc that you are going to mirror looks like a straight line going from the center of the circle to the lower-left corner of the polygon. At the Select objects: prompt, pick the arc that you just drew and press ↵ twice to complete the selection set. At the First point of mirror line: prompt, you can pick the center of the circle or the midpoint of the bottom arc. At the Second point: prompt, type **@1<90**↵.

2. At the Delete old objects? <N>: prompt, press ↵ to accept the default and not delete your original arc.

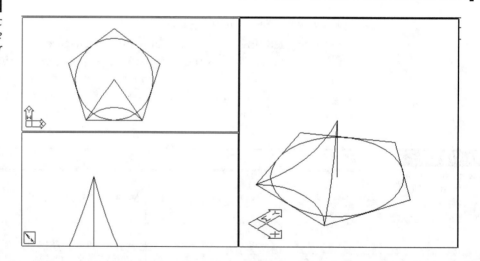

FIGURE 13.18

Mirroring the arc defining the side of the star

The top of the two-sided arc needs to be trimmed .2 units below the point and connected with a line to clip the point of the star.

1. Click in the lower-left viewport to change viewports. Click on the View UCS tool on the UCS toolbar to set a UCS parallel to this view.

2. Draw a horizontal line from the top of the arcs .5 units in the X direction. Click on the Line tool on the Draw toolbar, and at the From point: prompt, type **@.5<0**↵. Or, with the Ortho mode on, drag the mouse directly to the right and type **.5**↵ to use the Direct Distance Entry (DDE) feature.
3. Move the line .25 units in the –X direction and .2 units in the –Y direction. To do this, click on the line to highlight the grips and pick any of them to turn the Move tool on. Right-click to pop up the Grips menu and select Move. At the <Move to point>/Base point/Copy/Undo/eXit: prompt, type **@-.25,-.2**↵.
4. Use the Zoom Window tool on the Standard toolbar to enlarge the view of the point on the star in the lower-left viewport. Figure 13.19 shows several views of the clipped point.

NOTE

After you completed each command to draw, edit, or delete objects in one viewport, all of the other viewports were updated.

FIGURE 13.19

The line has been drawn and the lower-left viewport is highlighted and zoomed in on the point of the star.

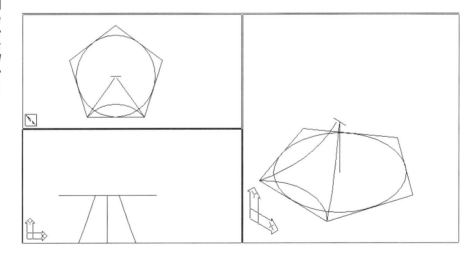

5. Click on the Trim tool from the Modify toolbar. At the `Select cutting edges: (Projmode = UCS, Edgemode = No extend) Select objects:` prompt, press ↵ to use the Automatic Trim feature. At the `<Select object to trim>/Project/Edge/Undo:` prompt, type **f**↵ to use the Fence selection set option, and draw a fence across all three objects (the two arcs and the line that we have been calling the Z-axis line) above the last line that you drew. Press ↵ when you are done. All three objects are trimmed.

6. Erase the last line that you drew. You only drew it as a construction line to help to trim the other objects.

7. Draw a line from the top end of one arc to the top end of the other arc. The Edgesurf command that you'll use to apply surfaces to your arcs requires four edges, and you have just drawn the fourth one.

NOTE

You can draw the line described in step 7 in any of the three viewports. The Isometric view will probably best demonstrate what is happening, but either of the other two viewports will allow you to pick the same points.

Finally, let's finish off this star by adding the surface representing the star's side and then arraying it around the star. Marketing has mentioned that they like a mesh texture.

1. Click on the Edge Surface button on the Surfaces toolbar, or enter **edgesurf**↵ at the command prompt. Change the viewport focus to the Isometric viewport (the one on the right).

NOTE

To display the Surfaces toolbar, choose Tools ➤ Toolbars ➤ Surfaces.

2. At the `Select edge 1:` prompt, pick the arc on the World UCS.
3. At the `Select edge 2:` prompt, pick either side arc.

In order for the Edgesurf command to work properly, the arcs—or any set of objects—used to define the boundary of a mesh with the Edge Surface option must be connected exactly end to end.

4. Change the viewpoint focus to the lower-left viewport (click once in the viewport to highlight it) for the most detailed view of the surface, and pick the last line that you drew. Change the viewport focus once more back to the Isometric view and pick the last unselected arc. (The arcs should be picked in a circular fashion, not crosswise.) A mesh will appear, filling the space between the four arcs. Your star is now complete.

TIP

Before starting the Array command, be sure that the UCS has been set to World.

5. Click on the Array tool on the Modify toolbar. At the `Select objects:` prompt, type **L↵** to select the last object drawn, then at the `Rectangular or Polar array (<R>/P):` prompt, type **p↵**.
6. At the `Base/<Specify center point of array>:` prompt, pick the center of the circle, and at the `Number of items:` prompt, type **5↵**. At the `Angle to fill (+=ccw, -=cw) <360>:` prompt, press ↵, and at the `Rotate objects as they are copied? <Y>:` prompt, press ↵ to accept the default.
7. Use View ➢ Hide to get a better view of the star. Be sure that the viewport is the Isometric view. You should have a view similar to Figure 13.20.
8. Save this file.

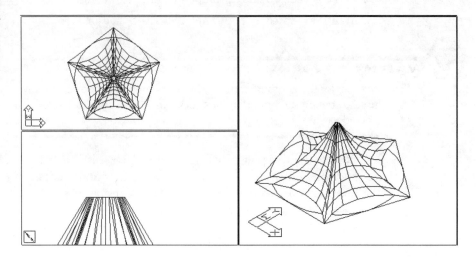

FIGURE 13.20
The completed star with meshes, in the Isometric viewport

At this point, you've tried a few of the options on the Surfaces toolbar. You'll get a chance to use a few more later in this chapter. Next, you'll learn how to edit mesh objects like the star.

Adjusting the Settings That Control Meshes

As you can see from Figure 13.20, the draping mesh of our star is made up of rectangular segments. If you want to increase the number of segments in the mesh (you can get a look like that in Figure 13.21), you can change the Surftab1 and Surftab2 system variables. Surftab1 controls the number of segments along edge 1, the first edge you pick in the sequence; and Surftab2 controls the number of segments along edge 2. Auto-CAD refers to the direction of edge 1 as *m* and the direction of edge 2 as *n*. These two directions can be loosely described as the X and Y axes of the mesh, with m being the X axis and n being the Y axis.

In Figure 13.21, the setting for Surftab1 is 25, and for Surftab2 the setting is 25. The default value for both settings is 6. If you would like to try different Surftab settings on the star mesh, you must erase the existing mesh, change the Surftab settings, and then use the Edge Surface tool again to define the mesh.

FIGURE 13.21
The star with different Surftab settings

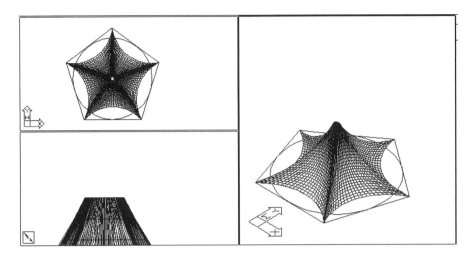

Creating a 3D Mesh by Specifying Coordinates

If you need to draw a mesh like the one in the previous example, but you want to give exact coordinates for each vertex in the mesh grid, you can use the 3D Mesh command. Suppose you have data from a survey of a part from a coordinate measuring machine; you can use 3D Mesh to convert your data into a graphic representation of its topography. Another use for 3D Mesh is to plot mathematical data to get a graphic representation of a formula.

Because you must enter the coordinate for each vertex in the mesh, 3D Mesh is better suited to scripts or AutoLISP programs, where a list of coordinates can be applied automatically to the 3D Mesh command in a sequential order. See Chapter 19 and the companion CD-ROM for more on AutoLISP.

Other Surface Drawing Tools

In the last exercise, you used the Edge Surface tool to create a 3D surface. There are several other 3D surface commands available which allow you to generate complex surface shapes easily.

All the surface drawing tools described in this section, along with the meshes you're already familiar with, are actually composites of 3D faces. This means these 3D objects can be exploded into their component 3D faces, which in turn can be edited individually.

Using Two Objects to Define a Surface

The Ruled Surface tool on the Surfaces toolbar draws a surface between two 2D objects, such as a line and an arc or a polyline and an arc. This command is useful for creating extruded forms that transform from one shape to another along a straight path. Let's see how this command works.

1. Open the file called Rulesurf.dwg from the companion CD-ROM. It looks like the top image of Figure 13.22. This drawing is of a simple half circle drawn using a line and an arc. Ignore the diagonal blue line for now.

2. Move the connecting line that is between the arc endpoints 10" in the Z axis. You already know how to use the Move tool from earlier in this chapter.

3. Now you are ready to use a 3D surface to connect the half circle with the line that used to connect the arc's ends. Click on the Ruled Surface button on the Surfaces toolbar, or choose Draw ➤ Surfaces ➤ Ruled Surface.

4. At the Select first defining curve: prompt, place the cursor toward the right end of the arc and click on it.

5. At the Select second defining curve: prompt, move the cursor toward the right end of the line and click on it, as shown in the bottom image of Figure 13.22. The surface will appear as shown in Figure 13.23.

The position you use to pick the second object will determine how the surface is generated.

FIGURE 13.22

Drawing two edges for the Ruled Surface option

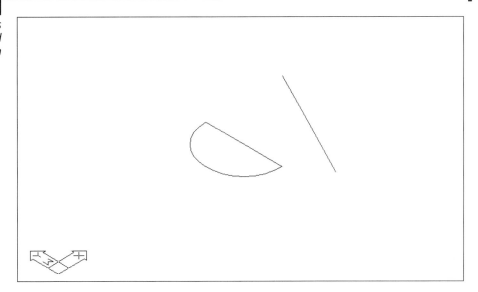
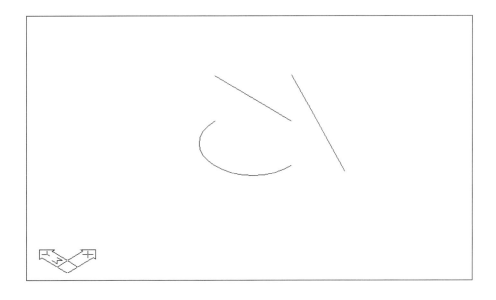

FIGURE 13.23

The Rulesurf surface

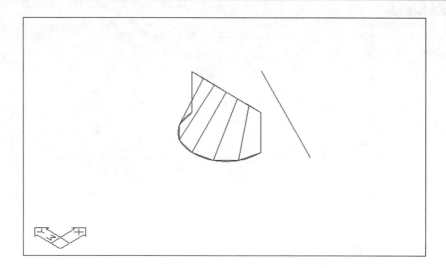

The location you use to select the two objects for the Ruled Surface is important. We asked you to select specific locations on the arc and line so that the ruled surface is generated properly. Had you selected the opposite end of the line, for example, your result would look more like Figure 13.24. Notice that the segments defining the surface cross each other. This crossing effect is caused by picking the defining objects near opposite endpoints. The arc was picked near its lower end, and the line was picked toward the top end. If you needed to produce a pinwheel or a propeller, you might actually want this twisted effect.

FIGURE 13.24

The ruled surface redrawn by using different points to select the objects

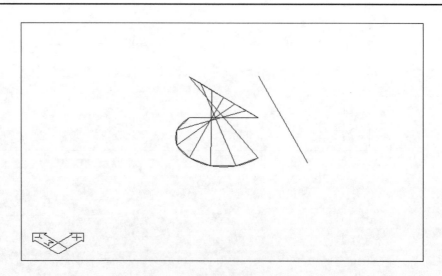

Extruding an Object along a Straight Line

The Tabulated Surface tool also uses two objects to draw a 3D surface, but instead of drawing the surface between the objects, Tabulated Surface extrudes one object in a direction defined by a direction vector. The result is an extruded shape the length and direction of the direction vector. To see what this means, try the following exercise.

1. While still in the Rulesurf drawing, click on the Undo button in the Standard toolbar to undo the Ruled Surface from the previous exercise.

2. Click on Tabulated Surface from the Surface toolbar, or choose Draw ➤ Surfaces ➤ Tabulated Surface.

3. At the Path curve: prompt, click on the arc.

4. At the Select Direction Vector: prompt, click on the lower end of the blue line farthest to the right. The arc is extruded in the direction of the blue line, as shown in Figure 13.25.

FIGURE 13.25

Extruding an arc using a direction vector

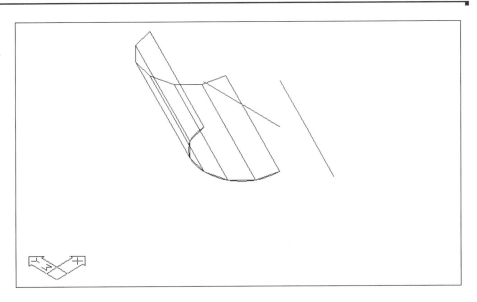

The direction vector can be any object, but AutoCAD will only consider the object's two endpoints when extruding the path curve. Just as with the Ruled Surface, the point at which you select the direction vector object affects the outcome of the extrusion. If you had selected a location near the top of the blue line, the extrusion would have gone in the opposite direction.

Since the direction vector can point in any direction, the Tabulated Surface tool allows you to create an extruded shape that is not restricted to a direction perpendicular to the object being extruded.

The path curve defining the shape of the extrusion can be an arc, circle, line, or polyline. You can use a curve-fitted polyline or a spline polyline to create more complex shapes, as shown in Figure 13.26.

FIGURE 13.26

Some samples of other shapes created using Ruled Surface and Tabulated Surface

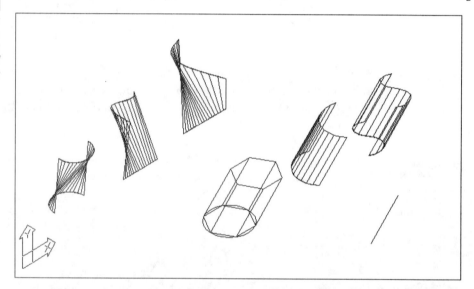

If you want to increase the number of facets in Ruled Surface or in Tabulated Surface, set the Surftab1 system variable to the number of facets you desire.

Extruding a Circular Surface

The Revolved Surface tool allows you to quickly generate circular extrusions. Typical examples are vases or tea cups. The following exercise shows how the Revolved Surface tool is used to draw a pitcher. You'll use an existing drawing that has a profile of the pitcher already drawn.

1. Open the `Pitcher.dwg` file from the companion CD-ROM. This file contains a polyline profile of a pitcher and a single line representing the center of the pitcher (see the top image of Figure 13.28, which we'll come back to in a moment). The profile and line have already been rotated to a position perpendicular to the WCS. The grid is turned on so you can better visualize the plane of the WCS.

2. Click on the Revolved Surface button on the Surfaces Toolbar.

3. At the `Select Path Curve:` prompt, click on the polyline profile, as shown in the top image of Figure 13.27.

4. At the `Select Axis of Revolution:` prompt, click near the bottom of the vertical line representing the center of the vase, as shown in the top image of Figure 13.27.

5. At the `Start angle <0>:` prompt, press ↵ to accept the 0 start angle.

6. At the `Included angle (+=ccw, -=cw) <Full circle>:` prompt, press ↵ to accept the Full Circle default. The pitcher appears, as shown in the bottom image of Figure 13.27.

FIGURE 13.27

Drawing a pitcher using the Revolved Surface tool

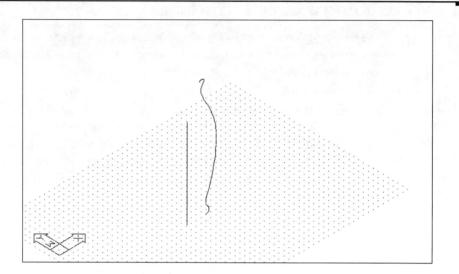

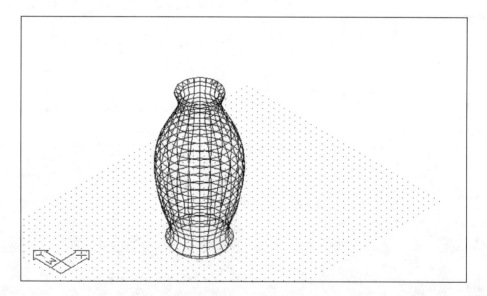

Notice that the pitcher is made up of a faceted mesh, like the mesh that is created by the Edge Surface tool. Just as with the Edge Surface tool, you can set the number of facets in each direction using the Surftab1 and Surftab2 system variable settings. Both Surftab1 and Surftab2 were already set to 24 in the `Pitcher.dwg` file, so the pitcher shape appears fairly smooth.

You may have noticed that in steps 5 and 6 of the previous exercise you have a few options. In step 5, you can specify a start angle. In this case, you accepted the 0 default. Had you entered a different value, 90 for example, the extrusion would have started in the 90° position relative to the current WCS. In step 6, you have the option of specifying the angle of the extrusion. Had you entered 180, for example, your result would have been half the pitcher. You can also specify the direction of the extrusion by specifying a negative or positive angle.

Editing a Mesh

Once you've created a mesh surface with either the Edge Surface or Revolved Surface tool, you can make modifications to it. For example, suppose you wanted to add a spout to the pitcher you created in the previous exercise. You can use grips to adjust the individual points on the mesh to reshape the object. Here, you must take care how you select points. The UCS will become useful for editing meshes, as shown in the following exercise.

1. Zoom into the area shown in the top image of Figure 13.28.
2. Click on the pitcher mesh to expose its grips.
3. Shift+click on the grips shown in the middle image of Figure 13.28.
4. Click on the grip shown in the bottom image of Figure 13.28 and slowly drag the cursor to the left. As you move the cursor, notice how the lip of the pitcher deforms.
5. When you have dragged the cursor to the approximate location indicated in the bottom image of Figure 13.28, select that point. The spout will be fixed in the new position.

FIGURE 13.28

Adding a spout to the pitcher mesh

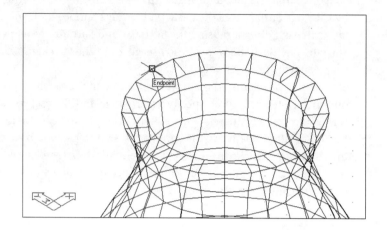

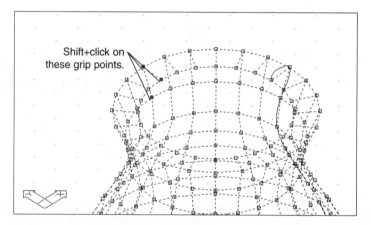

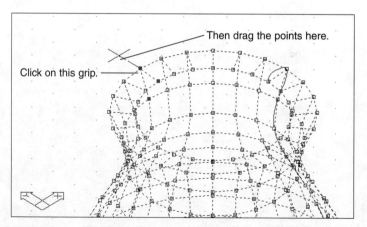

You can refine the shape of the spout by carefully adjusting the position of other grip points around the edge of the pitcher. Later, when you render the pitcher, you can apply a smooth shading value so that the sharp edges of the spout are smoothed out.

This exercise shows how easy it is to make changes to a mesh by moving individual grip locations. Be aware, however, that when you move mesh grips manually (rather than by entering coordinates), the grips' motion is restricted to a plane parallel to the current UCS. You can use this restriction to your advantage. For example, if you want to move the spout downward at a 30° angle, you can rotate the UCS so it is tipped at a 30° angle in relation to the top of the pitcher, and then edit the mesh grips as you did in the previous exercise.

Another option would be to specify a *relative* coordinate as opposed to selecting a point. By specifying a coordinate, such as @.5<50, you would not have to move the UCS. Using relative coordinates, however, removes the spontaneity of being able to select a point visually.

Other Mesh Editing Options

You can use Modify ➢ Object ➢ Polyline to edit meshes in a way similar to editing polylines. When you choose this option and pick a mesh, you get this prompt:

Edit vertex/Smooth surface/Desmooth/Mclose/Nclose/Undo/eXit <X>:

Here are descriptions of these options:

Edit vertex allows you to relocate individual vertices in the mesh.

Smooth surface is similar to the Spline option for polylines. Rather than having the mesh's shape determined by the vertex points, Smooth surface adjusts the mesh so that mesh vertices act as control points that pull the mesh—much as a spline frame pulls a spline curve.

You can adjust the amount of pull the vertex points exert on a mesh by using Smooth surface in conjunction with the Surftype system variable. If you'd like to know more about these settings, see Appendix D.

Desmooth reverses the effects of Smooth surface.

Mclose and **Nclose** allow you to close the mesh in either the m or n direction. When either of these options is used, the prompt line will change, replacing Mclose or Nclose with Mopen or Nopen, allowing you to open a closed mesh.

> **NOTE** The Edit Polyline tool in the Modify II toolbar performs the same function as Modify ➢ Object ➢ Polyline.

Moving Objects in 3D Space

AutoCAD provides two utilities for moving objects in 3D space: Align and 3D Rotate. Both of these commands are found on the Rotate flyout of the Modify toolbar. They help you perform some of the more common moves associated with 3D editing.

Aligning Objects in 3D Space

In mechanical drawing you often create the parts in 3D and then show an assembly of the parts. The Align option can greatly simplify the assembly process. The following exercise describes how Align works.

1. Choose Modify ➢ 3D Operation ➢ Align, or type **al**↵.
2. At the `Select objects:` prompt, select the 3D source object you want to align to another part. (The *source object* is the object you want to move.)
3. At the `1st source point:` prompt, pick a point on the source object that is the first point of an alignment axis—such as the center of a hole or the corner of a surface.
4. At the `1st destination point:` prompt, pick a point on the destination object to which you want the first source point to move. (The *destination object* is the object with which you want the source object to align.)
5. At the `2nd source point:` prompt, pick a point on the source object that is the second point of an alignment axis—such as another center point or other corner of a surface.
6. At the `2nd destination point:` prompt, pick a point on the destination object indicating how the first and second source points are to align in relation to the destination object.
7. At the `3rd source point:` prompt, you can press ↵ if two points are adequate to describe the alignment. Otherwise, pick a third point on the source object that, along with the first two points, best describes the surface plane you want aligned with the destination object.

8. At the 3rd destination point: prompt, pick a point on the destination object that, along with the previous two destination points, describes the plane with which you want the source object to be aligned. The source object will move into alignment with the destination object.

NOTE

The first source point will match the first destination point perfectly, assuming that Osnaps are used, the second and third points will be as close as possible. This technique can be used to make objects rotate in 3D space as they move to the desired alignment.

Figure 13.29 gives some examples of how the Align utility works.

FIGURE 13.29

Aligning two 3D objects

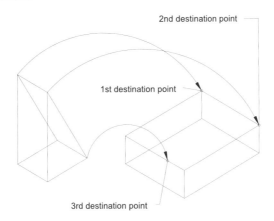

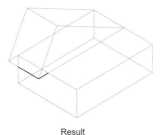

Result

Rotating an Object in 3D

If you just want to rotate an object in 3D space, the Modify ➢ 3D Operation ➢ Rotate 3D option on the menu bar can simplify the operation. Once you've selected this option and selected the objects you want to rotate, you get the following prompt:

```
Axis by Object/Last/View/Xaxis/Yaxis/Zaxis/<2points>:
```

This prompt is asking you to describe the axis of rotation. Here are descriptions of the options presented in the prompt:

Object allows you to indicate an axis by clicking on an object. When you select this option, you are prompted to pick a line, circle, arc, or 2D polyline segment. If you click on a line or polyline segment, the line is used as the axis of rotation. If you click on a circle, arc, or polyline arc segment, AutoCAD uses the line passing through the center of the circle or arc and perpendicular to its plane.

Last uses the last axis that was used for a 3D rotation. If no previous axis exists, you are returned to the Axis… prompt.

View uses the current view direction as the direction of the rotation axis. You are then prompted to select a point on the view direction axis to specify the exact location of the rotation axis.

Xaxis/Yaxis/Zaxis uses the standard X, Y, or Z axis as the direction for the rotation axis. You are then prompted to select points on the X, Y, or Z axis to locate the rotation axis.

<2points> uses two points that you provide as the endpoints of the rotation axis.

This completes your look at creating and editing 3D objects. You have had a chance to use nearly every type of object available in AutoCAD. You might want to experiment on your own with the predefined 3D shapes offered on the Surfaces toolbar. In the next section, you'll discover how you can generate perspective views.

Viewing Your Model in Perspective

So far, your views of 3D drawings have been in *parallel projection*. This means parallel lines appear parallel on your screen. Although this type of view is helpful while constructing your drawing, you will want to view your drawing in true perspective from time to time, to get a better feel for what your 3D model actually looks like.

There's really only one command that offers perspective views in AutoCAD, but that one command, called Dview, is a complex one. With this in mind, you may want

VIEWING YOUR MODEL IN PERSPECTIVE | 625

to begin these exercises when you know you have an hour or so to complete them all at one sitting.

If you are ready now, let's begin!

1. Open the `surfaces.dwg` file you created earlier in this chapter, or use the `ch13surfaces.dwg` file on the companion CD-ROM. Be sure you are in the World Coordinate System. Choose View ➢ 3D Viewpoint ➢ Plan View ➢ World UCS, or type **plan**↵↵. This will display a Plan view of the star (see Figure 13.30).

FIGURE 13.30

The Plan view of the star

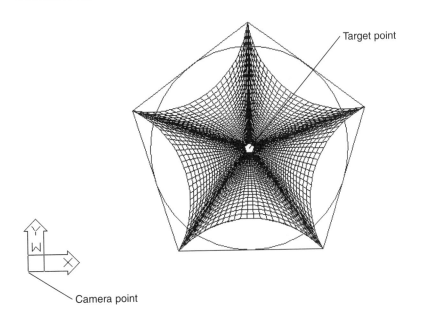

2. Use View ➢ Zoom ➢ All to get an overall view of the drawing.
3. Click on View ➢ 3D Dynamic View, or enter **dv**↵ at the command prompt.
4. At the Object Selection prompt, pick the bottom surface, the circle, and the polygon. You will use these objects as references while using the Dview command.
5. Next you will see a fairly lengthy prompt:

    CAmera/TArget/Distance/POints/PAn/Zoom/TWist/CLip/Hide/Off/Undo/<eXit>:

    at which you can select the appropriate option(These options are discussed in the remaining exercises in this chapter.) The screen will change to show only the objects you selected.

Dview allows you to adjust your perspective view in real time. For this reason, you are asked to select objects that will allow you to get a good idea of your view without slowing the realtime display of your model. If you had selected the whole star, view selection would be slower because the whole star would have to be dragged during the selection process. For a file as small as this star, view selection isn't a big problem, but for larger files, you will find that selecting too many objects can bog down your system.

Setting Up Your Perspective View

As with a camera, your perspective view in AutoCAD is determined by the distance from the object, camera position, view target, and camera lens type.

Follow these steps to determine the camera and target positions.

1. At the DVIEW: prompt, enter **po**↵ for the Points option.

2. At the Enter target point <current point>: prompt, pick the center of the circle. This will allow you to adjust the camera target point, which is the point at which the camera is aimed.

3. At the Enter camera point <current point>: prompt, pick the lower-left corner of the screen. This places the camera location (the position from which you are looking) below and to the left of the star, on the plane of the WCS (see Figure 13.31).

WARNING

When selecting views in this set of exercises using Dview and its options, be sure you click the mouse/pick button as indicated in the text. If you press the ↵ key (or click the left button on the mouse), your view will return to the default orientation, which is usually the last view selected.

Your view now changes to reflect your target and camera locations, as shown in Figure 13.31. The DVIEW: prompt returns, allowing you to further adjust your view.

FIGURE 13.31

The view with the camera and target positioned

If you like, you can press ↵ at the object-selection prompt without picking any object, and you will get the default image, a house, to help you set up your view (see Figure 13.32). Or you can define a block and name it **Dviewblock**. Dviewblock should be defined in a one-unit cubed space. AutoCAD will search the current drawing database, and if it finds Dviewblock, AutoCAD will use it as a sample image to help you determine your perspective view.

NOTE

Make Dviewblock as simple as possible, but without giving up the detail necessary to distinguish its orientation.

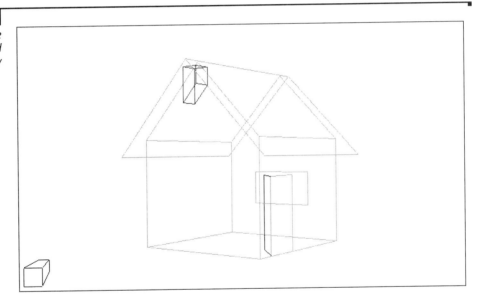

FIGURE 13.32
The default sample image used with Dview

Adjusting Distances

Next, you will adjust the distance between the camera and target.

1. At the DVIEW: prompt, enter **d**↵ for the Distance option. A slide bar appears at the top of the screen.

NOTE The Distance option actually serves two functions: Aside from allowing you to adjust your camera to target distance, it also turns on the Perspective View mode. The Off option of Dview changes the view back to a parallel projection.

2. At the `New camera/target distance <current distance>:` prompt, move your cursor from left to right. The star appears to enlarge and reduce. You can also see that the position of the diamond in the slide bar moves. The slide bar gives you an idea of the distance between the camera and the target point in relation to the current distance.

3. As you move the diamond, you see lines from the diamond to the 1x value (1x being the current view distance). As you move the cursor toward the 4x mark on the slide bar, the star appears to move away from you. Move the cursor toward 0x, and the star appears to move closer.

4. Move the cursor farther to the left; as you get to the extreme left, the star appears to fly off the screen. This is because your camera location has moved so close to the star that the star disappears beyond the view of the camera—as if you were sliding the camera along the floor toward the target point. The closer to the star you are, the larger and farther above you the star appears to be (see Figure 13.33).

FIGURE 13.33

The camera's field of vision

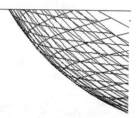

5. Adjust your view so it looks like Figure 13.34. To do this, move the diamond to a spot between 1x and 4x in the slide bar.

6. When you have the view that you want, click the mouse/pick button. The slide bar disappears and your view is fixed in place. Notice that you are now viewing the star in perspective.

FIGURE 13.34

The perspective view of the star using the Distance option

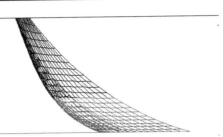

Adjusting the Camera and Target Positions

Next, you will adjust your view so that you can see the whole star. You are still in the Dview command.

 NOTE

Using Dview's Target option is like standing in one location while moving the camera around.

1. At the DVIEW: prompt, enter **ta**↵ for the Target option. The star will temporarily disappear from view.
2. When you see the following prompt:

    [T]oggle angle in/Enter angle from X-Y plane <.00>:

    move your cursor very slowly in a side-to-side motion. Keep the cursor centered vertically or you may not be able to find the star. The star moves in an exaggerated manner in the direction of the cursor. The sideways motion of the cursor simulates panning a camera from side to side across a scene (see Figure 13.35).

FIGURE 13.35
Adjusting the Target option is like panning your camera across a scene.

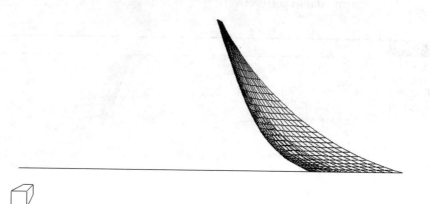

3. Center the star in your view, and then move the cursor slowly up and down. The star moves in the opposite direction to the cursor. The up-and-down motion of the cursor simulates panning a camera up and down.

TIP If you lose your view of the star but remember your camera angle, you can enter the angle at the Enter angle in X-Y plane: prompt to help relocate your view.

4. Enter **ta**↵. You will see the prompt:

   Toggle angle in/Enter angle from XY plane <-35.2644>:

   At this prompt, you can enter a value representing the vertical angle between the camera and the target point.

5. Press ↵ to accept the default. This fixes the vertical motion of the target location to its current default location. However, you can still move the target horizontally. Here you can enter an angle value representing the angle between the camera and target in a horizontal plane measured from AutoCAD's angle 0.

6. Position the view of the star so that it looks like Figure 13.36, and click the mouse/pick button. You've now fixed the target position.

FIGURE 13.36

While in the Dview Target option, set up your view to look like this.

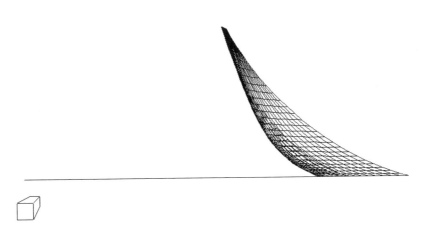

In steps 4 and 5, you could enter an angle value indicating either the vertical or horizontal angle to the target. Once you enter that value (or just press ↵), the angle becomes fixed in either the vertical or horizontal direction. Then, as you move your cursor, the view's motion will be restricted to the remaining unfixed direction. In the exercise, you fixed the vertical angle, and then visually selected a horizontal angle. If you would prefer to "fix" the horizontal angle and visually adjust the vertical, you can enter **ta↵t↵↵** at the Dview prompt.

Changing Your Point of View

Next, you will adjust the camera location to one that is higher in elevation.

1. At the DVIEW: prompt, enter **ca↵** to select the Camera option.

NOTE

Using Dview's Camera option is like changing your view elevation while constantly looking at the star.

2. At the following prompt:

[T]oggle angle in/Enter angle from XY plane <11>:,

move your cursor slowly up and down. As you move the cursor up, your view changes as if you were rising above the star (see Figure 13.37). If you know the vertical angle you want, you can enter it now.

FIGURE 13.37

In the Camera option, moving the cursor up and down is like moving your camera location up and down in an arc.

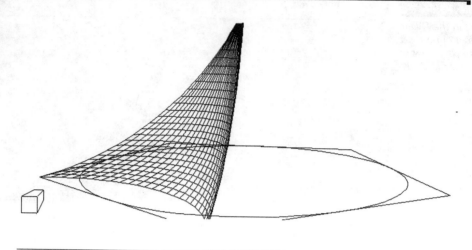

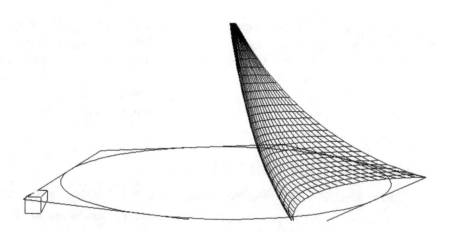

3. Move the cursor down so that you have a view roughly level with the star, and then move the cursor from side to side. Your view changes as if you were walking around the star, viewing it from different sides (see Figure 13.38).

FIGURE 13.38

Moving the cursor from side to side is like walking around the target position.

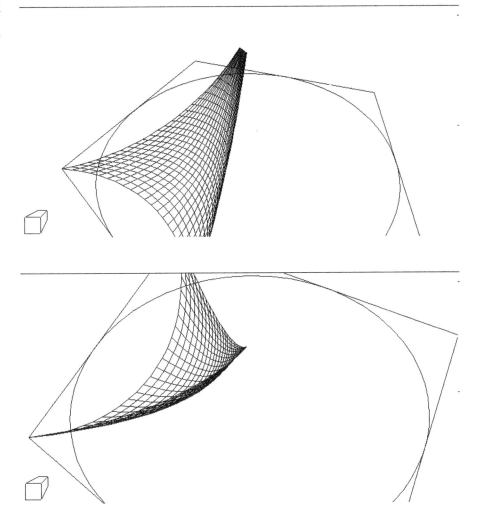

4. When you are ready, enter **t.⏎** to choose the [T]oggle angle in/Enter angle from XY plane option. The prompt changes to

   [T]oggle angle from/Enter angle in XY plane from x axis <-144>:

   If you know the horizontal angle you want, you can enter it now.

5. Position your view of the star so it is similar to the one in Figure 13.39, and click the mouse/pick button.

FIGURE 13.39

Set up your camera location so your view resembles this one.

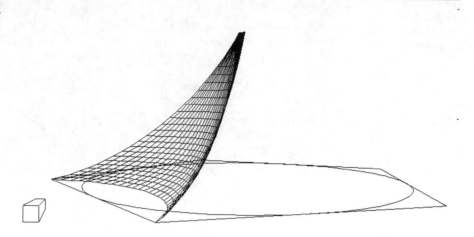

In steps 2 and 4, you could enter an angle value indicating either the vertical or horizontal angle to the camera. Once you indicate a value, either by entering a new one or by pressing ↵, the angle becomes fixed in either the vertical or horizontal direction. Then, as you move your cursor, the view's motion will be restricted to the remaining unfixed direction.

Using the Zoom Option as a Telephoto Lens

The Zoom option of Dview allows you to adjust your view's cone of vision, much like a telephoto lens in a camera. You can expand your view to include more of your drawing or narrow the field of vision to focus on a particular object.

1. At the Dview prompt, enter **z↵** for the Zoom option. Move your cursor from side to side, and notice that the star appears to shrink or expand. You also see a slide bar at the top of the screen, which lets you see your view in relation to the last Zoom setting, indicated by a diamond. You can enter a value for a different focal length, or you can visually select a view using the slide bar.

2. At the Adjust Lens Length <50.000mm>: prompt, press ↵ to accept the 50.000mm default value.

NOTE

When you don't have a perspective view (obtained by using the Distance option) and you use the Zoom option, you will get the prompt `Adjust zoom scale factor <1>:` instead of the `Adjust Lens Length:` prompt. The `Adjust zoom...` prompt acts just like the standard Zoom command.

Twisting the Camera

The Twist option lets you adjust the angle of your view in the viewport frame—like twisting the camera to make your picture fit diagonally across the frame.

1. At the `DVIEW:` prompt, enter **tw**↵ for the Twist option. Move your cursor, and notice that a rubber-banding line emanates from the view center; the star also appears to rotate, and the coordinate readout changes to reflect the twist angle.
2. At the `New view twist <0>:` prompt, press ↵ to keep the current 0° twist angle.
3. Press ↵ again to exit the Dview command, and the drawing regenerates, showing the star in perspective (see Figure 13.40).
4. Choose View ➤ Hide to see a hidden-line perspective view.

FIGURE 13.40

A perspective view of the star

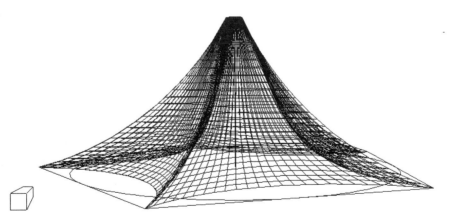

In the next section, you will look at one special Dview option—Clip—that lets you control what is included in your 3D views.

Using Clip Planes to Hide Parts of Your View

At times, you may want a view that would normally be obscured by objects in the foreground. For example, if you try to view the interior of an enclosure design, the objects closest to the camera obscure your view. The Dview command's Clip option allows you to eliminate objects in either the foreground or the background, so you can control your views more easily. In the case of the star, set the Clip/Front option to delete any surfaces in the foreground that might obscure your view of its interior (see Figure 13.41).

FIGURE 13.41

A view of the star interior using the Clip/Front plane

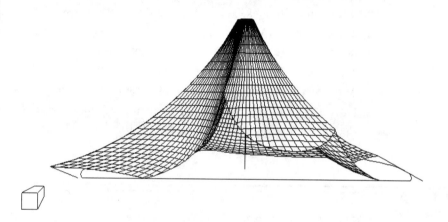

1. While still in the Dview command, enter **cl**↵ for the Clip option.
2. At the Back/front <off>: prompt, enter **f**↵ for the Front option. A slide bar appears at the top of the screen.
3. As you move the diamond on the slide bar from left to right, the star point's surfaces in the foreground begin to disappear, starting at the position closest to you. Moving the diamond the other way (right to left) brings the point's surfaces back into view. You can select a view either by using the slide bar or by entering a distance from the target to the Clip plane.

4. At the Eye/<distance from target> <current distance>: prompt, move the slide bar diamond until your view looks similar to Figure 13.41. Then click the mouse/pick button to fix the view.

5. To make sure the Clip plane is in the correct location, preview your perspective view with hidden lines removed. Enter **h↵** at the DVIEW: prompt. The drawing regenerates, with hidden lines removed.

There are several other Dview Clip options that let you control the location of the Clip plane:

Eye   places the Clip plane at the position of the camera itself.

Back   operates in the same way as the **Front** option, but it clips the view behind the view target instead of in front (see Figure 13.42).

Off   turns off any Clip planes you may have set up.

FIGURE 13.42

Effects of the Clip planes

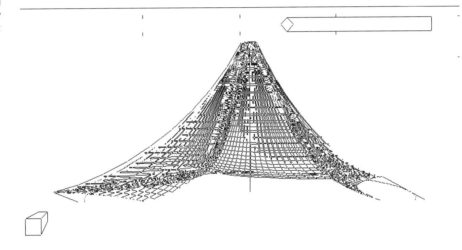

You've now completed the Dview command exercises, and you should have a better understanding of how Dview can be used to get exactly the image you want.

If you like, you can use View ➤ Named Views to save your perspective views. This is helpful when you want to construct several views of a drawing to play back later as part of a presentation. You can also use the Hide and Shade options on the Tools pull-down menu to help you visualize your model.

> The quickest and easiest way to establish a perspective view is to use the Viewpoint Presets dialog box to set up your 3D view orientation, then start Dview and use the Distance option right off the bat to set your camera-to-target distance. Once this is done, you can easily use the other Dview options as needed, or exit Dview to see your perspective view.

If You Want to Experiment...

You've worked with a lot of new commands in this chapter, so now is a good time to play with these commands to help you remember what you've learned. Draw a propeller beanie using curved surfaces and the Arc tool, the Hide tool, the twisting effect that you learned in the Rulesurf section, and the Mirroring tool. Remember to allow room for a head!

PART IV

Printing and Plotting as an Expert

LEARN TO:

- **Print and Plot Your Drawings**
- **Use Pre-existing Drawings**
- **Work with Raster Images**
- **Render in 3D**

Chapter 14

Printing and Plotting

FEATURING

Setting Plotter Origin

Selecting Paper Sizes

Controlling What Gets Printed

Setting Scale and Drawing Location

Setting Line Weights

Setting Plotter Speed

Previewing a Plot

Sending Your Drawings to a Service Bureau

Batch Plotting

Chapter 14

Printing and Plotting

Getting hard copy output from AutoCAD is something of an art. You'll need to be intimately familiar with both your output device and the settings available in AutoCAD. You will probably spend a good deal of time experimenting with AutoCAD's plotter settings and your printer or plotter to get your equipment set up just the way you want.

With the huge array of output options available, this chapter can only provide a general discussion of plotting. It's up to you to work out the details and fine-tune the way you and AutoCAD together work with your plotter. Here we'll take a look at the features available in AutoCAD and discuss some general rules and guidelines to follow when setting up your plots. We'll also discuss alternatives to using a plotter, such as plotter service bureaus and common desktop printers. There won't be much in the way of a tutorial, so consider this chapter more of a reference.

For more information on choosing between printers and plotters, see Appendix A.

Plotting the Plan

To see firsthand how the Plot command works, you'll plot the Base file using the default settings on your system. But first, take a preview of your plot using the new Release 14 Plot Preview option.

1. First, be sure your printer or plotter is connected to your computer and is turned on.
2. Start AutoCAD and open the Base.dwg file.
3. Click on View ➢ Zoom ➢ All to display the entire drawing.
4. Choose File ➢ Print Preview. The display changes to show you how the Plan file will look as output from your printer or plotter.

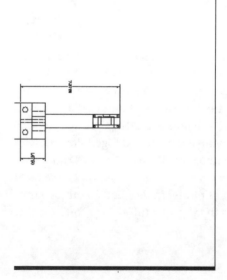

PLOTTING THE PLAN | 647

Your print preview will depend on the type of output device you chose when you installed AutoCAD or when you last selected a Plotter Device option (described in *Selecting an Output Device* in this chapter). It will also be affected by other settings in the Print/Plot Configuration dialog box such as those in the Additional Parameters group and Scale, Rotation, and Origin group. This example shows a typical preview view using the Windows default system printer in portrait mode.

Notice that the view also shows the Zoom Realtime cursor. You can use the Zoom/Pan Realtime tool to get a close-up look at your print preview. Now go ahead and plot the file.

5. Click on File ➢ Print or type **plot**↵ at the command prompt. The Print/Plot Configuration dialog box appears.

If you see prompts in the command window instead of the Print/Plot Configuration dialog box, then change the Cmddia system variable to 1. To do this, type **cmddia**↵**1**↵.

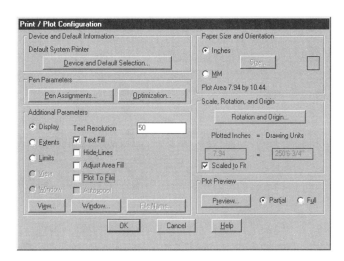

6. Click on OK. The dialog box closes and you see the following message at the command prompt:

```
PLOT Effective plotting area: 10.50 wide by 7.59 high
```

The width and height values shown in this message will depend on your system setup.

Your plotter or printer will print out the plan to no particular scale. You may see various messages while AutoCAD is plotting. When the plot is done, you see the following message:

Regeneration done 100% Plot complete.

You've just done your first printout. Now let's take a look at the wealth of settings available when you plot, starting with the selection of an output device.

Selecting an Output Device

Many AutoCAD users have more than one device for output. You may have a laser printer for your word-processed documents, in addition to a plotter. You may also require PostScript file output for presentations. AutoCAD offers you the flexibility of using several types of devices, quickly and easily.

When you configure AutoCAD, you have the option to specify more than one output device. Once you've configured several printers and plotters, you can select the desired device using the Device and Default Selection button. The current default device is shown just above this button. When you click on the button, you will see the Device and Default Selection dialog box.

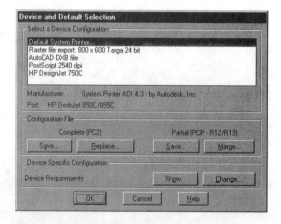

NOTE This graphic shows some sample plotter configurations in the Device Configuration list. Your system will show a different list, or only one plotter or printer.

> ### Using the Windows System Printer
>
> In Windows, you have even more output options than those provided by AutoCAD because Windows itself offers you the option to output to a wide range of plotters and printers. If you configure AutoCAD to use the System Printer, AutoCAD will then use the Windows output device. Set the Windows output device through the Printers option of the Windows Control Panel. Consult your Windows manual for more on the use of output devices and Windows.

To select a device, highlight the device name from the list and click on OK. The rest of the Print/Plot Configuration options will then reflect the requirements of the chosen device. For this reason, be sure you select the output device you want, *before* you adjust the other plotter settings.

Adding an Output Device

Before you can select more than one device from the list in the Device and Default Selection dialog box, you'll need to configure AutoCAD for multiple output devices. You have the opportunity to configure AutoCAD plotters or printers when you install AutoCAD. You can also configure additional printers through the Preferences dialog box. Here's how it's done.

1. Choose Tools ➤ Preferences.
2. Click on the Printer tab. The Printer options appear.

3. You'll see a list of the current printer drivers that have already been installed. To the right of the list, you should see several buttons.

4. Click on the button labeled New. The Add a Printer dialog box appears.

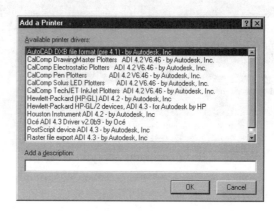

5. Select a printer from the list in the dialog box.

6. Add a description of your selection.

7. Click on OK; you'll then be presented with a series of prompts asking you for more detailed information on how you would like your device configured. You can generally accept the default answers to each prompt, since they are also presented to you in the Print/Plot Configuration dialog box when you plot. The exception is the Raster File and DXB output options, which will ask for a specific image size and resolution.

8. When you are done replying to the prompts, you are returned to the Preferences dialog box.

If you find that the settings you selected in step 7 are inappropriate, you can use the Modify button to change the printer configuration settings. Highlight the printer you want to reconfigure from the list, then click on Modify. The Reconfigure a Printer dialog box appears.

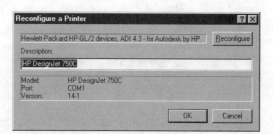

You can enter a new description for the printer in the Description input box. Click on Reconfigure to change the configuration settings. You will then see the prompts you saw in step 7.

Other options in the Printer tab of the Preferences dialog box let you store or load PC2 plot configuration files (see *Storing Plotter Settings*) or remove a configuration from the list. Here is a summary of the printer options:

Set Current makes the highlighted printer or plotter the current output device.

Modify lets you change the printer description shown in the list and also allows you to change configuration settings for the selected printer.

New lets you add a printer to the list of printers available at plot time.

Open lets you load a PC2 plot configuration file.

Save As lets you save the current configuration of a plotter as a PC2 file.

Remove removes a plotter from the list.

Plotting Image Files

If your work involves the production of manuals, reports, or similar documents, you may want to add the Raster File Export option to your list of plotter configurations. The Raster File Export option lets you plot your drawings as PCX, Targa, Tiff, or BMP files that you can later import into documents that accept bitmap images. Images can be up to 8000 × 8000 pixels and can contain 256 colors. If you need several different raster formats, you can use multiple instances of this or any plotter configuration.

If you would like to convert your 3D wire-frame models into 2D line drawings, add the AutoCAD DXB file format to your list of plotters. Then you can plot your hidden line 3D models to a DXB file format. Once you have a DXB file, choose File ➢ Import to import the DXB plot file. AutoCAD will create a 2D line drawing from the plot file data. For best results, use a value of 2900 for the plotter steps per drawing unit setting.

Another tool for 3D is the PostScript file output format. Like the DXB option, you can plot a 3D image to a file, then import the resultant EPS file using File ➢ Import. For best results, make sure that when you configure the PostScript printer you use the highest resolution available, and that you use the largest media size possible when you're plotting.

Storing Plotter Settings

In addition to the ability to set up multiple output devices, AutoCAD also lets you store setup information for each device, such as default paper sizes, sheet orientation, what to plot, and so on. This information is stored as a file in one of two formats: the older PCP format used in Releases 12 and 13, or the new PC2 format, which stores device-specific information as well as the general information in the PCP file.

If you find that you frequently plot your drawings using a particular plotter setup, the PCP and PC2 setup files will be especially helpful for saving time. They are also a crucial component for batch plotting, as you'll see later in this chapter.

Understanding Your Plotter's Limits

If you're familiar with a word-processor or desktop-publishing program, you know that you can set the margins of a page, thereby telling the program exactly how far from each edge of the paper you want the text to appear. With AutoCAD, you don't have that luxury. To accurately place a plot on your paper, you must know the plotter's *hard clip limits*. The hard clip limits are like built-in margins, beyond which the plotter will not plot. These limits vary from plotter to plotter (see Figure 14.1).

FIGURE 14.1

The hard clip limits of a plotter

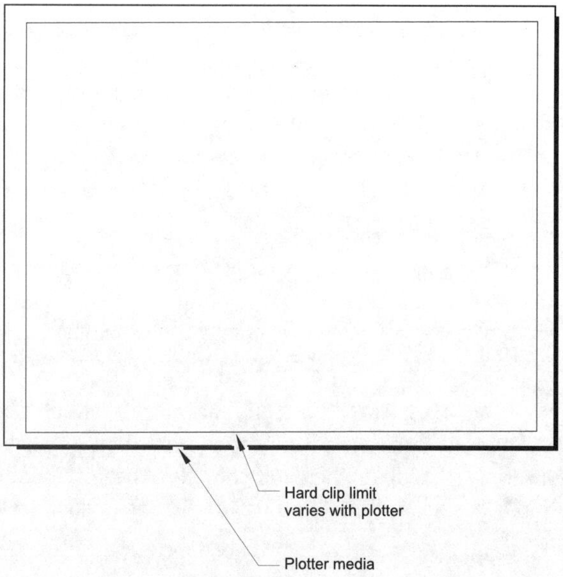

It's crucial that you know your printer/plotter's hard clip limits, so that you can place your drawings accurately on the sheet. Take some time to study your plotter manual and find out exactly what these limits are. Then make a record of them and store it somewhere, in case you or someone else needs to format a sheet in a special way.

Hard clip limits for printers are often dependent on the software that drives them. You may need to consult your printer manual, or use the trial-and-error method of plotting several samples to see how they come out.

Once you've established the limits of your plotter or printer, you can begin to lay out your drawings to fit within those limits (see *Setting the Output's Origin and Rotation* later in this chapter). You can then establish some standard drawing limits based on your plotter's limits. You'll also need to know the dimension of those hard clip limits to define your plotter sheet sizes. While AutoCAD offers standard sheet sizes in the Paper Size and Orientation button group of the Print/Plot Configuration dialog box, these sizes do not take into account the hard clip limits.

Knowing Your Plotter's Origins

Another important consideration is the location of your plotter's origin. For example, on some plotters the lower-left corner of the plot area is used as the origin. Other plotters will use the center of the plot area as the origin. When you plot a drawing that is too large to fit the sheet on a plotter that uses a corner for the origin, the image is pushed toward the top and to the right of the sheet (see Figure 14.2). When you plot a drawing that is too large to fit on a plotter that uses the center of the paper as the origin, the image is pushed outward in all directions from the center of the sheet.

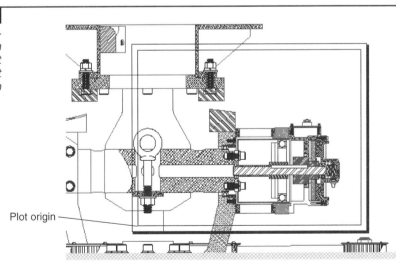

FIGURE 14.2

Plotting an oversized image on a plotter that uses the lower-left corner for its origin

In each situation, the origin determines a point of reference from which you can relate your drawing in the computer to the physical output. Once you understand this, you're better equipped to accurately place your electronic drawing on the physical media.

These origin placements also apply to plotter-emulation software that allows you to plot to a raster printer using the HPGL (Houston Instrument Graphic Language) format.

Selecting a Paper Size and Orientation

Next let's look at the Paper Size and Orientation button group in the Print/Plot Configuration dialog box. This is where you determine the size of your media and the standard unit of measure you will use. The Inches and MM radio buttons let you determine the unit of measurement you want to work with in this dialog box. This is the unit of measurement you will use when specifying sheet sizes and view locations on the sheet. If you choose MM, sheet sizes are shown in millimeters, and you must specify distances and scales in millimeters for other options.

If you need to convert from inches to millimeters, the scale factor is 1"=25.4 mm.

Once you've chosen a unit of measurement, you can click on the Size button to select a sheet size. This brings up the Paper Size dialog box.

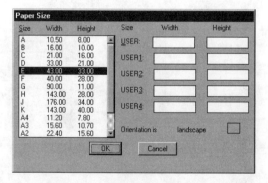

NOTE If you are using the System Printer, the Size option is not available. To set the size, you must use the Printers option in the Windows Control Panel, or click on the Device and Default Selection button, and with the Default System Printer selected in the list, click on Change in the Device Specific Configuration button group. This opens the Print Setup dialog box, where you can change your printer settings.

The range of sheet sizes will vary, depending on which plotter you have selected in the Device and Default Selection option (described later in this chapter). To select a sheet size, simply highlight it in the list box on the left and click on OK. If you choose, you can enter a nonstandard sheet size in the input boxes to the right. As you can see, you can store up to five custom sheet sizes.

Controlling the Appearance of Output

On the left side of the Print/Plot Configuration dialog box is the Additional Parameters button group. These options give you the most control over the appearance of your plot. From here, you can control what gets plotted and how.

Designating Hidden Lines, Fills, and Text Quality

Let's start by looking at the six options down the right side of this group. Using these check boxes, you can tell AutoCAD to store the plot in a file on disk, instead of sending the plot data to a plotter or printer for immediate output. You can specify whether to plot a 3D drawing with hidden lines removed, or whether your plotter is to compensate for pen widths when drawing solid fills.

Text Resolution

The Text Resolution input box controls the appearance of filled fonts such as Times New Roman TrueType or Helvetica PostScript. A higher number generates smoother edges around the font characters. A lower number causes the characters to appear rough. The default is 50, with a range from 0 to 100.

Text Fill

If you want your filled fonts to appear as outline fonts, make sure this option is *not* checked. Leave it checked for filled fonts.

Hide Lines

The Hide Lines check box is generally only used for 3D images. When this option is on, AutoCAD will remove hidden lines from a 3D drawing as it is plotted. This operation will add a minute or two to your plotting time.

Hide Lines is not required for Paper Space viewports that have been set to hide lines using the Hideplot option of the Mview command (View ➢ Floating Viewports ➢ Hideplot). The current setting for Hideplot can be found by using the List command and selecting the border of the Paper Space viewport.

Adjust Area Fill

Turning on Adjust Area Fill tells AutoCAD to compensate for pen width around the edges of a solid filled area in order to maintain dimensional accuracy of the plot. To understand this feature, you need to understand how most plotters draw solid areas.

Plotters draw solid fills by first outlining the fill area and then cross-hatching the area with a series of lines, much as you would do by hand. For example, if a solid filled area is drawn at a width of 0.090", the plotter will outline the area using the edge of the outline as the centerline for the pen, and then proceed to fill the area with a cross-hatch motion. Unfortunately, by using the outline as the centerline for the pen, the solid fill's actual width is 0.090" *plus* the width of the pen.

Generally, compensation for pen width is critical only when you are producing drawings as a basis for photo etchings or similar artwork, where close tolerances must be adhered to.

When you check the Adjust Area Fill box, AutoCAD pulls in the outline of the solid area by half the pen width. To determine the amount of offset to use, AutoCAD uses the pen-width setting you enter under the Pen Assignments dialog box (described later in this chapter). Figure 14.3 illustrates the operation of Adjust Area Fill.

CONTROLLING THE APPEARANCE OF OUTPUT | 657

FIGURE 14.3
A solid area shown without pen-width compensation and with pen-width compensation

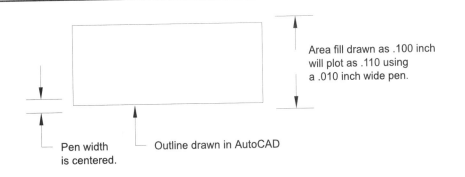

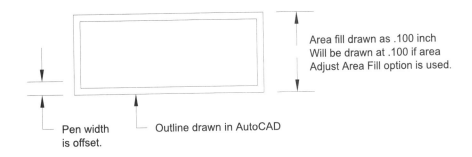

Plot to File

Turn this option on when you want to divert your printout to a file on disk and print it later. This can be useful if you are in an office in which a plotter is shared among several CAD stations. When the plotter is busy, or if a plotter buffer or spooler is full, plot to a file. Later you can download the plot file when your plotter is available.

Some plotters require that you plot to a file as a part of their configuration.

To use this option, do the following:

1. Click on the Plot To File check box.
2. Click on the File Name button (just below the check box). You'll see the Create Plot File dialog box, which is the same as the one used for most other file creation operations.

3. Enter the name for your plot file, or just accept the default file name, which is usually the same as the current file. The PLT file name extension is the default.

4. Click on OK to accept the file name.

Autospool

The Autospool option is only available for certain AutoCAD ADI plotter configurations and only when the Plot to File option is selected. With Autospool selected, AutoCAD will create a Plot spool file in the directory of your choice from the Files tab of the Preferences dialog box (Tools ➢ Preferences) under Print Spooler File Location. This option is usually associated with a Print Spooler application that reads the Print spool file and sends it to the appropriate device. You can also change devices via the Files tab of the Preferences dialog box.

If you want to learn more about Autospooling, check out the AutoCAD R14 Installation Guide, page 85, first paragraph for Autodesk's words of wisdom.

Determining What to Print

The radio buttons on the left side of the Additional Parameters button group let you specify which part of your drawing you wish to plot. You might notice some similarities between these settings and the Zoom command options.

The File ➢ Print Preview option will display a preview based on the settings in this section.

Display

This is the default option; it tells AutoCAD to plot what is currently displayed on the screen (see the top panel of Figure 14.4). If you let AutoCAD fit the drawing onto the sheet (that is, if you check the Scaled to Fit check box), the plot will be exactly the same as what you see on your screen (see the bottom panel of Figure 14.4).

FIGURE 14.4

The screen display and the printed output when Display is chosen and no Scale is used (the drawing is scaled to fit the sheet)

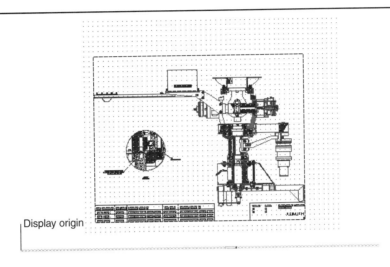

Display origin

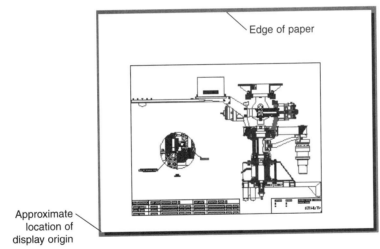

Edge of paper

Approximate location of display origin

Extents

The Extents option draws the entire drawing, eliminating any space that may border the drawing (see Figure 14.5). If you let AutoCAD fit the drawing onto the sheet (that is, if you check the Scaled to Fit check box), the plot will display exactly the same thing that you would see on the screen had you chosen View ➤ Zoom ➤ Extents.

FIGURE 14.5

The printed output when Extents is chosen

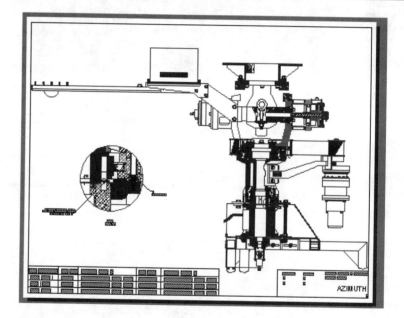

Limits

The Limits printing option uses the limits of the drawing to determine what to print (see Figure 14.6). If you let AutoCAD fit the drawing onto the sheet (by selecting Scaled to Fit), the plot will display exactly the same thing that you would see on the screen had you clicked on View ➤ Zoom ➤ All.

FIGURE 14.6

The screen display and the printed output when Limits is chosen

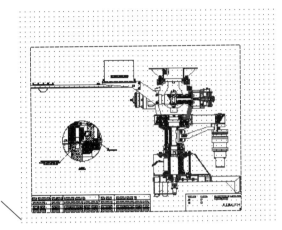

Drawing origin

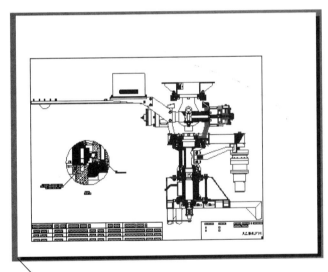

Plot origin is equal to drawing origin.

View

The View printing option uses a previously saved view to determine what to print (see Figure 14.7). To use this option, you must first create a view. Then click on the View button in the Print/Plot Configuration dialog box, and double-click on the desired view name in the drop-down list that appears.

FIGURE 14.7
A comparison of the saved view and the printed output

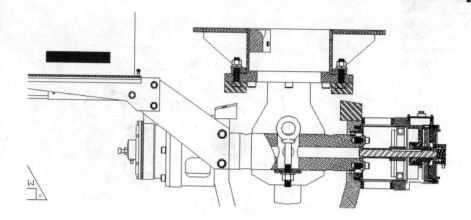

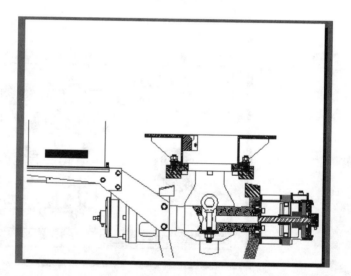

If you let AutoCAD fit the drawing onto the sheet (by selecting Scaled to Fit), the plot will display exactly the same thing that you would see on the screen if you recalled the view you are plotting.

Window

Finally, the Window option allows you to use a window to indicate the area you wish to plot (see the top image of Figure 14.8). Nothing outside the window will print.

FIGURE 14.8

A selected window and the resulting printout

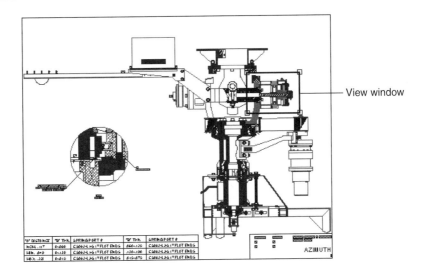

View window

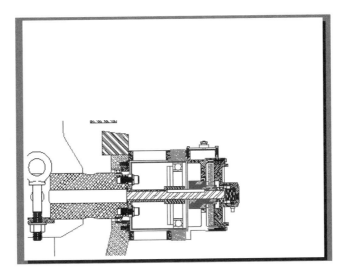

To use this option, click on the Window button. Then enter the coordinates of the window in the appropriate input boxes. Or you can click on the pick button to indicate a window in the drawing editor. The dialog box will temporarily close to allow you to select points. When you're done, click on OK.

If you let AutoCAD fit the drawing onto the sheet using the Scaled to Fit check box, the plot will display exactly the same thing that you enclose within the window.

TIP Do you get a blank print, even though you selected Extents or Display? Chances are the Scaled to Fit option is unchecked, or the Plotted Inches = Drawing Units setting is inappropriate for the sheet size and scale of your drawing. If you don't care about the scale of the drawing, then make sure the Scaled to Fit check box is checked. Otherwise, make sure the Plotted Inches = Drawing Units setting is set correctly. The next section describes how to set the scale for your plots.

Controlling Scale and Location

The Scale, Rotation, and Origin button group is where you tell AutoCAD the scale of your drawing, as well as how the image is to be rotated on the sheet, and the location of the drawing origin on the paper.

In the previous section, the descriptions of several options indicate that the Scaled to Fit check box must be checked. Bear in mind that when you apply a scale factor to your plot, it changes the results of the Additional Parameters settings, and some subtle problems can arise.

For example, the drawing fits nicely on the paper when you use Scaled to Fit. But if you tried to plot the drawing at a scale of 1"=1', you would probably get a blank piece of paper, because, at that scale, hardly any of the drawing would fit on your paper. AutoCAD would tell you that it was plotting and then that the plot was finished. You wouldn't have a clue as to why your sheet was blank.

If an image is too large to fit on a sheet of paper because of improper scaling, the plot image will be clipped differently depending on whether the plotter uses the center of the image or the lower-left corner for its origin (see Figure 14.1). Keep this in mind as you specify scale factors in this area of the dialog box.

Specifying Drawing Scale

To indicate scale, two input boxes are provided in the Scale, Rotation, and Origin button group: Plotted Inches (or Plotted MM if you use metric units) and Drawing Units.

You can specify a different scale from the one you chose while setting up your drawing, and AutoCAD will plot your drawing to that scale. You are not restricted in any way as to scale, but entering the correct scale is important: If it is too large, AutoCAD will think your drawing is too large to fit on the sheet, although it will attempt to plot your drawing anyway.

CONTROLLING SCALE AND LOCATION | 665

See Chapter 3 for a discussion on unit styles and scale factors.

If you plot to a scale that is different from the scale you originally intended, objects and text will appear smaller or larger than would be appropriate for your plot. You'll need to edit your text size to match the new scale. This can be done using a utility found on the companion CD-ROM. See Appendix C for details.

The Scaled to Fit check box, as already described, allows you to avoid giving a scale altogether and forces the drawing to fit on the sheet. This works fine if you are doing illustrations that are not to scale.

Setting the Output's Origin and Rotation

To adjust the position of your drawing on the media, enter the location of the view origin in relation to the plotter origin in X and Y coordinates (see Figure 14.9).

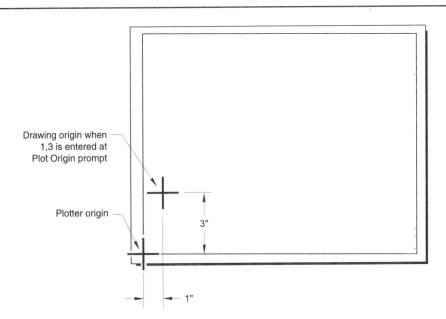

FIGURE 14.9
Adjusting the image location on a sheet

For example, suppose you plot a drawing, then realize that it needs to be moved 1" to the right and 3" up on the sheet. You would replot the drawing by making the following changes:

1. In the Print/Plot Configuration dialog box, click on the Rotation and Origin button in the Scale, Rotation, and Origin button group. The Plot Rotation and Origin dialog box appears.

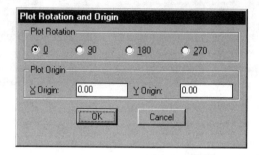

2. Double-click on the X Origin input box and type **1**.
3. Double-click on the Y Origin input box and type **3**.
4. Click on OK.

Now proceed with the rest of the plot configuration. With the above settings, when the plot is done, the image will be shifted on the paper exactly 1" to the right and 3" up.

> If you encounter problems with rotating plots, try setting up a UCS that is rotated 90° around the Z axis in the WCS. Then use the Plan command (View ➤ 3D Viewport Presets ➤ Plan View) to view your drawing in the new rotated orientation. Next, save the view using the View command (View ➤ Named Views). Once you do this, you can use the View button in the Additional Parameters group to plot the rotated view. See Chapter 11 for more on UCS.

The four radio buttons labeled 0, 90, 180, and 270 allow you to rotate the plot on the sheet. Each value indicates the number of degrees of rotation for the plot. The default is 0, but if you need to rotate the image to a different angle, click on the appropriate radio button.

Adjusting Pen Parameters and Plotter Optimization

The Pen Parameters group of the Print/Plot Configuration dialog box contains two buttons: Pen Assignments, which helps you control line weight and color, and Optimization, which lets you control pen motion in pen plotters.

Working with Pen Assignments and Line Weights

In most graphics programs, you control line weights by adjusting the actual width of lines in the drawing. In AutoCAD, you generally take a different approach to line weights. Instead of specifying a line weight in the drawing editor, you match plotter pen widths to colors in your drawing. For example, you might designate the color red to correspond to a fine line weight or the color blue to a very heavy line weight. To make these correlations between colors and line weights, tell AutoCAD to plot with a particular pen for each color in the drawing.

NOTE

You can also control line weights through the use of polylines. See Chapter 10 for details.

The Pen Assignments button in the Print/Plot Configuration dialog box is the entry point to setting the drawing colors for the pens on your plotter. When you click on this button, the Pen Assignments dialog box appears.

NOTE

This dialog box shows a listing for a Hewlett-Packard 750c plotter.

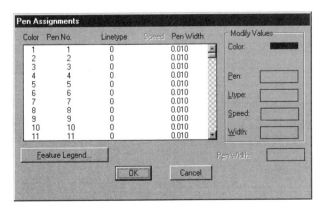

The predominant feature of this dialog box is the list of pen assignments. The color numbers in the first column of the list correspond to the colors in your drawing. The other columns show you what pen number, linetype, pen speed, and pen width are assigned to those colors.

To change these settings, click on the item in the list you want to change. The values for that selected item appear in the appropriate input boxes under Modify Values at the right. You can then change the values in the input boxes. You can highlight more than one color at a time to change the pen assignments of several colors at once.

Not all printers and plotters will offer all of the options presented here. System printers may have separate options in the Windows Control Panel for many of these options. Some plotters offer a "software mode" that will allow AutoCAD to have more control over the plots. Check the documentation for your particular output device for more detailed information.

ISO Pen Widths

You may have noticed a setting called ISO Pen Widths in the Linetype tab of the Layer & Linetype Properties dialog box discussed in Chapter 4 (Format ➢ Linetype). This setting is in the form of a drop-down list. When you select a pen width from that list, the linetype scale is updated to conform to the ISO standard for that width. This setting has no effect on the actual plotter output, however. If you are using ISO standard widths, it is up to you to match the color of the lines to their corresponding widths in the Print/Plot Configuration dialog box. Use the Pen Assignments dialog box to set the line colors to pen widths.

Pen No.

For pen plotters, you assign line weights to colors by entering a pen number in the Pen input box. The default for the color red, for example, is pen 1. If you have a pen plotter, you can then place a pen with a fine tip in the slot designated for pen 1 in your plotter. Then everything that is red in your AutoCAD drawing will be plotted with the fine pen.

Most ink-jet and laser printers allow you to set line widths through the Width option discussed later in this section, or in the Printers folder of the Windows Control Panel.

If you have a Hewlett-Packard ink-jet plotter which supports HP-GL/2, you can also control pen settings through the Hpconfig command by typing **hpconfig**↵ at the command prompt. This option allows you to store and retrieve advanced plotter configurations in addition to those found in the Print/Plot Configuration dialog box. These options include solid fill shading and area fill control. Other plotter manufacturers may offer similar configuration options. Consult your plotter manual for details.

Ltype

Some plotters offer linetypes independent of AutoCAD's linetypes. Using the Ltype input box, you can force a pen assignment to one of these hardware linetypes. If your plotter supports hardware linetypes, you can click on the Feature Legend button to see what linetypes are available and their designations. Usually linetypes are given numeric designations, so a row of dots might be designated as linetype 1, or a dash-dot might be linetype 5.

If your plotter can generate its own linetypes, you can also assign linetypes to colors. But this method is very seldom used, because it is simpler to assign linetypes to layers directly in the drawing.

Speed

The Speed setting lets you adjust the pen speed in inches per second for pen plotters. This is important because various pens have their own speed requirements. Refillable technical pens that use India ink generally require the slowest settings; roller pens are capable of very high speeds. Pen speeds will also be affected by the media you are using.

Selecting pens and media for your plotter can be a trial-and-error proposition. If you are in a high-production environment, you may want to consider purchasing or leasing a modern ink-jet plotter, as these are easier to maintain and considerably faster than pen plotters.

You can often consult a local dealer of plotting supplies for the best combination of pen and ink for your situation.

Width

If you have an ink-jet or laser plotter, this option usually allows you to control the line weights of your plots. Here you can set the line width for each AutoCAD color. The default line width is 0.010, though you can usually specify a finer width than this, depending on your plotter or printer's resolution (usually rated in dots per inch). Of course, you can specify a thicker line weight as well.

AutoCAD also uses the Width setting in conjunction with the Adjust Area Fill option under the Additional Parameters button group (discussed earlier). When AutoCAD draws solid fills, it draws a series of lines close together—much as you would do by hand (see Figure 14.10). To do this efficiently, AutoCAD must know the pen width. If the Width setting is too low, AutoCAD will take longer than necessary to draw a solid fill; if it is too high, solid fills will appear as cross-hatches instead of solids.

FIGURE 14.10

How solid fill areas are drawn by a plotter

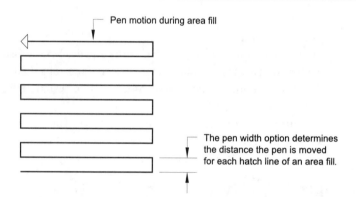

NOTE Technical drawings can have a beauty of their own, but they can also be deadly boring. What really sets a good technical drawing apart from a poor one is the control of line weights. Knowing how to vary and control line weights in both manual and CAD drawings can make a huge difference in the readability of the drawing. By emphasizing certain lines over others, visual monotony is avoided and the various components of the drawing can be seen more easily.

Optimizing Plotter Speed

AutoCAD does a lot of preparation before sending your drawing to the plotter. If you are using a pen plotter, one of the things AutoCAD does is to optimize the way it sends vectors to your plotter, so your plotter doesn't waste time making frequent pen changes and moving from one end of the plot to another just to draw a single line.

The Optimization button in the Pen Parameters group opens a dialog box that lets you control the level of pen optimization AutoCAD performs on your plots.

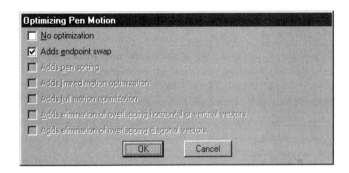

Here are brief descriptions of each setting:

No optimization   causes AutoCAD to plot the drawing just as it is regenerated on the screen.

Adds endpoint swap   forces the plotter to draw (as best it can) parallel lines in a back-and-forth motion, so that the pen moves the minimum distance between the end of one line and the beginning of another.

Adds pen sorting   sorts pens so that all of one color is plotted at once. This is fine if your plotter has a self-capping penholder, or if it is a single-pen plotter. You may not want to enable this option if you have a multipen plotter that does not cap the pens.

Adds limited motion optimization and **Adds full motion optimization**   further minimize pen travel over a plot.

Adds elimination of overlapping horizontal or vertical vectors   does just what it says: It eliminates overlapping lines. With pen plotters, overlapping lines can cause the weight of the line to increase noticeably. This setting helps reduce line weight build-up when a drawing contains numerous line overlaps. This setting does not affect raster plotters or printers.

Adds elimination of overlapping diagonal vectors performs a similar function as the previous option, but on diagonal lines.

None of these options, except the No Optimization and the Adds Endpoint Swap options, has any effect on dot-matrix printers, laser printers, or ink-jet plotters.

Other Plot Controls

Here are a couple of other handy, timesaving features available to you in the Print/Plot Configuration dialog box.

Previewing a Plot

If you're a seasoned AutoCAD user, you're probably all too familiar with the following scenario:

You're rushing to meet a deadline. You've got one hour to plot a drawing that you know takes 45 minutes to plot. You set up AutoCAD and start the plot, then run off to finish some paperwork. When you come back 45 minutes later, the plot image is half the size it is supposed to be.

You can avoid facing predicaments like this one with the Plot Preview feature. Once you've chosen all the settings you think you need for plotting your drawing, turn on the Full radio button in the Plot Preview group, and then click on Preview. AutoCAD will show you what your drawing will look like based on the settings you've chosen. Preview also lists any Warning messages that would appear during the actual plotting process.

You can press the Esc key to speed up the preview. Once the preview plot is done, you are immediately returned to the Print/Plot Configuration dialog box. You can also press Ctrl+c to terminate the preview generation when you've seen enough.

While in the Full Preview, you can zoom in on an area and pan around, just as you could with the File ≻ Print Preview option.

The Full Preview is worth using on a regular basis since it is considerably faster than an actual plot. If you're just interested in seeing how the drawing fits on the sheet, you can choose Partial instead of Full before clicking on Preview. The Partial option shows the sheet edge, image orientation triangle, and image boundary. The image itself is not shown. Using a small triangle in the corner of the drawing to indicate the lower-left corner of the drawing, AutoCAD shows you how the image is oriented on the sheet.

Reinitializing Your Input/Output Ports

If you are using two output devices (such as a printer and a plotter) on the same port, you may want to know about the Reinit command. This command lets you reinitialize a port for use by AutoCAD after another program has used the port.

To use Reinit, enter **reinit** at the command prompt or select Reinitialize from the Tools menu. You will see a dialog box containing check boxes labeled Digitizer, Plotter, Display, and PGP File. You can check them all, if you like, or just check the items you are most concerned with.

Reinit can also be accessed using the Reinit system variable.

Saving Your Settings

At times, drawings will require special plotter settings, or you may find that you frequently use one particular setting configuration. Instead of trying to remember the settings every time you plot, you can store settings as files that you can recall at any time.

In Release 14, you can now store plotter settings in two different formats. For compatibility with earlier versions of AutoCAD, you can store and load configuration files in the PCP format. This format stores the general plot configuration information you set up in the Print/Plot Configuration dialog box, such as sheet size, plot scale, and orientation. A new format called PC2 will store the same information as the PCP file, with additional configuration information regarding the specific plotter you are using. Here's a step-by-step description of how to store and recall plotter settings:

1. Set up the plotter settings exactly as you want them.
2. Click on the Device and Default Selection button.
3. In the Device and Default Selection dialog box, click under the Save button in either the Complete PC2 or Partial (PCP - R12/R13) heading. The Save to File dialog box appears.
4. Enter a name for the group of settings. The default name is the same as the current drawing name, with the .pcp or .pc2 file name extension.
5. Click on OK to create your plot file of settings.

To recall a settings file:

1. Click on File ➤ Print (or type **plot**↵ at the command line).
2. Click on the Device and Default Selection button.
3. In the Device and Default Selection dialog box, click on either the Replace button to recall a PC2 setup or on the Merge button to recall settings from a PCP file.
4. Click on the name of the desired plotter settings file.
5. Click on OK, and the settings will be loaded. You can then proceed with your plot.

The ability to store plotter settings in PC2 and PCP files gives you greater control over your output quality. It helps you to reproduce similar plots more easily when you need them later. You can store several different plotter configurations for one file, each for a different output device. In addition, PC2 files are required for performing batch plots to multiple devices (see the next section, *Batch Plotting*).

You can also open a plot configuration file with a text editor. You can even create your own PCP file by copying an existing one and editing it.

Batch Plotting

Release 14 includes a tool that enables you to plot several unattended drawings at once. This can be helpful when you've finished a set of drawings and would like to plot them during a break or overnight. Here's how to use the Batch Plot utility.

1. From the Windows Desktop, choose Start ➢ Programs ➢ AutoCAD R14 ➢ Batch Plot Utility. The Batch Plot Utility program opens, along with an AutoCAD session.

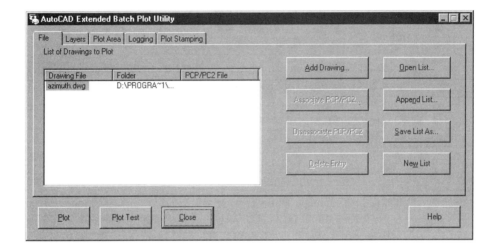

2. Click on the Add Drawing button. The Add Drawing dialog box opens. This is a typical Windows file dialog box.

3. Locate and select a file you wish to plot. The name of the file you select appears in the list box of the Batch Plot Utility window.

4. Click on the file name in the list. The three buttons labeled Associate PCP/PC2, Disassociate PCP/PC2, and Delete Entry become available. The Delete Entry option deletes the highlighted option.

5. Click on the Associate PCP/PC2 button. The Associate PCP/PC2 File dialog box appears. This is also a typical Windows file dialog box.

6. Locate and select a PCP or PC2 plot configuration file that contains the settings you want to use in conjunction with this file. If you want to plot to different devices for each drawing, you must use the PC2 plot configuration format.

7. Repeat steps 2–6 for additional files.

8. When you have selected all the files and associated a plot configuration file with each one, click on Plot Test to proceed with the plots.

We don't suggest that you try this on a 640 × 480 screen (like a laptop) because the Plot Test chops off 20% of the dialog box.

If you need to, you can store the list of files to plot so you can quickly recall them later. Click on the Save List As button and a file dialog box appears. You can then save the list under any name you choose. The file will have the .bp2 file extension. To restore a saved list, choose Open List; if you want to add a saved list to the current list, choose Append list.

If you have already set up a system using scripts for accessing the Plot command, you can restore the command line version of Plot by changing the Cmddia system variable to 0. Bear in mind that, due to minor changes in the way the command line plotting works, you may need to make changes to your plot script.

Other Extended Batch Plot Utility Options

In addition to the main File tab of the Extended Batch Plot Utility, four other tabs—Layers, Plot Area, Logging, and Plot Stamping—offer additional settings that allow greater control over your plots.

The Layers tab lets you set the layer on/off status for each file in the list of drawings to plot. The settings you make here don't affect the drawing, but they're stored with the BP2 batch plot file so they can be recalled during your batch plots. A Plot Test button is provided to test your batch plot settings. No plot is produced, but if errors are encountered, you will see messages describing them.

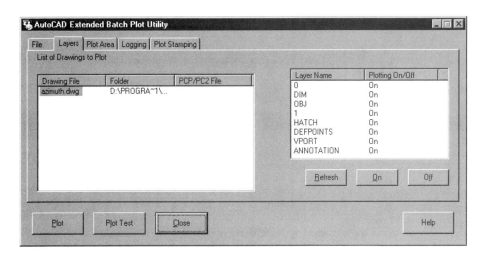

The Plot Area tab allows you to determine the portion of the drawing that is to be plotted. The options presented here are the same as those in the Print/Plot Configuration dialog box.

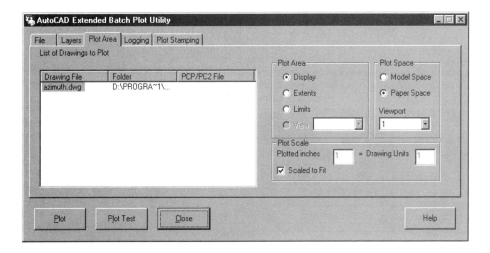

The Logging tab controls if and where a plot log file is generated at plot time. Log files are helpful in determining where problems are occurring when a batch plot fails. From here, you can set the location and name for the log file, or you can add user comments or header information.

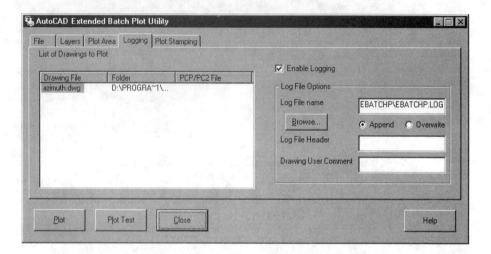

The Plot Stamping tab controls the plot-stamping feature that places a brief message on the plotter output to help identify the plot or otherwise add a stamp to the plot. This can be helpful in associating a plot with a drawing file, the creator of the plot, or other information. To add or change the text of a stamp, use the Change button. This opens another dialog box that allows you to edit the stamp's contents. The Change button also allows you to select an external text file for the contents of the stamp.

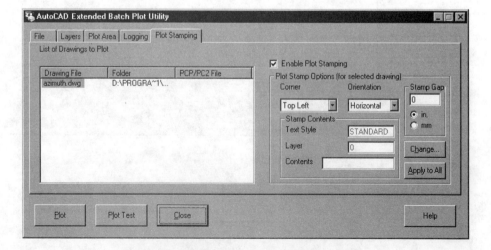

Sending Your Drawings to a Service Bureau

Using a plotting service can be a good alternative to purchasing your own plotter. Or you might consider using a low-cost plotter for check plots and then sending the files to a service for your final product. Many reprographic services, such as blueprinters, offer plotting in conjunction with their other services. Quite often these services include a high-speed modem that allows you to send files over phone lines, eliminating the need for using courier services or regular mail.

Service bureaus will often use an *electrostatic plotter*—it's like a very large laser printer, and is often capable of producing color plots. These plotters are costly, and you probably won't want to purchase one yourself. However, they are excellent for situations requiring high volume and fast turnaround. The electrostatic plotter produces high-quality plots, and it's fast: A 30" × 42" plot can take as little as two minutes. One limitation, however, is that these plotters require special media, so you can't use your preprinted title blocks. Ask your prospective plotter service organization about the media limitations.

Another device used by service bureaus is the *laser photo plotter*. This device uses a laser to plot a drawing on a piece of film. The film negative is later enlarged to the finished drawing size by means of standard reprographic techniques. Laser photo plotters yield the highest-quality output of any device, and they offer the flexibility of reproducing drawings at any size.

Finally, many service bureaus can produce full E-size plots of PostScript files. With AutoCAD's full PostScript support, you can get presentation-quality plots from any AutoCAD drawing. See Chapter 18 for more on PostScript and AutoCAD.

If You Want to Experiment...

At this point, since you aren't rushing to meet a deadline, you may want to experiment with some of the plotter and printer variables and see firsthand what each one does. Find a plotter in your company, through friends, or possibly at your local college. Configure AutoCAD for this plotter. Save a few plot files to a floppy disk and when you have the opportunity, copy them to the plotter.

Chapter 15

3D Rendering in AutoCAD

FEATURING

Things to Do Before You Start

Creating a Quick Study Rendering

Adding a Background

Effects with Lighting

Adding Reflections and Detail with Ray Tracing

Creating and Adjusting Texture Maps

Rendering Output Options

Editing Your Image

3D Rendering in AutoCAD

Just a few years ago, it took the power of a workstation to create the kind of images you will create in this chapter. Today, you can render not just a single image, but several hundred images to build computer animations. And with the explosion of game software, the Internet, and virtual reality, real-time walk-through sessions of 3D computer models are nearly as commonplace as word processors.

In this chapter, you'll learn how you can use rendering tools in AutoCAD to produce rendered still images of your 3D models. With these tools, you can add materials, control lighting, and even add landscaping and people to your models. You also have control over the reflectance and transparency of objects, and you can add bitmap backgrounds to help set the mood.

> **NOTE** Prior to Release 14, these rendering tools were sold as a separate, add-on product called AutoVision.

Things to Do Before You Start

You will want to take certain steps before you start working with the rendering tools so that you won't run into problems later. First, make sure you have a lot of free disk space on the drive where Windows is installed. Having 100 megabytes of free disk space will ensure that you won't exceed your RAM capacity while rendering. This may sound like a lot, but remember, you are attempting to do with your desktop computer what only workstations were capable of a few years ago. Also, make sure there is plenty of free disk space on the drive where your AutoCAD files are kept.

Creating a Quick Study Rendering

Throughout this chapter, you will work with a 3D model that was created using Auto-CAD's solid modeling tools (you learned about solid modeling in Chapter 11). You will use a version of your press model. The rest of the press parts have been added to the assembly and saved to the CD-ROM as Ch15press. You'll start by creating a basic rendering using the default settings in the Render dialog box.

1. Open the Ch15press.dwg file from the companion CD-ROM.
2. Open the Render toolbar from the Toolbars dialog box.
3. Choose View ➢ Render ➢ Render, or click on the Render tool in the Render toolbar.

The Render dialog box appears. In time, you will become intimately familiar with this dialog box.

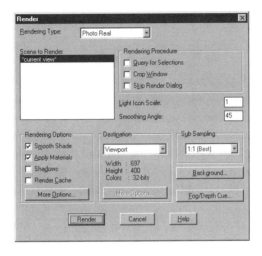

4. Click on the Render button. AutoCAD will take a minute or two to render the current view. While it's working, you will see messages in the Command window showing you the progress of the rendering. When AutoCAD is done, the model appears as a surface shaded model (see Figure 15.1).

FIGURE 15.1

The Press drawing rendered using all the default settings

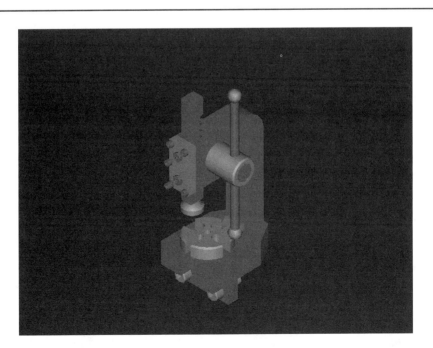

When you render a model without any special settings, you get what is called a *Z buffer shaded model*. The surfaces are shaded in their color and the light source is, by default, from the camera location. This view is much like a view with hidden lines removed and with color added to help distinguish surface orientation. You can actually get a similar view using the Shade tool in the standard AutoCAD Render toolbar.

Simulating the Sunlight Angle

The ability to add a sunlight source to a drawing is one of AutoCAD's key features for architects. For them, this is a frequently used tool in the design of buildings in urban and suburban settings. The Sun options let you accurately simulate the sun's location in relation to a model and its surrounding buildings. For mechanical users, the sun is a convenient and easy-to-use light source. AutoCAD also lets you set up multiple light sources other than the sun.

So let's add the sun to our model to give a better sense of the press assembly. In the following exercise, you'll explore most of the Sun options. Mechanical designers will seldom use all of these, but it is good to know that they are there. The day may come when you are asked to create something that goes outside and you will want to do a sun study.

1. Choose View ➤ Render ➤ Lights, or click on Lights from the Render toolbar.

Whenever you are creating a new light or other object with the Render tool, you usually have to give it a name first, before you can do anything else.

2. In the Lights dialog box, choose Distant Light from the drop-down list next to the New button toward the bottom left.

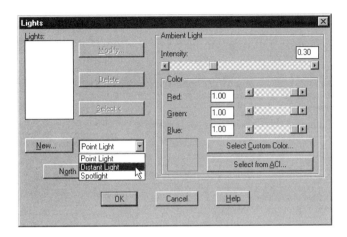

3. Click on the New button. The New Distant Light dialog box appears.

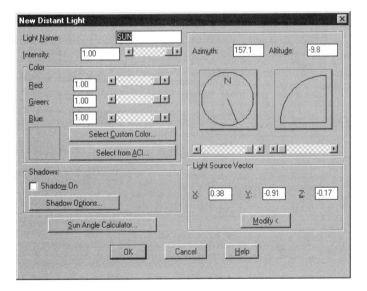

4. Type **SUN**. The word SUN appears in the Light Name input box toward the top of the dialog box. This dialog box lets you control various aspects of the light source, such as color and location.

5. Because we want to simulate the sun in this example, click on the button labeled Sun Angle Calculator. The Sun Angle Calculator dialog box appears.

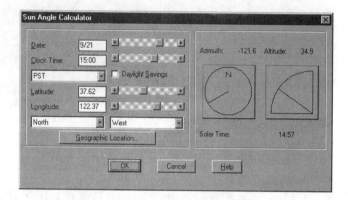

Notice that you have options for setting the date and time to determine the exact location of the sun. In addition, you have the option to indicate where true polar north is in relation to your model. AutoCAD assumes polar north is at the 90° position in the WCS.

6. One important factor for calculating the sun angle is finding your location on the earth. Click on the Geographic Location button. The Geographic Location dialog box appears.

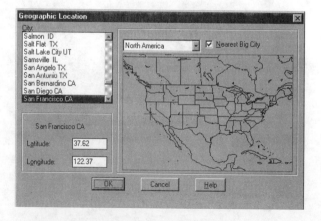

Here you can tell AutoCAD where your assembly is located in the world.

7. For the sake of this tutorial, let's suppose the Press model is located in San Francisco, California, USA. Select North America from the drop-down list above the map.
8. Locate and select San Francisco CA in the scrolling list to the left of the map. Notice that the Latitude and Longitude input boxes below the list change to reflect the location of San Francisco. For locations not listed, you can enter values manually in those input boxes.
9. Now click on OK to return to the Sun Angle Calculator dialog box. Set the Date for 9/21 and the time for 14:00 hours. Notice that the graphic to the right of the dialog box adjusts to show the altitude and azimuth angle of the sun for the time you enter.
10. Click on OK, and then click on OK again at the Lights dialog box.
11. Choose View ➢ Render ➢ Render, or click on Render from the Render toolbar. Then click on the Render button in the Render dialog box. Your model will be shaded to reflect the sun's location (see Figure 15.2).

FIGURE 15.2

The Press model with the Sun light source added

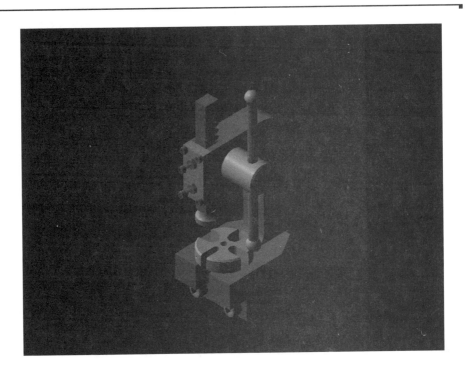

Notice that the press itself looks darker than before. Remember that in the first rendering, the light source is the same as the camera location, so the surface facing you receives more direct light. In this last rendering, the light source is at a glancing angle, so the surface appears darker.

We mentioned that you can set the direction of polar north. This is accomplished by clicking on the North Location button in the Lights dialog box. When selected, this button opens the North Location dialog box, shown in Figure 15.3. With this dialog box, you can set true north in any of three ways. You can click on the graphic to point to the direction, use the slide bar at the bottom to move the arrow of the graphic and adjust the value in the input box, or enter a value directly into the input box. You also have the option to indicate which UCS is used to set the north direction. For example, you may have already set a UCS to point to the true north direction. You only need to select the UCS from the list and leave the angle at 0.

FIGURE 15.3

The North Location dialog box

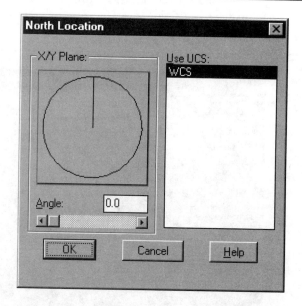

Improving the Smoothness of Circles and Arcs

You might notice that at times when using the Render, Hide, or Shade tools, solid or region arcs appear segmented rather than curved. This may be fine for producing layouts or backgrounds for hand-rendered drawings, but for final plots, you will want arcs and circles to appear as smooth curves. You can adjust the accuracy of arcs in your hidden, rendered, or shaded views through a setting in the Preferences dialog box.

The Rendered Object Smoothness setting in the Performance tab of the Preferences dialog box can be modified to improve the smoothness of arcs. Its default setting is 0.5, but you can increase this to as high as 10 to smooth out faceted curves. In the Press.dwg model, you can set Rendered Object Smoothness to 1.5 to render the curved surfaces as a smooth arc instead of a series of flat segments. This setting can also be adjusted using the Facetres system variable.

Adding Shadows

There is nothing like adding shadows to a 3D rendering to give the model a sense of realism. AutoCAD offers three methods for casting shadows. The default method is called Volumetric Shadows. This method takes a considerable amount of time to render more complex scenes. When using the second method—AutoCAD's Ray Trace option (described later in this chapter)—shadows will be generated using the Ray Trace method. The third method, called Shadow Map, offers the best speed but requires some adjustment to get good results. Shadow Map offers a soft-edge shadow. While Shadow Map is generally less accurate than the other two methods, its soft-edge option offers a level of realism not available in the other two methods.

In the following exercise, you will use the Shadow Map method. It requires the most adjustments and yields a fast rendering.

1. Choose View ➢ Render ➢ Lights, or click on Lights from the Render toolbar.

2. At the Lights dialog box, make sure Sun is highlighted, and then click Modify. The Modify Distant Light dialog box appears.

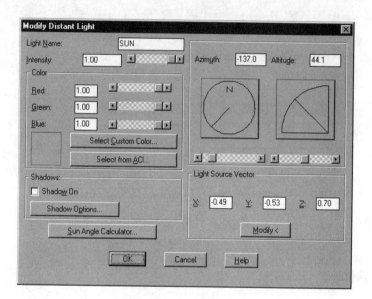

When adding shadows, remember that you must turn on the Shadow option for both the Render dialog box *and* for each light that is to cast a shadow.

3. Click on the Shadow On check box, and then click on the Shadow Options button. The Shadow Options dialog box appears.

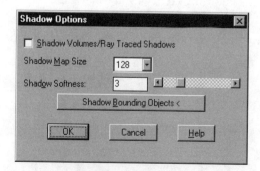

4. In the Shadow Map Size input box, select 512 from the drop-down list. This is the actual number of pixels used to create the shadow map.
 5. Click on the Shadow Bounding Objects button. The dialog box temporarily disappears to allow you to select objects from the screen. This option allows you to select the objects you want to cast shadows.
 6. Select the entire press assembly and press ↵. The Shadow Options dialog box reappears; click on OK to close the dialog box.
 7. In the Modify Distant Light dialog box, click on OK. It may take several seconds before the dialog box closes.
 8. When you get to the Lights dialog box, click on OK to close it.
 9. Click on the Render button on the Render toolbar.
 10. At the Render dialog box, check the Shadows check box, and then click on the Render button. After a minute or two, the model appears rendered with shadows (see Figure 15.4).

Don't panic if the shadows don't appear correct. The Shadow Map method needs some adjustment before it will give the proper shadows, because the default settings are appropriate for views of objects from a greater distance than our current view. The following exercise will show you what to do for "close-up" views.

FIGURE 15.4

The Press drawing rendered with shadows using the Shadow Map method

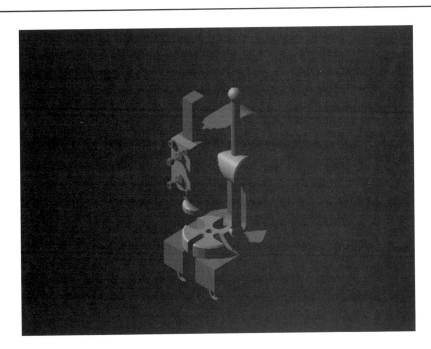

1. Open the Render dialog box again, and then click on the button labeled More Options. The Photo Real Render Options dialog box appears.

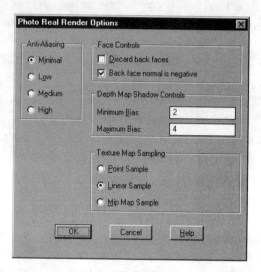

2. In the Depth Map Shadow Controls group, change the Minimum Bias value from 2 to 0.1.
3. In the same group, change the Maximum Bias value from 4 to 0.2.
4. Click on OK to close the dialog box, and then click on Render. Your next rendering will show more accurately drawn shadows (see Figure 15.5).

Consider placing a large yellow or cyan circle under the objects being rendered. Circles make convenient contrasting surfaces for Shade and Render, and will make the shadows more interesting.

The shadows still look a bit rough. You can further refine the shadows' appearance by increasing the Shadow Map Size to greater than 512. This setting can be found in the Shadow Options dialog box in step 4 of the exercise just before the last one. Figure 15.6 shows the same rendering with the Shadow Map Size set to 1024. As you increase the map size, you also increase render time and the amount of RAM required to render the view. If you don't have enough free disk space, you may find that AutoCAD will refuse to render the model. You will then either have to free up some disk space or decrease the map size.

CREATING A QUICK STUDY RENDERING | **695**

FIGURE 15.5
The rendered view with the Shadow bias settings revised

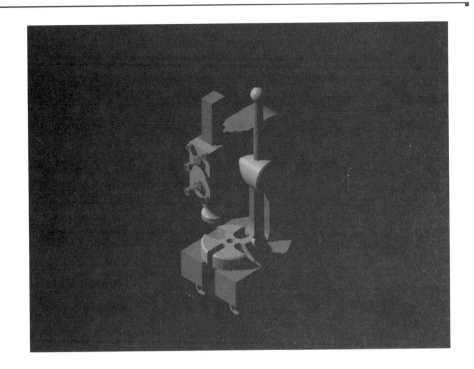

FIGURE 15.6
The rendered view with the Shadow Map Size set to 1024

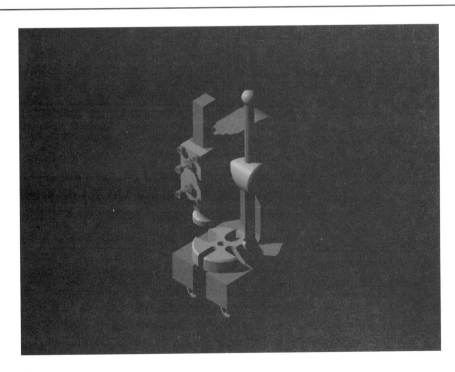

Notice that the shadow has a soft edge. You can control the softness of the shadow edge using the Shadow Options dialog box you saw in the exercise before the last one. The Shadow Softness input box and slide bar let you sharpen the shadow edge by decreasing the value or soften it by increasing the value. The soft shadow is especially effective for renderings of building interiors or scenes where you are simulating artificial light.

Adding Materials

The rendering methods you've learned so far can be an aid in your design effort. The Sun option offers a simple and effective light source for rendering. But the press still looks somewhat cartoonish. You can further enhance the rendering by adding materials to the objects in your model.

Let's suppose you want a granite-like finish on your press for purposes of illustration. You also want the press table and the knobs on the handle to look like molded plastic. The first step to adding materials is to acquire the materials from AutoCAD's Materials Library.

1. Choose View ➢ Render ➢ Materials, or click on Materials from the Render toolbar.

The Materials dialog box appears.

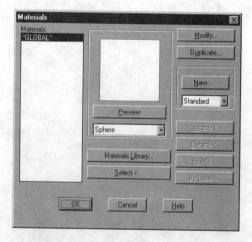

2. Click on the Materials Library button in the middle of the dialog box. The Materials Library dialog box appears.

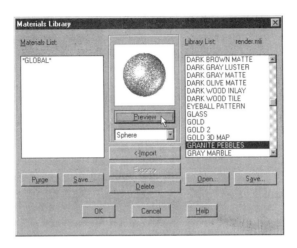

3. In the Library List box, find and select Granite Pebbles so it is highlighted. This is the material you will assign to the press.

4. Click on the Preview button in the middle of the dialog box. A view of the material appears on a sphere, giving you an idea of what the material looks like.

5. Click on the Import button. Notice that Granite Pebbles now appears in the Materials List box to the left. This list box shows the materials you've transferred to your drawing.

6. Now locate Molded Plastic in the Library list to the right and select it. Click on the Preview button again to see what it looks like. Notice that the preview shows a sphere reflecting some light. You may notice a textured effect caused by the low color resolution of the AutoCAD display.

7. Click on the Import button again to make Molded Plastic available in the drawing; then click on OK to exit the Materials Library dialog box.

Once you've acquired the materials, you will have to assign them to objects in your drawing.

1. In the Materials dialog box, highlight the Granite Pebbles item shown in the list to the left, and then click on the Attach button in the right half of the dialog box. The dialog box temporarily disappears, allowing you to select the objects you want to appear as Granite Pebbles.
2. Click on the blue parts of the model and then press ↵. After a moment, the Materials dialog box appears again.
3. Click on Molded Plastic from the Materials list, click on Attach, and choose the remaining parts of the model. Press ↵ when you are finished.

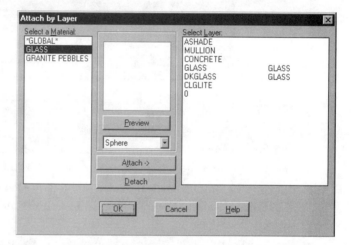

4. Click on OK to exit the Materials dialog box.
5. Now render your model. You may want to take a break at this point, as the rendering will take a few minutes. When AutoCAD is done, your rendering will look like Figure 15.7.

FIGURE 15.7
The press model with the molded plastic and granite pebbles materials added

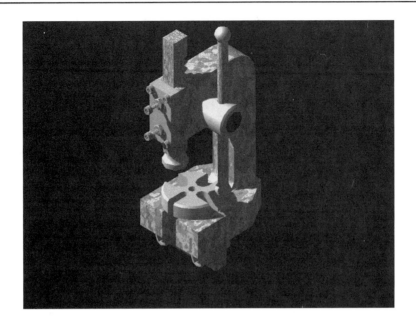

Adjusting the Materials' Appearance

The press looks more like it has an Army camouflage paint job than a granite finish. Fortunately, you can make several adjustments to the material. You will want to reduce the scale of the granite pebbles material so it is in line with the scale of the model.

1. Choose View ➣ Render ➣ Materials.
2. At the Materials dialog box, select Granite Pebbles from the Materials list, and then click on the Modify button. The Modify Granite Material dialog box appears.

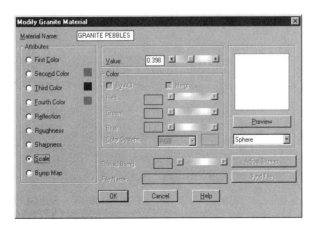

3. Click on the Scale radio button in the Attributes button group in the left-hand side of the dialog box.

4. Change the Value input box near the top of the dialog box from 0.398 to 0.010 to reduce the scale of the material.

5. Click on OK to return to the Materials dialog box.

The Modify Granite Material dialog box offers a variety of options that let you control reflectivity, roughness, color, transparency, and, of course, scale. The Help button on the Modify Granite Material dialog box will give a brief description of these options.

6. Render the view with the new material settings. After a few minutes, your view will look something like Figure 15.8.

There are four basic types of materials to choose from in the Materials list: Standard, Marble, Granite, and Wood. Each type has its own set of characteristics that you can adjust. You can even create new materials based on one of the four primary types of materials. Now let's continue by making another adjustment to the material settings.

The granite surface of the press is a bit too strong. You can reduce the graininess of the granite by further editing in the Modify Granite Material dialog box.

1. Click on Materials from the Render toolbar, and then select Granite Pebbles from the Materials list, and then choose Modify.

2. At the Modify Granite Material dialog box, click on the Sharpness attribute radio button, and then set the Value input box to 0.20.

3. Click on OK, and then click on OK again at the Materials dialog box.

4. Choose View ➣ Render ➣ Render, and then click on the Render button of the Render dialog box. After a few minutes, your rendering appears with a softer granite surface (see Figure 15.9).

CREATING A QUICK STUDY RENDERING | 701

FIGURE 15.8

The press model after modifying the material settings

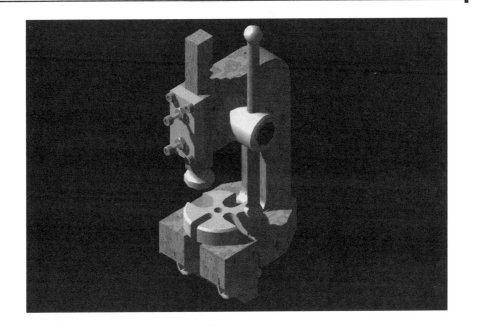

FIGURE 15.9

The rendered image with a softer granite surface

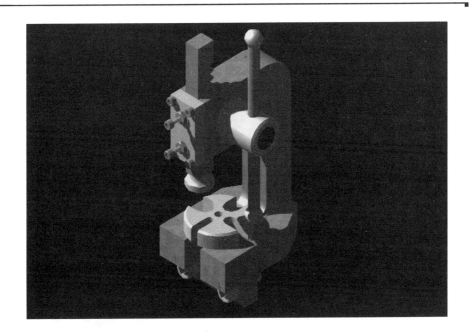

Adding a Background

You could continue by adding and adjusting materials to the other parts of model, but let's try dressing up our view by including a checkered background. To do so, we need to open the Background dialog box.

1. Open the Render dialog box, and then click on the button labeled Background. The Background dialog box appears.

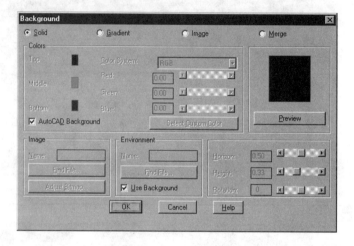

2. In the row of radio buttons across the top, find and click on Image. Notice that several of the options near the bottom of the dialog box are now available.

3. Click on the Find File button at the bottom left of the dialog box. The Background Image dialog box appears.

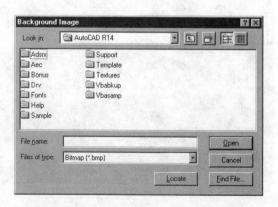

ADDING A BACKGROUND 703

4. Use the Background Image dialog box to locate the checkers.tga file. It can be found in the \Textures subdirectory of the \AutoCAD R14 directory.

NOTE

Notice that the Files of Type list box defaults to *.bmp, and you need to change it to *.tga.

5. Once back at the Background Image, click on Preview to see what the file looks like.
6. Click on OK. Then, at the Render dialog box, click on Render. The background appears behind the model, as shown in Figure 15.10.

FIGURE 15.10

The press model rendered with a checkered bitmap image for a background

We chose to add a bitmap image for a background, but you can use other methods to generate a background. For example, you might prefer to use a gradient shade or color for the background. This can help give a sense of depth to the image (see Figure 15.11). You can, of course, add a single color to the background if you prefer.

To create a gradient background, select the Gradient radio button at the top of the Background dialog box. You can then adjust the color for the top, middle, and bottom third of the background. AutoCAD automatically blends the three colors from top to bottom to create the gradient colors.

FIGURE 15.11

The Press model with a gradient color background

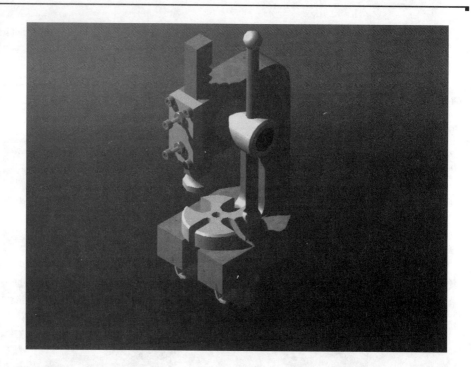

Effects with Lighting

Up to now, you've only used one light source, called a Distant Light, to create a sun. You have two other light sources available to help simulate light: point-light sources and spotlights. This section will show you some examples of how you can use these light sources, along with some imagination, to perform any number of visual tricks.

Spotlights

Spotlights are lights that are directed. They provide emphasis and are frequently used for product presentations. In this exercise, you'll use spotlights to illuminate the press.

You'll start by setting up a view to help place the spotlights. Once they are placed, you will make some adjustments to them to get a view you want.

1. Choose View ➢ 3D Viewpoint Presets ➢ SE Isometric; then zoom in to the press so that your view looks similar to Figure 15.12.
2. Choose View ➢ Render ➢ Lights. At the Lights dialog box, select Spotlight from the New drop-down list.
3. Click on New. Then at the New Spotlight dialog box, enter **Spot-L**. This designates a spotlight you will place on the left side of the press.
4. Enter **400** in the Intensity input box. Then click on the Modify button.
5. At the Enter Light Target Location prompt, use the Nearest Osnap and select the point on the window, as indicated in Figure 15.12.
6. At the Enter Light Location prompt, select the point indicated in Figure 15.12. Once you've selected the light location, return to the New Spotlight dialog box. You can, in the future, adjust the light location if you choose.
7. Click on OK. Then at the Lights dialog box, click on New again to create another spotlight.
8. This time, enter **Spot-R** for the name.
9. Click on the Modify button and select the target and light locations indicated in Figure 15.13.
10. Set the intensity to **400** as before. Click on OK to exit the New Spotlight dialog box, and then click on OK again at the Lights dialog box. You now have two spotlights.
11. Render the model (you should know how this is done by now). Your view will look similar to Figure 15.14.

FIGURE 15.12
Selecting the points for the first spotlight

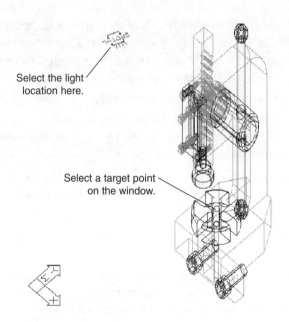

FIGURE 15.13
Selecting the points for the second spotlight

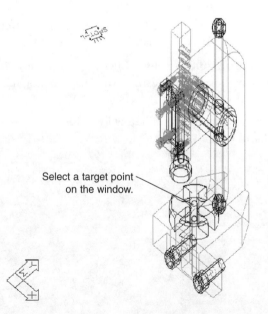

EFFECTS WITH LIGHTING | 707

FIGURE 15.14
The rendered view of the model with the spotlights

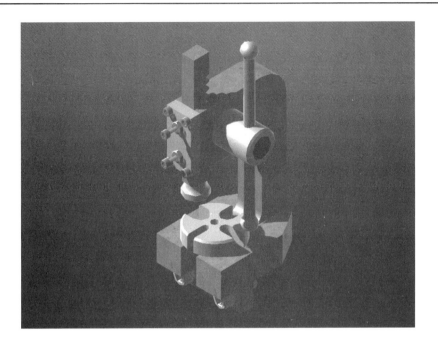

The rendered view has a number of problems. First, the sunlight source needs to be turned off. Second, the spotlights are too harsh. We'll solve these problems in the next section.

Controlling Lights with Scenes

The first problem we face is how to turn off the sun. You can set the sunlight intensity value to zero using the Modify Distant Light dialog box. Another way is to set up a scene. AutoCAD lets you combine different lights and views into named scenes. These scenes can then be quickly selected at render time so you don't have to adjust lighting or views every time you want a specific setup. Here's how it works.

1. Choose View ➢ Render ➢ Scenes, or click on Scenes from the Render toolbar.

You will see the Scenes dialog box.

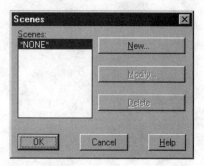

2. Click on New. The New Scene dialog box appears.

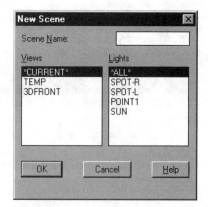

3. Enter **NIGHT** for the scene name. The name appears in the Scene Name input box.
4. Shift+click to select more than one item on SPOT-L and SPOT-R from the Lights list.
5. Click on OK. Notice that now you have NIGHT listed in the Scenes list in the Scenes dialog box.
6. Click on New again, and then type **DAY**.
7. Select SUN from the Lights list, and then click on OK. You now have two scenes set up.

8. Click on OK, and then open the Render dialog box. Notice that you have DAY and NIGHT listed in the Scene to Render list box in the upper left of the Render dialog box.

9. Select Night, and then click on the Render button. Your view will look like Figure 15.15.

FIGURE 15.15

Rendering the Night scene

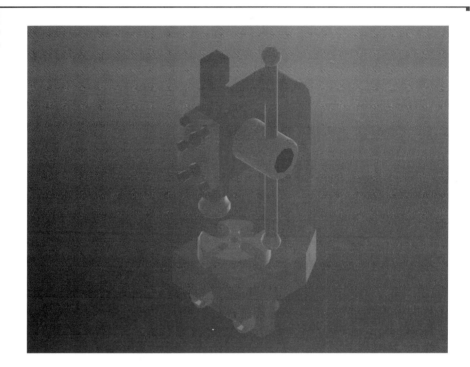

We now see that without the sunlight source, our view is considerably darker. Let's continue by adding a few more light sources and adjusting some existing ones.

1. Open the Lights dialog box, select Point Light from the New drop-down list, and click on New.

2. Enter the name **Point1**, and then give this new point light an intensity value of **300**.

3. Click on the Modify button, and place the Point1 light in the center, as shown in Figure 15.16.

4. Click on OK; then create another point-light source called **Point2** and give it an intensity of **150**.

5. Click on the Modify button, adjust your view so it looks similar to Figure 15.17, and then place the light on the Press, as shown in Figure 15.17. Use the .X, .Y, and .Z point filters to select the location of the light.

6. Click on OK. Then at the Light dialog box, select SPOT-L from the list and click on Modify.

7. At the Modify Spotlight dialog box, change the Falloff value in the upper right to 80. Click on OK.

8. Repeat step 6 for the SPOT-R spotlight.

9. Click on OK at the Modify Spotlight dialog box.

10. At the Light dialog box, increase the Ambient light intensity to **50**, and then click on OK.

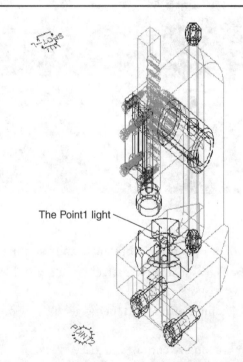

FIGURE 15.16
Adding another point-light source

EFFECTS WITH LIGHTING | 711

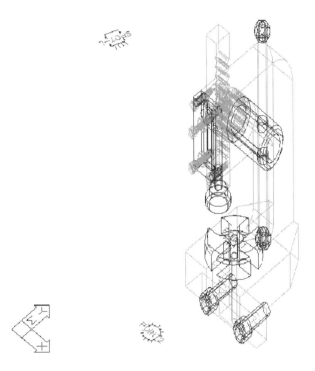

FIGURE 15.17
Adding a point-light source to the press

You've got the new lights installed and the spotlights adjusted. Before you render your scene, you need to include the new lights in the scene you set up for the night rendering.

1. Choose View ➢ Render ➢ Scenes.
2. Highlight NIGHT in the Scenes list, and then click on the Modify button.
3. Shift+click on Point1 and Point2 in the Lights list.
4. Click on OK at both the Modify Scene and Scenes dialog boxes.
5. Choose Render and make sure NIGHT is selected in the Scene to Render list.
6. Click on the Render button. Your view will look similar to Figure 15.18.

FIGURE 15.18
The Night rendering with added lights and an increased falloff area for the spotlights

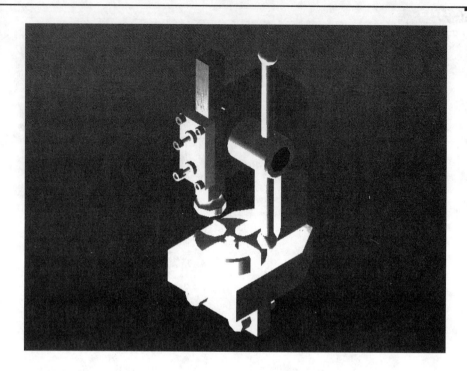

Adding Reflections and Detail with Ray Tracing

You've been gradually building up the detail and realism in your renderings by adding light and materials. In this section, you'll learn how using a different rendering method can further enhance your 3D models. Until now, you've been using the standard AutoCAD rendering method. Ray tracing can add interest to a rendering, especially where reflective surfaces are prominent in a model. In this section, you'll use the Ray Trace method to render your model after making a few adjustments to the material.

To make a long, complicated story short, ray tracing simulates the way light works. And it does it in somewhat of a reverse way. Ray tracing analyzes the light path to each pixel of your display, tracing the light or "ray" from the pixel to the light origin as it bounces off objects in your model. Ray tracing takes into account the reflectivity and, in the case of glass, the refraction of light as it is affected by objects in the model. Because more objects in a model offer more surfaces to reflect light, ray tracing becomes more time-consuming as the number of objects increases. Also, since each pixel is analyzed, a greater image size increases the render time geometrically. By doubling the width and height of the view size, for example, you are essentially increasing the number of pixels by four times.

AutoCAD offers ray tracing as an option for rendering both shadows and the entire scene. The Ray Trace options offer greater accuracy in exchange for slow rendering time. If you choose to select ray-traced shadows, for example, you can expect at least a fourfold increase in rendering time. Rendering an entire scene can increase rendering time by an order of magnitude.

Needless to say, if you are in a time crunch, you will want to save ray tracing just for the essential final renderings. Use Render or other rendering options for study rendering or for situations that don't require the accuracy of the Ray Trace method.

Getting a Sharp, Accurate Shadow with Ray Tracing

In *Adding Shadows*, earlier in this chapter, we showed you how to use the Shadow Map method for casting shadows. Shadow maps offer the feature of allowing a soft-edge shadow in exchange for accuracy. For exterior views, you may prefer a sharper shadow. The Press example loses some detail using the Shadow Map method; in particular, the grooves in the base of the press disappear. By switching to the Ray Trace method for casting shadows, you can recover some detail.

1. Choose View ➢ Render ➢ Lights. Then from the Lights list, select Sun and click on Modify.
2. At the Modify Distant Light dialog box, click on the Shadow Options button.
3. At the Shadow Options dialog box, click on Shadow Volumes/Ray Trace Shadows check box to place a check mark there.
4. Click on OK at all the dialog boxes to exit them and return to the AutoCAD view.
5. Render the view using the Photo Ray Trace for the Rendering Type and Day for the Scene to Render. Your view will look like Figure 15.19.

The new rendering is brighter. You can also see the effects of an increased falloff for the spotlights. They don't have the sharp edge they had in the first Night rendering, and the light is spread in a wider radius, illuminating more of the lower portion of the press. Notice that you can now see the press clearly. The shadows also appear sharper, especially around the surface detail of the press model.

FIGURE 15.19

The model using the Shadow Volumes/Ray Trace Shadows option

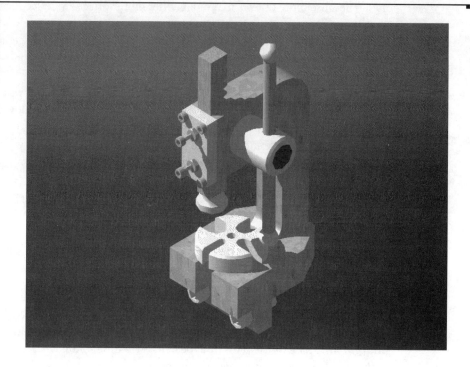

Creating and Adjusting Texture Maps

You've already seen how you can assign a material to an object by adding materials. Many of these materials make use of bitmap image files to simulate textures. You can create your own surface textures or use bitmaps in other ways to enhance your rendering. For example, you can include a photograph of machinery that may exist within the scene you are rendering.

Figure 15.20 shows a bitmap image that was scanned into the computer and edited using a popular paint program.

NOTE

AutoCAD uses two methods for refreshing your drawing display: *Regen* and *Redraw*. Regen meticulously regenerates all the information about a drawing as if recreating it from scratch. Even with the speed of your computer, this is a pretty lengthy process. Redraw grabs only the most relevant information to the view you have selected and produces a "virtual" reproduction without all the details, so it is considerably faster than Regen. For more information on using the Regen and Redraw commands, see Chapter 6.

FIGURE 15.20

A photographic image scanned into a computer and saved as a bitmap file

The following exercise will show you how to add a bitmap image to your rendering.

1. Click on Redraw from the Standard toolbar, and then adjust your view using Zoom or Dview so your press looks like the top image of Figure 15.21.
2. Draw a line 20 units long, as shown in the top image of Figure 15.21.
3. Change the thickness of the line to 20 units using the Properties tool.
4. Choose View ➤ Render ➤ Materials. Then at the Materials dialog box, click on New. Notice that the New Standard Materials dialog box is the same as the dialog box for the molded plastic material. The settings are not the same, however.
5. Enter **machine1** for the material name.
6. Make sure the Color/Pattern radio button is selected, and then click on the Find File button in the lower-right corner of the dialog box.
7. Click on the List Files of Type drop-down list. Notice that you have several file types from which to choose.
8. Choose TIF from the list, and then locate the Image01.tif file. This file comes with the other sample files from the companion CD-ROM.

9. Click on OK to exit this dialog box.
10. At the Materials dialog box, make sure Machine1 is selected in the materials list, and then click on the Attach button.
11. Select the line you added in step 1, and then press ↵.
12. Click on OK to exit the Materials dialog box, and then render the scene. Your view will look like the bottom image of Figure 15.21.

FIGURE 15.21
Adding a bitmap image to your rendering

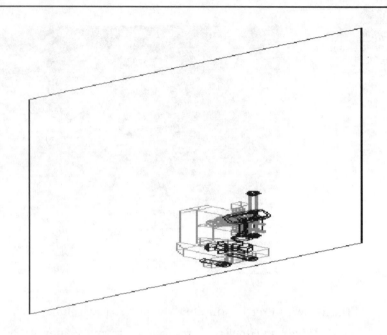

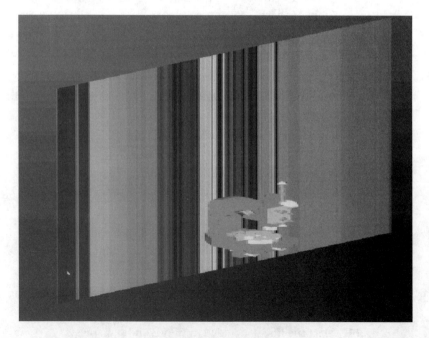

The bitmap image does not appear properly in the rendered view. Instead, it looks like vertical streaks of color. When you see this streaking, you know your bitmap image or material is not properly aligned with the object to which it is attached. The following exercise introduces you to the tools you need to properly align a bitmap image to an object.

1. Redraw the screen. Then choose View ➢ Render ➢ Mapping, or click on Mapping from the Rendering toolbar.

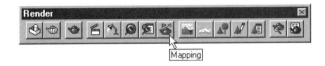

2. At the Select object: prompt, select the extruded line you created in the last exercise. The Mapping dialog box appears.

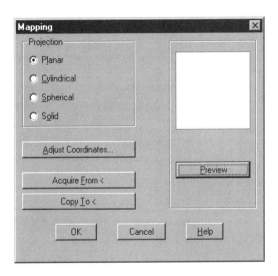

3. Click on the Adjust Coordinates button. The Adjust Planar Coordinates dialog box appears.

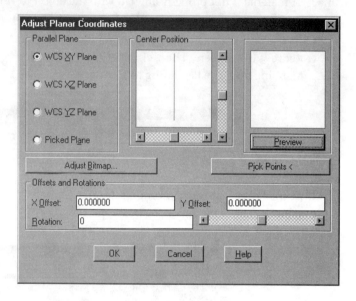

Notice the rectangle in the area labeled Center Position. This shows the relationship of the bitmap image to the object to which it has been assigned. All you can see is a vertical line.

4. Click on the WCS XZ Plane radio button. The plane defined by the X- and Z-axes is parallel to the surface on which you want the bitmap to appear.

5. Click on the Preview button. Now you can see how the bitmap will appear on the vertical surface.

6. Click on OK at each of the dialog boxes to close them, and then render the model. Your view will look like Figure 15.22.

Notice that now the image is correct and no longer looks like vertical streaks. There are no odd blank spaces either. As you have seen in the previous exercise, the Adjust Object Bitmap Placement dialog box allows you to stretch the image vertically or horizontally in case the image is distorted and it needs to be fitted to an accurately drawn object.

Another option is to use a paint program to refine the bitmap image before it is used in AutoCAD. AutoCAD attempts to place the bitmap accurately on a surface, so if the bitmap is fairly clean and doesn't have any extra blank space around the edges, you can usually place it on an object without having to make any adjustments other than its orientation.

FIGURE 15.22

The rendered view with the bitmap image adjusted

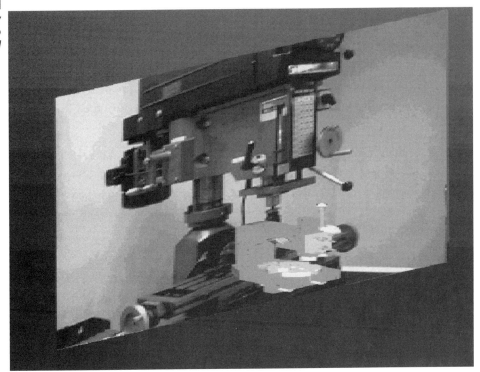

Rendering Output Options

All through this chapter, you have been rendering to the AutoCAD drawing area. You can also render to a file, which enables you to recall the image at any time in any application, or render to the Render window. From there, you have a number of options in dealing with the rendered image.

Rendering to the Render Window

The Render window lets you control the resolution and color depth of your image. It also lets you save the images you render in the Windows BMP format. Another advantage of the Render window is that you can render several views and then compare them before you decide which ones to save.

1. Open the Render dialog box, and then select Render Window from the Destination drop-down near the bottom of the dialog box.

2. Click on Render. After a moment the Render window will appear. It will then take a minute or two before the image finishes rendering and appears in the window.

Notice that the image is within its own window. If you render another view, that view will also appear in its own window, leaving the previous renderings undisturbed. You can use File ➢ Save in the Render Window to save the file as a BMP file for later editing or printing, or you can print directly from the Render window. You can also use the Render window to cut and paste the image to another application or to view other files in the BMP format.

To set the size of renderings, use the File ➢ Option tool in the Render window. This option opens the Windows Render Options dialog box (see Figure 15.23). Here you can choose from two standard sizes or enter a custom size for your rendering. You can also choose between 8-bit (256 colors) and 24-bit (16 million colors) color depth. Changes to these settings don't take effect until you render another view.

FIGURE 15.23

The Windows Render Options dialog box

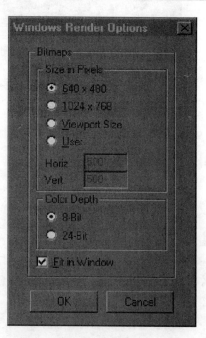

Rendering Directly to a File

Rendering to the Render window allows you to view and compare your views before you save them. However, you can only save your views in the BMP format. If you plan to further edit the image in an image-processing program, this may not be a problem. But if you want to use your image file with a program that requires a specific file format, you may want to render directly to a file. Here's how it's done:

1. Open the Render dialog box, and then select File from the Destination group in the upper right of the dialog box.
2. Choose More Options at the bottom of the Destination group. The File Output Configuration dialog box appears.

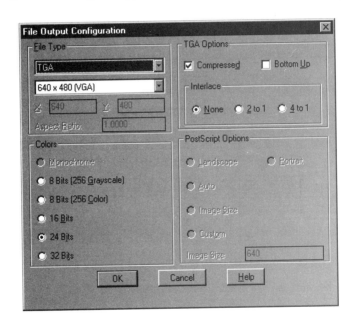

3. Click on the File Type drop-down list to see the options. You can save your image in BMP, TGA, TIF, PCX, or even PostScript format. You might also notice the other options available in the dialog box, such as color depth, resolution, and compression. Not all these options are available for all the file types. For example, PCX is limited to 256 colors, so the other color options will not apply.
4. Click on OK to return to the Render dialog box, and then click on the Render button. The Rendering File dialog box appears, prompting you for a file name for your image.
5. Enter **Press1**. AutoCAD will add the file name extension for you.

6. Click on OK and AutoCAD proceeds to render to the file.

As AutoCAD renders to the file, it tells you in the command line how much of the image has been rendered.

Improving Your Image and Editing

There will be times when you will be rushing to get a rendering done and won't want to wait for each trial rendering to become visible. AutoCAD offers several tools that can save you time by limiting the resolution or area being rendered.

1. Open the Render dialog box and set the Destination option to Viewport.

2. Click on the Crop Window check box to activate this option, and then click on the Render button.

3. You are prompted to `Pick Crop Window to Render`. Select the area shown in Figure 15.24, indicated by the rubber-banding square. Once you select the window, AutoCAD renders only the area you selected.

FIGURE 15.24

Selecting the Crop window

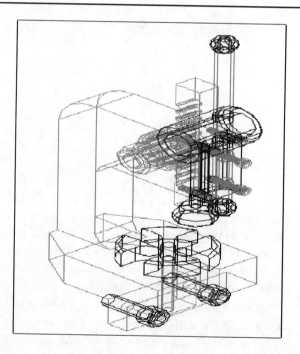

The Crop Window option is a working tool and is not available when File or Render Window are selected as destinations.

You can also select specific objects to be included in the rendering by checking the Query for Selections check box in the Render dialog box. This option asks you to select a set of objects before it proceeds to render. You can render to all three destination options with Query for Selections turned on.

If you want to get a quick rendering with a reduced resolution to check composition, you can use the Sub Sampling option drop-down list. Try the following exercise to see how it works.

1. Open the Render dialog box, and then open the Sub Sampling drop-down list.
2. Choose 3:1 from the list, make sure the Crop Window option is unchecked, and then click on the Render button. Your view will render faster but will look a bit crude (see Figure 15.25).

FIGURE 15.25
A rendered view with the Sub Sampling option set to 3:1

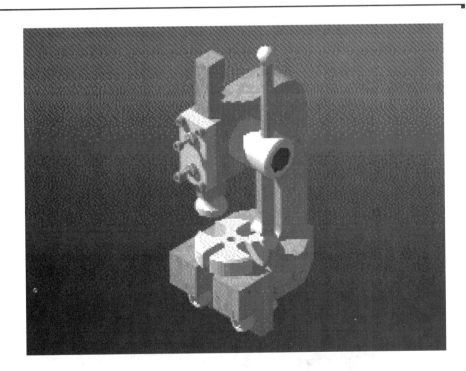

The different ratios in the Sub Sampling option tell you how many pixels are being combined to reduce the resolution of the image. For example, 3:1 will combine three pixels into one to reduce the resolution to a third of the original.

Smoothing Out the Rough Edges

The Sub Sampling option increases the jagged appearance of your rendering because of the reduced resolution. For your final rendering, you can actually improve the smoothness of edges and thereby increase the apparent resolution by using the Anti-Aliasing option in the Render dialog box. This option performs a kind of computer trick that reduces the jagged appearance of object edges. Anti-Aliasing blends the color of two adjacent contrasting colors. This gives the effect of smoothing out the "stair-step" appearance of a computer-generated image. The improvement to your rendering can be striking. Try the following exercise to see firsthand what Anti-Aliasing can do.

1. Open the Render dialog box, and then click on the More Options button.
2. In the Raytrace Rendering Options dialog box, click on the Medium radio button in the Anti-Aliasing button group, select Rendering Type: Photo Ray Trace, and then click on OK.
3. Select 1:1 from the Sub Sampling drop-down list, and then click on the Render button. The rendering will take several minutes, so you may want to take a break at this point. When the rendering is done, it will look similar to Figure 15.26.

FIGURE 15.26

A rendering with the Anti-Aliasing setting set to medium

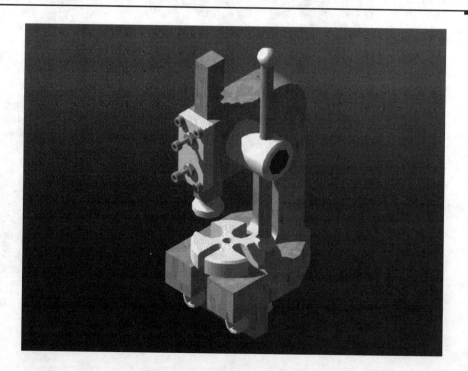

Notice that the edges are much smoother. You can also see that the vertical edges of the post are more clearly defined. One negative point is that the texture effect of the Press model has been reduced. You may have to increase the scale value for the Granite Pebbles material setting to bring the texture back.

As you can see from this exercise, you trade rendering speed for a cleaner image. You will want to save the higher Anti-Aliasing settings for your final output.

If You Want to Experiment...

In this chapter, you've taken a guided tour of AutoCAD's rendering tools and seen the main features of this product. Because of space considerations, we didn't go into the finer details of many of its features, but you do have the basic knowledge from which to build your rendering skills. Without too much effort, you can adapt much of what you've learned here to your own projects. If you need more detailed information, you can use the Help button found in all the Render dialog boxes.

Computer rendering of 3D models is a craft that takes some time to master. You will want to take some time to experiment with these rendering tools to see firsthand the types of results you can expect. You might want to try different types of views, like an Isometric or Elevation view.

Chapter 16

Working with Existing Drawings and Raster Images

FEATURING

Tracing, Scaling, and Scanning Drawings

Importing and Managing Raster Images

Importing PostScript Images

Working with Existing Drawings and Raster Images

At times you will want to turn a hand-drafted drawing into an AutoCAD drawing file. It may be that you are modifying a design you created before you started using AutoCAD, or that you are converting your entire library of drawings for future AutoCAD use. Or perhaps you want to convert a sketch into a formal drawing. This chapter discusses three ways to enter a hand-drafted drawing: tracing, scaling, and scanning. Each of these methods of drawing input has its advantages and disadvantages.

Tracing, Scaling, and Scanning Drawings

Tracing with a digitizing tablet is the easiest method, but a traced drawing usually requires some cleaning up and reorganization. If dimensional accuracy is not too important, tracing will do for entering existing drawings into AutoCAD. It is especially useful for drawings that contain irregular curves, such as the contour lines of a topographical map.

> Even if you don't plan to trace drawings into AutoCAD, be sure to read the tracing information anyway, because some of the information presented here will help your everyday editing tasks.

Scaling is recreating the paper drawing in AutoCAD using the data (dimensions) from the paper drawing. Scaling a drawing is the most flexible method because you don't need a tablet to do it and generally you are faced with less clean-up afterward. Scaling also affords the most accurate input of orthogonal lines because you can read dimensions directly from the drawing and enter those same dimensions into Auto-CAD. The main drawback with scaling is that if the drawing does not contain complete dimensional information, you must constantly look at the hand-drafted drawing and measure distances with a scale. Also, irregular curves are difficult to scale accurately.

Scanning offers some unique opportunities with Release 14, especially if you have a lot of RAM and a fast hard drive. Potentially, you can scan a drawing and save it on your computer as an image file, and then import the image into AutoCAD and trace over it. You still need to perform some clean-up work on the traced drawing, but because you can see your tracing directly on your screen, you have better control and you won't have quite as much cleaning up to do as you do when tracing from a digitizer.

There are also vectorizing programs that will automatically convert an image file into a vector file of lines and arcs. These programs may offer some help, but they will require the most cleaning up of the options presented here. Like tracing, scanning is best used for drawings that are difficult to scale, such as complex topographical maps containing more contours than are practical to trace on a digitizer, or non-technical line art, such as letterheads and logos.

Tracing a Drawing with a Digitizing Tablet

NOTE Most mechanical engineering offices don't use a digitizing tablet tool, so we'll give you just enough information to help you understand the concept. If you need more information on digitizing tablets, please spend some time with the manufacturer's manual for your tablet.

One way you can enter a hand-drafted drawing into AutoCAD is by tracing with a tablet digitizer. You may even find a digitizer the size of a drafting table in the back corner of the corporate office. These were used for some time to make drawings accessible to computers, but with the improvements in scanners and digital camera technology, such large digitizers do not see much use in the modern mechanical engineering office. Or, if you were working with a large drawing and you only had a small tablet, you would have had to cut the drawing into pieces that your tablet could manage, then trace each piece and reassemble the completed pieces into the large drawing.

The process of digitizing is very much like sketching, and you can use any of the tools that we will discuss either in a sketch or when using a tablet. If you do have a tablet, you can use either a stylus or a puck to trace drawings, but the stylus will offer the most natural feel because it is shaped like a pen. A puck has crosshairs that you have to center on the line you want to trace, and this requires a bit more dexterity.

Reconfiguring the Tablet for Tracing

If you had a tablet when you first installed AutoCAD, you configured the tablet to use most of its active drawing area for AutoCAD's menu template. Since you will need the tablet's entire drawing area to trace a drawing, you now need to reconfigure the tablet to eliminate the menu. Otherwise, you won't be able to pick points on the drawing outside the 4" × 3" pointing area AutoCAD normally uses (see Figure 16.1).

TIP You can save several different AutoCAD configurations that can be easily set using the Preferences dialog box. See Appendix B.

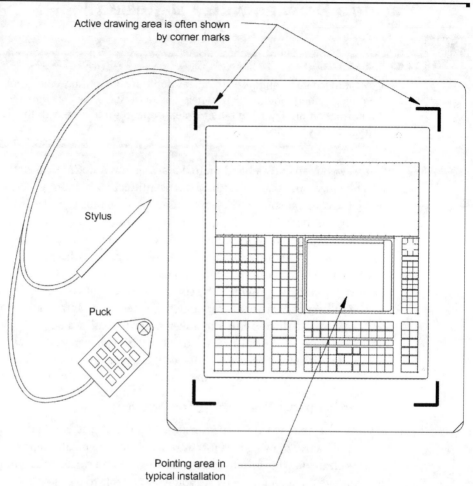

FIGURE 16.1
The tablet's active drawing area

Calibrating the Tablet for Your Drawing

Before you can trace anything into your computer, you must calibrate your tablet. This means you must give some points of reference so AutoCAD can know how distances on the tablet relate to distances in the drawing editor. For example, you may want to trace a drawing that was drawn at a scale of ½"=1". You will have to show AutoCAD two specific points on this drawing, as well as where those two points should appear in the drawing editor. This is accomplished by using the Tablet command's Cal option.

NOTE

When you calibrate a tablet, you are setting ratios for AutoCAD; for example, 2 inches on your tablet equals 1 inch in the drawing editor.

The word Tablet appears on the status bar to tell you that you are in Tablet mode. While in Tablet mode, you can trace the drawing but you cannot access the menus in Windows with some digitizers. (Check your digitizer manual for further information.) If you want to pick a menu item, you must toggle the Tablet mode off by using the F4 function key. Or you can enter commands through the keyboard. (If you need some reminders of the keyboard commands, type **help**↵ and click on Commands from the Help dialog box to get a list.)

Calibrating More Than Two Points

You can calibrate as many as 31 points. Why would anyone want to calibrate so many points? Often the drawing or photograph you are trying to trace will be distorted in one direction or another. For example, blueline prints are usually stretched in one direction because of the way prints are rolled through a print machine.

You can compensate for distortions by specifying several known points during your calibration. If you calibrate only two points, AutoCAD will scale X and Y distances equally. Calibrating three points causes AutoCAD to scale X and Y distances separately, making adjustments for each axis based on their respective calibration points.

Now suppose you want to trace a perspective view, but you want to "flatten" the perspective so that all the lines are parallel. You can calibrate four corners to stretch out the narrow end of the perspective view to be parallel with the wide end. This is a limited form of what cartographers call *rubber-sheeting*, where various areas of the tablet are stretched by specific scale factors.

When you select more than two points for calibration, you will get a message similar to that shown in Figure 16.2. Let's take a look at the parts of this message.

FIGURE 16.2

AutoCAD's assessment of the calibration

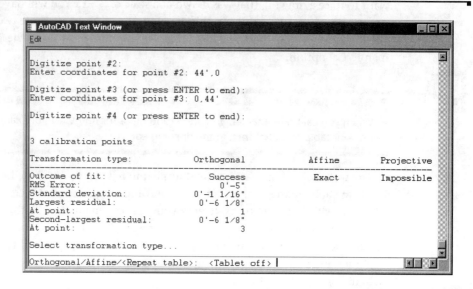

In the text window, you see the labels Orthogonal, Affine, and Projective. These are the three major *transformation types*, or types of calibrations. The orthogonal transformation scales the X and Y axes using the same values. Affine scales the X and Y axes separately and requires at least three points. The projective transformation stretches the tablet coordinates differently, depending on where you are on the tablet. It requires at least four calibration points.

Just below each of these labels you will see either Success, Exact, or Impossible. This tells you whether these transformation types are available to you. Because this example shows what you see when you pick three points, you get Impossible for the projective transformation.

The far-left column tells you what is shown in each of the other three columns.

Finally, the prompt at the bottom of the screen lets you select which transformation type to use. If you calibrate four or more points, the projective transformation is added to the prompt. The Repeat Table option simply refreshes the table.

Take care when you calibrate points on your tablet. Here are a few things to watch out for when calibrating your tablet:

- Use only known calibration points.
- Try to locate calibration points that cover a large area of your image.
- Don't get carried away. Try to limit calibration points to only those necessary to get the job done.

Tracing Lines from a Drawing

Once a tablet has been calibrated, you can trace your drawing from the tablet, even if the area you are tracing is not displayed in the drawing editor.

Tracing and sketching create inaccuracies. Some of the lines will be crooked, and others won't meet at the right points. These inaccuracies are caused by the limited resolution of your tablet, coupled with the lack of steadiness in the human hand. The best digitizing tablets have an accuracy of 0.001", which can't help you if the drawing that you are tracing is off by .03"—which was considered very accurate when the hand drawing was made.

Cleaning Up a Traced Drawing

In this section, we will discuss how to straighten and align the lines in your traced drawing. Then we'll look at adjusting the dimensions so that your new drawing reflects the dimensions in the original.

Straightening Lines

A problem in traced drawings is that some of the lines are not orthogonal. To straighten them, use the Change command, together with the Ortho mode. When used carefully, this command can be a real time saver.

WARNING

> The Change command's Change Point option changes the location of the endpoint closest to the new point location. This can cause erroneous results when you are trying to modify groups of lines. Also note that Change does not affect polyline line segments.

Aligning Lines

In addition to straightening lines, you can use Change to align a set of lines to another line. For example, when used with the Perpendicular Osnap, several lines can be made to align at a perpendicular angle to another line. However, this only works with the Ortho mode on.

You also can extend several lines to be perpendicular to a nonorthogonal line. To do so, you have to rotate the cursor to that line's angle (see the top image Figure 16.3), using the Snapang system variable. (You can also use the Snap Angle input box in the Tools ➤ Drawing Aids dialog box to rotate the cursor.) Then use the Change Point process to extend or shorten the other lines (see the bottom image of Figure 16.3).

FIGURE 16.3
You can use the Change command to quickly straighten a set of non-parallel lines and to align their endpoints to another reference line.

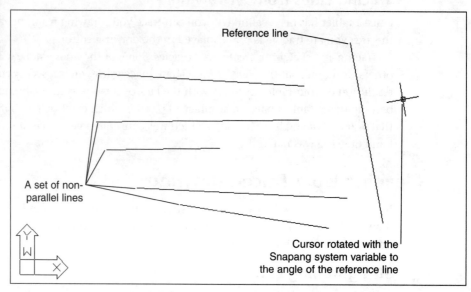

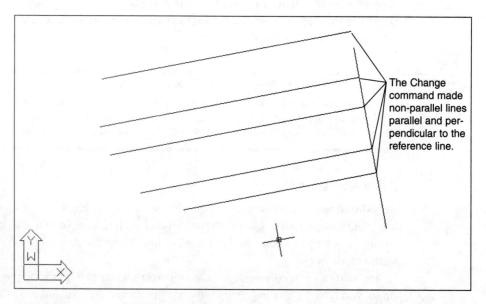

When changing several lines to be perpendicular to another line, you must carefully choose the new endpoint location. Figure 16.4 shows what happens to the same line work shown in the top image of Figure 16.3 when perpendicular reference is placed in the middle of the set of lines. Some lines are straightened to a perpendicular orientation, while others have the wrong endpoints aligned with the reference line.

FIGURE 16.4

The results of the Change command can be unpredictable if the endpoint location is too close to the lines being changed.

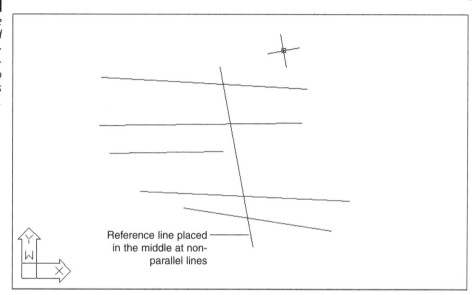

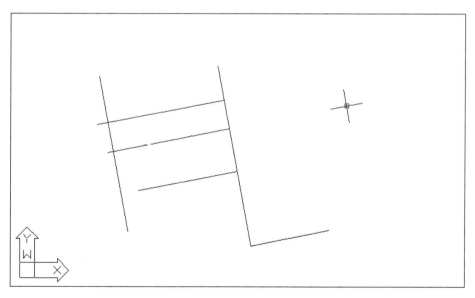

Adjusting Dimensions

Chances are, the dimensions of any drawing you trace will vary somewhat from the ones shown on the paper you are tracing. You will need to adjust your drawing to fit these dimensions. To make these adjustments, use the Stretch command and Mirror command. They and the other editing tools can be very useful in adding and correcting geometry that you have traced. Understanding how long it will take you to straighten and resize these objects will help you determine which method you want to use to convert them to CAD.

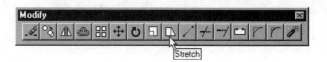

Scaling a Drawing

When a hand-drafted drawing is to scale, you can read the drawing's own dimensions or measure distances using an engineer's scale, and then enter the drawing into AutoCAD as you would a new drawing using these dimensions. Entering distances through the keyboard is slower than tracing, but you don't have to do as much clean-up because you are entering the drawing accurately. If the paper drawing has a column grid, input the grid first, and then use the grid as a reference for other dimensions.

When a drawing contains lots of curves, you'll have to resort to a different scaling method, which is actually an old drafting technique for enlarging or reducing a drawing. The following steps show you how it works.

1. First, draw a grid in AutoCAD to the same proportions as your hand-drafted drawing; that is, if your hand-drafted drawing is 40" × 30", make a 40" × 30" drawing in AutoCAD with grids spaced at ¼" intervals.
2. Plot this grid on translucent media and place it over the original paper drawing.
3. Place points on the plotted grid where the curves of the original drawing intersect the grid lines of the plot.
4. Finally, copy those intersection points from the plotted grid to the AutoCAD file of the grid.
5. Once you have positioned these points in your AutoCAD file, you can connect them to form the curves by using AutoCAD spline curves.

This method is somewhat time-consuming and not very accurate, but it will work in a pinch. If you plan to enter many drawings containing curves, it's best to purchase a tablet and trace them, or consider scanning or using a scanning service.

Scanning a Drawing

No discussion of drawing input can be complete without mentioning scanners. Imagine how easy it would be to convert an existing library of drawings into AutoCAD drawing files by simply running them through a scanning device. Unfortunately, scanning drawings is not quite that simple.

In scanning, the size of the drawing can be a problem. Desktop scanners are generally limited to an 8½" × 14" sheet size. Many low-cost, hand-held scanners will scan a 22" × 14" area. Larger-format scanners are available but more expensive.

Once the drawing is scanned and saved as a file, you have two paths to importing it into AutoCAD. One path is to convert the scanned image into AutoCAD objects such as lines, arcs, and circles. This requires special software and is usually a fairly time-consuming process. Finally, the drawing usually requires some clean-up, which can take even longer than cleaning up a traced drawing. The poorer the condition of the original drawing, the more clean-up you'll have to do.

Another path is to import a scanned image directly into AutoCAD and then use some or all of the scanned image in combination with AutoCAD objects. You can trace over the scanned image using the standard AutoCAD tools and discard the image when you are done, or you can use the scanned image as part of your AutoCAD file. With Release 14's ability to import raster images, you can, for example, import a scanned image of an existing paper drawing, and then mask off the area you wish to edit. You can then draw over the masked portions to make the required changes.

Whether a scanner can help you depends on your application. If you have drawings that would be very difficult to trace—large, multi-part space industry drawings, for example—a scanner may well be worth a look. You don't necessarily have to buy a scanner; some scanning services offer excellent value. And if you can accept the quality of a scanned drawing before it is cleaned up, you can save a lot of time. On the other hand, a drawing composed mostly of orthogonal lines and notes may be more easily entered directly by using the drawing's dimensions.

Scanning can be an excellent document management tool for your existing paper drawings. You might consider scanning your existing paper drawings for archiving purposes. You can then use portions or all of your scanned drawings later, without committing to a full-scale, paper-to-AutoCAD scan conversion.

Tips for Importing Raster Images

When you scan a document into your computer, you get a raster image file. Unlike AutoCAD files, raster image files are made up of a matrix of colors that form a picture. Vector files, like those produced by AutoCAD, are made up of lines, arcs, curves, and circles. The two formats, raster and vector, are so different that it is difficult to accurately convert one format to the other. It is easier to trace a raster file in AutoCAD than it is to try to have some computer program make the conversion for you.

But even tracing a raster image file can be difficult if the image is of poor quality. Here are a few points you should consider if you plan to use raster import for tracing drawings:

- Scan in your drawing using a grayscale or color scanner, or convert your black-and-white scanned image to grayscale using your scanner software.
- Use a paint program or your scanner software to clean up unwanted gray or spotted areas in the file before importing into AutoCAD.
- If your scanner software or paint program has a de-speckle or de-spot routine, use it. It can help clean up your image and ultimately reduce the raster image file size.
- Scan at a reasonable resolution. Remember that the human hand is usually not more accurate than a few thousandths of an inch, so scanning at 150 to 200 dpi may be more than adequate.
- If you plan to make heavy use of raster import, upgrade your computer to the fastest processor you can afford and don't spare the memory.

The raster import commands can incorporate paper plans into 3D AutoCAD drawings for presentations.

Importing Raster Images

If you have a scanner and you would like to use it to import drawings and other images into AutoCAD, you can use AutoCAD's raster image import capabilities. There are many reasons for wanting to import a scanned image. The old adage that states "a picture is worth a thousand words" is still true. You may want to add a photograph of an assembly to your assembly drawing or use an annotated photo as your assembly drawing. An alternative to scanning is the digital camera, which captures a raster image of the same sort that a scanner makes. You can save an incredible amount of time if you don't have to draft details that can be easily photographed with a digital camera and added to your work. Another reason for importing scanned images is to use the image as a reference to trace over. The first part of this section will show you how you might accomplish this import.

The Insert ➢ Raster Image option in the menu bar opens the Image dialog box, which in turn allows you to import a full range of raster image files.

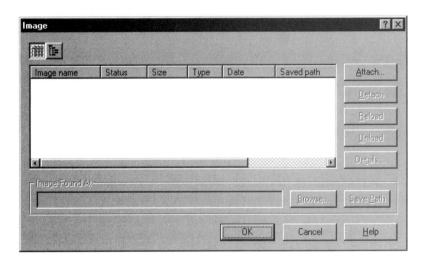

If you read Chapter 9, then this dialog box should look familiar. It looks and works just like the External Reference dialog box. The similarities are more than just cosmetic. Just like External References, raster images are loaded when the current file is open, but they are not stored as part of the current file when the file is saved. This helps keep file sizes down, but it also means that you need to keep track of inserted raster files. You will need to make sure that they are kept together with the AutoCAD files in which they are inserted.

> AutoCAD offers a "suitcase" utility that will collect AutoCAD files and their related support files, such as raster images, external references, and fonts, into any folder or drive that you specify. See Chapter 19 for details.

Another similarity between Xrefs and imported raster images is that you can clip a raster image so that only a portion of the image is displayed in your drawing. Portions of a raster file that are clipped will not be stored in memory, so your system won't get bogged down, even if the raster file is huge.

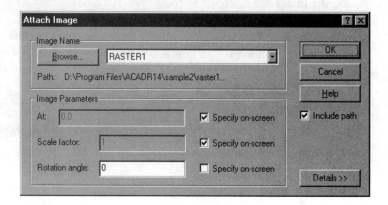

Controlling Object Visibility and Overlap with Raster Images

With the introduction of raster image support, AutoCAD inherits a problem fairly common to programs that use raster images. Raster images will obscure other objects that were placed before the raster image. In most cases, this overlap may not be a problem, but there will be situations where you will want AutoCAD vector objects to overlap an imported raster image. An example of this would be an assembly drawing which was scanned in from a photograph with a leader pointing out a significant feature of the assembly.

Paint and Page layout programs usually offer a "to front" and "to back" tool to control the overlap of objects and images. AutoCAD offers the Draworder command. Here's how it works.

1. Open a new drawing using the Versions template.

2. Draw a leader. Select the Leader tool from the Dimension toolbar. Start the leader at coordinates .5,.5 then drag the mouse to coordinates 1.25,.75. Press ↵ three times to get to the Text window, type **TEST**, then click on OK.

3. Choose Insert ➢ Raster Image or type **im**↵ to open the Image dialog box.

4. Click on the Attach button in the upper right of the dialog box. The Attach Image dialog box appears for you to select an image.
5. Use Browse to locate and select the Raster1.tif file from the companion CD-ROM. Notice that you can see a preview of the file in the right side of the dialog box.
6. Click on Open. The Attach Image dialog box appears. Click on OK.

The options in the Image dialog box perform the same functions as those for the External Reference dialog box. Refer to Chapter 9 for a detailed description of these options.

7. Press ↵ at the Insertion Point: prompt to accept the 0,0 coordinates.
8. At the Scale Factor: prompt, you can use the cursor to scale the image or you can enter a scale value. Press ↵ to accept the default of 1 and insert the raster image.

You can bypass the Image dialog box and go directly to the Attach Image dialog box by entering **iat**↵ at the command prompt.

The Image Dialog Box Options

The Attach Image dialog box you saw above in step 4 helps you manage your imported image files. It is especially helpful when you have a large number of images in your drawing. Like the External Reference dialog box, you can temporarily unload images (to help speed up editing of AutoCAD objects), reload, detach, and relocate raster images files. Refer to *Other External Reference Options* in Chapter 9 for a detailed description of these options.

Reordering a Raster Image

1. Choose View ➤ Zoom ➤ Extents to get an overall view of the image. Your drawing will look like Figure 16.5.

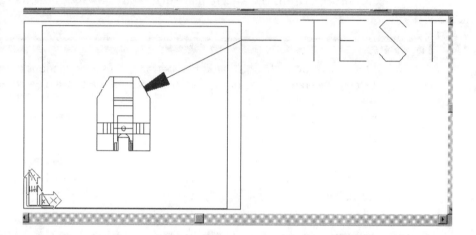

FIGURE 16.5

The raster image inserted over the leader

2. Choose Tools ➤ Display Order ➤ Bring Above Object.
3. At the `Select Objects:` prompt, select the leader line you drew when you first opened the file.
4. You could go on to select other objects. Press ↵ to finish your selection.
5. At the `Select Reference Object:` prompt, click on the edge of the raster image.

The drawing regenerates and the entire leader and arrowhead appear, no longer obscured by the raster image.

The Draworder command you just used actually offers four options. The pull-down menu options are as follows.

Tools ➤ Display Order ➤ Bring to Front places an object or set of objects at the top of the draw order for the entire drawing. The effect is that the objects are completely visible.

Tools ➤ Display Order ➤ Send to Back places an object or set of objects at the bottom of the draw order for the entire drawing. The effect is that the objects may be obscured by other objects in the drawing.

Tools ➢ Display Order ➢ Bring Above Object places an object or set of objects above another object in the draw order. This has the effect of making the first set of objects appear above the second selected object.

Tools ➢ Display Order ➢ Bring Under Object places an object or set of objects below another object in the draw order. This has the effect of making the first set of objects appear underneath the second selected object.

You can also use the **dr**↵ keyboard shortcut to issue the Draworder command. If you do this, you see the prompt:

    Above object/Under object/Front/<Back>:

You must then select the option by entering through the keyboard the capitalized letter of the option.

Although we've discussed the Display Order tools in relation to raster images, these tools can be also be invaluable in controlling visibility of line work in conjunction with hatch patterns and solid fills. See Chapter 10 for a detailed discussion of the Display Order tools and solid fills.

Under certain conditions, the draw order of external reference files may not appear properly. If you encounter this problem, open the Xref file and make sure the draw order is correct. Then use the Wblock command to export all its objects to a new file. Use the new file for the external reference instead of the original external reference.

Clipping a Raster Image

In Chapter 9, you saw how you can clip an External Reference object so that only a portion of it appears in the drawing. You can also clip imported raster images in the same way. Just as with Xrefs, you can create a closed outline of the area you want to clip, or you can specify a simple rectangular area. In the following exercise, you'll try out the Imageclip command to control the display of the raster image.

1. Choose Insert ➢ Raster Image from the menu bar.
2. Click on the Attach button in the upper right of the dialog box. The Select File to Attach dialog box appears.

3. Locate and select the `Raster2.tif` file. This is a file from the companion CD-ROM. Notice that you can see a preview of the file in the right side of the dialog box.

4. Click on Open. The Attach Image dialog box appears.

5. At the `Insertion Point:` prompt, enter the coordinates **3,0**↵.

6. At the `Scale Factor:` prompt, press ↵ to accept the default.

You have imported a picture of a bench with a press and other tools sitting on it. Next you will clip both images.

1. Choose Modify ➢ Object ➢ Image Clip, or type **icl**↵.

2. At the `Select Image to Clip:` prompt, click on the edge of the Raster1 image.

3. At the `ON/OFF/Delete/<New boundary>:` prompt, press ↵ to create a new boundary.

4. At the `Polygonal/<Rectangular>:` prompt, enter **p**↵ to draw a polygonal boundary.

5. Select the points shown in the top image of Figure 16.6; press ↵ when you are done. The raster image is clipped to the boundary you created, as shown in the bottom image of Figure 16.6.

6. Follow the same steps for the picture of the bench. See the top and bottom of Figure 16.6.

As the prompt in step 3 indicates, you can turn the clipping off or on, or delete an existing clipping boundary through the Imageclip option.

Once you have clipped a raster image, you can adjust the clipping boundary using its grips.

1. Click on the boundary edge of the raster image to expose its grips.

2. Click on a grip in the upper-right corner of the picture, as shown in the bottom image of Figure 16.6.

3. Drag the grip up and to the right, and then click on a point. The image adjusts to the new boundary.

FIGURE 16.6

Adjusting the boundary of a clipped image

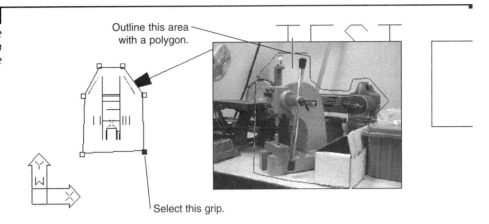

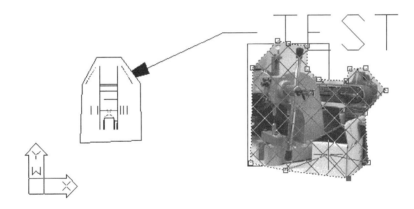

In addition to hiding portions of a raster image that may be unimportant to you, clipping an image file will reduce the amount of RAM the raster image uses during your editing session. AutoCAD will only load the visible portion of the image into RAM. The rest is ignored.

The Wipeout tool in the Bonus Standard toolbar can also mask out a portion of a raster image. This is useful if you want to hide portions of a raster image that may make it difficult to view overlapping AutoCAD objects such as text or dimensions. Wipeout can also be useful as a general masking tool for AutoCAD objects. See Chapter 19 for details on how this tool works.

Adjusting Brightness, Contrast, and Strength

AutoCAD offers a tool that allows you to adjust the brightness, contrast, and strength of a raster image. Try making some adjustments to the raster image of the picture of the press in the following exercise. Erase the extraneous lines, leaders, and raster drawing, then zoom the extents to see the press drawing.

1. Choose Modify ➤ Object ➤ Image ➤ Adjust, or type **iad**↵.

2. At the Select Image to Adjust: prompt, click on the edge of the raster image. The Image Adjust dialog box appears.

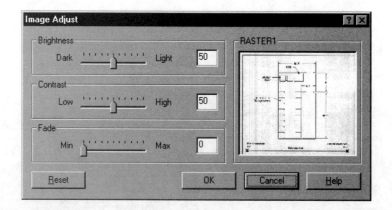

3. Click and drag the Fade slider to the right so that it is near the middle of the slider scale. Or enter **50** in the Fade input box that is just to the right of the slider. Notice how the sample image fades to the AutoCAD background color as you move the slider.

IMPORTING RASTER IMAGES | 749

4. Click on OK. The raster image appears faded.

You can adjust the brightness and contrast using the other two sliders in the Image Adjust dialog box. The Reset button resets all the settings to their default value.

By using the Image Adjust option in conjunction with image clipping, you can create special effects. Figure 16.7 shows two copies of the same raster image. One copy serves as a background, which was lightened using the same method demonstrated in the previous exercise. The second copy is the darker area of the image with a polygonal clip boundary applied. You might use this technique to bring focus to a particular area of a drawing you are preparing for a presentation.

Figure 16.7 also contains a leader drawn over the image. The text labels make use of the Wipeout bonus tool to hide their background, making them easier to read. The Draworder command is also instrumental in creating this type of image, allowing you to control which object appears on top of others.

If the draw order of objects appears incorrectly after opening a file or performing a pan or zoom, issue a Regen to recover the correct draw order view.

FIGURE 16.7

Two copies of the same image can be combined to create emphasis on a portion of the drawing.

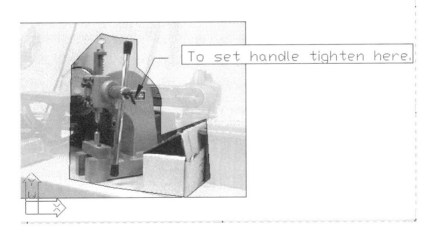

Turning off the Frame, Adjusting Overall Quality and Transparency

Three other adjustments can be made to your raster image: frame visibility, image quality, and image transparency.

By default, a raster image will display an outline or frame. In many instances, this frame may detract from your drawing. You can globally turn off image frames by choosing Modify ➢ Object ➢ Image ➢ Frame, then choosing On or Off depending on whether you want the frame visible or invisible (see Figure 16.8). You can also type **imageframe.↵off.↵** (the default is On so you could type **imageframe.↵↵** to get a frame).

FIGURE 16.8

A raster image with the frame off

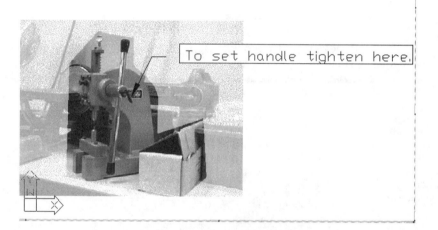

WARNING

If you turn off the frame of raster images, you will not be able to select them for editing. You can use this to your advantage if you don't want a raster image to be moved or otherwise edited. To make a raster image selectable, turn on the Image Frame setting. But beware: Erase All will erase all, even a raster image!

If your drawing doesn't require the highest-quality image, you can set the image quality to Draft mode. You might use Draft mode when you are tracing an image, or when the image is already of a high quality. To set the image quality, choose Modify ➢ Object ➢ Image ➢ Quality, and then enter **h** for High mode or **d** for Draft mode. In Draft mode, your drawing will regenerate faster.

The High mode softens the pixels of the raster image, giving the image a smoother appearance. The Draft mode displays the image in a "raw," pixelated state.

Finally, you can control the transparency of raster image files that allow transparent pixels. Some file formats, such as the CompuServe GIF 89a format, allow you to set a color in the image to be transparent (usually the background color). Most image-editing programs support this format because it is a popular one often used on Web pages.

By turning on the Transparency setting, objects normally obscured by the background of a raster image may show through. Choose Modify ➢ Object ➢ Image ➢ Transparency, and then select the raster image that you wish to make transparent. Choose On or Off depending on whether you want the image to be transparent or not. Unlike the Frame and Quality options, Transparency works on individual objects, rather than globally.

If you want quick access to the Transparency setting, along with other raster image settings, the Properties tool in the Object Properties toolbar offers many of the same adjustments described in this section. You can access the Image Adjust dialog box, hide or display clipped areas, or hide the entire raster image.

Importing PostScript Files

In addition to raster import, you can use the built-in PostScript import feature to import PostScript files. This feature can be accessed by choosing Insert ➢ Encapsulated PostScript or by typing **import**↵. The Import File dialog box lets you easily locate the file to be imported. Select Encapsulated PostScript (.EPS) from the List Files of Type drop-down list, and then browse to find the file you want. Once you select a file, the rest of the program works just like the raster import commands. You are asked for an insertion point, a scale, and a rotation angle. Once you've answered all the prompts, the image appears in the drawing.

AutoCAD supports drag and drop for many raster file formats, including PostScript EPS.

You can make adjustments to the quality of imported PostScript files by using the Psquality system variable (enter **psquality**↵). This setting takes an integer value in the range –75 to 75, with 75 being the default. The absolute value of this setting is taken as the ratio of pixels to drawing units. If, for example, Psquality is set to 50 or –50, then Psin will convert 50 pixels of the incoming PostScript file into 1 drawing unit. Use negative values to indicate that you want outlines of filled areas rather than the full painted image. Using outlines can save drawing space and improve readability on monochrome systems or systems with limited color capability.

Finally, if Psquality is set to 0, only the bounding box of the imported image is displayed in the drawing. Though you may only see a box, the image data is still incorporated into the drawing and will be maintained as the image is exported (using Psout).

Another system variable you will want to know about is Psdrag, which controls how the imported PostScript image appears at the insertion point prompt. Normally, Psdrag is set to 0, so Psin displays just the outline (bounding box) of the imported file as you move it into position. To display the full image of the imported PostScript file as you locate an insertion point, you can set Psdrag to 1. Access Psdrag from the menu bar by choosing File ➢ Options ➢ PostScript Display.

If you need PostScript fonts in your EPS output files, you can use AutoCAD fonts as substitutes. AutoCAD will convert fonts with specific code names to PostScript fonts during the EPS export process. See Appendix B for details.

If You Want to Experiment...

If you want to see firsthand how the PostScript Import and Export commands work, you can try the following exercise.

1. Open the Table drawing from the companion CD-ROM.

This sequence of steps is a way to convert a 3D model into a 2D line drawing.

2. Choose File ➢ Export to open the Export Data dialog box. Select Encapsulated PostScript (.EPS) in the List Files of Type drop-down list. You can also type

psout↵ at the command prompt. This will open the Create PostScript File dialog box and allow you to select or enter a file name.

3. Click on the Options button. The Export Options dialog box appears.

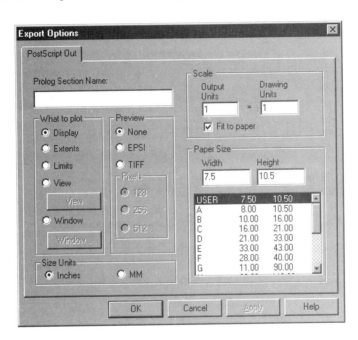

This dialog box offers options similar to the Print/Plot Configuration dialog box you saw in Chapter 14.

4. Select the desired area to plot, sheet size, scale, and units from the dialog box; then click on OK.
5. At the File dialog box, click on Save to accept the suggested PostScript file name of `table.eps`.
6. Click on Save in the Export Data box.

You may notice that arcs, circles, and curves are not very smooth when you use this method to export, and then import PostScript files. To improve the appearance of arcs and curves, specify a large sheet size at the `Enter the Size or Width,Height (in Inches) <USER>:` prompt.

7. Choose Insert ➤ Encapsulated PostScript to open the Select PostScript File dialog box.

8. Locate and select the `table.eps` file. You will be prompted for Insertion point and Scale factor. Position and size the image as required. The file then appears in your drawing.

PART V

Customization: Taking AutoCAD to the Limit

LEARN TO:

- *Use Attributes to Store Data*
- *Exchange CAD Data with Other Programs*
- *Create Keyboard Macros with AutoLISP*
- *Use ActiveX Automation*
- *Customize Toolbars*
- *Add Pull-Down Menus*
- *Create Custom Linetypes*

Chapter 17

Storing and Linking Data with Graphics

FEATURING

Using Attributes to Store Data

Defining an Attribute and Adding It to a Block

Entering Attribute Data

Editing Attribute Data, One Block at a Time

Making Global Changes to Attribute Values

Controlling Attribute Visibility

Exporting Attribute Data to Other Programs

Linking Objects to a Database

Using AutoCAD to View External Databases

Chapter 17

Storing and Linking Data with Graphics

Attributes are unique to computer-aided design and drafting; nothing quite like them exists in traditional drafting. Because of this, they are often poorly understood. Attributes enable you to store information as text that you later extract to use in database managers, spreadsheet programs, and word processors. By using attributes, you can keep track of virtually any object in a drawing, or maintain textual information, which can be queried.

Keeping track of objects is just one way of using attributes. You can also use attributes in place of text objects in situations where you must enter the same text, with minor modifications, in many places in your drawing. For example, if you are drawing a schedule that contains several columns of information, you can use attributes to help simplify your data entry.

Attributes can also be used where you anticipate global editing of text. For example, suppose a note referring to a part number occurs in several places. If you think you will want to change that part number in every note, you can make the part a block with an attribute. Later, when you know the new part number, you can use the global editing capability of the Attribute feature to change the old part number for all occurrences in one step.

In this chapter, you will use attributes for one of their more common functions: maintaining a list of parts. We will also describe how to import these attributes into a database management program. As you go through these exercises, think about the ways attributes can help you in your particular application.

Creating Attributes

Attributes depend on blocks. You might think of an attribute as a tag attached to a block, with the tag containing information about the block. For example, you could have included an attribute definition with the plate drawing you created in Chapter 2. If you had, then every time you subsequently inserted the plate you would have been prompted for a value associated with that plate. The value could be a part number, a height or width value, a name, or any type of textual information you want. When you insert the block, you are prompted for an attribute value. Once you enter a value, it is stored as part of the block within the drawing database. This value can be displayed as text attached to the plate, or it can be made invisible. The value can be changed at any time. You can even specify what the prompts say in asking you for the attribute value.

Linking Attributes to Parts

Using a *balloon and leader* is the most common way to identify objects in an assembly drawing. You place the attribute definition in a balloon (like a cartoon balloon) that identifies the part in the assembly and then draw a leader that links the attribute definition directly to the part. Linking an attribute to a part is a key concept for creating multiple drawings that use the same components.

You might use this linking technique if you have drawn multiple views of a screw that you intend to use in your 2D assembly drawing. If you showed a set of screw attributes in each view, when the screws to be used were counted, each instance of the screw and its attribute would be counted, possibly giving a variety of required screws that is too high. But if the screw and its attributes were stored as a separate file in the subdirectory you made to store your symbols, you could just refer to the existing instance each time that particular screw was required. This system allows you to get an accurate count of the number of each type of screw that is needed, and it saves the repeated entering of identical information on multiple drawings.

NOTE

A benefit of using 3D parts is that the attribute can be included directly with the object. Each instance of the solid represents one part in the assembly and cannot give false readings.

Adding Attributes to Blocks

In this exercise, you will use a screw drawing and add attributes to describe the size, pitch length, and other characteristics of any given screw.

Although in this exercise you will be creating a new file containing attribute definitions, you can also include such definitions in blocks you create using the Make Block tool (Block command) or in files you create using the Wblock command. Just create the attribute definitions, then include them with the Block or Wblock selections.

1. Create a new file and call it balloon.dwg.
2. Draw a circle with a radius of 0.188 and with its center at coordinate 7,5.

Since this is a new drawing, the circle is automatically placed on layer 0. Remember that objects on layer 0 that are in a block will take on the color and linetype assignment of the layer on which the block is inserted.

3. Next, zoom into the circle so it is about the same size as shown in Figure 17.1.
4. If the circle looks like an octagon, Choose View ➣ Regen or type **re**↵ to regenerate your drawing.
5. Choose Draw ➣ Block ➣ Define Attributes, or type **at**↵. The Attribute Definition dialog box appears.

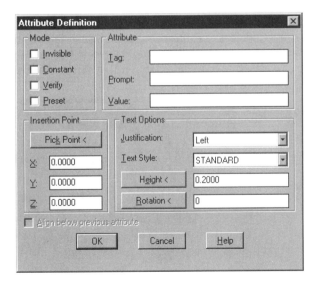

6. Click on the input box labeled Tag in the Attribute group. Enter **in**.

The Attribute Tag is equivalent to the field name in a database; it can be up to 31 characters long, but it cannot contain spaces. If you plan to use the attribute data in a database program, check that program's manuals for other restrictions on field names.

7. Press the Tab key or click on the input box labeled Prompt, and enter **ITEM NUMBER**. Here you enter the text for the prompt that will appear when you insert the block containing this attribute. Often the prompt is the same as the tag, but it can be anything you like. Unlike the tag, the prompt can include spaces.

Use a prompt that gives explicit instructions so that the user will know exactly what is expected. Consider including an example within the prompt. (Enclose the example in brackets to imitate the way AutoCAD prompts display defaults.)

8. Click on the input box labeled Value. This is where you enter a default value for the ITEM NUMBER prompt. Enter a hyphen.

If an attribute is to contain a number that will later be used for making sorts in a database, use a default such as 000 to indicate the number of digits required. The zeros may also serve to remind the user that values less than 100 must be preceded by a leading zero, as in 099.

9. Click on the Justification drop-down list, and then highlight Middle. This will allow you to center the attribute on the circle's center. You might notice several other options in the Text Options group. Since attributes appear as text, you can apply the same settings to them as you would to single-line text.

10. Double-click on the input box next to the button labeled Height, and then enter **0.125**. This will make the attribute text 0.125 inches high.

11. Check the box labeled Verify in the Mode group. This option instructs AutoCAD to verify any answers you give to the attribute prompts at insertion time (you'll see later in this chapter how Verify works).

12. Click on the button labeled Pick Point in the Insertion Point group. The dialog box closes momentarily to let you pick a location for the attribute.
13. Use the Center Osnap to pick the center of the circle. (You need to place the cursor on the circle, not the circle's center, to obtain the center using the Osnap.) The dialog box reappears.
14. Click on OK. You will see the attribute definition at the center of the circle (see Figure 17.1).

FIGURE 17.1

The attribute inserted in the circle

You have just created your first attribute definition. The attribute definition displays its tag in all uppercase letters to help you identify it. Later, when you insert this file into another drawing, you will see that the tag turns into the value you assign to it when it is inserted. If you only want one attribute, you can stop here and save the file. The next section shows you how you can quickly add several more attributes to your drawing.

Changing Attribute Specifications

Next, you will add a few more attribute definitions, but instead of using the Attribute Definition dialog box, you will make an arrayed copy of the first attribute, and then edit the attribute definition copies. This method can save you time when you want to create several attribute definitions that have similar characteristics. By making copies and editing them, you'll also get a chance to see firsthand how to make changes to an attribute definition.

1. Click on Array on the Modify toolbar or type **ar**↵.

2. At the Select objects: prompt, click on the attribute definition you just created, and then press ↵.
3. At the Rectangular or Polar array (<R>/P): prompt, type **r**↵.
4. At the Number of Rows: prompt, enter **7**↵.

5. At the `Number of Columns:` prompt, press ↵.

6. At the `Distance between Rows:` prompt, enter **–.18**↵. This is about 1.5 times the height of the attribute definition. Be sure to include the minus sign. This will cause the array to be drawn downward.

7. Issue a Zoom Extents command or use the Zoom Realtime tool to view all the attributes.

Now you are ready to modify the copies of the attribute definitions.

1. Click on the Properties button on the Object Properties toolbar.

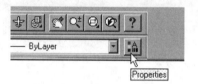

The Text Edit tool (Modify ➤ Object ➤ Text) lets you edit the tag, prompt, and default value of an attribute definition. However, it doesn't let you change an attribute definition's *mode*.

2. At the `Select Object to Modify:` prompt, click on the attribute definition just below the original and then press ↵. The Modify Attribute Definition dialog box appears.

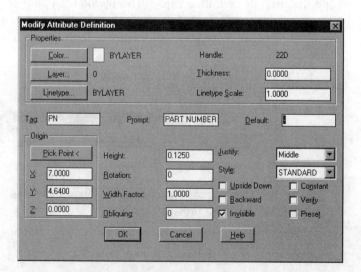

CREATING ATTRIBUTES

3. Click on the Invisible check box in the lower-right portion of the dialog box. This will cause this attribute to be invisible when the file is later inserted.
4. Double-click on the input box labeled Tag, and then enter **qty**.
5. Press Tab to move to the Prompt input box, and then type **QUANTITY**.
6. Press Tab a second time to move to the Default input box, and then enter **1**.
7. Click on OK. You will see the attribute definition change to reflect the new tag.
8. Continue to edit the rest of the attribute definitions using the attributes settings listed in Table 17.1. Make sure all but the original attribute have the Invisible option turned on.
9. After you have modified all the attributes, use the Draw ➢ Block ➢ Base option to change the base point of this drawing to the center of the circle. Use the Center Osnap to get the exact center.
10. Now you have finished creating your balloon symbol with attributes. Save the balloon. Your balloon drawing should look like Figure 17.2.

FIGURE 17.2
The finished balloon symbol drawing

TABLE 17.1: ADDITIONAL ATTRIBUTES FOR THE BALLOON SYMBOL (MAKE SURE THE INVISIBLE OPTION IS CHECKED)		
Tag	**Prompt**	**Default**
pn	PART NUMBER	-
description	DESCRIPTION	-
material	MATERIAL	-
finish	FINISH	-
notes	NOTES	-

When you later insert a file or block containing attributes, the attribute prompts will appear in the order that their associated definitions were created. If the order of the prompts at insertion time is important, you can control it by editing the attribute definitions so their creation order corresponds to the desired prompt order.

Understanding Attribute Definition Modes

In the Attribute Definition dialog box, you saw several check boxes in the Mode group. We've briefly described what two of these modes do, but below is a list describing all of the modes for your reference. You won't be asked to use any of the other modes in this tutorial, so the following set of descriptions is provided in case they are useful for your work.

Invisible controls whether the attribute is shown as part of the drawing.

Constant creates an attribute that does not prompt you to enter a value. Instead, the attribute simply has a constant, or fixed, value you give it during creation. Constant is used in situations where you know you will assign a fixed value to an object. Once they are set in a block, constant values cannot be changed using the standard set of attribute editing commands.

Verify causes AutoCAD to review the attribute values you enter at insertion time and asks you if they are correct.

Preset causes AutoCAD to automatically assign the default value to an attribute when its block is inserted. This saves time because a preset attribute will not prompt you for a value. Unlike the Constant option, you can edit an attribute that has the Preset option turned on.

You can have all four modes on, all four off, or any combination of modes. With the exception of the Invisible mode, none of these modes can be altered once the attribute becomes part of a block. Later in this chapter we will discuss how to make an invisible attribute visible.

Inserting Blocks Containing Attributes

In the last section, you created a balloon symbol at the desired size for the actual plotted symbol. This means that whenever you insert that symbol, you have to specify an X and Y scale factor appropriate to the scale of your drawing. This allows you to use the same symbol in any drawing, regardless of its scale. (You could have several balloon symbols, one for each scale you anticipate using, but this would be inefficient.)

1. Open the file named ch17assembly.dwg from the companion CD-ROM.
2. Click on Insert Block on the Draw toolbar or type **i**↵ to open the Insert dialog box.

3. At the Insert dialog box, click on the File button.
4. Locate the balloon file in the file list and double-click on it.
5. Click on OK.
6. Insert the symbol at the end of the leader near coordinate **12.26,6.70**.
7. At the X Scale Factor: prompt, enter ↵.
8. Press ↵ at the Y Scale Factor: prompt, and again at the Rotation Angle: prompt.
9. At the Enter attribute values ITEM NUMBER <->: prompt, enter **1**↵. Notice that this prompt is the prompt you created. Notice also that the default value is the hyphen you specified.

NOTE
Attribute data is case-sensitive, so text you enter in all capital letters will be stored in all capital letters.

10. At the QUANTITY <->: prompt, type **1**↵ because there is only one of this part in the assembly.
11. At the PART NUMBER <->: prompt, enter **36-120063-03 REV A**↵. Continue to enter the values for each prompt as shown in Table 17.2. You will use the values in Table 17.3–Table 17.5 in the next section.
12. When you have finished entering the values, the prompts repeat themselves to verify your entry (because you selected Verify from the Modes group of the Attribute Definition dialog box). You can now either change an entry or just press ↵ to accept the original entry.
13. Now you've finished and the balloon appears. The only attribute you can see is the one you selected to be visible: the Item Number.

TIP If the symbol does not appear, go back to the balloon.dwg file and make sure you have set the base point to the center of the circle.

14. Add the rest of the balloon symbols for the Assembly by copying the balloon block you just inserted. The balloon symbol can only have one base point. You need to place it with the quadrant of the circle on the end of the leader. See Figure 17.3 for the placement of the balloons.

   Don't worry that the attribute values won't be appropriate. We'll show you how to edit the attributes later in this chapter.

FIGURE 17.3
The location of the four balloons in the assembly

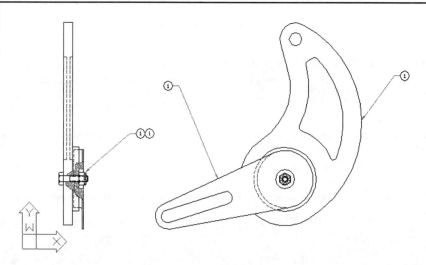

TABLE 17.2: ATTRIBUTE VALUES FOR ITEM 1

Prompt	Value
ITEM NUMBER	1
QUANTITY	1
PART NUMBER	36-120063-03 REV A
DESCRIPTION	CURVED LINK
MATERIAL	AL 6061-T6
FINISH	HARD BLACK ANODIZE
NOTES	

TABLE 17.3: ATTRIBUTE VALUES FOR ITEM 2

Prompt	Value
ITEM NUMBER	2
QUANTITY	1
PART NUMBER	36-120064-01 REV B
DESCRIPTION	SLOTTED LINK
MATERIAL	AL 6061-T6
FINISH	HARD BLACK ANODIZE
NOTES	

TABLE 17.4: ATTRIBUTE VALUES FOR ITEM 3

Prompt	Value
ITEM NUMBER	3
QUANTITY	1
PART NUMBER	72-160070-25
DESCRIPTION	BOLT #10-32 x .75 LG, HEX HEAD
MATERIAL	STEEL
FINISH	ZINC
NOTES	

TABLE 17.5: ATTRIBUTE VALUES FOR ITEM 4

Prompt	Value
ITEM NUMBER	4
QUANTITY	1
PART NUMBER	72-150070-25 REV A
DESCRIPTION	10-32 NUT, SLOTTED, THICK
MATERIAL	STEEL
FINISH	ZINC
NOTES	

Using a Dialog Box to Answer Attribute Prompts

You can set up AutoCAD to display an Enter Attributes dialog box (instead of individual prompts) for entering the attribute values at insertion time. Because this dialog box allows you to change your mind about a value before confirming your entry, the dialog box allows greater flexibility than individual prompts when entering attributes. You can also see all the attributes associated with a block at once, making it easier to understand what information is required for the block.

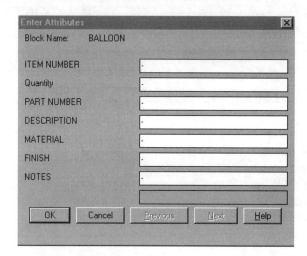

To turn this feature on, do the following:

1. Enter **attdia**↵ at the command prompt.
2. At the New value for ATTDIA <0>: prompt, type the number **1**↵.

Attributes set with the Preset mode on will also appear in the dialog box and are treated no differently from other non-constant attributes.

Editing Attributes

Because drawings are usually in flux even after manufacturing begins, you will eventually have to edit previously entered attributes.

Attributes can be edited *individually* (one at a time) or they can be edited *globally* (meaning you can edit several occurrences of a particular attribute tag all at one time). In this section you will make changes to the attributes you have entered so far, using both individual and global editing techniques, and you will practice editing invisible attributes.

Editing Attributes One at a Time

AutoCAD offers an easy way to edit attributes one at a time through a dialog box. The following exercise demonstrates this feature.

1. Choose Modify ➢ Object ➢ Attribute ➢ Single, or enter **ate**↵ at the command prompt.
2. At the Select Block: prompt, click on the balloon attached to the slotted link. A dialog box appears, showing you the value for the attribute in an input box. Note that the value is highlighted and ready to be edited.

See Table 17.3–Table 17.5 to find the information for items 2, 3, and 4. Use this data to fill in the input boxes in the Edit Attributes dialog box. Once you have finished, the assembly will look like Figure 17.4.

The Edit Attributes option is useful for reviewing attributes as well as editing them, because both visible and invisible attributes are displayed in the dialog box.

Editing Several Attributes in Succession

You can take advantage of the Tab key and the spacebar to quickly edit a series of attributes. In a dialog box, the Tab key moves you to the next option. In AutoCAD, the spacebar reissues the previous command. By combining these two tools, you can make quick work of editing attributes. The following steps describe the process.

1. Choose Modify ➢ Object ➢ Attribute ➢ Single, or enter **ate**↵ at the command prompt to start the attribute editing process. Select the first attribute you want to edit.
2. At the Edit Attributes dialog box, enter the new value for the attribute. The old attribute value should already be highlighted. For blocks with multiple attributes, you may need to press the Tab key until you get to the value you want to edit.
3. After entering the new attribute value, press the Tab key to advance to the OK button.
4. Press the spacebar twice; once to accept the OK button, and again to reissue the Attribute Edit command.
5. Click on the next attribute you want to edit and repeat steps 2–4. When you are finished, click on OK.

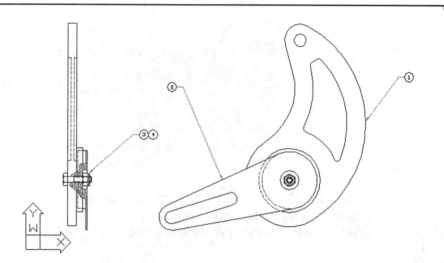

FIGURE 17.4
The assembly with corrected item numbers

If you are comfortable with the keyboard, you can get into a rhythm of selecting and editing attributes using these steps, especially if the block contains only one attribute.

Making Minor Changes to an Attribute's Appearance

Eventually, there will be situations where you will want to make a change to an attribute that doesn't involve its value, such as moving the attribute's location relative to the block it's associated with, or changing its color, its angle, or even its text style. To make such changes, you must use the Attedit command. Here's how to do it.

1. Choose Modify ➢ Object ➢ Attribute ➢ Global, or type **attedit**↵ at the command prompt.
2. At the `Edit Attributes One at a Time? <Y>:` prompt, press ↵ to accept the default of Yes.

If you just want to change the location of individual attributes in a block, you can move attributes using grips. Click on the block to expose the grips and then click on the grip connected to the attribute. Or if you've selected several blocks, Shift+click on the attribute grips; then move the attributes to their new location. They will still be attached to their associated blocks.

3. At the `Block Name Specification <*>:` prompt, press ↵. Optionally, you can enter a block name to narrow the selection to specific blocks.
4. At the `Attribute Tag Specification <*>:` prompt, press ↵. Optionally, you can enter an attribute tag name to narrow your selection to specific tags.
5. At the `Attribute Value Specification <*>:` prompt, press ↵. Optionally, you can narrow your selection to attributes containing specific values.
6. At the `Select Attributes:` prompt, you can pick the set of blocks that contain the attributes you wish to edit. Once you press ↵ to confirm your selection, one of the selected attributes becomes highlighted, and an X appears at its base point (see Figure 17.5).

FIGURE 17.5

Close-up of attribute with X

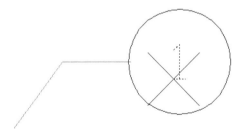

7. At the `Value/Position/Height/Angle/Style/Layer/Color/Next <N>`: prompt, you can enter the option that best describes the attribute characteristic you wish to change. After you make the change, the prompt returns, allowing you to make another change to the attribute. If you press ↵ to accept the default, No, another attribute highlights with an X at its base.

8. The `Value/Position/Height`: prompt appears again, allowing you to make changes to the next attribute.

9. This process repeats until all the attributes have been edited or until you press Esc.

At some point you may find that an attribute would function better as a single-line text object. A bonus tool called Explode Attributes to Text is available in the Bonus Text Tools toolbar. See Chapter 19 for more information on this bonus tool.

Making Global Changes to Attributes

There will be times when you'll want to change the value of several attributes in a file to be the same value. You can use the Edit Attribute Globally option to make any global changes to attribute values.

Suppose you decide you want to change all the notes to state "-02 VERSION ONLY". The following exercise demonstrates how this is done.

1. Choose Modify ➢ Object ➢ Attribute ➢ Global, or type **attedit**↵ at the command prompt.

2. At the `Edit Attributes One at a Time? <Y>`: prompt, enter **n**↵ for No. You will see the message `Global edit of attribute values`. This tells you that you are in the Global Edit mode.

3. At the `Edit Only Attributes Visible On Screen? <Y>` prompt, enter ↵. As you can see from this prompt, you have the option to edit all attributes, including those out of the view area.

4. At the `Block Name Specification <*>`: prompt, type **balloon**↵ to narrow the selection to specific blocks. Optionally, you can press ↵ and select the blocks later.

5. At the `Attribute Tag Specification <*>`: prompt, type **notes**↵. Optionally, you can enter an attribute tag name to narrow your selection to specific tags or press ↵ and be asked later to select tags.

6. At the Attribute Value Specification <*>: prompt, press ↵. Optionally, you can narrow your selection to attributes containing specific values.

7. At the String to Change: prompt, press ↵. You did not fill in this attribute when you edited the attributes.

8. At the New String: prompt, enter **-02 VERSION ONLY**↵. The balloon symbols all change to the new value.

When you are editing attributes, make sure you click on the attribute itself and not on other graphic components of the block containing the attribute.

If you only need to change a few visible attributes in a few of the balloons in your drawing, you can do so by answering the series of prompts in a slightly different way, as in the following exercise.

1. Try the same procedure again, but this time enter **n** at the Edit Only Attributes Visible On Screen: prompt (step 3 in the previous exercise). The message Drawing must be regenerated afterwards appears. The display will flip to text mode.

2. Once again, you are prompted for the block name; press ↵ instead of the block name. At the Attribute Tag Specification <*>: prompt, press ↵, and at the Attribute Value Specification <*>: prompt, press ↵.

3. At the String to Change: prompt, enter **1**↵ to indicate you want to change the rest of the balloon's attribute values.

4. At the New String: prompt, enter **2**↵. A series of 2s appears, indicating the number of strings that were replaced.

5. At the Select attributes: prompt, select the item number in each of the two balloons that are attached to the nut. Be sure that you select the text, not the balloon.

If the Regenauto command is off, you must regenerate the drawing to see the change.

TIP

If you find that this method for globally editing attributes is too complex, AutoCAD offers the Global Attribute Edit bonus tool in the Bonus Text Tools toolbar. See Chapter 19 for details.

You may have noticed in the last exercise that the Select Attribute: prompt is skipped and you go directly to the String to change: prompt. AutoCAD assumes that you want it to edit every attribute in the drawing, so it doesn't bother asking you to select specific attributes.

Making Invisible Attributes Visible

Invisible attributes, such as those in the balloon, can be edited globally using the tools just described. You may, however, want to be a bit more selective about which invisible attribute you want to modify. Or you may simply want to make them temporarily visible for other editing purposes. This section describes how you can make invisible attributes visible.

1. Enter **attdisp**↵.

TIP

You may also use the Windows menu to change the display characteristics of attributes. Choose Options ➢ Display ➢ Attribute Display, then click on the desired option on the cascading menu.

2. At the Normal/ON/OFF <Normal>: prompt, enter **on**↵. Your drawing will look like Figure 17.6. If Regenauto is turned off, you may have to issue the Regen command. At this point you could edit the invisible attributes individually, as in the first attribute-editing exercise. For now, set the attribute display back to Normal.

3. Enter **attdisp**↵ again; then at the Normal/ON/OFF prompt, enter **n**↵ for Normal.

NOTE

You get a chance to see the results of the On and Normal options. The Off option will make all attributes invisible, regardless of the mode used when they were created.

EDITING ATTRIBUTES 779

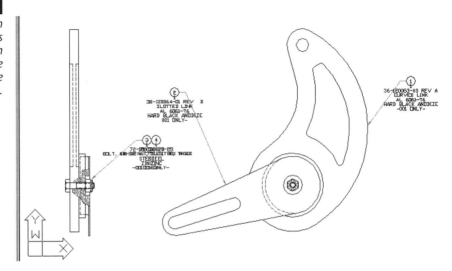

FIGURE 17.6
The drawing with all the attributes visible. The balloon symbols are so close together that the attributes overlap.

Because the attributes were not intended to be visible, they appear to overlap each other and cover other parts of the drawing when they are made visible. Just remember to turn them back off when you are done reviewing them.

Redefining Blocks Containing Attributes

Finally, you should be aware that attributes act differently from other objects when included in redefined blocks. Normally, blocks that have been redefined change their configuration to reflect the new block definition. But if a redefined block contains attributes, the attributes will maintain their old properties. This means that the old attribute position, style, and so on, do not change—even though you may have changed them in the new definition.

Fortunately, AutoCAD offers a tool specifically designed to let you update blocks with attributes. The following steps show how you would go about updating attribute blocks.

1. Before you can use the command to redefine an attribute block, you must first create the objects and attribute definitions that are going to make up the new, replacement attribute block. The simplest way to do this is to explode a copy of the attribute block you wish to update. This insures that you have the same attribute definitions in the updated block.

2. Make your changes to the exploded attribute block.

> Before you explode the attribute block copy, be sure that it is at a 1-to-1 scale. This is important—otherwise, you could end up with all of your new attribute blocks at the wrong size. Also be sure you use some marker device, such as a line, to locate the insertion point of the attribute block *before* you explode it.

3. Type **attredef**↵.
4. At the `Name of Block you wish to Redefine:` prompt, enter the appropriate name.
5. At the `Select Objects for New Block:` prompt, select all the objects, including the attribute definitions, you want to include in the revised attribute block.
6. At the `Insertion Base Point of New Block:` prompt, pick the same location used for the original block.

Once you pick the insertion point, AutoCAD will take a few seconds to update the blocks. The amount of time will vary depending on the complexity of the block and the number of times the block occurs in the drawing. If you include a new attribute definition with your new block, it too will be added to all the updated blocks, with its default value. Attribute definitions that are deleted from your new definition will be removed from all the updated blocks.

Common Uses for Attributes

Attributes are an easy way to combine editable text with graphic symbols without resorting to groups or separate text and graphic elements. One of the more common uses of attributes is in title blocks. You can create a number of title blocks and insert attributes for title, part number, drawn by, checked by, and so on. Each time the block is inserted, the drafter will be asked to fill in the data. This use of attributes means that all of your title blocks will have the same information at the same size and text font—it's a simple way to standardize this function.

Extracting and Exporting Attribute Information

Once you have entered the attributes into your drawing, you can extract the information contained in them and use it in other programs. You may, for example, want to send the parts list to be entered into the Material Requirements Planning (MRP) software, or you may want to create a bill of materials for your drawing.

The first step in extracting attribute information is to create a template file using a text editor like Windows Notepad. The template file used with attributes is an ASCII text file containing a list of the attributes you wish to extract and their characteristics. You can also extract information about the block that an attribute is associated with. The block's name, X and Y coordinates, layer, orientation, and scale are all available for extraction.

Don't confuse this attribute template file with the drawing template file you use to set up various default settings.

Determining What to Extract

In the template file, for every attribute you wish to extract, you must give the attribute's tag name followed by a code that determines whether the attribute value is numeric or text, and how many spaces to allow for the value. If the value is numeric, you must indicate how many decimal places to give the number. If you are familiar with database management programs, you'll know these are typical variables you determine when you set up a database.

You cannot have a blank line anywhere in the template file, or AutoCAD will reject it. Also, the last line in the file must end with ↵, or your data will not extract.

For example, to get a list of items with description balloons, you would create a text file with the following contents:

```
IN N003000
DESCRIPTION C032000
```

The first item on each line (IN and DESCRIPTION in this example) is the tag of the attribute you want to list. This is followed by at least one space, and then a code that describes the attribute. This code may look a little cryptic at first glance. The following list describes how the code is broken down from left to right:

- The first character of the code is always a *C* or an *N* to denote a character (C) or numeric (N) value.

- The next three digits are where you enter the number of spaces the value will take up. You can enter any number from 001 to 999, but you must enter zeros for null values. The balloon example shows the value of *003* for three spaces. The two leading zeros are needed because AutoCAD expects to see three digits in this part of the code.

- The last three digits are for the number of decimal places to allow if the value is numeric. For character values, these must always be zeros. Once again, AutoCAD expects to see three digits in this part of the code, so even if there are no decimal digits for the value, you must include *000*.

Using Spaces in Attribute Values

At times, you may want the default value to begin with a blank space. This enables you to specify text strings more easily when you edit the attribute. For example, you may have an attribute value that reads *3334333*. If you want to change the first 3 in this string of numbers, you have to specify **3334** when prompted for the string to change. If you start with a space, as in **_3334333** (we're only using an underline here to represent the space—it doesn't mean you type an underline character), you can isolate the first 3 from the rest by specifying **_3** as the string to change (again, typing a space instead of the underline).

You must enter a backslash character (\) before the space in the default value to tell AutoCAD to interpret the space literally, rather than as a press of the spacebar (which is equivalent to pressing ↵).

Now you will use the Windows Notepad application to create a template file. If you like, you can use any Windows word processor that is capable of saving files in the ASCII format.

1. In Windows 95/NT, locate and start up the Notepad application from the Accessories program group.
2. Enter the following text as it is shown. Press ↵ at the end of *each* line, including the last.

   ```
   IN          C003000
   QTY         C003000
   PN          C020000
   DESCRIPTION C050000
   MATERIAL    C050000
   FINISH      C050000
   NOTES       C050000
   ```

3. When you have finished entering these lines of text, click on File ➣ Save, then enter **Bill.txt** for the file name. For ease of access you should save this file to your current default directory, or the \ACADR14\ directory.
4. Close the Notepad, and return to AutoCAD.

You've just completed the setup for attribute extraction. Now that you have a template file, you can extract the attribute data.

Text Editor Line Endings

It is very important that the last line of your file end with a single ↵. AutoCAD will return an error message if you leave the ↵ off or if there is an extra ↵ at the end of the file. Take care to end the line with a ↵ and don't add an extra one. If the extraction doesn't work, check to see whether there is an extra ↵ at the end of the file.

Extracting Block Information Using Attributes

We mentioned that you can extract information regarding blocks, as well as attributes. To do this you must use the following format.

```
BL:LEVEL     N002000
BL:NAME      C031000
BL:X         N009004
BL:Y         N009004
BL:Z         N009004
BL:NUMBER    N009000
```

BL:HANDLE	C009000
BL:LAYER	C031000
BL:ORIENT	N009004
BL:XSCALE	N009004
BL:YSCALE	N009004
BL:ZSCALE	N009004
BL:XEXTRUDE	N009004
BL:YEXTRUDE	N009004
BL:ZEXTRUDE	N009004

WARNING

A template file containing these codes must also contain at least one attribute tag, because AutoCAD must know which attribute it is extracting before it can tell what block the attribute is associated with. The code information for blocks works the same as for attributes.

We have included some typical values for the attribute codes in this example. The following list describes what each line in the above example is used for.

LEVEL returns the nesting level.

NAME returns the block name.

X returns the X coordinate of the insertion point.

Y returns the Y coordinate of the insertion point.

Z returns the Z coordinate of the insertion point.

NUMBER returns the order number of the block.

HANDLE returns the block's handle. If no handle exists, a 0 is returned.

LAYER returns the layer the block is inserted on.

ORIENT returns the insertion angle.

XSCALE returns the X scale.

YSCALE returns the Y scale.

ZSCALE returns the Z scale.

XEXTRUDE returns the block's X extrusion direction.

YEXTRUDE returns the block's Y extrusion direction.

ZEXTRUDE returns the block's Z extrusion direction.

Performing the Extraction

AutoCAD allows you to extract attribute information from your drawing as a list in one of three different formats:

- CDF (comma-delimited format)
- SDF (space-delimited format)
- DXF (data exchange format)

Using the CDF Format

The *comma-delimited format* (CDF) can be read by many popular database management programs, as well as programs written in BASIC. This is the format you will use in this exercise.

1. Type **ddattext**↵. The Attribute Extraction dialog box appears.

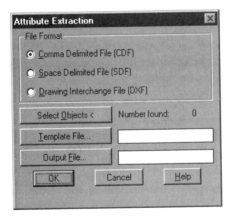

2. If it isn't already selected, click on the radio button labeled Comma Delimited File (CDF).
3. Click on the Template File button. Then, using the Template File dialog box, locate and select the Bill.txt file you created earlier.

You can select file names of existing template and output files by clicking on the Template File or Output File buttons in the Attribute Extraction dialog box. The File dialog box appears, allowing you to select files from a list.

4. Click on the Output File button. Then, using the File dialog box, enter the name Bom.txt for your output file and place it in the \Program Files\Acadr14 directory.

5. Click on OK at the Attribute Extraction dialog box. The message 4 records in extract file appears.

NOTE You may have noticed the Select Objects button in the Attribute Extraction dialog box. When you click on this button, the dialog box temporarily closes to let you single out attributes to extract by picking their associated blocks from the display.

AutoCAD has created a file called Bom.txt that contains the extracted list. Let's take a look at its contents.

1. Open the Windows Notepad application (Start ➤ Programs ➤ Accessories ➤ Notepad).

2. Choose File ➤ Open to open Bom.txt from the current directory. (In this exercise, the template file is located in the \Program Files\ACADR14 directory.) You will get the following list:

```
'1','1','36-120063-03 REV A','C-LINK','AL 6061-T6','HDBLK AN','-001 ONLY-'
'2','1','36-120064-01 REV B','S-LINK','AL 6061-T6','HDBLK AN','001 ONLY-'
'3','1','72-160070-25','BOLT, #10-32x.75 LG, HX HD','ST','ZINC','-001 ONLY-'
'4','1','71-150070-25','10-32 NUT, SLOTTED, THICK','STEEL','ZINC','001 ONLY-'
```

Since you picked the comma-delimited format (CDF), AutoCAD placed commas between each extracted attribute value (or *field,* in database terminology).

NOTE Notice that the individual values are enclosed in single quotes. These quotes delimit character values. If you had specified numeric values, the quotes would not have appeared. Also notice that the fields are in the order they appear in the template file.

The commas are used by some database management programs to indicate the separation of fields in ASCII files. This example shows everything in uppercase letters because that's the way they were entered when we inserted the attribute blocks in our own working sample. The extracted file maintains the case of whatever you enter for the attribute values.

Using Other Delimiters with CDF Some database managers require the use of other symbols, such as double quotes and slashes, to indicate character values and field separation. AutoCAD allows you to use a different symbol in place of the single quote or comma. For example, if the database manager you use requires double-quote delimiters for text in the file to be imported, you can add the statement

```
c:quote "
```

to the template file to replace the single quote with a double quote. A line from an extract file using *c:quote "* in the template file would look like this:

```
"1","1","36-120063-03 REV A","C-LINK","AL 6061-T6","HD BLK AN","-001 ONLY-"
```

Notice that the single quote (') is replaced by a double quote ("). You can also add the statement

```
c:delim /
```

to replace the comma delimiter with the slash symbol. A line from an extract file using both *c:quote "* and *c:delim /* in the template file would look like this:

```
"1"/"1"/"36-120063-03 REV A"/"C-LINK"/"AL 6061-T6"/"HD BLK AN"/"-001 ONLY-"
```

Here the comma is replaced by a forward slash. You can add either of these statements to the beginning or end of your template file.

Using the SDF Format

Like the CDF format, the *space-delimited format* (SDF) can be read by most database management programs. This format is the best one to use if you intend to enter information into a word-processed document because it leaves out the commas and quotes. You can even import it into an AutoCAD drawing using the method described in Chapter 7. Now let's try using the SDF option to extract the same list we extracted a moment ago using CDF.

1. Type **ddattext**↵ at the command prompt.
2. At the Attribute Extraction dialog box, use the same template file name, but for the attribute extract file name, use **Bill-SDF.txt** to distinguish this file from the last one you created.
3. Click on the SDF radio button, and then click on OK. You will see a message that reads 4 records in extract file.
4. After AutoCAD has extracted the list, use the Windows Notepad to view the contents of the file. You will get a list similar to this one (we've removed the spaces for clarity):

```
1 1 36-120063-03 REV A C-LINK AL 6061-T6 HD BLK AN -001 ONLY
2 1 36-120064-01 REV B S-LINK AL 6061-T6 HD BLK AN  001 ONLY
3 1 72-160070-25 BOLT, #10-32x.75 LG, HX HD ST ZINC -001 ONLY
4 1 71-150070-25 10-32 NUT, SLOTTED, THICK STEEL ZINC  001 ONLY
```

This format shows text without any special delimiting characters.

5. Save your work.

Using the DXF Format

The third file format is the *data exchange format* (DXF). There are actually two methods for DXF extraction. The Attribute Extraction dialog box you saw earlier offers the DXF option. This option extracts only the data from blocks containing attributes. Choose the File ➢ Export option then select .DXF from the List Files of Type drop-down list to convert an entire drawing file into a special format for data exchange between AutoCAD and other programs (for example, other PC CAD programs). We will discuss the DXF format in more detail in Chapter 18.

NOTE As an alternative, you can choose File ➢ Export, then at the Export Data dialog box, choose DXX Extract (*.DXX) from the List File of Type drop-down list. Enter a name for the extracted data file in the File Name input box, and then click on OK. Finally, in the drawing editor, select the attributes you want to extract. The DXX format is a subset of the DXF format.

Using Extracted Attribute Data with Other Programs

You can import any of these lists into any word-processing program that accepts ASCII files. They will appear as shown in our examples.

As we mentioned earlier, the extracted file can also be made to conform to other data formats.

Microsoft Excel

Excel has no specific requirements for the formatting of AutoCAD extracted data. Use File ➢ Open from the Excel menu bar, and then select the .TXT file type from the File dialog box. Excel then opens a very simple-to-use import tool that steps you through import process. You can choose the delimiting method—whether each field is a character, date, or other type of data—and other formatting options.

If you are a database user familiar with Microsoft Visual Basic, you will want to look at the sample Visual Basic utility supplied with AutoCAD. It shows how you can use VBA to export attribute data directly from AutoCAD to Excel without having to create a template file or perform an export using the Ddattext command. You can find the sample program in the \sample\axtiveX\ExtAttr directory. See Chapter 20 for more on this and other useful VBA-related utilities.

Microsoft Access

Like Excel, you can open the exported attribute data directly in Access. First create a new database table. You can then use the File ➢ Get External Data ➢ Import option from the Access menu bar, and then choose Text Files from the Files of Type dropdown list. After you've selected the attribute extracted file, Access offers an easy-to-use import tool that steps you through the process of importing the file. You can specify field and record names, as well as the delimiting method.

If you plan to update the attribute data file on a regular basis, you can have the file linked, instead of imported (File ➢ Get External Data ➢ Link Tables). A linked file is best suited for situations where you want to read the data without changing it. You can generate reports, link the attribute data to other databases, perform searches, or sort the attribute data.

As with Excel, if you are an Visual Basic user, you can add controls to enhance your attribute data link to Access.

Accessing External Databases

AutoCAD offers a way to access an external database from within AutoCAD—the *AutoCAD SQL Extension* (*ASE*). (SQL stands for Structured Query Language. It is a standard by which databases are organized.) With the ASE, you can read and manipulate data from external database files. You can also use ASE to *link* parts of your drawing to an external database. There are numerous reasons for doing this. The most obvious is to keep inventory on parts of your drawing. You can also keep track of the title block data and notes in your drawings.

In this section we specifically avoid the more complex programming issues of database management systems, and we do not discuss the SQL language on which much of ASE is based. Still, you should be able to make good use of ASE with the information provided here.

In these exercises, we do assume that you are somewhat familiar with databases. For example, we will frequently refer to something called a *table*. A table is a SQL term referring to the row-and-column data structure of a typical database file. For the sake of this tutorial, you can think of a table as a dBASE database file. Other terms we'll use are *rows*, which are the records in a database, and *columns*, which refer to the database fields.

Finally, it is *very* important that you follow the instructions in these beginning exercises carefully. If you make mistakes in the beginning, later exercises will not work properly.

Database Managers and AutoCAD

Chances are, you are already familiar with at least one database program. ASE provides support for dBASE, ODBC, and Oracle databases. Many other database and spreadsheet applications can read and write to at least one of these database formats. If you are a Microsoft Office user, you can use Access or Excel to create database files for use with AutoCAD ASE.

Setting Up ASE to Locate Database Files

ASE doesn't create new database files. You must use existing files or create them before you use ASE. In the first set of exercises below, you will use a file from the CD-ROM named `Partslist.dbf`, located in the `\Acadr14\sample\Dbf` subdirectory. The contents of this database file are shown in Figure 17.7.

FIGURE 17.7

The contents of the Partslist.dbf dBASEIII file as it would appear in Microsoft Excel

ITEM	QTY	PART	REVISION	NAME	MATERIAL	FINISH	PRICE	VENDOR	ASSEMBLY	NOTES
1	1	36-12006303	A	CURVED LINK	AL 6061-T6	HARD BLACK ANODIZE	26.050	FAB MART	02-143786	
2	1	36-12006401	B	SLOTTED LINK	AL 6061-T6	HARD BLACK ANODIZE	17.400	FAB MART	02-143786	
3	1	72-160070-25	A	BOLT,#10-32x .75 LG, HEX HEAD	STEEL	ZINC	0.050	EXPERT FASTENERS	02-143786	REVISION
4	1	71-190070-25	A	10-32 NUT, SLOTTED, THICK	STEEL	ZINC	0.025	EXPERT FASTENERS	02-143786	

Before you start to work with databases, you must tell AutoCAD about your database *environment*. The environment consists of a database management application and its associated directory paths, user access information, and the database tables for that particular database.

If you have installed the full version of AutoCAD, then you are already set up to proceed with the tutorial in the next section. AutoCAD has already included information about the sample database environment. But you will still want to know the method for setting up AutoCAD to recognize a database environment, so that in the future you can access other databases besides those in this sample. To do this, use the External Database Configuration application that comes with AutoCAD.

Knowing Your Database Structure

Before you actually use the External Database Configuration program, you will want to make sure you are familiar with the way your database files are organized and how that organization relates to the database nomenclature. You will need to know three terms: catalog, schema, and tables.

To help explain these terms, we'll set up a database environment using the database file that is located on the companion CD-ROM. The file is located on the CD-ROM in the AutoCAD subdirectory called \DBF. It is known as the *table* of the database, while the directory that contains it is the *schema*. The directory path that leads to the schema, \Autocad\sample, is known as the *catalog*. As you will see, catalogs and schemas need names, so if you are setting up a database to work with AutoCAD, you may want to start thinking of a meaningful name for these components of your database.

Configuring AutoCAD for the External Database

Now let's use this example to set up AutoCAD's database configuration.

1. From the Windows 95/NT desktop, choose Start ➢ Programs ➢ AutoCAD R14 ➢ External Database Configuration.
2. The External Database Configuration dialog box appears.

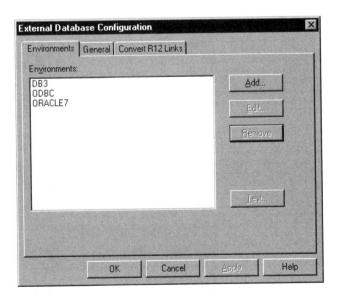

3. Click on Add to create a new environment. (The environment you will be working with in the following tutorial is the DB3 environment that already appears in the Environment list.) The Select DBMS for New Environment dialog box appears.

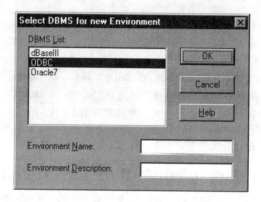

4. Select the type of database you are using from the list. In this example, click on dBaseIII (DB3).
5. Enter **Tutorial** in the Environment Name input box.
6. Click on OK. The Environment dialog box appears.

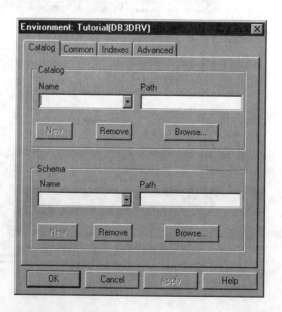

WARNING

The configuration of the Environment dialog box will be different depending on the type of database you select in step 4. For more information on the ODBC environment settings, see Appendix A.

7. In the Catalog button group, click on Browse; then in the Browse dialog box, locate and select the \Program Files\Acadr14\Support directory that is the parent to the actual subdirectory that contains the database tables. Your path may be different, depending on your AutoCAD installation. Click on OK to continue.

8. Enter **Cat** in the Name input box of the Catalog button group, and then click on the New button.

9. In the Schema button group, click on the Browse button; then in the Browse dialog box, locate and select the \Program Files\Acadr14\Sample\Dbf directory that contains the tables. Click on OK to continue.

10. Enter **Catfiles** in the Name input box of the Schema button group, then click on New. Remember these names because you will use them when you open the database from AutoCAD.

11. Click on New, then click on OK. You will see the name of the database environment you just created in the Environments list box.

12. Click on OK to exit the External Database Configuration dialog box.

13. Use Windows Explorer or any file management tool that you like to copy the Partslist.dbf file located on the companion CD-ROM under \Autocad\Sample\DBF to your hard drive's \Acadr14\Sample\Dbf directory.

Now you can access your database from AutoCAD using the environment, catalog, and schema you have just set up. Other options in the External Database Configuration application let you set up user names and passwords so you can restrict access to the databases, or include indexes for the databases.

Loading the External Database Toolbar

Before you proceed with the following exercises, open the External Database toolbar.

1. Start AutoCAD if it isn't already open.

2. Right-click on any toolbar, then click on External Database from the Toolbars dialog box. Click on Close to close the Toolbars dialog box.

Opening a Database from AutoCAD

Now you're ready to access your database files directly from AutoCAD. In the following exercise, you'll take the first step by making a connection between a database table and AutoCAD.

1. Open the assembly drawing that you were working on earlier in this chapter, if it is not still open. You can also find a copy of this file on this book's companion CD-ROM.

If ASE does not initialize, be sure that the \Program Files\acadr14\Support directory is included in the Support option of the Environment settings. This option can be found by choosing Tools ➢ Preferences; then at the Preferences dialog box, click on the tab labeled Files. The Support subdirectory should be listed in the Support Files Search Path listing. Be sure to close and restart AutoCAD after making the environment change.

2. Click on the Administration button on the External Database toolbar, or type **aad**↵.

The Administration dialog box appears.

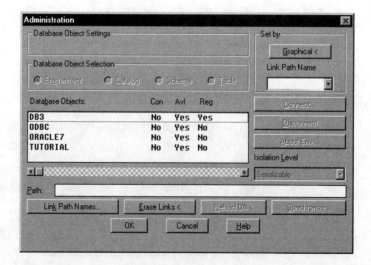

3. Click on TUTORIAL from the Database Objects list. Notice that the Environment button is automatically selected.
4. Click on the Connect button to the right of the list. The Connect to Environment dialog box appears.

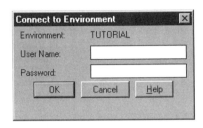

5. Because the user name and password are not important to this tutorial, click on OK.
6. Click on the Catalog button in the Database Object Selection group, and then click on CAT, which appears in the Database Objects list.
7. Click on the Schema button, and then click on CATFILES from the list.
8. Click on the Table button, and then click on PARTSLIST from the list.
9. Finally, click on OK. You've just linked to the Partslist.dbf file.

Finding a Record in the Database

Now that you are connected to the database, suppose you want to find the record for a specific part. You might already know that the part you're looking for is made by Fabmart.

1. Click on the Rows button on the External Database toolbar, or type **aro**↵.

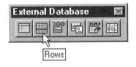

The Rows dialog box appears.

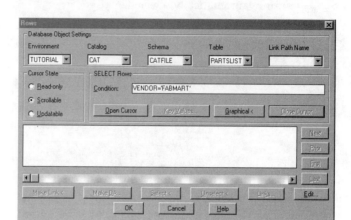

2. Click on the Condition input box and type **VENDOR='FABMART'**. Make sure you include the single quotes and use upper case.
3. Click on the Scrollable radio button in the Cursor State group, and then click on the Open Cursor button below the Condition input box. The first item that fulfills the Condition criterion appears in the list box.

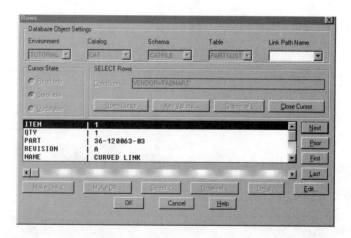

4. Click on the Next button to view the next row that meets the search criterion.
5. Click on Last to view the last row.
6. Click on OK to close the Rows dialog box.

ACCESSING EXTERNAL DATABASES 797

You can also click on Open Cursor without entering anything in the input box in step 2. This essentially tells ASE to select all the rows in the current database table. You would then be able to scan the entire table. To exit the list, click on Close Cursor or simply exit the Rows dialog box.

You've just seen how you can locate and view a record in a database. If you wanted to edit or delete that record, you would have clicked on the Updatable radio button in step 3, instead of the Scrollable button. You could then make changes to the database item shown in the list box by clicking on the Edit button. The Edit button then opens another dialog box which allows you to edit individual items in a record. In the following section, you'll use the Edit Row dialog box, not to change a record, but to add one.

The ability to access databases in this way can help you connect AutoCAD graphic data with database information. Database access works in the other direction as well; you can store graphical objects in most databases that are Windows 95 compliant. Polished final drawings could be available for Marketing or Sales purposes, and they could be quickly amended to show customization by allowing a dialog between the drawing and a database of variations.

Adding a Row to a Database Table

Now let's get back to our assembly example. Suppose you are adding a new part to your assembly. The first thing you'll want to do is add the part to the database. Here's how it's done.

1. Click on the Rows button on the External Database toolbar, or enter **aro**↵.
2. Click on the Updatable button in the Cursor State group, and then click on the Edit button toward the lower-right corner of the dialog box. The Edit Row dialog box appears.

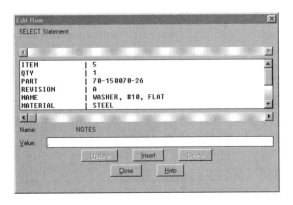

3. Enter the following data in the dialog box. To add an item, highlight the item in the list, and then add the information in the input area at the bottom of the dialog box. After you type in each item, press ↵ to move to the next item.

Data Type	You Enter
ITEM	**5**
QTY	**1**
PART	**70-150070-26**
REVISION	**A**
NAME	**WASHER, #10, FLAT**
MATERIAL	**STEEL**
FINISH	**ZINC**
PRICE	**0.001**
VENDOR	**EXPERT FASTENERS**
ASSEMBLY	**02-143785**
NOTES	

4. When you've finished entering the list, click on Insert, and then click on Close.
5. Click on OK at the Rows dialog box. You've just added a row to the database, and it is the current one.

Linking Objects to a Database

So far, you've looked at ways you can access an external database file. You can also *link* specific drawing objects to elements in a database. But before you can link your drawing to data, you must *register* the table (that is, the database file) to which you want to link, and then you must create a *link path*.

Registering a table is a way of naming a group of links between your drawing and the database file. You can register a table numerous times, allowing you to set up several different sets of links. For example, you may want to create a set of links between the Item column of your database table and the balloons in your drawing. In another instance, you may want to link the NAME field to a parts listing in the drawing. To identify these different sets of links, create a link path name, which is really just a name you give to each set of links.

Now let's see how you can create a link to the database by linking your new part to an empty balloon.

Creating a Link

In the following set of exercises, you will link an AutoCAD object to a record in the Partslist database table.

1. Click on the Administration button on the External Database toolbar.
2. With the Partslist table selected, click on the Link Path Names button in the lower-left corner of the dialog box. The Link Path Names dialog box appears.

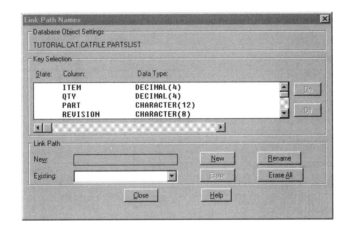

3. Click on ITEM in the Key Selection group; then click on the On button to the right. The word ON appears next to the name.
4. Repeat step 3 for the item labeled NAME, which is farther down the list.
5. Type **r-parts** in the New input box in the Link Path group.
6. Click on New to the right of the New input box. The name R-PARTS appears in the Existing drop-down list and the message Registered Successfully appears in the lower-left corner. At this point, you could select another set of items from the list and create another link path name.
7. Click on Close to return to the Administration dialog box. Click on OK to exit this dialog box.

You've just registered the Parts list and created a link path name for the ITEM and NAME fields. You can create other link paths that include other fields.

Now you are ready to add a link to the assembly. The first step is to locate the record that is associated with the washer.

1. Zoom into the area shown in Figure 17.8.

2. Draw a circle the same size as the balloon for Item 4 and place it next to the Item 4 balloon.

3. Click on Rows on the External Database toolbar.

4. Click on the Conditions input box, and then type **NAME='WASHER, #10, FLAT'**.

5. Click on Open Cursor. The record for WASHER appears in the list.

6. Click on the button labeled Make Link. The dialog box temporarily disappears.

7. Click on the circle that you just created. Press ↵ to return to the Rows dialog box.

8. Click on OK to exit the Rows dialog box.

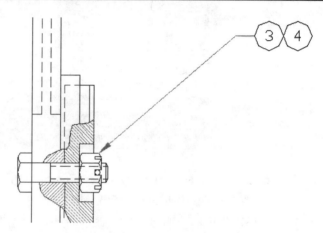

FIGURE 17.8

Enlarging your view before adding the washer

Now you have a link established between the row for "washer" in the database and an object in your drawing. Later, you can continue to build links between objects and database records, but for now, let's look at other things you can do with the link.

Adding Labels with Links

Once you've got a link established, you can use it to perform a variety of editing tasks. For example, you can add labels to your drawing based on information in the database. The following exercise shows you how you can add the Item Number to the drawing.

1. Click on Rows on the External Database toolbar. The Rows dialog box appears.

2. Click on the Graphical button. The dialog box disappears momentarily, allowing you to select an object.

LINKING OBJECTS TO A DATABASE | **801**

3. At the Select Objects: prompt, click on the circle that you just created. The Links dialog box appears.

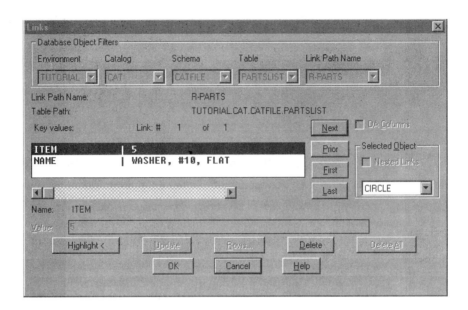

4. Click on the Link Path Name drop-down list and select R-PARTS.

5. Click on OK. The Rows dialog box appears again and displays the record information that is linked to the number.

6. Now click on Make DA. The Make Displayable Attribute dialog box appears.

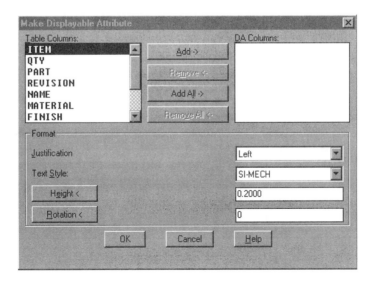

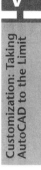

PART
V

Customization: Taking
AutoCAD to the Limit

The *DA* in *Make DA* stands for Displayable Attribute. Don't confuse this with the attributes we've discussed earlier in this chapter. They are not the same.

This dialog box shows two lists. The list on the left, Table Columns, shows the items in the database record. The list on the right, DA Columns, holds the records you want displayed in the drawing as labels. Right now the DA Columns list is empty.

7. Highlight ITEM in the Table Columns list, and then click on the Add button. The item is copied into the DA Columns list. In the Format button group, change the Justification to Middle Center and the Height to 0.125.

8. Use the Center Osnap to locate and pick the center of the circle that you just drew. The Rows dialog box appears.

9. Click on OK. The Rows dialog box closes, and you see a new label with the item number in the circle. Unlike the attributes of the adjacent balloons, this new label is linked to the record in your database.

You used the Format group in the Make Displayable Attribute dialog box in step 7. These items allow you to control the graphic characteristics of the label, such as Justification, Text Style, Height, and Rotation. The dialog box also has a button labeled Add All. This copies all the columns from the Table Columns list to the DA Columns list.

If a table is modified outside of AutoCAD, you must use the Synchronize button in the Administration dialog box to re-synchronize the links between your drawing and the database the next time you open the linked drawing file.

Updating Rows and Labels

Chances are the contents of your database will change frequently. In the following exercise, you will get to change a database row and update the label you just placed in your drawing. You will start by quickly setting the current row by selecting an object.

1. Click on Rows on the External Database toolbar; and at the dialog box, click on the Updatable radio button in the Cursor State group.

2. Click on Graphical. The dialog box closes and the Select Object: prompt appears.
3. Click on the label you just entered. The current row is now set to the one for the new item.
4. Click on the Edit button; then, at the Edit Row dialog box, highlight ITEM.
5. Change the Value input box to read **6**, then press ↵. The ↵ is important; without it the Update button is not activated.
6. Click on Update; then click on Close to return to the Rows dialog box.

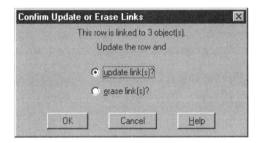

7. Click on OK at the Rows dialog box. The label now reflects the change you made in the database.

You have just edited a row in the database and updated the ITEM label at the same time.

Finding and Selecting Graphics through the Database

You've just seen how you can use links to add and update labels in your drawing. Links can also help you find and select objects in a drawing that are linked to a database. The next exercise shows, in a simplified way, how this works.

1. Click on Rows on the External Database toolbar; then type **VENDOR='Expert Fasteners'** into the Condition input box in the Select Rows group.
2. Click on Open Cursor. Once again, you see the familiar record we've been working with.
3. Click on Select. The dialog box temporarily disappears, and you see all the items linked to this record highlighted in the drawing.
4. Press ↵ to return to the Rows dialog box, and then click on OK. The objects linked to the selected record remain highlighted.

Once these steps have been taken, you can use the Previous Object Selection option to select those objects that were highlighted in step 3.

Deleting a Link

Earlier, you had to change a link that connected a circle object in the drawing to an Item and Name in the database record. This next exercise shows you how you can delete the link to the name.

1. Click on Rows on the External Database toolbar; then, at the dialog box, click on Graphical.

2. Click on the circle for item 6. The row assigned to that label is now the current row. The Links dialog box appears.

3. Click on the Environment drop-down list and select TUTORIAL.

4. Click on OK.

5. At the Rows dialog box, click on Links.

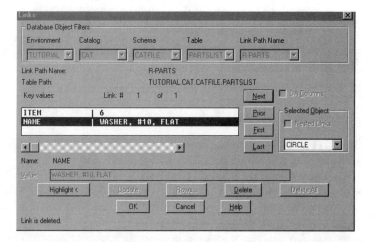

The Key Values list box shows the key values that were selected when you first created the link path name. Above the list, you see the statement Link: # 1 of 1. This tells you which object link in the drawing you are currently working with.

1. To see which object #1 is, click on the Highlight button. The dialog box disappears and the drawing shows the circle with item 5 highlighted.

2. Press ↵ to return to the Links dialog box.

3. Click on the Delete button at the bottom of the dialog box. The link between the record and the circle is deleted.

4. Click on OK, then click on OK again at the Rows dialog box.

Filtering Selections and Exporting Links

There are two options you didn't get a chance to try. They are Object Selection and Export Links. Object Selection offers a way to select objects based on a combination of graphical and database criteria. For example, suppose you want to select all the items of a certain description from one region of your assembly. Assuming you have already created links between your database table and your drawing, you could do the following steps.

1. Click on the Select Objects button on the External Database toolbar, or type **ase**↲.

The Select Objects dialog box appears.

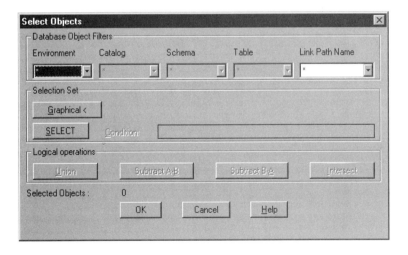

2. Select the appropriate Environment, Catalog, Schema, and Table. In the Condition input box, enter the search criteria for the washer.

3. Under the Logical Operations group, click on the Intersect button, and then click on Graphical. The dialog box temporarily disappears to allow you to select objects.

4. Using a window, select the area in the drawing that includes the item you want, and then press ↵.

5. Click on OK to exit the dialog box. You can now use the Previous Selection option to select the specified washer in the area you selected.

The Export Links feature lets you export information about the links between your drawing and your database. This can be useful when you are preparing reports from your database application, because database applications are unaware of the number of links that occur to your drawing. Once you export link information, you can then incorporate that information into your report. The exported information lists the object handles and the associated linked database value. Here's a description of how the Export Link function works.

1. Click on the Export Links button on the External Database toolbar, or type **aex**↵. You are then asked to select an object.

2. Select the objects that contain the links you wish to export. The Export Links dialog box appears.

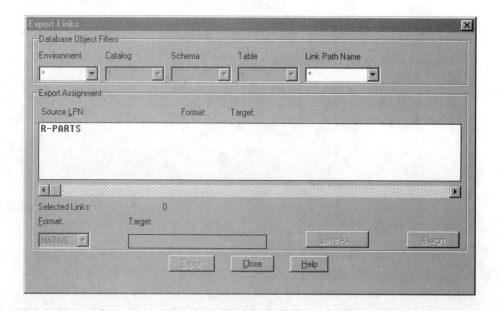

LINKING OBJECTS TO A DATABASE | **807**

3. You can, at this point, reduce the scope of the link information you will export, by selecting a specific Environment, Catalog, Schema, Table, or Link Path Name.

4. Select a Link Path Name from the list box.

5. Select NATIVE from the Format drop-down list, and then enter a file name in the Target input box. This will be the name given to the table you want to create.

6. Click on Assign to assign the format and target to the link path name.

7. Choose Export to complete the export process.

You can export multiple link path names by repeating steps 3–6, before completing step 7.

Using SQL Statements

For those who are more familiar with SQL, AutoCAD offers a way that lets you access and manipulate database information with a greater degree of control. The SQL Editor can be accessed by clicking on the SQL Editor tool on the External Database toolbar, or by typing **asq**↵.

The SQL dialog box appears.

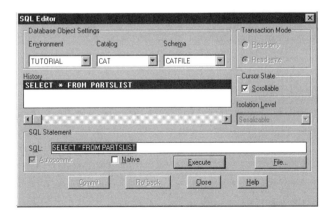

You can then enter an SQL statement in the edit box and click on Execute. Results of your query are displayed in the SQL Cursor dialog box.

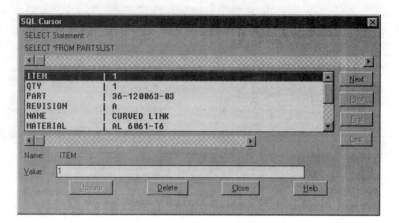

Where to Go from Here

You've seen how you can access and link your drawing to a database. We hope that in this brief tutorial, you can find the information you will need to develop your own database.

If you understand SQL, you can take advantage of it to perform more sophisticated searches. You can also expand the functionality of the basic ASE package included with AutoCAD. These topics are, unfortunately, beyond the scope of this book. For more detailed information about ASE and SQL, refer to the *AutoCAD SQL Extension* reference manual.

If You Want to Experiment...

Attributes can be used to help automate data entry into drawings. To demonstrate this, try the following exercise.

Create a drawing file called **Record** with the attribute definitions shown in Figure 17.9. Note the size and placement of the attribute definitions as well as the new base point for the drawing. Save and exit the file, and then create a new drawing called **Schedule** containing the schedule shown in Figure 17.10. Use the Insert command and insert the Record file into the schedule at the point indicated. Note that you are prompted for each entry of the record. Enter any value you like for each prompt. When you are done, the information for one record is entered into the schedule.

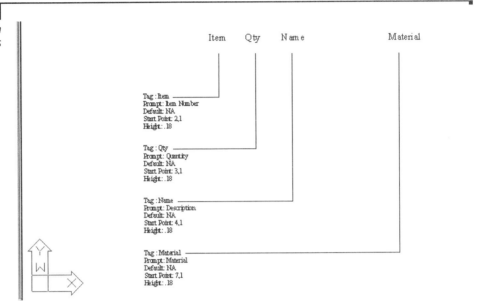

FIGURE 17.9

The Record file with attribute definitions

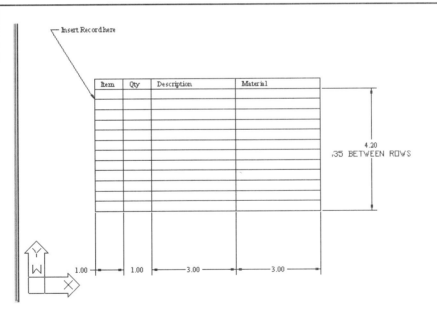

FIGURE 17.10

The Schedule drawing with Record inserted

Chapter 18

Getting and Exchanging Data from Drawings

FEATURING

Getting Information about a Drawing

Exchanging CAD Data with Other Programs

Using AutoCAD Drawings in Desktop Publishing

Combining Data from Different Sources

Chapter 18

Getting and Exchanging Data from Drawings

AutoCAD drawings contain a wealth of data. In them you can find graphic information such as distances and angles between objects, as well as precise areas and the properties of objects. But as you become more involved with AutoCAD, you will find that you also need data of a different nature. For example, as you begin to work in groups, the various settings in a drawing become important. Statistics on the amount of time you spend on a drawing are needed if you are consulting and you are billing computer time. As your projects become more complex, file maintenance requires a greater degree of attention. To take full advantage of AutoCAD, you will want to exchange much of this data with other people and other programs.

In this chapter, you will explore the ways in which many types of data can be extracted from AutoCAD and made available to you, your coworkers, and other programs. First, you will discover how to get specific data on your drawings. Then you will look at ways to exchange data with other programs—such as word processors, desktop publishing software, and even other CAD programs.

Getting Information about a Drawing

AutoCAD can instantly give you precise information about your drawing, such as the area, perimeter, and location of an object; the base point, current mode settings, and space used in a drawing; and the time at which a drawing was created and last edited. In this section you will practice extracting this type of information from your drawing, using the tools found in the Tools ➤ Inquiry pull-down menu.

To find absolute coordinates in a drawing, use the ID command. Choose Tools ➤ Inquiry ➤ ID point, or type **id**↵. At the prompt ID Point, use the Osnap overrides to pick a point, and its X, Y, and Z coordinates will be displayed on the prompt line. The last point located with ID becomes the default point location in the next command that requires a point. You can examine your drawing, locate a point and use ID to place the point temporarily into memory.

Finding the Area or Location of an Object

AutoCAD simplifies the task of finding areas, locations, and perimeters. In this section you will practice determining the areas of both regular and irregular objects.

First you will find out the area of a simple shape.

1. Start AutoCAD and open the ch18shapes.dwg file from the companion CD-ROM.

2. Don't zoom out yet to see the whole drawing. Your view will be the same as Figure 18.1.

3. Choose Tools ➤ Inquiry ➤ Area, or type **area**↵ at the command prompt. You can also click and drag List from the Standard toolbar, and then select Area.

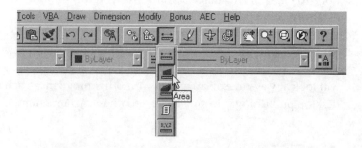

GETTING INFORMATION ABOUT A DRAWING | 815

4. Using the Endpoint Osnap, start with the lower-left corner of the shape and select the points indicated in Figure 18.1. You are indicating the boundary of the shape.

5. When you have come full circle to the eighth point shown in Figure 18.1, press ↵. You get the message:

```
Area = 3.8602 sq units, Perimeter = 8.8887 units
```

FIGURE 18.1

Selecting the points to determine the area of a simple shape

There is no limit to the number of points you can pick to define an area, so you can obtain the areas of very complex shapes.

Using Boundary

Using the Object option of the Area command, you can also select circles and polylines for area calculations. Using this option in conjunction with another AutoCAD utility called Boundary, you can quickly get the area of a bounded space. Here's how to use it.

1. Set the current layer to Phantom.

2. Choose Draw ➤ Boundary, or type **bo**↵. The Boundary Creation dialog box appears.

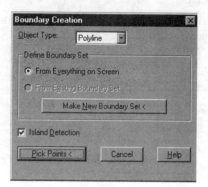

There is one caveat to using Boundary: You must be sure the area you are trying to define has a continuous border. If there are any gaps at all, no matter how small, Boundary will give you an error message.

3. Click on the Pick Points button. The dialog box closes.
4. At the `Select internal point:` prompt, click on the interior of the shape. The outline of the shape is highlighted (see Figure 18.2).

FIGURE 18.2

Once you select a point on the shape using Boundary, an outline of the area is highlighted and surrounded by a dotted line.

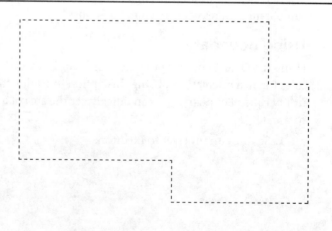

5. Press ↵. Boundary draws an outline of the shape area using a polyline. Since the current layer is Phantom, the boundary is drawn on the Phantom layer and given the default cyan color of the layer.

6. Choose Tools ➢ Inquiry ➢ Area again, or type **area**↵ at the command prompt; and then enter **o**↵ for the Object option.

7. Click on the boundary; when it is highlighted, press ↵. You get the same Area… message you got in the previous exercise.

If you need to recall the last area calculation value you received, you can enter '**setvar**↵ **area**↵. The area will be displayed in the prompt. Enter '**perimeter**↵, and you get the last perimeter calculated.

The Boundary command creates a polyline that conforms to the boundary of an area. This feature, combined with the ability of the Area command to find the area of a polyline, makes short work of area calculations.

Finding the Area of Complex Shapes

The Boundary command works fine as long as the area does not contain *islands* that you do not want included in the area calculation. An island is a closed area within a larger area within which you are attempting to hatch or create a boundary. In the case of the flange part, the islands are the two circles at the lower end of the part.

For areas that do contain islands, you must enlist the aid of the other Area command options: Object, Add, and Subtract. Using Add and Subtract, you can maintain a running total of several separate areas being calculated. This gives you flexibility in finding areas of complex shapes.

The exercise in this section guides you through the use of these options. First, you'll look at how you can keep a running tally of areas. For this exercise, you will use a flange shape that contains circles. This shape is composed of simple arcs, lines, and circles.

1. Zoom the extents and then zoom into the U-shaped object on the lower-right side (see Figure 18.3).

2. Choose Draw ➢ Boundary.

3. At the Boundary Creation dialog box, click on Pick Points.

4. Click on the interior of the flange shape. Notice that the entire shape is highlighted, including the circle islands.

5. Press ↵.

FIGURE 18.3

A flange to a mechanical device

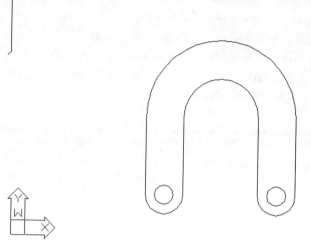

You now have a polyline outline of the shape. As you saw in the previous exercise, the polyline will aid you in quickly obtaining the area. Now let's continue by using the Area command's Add and Subtract options.

1. Choose Tools ➢ Inquiry ➢ Area.
2. Type **a**↵ to enter the Add mode, and then **o**↵ to select an object.
3. Click on the outline of the flange. You'll see the message:

   ```
   Area = 2.3744, Perimeter = 10.2832
   Total area = 2.3744
   ```

4. Press ↵ to exit the Add mode.
5. Type **s**↵ to enter the Subtract mode, and then type **o**↵ to select an object.
6. Click on one of the circles. You'll see the message:

   ```
   Area = 0.0491, Perimeter = 0.7854
   Total area = 2.3254
   ```

   This shows you the area and perimeter of the selected object, and a running count of the total area of the flange outline minus the circle.

7. Click on the other circle. You'll see the message:

   ```
   Area = 0.0491, Perimeter = 0.7854
   Total area = 2.2763
   ```

Again, you see a listing of the area and perimeter of the selected object along with a running count of the total area, which now shows a value of 2.2763. This last value is the true area of the flange.

8. Press ↵ twice to exit the Area command.

In the exercise, you first selected the main object outline, and then subtracted the island objects. You don't have to follow this order; you can start by subtracting areas to get negative area values, and then add other areas to come up with a total. You can also alternate between Add and Subtract modes, in case you forget to add or subtract areas.

You may have noticed that the Area command prompt offered <first point> as the default option for both the Add and Subtract modes. Instead of using the Object option to pick the circles, you could have started selecting points to indicate a rectangular area, as you did in the first exercise of this chapter.

It is important to remember that whenever you press ↵ while selecting points for an area calculation, AutoCAD automatically connects the first and last points and returns the area calculated. If you are in the Add or Subtract mode, you can then continue to select points, but the additional areas will be calculated from the *next* point you pick.

As you can see from these exercises, it is simpler to first outline an area with a polyline, whenever possible, and then use the Object option to add and subtract area values of polylines.

In this example, you obtained the area of a mechanical object. However, the same process works for any type of area you want to calculate. It can be the area of a piece of property on a topographical map or the area of a floor plan. For example, you can use the Object option to find an irregular area like the one shown in Figure 18.4, as long as it is a polyline.

FIGURE 18.4

A complex shape with the area to be calculated shown as hatched

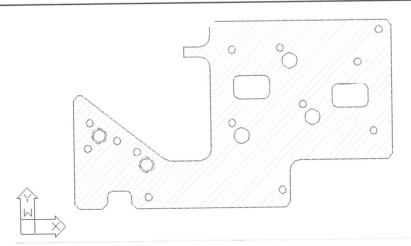

As an alternative, use the Region command to create a Boolean object from the boundary polyline of the shape. Then use the Subtract command to remove any islands composed of circles or closed polyline boundaries. The resulting region is an object, which can be easily checked with the Area command's Object option.

Recording Area Data in a Drawing File

Once you find the area of an object, you'll often need to record it somewhere. You can write it down in a project log book, but this is easy to overlook. A more dependable way to store area information is to use *attributes*.

For example, you can create a block that contains attributes for the area, and the date when the area was last taken. You might make the area and date attributes invisible. This block could then be inserted into every object you wish to measure. Once you find the area, you can easily add it to your block attribute with the Ddatte command. In fact, such a block could be used with any drawing in which you wished to store area data. See Chapter 17 for more on attributes.

Appendix C describes an AutoLISP program on your companion CD-ROM that lets you easily record area data in a drawing file.

Determining the Drawing's Status

When you work with a group of people on a large project, keeping track of a drawing's setup becomes important. The Status command enables you to obtain some general information about the drawing you are working on, such as the base point, current mode settings, and workspace or computer memory use. Status is especially helpful when you are editing a drawing someone else has worked on, because you may want to identify and change settings for your own style of working. When you select Tools ➢ Inquiry ➢ Status, you get a list like the one shown in Figure 18.5.

GETTING INFORMATION ABOUT A DRAWING

NOTE

If you have problems editing a file created by someone else, the difficulty can often be attributed to a new or different setting you are not used to working with. If you find that AutoCAD is acting in an unusual way, use the Status command to get a quick glimpse of the file settings before you start calling for help.

FIGURE 18.5

The Status screen of the AutoCAD Text Window

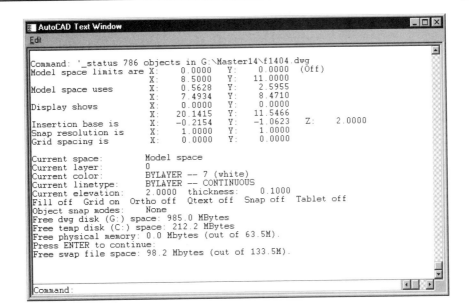

Here is a brief description of each item on the status screen. Note that some of the items you see listed on your screen will vary somewhat from what we've shown here, but the information applies to virtually all situations, except where noted.

(Number) objects in G:\Master14\f1404.dwg: the number of entities or objects in the drawing

Model space limits are: the coordinates of the Model Space limits (see Chapter 3 for more details on limits)

Model space uses: the area the drawing occupies; equivalent to the extents of the drawing

**Over: if present, means that part of drawing is outside the limit boundary

Display shows: the area covered by the current view

Insertion base is, Snap resolution is, and Grid spacing is: the current default values for these mode settings

Current space: Model Space or Paper Space

Current layer: the current default layer

Current color: the color assigned to new objects

Current linetype: the linetype assigned to new objects

Current elevation/thickness: the current default Z coordinate for new objects, plus the default thickness of objects; these are both 3D-related settings (see Chapter 11 for details)

Fill, Grid, Ortho, Qtext, Snap, and Tablet: the status of these options

Object snap modes: the current default Osnap setting

Free dwg disk (drive:): the amount of space available to store drawing-specific temporary files

Free temp disk (drive:): the amount of space you have left on your hard drive for AutoCAD's resource temporary files

Free physical memory: the amount of available RAM

Free swap file space: the amount of Windows swap file space available

NOTE

When you are in Paper Space, the Status command displays information regarding the Paper Space limits. See Chapter 9 for more on Model Space and Paper Space.

In addition to being useful in understanding a drawing file, Status is an invaluable tool for troubleshooting. Frequently, problems can be isolated by a technical support person using the information provided by the Status command.

NOTE For more information on memory use, see Appendix A.

Keeping Track of Time

The Time command allows you to keep track of the time spent on a drawing, for billing or analysis purposes. You can also use Time to check the current time and find out when the drawing was created and most recently edited. Because the AutoCAD timer uses your computer's time, be sure the computer's clock is set correctly.

To access the Time command, enter **time**↵ at the command prompt, or select Tools ➢ Inquiry ➢ Time. You get a message like the one in Figure 18.6.

FIGURE 18.6

The Time screen in the AutoCAD Text Window

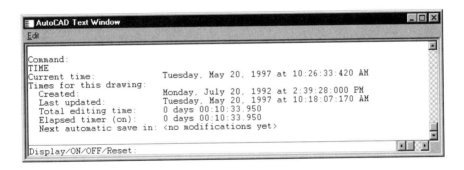

The first three lines tell you the current date and time, the date and time the drawing was created, and the last time the drawing was saved or ended.

The fourth line shows the total time spent on the drawing from the point that the file was opened. This elapsed timer lets you time a particular activity, such as changing the dimension of all the screws in a drawing or redesigning a piece of machinery. You can turn the elapsed timer on or off, or reset it, by entering **on, off,** or **reset** at the prompt shown. The last line tells you when the next automatic save will be.

Getting Information from System Variables

System variables store information about commands, drawing status, and defaults. You can check the status or change the setting of any system variable while you are in the middle of another command. To do this, you simply type an apostrophe ('), followed by the name of the system variable, at the command prompt.

For example, if you have started to draw a line, and you suddenly decide you need to rotate your cursor 45°, you can do the following:

1. At the To point: prompt, enter '**snapang**↵.
2. At the New value for Snapang: prompt, enter a new cursor angle. Once you have entered an angle value, you are returned to the Line command with the cursor in its new orientation.

You can also recall information such as the last area or distance calculated by Auto-CAD. Because the Area system variable duplicates the name of the Area command, you need to choose Tools ➣ Inquiry ➣ Set Variables, and then type **area**↵ to read the last area calculation. You can also type '**setvar**↵**area**↵. The Tools ➣ Inquiry ➣ Set Variables option also lets you list all the system variables and their status, as well as access each system variable individually by entering a question mark (?).

Many of the system variables give you direct access to detailed information about your drawing. They also let you fine-tune your drawing and editing activities. In Appendix D you'll find all the information you need to familiarize yourself with the system variables available. Don't feel that you have to memorize them all at once; just be aware that they are available.

NOTE Many of the dialog box options you have been using throughout this book, such as the options found in the Preferences dialog box, are actually system variable settings.

Keeping a Log of Your Activity

At times you may find it helpful to keep a log of your activity in an AutoCAD session. A log is a text file containing a record of your activities in AutoCAD. It may also contain notes to yourself or others about how a drawing is set up. Such a log can help you determine how frequently you use a particular command, or it can help you construct a macro for a commonly used sequence of commands.

The following exercise demonstrates how you can save and view a detailed record of an AutoCAD session using the Log feature.

1. Click on Tools ➢ Preferences; then at the Preferences dialog box, click on the tab labeled General at the top of the dialog box. A new set of options appears.

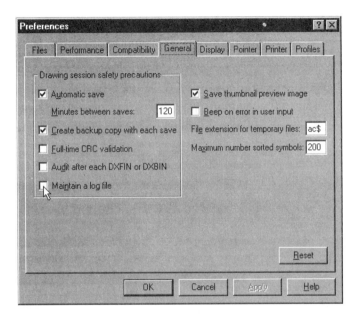

As a shortcut, you can quickly turn the Log File feature on and off by typing **logfileon**↵ and **logfileoff**↵ at the command prompt in AutoCAD.

2. In the Drawing Session Safety Precautions group, click on the check box labeled Maintain a Log File and then click on OK. You must turn off the Log File feature before you can actually view this file in AutoCAD. To do this, Click on Tools ➢ Preferences.

3. Return to the General tab of the Preferences dialog box, and then click on the Maintain a Log File check box again remove the check mark.

4. Click on OK to exit the dialog box.

5. Switch over to the Program Manager and start the Notepad application, or any text editor.

6. With the text editor, open the file called Acad.log in the \Program Files\ Acadr14 directory. This is the file that stores the text data from the command prompt whenever the Log File option is turned on.

Since Acad.log is a standard text file, you can easily send it to other members of your workgroup or print it out for a permanent record.

Easy Access to the Acad.log File

If you want to have easy access to the Acad.log file, place a shortcut to it in your AutoCAD Release 14 program group. Here's how it's done:

1. Open the Windows Explorer (Start ➢ Explorer) and locate and highlight the Acad.log file.

2. Right-click on the Acad.log file.

3. At the pop-up menu, select Create Shortcut. A new file named Shortcut to acad.log is created.

4. Shift+click and drag this Shortcut file to the Desktop. You can Shift+click on a file to move the file instead of making a copy. (Note: Early versions of Windows 95 did not allow you to make a shortcut anywhere except the Desktop. This step may not be necessary.)

5. Right-click on the Start button.

6. Click on the Open option in the pop-up menu. The Start Menu window appears.

7. Double-click on the Programs icon in the Start Menu window. The Programs Window appears.

8. Adjust your view of the Programs window so you can see the AutoCAD Release 14 program group icon.

9. Click and drag the Shortcut to Acad.log file into the AutoCAD Release 14 program group icon.

Once this is done, you can open the Acad.log file by choosing Start ➢ Programs ➢ AutoCAD R14 ➢ Shortcut to Acad.LOG from the Windows 95 or NT 4 Start menu. Note that you cannot access this file while AutoCAD is open and the Log File feature is turned on.

Capturing and Saving Text Data from the AutoCAD Text Window

If you are working in groups, it is often quite helpful to have a record of the status, editing time, and system variables for particular files, readily available to other group members. It is also convenient to keep records of block and layer information, so you can see if a specific block is included in a drawing or what layers are normally on or off.

You can use the Windows Clipboard to capture and save such data from the AutoCAD Text window. The following steps show you how it's done.

1. Move the arrow cursor to the command prompt at the bottom of the AutoCAD window.
2. Right-click on the mouse. A pop-up menu appears.
3. Click on Copy History. This copies the entire contents of the Text window to the Clipboard.

By default, the Text window stores 400 lines of text. You can change this number by changing the options in the Text Window group in the Display tab of the Preferences dialog box.

If you only want to copy a portion of the Text window to the Clipboard, perform the following steps.

1. Press the F2 function key to open the Text window.
2. Using the I-beam text cursor, highlight the text you wish to copy to the Clipboard.
3. Right-click on your mouse, and then click on Copy from the pop-up menu. You can also choose Edit ≻ Copy from the Text window's menu bar. The highlighted text is copied to the clipboard.
4. Open a Notepad or other word processor file and paste the information.

You may notice three other options on the pop-up menu: Paste to CmdLine, Paste, and Preferences. Paste to CmdLine offers a way to capture data, such as layer, linetype, block names, or even commands, and paste them into the command prompt window. The Paste option will paste the first line of the contents of the Clipboard into the command line or input box of a dialog box. This can be useful for entering repetitive text or for storing and retrieving a frequently used command. The Preferences menu item will open the Preferences dialog box.

> **TIP** Items copied to the Clipboard from the AutoCAD Text window can be pasted into dialog box input boxes. This can be a quick way to transfer layer, linetype, or other named items from the Text window into dialog boxes.

Recovering Corrupted Files

No system is perfect. Eventually, you will encounter a file that is corrupted in some way. AutoCAD offers two tools that can sometimes salvage a corrupted file: Audit and Recover. *Audit* allows you to check a file that you are able to open, but you suspect has some problem. *Recover* allows you to open a file that is so badly corrupted that AutoCAD is unable to open it in a normal way. You can access these tools from the File ➣ Drawing Utilities on the menu bar. The functions of these options are described here.

Audit checks the current file for any errors and displays the results in the Text window.

Recover attempts to recover damaged or corrupted AutoCAD files. The current file is closed in the process as Recover attempts to open the file to be recovered.

More often than not, these tools will do the job, although they aren't a panacea for all file corruption problems. In the event that you cannot recover a file, even with these tools, make sure your computer is running smoothly and that other systems are not faulty.

Exchanging CAD Data with Other Programs

AutoCAD offers many ways to share data with other programs. Perhaps the most common type of data exchange is to simply share drawing data with other CAD programs. In this section, you'll look at how you can export and import CAD drawings using the DXF file format. You'll also look at how you can share bitmap graphics, both to and from AutoCAD, through the Windows Clipboard.

Other types of data exchange involve text, spreadsheets, and databases. We cover database links in Chapter 17, but here we'll look at how you can include text, spreadsheet, and database files in a drawing or include AutoCAD drawings in other program files using a Windows feature called Object Linking and Embedding.

Using the DXF File Format

A DXF file is a DOS text file containing all the information needed to reconstruct a drawing. It is often used to exchange drawings created with other programs. Many micro-CAD programs, including some 3D perspective programs, can generate or read files in DXF format. You may want to use a 3D program to view your drawing in a perspective view, or you may have a consultant who uses a different CAD program that accepts DXF files. There are many 3D rendering programs that read DXF files. Most 2D drafting programs also read and write DXF files.

You should be aware that not all programs that read DXF files will accept all the data stored therein. Many programs that claim to read DXF files will "throw away" much of the DXF file's information. Attributes are perhaps the most commonly ignored objects, followed by many of the 3D objects, such as meshes and 3D faces. But DXF files, though not the perfect medium for translating data, have become something of a standard.

> The IGES (Initial Graphics Exchange Specification) standard for CAD data translation is only supported in the Mechanical Desktop add-on, not AutoCAD.

Exporting DXF Files

To export your current drawing as a DXF file, try the following steps.

  1. Choose File ➢ Export. The Export Data dialog box appears.

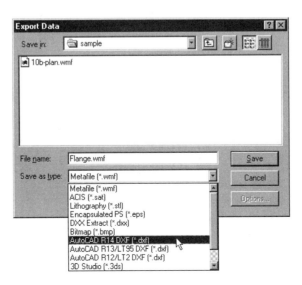

You can also click on File ➢ Export, and then enter the name of your export file, including the .dxf extension. AutoCAD will skip the prompt in step 3 and proceed to create the DXF file.

2. Click on the Save as Type drop-down list. You then have the option to export your drawing under a number of formats, including three different DXF formats.

3. Select the appropriate DXF format, and then enter a name for your file. You do not have to include the .dxf file name extension.

4. Select a Folder for the file, and then click on Save.

In step 2, you can select from the following DXF file formats:

- AutoCAD R14 DXF
- AutoCAD R13/LT 95 DXF
- AutoCAD R12/LT 2 DXF

Choose the format appropriate to the program you are exporting to. In most cases, the safest choice is AutoCAD R12/LT 2 DXF if you are exporting to another CAD program, although AutoCAD will not maintain the complete functionality of Release 14 for such files.

Once you've selected a DXF format from the Save as Type drop-down list, you can set more detailed specifications by clicking on the Options button in the Export Data dialog box. Doing so opens the Export Options dialog box.

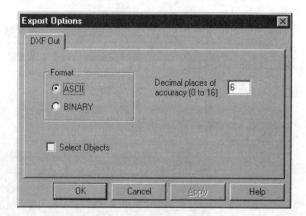

Here you have the following options:

Format lets you choose between ASCII (plain text) or binary file formats. Most other programs will accept ASCII, so it is the safest choice. Some programs will accept binary DXF files, which have the advantage of being more compact than the ASCII format.

Select Objects lets you select specific objects within the drawing for export. You are given the option to select objects after you have closed the Export Options dialog box and have selected Save from the Export Data dialog box.

Decimal places of accuracy allows you to determine the accuracy of the exported file. Keeping this value low will help reduce the size of the export file, particularly if it is to be in ASCII format. Some CAD programs do not support the high accuracy of AutoCAD, so using a high value here may have no significance.

You can also type **dxfout**↵ at the command prompt to open the Create DXF File dialog box. This is a standard Windows file dialog box that includes the button options described here. This dialog box displays only DXF file formats in the Save as Type drop-down list.

Opening or Importing DXF Files

Some companies have standardized their CAD drawings on the DXF file format. This is most commonly seen in companies that use a variety of CAD software besides Auto-CAD. Release 14 lets you open DXF files directly, as if they were DWG files. Here's how it's done.

1. Choose File ➢ Open.
2. At the Select File dialog box, choose .DXF from the Files of Type drop-down list.
3. Locate and select the DXF file you wish to open.

You can also import DXF files into a new open file.

If you attempt to Dxfin a *.dxf file into an existing drawing, you get the following error message: `AutoCAD cannot DXFIN this file. You can DXFIN the file into a new drawing, save the drawing, and use INSERT* to include the entities from that drawing into the current drawing.`

1. Type **dxfin**↵ at the command prompt. The Select DXF File dialog box appears. This is a typical Windows-style file dialog box.
2. Locate and select the DXF file you wish to import.
3. Double-click on the file name to begin importing it.

If the import drawing is large, AutoCAD may take several minutes.

Exchanging Files with Earlier Releases

One persistent dilemma that has plagued AutoCAD users is how to exchange files between earlier versions of the program. In the past, if you upgraded your AutoCAD, you would find that you were locked out from exchanging your drawings with people using earlier versions. Release 12 alleviated this difficulty by making Release 12 files compatible with Release 11 files.

With Release 13, we had a file structure that was radically different from earlier versions of AutoCAD. Fortunately, you can still exchange files with earlier versions of the program, using the Save as Type drop-down list in the Save Drawing As dialog box.

You don't get something for nothing, however. While you can make a "round trip" from Release 14 to Release 13 and back without losing any drawing data, there are certain things you will lose when you save a Release 14 file in Release 12 format. Bear these considerations in mind when you use Save As to save a file to Release 12:

- Splines become polyline splines.
- 3D solids become polylines representing the wire frame of the solid.
- Multilines become polylines.
- Linetypes with embedded shapes are separated into lines and shapes.
- Dimension styles are not completely translated.
- Linetypes are not completely translated.
- TrueType fonts are not supported in Release 12 or earlier versions.

When you save a drawing as a Release 12 file, you will receive a message telling you how the file is being modified to accommodate the limited Release 12 format. You may want to use the Logfileon command to store this conversion message for future reference.

Using AutoCAD Drawings in Desktop Publishing

As you probably know, AutoCAD is a natural for creating line art, and because of its popularity, most desktop publishing programs are designed to import AutoCAD drawings in one form or another. Those of you who employ desktop publishing software to generate user manuals or other technical documents will probably want to be able to use AutoCAD drawings in your work. In this section, we will examine ways to output AutoCAD drawings to formats that most desktop publishing programs can accept.

There are two methods for exporting AutoCAD files to desktop publishing formats: *raster export* and *vector file export*.

Exporting Raster Files

In some cases, you may only need a rough image of your AutoCAD drawing. You can export your drawing as a raster file that can be read in virtually any desktop publishing or word processing program. Here are the steps for accessing this feature:

1. Click on Tools ➢ Preferences to open the Preferences dialog box.
2. Click on the Printer tab. You will see a list of printer configuration options.

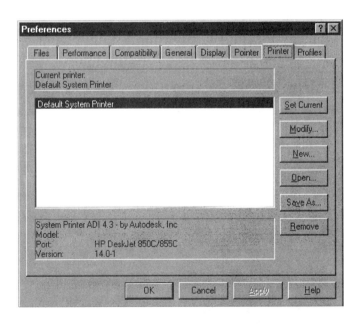

3. Click on New to see a listing of printer options in the Add a Printer dialog box.

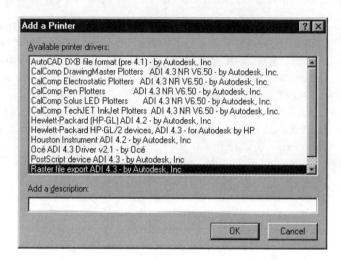

4. Click on the option that reads Raster file export ADI 4.3 - by Autodesk, Inc. You may also add a description for this selection in the Add a Description input box.

5. Click on OK. The AutoCAD Text window appears with a display of raster size options.

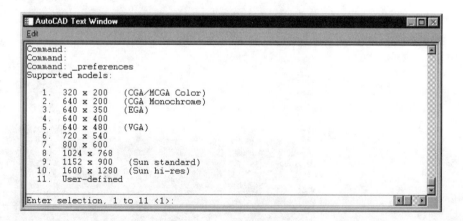

6. Select a size you want to work with by typing its number and pressing ↵. If you are uncertain about which size you want, you can select any size at first, and then come back and add another version of the Raster File Export option for a different raster image size.

7. Once you've selected a size, you are shown a list of file formats:

    ```
    1.   Microsoft Windows Device-independent Bitmap (.BMP)
    2.   TrueVision TGA Format
    3.   Z-Soft PCX Format
    4.   TIFF (Tag Image File Format)
    ```

8. Select a format by entering its number followed by ↵. You are then presented with a list of color options. The following list is offered when you select option 4, the TIFF format:

    ```
    1.   Monochrome
    2.   Color -   16 colors
    3.   Color - 256 colors
    ```

9. Select the color option you desire. If you select an option other than monochrome, you then get the following message:

    ```
    You can specify the background color to be any of AutoCAD's 256 standard
    colors. The default of 0 selects a black screen background.
    Enter selection, 0 to 255 <0>:
    ```

10. As the message tells you, you can determine the background color of your raster images. The default 0 gives you a black background. Enter **7** for a white background, or enter another color number. Once you've entered a color, you'll see the following message:

    ```
    Sizes are in Inches and the style is landscape
    Plot origin is at (0.00,0.00)
    Plotting area is 640.00 wide by 480.00 high (MAX size)
    Plot is NOT rotated
    Hidden lines will NOT be removed
    Plot will be scaled to fit available area

    Do you want to change anything? (No/Yes/File) <N>:
    ```

NOTE

The color numbers are based on the standard AutoCAD color index that you see when you select colors for objects or layers.

11. The options presented here are the same as those presented in a more understandable way in the Print/Plot Configuration dialog box. You have the opportunity to change them when you are ready to create your raster file, so press ↵ to accept the default, No. You will return to the Preferences dialog box.

12. Click on OK. You now have a printer option that will generate a raster file of your drawing (rather than print or plot your drawing).

When you are ready to create a raster file, proceed as if you were to plot your file. Then at the Print/Plot Configuration dialog box, select the raster output option from the Device and Default Selection dialog box (see Chapter 14 for details on using the Print/Plot Configuration dialog box).

The range of raster output options is quite comprehensive. Chances are, if you need raster output from AutoCAD, at least one of the available options should fill your needs.

Exporting Vector Files

If you need to preserve the accuracy of your drawing, or if you wish to take advantage of TrueType or PostScript fonts, you really have no choice but to use either the DXF or PostScript vector formats.

The DXF format is the easiest to use for vector format files, and with TrueType support, DXF can preserve font information between AutoCAD and desktop publishing programs that support the DXF format. PostScript is a raster/vector hybrid file format that AutoCAD supports; unfortunately, Release 14 has dropped direct PostScript font support. However, you can still use substitute fonts to stand in for PostScript fonts. These substitute fonts will be converted to true PostScript fonts when AutoCAD exports the drawing. You won't actually see the true results of your PostScript output until you print your drawing out on a PostScript printer.

We've already covered DXF file export in *Using the DXF File Format*, so we'll concentrate on PostScript in this section.

If you are a circuit board designer or drafter, you may want to use the PostScript Out option to output your layout to a PostScript typesetting devices. This will save time and file size since this option converts AutoCAD entities into true PostScript descriptions.

PostScript Output

AutoCAD is capable of exporting to the Encapsulated PostScript file format (.EPS). You actually have two ways of obtaining PostScript output. You can use the File ➢ Export option from the menu bar, or you can install a PostScript printer driver and plot your drawing to an EPS file. In the *If You Want to Experiment* section of Chapter 16, we described the steps for using File ➢ Export to export EPS files. To set up AutoCAD to plot your drawing to an EPS file, follow the same steps described in the previous *Exporting Raster Files* section, but in step 4, select the option that reads `PostScript device ADI 4.3 - by Autodesk, Inc` instead of the raster file driver.

AutoCAD does not preserve font information when creating EPS files from the printer option.

PostScript Font Substitution

We mentioned earlier that AutoCAD will replace its own fonts with PostScript fonts when a file is exported to an EPS file using the Export Data dialog box. If your work involves PostScript output, you will want to know these font names in order to make the appropriate substitution. Table 18.1 shows a listing of AutoCAD font names and their equivalent PostScript names.

To take advantage of AutoCAD's ability to translate fonts, you need to create AutoCAD fonts that have the names listed in the first column of Table 18.1. You then need to use those fonts when creating text styles in AutoCAD. AutoCAD will convert the AutoCAD fonts into the corresponding PostScript fonts.

Creating the AutoCAD fonts can be simply a matter of copying and renaming existing fonts to those listed in Table 18.1. For example, you could make a copy of the `Romans.shx` font and name it `Agd.shx`. Better yet, if you have the `PostScript .pfb` file of the font, you can compile it into an AutoCAD font file and re-name the compiled file appropriately. By compiling the PFB file, you will get a close approximation of its appearance in AutoCAD. See the *Using PostScript Fonts* sidebar in Chapter 7 for a description on how to compile PostScript fonts.

TABLE 18.1: A LISTING OF AUTOCAD FONT FILE NAMES AND THEIR CORRESPONDING POSTSCRIPT FONTS

AutoCAD Font Name	PostScript Font Name	AutoCAD Font Name	PostScript Font Name
agd	AvantGarde-Demi	agdo	AvantGarde-DemiOblique
agw	AvantGarde-Book	agwo	AvantGarde-BookOblique
bdps	Bodoni-Poster	bkd	Bookman-Demi
bkdi	Bookman-DemiItalic	bkl	Bookman-Light
bkli	Bookman-LightItalic	c	Cottonwood
cibt	CityBlueprint	cob	Courier-Bold
cobo	Courier-BoldOblique	cobt	CountryBlueprint
com	Courier	coo	Courier-Oblique
eur	EuroRoman	euro	EuroRoman-Oblique
fs	FreestyleScript	ho	Hobo
hv	Helvetica	hvb	Helvetica-Bold
hvbo	Helvetica-BoldOblique	hvn	Helvetica-Narrow
hvnb	Helvetica-Narrow-Bold	hvnbo	Helvetica-Narrow-BoldOblique
hvno	Helvetica-Narrow-Oblique	hvo	Helvetica-Oblique
lx	Linotext	ncb	NewCenturySchlbk-Bold
ncbi	NewCenturySchlbk-BoldItalic	nci	NewCenturySchlbk-Italic
ncr	NewCenturySchlbk-Roman	par	PanRoman
pob	Palatino-Bold	pobi	Palatino-BoldItalic
poi	Palatino-Italic	por	Palatino-Roman
rom	Romantic	romb	Romantic-Bold
romi	Romantic-Italic	sas	SansSerif
sasb	SansSerif-Bold	sasbo	SansSerif-BoldOblique
saso	SansSerif-Oblique	suf	SuperFrench
sy	Symbol	te	Technic
teb	Technic-Bold	tel	Technic-Light
tib	Times-Bold	tibi	Times-BoldItalic
tii	Times-Italic	tir	Times-Roman
tjrg	Trajan-Regular	vrb	VAGRounded-Bold
zcmi	ZapfChancery-MediumItalic	zd	ZapfDingbats

If you are using PostScript fonts not listed in Table 18.1, you can add your own AutoCAD-to-PostScript substitution by editing the Acad.psf file. This is a plain text file that contains the font substitution information as well as other PostScript translation data. See Appendix A for more information on this file.

HPGL plot file format is another vector format you can use to export your AutoCAD drawings. Use the method described earlier in *Exporting Raster Files* to add the HPGL plotter driver to your printer/plotter configuration.

Combining Data from Different Sources

Imagine being able to import spreadsheet data into an AutoCAD drawing and display it there. Further imagine that you could easily update that spreadsheet data, either from directly within the drawing or remotely by editing the source spreadsheet document. With a little help from a Windows feature called *Object Linking and Embedding* (OLE), such a scenario is within your grasp. The data is not limited to spreadsheets; it could be a word-processed document, a database report, or even a sound or video clip.

To import data from other applications, you use the Cut and Paste feature found in virtually all Windows programs; cut the data from the source document, then paste it into AutoCAD.

When you paste data into your AutoCAD file, you have the option to *link* the data to the source file or to *embed* the data. If you link data to the source file, then the pasted data will be updated whenever the source file is modified. This is similar to an AutoCAD cross-referenced file (see Chapter 9 for more on cross-referenced files).

You can also paste data into AutoCAD without linking it; such data is considered an embedded object. You can still open the application associated with the data by double-clicking on it, but the data is no longer associated with the source file. This is similar to a drawing inserted as a block where changes in the source drawing file have no effect on the inserted block.

Let's see firsthand how OLE works. The following exercise shows how to link an Excel spreadsheet to AutoCAD. You will need a copy of Excel for Windows 95/NT, but if you have another application that supports OLE, you can follow along.

840 CHAPTER 18 • GETTING AND EXCHANGING DATA FROM DRAWINGS

1. Open the file called 18assembly.dwg from the companion CD-ROM. This is a copy of the drawing you worked with in Chapter 17.

2. Open the Excel spreadsheet called 18assembly.xls, also from the companion CD-ROM.

3. In Excel, highlight the data, as shown in Figure 18.7, by clicking on cell A1 and dragging to cell G6.

FIGURE 18.7

The Excel spreadsheet

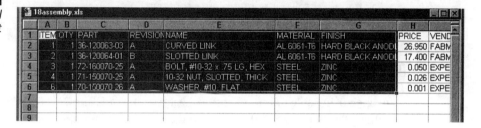

4. Choose Edit ➢ Copy. This places a copy of the selected data into the Windows Clipboard.

5. Switch to AutoCAD, either by clicking on a visible portion of the AutoCAD window, or by clicking on the AutoCAD button in the Taskbar at the bottom of the Windows Desktop.

6. Choose Edit ➢ Paste Special. The Paste Special dialog box appears.

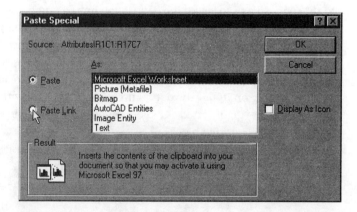

7. Click on the Paste Link radio button to tell AutoCAD you want this paste to be a link. Notice that the list of source types changes to show only one option: Microsoft Excel Worksheet.

8. Click on OK. The spreadsheet data appears in the drawing (see Figure 18.8).

FIGURE 18.8

The AutoCAD drawing with the spreadsheet pasted

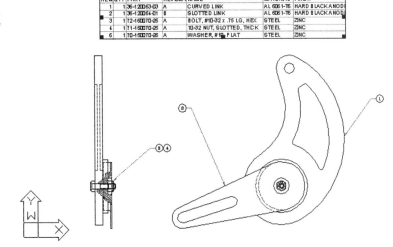

9. Place the cursor on the upper-left corner of the spreadsheet so that a double-headed diagonal arrow appears, and then click and drag the corner downward and to the right to make the spreadsheet the size shown in Figure 18.9.

10. Place the cursor over the spreadsheet data so that the cursor changes to a cross. The spreadsheet no longer has spreadsheet qualities, but is now an object. All the grips along the outside perimeter are for sizing. Click anywhere inside the spreadsheet and you can drag it to another location.

11. Zoom into the spreadsheet so you can read its contents clearly.

FIGURE 18.9
Resizing the spreadsheet within AutoCAD

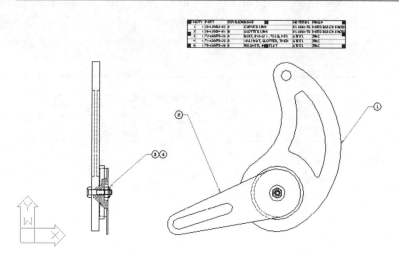

As you saw in steps 8 and 9, you can resize a pasted object using the corner or side grips. The corner grips will maintain the original proportion of the inserted object.

You now have a linked object inserted into the AutoCAD drawing. You can save this file and send it off to someone else, along with the pasted document `18assembly.xls`, and the other person will be able to open the AutoCAD file and view the drawing with the spreadsheet.

WARNING
Objects that are pasted into AutoCAD are maintained within AutoCAD until you use Erase to delete them. They act like other AutoCAD objects where layers are concerned. One limitation of pasted objects is that they will not appear in prints or plots unless you use the Windows System printer or plotter.

Now let's see how the Link feature works by making some changes to the spreadsheet data.

1. Go back to Excel by clicking on the Excel button in the Windows toolbar.

2. Click on the cell just below the column heading REVISION.

3. Change the cell's contents by typing **c**↵.
4. Go back to AutoCAD and notice that the corresponding cell in the inserted spreadsheet will change, after a moment or two, to reflect the change you made to the original document. Since you inserted the spreadsheet as a linked document, OLE updates the pasted copy whenever the original source document changes.
5. Now close both the Excel spreadsheet and the AutoCAD drawing.

Editing Links

Once you've pasted an object with links, you can control the link by selecting Edit ➢ OLE Links (Olelinks). If there are no linked objects in the drawing, this option does nothing; otherwise it opens the Links dialog box.

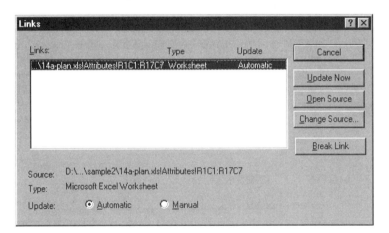

The following list describes the options available in this dialog box:

Cancel does just that. It cancels the link between a pasted object and its source file. Once this option is used, changes in the source file have no effect on the pasted object. This is similar to using the Bind option in the Xref command.

Update Now updates an object's link when the Manual option is selected.

Open Source opens the application associated with the object and lets you edit it.

Change Source lets you change the object's link to different file. When you select this option, AutoCAD opens the Change Link dialog box, which lets you select another file of the same type. For example, if you are editing the link to a sound file, the Change Link dialog box will display files with the .wav file extension.

Automatic and **Manual** radio buttons control whether linked objects are updated automatically or manually.

Break Link disconnects the link between the inserted data and the source document. The inserted data then becomes embedded, rather than linked.

Adding Sound, Motion, and Photos to Your Drawings

You've already seen how you can include scanned images in AutoCAD drawings through the Raster Image tools. Through Object Linking and Embedding, you can also include sound files, video clips, and animation. Imagine how you might be able to enhance your AutoCAD files with these types of data. You can include voice annotation or, if the file is to go to a client, an animated walk-through of your building or mechanical design. The potential for this feature is enormous.

Options for Embedding Data

If you don't need to link the imported data to its source, the Paste Special dialog box lets you convert the imported data to a number of other formats. Here is a brief description of each available format.

Picture (Metafile) imports the data as vector or bitmap graphics, whichever is appropriate. If applicable, text is also maintained as text, though not editable within AutoCAD.

Bitmap imports the data as a bitmap image, closely reflecting the appearance of the data as it appears on your computer screen in the source application.

AutoCAD Entities converts the data into AutoCAD objects such as lines, arcs, and circles. Text is converted into AutoCAD single-line text objects.

Image Entity converts the data into an AutoCAD raster image. You can then edit it using the raster image-related tools found by selecting Modify ➢ Object ➢ Image from the menu bar. See Chapter 16 for more on how to use raster images.

Text converts the data into AutoCAD multiline text objects.

The options you see in the Paste Special dialog box will depend on the type of data being imported. You saw how the Microsoft Excel Worksheet option maintains the imported data as a Microsoft worksheet. If the contents of the Clipboard come from another program, you will be offered that program as a choice in place of Excel.

NOTE

The Edit ➣ Paste option will embed OLE data objects into AutoCAD, as will the Paste From Clipboard tool in the Standard toolbar.

Using the Clipboard to Export AutoCAD Drawings

Just as you can cut and paste data into AutoCAD from applications that support OLE, you can also cut and paste AutoCAD images into other applications. This can be useful as a way of including AutoCAD images into word-processed documents, spreadsheets, or desktop-publishing documents. It can also be useful in creating background images for visualization programs such as 3D Studio, or paint programs such as Fractal Painter.

NOTE

If you cut and paste an AutoCAD drawing to another file using OLE, then send the file to someone using another computer, the recipient must also have AutoCAD installed before the pasted AutoCAD drawing can be edited.

The receiving application does not need to support OLE, but if it does, then the exported drawing can be edited within AutoCAD and will maintain its accuracy as a CAD drawing. Otherwise, the AutoCAD image will be converted to a bitmap graphic.

To use the Clipboard to export an object or set of objects from an AutoCAD drawing, use the Edit ➣ Copy option. You are then prompted to select the objects you want to export. If you want to simultaneously export and erase objects from AutoCAD, choose Edit ➣ Cut.

If you want the AutoCAD image to be linked to AutoCAD, use Edit ➣ Copy Link. You won't be prompted to select objects. The current visible portion of your drawing will be exported. If you want the entire drawing to be exported, use View ➣ Zoom ➣ Extents before using the Copy Link option. Otherwise, set up AutoCAD to display the portion of your drawing you want exported before using Copy Link.

In the receiving application, choose Edit ➣ Paste ➣ Special. You'll see a dialog box similar to AutoCAD's Paste Special dialog box. Select the method for pasting your AutoCAD image, and then click on OK. If the receiving application does not have a

Paste Special option, choose Edit ➢ Paste. The receiving application will convert the image into a format it can accept.

You can copy multiple viewport views from Paper Space to the Clipboard using the Edit ➢ Copy Link option.

If You Want to Experiment...

With a little help from a Visual Basic macro and OLE, you can have Excel extract attribute data from a drawing, and then display that data in a spreadsheet imported into AutoCAD. To demonstrate, the following exercise uses a Visual Basic macro embedded in the 18assembly.xls file you used in an earlier exercise.

1. Open the 18assembly.dwg file in AutoCAD again.
2. Open the 18assembly.xls file in Excel.
3. Repeat the exercise in the *Combining Data from Different Sources* section of this chapter, but stop before exiting the two files.
4. In AutoCAD, choose Modify ➢ Object ➢ Attribute ➢ Single, and then click on the balloon block for the slotted link.
5. Change the NOTES Attribute value to –005 ONLY, and then click on OK.
6. Go to Excel, and then choose Tools ➢ Macro ➢ Macros.
7. At the Macros dialog box, highlight the Extract macro, and then click on Run. Excel will take a moment to extract the attribute data from the open file; then it will display the data in the spreadsheet.
8. Return to AutoCAD and check the NOTES in the imported spreadsheet. The imported spreadsheet reflects the change you made in the attribute in step 5.
9. Close both files.

The macro you used in the Excel file is a small sample of what can be done using AutoCAD's implementation of Visual Basic Automation. You'll learn more about VBA in Chapter 20.

In this chapter, you have seen how AutoCAD allows you to access information ranging from the areas of objects to information from other programs. You may never use some of these features, but knowing they are there may at some point help you to solve a production problem.

Chapter 19

Introduction to Customization

FEATURING

Enhancements Straight from the Source

Utilities Available from Other Sources

Putting AutoLISP to Work

Loading AutoLISP Programs Automatically

Creating Keyboard Macros with AutoLISP

Using Third-Party Software

Getting the Latest Information from Online Services

Posting and Accessing Drawings on the Web

Introduction to Customization

AutoCAD offers a wealth of features that you can use to improve your productivity. But even with these aids to efficiency, there are always situations that can use some further automation. In this chapter, you'll be introduced to the different ways AutoCAD can be customized and enhanced with add-on utilities.

First, you'll discover how the AutoCAD Bonus tools can help boost your productivity. Many of these utilities were created using the programming tools that are available to anyone, namely AutoLISP and VBA (Visual Basic for Applications).

Next, you'll learn how to load and run AutoLISP utilities that are supplied on this book's companion CD-ROM. By doing so, you'll be prepared to take advantage of the many utilities available from user groups and online services. Finally, you'll finish the chapter by taking a look at how third-party applications and the Internet can enhance AutoCAD's role in your workplace.

Enhancements Straight from the Source

If you've followed the tutorials in this book, you've already used a few add-on programs that come with AutoCAD, perhaps without even being aware that they were not part of the core AutoCAD program. In this section we'll introduce you to the AutoCAD Bonus tools: a set of AutoLISP, ARX, and VBA tools that showcase these powerful customization environments. The best part about the Bonus tools is that you don't have to know a thing about programming to take advantage of them

There are so many of these Bonus tools that we can't provide step-by-step instructions for all of them. Instead, we will offer a detailed look at some of the more complicated tools and provide shorter descriptions for others. We'll start with the Bonus Layer tools.

Loading the Bonus Tools

If you installed AutoCAD using the Typical Installation option, you may not have installed the AutoCAD Bonus tools yet. Fortunately, you can install these utilities separately without having to reinstall the entire program.

Proceed as though you are installing AutoCAD for the first time. When you see the Setup Choices dialog box, choose the Add button to add new components to the current system. You will see an item called Bonus in the Custom Components dialog box that appears next. Place a check in the Bonus check box, and then proceed with the installation. When Setup is finished, open AutoCAD and load the Bonus menu. See Chapter 21 for more detailed information on loading menus.

Tools for Managing Layers

In a recent survey of AutoCAD users, Autodesk discovered that one of the most frequently used features in AutoCAD was the Layer command. As a result, the layer controls in Release 14 have been greatly improved. Still, there is room for some improvement. The Bonus Layer tools offer some shortcuts to controlling layer settings, as well as one major layer enhancement called the Layer Manager.

TIP
All of the Bonus tools discussed in this section have keyboard command equivalents. Check the status bar for the keyboard command name when selecting these tools from the toolbar or the menu bar.

Saving and Recalling Layer Settings

The Layer Manager lets you save layer settings. For example, you have been creating a complex assembly with many parts. You have been careful to name the layers so that they are easily managed, but it now takes some time to turn off all of the dimensions and hidden lines in each of the parts. The Layer Manager tool can assist you with this tedious task. You can turn layers on and off and then save the layer settings. Later, when you need to see or not to see the information on these layers, you can recall the layer setting to view the data. Here's how the Layer Manager works:

1. In AutoCAD, while you are working in a file, open the Layer & Linetype Properties dialog box and turn on all the layers except the Dimensions and Hidden Line layers.

2. Click on the Layer Manager tool in the Bonus Layer Tools toolbar.

The Layer Manager dialog box appears.

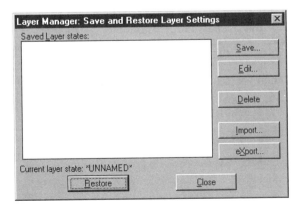

3. Click on the Save button. The Layer State Name dialog box appears.
4. Enter the name **no dims or hidden lines**, and then click on OK. You return to the Layer Manager dialog box. Notice that the name you entered for the layer state appears in the list box.
5. Click on the Close button.
6. Now open the Layer & Linetype Properties dialog box again, turn on the Dimensions and Hidden Lines layers, and turn off the Center layer.
7. Click on the Layer Manager tool again.
8. Click on NO DIMS OR HIDDEN LINES from the list, and then click on Restore.
9. Click on Close. Your drawing reverts to the previous view, with the Dimensions layer and Hidden Lines layer turned off.

The layer states are saved with the file so you can retrieve them at a later date. As you can see from the Layer Manager dialog box, you have a few other options. Here is a listing of those options and what they do:

Edit opens the Layer & Linetype Properties dialog box to let you edit the settings for a layer state. Highlight the layer state from the list, and then choose Edit.

Delete deletes a layer state from the list.

Import imports a set of layer states that have been exported using the Export option of this dialog box. It will create any layers not already in the drawing.

Export saves a set of layer states as a file. By default, the file is given the name of the current file with the .lay file name extension. You can import the layer state file into other files.

Changing the Layer Assignment of Objects

In addition to the Layer Manager, the Bonus Layer Tools toolbar offers two tools that change the layer assignments of objects. The Match Objects Layer tool is similar to the Match Properties tool, but is streamlined to just operate on layer assignments. After clicking on this tool, you first select the object or objects you wish to change, and then you select an object whose layer you wish to match.

The Change to Current Layer tool changes an object's layer assignment to the current layer. This tool has long existed as an AutoLISP utility and you'll find that you'll get a lot of use from it.

Controlling Layer Settings through Objects

The remaining set of Bonus Layer tools lets you make layer settings by selecting objects in the drawing. They are simple to use: Just click on the tool then select an object. The following list describes what each tool does.

Isolate Object Layer will turn off all the layers except for the layer of the selected object.

Freeze Object Layer will freeze the layer of the selected object.

Turn Object Layer Off will turn off the layer of the selected object.

Lock Object Layer will lock the layer of the selected object. A locked layer is one that is visible but cannot be edited.

Unlock Object Layer will unlock the layer of the selected object.

Tools for Editing Text

It seems that we can never have enough text-editing features. Even the realm of word processors contains seemingly innumerable tools for setting fonts, paragraphs, tabs, and tables. Some programs even check grammar. While we're not trying to write the Great American Novel in AutoCAD, we are interested in getting our text in the right location, at the right size, with some degree of style. This often means using a mixture of text and graphics editing tools.

Release 14 has brought us some great improvements in text handling through an improved text editor. Here are some additional tools that will help ease your way through some otherwise difficult editing tasks.

Masking Text Backgrounds

One problem AutoCAD users frequently face is how to get text to read clearly when it is placed over a hatch pattern or other graphic. The Hatch command will hatch around existing text, leaving a clear space behind it. But what about those situations where you must add text *after* a hatch pattern has been created? Or what about those instances where you need to mask behind text that is placed over a non-hatch object, such as dimension leaders or raster images?

The Text Mask tool addresses this problem by masking the area behind text with a special masking object called a Wipeout. Try the following exercise to see firsthand how it works.

1. Start a new drawing with the Versions template and save it as `raster.dwg`. Use the Tools menu and use the Raster Image tool. Insert the `raster1.tif` file from the companion CD-ROM. See Chapter 16 if you need to refresh your memory on this process. Make the image just about as large as the screen.

2. Adjust your view so that you see the entire photo. Use the Text tool from the Draw toolbar to place the words **Image is not Actual Size** over the press. Your drawing should look like the top image of Figure 19.1. Notice that the text is obscured by the parts of the photo.

3. Choose Text Mask from the Bonus Text Tools toolbar.

You'll see the following message:

```
Initializing...
Loading WIPEOUT for use with TEXTMASK...
Enter offset factor relative to text height <0.35>:
```

4. Here you can enter the amount of space you want around the text as a percentage of the text height. Press ↵ to accept the default.

5. At the `Select Text to MASK` prompt, use a window to select the image text. When you're done selecting text, press ↵. You'll see the message `Wipeout created` for each object selected, and the text will appear on a white background, as shown in the bottom image of Figure 19.1.

Text Mask creates an object called a Wipeout that masks other objects behind the text. Wipeout is not a standard geometry AutoCAD object; it is a new object created through AutoCAD's programming interface.

The Wipeout object has its own little quirks that you will want to know about. To get a bit more familiar with Wipeout objects, try the following exercise.

1. Click on the Image text. Notice that both the text and the Wipeout object are selected.

2. Click on Move on the Modify toolbar.

3. Move the text and Wipeout object to the right about 1/2 inch. The text seems to disappear.

4. Type **re**↵ to issue a Regen. The text appears once again. Wipeout color will always be same as the display screen and plotted sheet.

FIGURE 19.1

Creating a mask behind text

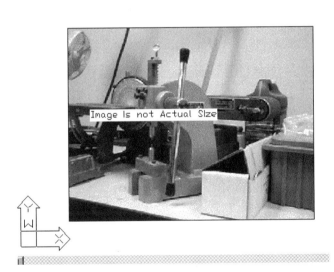

The text and Wipeout objects are linked so that if you select the text, you automatically select the Wipeout object. Also, the display order of the two objects gets mixed up when you move them, so you need to issue a Regen to restore the text's visibility. You can also edit or erase the Wipeout object. There is a description of how to edit Wipeout objects in the *Bonus Standard Tools* section later in this chapter.

Next, you'll look at ways to globally change text objects.

Making Global Changes to Text Objects

Technical drawings often contain repetitive text. Frequently, an AutoCAD user will create one text object and copy it several times, saving the time it takes to type the text in multiple times. This is a great method for adding text to a drawing, but it also tends to multiply mistakes. Here are two tools that will help you make changes to multiple sets of text objects with a minimum amount of pain.

Changing Multiple Text Items   The Change Multiple Text Items tool lets you change the height, justification, rotation angle, style, and width of a selected set of text. You can also change the text itself if you need to. It's fairly straightforward to use. Here's how.

1. Click on Change Multiple Text Items on the Bonus Text Tools toolbar.

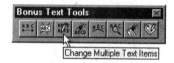

2. Select the text you want to edit, and then press ↵. You will then see the prompt:

   Height/Justification/Location/Rotation/Style/Text/Undo/Width:

3. Type in the letter of the option you want, and then proceed to answer the prompts that follow. For example, if you enter **h**↵ at the prompt, AutoCAD will ask you for a new text height.

4. After entering a new height value, the selected text will change to the new height and the Height/Justification/Location prompt will return, allowing you to make further changes.

NOTE   If you've used the Text Mask tool on the text you are changing, the masking Wipeout object will not be affected. See the *Wipeout* section later in this chapter for information on editing Wipeout objects.

The following list describes the options offered by the Change Multiple Text Items tool:

Height   lets you change the height of selected text either individually or all at once.

Justification   lets you change the justification of the text.

Location   lets you move individual text objects to a new location.

Rotation lets you change the rotation angle of the selected text. You can rotate all the text to the same angle or set the text angle individually.

Style lets you assign a new text style to the selected text. You can change the style for all the selected text at once or for individual occurrences.

Text lets you perform a search and replace on the selected text, or replace each individual occurrence of text with new text.

Undo will undo the last option you used.

Width lets you change the width factor of the selected text, either all at once or for each individual occurrence.

Finding and Replacing Text The Find and Replace Text tool does just what its name says. It locates a particular string of text that you specify and replaces it with another string. Unfortunately, it only works for single-line text objects.

When you click on the Find and Replace Text tool, you are presented with the Find and Replace dialog box.

After you've entered Find and Replace With values, and clicked on OK, you are prompted to select objects. You can select any set of single-line text objects. Once you've made your selection, you'll see the second Find and Replace dialog box.

Along with this dialog box, you will see one of the selected text objects highlighted. You can click on any of the options presented in this dialog box to replace the highlighted text, proceed to replace all the text (Auto), skip the highlighted text and move on to the next occurrence, or cancel the Find and Replace tool altogether. If you know you want to replace all the selected text, you can click on the Global Change check box in the first dialog box and forego the second dialog box.

Other Bonus Text Tools

We've shown you three of the main text editing tools in the Bonus Text Tools toolbar. There are several more that you may find useful. By now, you should feel comfortable in exploring these tools on your own. The following is a brief description to get you started:

> **Global Attribute Edit** simplifies the global editing of Attribute text.
>
> **Text Fit** lets you visually stretch or compress text to fit within a given width.
>
> **Arc Aligned Text** creates text that follows the curve of an arc. If the arc is stretched or changed, the text follows the arc's shape. This is one of the more interesting Bonus Text tools, offering a wide range of settings presented in a neat little dialog box.

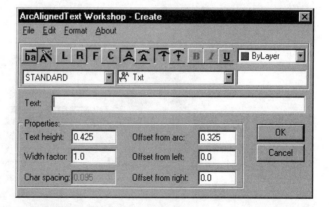

> **Explode Attributes to Text** explodes blocks containing attributes so that the attribute values are converted into plain, single-line text.
>
> **Explode Text** converts the individual characters in a text object into polylines. Beware! This tool can take some time while it works.

ENHANCEMENTS STRAIGHT FROM THE SOURCE | 859

If you want text to follow a curved path, take a look at the Txtpath.lsp utility from the companion CD-ROM. It draws text on a spline curve to follow virtually any contour you want.

Bonus Standard Tools

The Bonus Standard toolbar seems to be the answer to most AutoCAD users' wish lists. Like many of the Bonus tools discussed so far, some of these tools have been floating around in the AutoCAD user community as AutoLISP utilities. However, some are completely new. We'll start with a look at one tool that has been on our wish list for quite some time.

Wipeout

One method for masking hatch patterns is to use the Wipeout tool. Wipeout creates an object called Wipeout, which acts like a mask. If you read the previous section on the Text Mask tool, you've gotten a glimpse at how Wipeout works, because the Text Mask tool uses the Wipeout object. The following exercise demonstrates how to use this tool in another application.

Imagine that you've set up a Paper Space layout showing an enlarged view of an assembly such as the Azimuth drawing. You want to show dimensions and notes, but there are too many other objects in the way. The Wipeout tool can be of great help in this situation. Here's how.

1. Open the Azimuth.dwg file located in the Sample sub-directory of your AutoCAD R14 directory. There is also a copy of this file on the companion CD-ROM. When you open this file you will be in Paper Space.

2. While still in Paper Space, zoom into the bearing seal area, so that your view looks similar to Figure 19.2.

3. Create a layer called Wipeout and make it current.

4. Switch to Floating Model Space by choosing View ➣ Model Space (Floating), or by double-clicking on the PAPER label in the status bar.

5. Draw the closed polyline shown in Figure 19.2. You don't have to be exact about the shape; you can adjust it later.

6. Click on Wipeout from the Bonus Standard toolbar.

7. At the `Wipeout Frame/New <New>:` prompt, press ↵ to accept the default New option.

8. At the `Select a polyline:` prompt, select the polyline you just drew.

9. At the `Erase polyline? Yes/No <No>:` prompt, enter **y**↵ to erase the polyline. The area enclosed by the polyline will be masked out.

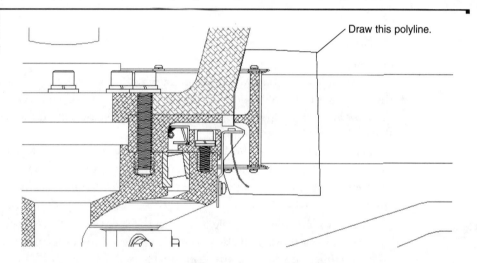

FIGURE 19.2

Adding a polyline to the enlarged view

The Wipeout object has a border that can be turned on and off. When it is visible, you can click on the Wipeout border and use its corner grips to reshape the area that it covers. You can also erase, move, or copy the Wipeout object using its border. In the example of the Azimuth drawing, you will want to hide the Wipeout border. Take the following steps to turn off the Wipeout border's visibility.

1. Click on the Wipeout button in the Bonus Standard toolbar.

2. At the `Frame/New <New>:` prompt, type **f**↵.

3. At the OFF/ON <ON>: prompt, type **off**.↵. The frame disappears.

When the frame is off, you cannot edit the Wipeout object. Of course, you can turn it back on using the Frame option you used in step 2 of the previous exercise.

If you need to edit the Text Mask tool described earlier in this chapter, use the Frame option presented in the above exercise to turn on the Text Mask border.

With the Wipeout object in place and its border turned off, you can add dimensions and notes around the image without having the adjoining graphics interfere with the visibility of your notes. Figure 19.3 shows the Azimuth drawing with the dimensions inserted.

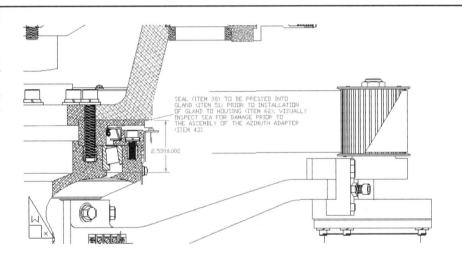

FIGURE 19.3

The Azimuth drawing with dimensions added and the viewport border adjusted to hide the graphics beyond the Wipeout object

If you switch back to Paper Space and zoom out to view the entire Paper Space drawing, you'll notice that the Wipeout object appears in the Overall view at the top of the screen (see Figure 19.4). Fortunately, you can freeze the Wipeout layer in the viewport with the Overall view to hide the Wipeout object.

FIGURE 19.4
The Wipeout object as it appears in the Overall view

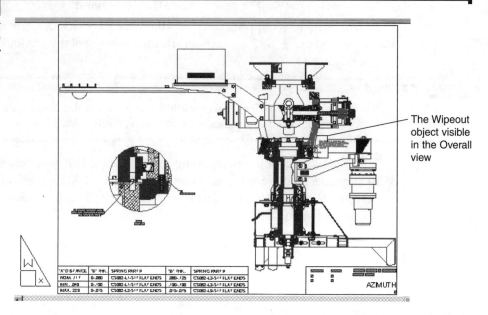

Revision Cloud

A *revision cloud* is a cloud-like outline drawn around parts of a drawing that have been revised. Revision clouds are used to alert the viewer to any changes that have occurred in the design of a project since the drawings were last issued. Revision clouds are fairly common in most types of technical drawings, including mechanical, civil, and architectural drawings.

As simple as they might appear, revision clouds are difficult to draw using the standard tools offered by AutoCAD. But now we have a single tool that makes them easy to draw. Try using the Revision Cloud tool on the Azimuth.dwg file by following these steps.

1. If you haven't already done so, switch your drawing to Paper Space.
2. Click on the Revision Cloud tool on the Bonus Standard toolbar. Then click on a point near the top right side of the Azimuth assembly, as shown in Figure 19.5.

3. Move the cursor in a counterclockwise direction to encircle the bolts. As you move the cursor, the cloud is drawn.

4. Bring the cursor back full circle to the point from which you started. When you approach the beginning of the cloud, the revision cloud closes and you exit the Revision Cloud tool.

FIGURE 19.5

Drawing a revision cloud

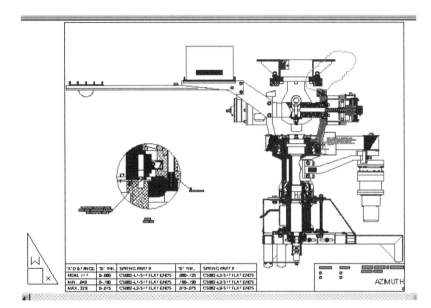

If you need to change the size of the arcs in the revision cloud, you can do so in step 2 by entering **a**↵. You can then enter an arc length. Also note that you must draw the cloud in a counterclockwise direction, otherwise the arcs of the cloud will point in the wrong direction.

Pack 'n Go

When it comes time to send the AutoCAD file and its Xrefs (external reference files) to other team members, vendors, and consultants, you have to figure out which files are external references for other files. In a large project, management of Xrefs can become a major headache.

The Pack 'n Go utility is designed to help you manage Xrefs, as well as most other external resources that an AutoCAD drawing may depend on, such as linetype definitions and text fonts. Here's how it works.

1. Click on the Pack 'n Go tool on the Bonus Standard toolbar.

The Pack & Go dialog box appears.

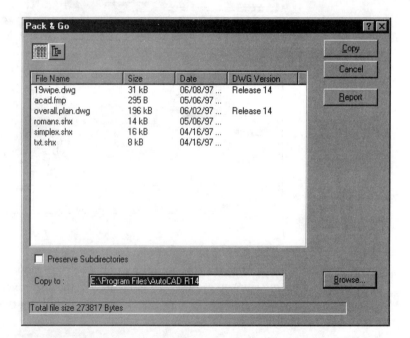

This dialog box shows all external references and resources the current file is using. It also allows you to move all of these resources into one location, such as a directory you've set up to collect a set of files to send.

2. To choose a location for the copies of your files, click on the Browse button in the lower-right corner of the dialog box. A Browse for Folder dialog box lets you locate and select a folder in which to place your copies.

3. Click on Copy to copy all the drawing files and resources to the selected directory.

In addition to the files and resources, Pack 'n Go generates three script files designed to convert the drawing files into any format from Release 12 to Release 14.

Another helpful feature of the Pack 'n Go tool is the Report generator. If you click on the Report button in the Pack 'n Go dialog box, a Report dialog box opens, providing a written description of the current file and its resources.

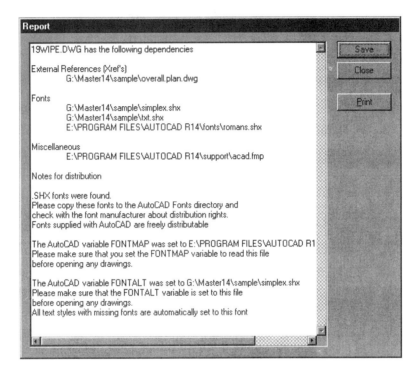

You can save this report as a text file by clicking on the Save button. Such a report can be used as a readme file when sending drawings to clients or consultants.

Extended Change Properties

When you use the Properties tool with multiple objects, you see a dialog box that lists the properties common to all objects: color, layer, linetype, and thickness. At times, this limited range of property options seems restrictive. Now, Release 14 offers the Extended Change Properties tool, which adds a few more properties to those found in the standard Properties dialog box.

When you click on the Extended Change Properties tool and select several objects, you see a slightly different version of the Change Properties dialog box.

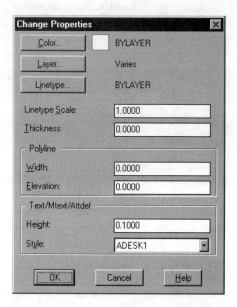

This dialog box offers a few extra options that relate to polylines and text. If you happened to have included either polylines or text in the selection set, you can enter a value to set the width and elevation of polylines or the height and style of text.

Multiple Pedit

If you only want to change the properties of polylines, you may want to use the Multiple Pedit tool.

This tool works exactly like the standard Pedit command (found by choosing Modify ➤ Object ➤ Polyline) with two exceptions: It does not offer the Edit Vertex option and you are not limited to a single polyline. This means that you can select multiple polylines to change their width, curvature, or open/close status. Multiple Pedit also lets you easily convert multiple lines and arcs into polylines.

One of the most common uses for this tool is to change the width of a set of lines, arcs, and polylines. If you include lines and arcs in a selection set with this tool, they will be converted into polylines and the specified width will be applied.

Multiple Entity Stretch

The Stretch command has always been limited by the fact that you can only select one set of vertices. The Multiple Entity Stretch tool removes that limitation and makes stretching multiple objects a simpler task. Here's how it works.

1. Click on Multiple Entity Stretch on the Bonus Standard toolbar.

You'll see the following message:

```
Define crossing windows or crossing polygons...
CP(crossing polygon)/<Crossing First point>:
```

2. Place crossing windows around the vertices you want to stretch. You may also enter **cp**↵ and proceed to place crossing polygons around the vertices.

3. When you are done selecting vertices, press ↵.

4. Go ahead and select a base point and second point to move the vertices.

Quick Multiple Trims with Extended Trim

The Extended Trim tool is actually best described by its title in the Bonus pull-down menu: Cookie Cutter Trim. It is capable of trimming a set of objects to a closed shape, such as a circle or closed polyline. You can, for example, use it to cut out a star shape in a crosshatch pattern. You can also trim multiple objects to a line or arc. To use it, do the following.

1. Click on Extended Trim from the Bonus Standard toolbar, or select Bonus ➤ Cookie Cutter Trim.

2. Select an object that is to be the trim boundary: that is, the object you want to trim to.

3. Click on the side of the selected object that you want trimmed.

Extended Trim only allows you to select a single object to trim to, but it trims multiple objects quickly and with fewer clicks of the mouse.

Clipping a Curved Shape with Extended Clip

In Chapters 9 and 16, you saw how you can clip portions of an Xref or raster image so that only a portion of these objects was visible. One limitation to the Xclip command and the Image Clip option is that you can only clip areas defined by straight lines. You cannot, for example, clip an area defined by a circle or ellipse.

Extended Clip is designed for those instances where you absolutely need to clip an Xref, raster image, or block to a curved area. Here's how it works.

1. Open a new file using the versions template. Choose Insert ➢ Raster Image from the menu bar to insert the `Raster1.tif` file from the companion CD-ROM. Window most of the screen at the `Base image size: Width: 1.00, Height: 0.76 <Unitless> Unit/Scale factor <1>:` prompt.

2. Create a clip boundary using a curved polyline or circle.

3. Click on Extended Clip from the Bonus Standard toolbar.

4. Click on the boundary.

5. Click on the Xref, block, or image you wish to clip.

6. At the `Enter max error distance for resolution of arcs <7/16">:` prompt, press ↵. The Xref, block, or image will clip to the selected boundary.

7. You may erase the boundary you created in step 2 or keep it for future reference.

Extended Clip really doesn't clip to the boundary you created; instead, it approximates that boundary by creating a true clip boundary with a series of very short line segments. In fact, the prompt in step 6 lets you specify the maximum allowable distance between the straight line segments it generates and the curve of the boundary you create (see Figure 19.6).

Once you've created a boundary using Extended Clip, you can edit the properties of the boundary using Modify ➢ Object ➢ Clip for Xrefs and blocks, or Modify ➢ Object ➢ Image Clip for raster images.

FIGURE 19.6
Extended Clip allows you to set the maximum distance from the your clip boundary and the one it generates.

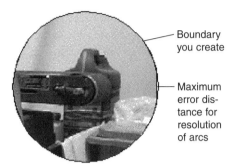

Boundary you create

Maximum error distance for resolution of arcs

Clip boundary created by Extended Clip

Block-Related Tools

Every now and then, you run into a situation where you want to use objects within a block to trim or extend to, or perhaps you want to copy a part of a block to another part of your drawing. Here are three tools that will let you do these things. They're fairly simple to use, so the following descriptions should be enough to get you started.

Copy Nested Entities lets you copy single objects within a block. You are only allowed to select objects individually—one click at a time. The copied objects will be placed on the current layer.

Trim to Block Entities lets you trim to objects in a block. It works just like the standard Trim command with the exception that you must select the objects to trim to individually.

Extend to Block Entities lets you extend to objects in a block. It also works like its standard counterpart with the exception that you must select the objects you wish to extend to individually.

Miscellaneous Bonus Standard Tools

This last set of tools offers a mixed bag of functions. Try them out at least once. They might fill a need on your wish list.

Quick Leader draws a leader with text, just like the standard Leader tool on the Dimension toolbar. The main difference is that Quick Leader doesn't ask you as many questions.

Move Copy Rotate combines these three functions into one tool. It's like a streamlined Grip Edit tool without the grips.

List Xref/Block Entities will display basic information about an Xref or block.

Tools on the Bonus Pull-Down Menu

All the Bonus tools we've discussed so far are available as options in the Bonus pull-down menu. There are some additional options on the pull-down menu you won't see in any of the toolbars. You won't want to miss these additional tools. They can greatly enhance your productivity on any type of project.

Popup Menu

Pop-up menus are great for providing quick access to frequently used functions. The Popup Menu option in the Bonus pull-down menu lets you turn any pull-down menu into a pop-up menu. So if you find that you use a particular pull-down frequently, give this handy utility a try. Here's how it works.

1. Choose Bonus ➤ Tools ➤ Popup Menu. This places a check mark next to this option.

2. Now Ctrl+right-click on your mouse. The View menu pops up. You can then select from the View menu as you would if it were open from the menu bar.

3. To select another menu as a pop-up candidate, Alt+right-click. The Pick a Popup Menu dialog box appears.

4. Select the menu you want from the list, and then click on OK. From then on, the menu you select will pop up when you Ctrl+right-click.

If you don't care for any of the existing pull-down menus, you can create your own pull-down menu, and then turn it into a pop-up menu. See Chapter 21 for more on creating pull-down menus.

Command Alias Editor

Throughout this book, we've been showing you the keyboard shortcuts to the commands of AutoCAD. All of these shortcuts are stored in a file called Acad.pgp in the \Program Files\AutoCAD R14\Support directory. In the past, you had to edit this file with a text editor to modify these command shortcuts (otherwise known as *command aliases*). But to make our lives simpler, Autodesk has supplied the Command Alias Editor, which automates the process of editing, adding, or removing command aliases from AutoCAD.

In addition, the Command Alias Editor lets you store your own alias definitions in a separate file. You can then recall your file to load your own command aliases. Here's how the Command Alias Editor works.

1. Choose Bonus ➤ Tools ➤ Command Alias Editor. The AutoCAD Alias Editor dialog box appears.

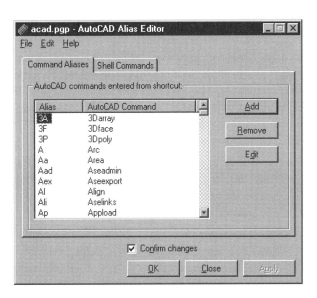

2. As you can see from the button options, you can add a new alias or delete or edit an existing alias. If you click on the Add button, you see the New Command Alias dialog box.

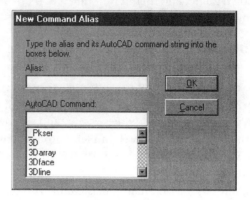

In this dialog box, enter the desired alias in the Alias input box, and then select the command from the list box below. You can also enter a command or macro name, such as Wipeout, in the input box. When you click on the Edit option in the AutoCAD Alias Editor dialog box, you see a dialog box identical to this one with the input boxes already filled in.

3. When you are done creating or editing an alias, click on OK. You will return to the AutoCAD Alias Editor dialog box.

4. Click on OK to exit the dialog box. You will see a warning message.

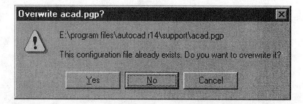

This message is telling you that you are about to overwrite the Acad.pgp file.

5. Click on No to leave the Acad.pgp file untouched. You will then see the Save As dialog box. You can enter an alternate file name, such as Myalias.pgp, to store your personal set of command aliases.

6. Once you've entered a name and saved your settings, you will see a message telling you that your new settings have taken effect. Click on OK to return to AutoCAD.

If you're a veteran AutoCAD user, you may have become accustomed to your own set of command aliases. If so, you might want to leave the original Acad.pgp file alone and create your own PGP file as we suggested in step 5. Then, whenever you use AutoCAD, you can open the AutoCAD Alias Editor, and then choose File Open and load your personal PGP file. From then on, the aliases in your file will supersede those of the standard Acad.pgp file.

System Variable Editor

Depending on the type of drawing you are working on, certain AutoCAD settings work better than others. For example, a 3D modeling project may work more smoothly with the Performance settings in the Preferences dialog box set a certain way, while a large 2D plan layout requires entirely different settings. If you find yourself wishing you could store all of AutoCAD's dialog box settings at a given moment in time, then here's the tool for you.

Nearly all of AutoCAD's settings are controlled through *system variables*. Dialog box options are frequently tied to system variables as well as some command option default settings. The System Variable Editor is a tool that lets you save the current system variable settings to a file for later recall. With this tool, you can save and restore a carefully tuned AutoCAD setup with a few clicks of the mouse, instead of reconstructing individual dialog box or system variable settings. Here's how it works.

1. Choose Bonus ➢ Tools ➢ System Variable Editor, or type **sd**↵. The System Variables dialog box appears.

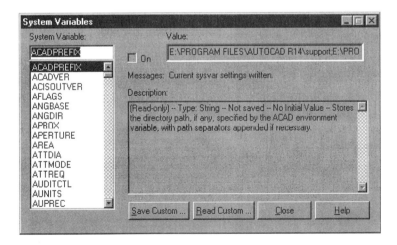

2. You can enter a system variable name or select one from the list box on the left. A description of the system variable is displayed in the Description box.

3. Once you've selected a system variable, you can enter a value for it in the Value input box. An On check box is provided for system variables that have an On/Off state instead of a value. Changes you make to system variables take effect immediately.

4. Finally, the buttons labeled Save Custom and Read Custom allow you to save the current state of the system variables to a file. The file will have the .svf file extension.

If you create a custom SVF file, AutoCAD will automatically save the system variable settings prior to your changes in an SVF file with the name of the current drawing. This is a safety measure in case you've made a terrible blunder in your settings. You can use the Read Custom button to restore your saved settings. AutoCAD also provides a special SVF file called Defaults.svf in the AutoCADr14\Bonus subdirectory. You can open this file from the System Variable Settings dialog box (Bonus ➢ Tools ➢ System Variable Editor, or type **sd**↵) to restore the "factory" default AutoCAD system variable settings.

Dimstyle Export and Dimstyle Import

Most AutoCAD users really only need to set up their dimension styles once, then make minor alterations for drawing scale. You can set up your dimension styles in a template file, then use that template whenever you create new drawings. That way, your dimension styles will already be set up the way you want them.

Frequently, however, you will receive files that were created by someone else who may not have the same ideas about dimension styles as you do. Normally, this would mean that you must re-create your favorite settings in a new dimension style. Now, with Release 14's Bonus tools, you can export and import dimension styles at any time, saving yourself the effort of re-creating them. Here's how it works.

1. Open a file from which you wish to export a dimension style.

2. Choose Bonus ➢ Tools ➢ Dimstyle Export, or enter **dimex**↵. The Dimension Style Export dialog box appears.

3. Click on the Browse button at the top of the dialog box to locate and name a file for storing your dimension style. AutoCAD appends the .dim file name extension.

4. Click on Open in the Open dialog box. If the file you specified does not exist, AutoCAD will ask you if you want to create it. Click on OK to create a new DIM file.

5. Select the name of the dimension style you want to export from the Available Dimension Styles list box.

6. Click on the Full Text Style Information radio button to include all the information regarding the associated text style.

7. Click on OK. You will see a message in the Command widow telling you that your dimension style was successfully exported.

To import a style you've exported, take the following steps.

1. Open a file into which you want to import a dimension style.

2. Choose Bonus ➤ Tools ➤ Dimstyle Import, or type **dimim**↵. The Dimension Style Import dialog box appears.

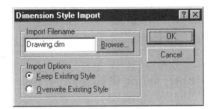

3. Click on the Browse button to open the Open dialog box.

4. Locate and select the dimension style file you saved earlier, and then click on Open.

5. Click on either the Keep Existing Style or the Overwrite Existing Style radio button to choose which action to take.

6. Click on OK.

Options for Selecting Objects, Attaching Data to Objects, and Updating Polylines

This last set of options is less likely to get as much use as the others we've discussed so far, so we've included a brief description of them here without going into too much detail. They're actually fairly easy to use, and you shouldn't have trouble trying them out.

Get Selection Set lets you create a selection set based on layer and type of object. You can either enter a layer or object type when prompted to do so or select a representative object from the screen.

Xdata Attachment lets you attach extended data to objects. Extended data is usually only used by AutoLISP, ADS, or ARX applications. You are asked to select the object that will receive the data, and then for an application name that serves as a tag to tell others to whom the data belongs. You can then select a data type. Once this is done, you can enter your data.

List Entity Xdata will display extended data that has been attached to an object.

Pline Converter converts all polylines, except for curve fitted or spline polylines, into the Release 14 lightweight polyline. This only applies to drawings created in an earlier version of AutoCAD. Generally, old polylines are automatically converted when opened by Release 14, but there are some situations where old polylines will persist:

- If the Plinetype system variable is set to any value other than 2
- If a drawing containing polylines from an older version of AutoCAD is inserted as a block and then exploded
- If a third-party application constructs an old-style polyline

Utilities Available from Other Sources

The utilities listed in the previous section are just a few samples of the many available for AutoCAD. Other sources for AutoLISP utilities are the AutoCAD journals *Cadence* and *Cadalyst*. Both offer sections, written by readers and editorial staff, that list utilities. If you don't already have a subscription to one of these publications and want to know more about them, their addresses are:

Cadence
Miller Freeman, Inc.
600 Harrison Street
San Francisco, CA 94107

Cadalyst
Advanstar Communications
859 Willamette Street
P.O. Box 10460
Eugene, OR 97440-2460

Finally, the companion CD-ROM included with this book contains some freeware and shareware utilities.

If you're using AutoCAD's 3D features, you'll want to check out the Eye2eye replacement for the Dview command. Eye2eye lets you create Perspective views easily using a camera and target object. For more information on what is included on the companion CD-ROM, see Appendix C.

Putting AutoLISP to Work

Most high-end CAD packages offer a macro or programming language to help users customize their systems. AutoCAD has *AutoLISP*, which is a pared-down version of the popular Common LISP artificial intelligence language.

Don't let AutoLISP scare you. In many ways, an AutoLISP program is just a set of AutoCAD commands that help you build your own features. The only difference is that you have to follow a different set of rules when using AutoLISP. But this isn't so unusual. After all, you had to learn some basic rules about using AutoCAD commands, too—how to start commands, for instance, and how to use command options.

If the thought of using AutoLISP is a little intimidating to you, bear in mind that you don't really need substantial computer knowledge to use this tool. In this section, you will see how you can get AutoLISP to help out in your everyday editing tasks, without having to learn the entire programming language.

Other Customization Options

If you are serious about customization, you'll want to know about the Autodesk's ObjectARX programming environment that allows Microsoft Visual C++ programmers to develop full applications that work within AutoCAD. ObjectARX allows programmers to create new objects within AutoCAD as well as add functionality to existing objects. ObjectARX is beyond the scope of this book, so to find out more, contact your AutoCAD dealer or visit Autodesk's Web site at www.autodesk.com.

If you are familiar with Visual Basic, Release 14 offers Visual Basic ActiveX Automation as part of its set of customization tools. ActiveX Automation offers the ability to create macros that operate across different applications. It also gives you access to AutoCAD objects through an object-oriented programming environment. Automation is a broad subject, so Chapter 20 is devoted to this topic.

Finally, you'll see references to *ADS* as you read through this section. ADS stands for the AutoCAD Development System. It is an older AutoCAD programming environment for C programmers. While it is still supported, it is being phased out, and developers using ADS are encouraged to move to ObjectARX.

Loading and Running an AutoLISP Program

Many AutoCAD users have discovered the usefulness of AutoLISP through the thousands of free AutoLISP utilities that are available from bulletin board and online services. In fact, it's quite common for users to maintain a "toolbox" of their favorite utilities on a diskette. But before you can use these utilities, you need to know how to load them into AutoCAD. In the following exercise, you'll load and use a sample AutoLISP utility found on the companion CD-ROM.

1. Start AutoCAD and open the Azimuth file.

2. Click on Tools ➣ Load Application. The Load AutoLISP, ADS, and ARX Files dialog box appears.

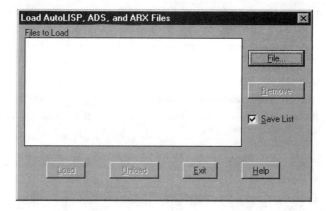

3. Click on the File button.

4. In the Directories list, locate and select the Getarea.lsp file from the companion CD-ROM. It is included among the sample drawing files.

5. Highlight Getarea.lsp and click on the Load button. You will see the message C:Getarea Loaded.

6. Now enter **getarea**↵.

7. At the Pick point inside area to be calculated: prompt, click inside the cloud.

8. At the select location for area note: prompt, pick a point just above cloud. A label appears displaying the area of the cloud less the Azimuth assembly in square feet.

You have just loaded and used an AutoLISP utility. As you saw in the File dialog box, there are several other utilities you can load and try out. You'll be introduced to a few more of these utilities later on in Appendix C, but for now, let's look more closely at the Load AutoLISP, ADS, and ARX Files dialog box.

NOTE   Some of the more popular AutoLISP utilities have become part of the core AutoCAD Program. Two new tools in Release 14—Match Properties and Make Objects Layer Current—have been around as AutoLISP utilities since the earliest releases of AutoCAD.

Working with the Load AutoLISP, ADS, and ARX Files Dialog Box

The Load AutoLISP, ADS, and ARX Files dialog box gives you plenty of flexibility in managing your favorite AutoLISP utilities. As you saw from the previous exercise, you can easily find and select utilities using this dialog box. Once you locate a file, it becomes part of the list, saving you from having to hunt down your favorite utility every time you want to use it.

Even when you exit AutoCAD, the dialog box retains the name of any AutoLISP file you select. This is because, by default, the Save List check box in the dialog box is checked. (If you don't want to retain items in the list, turn this option off.)

To remove an item from the list, just highlight it and click on the Remove button. As with any other list box, you can select multiple items and load them all at once.

Loading AutoLISP Programs Automatically

As you become more familiar with the AutoCAD user community, either online or through user groups, you may find yourself building a library of AutoLISP utilities of your own. If that library becomes extensive, you may find it tedious to load each utility one by one from the Load AutoLISP, ADS, and ARX Files dialog box.

If you find yourself in this situation, you can have AutoCAD automatically load all of your utilities for you at start-up time. To do this, you will need to create a file called Acad.1sp, and include some special code in that file. Here's how to create your Acad.1sp file.

WARNING   Some third-party AutoCAD products use an Acad.1sp file of their own. Before you try to append any of these products to an existing Acad.1sp file, be sure you have a backup copy of that file in case anything goes wrong.

Use the Windows Notepad text editor to create a file called **Acad.lsp** in your AutoCAD directory. For example, suppose you have three AutoLISP files named Edge.lsp, Xplode.lsp, and Xdata.lsp. To have AutoCAD automatically load them, you would create an Acad.lsp file that contains the following lines:

```
(load ~"Xdata~")
(load ~"Xplode~")
(load ~"Edge~")
```

NOTE

In the Acad.lsp file, each line is entered in the same way you would enter it when you load the utilities from the command prompt. Note that the .lsp extension is not required.

As a final step, you must make sure that AutoCAD can find these files. This means you must include their location in the Files tab of the Preferences dialog box. In particular, you should include them in the Support File Search Path listing.

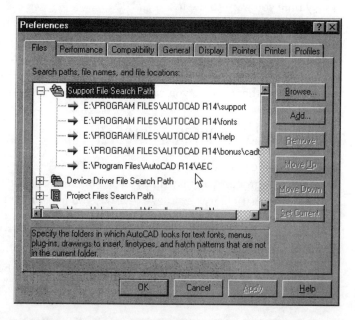

Creating Keyboard Macros with AutoLISP

You can write some simple AutoLISP programs of your own that create *keyboard macros*. Macros—like Script files—are strings of predefined keyboard entries. They are invaluable for shortcuts to commands and options you use frequently. For example, you might find that, while editing a particular drawing, you often use the Break command to break an object at a single point. Here's a way to turn this operation into a macro.

1. Open the Azimuth file and, at the command prompt, enter the following text. Be sure you enter the line exactly as shown here. If you make a mistake while entering this line, you can use the I-beam cursor or arrow keys to go to the location of your error to fix it.

    (defun C:breakat () (command ~"_break~" pause ~"f~" pause ~"@~")) ↵

2. Next, enter **breakat**↵ at the command prompt. The Break command will start and you will be prompted to select an object.

3. Click on a line.

4. At the Enter First Point: prompt, click on a point on that line where you want to create a break.

5. To see the result of the break, click on the line again. You will see that it has been split into two lines, as shown in Figure 19.7.

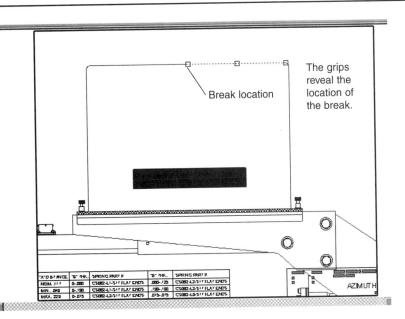

FIGURE 19.7

With the grips exposed, you can see that the line is cut into two lines.

You've just written and run your first AutoLISP macro! Let's take a closer look at this very simple program (see Figure 19.8). It starts out with an opening parenthesis, as do all AutoLISP programs, followed by the word *defun*. Defun is an AutoLISP function that lets you create commands; it is followed by the name you want to give the command (*breakat*, in this case). The command name is preceded by *c:*, telling Defun to make this command accessible from the command prompt. If the c: were omitted, you would have to start Breakat using parentheses, as in **(Breakat)**.

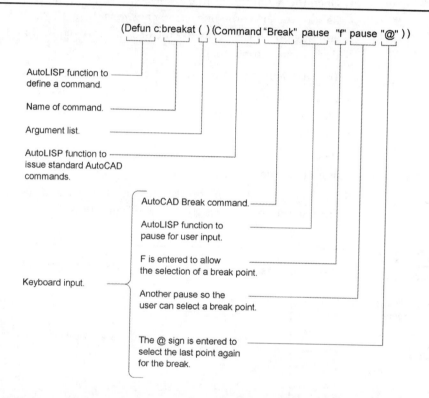

FIGURE 19.8

Breakdown of the Breakat macro

After the command name is a set of opening and closing parentheses. This encloses what is called the *argument list*. We won't go into detail about it here; just be aware that these parentheses must follow the command name.

Finally, a list of words follows, enclosed by another set of parentheses. This list starts with the word *command*. Command is an AutoLISP function that tells AutoLISP that whatever follows should be entered just like regular keyboard input. Only one item in the Breakat macro—the word *pause*—is not part of the keyboard input series. Pause is an AutoLISP function that tells AutoLISP to pause for input. In this particular macro, AutoLISP pauses to let you pick an object to break.

Notice that most of the items in the macro are enclosed in quotation marks. Literal keyboard input must be enclosed in quotation marks in this way. The Pause function, on the other hand, does not require quotation marks because it is a proper function, one that AutoLISP can recognize.

Finally, the program ends with two closing parentheses. All parentheses in an AutoLISP program must be in balanced pairs, so these two parentheses close the opening parenthesis at the start of the Command function as well as the opening parenthesis back at the beginning of the Defun function.

Storing AutoLISP Macros as Files

When you create a program at the command prompt like the Breakat macro (unless you include the code persistent AutoLISP, in the Tools, Preferences, compatibility, dialog, 'Reload AutoLISP between drawings') AutoCAD remembers it only until you exit the current file. Unless you want to re-create this macro the next time you use AutoCAD, you can save it by copying it into an ASCII text file with an .lsp extension, as shown in the following example, where we've saved the Breakat macro along with some other frequently used macros.

Figure 19.9 shows the contents of a file we've named Keycad.lsp. This file contains the macro you used above, along with several others. The other macros are commands that include optional responses. For example, the third item, defun c:corner, would cause AutoCAD to start the Fillet command, enter an **r**↵ to issue the Radius option, and finally enter a **0**↵ for the fillet radius. Table 19.1 shows the command abbreviations and what they do.

FIGURE 19.9

The contents of Keycad.lsp

```
(defun c:breakat () (COMMAND "break" PAUSE "f" PAUSE "@"))
(defun c:arcd    () (COMMAND "arc" pause "e" pause "d"))
(defun c:corner  () (COMMAND "fillet" "r" "0" "fillet"))
(defun c:ptx     () (COMMAND "pdmode" "3"))
```

TABLE 19.1: THE KEYBOARD MACROS PROVIDED BY THE KEYCAD.LSP FILE

Abbreviation	Command or Action Taken
Breakat	Breaks an object at a single point
Arcd	Draws an arc using the Start, End, Direction sequence
Corner	Sets the Fillet radius to 0, then starts the Fillet command
ptx	Sets the point style to be in the shape of an X

Use the Windows Notepad application and copy the listing in Figure 19.9. Give this file the name **Keycad.lsp**, and be sure you save it as an ASCII file. Then, whenever you want to use these macros you don't have load each one individually; instead, load the Keycad.lsp file the first time you want to use one of the macros, and they're all available for the rest of the session.

Once the Keycad.lsp file is loaded, you can use any of the macros contained within it just by entering the macro name. For example, entering **ptx**↵ will set the point style to the shape of an X.

Macros loaded in this manner will be available to you until you exit AutoCAD. Of course, you can have these macros loaded automatically every time you start AutoCAD by including the statement **(load "keycad")** in your Acad.lsp file. That way, you don't have to remember to load the file in order to use the macros.

Now that you have some firsthand experience with AutoLISP, we hope these examples will encourage you to try learning more about this powerful tool. If you would like to learn more about AutoLISP, see *The ABCs of AutoLISP* on the companion CD-ROM. This 400-page book, converted into an electronic document, is a complete resource for AutoLISP, including tutorials and example programs.

Using Third-Party Software

One of the most significant reasons for AutoCAD's popularity is its strong support for third-party software. AutoCAD is like a chameleon; it can change to suit its environment. Out of the box, AutoCAD may not fulfill the needs of some users. But by incorporating any of the over 300 third-party add-ons, you can tailor AutoCAD to suit your specific needs.

This section discusses a few of the third-party add-ons that are popular today, giving you an idea of the scope of third-party software. For more information on the myriad third-party tools out there, check out the Autodesk Web site at http://www.autodesk.com.

Custom-Tailoring AutoCAD

The needs of a mechanical designer are far different from those of an architect or a civil engineer. Third-party developers have created some specialized tools that help users of specific types of AutoCAD applications.

Many of these tools come complete with libraries of parts or symbols, AutoLISP, ADS, or ARX programs, and menus—all integrated into a single package. These packages offer added functions to AutoCAD that simplify and speed up the AutoCAD user's work. For example, mechanical add-ons offer utilities for drawing weld and surface finish symbols, hardware, improved dimensioning, and parts lists. These functions can be performed with the stock AutoCAD package but usually require a certain amount of effort. Certainly, if you have the time, you can create your own system of symbols, AutoLISP programs, and menus, and often this is the best way of molding AutoCAD to your needs. However, if you want a ready-made solution, these add-ons are invaluable.

> The companion CD-ROM includes a good mechanical add-on called *SI-Mechanical*. This add-on provides many tools for creating mechanical CAD drawings, some reusable shapes, and some great utilities for your everyday use.

Specialized third-party add-ons are available for mechanical, AEC, civil engineering, piping, mapping, finite element analysis, numeric control, GIS, and many other applications. They can save you a good deal of frustration and time, especially if you find just the right one for your environment. Like so many things, however, third-party add-ons can't be all things to all people. It is likely that no matter which add-on you purchase, you will find something lacking. When you're considering custom add-ons, make sure that there is some degree of flexibility in the package, so that if you don't like something, you can change it or add to it later.

Check with your AutoCAD dealer for information on third-party add-ons. Most AutoCAD dealers carry the more popular offerings. You might also want to get involved with a user group in your area.

Third-Party Product Information on the World Wide Web

The World Wide Web is another good place to start looking for the third-party add-ons. In particular, take a look at the Autodesk Registered Developers Web site at www.ipc.com/autodesk/. There you'll find the complete *AutoCAD Resource Guide* from ICP, Inc. As of this book's publication, this guide contains listings and information on virtually all the third-party products available for AutoCAD. You can search for products using a variety of criteria.

Autodesk's Own Offerings

Autodesk also offers a wide variety of add-ons to AutoCAD from simple symbols libraries to full-blown, industry-specific applications. There are offerings for mechanical, civil, architecture, mapping, data management, and 3D Visualization. Check out the Autodesk Web site for full details.

Getting the Latest Information from Online Services

Many resources are available for the AutoCAD user. Perhaps the most useful can be found on today's popular online services and in AutoCAD-related newsgroups. If you don't already subscribe to one, you would do well to get a modem and explore the AutoCAD newsgroups, departments, or forums on online services.

To start with, check out the two Internet newsgroups devoted to AutoCAD users:

- alt.cad.autocad
- comp.cad.autocad

Both offer a forum for you to discuss your AutoCAD questions and problems with other users. Most Internet browsers let you access newsgroups. For example, you can open a News window from Netscape Communicator by choosing Window ➢ Netscape News. From the Netscape News window, choose File ➢ Add Newsgroup, and then enter the name of the newsgroup into the input box that pops up. From then on, you can read messages, reply to posted messages, or post your questions.

The Autodesk forum on CompuServe is another excellent source of useful AutoCAD-related information, as well as utilities and troubleshooting tips. You can often get the latest information on new products, updates, bug fixes, and more. To get to the Autodesk forum, click on the GO button on the Compuserve button bar, and then enter **ACAD** at the Go dialog box.

Another online service that offers help to AutoCAD users is America Online (AOL). Although it doesn't offer a direct line to Autodesk, there is a forum for AutoCAD users to exchange ideas and troubleshooting tips. AOL also offers a library of AutoCAD-related utilities. To get to the AutoCAD folder in AOL, choose Go To ➢ Keyword. Then at the Keyword dialog box, enter **CAD** and click on GO.

Cadalyst and *Cadence*, the two North American magazines devoted to AutoCAD that we mentioned earlier, both have their own Web sites. *Cadalyst* is at http:\\www.cadonline.com; *Cadence* is at http:\\www.cadence.com. Also check out the Sybex Web site at http:\\www.sybex.com for the latest information on more great books on AutoCAD. Finally, you might check out George's site—http:\\www.omura.com—for information concerning this and other books, files, and links to other AutoCAD resources.

Posting and Accessing Drawings on the World Wide Web

The World Wide Web offers AutoCAD users some real, practical benefits through its ability to publish drawings and other documents online. AutoCAD gives you tools that allow you to post drawings on the Web that others can view and download. For the mechanical designer in particular, this can mean easier access to current catalogs and services, price and availability of purchased parts, and general libraries like those at the Autodesk Parts Spec and Material Spec Web sites. Some suppliers use the Web to post 3D solid models of their products. You could use the World Wide Web or an intranet to share data with others within your own organization.

In this section, you'll learn about the tools AutoCAD provides for publishing and accessing drawings on the Web. This section is for those with some knowledge of the HTML format used to create Web pages, and a basic familiarity with browsing the Web and using FTP sites. If you need to learn more about these topics, see the listing of relevant books from the Sybex Web site at sybex.com.

Let's start by creating a Web-viewable drawing.

Using Your Web Space

Nearly all Internet service providers offer *Web space* as part of their basic package. Web space is an area on your Internet provider's computer reserved for you alone. You can place your Web page documents there for others to view. Of course, along with the Web space you get your own Web address. If you are using the Internet now, but aren't sure whether you can use your Internet account to post Web pages, check with your Internet service provider. You may already have the Web space and not even know it.

Once you've established your Web space, you'll need to know how to post your Web pages. If you need help in this area, Sybex offers *Mastering Web Design* (1997).

Creating a Web-Compatible Drawing

AutoCAD users have been looking for ways to publish their drawings on the Web for some time. Early efforts involved capturing bitmap images of drawings and adding them to Web pages. While this is fairly simple to do, it allowed for only the crudest of images to be displayed. Drawings had to be limited in size and resolution to make them easily accessible. If you wanted to add URL links (clickable areas on an image that open other documents), you had to delve into the inner workings of Web-page design.

URL stands for Uniform Resource Locator and is a standard system for addressing Internet locations on the World Wide Web.

Fortunately, Autodesk has come up with the DWF drawing file format. DWF allows you to easily add vector format images to your Web pages. DWF images can be viewed using the same Pan and Zoom tools available in AutoCAD, thereby allowing you to present greater detail. In addition, you can embed URL links that can open other documents with a single mouse click. These links can be embedded in objects or areas in the drawing.

Opening the Internet Utilities Toolbar

Before you get started, make sure the Internet Utilities toolbar is open.

If it isn't, do the following:

1. Right-click on any toolbar.
2. In the Toolbars dialog box, click on the Menu Group drop-down list.
3. Select Inet from the list. The toolbar list will change to show the Internet Utilities.
4. Click on the Internet Utilities check box. The Internet Utilities toolbar appears.

If you can't find the Inet menu group in step 3, you need to install the Internet Utilities. You can do so by running the AutoCAD Setup program from the AutoCAD Release 14 CD-ROM.

Installing the Internet Utilities

Start the AutoCAD installation. At the Welcome screen, click on Next. At the Setup Choices screen, click on Add. At the Custom Components dialog box, locate Internet in the Components list and click on its check box. Click on Next. You will see a message telling you where Setup is installing the Internet Utilities and the amount of space it will take. Click on Next to proceed with the installation.

Creating a DWF file

Now you're ready to move ahead. Take the following steps to see how you can save a drawing in the DWF format.

1. Open the CH19brkt.dwg file from the companion CD-ROM.
2. Choose File ➤ Export.
3. At the Export Data dialog box, open the Save as Type drop-down list and choose Drawing Web Format (*.dwf). You may need to scroll down the list, as this option is at the bottom.
4. Click on Save. You've just created a DWF file.

NOTE You can also type **dwfout**⏎ to open the Create DWF File dialog box. This dialog box is similar to the one you saw in step 3 but automatically limits the output file to the DWF format.

In step 3, you may have noticed that the Options button becomes available after you select the DWF format. When you click on Options, the DWF Export Options dialog box appears.

This dialog box lets you set the level of accuracy your exported drawing will offer. The higher the accuracy, the greater the file size, so you don't want to go for the highest level of accuracy unless you really need it. In most cases, you may get by with the Low setting. Where your drawing covers a large object, an airplane for example, you may want to use Medium or High. Make a few sample files and view them through your Web browser to see firsthand which quality level works best.

Now let's continue by adding the DWF file to a Web page.

Adding a DWF file to a Web Page

Creating a DWF file is simple. Adding it to a Web page is a bit more work. For this section, we'll assume that you are familiar with the creation and editing of HTML documents. HTML stands for Hypertext Markup Language, but don't let the fancy title scare you. The fundamentals of HTML are really quite easy to learn. If you need to know more about HTML, check out *HTML 4.0: No Experience Required* (Sybex, 1997). Now let's proceed with the instructions.

Open your HTML document, either in a word processor or in a Web page creation program, and then insert the following code in the location where you want the DWF file to appear.

```
<object
 classid ="clsid:B2BE75F3-9197-11CF-ABF4-08000996E931"
 codebase = "ftp://ftp.autodesk.com/pub/autocad/plugin/whip.cab#version=2,0,0,0"
```

```
    width=400
    height=300 >
<param name="Filename"   value="drawingname.dwf">
<param name="View"       value="10000,20000 30000,40000">
<param name="Namedview" value="viewname">
<embed name= drawingname src=" drawingname.dwf"
    pluginspage=http://www.autodesk.com/products/autocad/whip/whip.htm
width=400
height=300
view="10000,20000 30000,40000"
namedview="INITIAL">
</object>
```

This code includes all the data required for both Netscape Communicator and Microsoft Internet Explorer. You'll want to keep the code for both browsers in your HTML file so that users of either browser can view your drawings.

We're showing some generic code in this example. You'll want to make a few changes to it to make it applicable to your specific DWF file. The items in italics are the ones you will want to change. Let's look at them one by one so you know exactly how to replace them in your file.

Determining Files and Opening Views

As you scan down the code, you'll see two lines labeled width and height.

```
    width=400
    height=300 >
```

These two lines are for the benefit of Microsoft Internet Explorer. The numeric values in this bit of code determine the size of your drawing in the Explorer window. They describe the width and height in pixels. You can replace the numbers here to whatever value you want, but keep them in a range that will fit neatly in a typical Web page format.

The next set of lines determines the actual file name and opening view for the DWF file.

```
<param name="Filename"   value="drawingname.dwf">
<param name="View"       value="10000,20000 30000,40000">
<param name="Namedview" value="viewname">
<embed name= drawingname src=" drawingname.dwf"
```

Again, the first three lines are for Microsoft Internet Explorer. In the first line, replace the italicized letters with your own DWF file name. You can also specify an URL for a DWF file at another location, such as http://www.omura.com/sample.dwf.

The next two lines describe the view name to which Explorer will open the DWF file, or the actual coordinates for the opening view. Use one or the other, but don't use both view parameters.

With regard to the view name in the HTML code, AutoCAD will create a view called *Initial* when you create the DWF file. This view will be of the drawing at the time that the DWF file is created. You may want to use the Initial view name just to simplify your page creation efforts.

The fourth line is for the benefit of Netscape Communicator. Again, you enter the name of the DWF file here in place of the *drawingname.dwf* letters. You'll also see "name = *drawingname*". This lets you identify the drawing with Java and JavaScript applications. You can replace the italicized name with your own name. The italicized name doesn't have to be the drawing name.

Farther down the listing you will see the width, height, and view parameters repeated.

```
width=400
 height=300
 view="10000,20000 30000,40000"
 namedview="viewname">
```

This set of parameters is intended for Netscape Communicator and should match the data you provide for Microsoft Internet Explorer in the previous set of lines.

Your Internet Provider Needs to Know…

Finally, whether you are using an Internet provider or an in-house Web server, your server will need to be able to recognize the DWF file type. You will need to inform your Webmaster that you intend to use the DWF file in your Web page. The Webmaster in turn will need to add the MIME type of "drawing/x-dwf" to your Internet server. This registers the DWF file type with the Internet server software, enabling others to view your drawings online.

Viewing Your Web Page

Once you have your HTML file completed, you are ready to view it with a browser. You will need version 3.0 of either Microsoft Internet Explorer or Netscape Navigator *and* the Autodesk Whip 2.0 driver. This driver can be obtained from the Autodesk Web page. There are separate drivers for the Internet Explorer and Netscape Navigator, so make sure you locate the Whip driver for your browser. These drivers come in the form of fairly large installation files. Set aside at least a half an hour of download time.

Once the installation file is downloaded, double-click on it and follow the instructions. It will locate your browser automatically, then prompt you for a location for the Whip driver software. Once this is done, you're ready to view your page!

NOTE You may have noticed that the sample HTML code you saw in the previous section included some Web addresses. These addresses allow a viewer without the Whip driver to automatically locate and download the driver when they attempt to view your page.

We've provided a sample page on the companion CD-ROM in case you are in a hurry to see how DWF files look on a Web page. The following exercise steps you through the opening and viewing of that sample page. This will give you a chance to see just how useful a DWF file can be.

1. Using the Windows Explorer, locate the `bracket.html` file from the companion CD-ROM and double-click on it. Your Web browser will open with a view of the bracket drawing.

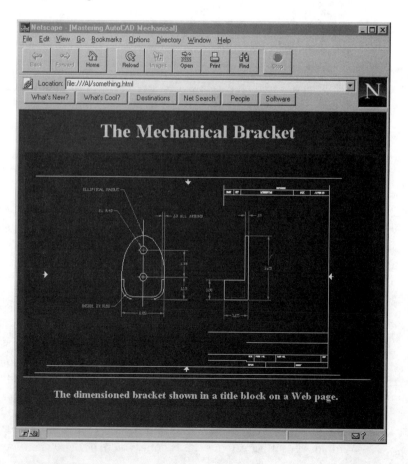

As you move the mouse over the image, you'll see the Pan Realtime hand cursor.

2. Click and drag the Pan Realtime cursor over the image. Notice that it works just like the Pan Realtime tool in AutoCAD.

3. Right-click on the mouse. A pop-up menu appears.

Notice that this menu is also similar to the one you see in AutoCAD, with some additions.

4. Select Zoom from the menu. The cursor changes to the Zoom Realtime cursor.

5. Zoom in on the view as you would in AutoCAD.

6. Adjust your view so you see the lower-left corner of the bracket, similar to Figure 19.10.

7. Double-click on the revision block shown in Figure 19.10. A new page appears, showing a revision document associated with the revision you clicked on.

8. Click on the Back button on the Netscape toolbar to return to the previous page.

In this example, you saw how you can click on an area of the drawing to get to another drawing. These clickable objects are called *URL links*. It is possible to set up these URL links to open the actual revision document associated with the document. In the next section, you'll learn how these links were created.

FIGURE 19.10

Exploring a Web page containing a DWF file

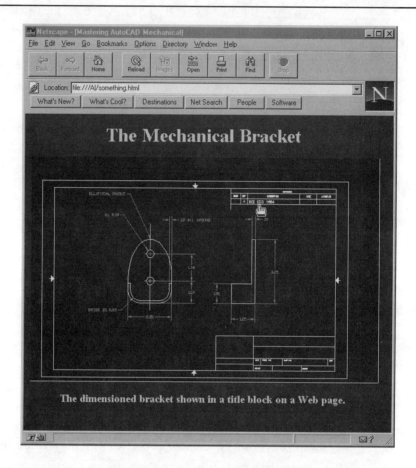

Adding Links to Other Drawings or Web Pages

In the previous exercise, the first file you opened in your browser was really not much more than the HTML file you looked at earlier. Only the names of files were modified to include the bracket. The ability to have a clickable object in the drawing comes from the DWF file itself. You add links to other HTML files while in AutoCAD before you save your drawing in the DWF format. The following steps show how links were added to the plan you saw in Figure 19.10.

   1. In AutoCAD, open the file called ch19brkt.dwg from the CD-ROM.

2. Click on Attach URL from the Internet Utilities toolbar.

3. At the URL by (Area/<Objects>): prompt, press **a**↵ to assign URLs to an area in the drawing.
4. Drag a rectangle around the revision block as shown in Figure 19.11. When you're done, press ↵. (In the sample file you looked at previously, we added a link to all of the hexagonal symbols.)
5. At the Enter URL: prompt, enter **change.html**, which is the name of the HTML file that contains the revision document.
6. Choose File ➢ Export and export the current file as a DWF file.

FIGURE 19.11
Adding URL links to the bracket revision block drawing

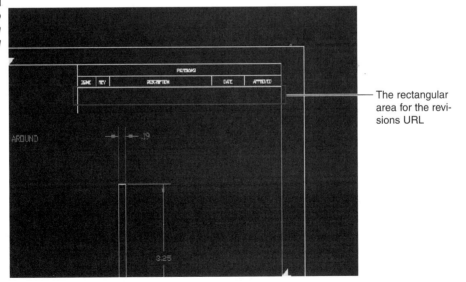

The rectangular area for the revisions URL

As indicated by the prompt in step 3, you can assign an URL to an object instead of an area by entering ↵, and then selecting objects in the drawing.

If you place anchors in your HTML files, you can include them in your URL address in step 5. For example, if you place the anchor <A name = "brkt-1"> </A> in a particular place in the HTML document, you can have your link go directly to that location by including the number sign followed by the anchor name in your URL, as in Brktsch.html#door-1.

You don't have to limit your links to HTML files containing AutoCAD drawings. You can link to all sorts of Web documents, even to other sites. For example, in step 5 you could have entered **http://www.autodesk.com** to link the hexagonal symbols to the Autodesk Web site.

Viewing and Removing Links

A couple of other tools on the Internet Utilities toolbar allow you to view or remove the URLs associated with objects and areas. The List URL tool will display the URL attached to an object or area. The Detach URL tool will clear the URL that is attached to an area or object. If you have difficulty remembering which objects have URLs assigned to them, you can use the Select URL tool to highlight all the objects and areas that have URLs attached to them.

Opening, Inserting, and Saving DWG Files over the Web

If you find you need to share your AutoCAD drawings with a lot of people, you can post your files on your Web site and allow others to download it. This is typically done through a Web page by assigning a graphic or a string of text to a file. A person viewing your page can then click on the graphic or text to start the download process.

AutoCAD also offers a way to open files directly from a Web page. You'll need to know the name of the file you are downloading, but beyond that, the process is quite simple. Try downloading a sample drawing from my Web page at www.omura.com to see how this process works.

1. Make the connection to your Internet service provider.
2. With AutoCAD open, click on Open from URL from the Internet Utilities toolbar. The Open DWG from URL dialog box appears.

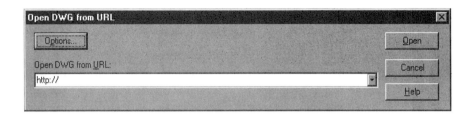

3. Click on the Options button. The Internet Configuration dialog box appears.

4. Make sure the Anonymous Login check box is checked. The anonymous user name is a standard way of accessing Web sites for file transfer where no password is assigned or needed.

5. Click on OK to return to the Open DWG from URL dialog box.

6. Enter **www.omura.com/sample.dwg** in the input box. The input box should read http://www.omura.com/sample.dwg when you are done.

7. Click on Open. The Remote Transfer in Progress dialog box appears as the file is downloaded. After a minute or two, the file appears in the drawing editor.

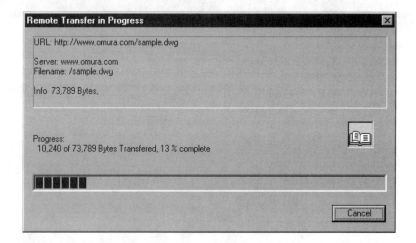

The Insert from URL tool works in exactly the same way as the Open from URL tool. The only difference is that AutoCAD will insert the downloaded file into the current file instead of closing the current file and opening the downloaded file.

You can also click and drag a DWF file from a Web page into AutoCAD to download and open a DWG file, provided a DWG file with the same name as the DWF file exists on the Web site.

If you've already downloaded this file once before, you will get a message telling you that the file already exists in the \Windows\Temp directory.

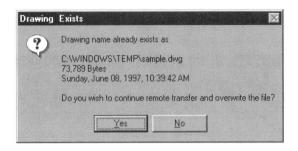

Regardless of whether you use the Open from URL or the Insert from URL tool, AutoCAD saves the downloaded file in the \Windows\Temp directory first before opening it.

Saving a File to a Web or FTP Site

If you have both read and write privileges to a Web or FTP site, you can save files to that site directly from AutoCAD. The first step in doing this is to set up AutoCAD to provide your user name and password. Take the following steps to set this up.

1. While in AutoCAD, click on the Configure Internet Host tool in the Internet Utilities toolbar. The Internet Configuration dialog box appears. Notice that this is the same dialog box you opened in step 3 of the previous exercise.

2. Click on the Anonymous Login check box to remove the check mark. This allows you to enter values in the User Name and Password input boxes in the FTP Login group.

3. Enter your user name in the User Name input box and your password in the Password input box, then click on OK.

Let's take a moment to review the Internet Configuration dialog box because this is one place where problems can occur if settings are incorrect.

If you are an individual user on a single computer connected to an Internet service provider, you would most likely use the Direct Connection option in the Connection group. The Proxy Server option is intended for systems on a company-wide network whose access to the Internet is controlled through a single computer called a *proxy server*. If you find yourself in this situation, you will have to consult your network administrator to gain access to the Internet. You can use the Help button in this dialog box to gain further information and to fill out the Proxy Information.

The HTTP Secure Access group should be filled in if you find yourself logging on frequently to Web sites that require password access.

Once your Internet configuration is up and working, you are ready to use the Save to URL tool in the Internet Utilities toolbar to upload a file to your Web site.

1. Choose Save to URL from the Internet Utilities toolbar. The Save DWG to URL dialog box appears.

2. Enter the File Transfer Protocol (FTP) name of your Web site in the Save DWG to URL input box. Typically, this would the same as your Web site with an *ftp* prefix instead of the usual *www* prefix, and would include the name of the directory where your files are stored. An example might be `ftp://ftp.myips.com/mypublic_dir/sample.dwg`. Remember to include the drawing name at the end.

3. Click on OK to save the file to your Web site. You'll see the Remote Transfer in Progress dialog box as the file is being saved.

Getting Help

These Web access features open up a whole new world of possibilities for AutoCAD users. But there are many variables and places where you can encounter problems. AutoCAD offers the Internet Help button on the Internet Utilities toolbar to answer your questions about the Internet Utilities. There you'll find detailed information about the topics covered in this section.

In addition, you will want to check out the `Aipk.html` and `Opening.html` files that come with the Whip2 plug-ins. These documents—located in the `\Program Files\AutoCAD Internet` directory—provide tips, tutorials, and ideas on how to use the Internet and World Wide Web with AutoCAD. They even offer animated GIF images and sample pages. Check it out!

If You Want to Experiment...

If you haven't already done so, take a careful look at the readme HTML document that comes with the Whip2 plug-in. It offers some additional information that can be helpful to the first-time Web author.

1. Choose Start ➤ Programs ➤ Whip Netscape Plug-in ➤ ReadMe. Or, if you are using the Microsoft Internet Explorer, choose Start ➤ Programs ➤ Whip Explorer Plug-in ➤ ReadMe. Your Internet browser opens the readme file.

2. Scroll down the file until you see the Publish DWF file for Whip listing.
3. Double-click on the HREF tag topic to find out how to add an URL link to a Web page.

Chapter 20

Using ActiveX Automation with AutoCAD

FEATURING

Understanding ActiveX Automation

The AutoCAD Object Model

Automation Techniques

The AutoCAD VBA Preview

Chapter 20

Using ActiveX Automation with AutoCAD

One of the major leaps forward in AutoCAD 14 is the addition of ActiveX Automation server capabilities to AutoCAD itself. This addition makes it possible to automate AutoCAD operations from a wide variety of other applications, including Microsoft Visual Basic, Microsoft Excel, and Microsoft Word. In this chapter, we'll review the basics of ActiveX Automation and investigate the integration of this powerful technique with AutoCAD. We'll also see some examples of using ActiveX Automation, and touch on the AutoCAD Release 14 Preview, VBA Edition.

Unlike the other skills we've covered in this book, ActiveX Automation programming is not done within AutoCAD, but within other programs such as Visual Basic or Excel. So we've departed from our usual tutorial style to provide reference information on this exciting new topic. You'll find all of the examples discussed in this chapter on the companion CD-ROM. The Visual Basic samples are on the CD-ROM in both source and compiled format, so you can test them even if you don't own a copy of Visual Basic.

What Is ActiveX Automation?

You've seen by now that AutoCAD is an immensely flexible and programmable drawing system. You can control AutoCAD from the command prompt, from menus and toolbars, and from AutoLISP macros and programs. Autodesk also supplies a set of extensions for controlling AutoCAD from the C++ language. These extensions, called ObjectARX, are designed for use from C++ only, and are beyond the scope of this book.

Now, with AutoCAD 14, there's an entirely new way to control AutoCAD: through ActiveX Automation. ActiveX Automation is a Microsoft-created standard, formerly called OLE Automation, that allows one Windows application to control another Windows application through exposed objects. In this chapter, we'll see how you can use an external application to control AutoCAD, and get familiar with the ActiveX Automation capabilities of AutoCAD 14.

If you want to see some immediate examples of ActiveX Automation programming, skip forward to the *Automation Techniques* section. But be sure to come back and read other material in this chapter for an overview of the potential of ActiveX Automation with AutoCAD.

Integrating Applications

AutoCAD is a marvelously flexible application, but like all computer applications, it's specialized in one particular area. AutoCAD's specialty is producing drawings. Despite some capabilities for handling text, numbers, and data storage, AutoCAD isn't a word processor, spreadsheet, or database program. Nor should it be. By concentrating their efforts on drawing functionality, the Autodesk developers can produce the best possible drawing application and let other teams produce the best possible word processors, spreadsheets, and databases.

However, there are times when it would be nice to combine the capabilities of multiple programs. Imagine that you have a complex drawing for an entire automobile, and you need to produce some custom reports on the engine size, powertrain, and transaxle variations that the plan includes. Yes, you could figure out a way to do this within AutoCAD itself. But wouldn't it be much easier to somehow use the AutoCAD data within a database program that has a very strong report generator? Or suppose you need to include a summary of windshield requirements for a one-month build-cycle in a document you're preparing in a word processor. Wouldn't you like to be able to read the necessary information from the drawing file you've been working with?

ActiveX Automation was developed to exploit these synergies between different applications. Windows itself allows multiple applications to execute at the same time; ActiveX Automation allows those applications to communicate among themselves. Each application can decide what information and capabilities it wants to publicize, or expose, to other applications on the system.

NOTE

You'll find references to "OLE Automation" in books about Windows programming. As it's expanded beyond its original roots in the Object Linking and Embedding standard, this technology has been renamed "ActiveX Automation." Whatever you call it, the ideas are the same.

Clients and Servers

Although ActiveX Automation always involves a conversation between two applications, it's not a two-way conversation between equal partners. Every piece of ActiveX Automation programming involves two programs with different roles. The *client* is the application that initiates the conversation. The *server* is the application that responds to the client. ActiveX Automation code runs in the client, while the actions that this code controls are executed by the server. Figure 20.1 shows the relationship between client and server in a typical ActiveX Automation exchange.

FIGURE 20.1

ActiveX Automation in action

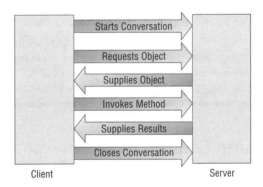

As the figure shows, there are three steps in every ActiveX Automation operation. First, one application decides to start the ActiveX Automation operation. This application automatically becomes the client, and the application it is calling, or *invoking*, becomes the server. We'll see how to program this part of the exchange in a few moments.

After the conversation has started, the client application runs code that contains server commands. It's up to the server application to decide which commands to expose, but the client application chooses which ones to actually use at any given time. In the figure, we've used the single line of code, obj.Update, to call the Update method of the AutoCAD Application object.

ActiveX Automation commands are then passed to the server, which responds to them appropriately. In the case of this particular command, the response of AutoCAD is to regenerate the current drawing.

The client may continue sending commands to the server as long as it cares to do so. The server will faithfully execute each of these commands in turn (assuming that the commands don't contain syntactical errors). When it's done controlling the server, the client can explicitly end the conversation, or it can simply stop sending commands.

There are many more ActiveX servers than ActiveX clients available. That's because programming ActiveX server support into an application is a much simpler task than making that application an ActiveX client. However, there is still a wide variety of ActiveX clients from which to choose. Here are a few of the applications you can use to control ActiveX servers, including AutoCAD:

- Visual Basic 4.0 or 5.0
- Excel 95 or 97
- Word 95 or 97
- Access 95 or 97
- PowerPoint 97

Although the code in this chapter should work in any of these clients, it's all taken from the chapter examples, which are written in Excel 95 or Visual Basic 5.0. The sample code is all included on the companion CD-ROM, but of course you'll have to provide your own copy of one of these ActiveX clients to test it.

Automation Objects

An ActiveX Automation server (such as AutoCAD 14) exposes its functionality by way of *objects*. An object is simply an abstract representation of some piece of the server application. It can be the application itself, a portion of a document managed by the application, or a part of the application's interface such as a toolbar. An object is distinguished from other objects by three things:

- The *properties* of the object
- The type, or *class*, of the object
- The *methods* of the object

The *properties* of an object are the various characteristics that describe that object. The *class* of an object is the group of objects with similar properties to which the object belongs. The *methods* of an object are the various operations that the object can perform. A server application can choose which of the properties and methods of its objects it chooses to make available by way of ActiveX Automation.

For example, one of the objects supplied by AutoCAD as an ActiveX Automation server is a Line object. Not surprisingly, this represents a single line somewhere on an AutoCAD drawing. The properties of a line include a number of things you can set about that line if you're working directly in AutoCAD. A few of the properties of the Line object are

- Color
- Layer
- StartPoint
- EndPoint
- Thickness

The methods of a line include things you can do with a line in the AutoCAD user interface. Some of the methods of a Line object are

- Copy
- Erase
- Mirror
- Move
- Rotate

Automation and AutoCAD

Applications vary widely in the extent to which they expose their capabilities via ActiveX Automation. Some applications, such as Microsoft Excel, allow client applications to manipulate both their data and their interface. You can use ActiveX Automation, for example, to change the contents of a cell on an Excel worksheet or to add an item to an Excel menu. Other applications, such as Microsoft Visual Basic, expose only a few bits of their interface to ActiveX Automation clients.

AutoCAD takes a middle road here. All of the objects in your drawings are available through ActiveX Automation, as are the major factors that control your view of the drawing, such as the size and placement of viewports. But the parts of AutoCAD that do not have direct impact on drawings, such as the location and contents of toolbars, are not exposed to ActiveX Automation clients.

This decision on the part of Autodesk means that you can use ActiveX Automation to manipulate drawings in almost any imaginable way, but you can't use ActiveX Automation to extend the capabilities of the AutoCAD environment itself. You can't, for example, write a utility that stores the locations of AutoCAD toolbars in a database and then restores these locations on demand. The information that would be needed to do this simply isn't available to ActiveX clients.

An Automation Sample

Before we go further into a systematic exploration of the ActiveX Automation capabilities of AutoCAD 14, let's take a look at an example of using these capabilities. The file census1.xls, on the companion CD-ROM, demonstrates the basic principles of using Automation with AutoCAD. This is an Excel 95 workbook that contains two worksheets. The worksheet called Module1 has code that uses ActiveX Automation to return a list of the objects in an open AutoCAD drawing. Figure 20.2 shows such a census for the r300-20.dwg file that ships with AutoCAD.

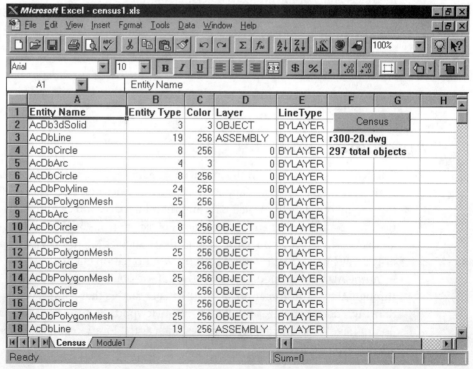

FIGURE 20.2

Census for the r300-20.dwg sample

ActiveX Automation code won't work if AutoCAD is in the middle of a command. For example, if you've chosen Linetype from the Format menu, which opens the Layer & Linetype Properties dialog box, you'll find that none of the code in this chapter will function. Make sure that all dialog boxes are closed when you're experimenting with ActiveX Automation code.

Let's review some of the Excel code that created this census. Although it's not our intent to try to teach Excel programming in this book, the basic principles of using ActiveX Automation are the same for any application that uses a dialect of Visual Basic as its programming language.

To view the code within the Excel workbook, just click on the Module1 tab at the bottom of the workbook.

All of the code for this simple example is contained within a single procedure named GetCensus. The procedure starts out by defining some variables to use:

```
Dim oAutoCad As Object
Dim oModelSpace As Object
Dim wksCensus As Worksheet
Dim intI As Integer
```

Here oAutoCad and oModelSpace are variables that will be used to represent objects supplied by AutoCAD (respectively, the AutoCAD application itself and the Model Space of the current drawing), while wksCensus is a variable that will be used to represent the Excel worksheet (an object within the same application as the code), and intI is a simple loop counter. You'll see that every variable has a type: Object, Worksheet, or Integer, in this case. Worksheet is an example of an *early-bound* variable: one whose type is specified when the variable is defined. Object is a *late-bound* variable: one that can hold any sort of object from any application.

Although early-bound variables are generally faster than late-bound variables, AutoCAD 14 does not support early-bound variables as an Automation server. You'll have to declare all variables that will hold AutoCAD objects "As Object."

In order to start the conversation with AutoCAD, Excel uses a `GetObject` statement:

```
Set oAutoCad = GetObject(, "AutoCAD.Application")
```

This tells Excel to open an ActiveX Automation session with an active instance of AutoCAD; that is, one that was opened manually or by some other application before this code was executed. `AutoCAD.Application` is the class name for AutoCAD, which uniquely identifies AutoCAD to ActiveX Automation clients. There are other ways to open an Automation session, which we'll see later.

Once Excel is talking to AutoCAD, it retrieves some simple information and places it on the Census worksheet:

```
Set wksCensus = Worksheets("Census")
wksCensus.Range("A2", "E1000").Clear
Set oModelSpace = oAutoCad.ActiveDocument.ModelSpace
wksCensus.Cells(3, 6) = oAutoCad.ActiveDocument.Name
wksCensus.Cells(4, 6) = oModelSpace.Count & " total objects"
```

The first two lines of this block of code create a reference to Excel's own worksheet object and use its `Clear` method to remove any data that is already on the worksheet. The next line initializes the `oModelSpace` variable to refer to the Model Space of the active document within AutoCAD; that is, to whatever drawing is loaded. It does this by locating the Model Space within a hierarchy of AutoCAD objects. We'll learn more about this hierarchy in the next section.

The next two lines write the values of AutoCAD properties to particular cells on the worksheet. `oAutoCad.ActiveDocument.Name` retrieves the Name property of the active document, while `oModelSpace.Count` retrieves the number of objects within the Model Space. You'll see that properties are referred to by separating them from the name of their parent objects with a single dot.

The main work of the code is done within a loop that visits every object in the drawing's Model Space:

```
For intI = 0 To oModelSpace.Count - 1
    With oModelSpace.Item(intI)
        wksCensus.Cells(intI + 2, 1) = .EntityName
        wksCensus.Cells(intI + 2, 2) = .EntityType
        wksCensus.Cells(intI + 2, 3) = .Color
        wksCensus.Cells(intI + 2, 4) = .Layer
        wksCensus.Cells(intI + 2, 5) = .Linetype
    End With
Next intI
```

TIP If you know AutoLISP, but not Visual Basic, look up "AutoLISP and Automation Comparison" in the AutoCAD Help file. You'll find a handy list of equivalents between the two languages.

The `oModelSpace` object is a *collection* object: one that contains many other objects. The Count property of the collection tells you how many objects are in that collection. These objects are each assigned a number, starting at zero, so the highest-numbered object is at `Count-1`. The loop here goes through the entire collection, referring in turn to each object within it. The line of code `With oModelSpace.Item(intI)` tells Excel that everything until the corresponding `End With` refers to this particular numbered object within the collection. Each of these objects has properties such as `EntityName` and `EntityType`, and Excel retrieves these and writes them out to the worksheet. Our census is complete!

WARNING This is not production-quality code! In particular, there's no provision for unexpected errors.

The AutoCAD Object Model

Almost every ActiveX Automation server supplies more than one object to clients, and AutoCAD is no exception. AutoCAD exposes about 70 objects to Automation clients, with a total of about 500 methods and properties. The developer needs some organizing principle to help make sense of this overwhelming complexity. This principle is provided by the AutoCAD *object model*. An object model provides a simple framework for understanding the relation of objects to each other and to the overall functionality of the application. In this section we'll investigate this object model to see some of the capabilities of ActiveX Automation in AutoCAD.

TIP The lists of objects, methods, and properties in this chapter are only partial. For the full AutoCAD object model, refer to the `acadauto.hlp` file that ships with the product.

The Object Hierarchy

It's convenient to think of the AutoCAD objects as arranged in a hierarchy. Figure 20.3 shows the top portion of this hierarchy, beginning with the AutoCAD Application object.

FIGURE 20.3

Top part of the AutoCAD object hierarchy

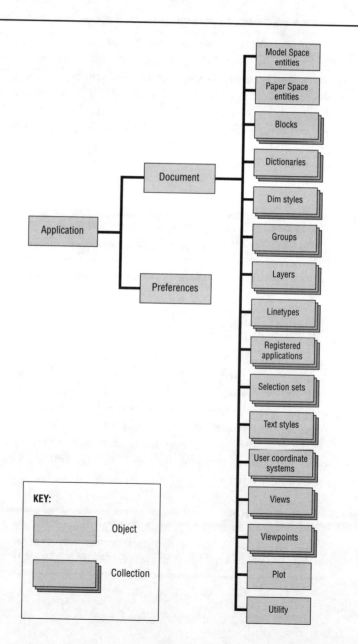

Each object in the hierarchy contains the objects beneath it. For example, the Application object contains a Preferences object and a Document object. These are objects themselves, but they may also be referred to by using properties of their parent object. This is a general principle of ActiveX Automation: You retrieve objects by working your way down from the top of the object hierarchy, using the Visual Basic Set keyword to refer to each object as you go. If you need to use both the Document object and the Preferences object in your code, you could do so like this:

```
Dim oAutoCad As Object
Dim oDocument As Object
Dim oPreferences As Object
Set oAutoCad = GetObject(, "AutoCAD.Application")
Set oDocument = oAutoCad.ActiveDocument
Set oPreferences = oAutoCad.Preferences
```

Here, `ActiveDocument` is the property of the Application object that returns the Document object, and `Preferences` is the property of the Application object that returns the Preferences object.

If you're experienced with ActiveX Automation with other servers, you might have expected the properties to have the same name as the objects they return; that is, oAutoCad.Document would be the correct syntax. Unfortunately, AutoCAD doesn't always follow the usual conventions in this area, so you'll have to check Help carefully as you return objects.

You may have noticed that none of the objects in Figure 20.3 refer to drawing objects. You'll find these listed under the *Block* section below, as all drawing objects are only available via a Block object.

Application

The Application object represents AutoCAD itself. It's always the first object your code will use in any ActiveX Automation application, because it's the single object that AutoCAD allows other objects to access directly (without going through the object hierarchy). You can use either the `GetObject` statement or the `CreateObject` statement in your Visual Basic code to connect to the Application object. GetObject is used to connect to an existing instance of AutoCAD, while CreateObject launches a new instance of AutoCAD.

Often your client application won't know whether AutoCAD is running or not, and whether the drawing you want to work with is loaded or not. In such circumstances, you can use code similar to this (taken from the AutoCAD Drawing Explorer sample on the companion CD-ROM) to put AutoCAD into the state you want:

```
' Connect to AutoCAD and load the requested drawing.
' Start by trying to connect to a running instance
On Error Resume Next
Set oAcad = GetObject(, "AutoCAD.Application")
If Err.Number Then
    ' We couldn't find a running instance, so try to launch
        ' a new instance of AutoCAD
    Err.Clear
     Set oAcad = CreateObject("AutoCAD.Application")
     If Err.Number Then
        ' Couldn't find it, couldn't create it: give up
        MsgBox "Unable to launch AutoCAD session", _
        vbCritical, "AutoCAD Drawing Explorer"
        End
    Else
        ' CreateObject launches an invisible session.
        ' Make it visible.
        oAcad.Visible = True
    End If
End If
On Error GoTo HandleErr
' If we get this far, AutoCAD is open. Now see what
' drawing (if any) is loaded
If oAcad.ActiveDocument.FullName <> txtDrawing Then
    ' Some other drawing is loaded. Save it, and
    ' load the one we want
    oAcad.ActiveDocument.Save
    oAcad.ActiveDocument.Open txtDrawing
End If
```

Although CreateObject *should* create a completely invisible instance of AutoCAD, there's a bug in AutoCAD 14: Floating toolbars in an instance launched in this fashion are visible, whether the application itself is visible or not. Since the AutoCAD object model doesn't allow you to manipulate toolbar visibility, you really have no choice but to make the whole application visible.

Table 20.1 shows some of the methods and properties of the Application object. You shouldn't try to memorize this table (or the others later in the chapter). Rather, they're provided to allow you to familiarize yourself with some of the capabilities that AutoCAD makes available as an ActiveX Automation server.

TABLE 20.1: APPLICATION METHODS AND PROPERTIES

Name	Type	Description
LoadARX	Method	Loads an AutoCAD ARX application
Quit	Method	Closes the current drawing and close AutoCAD
Update	Method	Updates the entire drawing
ActiveDocument	Property	Returns the Document object
Caption	Property	The caption of the AutoCAD window
Top	Property	Top position of the AutoCAD window (pixels from the top of the screen)
Left	Property	Left position of the AutoCAD window (pixels from the left edge of the screen)
Height	Property	Height of the AutoCAD window (pixels)
Width	Property	Width of the AutoCAD window (pixels)
Visible	Property	True if AutoCAD is currently visible

You can use the AutoCAD Drawing Explorer on the companion CD-ROM to view typical values of many AutoCAD properties.

TIP

Unlike many ActiveX Automation servers, AutoCAD does not attach miscellaneous methods (such as prompting for user input) to either the Application or the Document object. Instead, there's a separate Utility object, described later in this chapter, for this purpose.

Preferences

The Preferences object provides your ActiveX Automation applications with access to just about every option in the AutoCAD Tools ➤ Preferences dialog box. Just as there's only one such dialog box in a session of AutoCAD, each instance of the AutoCAD Application object supplies precisely one Preferences object. This object can be retrieved by using the `Application.Preferences` property. The properties of the Preferences object tell you what preferences the user has chosen in the AutoCAD Preferences dialog box. By assigning new values to these properties, you can actually change those preferences.

TIP

If you change any of the user's preferences, store the old values first and provide a way to reset the changes. This way you can avoid destroying customizations that might have taken the user a long while to set up.

Table 20.2 shows some of the methods and properties of the Preferences object. This object in particular has many properties. The table shows only a few of them to give you some sense of what's available.

TABLE 20.2: PREFERENCES METHODS AND PROPERTIES

Name	Type	Description
DeleteProfile	Method	Deletes an AutoCAD user profile
ExportProfile	Method	Exports an AutoCAD user profile
ImportProfile	Method	Imports an AutoCAD user profile
ActiveProfile	Property	Current user profile
CustomDictionary	Property	Name of the custom dictionary to use
TempFilePath	Property	Directory AutoCAD uses to store temporary files
DisplayDraggedObjects	Property	Controls whether objects are updated on the screen while being dragged

Continued

TABLE 20.2: PREFERENCES METHODS AND PROPERTIES (CONTINUED)

Name	Type	Description
CreateBackup	Property	True if AutoCAD should always create backup files
TextFont	Property	Font to use for new text
TextFontSize	Property	Size of new text in points
CursorSize	Property	Cursor size as a percentage of the screen size

Although the Preferences object provides ways to create, save, and change profiles, the AutoCAD object model has no way to enumerate all of the available profiles. If you need this information, you'll have to read it from the Windows Registry, under HKEY_CURRENT_USER\Software\Autodesk\AutoCAD\R14.0\<your serial number>\Profiles.

Document

Unlike most other ActiveX servers (for example, Microsoft Word, Excel, or PowerPoint), AutoCAD can only open a single document (drawing) at a time. Thus, the object hierarchy includes only one Document object, which represents the loaded drawing (assuming that one is open). This object is returned by the Application.ActiveDocument property.

As you might expect, most of the methods of the Document object relate to file manipulation. There are two major classes of properties for this object: defaults and collections. *Defaults*, such as ActiveLayer, control the results of subsequent drawing actions. To create new objects on a specific layer, you first need to set the Document.ActiveLayer property to that layer, and then create the objects. *Collections* include groups of other objects that the document keeps track of. For example, there is a Layers collection that includes Layer objects, one for each layer in the drawing. Layers is a collection, but it's also a property of the Document object. To access the collection, you need to use the property:

```
Set oLayers = oAutoCad.ActiveDocument.Layers
```

Table 20.3 shows some of the methods and properties of the Document object.

TABLE 20.3: DOCUMENT METHODS AND PROPERTIES

Name	Type	Description
AuditInfo	Method	Checks the integrity of a drawing file and optionally attempt to fix errors
Export	Method	Exports the drawing to a non-AutoCAD format
GetVariable	Method	Gets the value of an AutoCAD system variable
HandleToObject	Method	Returns an object, given its handle
Import	Method	Imports a non-AutoCAD drawing
New	Method	Creates a new drawing
Open	Method	Opens an existing drawing
Regen	Method	Regenerates the entire drawing
Save	Method	Saves the drawing
SaveAs	Method	Saves the drawing under a new name
SetVariable	Method	Sets the value of a system variable
WBlock	Method	Writes a selection set out as a new drawing file
ActiveDimStyle	Property	Default dimension style for new objects
ActiveLayer	Property	Default layer for new objects
ActiveLineType	Property	Default line type for new objects
ActiveSpace	Property	Toggles between Model Space and Paper Space
FullName	Property	Path and file name of the current drawing
Name	Property	File name (only) of the current drawing
ObjectSnapMode	Property	True if Object Snap is on
Saved	Property	True if the document has no unsaved changes

The HandleToObject method provides an easy way to retrieve objects that you've already worked with, but which are located deep in the hierarchy of the drawing, without going through all of the intermediate objects. Handles are discussed in the *Block* section below.

The AutoCAD Automation Help file reverses the meaning of Document.Saved.

Block

The Block object represents a single block within the AutoCAD drawing. A block can consist of a single entity or any number of entities. Each block has a name that can be retrieved, and a count of the entities under the block. Figure 20.4 shows the bottom portion of the AutoCAD object hierarchy, starting with the Block object.

FIGURE 20.4

Lower part of the AutoCAD object hierarchy

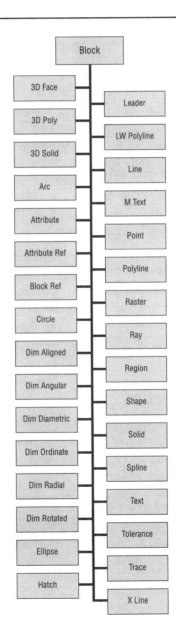

Note that although the Block object contains many types of entities, there is not a separate collection for each of these types. Rather, the different objects are put together into one heterogeneous collection of entities. You use the Block object's item method, or just the index of the collection, to retrieve any given object that belongs to the Block. For example, given a block containing a single line, you can retrieve the line this way:

```
Set oLine = oBlock.Item(0)
```

The next line of code is equivalent to the last one, since Item is the default property of the Block object:

```
Set oLine = oBlock(0)
```

The different entities within the block can be identified by means of their unique *handle* values. Each of the objects that can be contained in a block has a Handle property, and these properties are unique across all objects in the drawing. You can use the Document object's HandleToObject method (see Table 20.3) to retrieve an object, given its handle.

In addition to any named blocks you might have created, there are two special blocks in every drawing: the ModelSpace and PaperSpace objects. The ModelSpace object contains every entity in the drawing's Model Space, while the PaperSpace object contains every entity in the drawing's Paper Space. You can retrieve these special blocks by using properties of the document object:

```
Set oModelSpace = oDocument.ModelSpace
Set oPaperSpace = oDocument.PaperSpace
```

In addition, if you go through all the members of the Document's Blocks collection, you'll find that two of them have the special names *PAPERSPACE and *MODELSPACE. These are copies, of course, of the PaperSpace and ModelSpace objects. If you're trying to carry out some operation on all user-created blocks, you'll want to skip these two blocks.

Entity Properties

Each entity you might find in a Block's collection has its own properties. A Line object, for example, has StartPoint and EndPoint properties, while an Arc object has StartAngle, EndAngle, Center, and Radius properties—to name a few. There are some properties common to nearly all entities:

- Color
- EntityName
- EntityType
- Handle
- Layer

- LineType
- Visible

All the other properties are specific to their entity types. If your ActiveX Automation code acts on all entities in a particular Block, you'll need to check the EntityName or EntityType properties of each entity to determine whether it's an appropriate target for your actions.

Covering all of the properties of every entity is beyond the scope of this book. The *Automation Techniques* section later in this chapter shows what you can do with a few selected properties. The AutoCAD Drawing Explorer on the companion CD-ROM will display a small selection of these properties, as well. Beyond that, you'll need to refer to the `acadauto.hlp` file.

Dictionary

A Dictionary object represents a single AutoCAD dictionary: an arbitrary collection of objects, each associated with a keyword. Don't confuse a Dictionary object with the sort of dictionary you'd use for spell checking; it's a completely different animal. (If you're familiar with Visual Basic, you'll recognize a similarity between AutoCAD dictionaries and VBA collections.)

You can retrieve the objects in a dictionary if you know their keywords. You can also create new dictionaries and add and delete objects from dictionaries. What you can't do is get a complete list of all the objects in a dictionary. You have to know what's in there. Thus, dictionaries are most useful to applications that need to organize their own data, not to applications that are trying to work with data left by other applications.

Table 20.4 shows some of the methods and properties of the Dictionary object.

TABLE 20.4: DICTIONARY METHODS AND PROPERTIES

Name	Type	Description
AddObject	Method	Adds an object to an existing dictionary
Delete	Method	Deletes an entire dictionary
GetName	Method	Given an object, retrieves its keyword
GetObject	Method	Given a keyword, retrieves an object
GetXData	Method	Gets the extended data for an object
Remove	Method	Removes an object from a dictionary
Rename	Method	Renames an object in a dictionary
Replace	Method	Replaces an object in a dictionary with a new object
SetXData	Method	Sets the extended data for an object
Name	Property	The name of the dictionary

DimStyle

A DimStyle object represents a set of dimensioning properties that control the appearance of dimensions in your drawing. You can add new DimStyles by using the Add method of the DimStyles collection, but you can't set their properties directly by manipulating the DimStyle object. Rather, you need to change the dimensioning system variables, by using the methods of the Document object. So creating and saving a new DimStyle object follows this outline:

1. Use `Document.DimStyles.Add` to create the new object.
2. Use `Document.SetVariable` to change the appropriate system variables.
3. Use `Document.ActiveDimStyle` to make your new object the default for new dimensions.

If you need to change the style on an existing dimension, set the StyleName property of the dimension to the name of a DimStyle object with the styles you desire.

Group and SelectionSet

Both Groups and SelectionSets are collections of entities from a drawing (just as a block is a collection of entities). The difference between the two is mainly in the methods you use to manipulate the objects in the collection. You can think of a Group object as representing the collection created with the AutoCAD Group command, and a SelectionSet object as representing a group of objects selected with a selection window or a selection fence in the AutoCAD user interface.

Generally, you won't find Group objects to be of very much use in ActiveX Automation applications because there's nothing you can do to the group as a whole. SelectionSet objects provide a nice interface to allow the user to choose entities on screen for your program to work with. There's an example of using a SelectionSet object for this purpose later in this chapter, in the *User Interaction* section.

Layer

A Layer object represents a single layer on your AutoCAD drawing. Once you've added some Layer objects to your drawing, you can control the layer for new objects by setting the ActiveLayer property of the Document object. You can change the layer of a drawing entity simply by assigning the name of a different Layer object to the entity's Layer property.

The properties of a layer (shown in Table 20.5) provide you with the same control you get from the Layer Control on the AutoCAD Object Properties toolbar.

TABLE 20.5: LAYER METHODS AND PROPERTIES

Name	Type	Description
Delete	Method	Deletes an entire layer
Color	Property	Color used to plot objects on the layer
Freeze	Property	True if the layer is frozen
LayerOn	Property	True if the layer is on
LineType	Property	Default line type for the layer
Lock	Property	True if the layer is locked
Name	Property	Name of the layer

Layers are among the objects with a Color property (drawing entities and Groups also have a Color property). AutoCAD supplies constants for some of the most common colors, but for the most part, you use the number that you'd see in the Select Color dialog box to specify a color. Here's a Visual Basic procedure you can use to translate the common colors into friendlier names:

```
Function TranslateColor(intColor As Integer) As String
    ' Translate an AutoCAD color value to a string
    Select Case intColor
        Case acByBlock
            TranslateColor = "By Block"
        Case acByLayer
            TranslateColor = "By Layer"
        Case acRed
            TranslateColor = "Red"
        Case acYellow
            TranslateColor = "Yellow"
        Case acGreen
            TranslateColor = "Green"
        Case acCyan
            TranslateColor = "Cyan"
        Case acMagenta
            TranslateColor = "Magenta"
        Case acBlue
            TranslateColor = "Blue"
        Case acWhite
            TranslateColor = "White"
```

```
            Case 8
                TranslateColor = "Light Gray"
            Case 9
                TranslateColor = "Dark Gray"
            Case Else
                TranslateColor = "Custom Color #" & CStr(intColor)
        End Select
    End Function
```

LineType

A LineType object represents a single style of line used in objects, and its combination of dots, dashes, and symbols. The Document.ActiveLineType property sets the type of line to be used for new objects in your drawing. A LineType object has only two properties of interest: its Name and its Description, which is a series of characters representing the line.

WARNING

With most AutoCAD objects, you can use the Add method of the parent collection to create new instances. Unfortunately, because there is no way to edit the properties of a LineType object, this doesn't really do you any good. To add new linetypes to a drawing, you need to use the LineTypes.Load method to load them from an AutoCAD linetype definition file. To create new linetypes, you'll need to use the Visual Basic file and string manipulation capabilities to modify a linetype definition file, and then load your new linetype from the file—not a task for a novice.

RegisteredApplication

The RegisteredApplications collection provides an ActiveX Automation equivalent of the AutoLISP regapp function. Before an external application can use AutoCAD extended data, it needs to be registered in the drawing. From an Automation client you can use the RegisteredApplications.Add method to accomplish this. You can also use this collection to list all the applications that are already registered for the current drawing.

TextStyle

A TextStyle object represents a set of formatting properties for a piece of text (which itself is represented by a Text object in the Drawing's collection of entities). You can use the Add method of the TextStyles collection to create a new TextStyle object, and then set its properties. To control the style of newly created text, use the Document.ActiveTextStyle property. To change the style of existing text, set the StyleName property of the Text object to the name of a TextStyle in the drawing.

Table 20.6 shows some of the methods and properties of the TextStyle object.

TABLE 20.6: TEXTSTYLE METHODS AND PROPERTIES

Name	Type	Description
Delete	Method	Deletes the TextStyle from the drawing
FontFile	Property	Path and file name of the font to use
Height	Property	Height of the text in drawing units
LastHeight	Property	Previous value of the Height property
ObliqueAngle	Property	Slant of the text
TextGenerationFlag	Property	Set to acTextFlagBackward for backward text, or acTextFlagUpsideDown for inverted text

UserCoordinateSystem

A UserCoordinateSystem object represents a User Coordinate System (UCS). You define a new UCS by specifying its Origin, X vector, and Y vector; these properties together are sufficient to determine the Z vector of the UCS. You can also retrieve the UCS Matrix for any given UCS, which provides the necessary 4 × 4 transformation matrix to transform values from the UCS back to the World Coordinate System. Table 20.7 shows some of the methods and properties for the UCS object.

TABLE 20.7: USERCOORDINATESYSTEM METHODS AND PROPERTIES

Name	Type	Description
GetUCSMatrix	Method	Retrieves the transformation matrix for the UCS
Origin	Property	Origin of the UCS
XVector	Property	X vector of the matri
YVector	Property	Y vector of the matrix

TIP All coordinates entered through ActiveX Automation are entered in the World Coordinate System (WCS).

View

A View object represents a single 3D view of your drawing. Views don't contain drawing objects. Rather, they are a set of instructions that tell AutoCAD how to portray the drawing objects. Views are used only as input to other objects. You can use a view as an argument to the PlotView method of the Plot object or the SetView method of the Viewport object. This has the effect of defining the view that those other objects use for their own operations.

Table 20.8 shows some of the methods and properties for the View object.

TABLE 20.8: VIEW METHODS AND PROPERTIES

Name	Type	Description
Delete	Method	Deletes the view from the drawing
Center	Property	Starting point of the view line of sight
Direction	Property	Direction vector of the view
Height	Property	Height of the view
Target	Property	Ending point of the view line of sight
Width	Property	Width of the view

Viewport

A Viewport object defines the properties of the on-screen view of your drawing. You select a viewport by setting the `Document.ActiveViewport` property to the name of a Viewport object. In addition to defining the portion of a drawing that you can see on the screen, a Viewport object provides methods to allow you to zoom in on a portion of the drawing. Table 20.9 lists some of the methods and properties of the Viewport object.

TABLE 20.9: VIEWPORT METHODS AND PROPERTIES

Name	Type	Description
GetGridSpacing	Method	Gets the grid spacing for this viewport
GetSnapSpacing	Method	Gets the snap spacing for this viewport
SetGridSpacing	Method	Sets the grid spacing for this viewport
SetSnapSpacing	Method	Sets the snap spacing for this viewport
SetView	Method	Chooses the View to associate with this viewport
ZoomAll	Method	Zooms to show the entire drawing
ZoomScaled	Method	Zooms to a particular magnification
Center	Property	Starting point of the viewport line of sight
Direction	Property	Direction vector of the viewport
GridOn	Property	True to show the grid
Height	Property	Height of the viewport
OrthoOn	Property	True to turn Ortho mode on
SnapOn	Property	True to turn Snap mode on
Target	Property	Ending point of the viewport line of sight
UCSIconOn	Property	True to show the UCS icon
Width	Property	Width of the viewport

Plot

Each AutoCAD drawing has precisely one Plot object, retrieved with the PlotObject property of the Document Object. This object is a repository for the settings that control the plotted (or printed) output of the drawing. The methods of the Plot object allow you to trigger a plot from your ActiveX Automation code. Table 20.10 shows some of the methods and properties of the Plot object.

TABLE 20.10: PLOT METHODS AND PROPERTIES

Name	Type	Description
PlotExtents	Method	Plots the portion of the drawing that contains objects
PlotPreview	Method	Displays the Plot Preview dialog box
PlotToDevice	Method	Sends the plot to the specified device
PlotToFile	Method	Sends the plot to the specified file
PlotView	Method	Specifies the view to use for the plot
PlotWithConfigFile	Method	Plots according to the specified configuration file
HideLines	Property	Set to True to remove hidden lines
Origin	Property	Origin of the plot on the paper
PaperSize	Property	Paper size for the plot

WARNING Because the PlotPreview method displays a dialog box within the AutoCAD interface, it halts all of your code until the user responds. Don't call PlotPreview if the Visible property of the Application object is set to False!

Utility

Designers of Automation models often face the problem of deciding how to expose useful functionality that doesn't seem to be tightly associated with a particular object in the application. AutoCAD's answer to this problem is the Utility object. This is an object with no properties (except for the standard reference back to the Application object) but a host of methods. Table 20.11 lists these methods. You might think of this as the "Miscellaneous" object of AutoCAD.

TABLE 20.11: METHODS OF THE UTILITY OBJECT

Name	Description
AngleFromXAxis	Calculates the angle of a given line from the X axis
AngleToReal	Converts an angle as a string to a real number in a specified unit of measurement
AngleToString	Converts an angle as a real number to a string in a specified unit of measurement
DistanceToReal	Converts a distance as a string to a real number in a specified unit of measurement
EndUndoMark	Marks the end of a group of operations (see StartUndoMark)
GetAngle	Interactively gets an angle from the user (considering the Angbase system variable)
GetCorner	Interactively gets the corner of a rectangle from the user
GetDistance	Interactively gets a distance from the user
GetInput	Interactively gets a keyword index from the user (see InitializeUserInput)
GetInteger	Interactively gets an integer from the user
GetKeyword	Interactively gets a keyword from the user (see InitializeUserInput)
GetOrientation	Interactively gets an angle from the user (ignoring the Angbase system variable)

Continued

TABLE 20.11: METHODS OF THE UTILITY OBJECT (CONTINUED)	
Name	Description
GetPoint	Interactively gets a point from the user
GetReal	Interactively gets a real number from the user
GetString	Interactively gets a string from the user
InitializeUserInput	Sets up a GetXXXX method that can accept keywords
PolarPoint	Gets a point at a specified distance and angle from a given point
RealToString	Converts a distance as a real number to a string in a specified unit of measurement
StartUndoMark	Marks the start of a group of operations (see EndUndoMark)
TranslateCoordinates	Translates a point from one set of coordinates to another

The GetXXXX methods of the Utility object prompt for their input in the Command window. To prompt the user to select entities from a drawing, use the `SelectionSet .SelectOnScreen` method.

Automation Techniques

As you've seen, the AutoCAD object model is fairly complex. It's also oriented toward drawing, rather than control of the AutoCAD user interface. While the ActiveX Automation control over a drawing is limited mainly by your imagination, it's sometimes difficult to know where to start. In this section, we'll explore a couple of simple ActiveX Automation operations. All of these samples are on the companion CD-ROM. If you're new to ActiveX Automation, you may want to open the samples and single-step through them so you can trace the full flow of the programs.

The samples are supplied both as executable programs and as Visual Basic 5.0 source code. You can run the samples and see them in action without having Visual Basic on your computer, but to inspect the source code you'll need Visual Basic.

Creating a Drawing

We've already seen (in the section on the Application object) how to open an existing drawing in AutoCAD. What if you'd like to create a new drawing? This ability comes in handy if your program is designed to be a sort of "Wizard" program, automating the process of getting started with AutoCAD. For example, you might prompt the user for a new drawing name and for a type of drawing, and automatically generate the drawing for them. This is similar to the capability already provided by the Create New Drawing ➤ Use a Template feature of AutoCAD. The difference is that you can take total control of the drawing created by code, and possibly change settings based on user interaction.

This code is extracted from the GridMaker sample application on the companion CD-ROM:

```
' Make sure we have enough information to proceed
    If Len(txtDrawing & "") = 0 Or Len(txtTemplate & "") = 0 Then
        GoTo ExitHere
    End If
    ' Connect to AutoCAD and load the requested drawing.
    ' Start by trying to connect to a running instance
    On Error Resume Next
    Set moAcad = GetObject(, "AutoCAD.Application")
    If Err.Number Then
        ' We couldn't find a running instance, so try to launch
        ' a new instance of AutoCAD
        Err.Clear
        Set moAcad = CreateObject("AutoCAD.Application")
        If Err.Number Then
            ' Couldn't find it, couldn't create it: give up
            MsgBox "Unable to launch AutoCAD session", _
                vbCritical, "AutoCAD Drawing Explorer"
            End
        Else
            ' CreateObject launches an invisible session.
            ' Make it visible.
            moAcad.Visible = True
        End If
    End If
    On Error GoTo HandleErr
    ' If we get this far, AutoCAD is open. Save whatever
```

```
' drawing is open and create a new drawing. If there
' is no drawing open, this command is harmless
moAcad.ActiveDocument.Save
' Create a new document based on the selected template
Set moDocument = moAcad.ActiveDocument.New(txtTemplate)
' Give the drawing a name by calling the SaveAs method
moDocument.SaveAs (txtDrawing)
```

In this sample, the txtDrawing control contains the name of the drawing to create, and the txtTemplate control contains the name of the template to use for it. Conceptually, the process is fairly simple. First, the code gets a pointer to the running instance of AutoCAD or if none can be found, it launches AutoCAD. Then, it saves any document that might be open and uses the New method of the ActiveDocument to create a new document. Finally, it uses the SaveAs method to assign the selected name to the document.

Because the Name and FullName properties of the Document object are read-only, the only way to assign a name to a drawing is to use the Document.SaveAs method.

AutoCAD Help claims that supplying a template name to the New method is optional. We did not find this to be the case. The New method will only work if you supply a template name.

Drawing a Line

Once you've created or retrieved a drawing using ActiveX Automation, you'll probably want to modify the drawing. We'll demonstrate this process with the GridMaker sample, which generates a series of lines. Figure 20.5 shows GridMaker in action, atop an AutoCAD drawing that it has just created. It doesn't look like much, does it? But when you realize that this entire drawing was created automatically, without having to enter the coordinates for any points or touch the AutoCAD interface with your mouse, it takes on new significance.

FIGURE 20.5
Drawing created with GridMaker

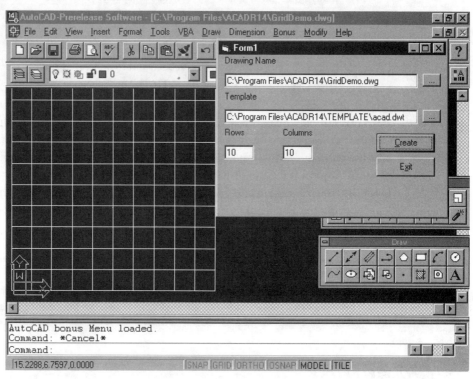

Drawing a line is simple. First, you need to define the starting and ending points of the line. In AutoCAD, a point is represented by a three-element array of double-precision floating-point numbers. As you might expect, these three elements represent the X, Y, and Z positions of the point.

Once you've defined two points, you draw the line by calling the AddLine method of the Model Space object (assuming you haven't switched to Paper Space view; the PaperSpace object also contains an AddLine method). This draws the line, and returns the Line object to your code. To actually show the line, you need to either regenerate the drawing (using the `Document.Regen` method), or tell AutoCAD to display the line using its own Update method.

The code in GridMaker simply runs a pair of loops, one to draw the horizontal lines (by holding X constant and varying Y) and one to draw the vertical lines (by holding Y constant and varying X):

```
' Now create the lines
    For intI = 0 To txtRows
```

```
            ptStart(0) = 0
            ptStart(1) = intI
            ptStart(2) = 0
            ptEnd(0) = txtColumns
            ptEnd(1) = intI
            ptEnd(2) = 0
            Set oLine = moDocument.ModelSpace.AddLine(ptStart, ptEnd)
            oLine.Update
        Next intI
        For intI = 0 To txtColumns
            ptStart(0) = intI
            ptStart(1) = 0
            ptStart(2) = 0
            ptEnd(0) = intI
            ptEnd(1) = txtRows
            ptEnd(2) = 0
            Set oLine = moDocument.ModelSpace.AddLine(ptStart, ptEnd)
            oLine.Update
        Next intI
```

This code uses the Update method on each line as it's created, so you can actually watch the drawing being created. In a real application, you'd probably want to wait until the end of the process and call Document.Regen instead, so all the drawing is done at once.

User Interaction

The Highlite sample application demonstrates a technique that lets the user of your application select objects on the AutoCAD drawing. This sample shows you how to work with an arbitrary set of AutoCAD drawing objects. When you press the Start Cycling button, AutoCAD prompts you to select objects, and then varies the color of the selected objects so that you get strong visual feedback as to what you selected. When you press Stop Cycling, the objects are returned to their original colors.

You might think you would have to write a tremendous amount of code to handle all of the standard ways of selecting objects: clicking on them, boxing them, and so on. But one of the strengths of ActiveX Automation is that applications can expose such complex behavior as a single method—in this case, the SelectOnScreen method of the SelectionSet object.

To use this method, your code first needs to create a new SelectionSet object, and then call the SelectOnScreen method. You can name your object anything you wish when you create it:

```
' Create a new selection set
    Set moSelSet = moAcad.ActiveDocument. _
    SelectionSets.Add("Highlite")
' Get some objects in it
    moSelSet.SelectOnScreen
```

When you call this method, control leaves your code and enters the hands of the end user (your application will wait until the user is done). AutoCAD places the standard Select objects: prompt on the command line, and the user can use any of the standard AutoCAD methods to choose the objects to work with. When the user is done, you can query the SelectionSet to see what he or she chose. For example, here's the code from Highlite that stores the color of each object the user selected so that we can restore it later:

```
' Save the colors of all selected objects
    ReDim maSaveColor(0 To moSelSet.Count - 1)
    For intI = 0 To moSelSet.Count - 1
        maSaveColor(intI) = moSelSet(intI).Color
    Next intI
```

Setting Active Properties

Your application may need to insure that AutoCAD is in a particular state before you start operating on the contents of the drawing. Conversely, you might want to sense the state of AutoCAD to determine whether it's safe to proceed with some operation. The Active sample application on the companion CD-ROM (see Figure 20.6) demonstrates these techniques by synchronizing its user interface with the selections in AutoCAD.

FIGURE 20.6

The Active sample

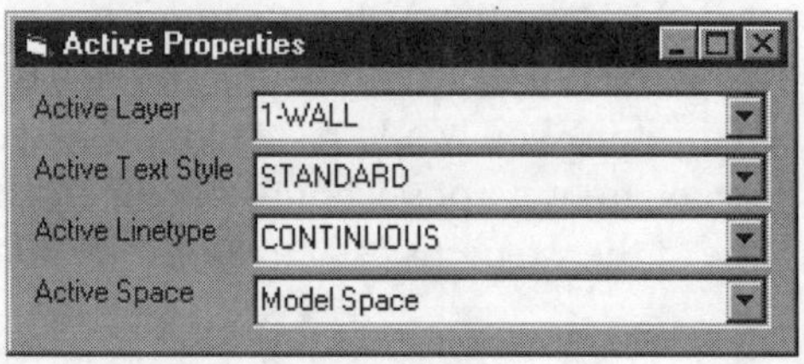

This sample starts out by retrieving lists of available layers, linetypes, and text styles from AutoCAD, and using them to fill combo boxes. For example, this code provides Active with a list of all of the available AutoCAD layers:

```
' Retrieve the list of layers
    cboLayer.Clear
    For intI = 0 To moDocument.Layers.Count - 1
        cboLayer.AddItem moDocument.Layers(intI).Name
    Next intI
```

To keep the Active interface synchronized with AutoCAD, we must employ a Visual Basic timer. This lets us run some code once a second, so that we can check the current values of the appropriate AutoCAD properties. Using a timer for this is necessary because AutoCAD doesn't notify our program by means of events when the user makes some change. To set the combo box on the Visual Basic form to the current value of the Document.ActiveLayer property in AutoCAD, you need this code:

```
With cboLayer
    For intI = 0 To .ListCount - 1
        If .List(intI) = moDocument.ActiveLayer.Name Then
            .ListIndex = intI
            Exit For
        End If
    Next intI
End With
```

Finally, to update the property in AutoCAD, you wait until the user selects a new value in the combo box, and use that value to set the property. To do so, you must retrieve an appropriate object and set it into the Document.ActiveLayer property:

```
' Retrieve the layer object we selected
    Set oLayer = moDocument.Layers(cboLayer.Text)
' And set the ActiveLayer property
    Set moDocument.ActiveLayer = oLayer
```

NOTE

As it's written, the Active sample application won't know if you add a new layer, linetype, or text style to the AutoCAD drawing. To add this capability, you could add a second timer to the form and use it to refresh the list of available items in the combo boxes, perhaps once every ten seconds.

The AutoCAD VBA Preview

AutoCAD R14 actually ships with two versions on the same CD-ROM. The first is the regular version that this whole book is about. The second is the AutoCAD Release 14 Preview, VBA Edition (we'll call it *AutoCAD VBA*). AutoCAD VBA includes, in addition to the usual AutoLISP programming language, a second language: Microsoft Visual Basic for Applications.

To install AutoCAD VBA, run the \acad\vbainst\setup.exe program from your AutoCAD CD-ROM. When you've done this, there's a new command available: Vbaide. When you enter this command, you'll be placed in the VBA development environment, shown in Figure 20.7.

FIGURE 20.7

The VBA development environment

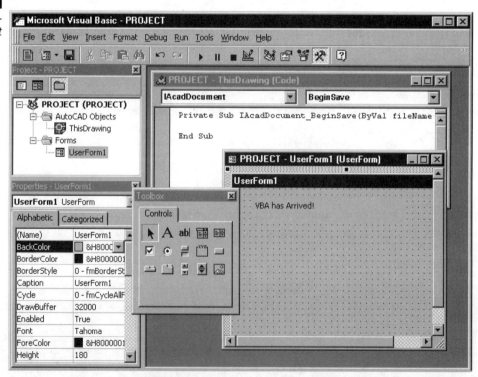

If you've used any other product with VBA 5.0, such as Word 97, Excel 97, or Visio 5.0, you'll feel right at home in this window. If not, don't panic! Because this feature was added relatively late during the creation of AutoCAD R14, we're not able to cover it in depth in the printed version of this book. But by the winter of 1997, you'll find two additional chapters on VBA on the Sybex Web site at http://www.sybex.com under the downloads for the architectural version of this book, called *Mastering AutoCAD 14*, by George Omura (ISBN # 0-7821-2109-8).

AutoCAD VBA has not yet been localized. Even if you've purchased a non-English version of AutoCAD, you'll find that the VBA development environment is presented in English.

AutoCAD VBA is currently a preview, not a supported product! If you've installed this version and run into problems, Autodesk technical support won't help you. However, if you're an AutoLISP developer, you should definitely look into VBA: You may find that your applications can be moved to this new environment with relatively little trouble. Autodesk is anxious to receive customer feedback on the potential of VBA in AutoCAD. You can send your comments to vba.support@autodesk.com, with a subject line starting with "VBA:".

If You Want to Experiment...

The code examples in this chapter are designed to show you just some of the possibilities of using ActiveX Automation with AutoCAD. To go further, take a look at the acadauto.hlp file on the AutoCAD CD-ROM, which is peppered with short code examples. You'll also need an ActiveX client program such as Visual Basic. Here are a few ideas to start you thinking:

- Allow the user to specify a group of objects and a layer to which to move them.
- Create a list showing the location of all screws on your arbor press.
- Select a group of drawings and send them to the plotter one by one.
- Extract the blocks from one drawing and re-create them in another.
- Create a new drawing layer and use it for backup copies of selected objects.

Chapter 21

Integrating AutoCAD into Your Projects and Organization

FEATURING

Customizing Toolbars

Adding Your Own Pull-Down Menus

Creating Custom Linetypes and Hatch Patterns

Supporting Your System and Working in Groups

Establishing Company Standards

Maintaining Files

Using Networks and Keeping Records

Understanding What AutoCAD Can Do for You

Chapter 21

Integrating AutoCAD into Your Projects and Organization

AutoCAD offers a high degree of flexibility and customization, allowing you to tailor the software's look and feel to your requirements. In this final chapter, you will examine how you can fit AutoCAD into your workgroup and company environment.

The first part of the chapter shows how you can adapt AutoCAD to fit your particular needs. You will learn how to customize AutoCAD by modifying its menus, and how to create custom macros for commands that your work group uses frequently.

Then we'll examine some general issues of using AutoCAD in a company. In this discussion you may find help with some problems you have encountered when using AutoCAD in your particular work environment. We'll also discuss the management of AutoCAD projects.

Customizing Toolbars

The most direct way to adapt AutoCAD to your way of working is to customize the toolbars. AutoCAD offers new users an easy route to customization. You can create new toolbars, customize tools, and even create new icons. In this section, you'll discover how easy it is to add features to AutoCAD.

Taking a Closer Look at the Toolbars Dialog Box

Throughout this book, you've used the Toolbars dialog box to open toolbars that are specialized for a particular purpose. The Toolbars dialog box is one of the many entry points to customizing AutoCAD. Let's take a closer look at this dialog box to see what other options it offers besides just opening other toolbars.

1. Right-click on the Draw toolbar. The Toolbars dialog box opens.

The Toolbars list box shows a listing of all the toolbars available in AutoCAD.

2. Scroll the list box up until you see Inquiry. Highlight it and then click on the Properties button. The Toolbar Properties dialog box appears.

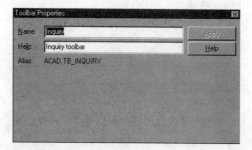

The Name input box controls the title that appears in the toolbar's title bar. The Help input box controls the Help message that appears in the status line.

3. Click on Close in the Toolbars dialog box to close both dialog boxes.

As you can see from the Toolbar Properties dialog box, you can rename a toolbar and alter its Help message, if you choose. Here is a brief description of some of the other options in the Toolbars dialog box.

Close closes the dialog box.

New lets you create a new toolbar.

Delete deletes a toolbar from the list.

Customize opens the Customize Toolbar dialog box from which you can click and drag pre-defined buttons.

Properties opens the Toolbar Properties dialog box.

Help displays helpful information about the Toolbars dialog box.

Large Buttons changes all the tools to a larger format.

Show Tooltips controls the display for the tool tips.

You'll get to use most of these options in the following sections.

NOTE

Typically, AutoCAD stores new toolbars and buttons in the Acad.mns file (see *The Windows Menu Files* sidebar later in this chapter). You can also store them in your custom menu files. Once you've created and loaded your menu file (as described in *Adding Your Own Pull-Down Menu*), choose your menu from the Menu Group drop-down list in the New Toolbar dialog box described below.

Creating Your Own Toolbar

You may find that instead of using one toolbar or flyout, you are moving from flyout to flyout on a variety of different toolbars. If you keep track of the tools you use most frequently, you can create your own custom toolbar containing your favorite tools. Here's how it's done.

1. Right-click on a button in any toolbar. The Toolbars dialog box opens.

2. Click on the New button. The New Toolbar dialog box appears.

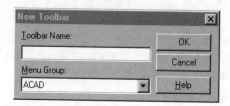

3. Enter **My Toolbar** in the Toolbar Name input box, and then click on OK. A small, blank toolbar appears in the AutoCAD window.

Notice that My Toolbar now appears in the ACAD Menu Group drop-down list of the Toolbars dialog box. You can now begin to add buttons to your toolbar.

4. Click on Customize in the Toolbars dialog box. The Customize Toolbars dialog box appears.

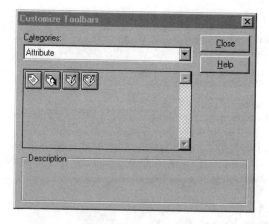

5. Open the Categories drop-down list. Notice that the list contains the main categories of commands.

6. Choose Draw from the list. The list box displays all the tools available for the Draw category. Notice that the dialog box offers several additional Arc and Circle tools not found in the Draw toolbar.

7. Click on the first tool in the top row: the Line tool. You'll see a description of the tool in the Description box at the bottom of the dialog box.

8. Click and drag the Line tool from the Customize Toolbars dialog box into the new toolbar you just created. The Line tool now appears in your toolbar.

CUSTOMIZING TOOLBARS 947

9. Click and drag the Arc Start End Direction tool to your new toolbar.

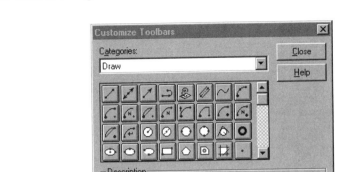

10. Exit the Customize Toolbars dialog box and the Toolbars dialog box

You now have a custom toolbar with two buttons. You can add buttons from different categories if you like. You are not restricted to buttons from one category.

NOTE

If you need to remove a tool from your toolbar, click and drag it out of your toolbar into the blank area of the drawing. Do this while the Customize Toolbars dialog box is open.

AutoCAD will treat your custom toolbar just like any other toolbar. The toolbar will appear when you start AutoCAD and will remain until you close it. You can recall your toolbar by the same method described in the first exercise.

Opening Toolbars from the Command Line

You may want to know how to open toolbars using the command line. This can be especially helpful if you want to create toolbar buttons that open other toolbars.

1. Type **-toolbar**↵ at the command prompt (don't forget to include the minus sign at the beginning of the Toolbar command).

2. At the Toolbar Name <All> prompt, enter the name of the toolbar you want to open.

Continued

CONTINUED

3. At the Show/Hide/Left/Right/Top/Bottom/Float: <Show>: prompt, press ↵. The toolbar appears on the screen.

A typical button macro for opening a toolbar might look like this:

```
^c^cToolbar[space]ACAD.Arc[space][space]
```

Here, the [space] is added for clarity. You would press the spacebar in its place. This example shows a macro that opens the Arc toolbar (ACAD.Arc).

As the prompt in step 3 indicates, you can specify the location of the toolbar by left, right, top, or bottom. Float lets you specify the location and number of rows for the toolbar.

The following list shows the toolbar names available in the standard AutoCAD system. Use these names with the Toolbar command.

ACAD.TB_OBJECT_PROPERTIES

ACAD.TB_STANDARD

ACAD.TB_DIMENSIONING

ACAD.TB_DRAW

ACAD.TB_EXTERNAL_DATABASE

ACAD.TB_INQUIRY

ACAD.TB_INSERT

ACAD.TB_MODIFY

ACAD.TB_MODIFYII

ACAD.TB_OBJECT_SNAP

ACAD.TB_REFERENCE

ACAD.TB_RENDER

ACAD.TB_SOLIDS

ACAD.TB_SURFACES

ACAD.TB_UCS

ACAD.TB_VIEWPOINT

ACAD.TB_ZOOM

You may notice that the toolbar names in the Toolbars dialog box do not match those shown in this list. You can find the "true" name of a toolbar by highlighting it in the Toolbars dialog box, and then clicking on Properties. The toolbar's true name is displayed under Alias.

Customizing Buttons

Now let's move on to some more serious customization. Suppose you want to create an entirely new button with its own functions. For example, you might want to create a set of buttons that will insert your favorite symbols. Or you might want to create a toolbar containing a set of tools that open some other toolbars which are normally "put away."

Creating a Custom Button

In the following exercises, you'll create a button that inserts a socket-head cap screw symbol. You'll add your custom button to the toolbar you just created.

1. Open the Toolbars dialog box again, and then click on Customize.
2. Select Custom from the drop-down list. The list box now shows two blank buttons, one for a single command and another for flyouts. (The one for flyouts has a small triangle in the lower-right corner.)
3. Click and drag the single command blank button to your new toolbar.
4. Right-click on the blank button in your new toolbar. The Button Properties dialog box appears. This dialog box lets you define the purpose of your custom button.

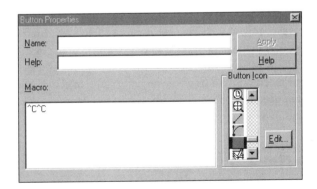

Let's pause for a moment to look at this dialog box. The Name input box lets you enter a name for your button. This name will appear as a tool tip. You must enter a name before AutoCAD will create the new button definition.

The Help input box just below the name lets you add a help message. This message will appear in the lower-left corner of the AutoCAD window when you point to your button.

The Macro area is the focus of this dialog box. Here, you can enter the keystrokes that you want to "play back" when you click on this button.

Finally, to the right, you see a scroll bar that lets you scroll through a set of icons. You also see a button labeled Edit. When you highlight an icon in the scroll box, and then click on Edit, an Icon Editor tool appears, allowing you to edit an existing icon or create a new icon.

Now let's go ahead and add a macro and new icon to this button.

1. In the Name input box, enter **Socket Head Cap Screw**. This will be the tool tip for this button.

2. In the Help input box, enter **Inserts a socket-head cap screw**. This will be the Help message for this button

3. In the Macro input box, enter the following:

    ^c^cinsert shcs

You can put any valid string of keystrokes in the Macro input box, including AutoLISP functions. See Chapter 19 or the *ABCs of AutoLISP* on the companion CD-ROM for more on AutoLISP. You can also include pauses for user input using the backslash (\) character. See the *Pausing for User Input* section later in this chapter.

Note that the two ^Cs already appear in the Macro input box. These represent two Cancels being issued. This is the same as pressing the Esc key twice. It ensures that when the macro starts, it cancels any unfinished commands.

Follow the two cancels with the Insert command as it is issued from the keyboard. If you need help finding the keyboard equivalent of a command, consult the *AutoCAD Instant Reference* on the companion CD-ROM, which contains a listing of all the command names.

It is important that you enter the exact sequence of keystrokes that follow the command, otherwise your macro may get out of step with the command prompts. This will take a little practice and some going back and forth between testing your button and editing the macro.

After the Insert command, there is a space, and then the name shcs. This is the same sequence of keystrokes you would enter at the command line to insert the socket-head cap screw drawing you created in Chapter 11. You could go on to include an insertion point, scale factor, and rotation angle in this macro, but these options are better left for the time when the macro is actually inserted.

Creating a Custom Icon

You have all the essential parts of the button defined. Now you just need to create a custom icon to go with your Socket Head Cap Screw button.

1. In the Icon scroll box, scroll down the list until you see a blank icon.
2. Click on the blank icon, and then click on Edit. The Button Editor appears.

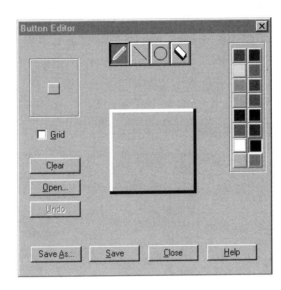

NOTE

If you prefer, you can use any of the predefined icons in the scroll box. Just click on the icon you want to use, and then click on Apply.

The Button Editor is like a very simple drawing program. Across the top are an eraser and the tools to draw lines, circles, and points. Along the right side, you see a Color toolbar from which you can choose colors for you icon button. In the upper left, you see a preview of your button. The following describes the other options.

Grid turns a grid on and off in the drawing area. This grid can be an aid in drawing your icon.

Clear erases the entire contents of the drawing area.

Open opens a BMP file to import an icon. The BMP file must be small enough to fit in the 16 × 16 pixel matrix provided for icons (24 × 24 for large format icons).

Save As saves your icon as a BMP file under a name you enter.

Save saves your icon under a name that AutoCAD provides, usually a series of numbers and letters.

Close exits the Button Editor.

Help displays helpful information about the features of the Button Editor.

Undo undoes the last operation you performed.

Now let's continue by creating a new icon.

3. Draw the Socket Head Cap Screw icon shown here. Don't worry if it's not perfect; you can always go back and fix it.

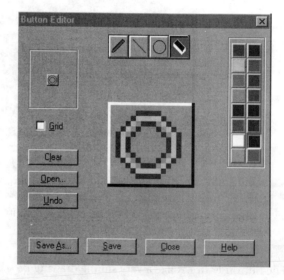

4. Click on Save, and then on Close.

5. In the Button Properties dialog box, click on Apply. You'll see the icon appear in the button in your toolbar.

6. Now click on the Close button of the Toolbars dialog box.

WARNING

The shcs drawing must be in the default directory, or in the Acad search path, before the Socket Head Cap Screw button will be inserted.

7. Click on the Socket Head Cap Screw button of your new toolbar. The socket-head cap screw appears in your drawing, ready to be placed.

You can continue to add more buttons to your toolbar to build a toolbar of symbols. Of course, you're not limited to a symbols library. You can also incorporate your favorite macros or even AutoLISP routines you may accumulate as you work with AutoCAD. The possibilities are endless.

Setting the Properties of Flyouts

Just as you added a new button to your toolbar, you can also add flyouts. Remember that flyouts are really just another kind of toolbar. This next example shows how you can add a copy of the Zoom toolbar to your custom toolbar, and then make adjustments to the properties of the flyout.

1. Right-click on the Socket Head Cap Screw button in the toolbar you just finished. The Toolbars dialog box opens with My Toolbar already highlighted.
2. Click on Customize, and then at the Customize Toolbars dialog box, open the drop-down list and select Custom.
3. Click and drag the Flyout button from the list box into the My Toolbar toolbar.

To delete a button from a toolbar, open the Toolbars dialog box, and then click on Customize. When the Customize Toolbars dialog box appears, click and drag the button you want to delete out of the toolbar and into the drawing area.

You could have simply clicked and dragged an existing flyout from the Customize Toolbars dialog box. You now have a blank flyout to which you can add your own icon. Let's see what options are available for flyout buttons.

4. Close the Customize Toolbars dialog box, and then right-click on the new, blank flyout button you just added to your toolbar. The Flyout Properties dialog box appears.

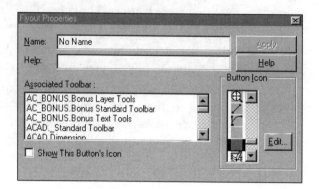

Notice how this dialog box resembles the Button Properties dialog box you used in the previous exercise. But instead of the Macro input box, you see a list of button toolbars.

5. Scroll down the list of toolbar names until you find ACAD.Zoom. This is a predefined toolbar, although it can be a toolbar you define yourself.

6. Highlight ACAD.Zoom, and then enter **My Zoom Flyout** in the Name input box. Enter **My very own flyout** in the Help input box.

7. In the Button Icon scroll box, locate an icon that looks like a magnifying glass, and click on it.

8. Click on the check box labeled Show This Button's Icon so that a check mark appears in the box.

9. Click on Apply. The flyout button in the My Toolbar toolbar shows the icon you selected.

10. Close the Toolbars dialog box, and then place the arrow cursor on the flyout to display its tool tip. Notice that your new tool tip and Help message appear.

11. Click and drag the Zoom icon, and then select Zoom In from the flyout. Notice that even though you selected Zoom In, the original Zoom icon remains as the icon for the flyout.

Step 11 demonstrates that you can disable the feature that causes the last flyout to appear as the default on the toolbar. You disabled this feature in step 8 by checking the Show This Button's Icon check box.

Editing Existing Buttons

If you want to edit an existing button or flyout, you can go directly to either the Button Properties or the Flyout Properties dialog box by double-right-clicking on a button. Once one of these dialog boxes is open, you can make changes to any component of the button definition.

The Windows Menu Files

As you create and modify icon buttons and toolbars, you'll see messages appear momentarily in the status line at the bottom of the AutoCAD window. These messages are telling you that AutoCAD is creating new menu files. The Windows version of AutoCAD creates several menu files it uses in the course of an editing session. Here's a brief rundown of what those different menu files are:

Acad.mnu is the source text file that contains the information required to build the AutoCAD menu. If you are a programmer, you would use this file to do detailed customization of the AutoCAD menu. Here, you can edit the pull-downs, image tiles, buttons, etc. Most users won't need to edit this file.

Acad.mnc is AutoCAD's translation of the Acad.mnu file. AutoCAD translates, or "compiles," the Acad.mnu file so that it can read the menu faster.

Acad.mns is a text file created by AutoCAD containing the source information from the MNU file plus additional comments. This file is rewritten whenever an MNU file is loaded.

Acad.mnr is the menu resource file. It is a binary file that contains the bitmap images used for buttons and other graphics.

As you create or edit icon buttons and toolbars, AutoCAD first adds your custom items to the Acad.mns file. It then compiles this file into the Acad.mnc and Acad.mnr files for quicker access to the menus. If you reload the MNU version of your menu, AutoCAD recreates the MNS file, thereby removing any toolbar customization you may have done.

Note that the Acad.mns file is the file you will want to copy to other computers to transfer your custom buttons and toolbars. Also, if you create your own pull-down menu files, as in the Mymenu.mnu example in this chapter, AutoCAD will create the source, compiled, and resource files for your custom menu file. You have the option to store new toolbars in your custom menu through the Menu Group pull-down list of the New Toolbar dialog box.

Adding Your Own Pull-Down Menu

Besides to adding buttons and toolbars, AutoCAD lets you add pull-down menu options. This section looks at how you might add a custom pull-down menu to your AutoCAD environment.

Creating Your First Pull-Down Menu

Let's start by trying the following exercise to create a simple pull-down menu file called My Menu:

1. Using a text editor, like the Windows Notepad, create a file called **Mymenu .mnu** containing the following lines:

   ```
   ***POP1
   [My 1st Menu]
   [Line]^c^c_line
   [–]
   [->More]
   [Arc-SED]^c^c_arc \_e \_d
   [<-Break At]^c^c(defun c:breakat ()+
   (command "break" pause "f" pause "@")+
   );breakat
   [Fillet 0]^c^c_fillet r 0;;
   [Point Style X]'pdmode 3
   ***POP2
   [My 2nd Menu]
   [door]^c^cInsert door
   [Continue Line]^C^CCLINE;;
   ```

2. Save this file, and be sure you place it in your \AutoCAD R14\support directory.

WARNING

Pay special attention to the spaces between letters in the commands described in this chapter (although you need not worry much about whether to type upper- or lowercase letters).

Once you've stored the file, you've got your first custom pull-down menu. You may have noticed some familiar items among the lines you entered. The menu contains the Line and Arc commands. It also contains the Breakat macro you worked on in Chapter 19; this time, that macro is broken into shorter lines.

Now let's see how My Menu works in AutoCAD.

Loading a Menu

In the following exercise, you will load the menu you have just created and test it out. The procedure described here for loading menus is the same for all menus, regardless of their source.

1. Click on Tools ➤ Customize Menus. The Menu Customization dialog box appears.

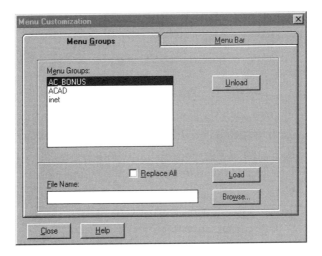

2. Click on Browse at the bottom of the dialog box. The Select Menu file dialog box appears.
3. Click on the Files of Type drop-down list, and then select Menu Template (*.mnu).
4. Locate the Mymenu.mnu file, highlight it, and then click on Open. You return to the Menu Customization dialog box.
5. Click on Load. You'll see a warning message telling you that you will lose any toolbar customization you have made. This only refers to the specific menu you are loading. Since you haven't made any toolbar customization changes to your menu, you won't lose anything.
6. Click on Yes. The warning dialog box closes and you'll see the name of your menu group listed in the Menu Group list box.

7. Highlight Mymenu.mnu in the list box, and then click on the Menu Bar tab at the top of the dialog box. The dialog box changes to show two lists.

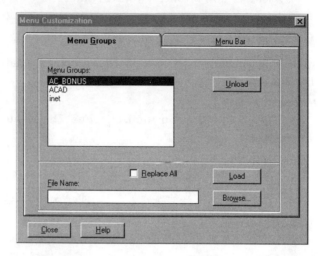

On the left is the name of the menus available in your menu file. The list on the right shows the currently available pull-down menus.

8. Highlight Help in the right-hand column. This tells AutoCAD you want to add your pull-down in front of the Help pull-down.

9. Highlight My 1st Menu from the list on the left, and then click on the Insert button. My 1st Menu moves into the right-hand column, and it appears in the AutoCAD menu bar.

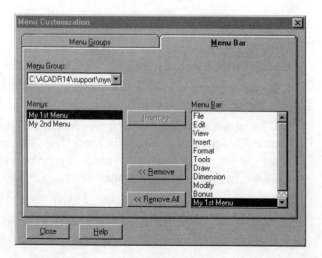

10. Highlight My 2nd Menu on the list on the left, and then click on Insert again. My 2nd Menu is copied to the right-hand column, and it also appears in the menu bar.

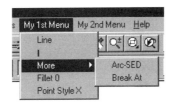

11. Close the Menu Customization dialog box.
12. Draw a line on the screen, and then try the My 1st Menu ➤ More ➤ Break At option.

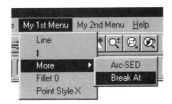

With just an 11-line menu file, you created a menu that contains virtually every tool used to build menus. Now let's take a more detailed look at how menu files work.

How the Pull-Down Menu Works

Let's take a closer look at the Mymenu.mns file. The first item in the file, ***POP1, identifies the beginning of a pull-down menu. The item just below this is the phrase My 1st Menu in square brackets. This is the title of the pull-down menu; it is what appears in the toolbar. Every pull-down menu must have this title element.

Following the title, each item on the list starts with a word enclosed in brackets; these words are the options that actually appear when you open the pull-down menu. If you were to remove everything else, you would have the menu as it appears on the screen. The text that follows the item in brackets conveys instructions to AutoCAD about the option.

Finally, in the My Menu sample, you see ***POP2. This is the beginning of a second pull-down menu. Again, you must follow this with a pull-down menu title in square brackets. Below the title, you can add other menu options.

Calling Commands

Now look at the Line option in the Mymenu.mnu listing. The two Ctrl+C (^C) elements that follow the square brackets will cancel any command that is currently operative. The Line command follows, written just as it would be entered through the keyboard. Two Cancels are issued in case you are in a command that has two levels, such as the Edit Vertex option of the Pedit command (Modify ➢ Object ➢ Polyline).

The underline that precedes the Line command tells AutoCAD that you are using the English-language version of this command. This feature lets you program non-English versions of AutoCAD using the English-language command names.

You may also notice that there is no space between the second ^C and the NEW command. A space in the line would be the same as a ↵. If there were a space between these two elements, a ↵ would be entered between the last Ctrl+C and the New command, causing the command sequence to misstep. Another way to indicate a ↵ is by using a semicolon, as in the following example:

    [Continue Line]^C^CLINE;;

When you have many ↵s in a menu macro, using semicolons instead of spaces can help make your macro more readable.

In this sample menu option, the Line command is issued, and then an additional ↵ is added. The effect of choosing this option would be a line that continues from the last line entered into your drawing. The two semicolons following the word LINE tell AutoCAD to start the Line command, and then issue ↵ twice to begin a line from the endpoint of the last line entered. (AutoCAD automatically issues a single ↵ at the end of a menu line. In this case, however, you want two ↵s, so they must be represented as semicolons.)

Pausing for User Input

Another symbol used in the menu file is the backslash (\); it is used when a pause is required for user input. For example, when you selected the Arc-SED option in My Menu, it started the Arc command and then paused for your input.

    [Arc-SED]^c^c_arc _e _d

NOTE The underscore that precedes the command name and option input tells AutoCAD that you are entering the English-language versions of these commands.

The space between ^c^c_arc and the backslash (\) represents the pressing of the spacebar. The backslash indicates a pause to allow you to select the starting endpoint for the arc. Once you have picked a point, the _e represents the selection of the Endpoint option under the Arc command. A second backslash allows another point selection. Finally, the _d represents the selection of the Direction option. Figure 21.1 illustrates this.

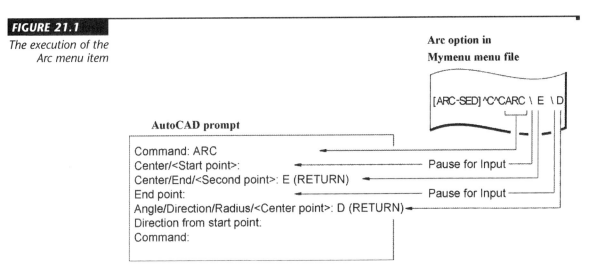

FIGURE 21.1
The execution of the Arc menu item

If you want the last character in a menu item to be a backslash, you must follow the backslash with a semicolon.

Using the Plus Sign for Long Lines

As you browse through the Acad.mnu file, you'll notice that many of the lines end with a plus sign (+). The length of each line in the menu file is limited to about

80 characters, but you can break a line into two or more lines by adding a plus sign at the end of the line that continues, like this:

```
[<-Break At]^c^c(defun c:breakat ()+
(command "break" pause "f" pause "@")+
);breakat
```

It's okay to break an AutoLISP program into smaller lines. In fact, it can help you read and understand the program more easily.

This example shows how you would include the Breakat AutoLISP macro in a menu. Everything in this segment is entered just as it would be with the keyboard. The plus sign is used to indicate the continuation of this long item to the subsequent lines, and the semicolon is used in place of ↵.

Creating a Cascading Menu

Look at the More option in the File pull-down menu group; it starts with these characters: ->. This is the way you indicate a menu item that opens a cascading menu. Everything that follows the [->More] menu item will appear in the cascading menu. To indicate the end of the cascading menu, use the <- characters, as in the [<-Rotate90] menu item farther down. Anything beyond this <- item appears in the main part of the menu. If the last item in a cascading menu is also the last item in the menu group, you must use <-<-, as in [<-<-.XZ].

Placing Division Lines and Dimmed Text in Pull-Down Menus

Two symbols are used to place dividing lines in your pull-down menus. One is the *double-hyphen* symbol (--). This is used to divide groups of items in a menu; it will expand to fill the entire width of the pull-down menu with a line of hyphens. The other option is the *tilde* symbol (~). If the tilde precedes a bracketed option name, that option will be dimmed when displayed; when clicked on, it will have no effect. You have probably encountered these dimmed options on various pull-down menus in the programs you use. When you see a dimmed menu item, it usually means that the option is not valid under the current command.

Loading AutoLISP Macros with Your Submenu

As you become a more advanced AutoCAD user, you may find that you want to have many of your own AutoLISP macros load with your menus. This can be accomplished by combining all of your AutoLISP macros into a single file. Give this file the same name as your menu file with the .mnl file extension. Such a file will be automatically loaded with its menu counterpart. For example, say you have a file called Mymenu.mnl containing the Breakat AutoLISP macro. Whenever you load the Mymenu.mns, Mymenu.mnl will automatically load along with it, giving you access to the Breakat macro. This is a good way to manage and organize any AutoLISP program code you want to include with a menu.

Adding Help Messages to Pull-Down Menu Items

Earlier in this chapter, we showed you how you can include a Help message with an icon button. The Help message appears in the status bar of AutoCAD window when you highlight an option. You can also include a Help message with a pull-down menu item. Here's how.

First, you must give your pull-down menu file a menu group name. This helps AutoCAD isolate your file and its Help messages from other menus that might be loaded along with yours. To give your menu file a group name, add the following line at the top of the file:

```
***MENUGROUP=MYMENU
```

where *MYMENU* is the name you want for your menu group name.

Next, you will have to add an ID name to each menu item that requires a Help message. The following shows a sample of how this might be done for the My 1st Menu example you used earlier:

```
***MENUGROUP=MYMENU
***POP1
[My 1st Menu]
ID_1line     [Line]^c^c_line
[--]
[->More]
ID_1arc-sed  [Arc-SED]^c^c_arc \_e \_d
```

```
ID_1breakat   [<-Break At]^c^c(defun c:breakat ()+
(command "break" pause "f" pause "@")+
);breakat
ID_1fillet0   [Fillet 0]^c^c_fillet r 0;;
ID_1pointx    [Point Style X]'pdmode 3
***POP2
[My 2nd Menu]
[door]^c^cInsert shcs
[Continue Line]^C^CLINE;;
```

The ID name starts with the characters ID followed by an underline, and then the name for the menu item. Several spaces are added so the menu items align for clarity. Each menu item must have a unique ID name.

Finally, you add a section at the end of your file called ***HELPSTRINGS. For this example, it would look like this:

```
***HELPSTRINGS
ID_1line      [Draws a line]
ID_1arc-sed   [Draws an arc with start, end, direction]
ID_1breakat   [Breaks an object at a single point]
ID_1fillet0   [Sets the Fillet radius to zero]
ID_1pointx    [Sets the Point style to an X]
```

The menu item ID names are duplicated exactly, followed by several spaces, and then the actual text you want to have appear in the status line, enclosed in brackets. The spaces between the ID and the text are for clarity.

WARNING

The ID names are case-sensitive, so make sure they match up in both the HELPSTRINGS section and in the menu section.

Once you've done this, and then loaded the menu file, you will see these same messages appear in the status bar when these menu options are highlighted. In fact, if you browse your Acad.mnu file, you will see similar ID names. If you prefer, you can use numbers in place of names.

Creating Accelerator Keys

Perhaps one of the more popular methods for customizing AutoCAD has been the keyboard *accelerator keys*. Accelerator keys are Ctrl or Shift key combinations that invoke commonly used commands or tools in AutoCAD. The Osnaps are a popular candidate for accelerator keys, as are the display commands.

To add accelerator key definitions to AutoCAD, you need to add some additional code to your menu file. The following is an example of what you can add to the `Mymenu.mnu` file to define a set of accelerator keys.

```
***ACCELERATORS
[CONTROL+SHIFT+"E"]endp
[CONTROL+SHIFT+"X"]int
ID_1breakat    [CONTROL+SHIFT+"B"]
```

The `***ACCELERATORS` line at the top of the list is the group heading, similar to the `***HELPSTRINGS` heading in that it defines the beginning of the section. This is followed by the accelerator descriptions.

There are two methods shown here for defining an accelerator key. The first two follow the format you've already seen for the pull-down menu. But instead of the menu text in square brackets, you see the keys required to invoke the action that follows the brackets. So in the first line,

```
[CONTROL+SHIFT+"E"]endp
```

the `CONTROL+SHIFT+"E"` tells AutoCAD to enter the Endpoint Osnap (endp) whenever the Ctrl+Shift+E key combination is pressed. Notice that the E is in quotation marks and that all the characters are uppercase. Follow this format for quotation marks and capitalization when you create your own accelerator keys.

A second method is shown in the third line:

```
ID_1breakat    [CONTROL+SHIFT+"B"]
```

Here, the `ID_1breakat` is used to associate a keystroke combination with a menu option not unlike the way it is used to associate a menu item with a helpstring. This line tells AutoCAD to issue the Breakat macro listed earlier in the menu whenever the Ctrl+Shift+B key combination is pressed.

You can use Ctrl or Shift individually or together in combination with most keys on your keyboard, including the function keys. You can also assign keys without the Ctrl or Shift options. Note that the function keys and the Esc key are already defined, so take care that you don't redefine them unless you really want to. Table 21.1 contains a brief listing of some of the special keys you can define and how you need to specify them in the menu file.

TABLE 21.1: KEY NAMES TO USE IN YOUR ACCELERATOR KEY DEFINITIONS

Key	Format used in AutoCAD Menu
Numeric keypad	"NUMPAD0" through "NUMPAD9"
Ins	"INSERT"
Del	"DELETE"
Function keys	"F2" through "F12"
Up arrow	"UP"
Down arrow	"DOWN"
Left arrow	"LEFT"
Right arrow	"RIGHT"

WARNING

Although you can use the F1 and Esc keys for accelerator keys, their use is discouraged because they serve other functions in both Windows and AutoCAD.

Creating Custom Linetypes and Hatch Patterns

As your drawing needs expand, you may find that the standard linetypes and hatch patterns are not adequate for your application. Fortunately, you can create your own. This section explains how to go about creating custom linetypes and patterns.

Viewing Available Linetypes

Although AutoCAD provides the linetypes most commonly used (see Figure 21.2), the dashes and dots may not be spaced the way you would like, or you may want an entirely new linetype.

NOTE

AutoCAD stores the linetypes in a file called Acad.lin, which is in ASCII format. When you create a new linetype, you are actually adding information to this file. Or, if you create a new file containing your own linetype definitions, it, too, will have the extension .lin. You can edit linetypes as described here, or you can edit them directly in these files.

FIGURE 21.2
The standard AutoCAD linetypes

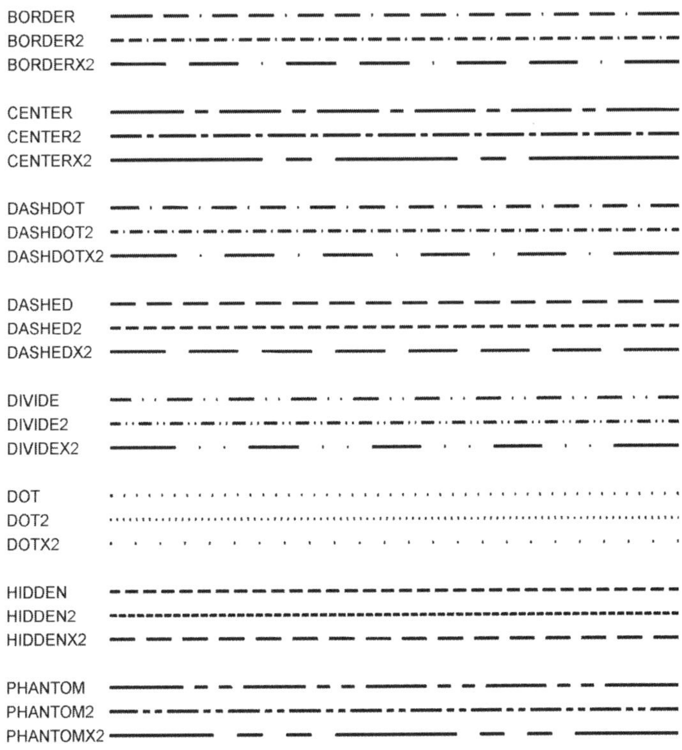

Linetype	Pattern
BORDER	— — · — — · — — · — — · — — ·
BORDER2	– – · – – · – – · – – · – – ·
BORDERX2	—— —— · —— —— · —— —— ·
CENTER	—— — —— — —— — ——
CENTER2	— - — - — - — -
CENTERX2	———— —— ———— ——
DASHDOT	— · — · — · — · — ·
DASHDOT2	- · - · - · - · - ·
DASHDOTX2	—— · —— · —— · —— ·
DASHED	— — — — — — — —
DASHED2	- - - - - - - -
DASHEDX2	—— —— —— —— ——
DIVIDE	— · · — · · — · · —
DIVIDE2	- · · - · · - · · -
DIVIDEX2	—— · · —— · · —— · ·
DOT	· · · · · · · · · · · ·
DOT2	············
DOTX2	· · · · · ·
HIDDEN	- - - - - - - - - - -
HIDDEN2	-------------
HIDDENX2	— — — — — — —
PHANTOM	—— — — —— — — ——
PHANTOM2	— - - — - - — - -
PHANTOMX2	———— —— —— ————

To create a custom linetype, use the Linetype command. Let's see how this handy command works, by first listing the available linetypes.

1. Open a new AutoCAD file.
2. Enter **-linetype**↵ at the command prompt. (Don't forget the minus sign at the beginning of the word *linetype*.)
3. At the ?/Create/Load/Set: prompt, enter **?**↵.
4. In the file dialog box that appears, locate and double-click on ACAD in the listing of available linetype files. You will get the listing shown in the top and bottom images of Figure 21.3, which shows the linetypes available in the Acad.lin file, along with a simple description of each line. The top image of Figure 21.3 shows the standard linetypes; the bottom image shows the ISO and complex linetypes.

FIGURE 21.3

The lines in this listing of standard linetypes were generated with the underline key and the period, and are only rough representations of the actual lines.

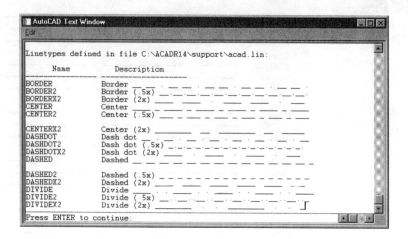

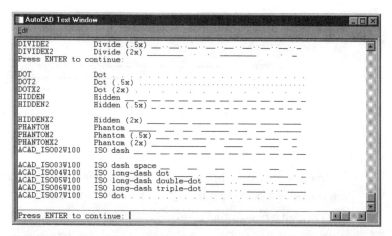

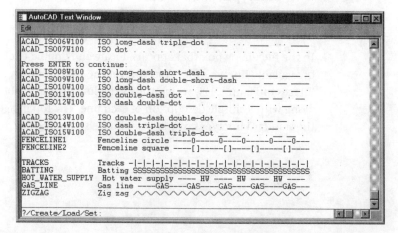

Creating a New Linetype

Next, try creating a new linetype.

1. At the ?/Create/Load/Set: prompt, enter **c**↵.
2. At the Name of linetype to create: prompt, enter **custom**↵ as the name of your new linetype.
3. Notice that the file dialog box you see next is named Create or Append Linetype File. You need to enter the name of the linetype file you want to create or add to. If you pick the default linetype file, ACAD, your new linetype will be added to the Acad.lin file. If you choose to create a new linetype file, AutoCAD will open a file containing the linetype you create and add the .lin extension to the file name you supply.
4. Let's assume you want to start a new linetype file; enter **newline**↵ at the File Name input box.

NOTE

If you had accepted the default linetype file, ACAD, the prompt in step 5 would say Wait, checking if linetype already defined.... This protects you from inadvertently overwriting an existing linetype you may want to keep.

5. At the Creating new file... Descriptive text: prompt, enter a text description of your linetype. You can use any keyboard character as part of your description, but the actual linetype can be composed only of a series of lines, points, and blank spaces. For this exercise, enter

   Custom - My own center line _____ _ _____ ↵

   using the underline key to simulate the appearance of your line.

6. At the Enter pattern (on next line): prompt, enter the following numbers, known as the linetype code (after the a that appears automatically):

   1.0,-.125,.25,-.125↵

WARNING

If you use the Set option of the -Linetype command to set a new default linetype, you will get that linetype no matter what layer you are on.

7. At the New definition written to file. ?/Create/Load/Set: prompt, press ↵ to exit the -Linetype command.

Remember, once you've created a linetype, you must load it in order to use it, as discussed in the *Assigning Linetypes to Layers* section of Chapter 4.

You may also open the Acad.lin or other LIN file with the Windows Notepad and add the descriptive text and linetype code directly to the end of the file.

The Linetype Code

In step 6 of the previous exercise you entered a series of numbers separated by commas. This is the linetype code, representing the different lengths of the components that make up the linetype. The separate elements of the linetype code are explained as follows:

- The 1.0 following the a is the length of the first part of the line. (The a that begins the linetype definition is a code that is applied to all linetypes.)

- The first -.125 is the blank or broken part of the line. The minus sign tells AutoCAD that the line is *not* to be drawn for the specified length, which is .125 units in this example.

- Next comes the positive value of 0.25. This tells AutoCAD to draw a line segment .25 units long after the blank part of the line.

- Finally, the last negative value, -.125, again tells AutoCAD to skip drawing the line for the distance of .125 units.

This series of numbers represents the one segment that is repeated to form the line (see Figure 21.4).

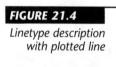

FIGURE 21.4

Linetype description with plotted line

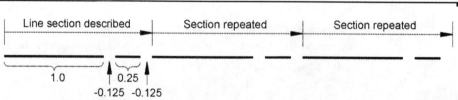

You may be wondering what purpose the a serves at the beginning of the linetype code. A linetype is composed of a series of line segments and points. The a, which is supplied by AutoCAD automatically, is a code that forces the linetype to start and end on a line segment rather than a blank space in the series of lines. At times, AutoCAD stretches the last line segment to force this condition, as shown in Figure 21.5.

FIGURE 21.5
AutoCAD stretches the beginning and end of the line as necessary

NOTE The values you enter for the line-segment lengths are multiplied by the Ltscale factor, so be sure to enter values for the *plotted* lengths.

As mentioned in the beginning of this section, you can also create linetypes outside AutoCAD by using a word processor or text editor such as Windows Notepad. The standard Acad.lin file looks like Figure 21.5. This is the same file you saw earlier, with the addition of the code used by AutoCAD to determine the line segment lengths.

Normally, to use a linetype you have created, you have to load it, through either the Layers or the Linetype dialog box (Data ➤ Layers, or Data ➤ Linetype). If you use one of your own linetypes frequently, you may want to create an icon button macro, so it will be available as an option on a menu.

Creating Complex Linetypes

A complex linetype is one that incorporates text or special graphics. For example, if you want a gas line in a pneumatic schematic, you could show a line with an intermittent *G*, as shown in Figure 21.6.

For the graphics needed to compose complex linetypes, you can use any of the symbols found in the AutoCAD font files discussed in Chapter 7. Just create a text style using these symbols fonts, and then specify the appropriate symbol by using its corresponding letter in the linetype description.

To create a linetype that includes text, use the same linetype code described earlier, with the addition of the necessary font file information in brackets. For example, say you want to create the linetype for the gas line mentioned above. You would add the following to your Acad.lin file:

```
*Gas-Line,Gas-Line --- G --- G --- G --- G ---
A,1.0,-0.25,[" ",STANDARD,S=.2,R=0.0,X=-0.1,Y=-.125],-.25".
```

FIGURE 21.6

Samples of complex linetypes

Fenceline1

Fenceline2

Tracks

Batting

Hot_water_supply

Gas_line

Zigzag

Just_G

FencelineX

The repeat of *Gas-Line* is required if you enter the custom linetype in an editor such as Wordpad.

The information in the square brackets describes the characteristics of the text. The actual text that you want to appear in the line is surrounded by quotes. Next are the text style, scale, rotation angle, X displacement, and Y displacement.

You cannot use the -Linetype command to define complex linetypes. Instead, you must open the Acad.lin file using a text editor, such as the Windows Notepad, and add the linetype information to the end of the file. Make sure you don't duplicate the name of an existing linetype.

You can replace the rotation angle (the R value) with an A, as in the following example:

```
a,1.0,-0.25, ["G", standard, S=.2 A=0, X=.1, Y=.1], -0.25
```

This has the effect of keeping the text at the same angle, regardless of the line's direction. Notice that in this sample, the X and Y values are .1; this will center the Gs on the line. The scale value of .2 will cause the text to be .2 units high, so the .1 is half the height.

In addition to fonts, you can also specify shapes for linetype definitions. Instead of letters, shapes display symbols. Shapes are stored not as drawings, but as definition files, similar to text-font files. In fact, shape files have the same .shx extension as text and are also defined similarly. Figure 21.7 shows some symbols from shape files supplied on the companion CD-ROM.

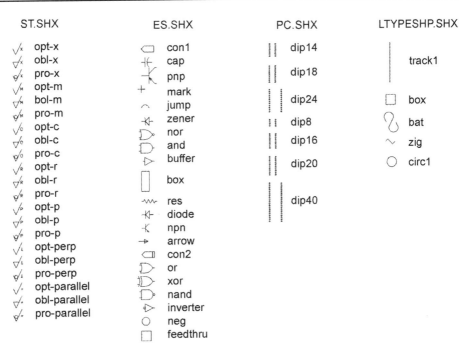

FIGURE 21.7

Samples of shapes available on the companion CD-ROM

To use a shape in a linetype code, use the same format as shown previously for text. However, instead of using a letter and style name, you use the shape name and the shape file name, as in the following example:

```
*Capline, ====
a,1.0,-0.25,[CAP,ES.SHX,S=.2,R=0,X=-.1,Y=-.1],-0.25
```

This example uses the Cap symbol from the Es.shx shape file. The symbol is scaled to .2 units with 0 rotation and an X and Y displacement of –.1.

Creating Hatch Patterns

AutoCAD provides several predefined hatch patterns you can choose from (see Figure 21.8), but you can also create your own. This section demonstrates the basic elements of pattern definition.

FIGURE 21.8

The standard hatch patterns

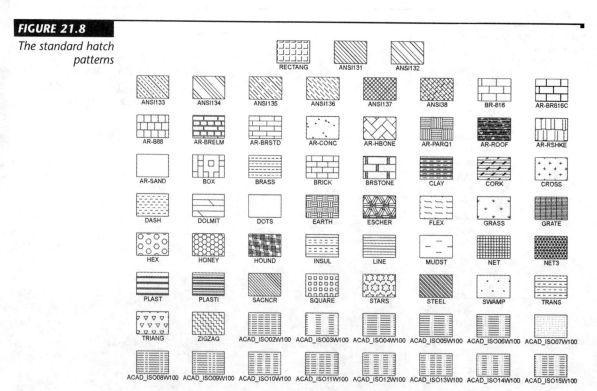

Unlike linetypes, hatch patterns cannot be created while you are in an AutoCAD file. The pattern definitions are contained in an external file named Acad.pat. This file can be opened and edited with a text editor that can handle ASCII files, such as the Windows Notepad. Here is one hatch pattern definition from that file:

```
*square,Small aligned squares
0,  0,0, 0,.125,  .125,-.125
90, 0,0, 0,.125,  .125,-.125
```

You can see some similarities between pattern descriptions and linetype descriptions. They both start with a line of descriptive text, and then give numeric values defining the pattern. However, the numbers in pattern descriptions have a different meaning. This example shows two lines of information. Each line represents a line in the pattern. The first line determines the horizontal line component of the pattern, and the second line represents the vertical component. Figure 21.9 shows the hatch pattern defined in the example.

FIGURE 21.9

Square pattern

A pattern is made up of *line groups*. A line group is like a linetype that is arrayed a specified distance to fill the area to be hatched. A line group is defined by a line of code, much as a linetype is defined. In the square pattern, for instance, two lines—one horizontal and one vertical—are used. Each of these lines is duplicated in a fashion that makes the lines appear as boxes when they are combined. Figure 21.10 illustrates this point.

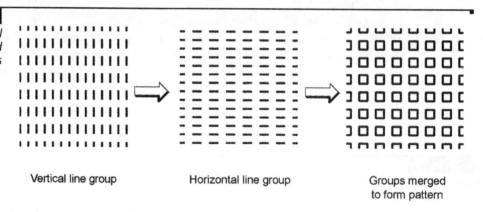

FIGURE 21.10
The individual and combined line groups

Vertical line group   Horizontal line group   Groups merged to form pattern

Look at the first line in the definition:

0, 0,0, 0,.125, .125,-.125

This example shows a series of numbers separated by commas, and it represents one line group. It actually contains four sets of information, separated by blank spaces:

- The first component is the 0 at the beginning. This value indicates the angle of the line group, as determined by the line's orientation. In this case it is 0 for a horizontal line that runs from left to right.

- The next component is the origin of the line group, 0,0. This does not mean that the line actually begins at the drawing origin (see Figure 21.11). It gives you a reference point to determine the location of other line groups involved in generating the pattern.

NOTE

If you have forgotten the numeric values for the various directions, refer to Figure 2.4 in Chapter 2, which shows AutoCAD's system for specifying angles.

CREATING CUSTOM LINETYPES AND HATCH PATTERNS

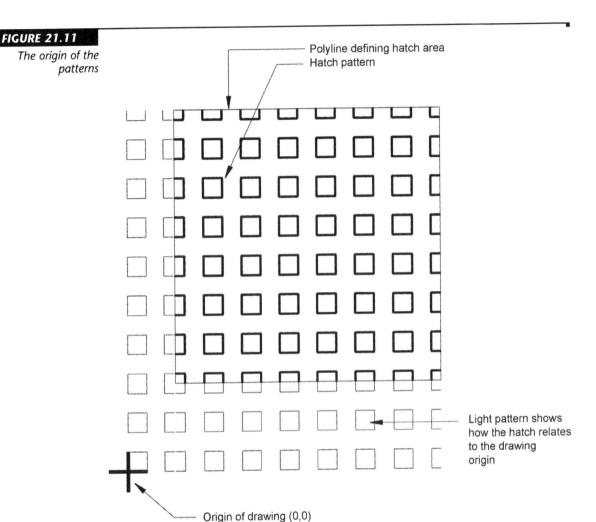

FIGURE 21.11
The origin of the patterns

- The next component is 0,.125. This determines the distance for arraying the line and in what direction, as illustrated in Figure 21.12. This value is like a relative coordinate indicating X and Y distances for a rectangular array. It is not based on the drawing coordinates, but on a coordinate system relative to the orientation of the line. For a line oriented at a 0° angle, the code 0,.125 indicates a precisely vertical direction. For a line oriented at a 45° angle, the code 0,.125 represents a 135° direction. In this example, the duplication occurs 90° in relation to the line group, because the X value is 0. Figure 21.13 illustrates this point.

FIGURE 21.12

The distance and direction of duplication

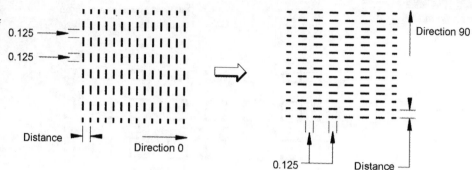

Result

FIGURE 21.13

How the direction of the line group copy is determined

The X and Y coordinate values given for the array distance are based on the orientation of the line group.

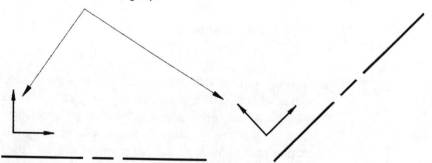

- The last component is the actual description of the line pattern. This value is equivalent to the value given when you create a linetype. Positive values are line segments, and negative values are blank segments. This part of the line group definition works exactly as in the linetype definitions you studied in the previous section.

This system of defining hatch patterns may seem somewhat limiting, but you can actually do a lot with it. Autodesk managed to come up with 53 patterns—and that was really only scratching the surface.

Supporting Your System and Working in Groups

So far in this book you have worked with AutoCAD as an individual learning a program. However, chances are you are usually not alone when you use AutoCAD on a project. Your success with AutoCAD may depend as much on the people you work with as on your knowledge of the program. In the last part of this chapter, we'll look at some of the issues you may face as a member of an interactive group: selecting a system and obtaining support for it; what happens once that system arrives in the company; and some ways you can manage your system.

Getting Outside Support

It helps to have knowledgeable people to consult when questions arise. Most often, the vendor who sells the CAD system is also the source for technical support. Another source is an independent CAD consultant. And don't overlook colleagues who have had some solid experience with AutoCAD. Most likely, you will tap all three of these sources at one point or another as you start to implement an AutoCAD system in your work.

It is well worth searching for a knowledgeable vendor. Your vendor can save you several times the sales commission in work hours your company might otherwise spend trying to solve hardware and software problems. Your vendor can also help you set up a system of file organization and management—something that can become a nightmare if left unattended.

It can be difficult to find a vendor who understands your special needs. This is because the vendor must have specialized knowledge of computers as well as design or production. A good vendor should offer training and phone support, both of which are crucial to productive use of a program with as much complexity as AutoCAD. Some vendors even offer user groups as a means of maintaining an active and open communication with their clients. Here are some further suggestions:

- In addition to your vendor, you might want to find an independent consultant who is familiar with AutoCAD. Although it may be harder to find a good CAD consultant than to find a good vendor, the consultant's view of your needs is unbiased by the motivation to make a sale. The consultant's main goal is to help you gain productivity from your CAD system, so he or she will be more helpful in these areas than the average vendor.

- Get references before you use anyone's services. Your own colleagues may be the best source of information on vendors, consultants, and even hardware. You can learn from their good fortune or mistakes.

- Don't rely on magazine reviews and product demonstrations at shows. These can often be misleading or offer incomplete information. If you see something you like, test it before you buy it. For some products this may be difficult, but because of the complex nature of computer-aided design and drafting, it is important to know exactly what you are getting.

Choosing In-House Experts

Perhaps even more important than a good vendor is an individual within your company who knows your CAD system thoroughly. Ideally, everyone closely involved in your drafting and design projects should be a proficient AutoCAD user, but it is impractical to expect everyone to give time to system management. Usually one individual is chosen for this task. It can sometimes be a thankless job, but when the going gets rough, an in-house expert is indispensable.

NOTE

If you find that a task can be automated, a knowledgeable in-house person can create custom macros and commands on the spot, saving your design or production staff hundreds of work hours—especially if several people are performing that task. Your in-house authority can also train new personnel and answer users' questions.

In a smaller company, the in-house authority may need to be an expert on design, production, and computers—all rolled into one. The point is that to really take advantage of AutoCAD or any CAD system, you should provide some in-house expertise. AutoCAD is a powerful tool, but that power is wasted if you don't take advantage of it.

However, you must be aware that the role of the in-house expert will require significant time spent away from other tasks. Dealing with questions about the program's use can be disruptive to the expert's own work, and writing custom applications is often at least a part-time job in itself. Keep this under consideration when scheduling work or managing costs on a project. It also pays to keep the following in mind:

- The in-house expert should be a professional trained in your firm's field of specialization, rather than someone with a computer background. Every engineer on your staff represents years of training, while learning AutoCAD can take a matter of weeks or months. The expert-to-be, however, should be willing to develop some computer expertise.

- Running a CAD system is not a simple task. It takes clear thinking and good organizational skills, as well as good communication skills. The in-house authority should have some interest in teaching, and the ability to be patient when dealing with interruptions and "stupid" questions. He or she may well be a manager and will need access to the same information as any key player on your design team.

- If you have several computers, you may also want to obtain some general technical support. Especially if your company is implementing its first computer environment, many questions will arise that are not directly related to AutoCAD. The technical support person will be able to answer highly technical questions for which a computer background is a more essential requirement than familiarity with your professional specialty.

- Consider contracting with an outside consultant to provide occasional additional support. This will allow your company to develop custom applications without waiting for staff to develop the necessary skills. A consultant can also help train your staff and might even fill in from time to time when production schedules become too tight.

Acclimatizing the Staff

Once an AutoCAD system is installed and operational, the next step is to get the staff acquainted with that system. This can be the most difficult task of all. In nearly every company, there is at least one key person who resists the use of computers. This can be a tremendous obstacle, especially if that individual is at management level—although nearly anyone who is resisting the project goals can do damage. Unfortunately, there is no easy solution to this problem aside from fostering a positive attitude toward the CAD system's capabilities and its implementation.

AutoCAD has a way of adding force to everything you do, both good and bad. Because it is capable of reproducing work rapidly, it is very easy to unintentionally multiply errors until they are actually out of hand. This also holds true for project management. Poor management tends to be magnified when AutoCAD comes into the picture. You are managing yet another dimension of information—blocks, symbols, layers, and so on. If the users cannot manage and communicate this information, problems are sure to arise.

On the other hand, a smooth-running, well-organized project is reflective of the way AutoCAD enhances your productivity. In fact, good management is essential for realizing productivity gains with AutoCAD. A project on AutoCAD is only as good as the information you provide and the manner in which the system is administered. Open communication and good record keeping are essential to the development and integrity of a design or a set of drawings. The better managed a project is, the fewer problems arise, and the less time is required to get results.

Discussing CAD management procedures in your project kickoff meetings will help get people accustomed to the idea of using the system. Exchanging information with your consultants concerning your CAD system standards is also an important step in keeping a job running smoothly from the start, especially if they are also using AutoCAD.

Learning the System

Learning AutoCAD can be time consuming. If you are the one who is to operate the AutoCAD system, at first you won't be as productive as you were when you were doing everything manually, and don't expect to perform miracles overnight. Once you have a good working knowledge of the program, you still have to integrate it into your day-to-day work. It will take you a month or two, depending on how much time you spend studying AutoCAD, to get to a point where you are entering drawings with any proficiency. It also helps to have a real project you can work on while you are in training. Choose a project that doesn't have a tight schedule, so that if anything goes wrong you have enough time to make corrections.

Remember that it is important to communicate to others what they can expect from you. Otherwise, you may find yourself in an awkward position because you haven't produced the results that someone anticipated.

Making AutoCAD Use Easier

Not everyone in your organization needs to be an AutoCAD expert, but in order for your firm to obtain maximum productivity from AutoCAD, almost everyone involved in design or production should be able to use the system. Designers especially should be involved, since AutoCAD can produce significant time savings in the design phase of a project.

You may want to consider an add-on software package to aid those who need to use AutoCAD but who are not likely to spend a lot of time learning it. These add-ons automate some of the typical functions of a particular application. They can also provide ready-made company standards for symbols and layers. Add-ons are available for mechanical designers, architects, circuit board designers, electrical engineers, and civil engineers, to name a few.

If you are serious about being productive with AutoCAD, you will want to develop custom applications. See Chapter 19 to find out more about customization and third-party software.

Add-ons shouldn't be viewed as the only means of using AutoCAD within your company, but rather as aids to casual users, and partners to your own custom applications. No two companies work alike and no two projects are exactly the same, so add-ons cannot be all things to all people. Remember, AutoCAD is really a graphics tool whose commands and interface can be manipulated to suit any project or company. And that is the way it should be.

Managing an AutoCAD Project

If you are managing a project that is to be put on AutoCAD, be sure you understand what it can and can't do. If your expectations are unreasonable, or if you don't communicate your requirements to the design or production team, friction and problems may occur. Open and clear communication is of the utmost importance, especially when using AutoCAD or any CAD program in a workgroup environment. Here are some further points to consider:

- If your company is just beginning to use AutoCAD, be sure you allow time for staff training. Generally, an individual can become independent on the program after 24 to 36 hours of training. ("Independent" means able to produce drawings without having to constantly refer to a manual or call in the trainer.) This book should provide enough guidance to accomplish this level of skill.

- Once at the point of independence, most individuals will take another month or so to reach a work rate comparable to hand drafting. After that, the individual's productivity will depend on his or her creativity and problem-solving ability. These are very rough estimates, but they should give you an idea of what to expect.

- If you are using a software add-on product, the training period may be shorter, but the user won't have the same depth of knowledge as someone who isn't using the enhancements. As mentioned earlier, this kind of education may be fine for casual users, but you will reach an artificial upper limit on productivity if you rely too heavily on add-ons.

As you or your staff members are learning AutoCAD, you will need to learn how to best utilize this new tool in the context of your company's operations. This may mean rethinking how you go about running a project. It may also mean training coworkers to operate differently.

For example, one of the most common production challenges is scheduling work so that check plots can be produced on a timely basis. Normally, project members have grown accustomed to looking at drawings at convenient times as they progress, even when there are scheduled review dates. With AutoCAD, you won't have that luxury. You will have to consider plotting time when scheduling drawing review dates. This means that the person doing the drawings must get accurate information in time to enter last-minute changes and to plot the drawings.

Establishing Company Standards

Communication is especially important when you are one of many people working on the same project on separate computers. A well-developed set of standards and procedures helps to minimize problems that might be caused by miscommunication. In this section, you'll find some suggestions on how to set up these standards.

Establishing Layering Conventions

You have seen how layers can be a useful tool. But they can easily get out of hand when you have free rein over their creation and naming. This can be especially troublesome when more than one person is working on a set of drawings. The following scenario illustrates this point.

One day the drawing you are working on has 20 layers. Then the next day, you find that someone has added six more, with names that have no meaning to you whatsoever. You don't dare delete those layers, or modify the objects on them, for fear of retaliation from the individual who put them there. You ask around, but no one seems to know anything about these new layers. Finally, after spending an hour or two tracking down the culprit, you discover that the layers are not important at all.

With an appropriate layer-naming convention, you can minimize this type of problem (though you may not eliminate it entirely). A too-rigid naming convention can cause as many problems as no convention at all, so it is best to give general guidelines rather than force everyone to stay within narrow limits. As mentioned in Chapter 4, you can create layer names in a way that allows you to group them using wildcards. AutoCAD allows up to 31 characters in a layer name, so you can use descriptive names.

Line weights should be standardized in conjunction with colors. If you intend to use a service bureau for your plotting, check with them first; they may require that you conform to their color and line-weight standards.

Maintaining Files

As you use your computer system, you will generate files rapidly. Some files will be garbage, some will be necessary but infrequently used, and others will be current job files or system files that are constantly used. In addition, AutoCAD automatically creates backup files (they have the file name extension .bak). If something is wrong with the most-recent version, you can restore the BAK file to a drawing file by simply changing the .bak extension to .dwg. You may or may not want to keep these backup files.

An AutoCAD backup file is a copy of the most-recent version of the file before you issued the last Save or End command. If you save the BAK files, you will always have the next-to-most recent version of a file.

All these files take up valuable space, and you may end up with insufficient room for your working files. Because of this, you will want to regularly clear the unused files from the hard disk, by erasing unwanted files and archiving those that are used infrequently. If you are connected to a network you may back up files there; otherwise you will need to back up on floppy disk or tape. It's wise to do this at every editing session, so you don't confuse meaningful files with garbage. You may want to erase all your BAK files, as well, if you don't care to keep them.

You can set up AutoCAD so the it does not create the BAK files. Open the Preferences dialog box, click on the General tab, and then remove the check mark from the Create Backup Copy with Each Save check box.

Backing Up Files

Besides keeping your hard disk clear of unused and inactive files, consider backing it up daily or at least once a week. This might be done by the in-house expert or the technical support staff. You needn't back up the entire hard disk, just those crucial drawing files you've been slaving over for the past several months. When you do back up your entire hard disk, you save the configuration of all your programs and the directory structure. In the event of a hard disk failure, you won't have to reinstall and reconfigure all your software—just restore the backups once the problem with the hard disk is remedied.

> In the event that you do have a system failure and some files become corrupted, AutoCAD has a built-in file-recovery routine. Whenever AutoCAD attempts to load a corrupted file, it will tell you the file is corrupted and proceed to fix the problem as best it can. See Appendix A for more on this feature.

A good policy to follow is to back up your entire hard disk every week, and your data files every day, as a compromise to backing up everything daily.

There are two methods for backing up a hard disk. One is through software, using your existing floppy-disk drive and removable disks as the backup media. DOS provides software for this purpose, but it is slow and difficult to use. Tape backup systems are the preferred backup method for a hard disk. These systems transfer the contents of your hard disk onto a tape cartridge or cassette. The software to operate the tape system is usually provided as part of the system.

Tape backup systems cost a bit more than the software-only alternative, but tape systems offer more flexibility and ease of use. For example, with backup software, you must constantly insert and remove disks from the computer as they are filled up. A tape backup system allows you to start the backup process and then walk away to do something else. Some systems offer timed backup so you don't even have to think about starting them—it simply happens at a predetermined time, usually in the evening.

Tape backup systems can back up a hard disk at the rate of about 8MB to 20MB per minute. The price of these systems ranges from $150 for a system installed inside your PC to $2,000 for an external multiuser unit, depending on tape capacity. The cartridges or cassettes used for storage cost around $15. Although you initially spend more on a tape backup system, you can save time and pay less for storage media.

There are other, faster, more flexible options, such as rewritable optical drives that can store 500MB to 1GB of data. Such a system usually costs twice as much as a tape system. Rewritable optical drives can be invaluable, however, in situations where large amounts of data must be archived and readily retrieved. You might also consider recordable CD-ROM drives as a backup tool. They can also double as a means to distribute drawings or other project files to clients and consultants.

Labeling Hard Copies

A problem you will run into once you start to generate files is keeping track of which AutoCAD file goes with which hard copy drawing. It's a good idea to place an identifying tag on the drawing (that will plot with the drawing) in some inconspicuous place. As well as the file name, the tag should include such information as the date and time the drawing was last edited, and who edited it. All these bits of information can prove helpful in the progress of a design or production project.

The Batch Plot utility described in Chapter 14 allows you to add a descriptive label to your drawings. Also, many of the modern ink-jet plotters feature a plot-stamping function.

Using Networks with AutoCAD

In an effort to simplify your file maintenance, you may want to consider installing a network to connect your computers electronically. Networks offer a way to share files and peripherals among computers in different locations.

Two basic types of network can be used with AutoCAD: dedicated file server/client systems and peer-to-peer systems. The *dedicated file server* offers you a way to maintain files on a single computer. The computers connected to the server are referred to as *clients* or *nodes*. You can store all of your common symbols, AutoLISP programs, custom menus, and working files on the server, thus reducing the risk of having dupli-cate files. The client computers are simpler and less powerful. They use the server as their main storage device, accessing the programs and data stored on the server through the network. Networks with servers also have all the peripheral output devices connected to the server. This centralized type of system offers the user an easier way of managing files.

> In simplified terms, a *dedicated file server* is a storage device, often a computer with a large-capacity hard disk, that acts as a repository of data. A server will often have a tape-backup device to facilitate regular system backup.

A *peer-to-peer network* does not use a server. Instead, each computer has equal status and can access files and peripherals on other computers on the network. Generally, this type of network is less expensive because you don't need to dedicate a computer to the single task of server. Peripherals such as plotters and printers are shared among computers. Even hard disks are shared, although access to directories can be controlled at each computer.

Networks can be useful tools in managing your work, but they can also introduce new difficulties. For some network users, file-version control becomes a major concern. Speed of file access can be another problem. No matter what form of network you have or install, you will need a system manager whose duties include backing up files in the server and making sure the network's output devices are operating as they should. Used properly, a network can save time by easing the flow of information between computer users; it must, however, be managed carefully.

Here are some tips on using AutoCAD on a network:

- If you can afford it, use a star topology for your network, with an active hub and a dedicated server.
- Configure AutoCAD to store temporary files on the client or node computer.
- Configure your network version of AutoCAD for multiple swap files and local code pages, and set the AutoCAD preferences to use the client or node computer for swap files (see Appendix A for more on these tasks).

Keeping Records

Computers are said to create "the paperless office." As you work more and more with them, however, you may find that quite the opposite is true. Although you can store more information on magnetic media, you will spend a good deal of time reviewing that information on hard copy because it is very difficult to spot errors on a computer monitor. When you use AutoCAD on large projects, another level of documentation must also exist: a way of keeping track of the many elements that go into the creation of drawings.

Because job requirements vary, you may want to provide a layer log to keep track of layers and their intended uses for specific jobs, or better yet, use the Layer Manager bonus tool described in Chapter 19. Also, to manage blocks within files, a log of block names and their insertion values can help. Finally, plan to keep a log of symbols. You will probably have a library of symbols used in your work group, and this library will grow as your projects become more varied. Documenting these symbols will help to keep track of them.

Another activity that you may want to keep records for is plotting—especially if you bill your clients separately for computer time or for analyzing job costs. A plot log might contain such information as the time spent on plotting, the type of plot done, the purpose of the plot, and even a list of plotting errors and problems that arise with each drawing.

Although records may be the last thing on your mind when you are working to meet a deadline, in the long run they can save time and aggravation for you and the people you work with.

Understanding What AutoCAD Can Do for You

Many of us have only a vague idea of what AutoCAD can contribute to our work. We think it will make our drafting tasks go faster, but we're not sure exactly how; or we believe it will make us produce better-quality drawings. Some people expect AutoCAD will make them better designers or allow them to produce professional-quality drawings without having much drawing talent. All these things are true to an extent, and AutoCAD can help you in some ways that are less tangible than speed and quality.

Seeing the Hidden Advantages

We have discussed how AutoCAD can help you meet your drafting and design challenges by allowing you to visualize your ideas more clearly and by reducing the time it takes to do repetitive tasks. AutoCAD also forces you to organize your drawing process more efficiently. It changes your perception of problems and, although it may introduce new ones, the additional accuracy and information AutoCAD provides will minimize errors.

AutoCAD also provides drawing consistency. A set of drawings done on AutoCAD is more legible and consistent, reducing the possibility of errors caused by illegible handwriting or poor drafting. In our litigious culture, this is a significant feature.

Finally, because AutoCAD systems are becoming more prevalent, it is easier to find people who can use AutoCAD proficiently. As this group of experts grows, training will become less of a burden to your company.

Taking a Project Off AutoCAD

As helpful as AutoCAD can be, there are times when making revisions on AutoCAD is simply not worth the effort. Last-minute changes that are minor but pervasive throughout a set of drawings are best made by hand on the most up-to-date hard copy. That way, you don't waste time and drawing media in plotting drawings.

Also, it is a good idea to have a label on every hard copy of a file so that users know which file to edit when the time comes to make changes.

To keep your AutoCAD files up to date, once a project is done, you should go back and enter any final changes you made on the hard copy into the AutoCAD version of the file.

Only your experience can help you determine the best time to stop using AutoCAD and start making changes by hand. Many factors will influence this decision: the size and complexity of the project, the people available to work on it, and the nature of the revisions—to name just a few.

As you have seen, AutoCAD is a powerful software tool, and like any powerful program, it is difficult to master. We hope this last chapter has given you the incentive to take advantage of AutoCAD's full potential. Remember: Even after you've mastered AutoCAD, there are many other issues that you must confront while using AutoCAD in the company environment.

Unlike words, drawings have few restrictions. The process of writing requires adherence to the structures of our language. The process of drawing, on the other hand, has no fixed structure. For example, there are a million ways to draw a face or a machine part. For this reason, your use of AutoCAD is much less restricted, and there are certainly potential uses for it that are yet to be discovered. We encourage you to experiment with AutoCAD and explore the infinite possibilities it offers for solving your design problems and communicating your ideas.

We hope *Mastering AutoCAD 14 for Mechanical Engineers* has been of benefit to you and that you will continue to use it as a reference. If you have comments, criticisms, or ideas about how we can improve the next edition of this book, write to one of us at the address below. And thanks for choosing *Mastering AutoCAD 14 for Mechanical Engineers*.

George Omura
Omura Illustration
P.O. Box 357
Albany, CA 94706-0357
E-Mail: `gomura@sirius.com`

Steven Keith
San Jose, CA
E-Mail: `steven.keith@home.com`

Appendix A

Hardware and Software Tips

Hardware and Software Tips

Because some of the items that make up an AutoCAD system are not found on the typical desktop system, we have provided this appendix to help you understand some of the less common items you may need. This appendix also discusses ways you can improve AutoCAD's performance through software and hardware. Note, however, that when you first install AutoCAD R14, it is automatically configured to the Windows devices already installed on your computer. If you need more information of the installation process, see Appendix B, where installation is discussed in detail.

The Graphics Display

There are two issues to consider concerning the graphics display: *resolution* and *performance*.

If you are running Windows 95 or Windows NT 4 software, you probably already have a high-resolution display card for your monitor. If you haven't set up your system yet, then a high-resolution display card and monitor is a must. SVGA display cards are inexpensive and offer high-quality display.

Graphics performance comes down to speed. To enhance graphics performance, consider the following resources:

- **PCI Bus:** Contemporary motherboards use high-speed PCI expansion slots that offer the fastest video throughput available. Check that yours is a system of this type. If you are shopping for a graphics card, make sure that it is a "plug-and-play" card for ease of installation.

- **Video RAM:** If you are shopping for a display system, you should also make sure that it has 2–8 megabytes of *video RAM*. Standard video cards typically have only 2MB of video RAM. Greater amounts, of up to 8MB, will give you noticeably better performance with AutoCAD.

In earlier versions of AutoCAD it was advisable to obtain *display list* software to improve graphics performance. With Release 14 this is no longer necessary (although some third-party add-on video drivers may offer other features that enhance AutoCAD). AutoCAD comes equipped with display list software. The display list gives you nearly instantaneous pans and zooms.

Pointing Devices

Our most basic means of communicating with computers is the keyboard and the pointing device. Most likely, you will use a mouse, but if you are still in the market for a pointing device, choose an input device that generates smooth cursor movement. Some of the lesser-quality input devices cause erratic movement. When looking for an input device other than a mouse, choose one that provides positive feedback, such as a definitive button click when you pick an object on the screen. Many low-cost digitizers have a very poor button feel, which can cause errors when you are selecting points or menu options. The Windows or System Pointing Device is automatically available when you first load AutoCAD.

In general, use a high-resolution mouse if you do not plan to do any tracing. If you must use a tablet menu, or if you know you are going to trace drawings, then get a digitizer, but be sure it is of good quality.

The Digitizing Tablet

If you need to trace drawings, you should consider a *digitizing tablet*. This is usually a rectangular object with a penlike *stylus* or a device called a *puck*, which resembles a mouse. It has a smooth surface on which to draw. The most popular size is 11" × 11", but digitizing tablets are available in sizes up to 60" × 70". The tablet gives a natural feel to drawing with the computer because the movement of the stylus or puck is directly translated into cursor movement. While many digitizers come with a stylus, you will want a multi-button puck to work with AutoCAD.

A digitizing tablet's puck often has *function buttons*. These buttons can be programmed through the AutoCAD menu file system to start your most frequently used commands, which is much faster than searching through an on-screen menu. You can also select commands from the tablet's surface if you install the *menu template* supplied with AutoCAD. A menu template is a flat sheet of plastic with the AutoCAD commands printed on it. You can select commands simply by pointing at them on the template. If you have a digitizing tablet, refer to Appendix B, which tells you how to install a template.

AutoCAD supports Wintab-compatible digitizers. If your digitizer has a Wintab driver, you can use your digitizer as both a *tracing device* (to trace drawings on a tablet) and a *pointing device* (to choose AutoCAD or Windows 95 menu items).

Your Wintab digitizer must be installed and configured under Windows. Check that it is working in Windows before enabling it in AutoCAD, otherwise you will not be able to use the digitizer as a pointing device (mouse). To enable the digitizer in AutoCAD, choose Tools ➢ Preferences and click on the Pointer tab. Then select the Digitizer or Digitizer & Mouse option in the Accept Input section.

Output Devices

Output options vary greatly in quality and price. Quality and paper size are the major considerations for both printers and plotters. Nearly all printers give accurate drawings, but some produce better line quality than others. Some plotters give merely acceptable results, while others are quite impressive in their speed and accuracy.

AutoCAD Release 14 can use the Windows 95 or NT system printer, so any device that Windows supports is also supported by AutoCAD. AutoCAD also gives you the option of plotting directly to an output device. By plotting directly to the output device, instead of going through Windows 95, AutoCAD can offer more control over the final output.

Printers

There are so many kinds of printers available these days that it has become more difficult to choose the right printer for your application. Here, we offer a description of the broad categories of printers available and how they relate to AutoCAD. You will also want to consider the other uses for your printer, such as word processing or color graphics. Here are a few printing options:

Laser Printers produce high-quality line work output. The standard office laser printer is usually limited to 8½" × 11" paper; however, 11" × 17" laser printers for graphics and CAD work are now becoming affordable and are commonly used for proof plots. Resolution and speed are the major considerations if you are buying a laser printer. An output of 300 dpi (dots per inch) produces very acceptable plots, but 600 dpi is fast becoming the standard. You should look also for a laser printer with sufficient built-in memory to improve spooling and plotting speeds.

Ink-Jet or Bubble-Jet Printers offer speed and quality output. Some ink-jet printers even accept 17" × 22" paper and offer PostScript emulation at up to 720 dpi. Since ink-jet printers are competitively priced, they can offer the best solution for low-cost check plots. And the 17" × 22" paper size is quite acceptable for half-size plots, a format that most architects and engineers are using.

PostScript Printers offer control of the nuances of your printout. Use a PostScript device to output your drawings. The best method is to use File ➢ Export or the Psout command. These options convert your drawing into a true PostScript file. You can then send your file to a PostScript printer or typesetting machine. This can be especially useful for PCB layout where you require photo negatives for output. If you need presentation-quality drawings, you may want to consider using the Encapsulated PS (*.eps) option in the Export Data dialog box. Often service bureaus which offer a raster plotter service can produce E-size Post-Script output from a PostScript file.

Plotters

A *plotter* is a mechanical drafting device used to draw a computer image on sheets of paper, vellum, or polyester film. In the early days of CAD, most plotters used ink pens. Now, ink-jet technology has taken over the plotter market. You can also find laser, thermal, or electrostatic plotters at much higher prices. Black-and-white ink-jet plotters offer the best value; they are fast and fairly inexpensive compared to the older pen plotters they supersede. For a bit more money, you can step up to a color plotter. Black-and-white ink-jet plotters are capable of printing raster images in larger formats. And they can print patterns and screens for highlight effects.

If you need large plots but can't afford a large plotter, many service bureaus (or blueprint companies) offer plotting as a service. This can be a very good alternative to purchasing your own plotter. Check with your local service bureau.

Fine-Tuning PostScript File Export

AutoCAD provides the PostScript user with a great deal of control over the formatting of the PostScript file that is output with the File ➢ Export and the Psout command. The options range from font-substitution mapping to custom PostScript Prologue data.

However, AutoCAD does require you to master the PostScript programming language to take full advantage of AutoCAD PostScript output. Most of this control is offered through a file named Acad.psf. This is the master support file for the Psin and Psout commands. You can customize the PostScript file created by Psout by making changes in the Acad.psf file. Acad.psf is divided into sections that affect various parts of the PostScript output. Each section begins with a title preceded by an asterisk. The following list briefly describes these sections.

*fonts lets you control font substitution. You can assign a PostScript font to an AutoCAD font file or PostScript PFB file used in the drawing.

*figureprologue defines the procedures used for embedding figures included with Psin PostScript images.

*isofontprologue defines the procedures used to re-encode fonts in order to be compatible with the ISO 8859 Latin/1 character set.

*fillprologue defines the code used in the Psout file to describe area fills.

*fill allows you to include your own custom fill patterns.

Most of these sections, with the exception of the *fonts section, will be of little use to the average user, but if you are a PostScript programmer, you can take advantage of these sections to customize your PostScript output.

You will also want to know about the *Psprolog* system variable. This system variable instructs Psout to include your custom prolog statement in its PostScript output. (See Chapter 18 for details on using Psout.) You add your custom prolog to the Acad.psf file using a text editor. The prolog should begin with a section heading that you devise. The heading can say anything, but it must begin with an asterisk like all the other section headings. Everything following the heading, up to the next heading or the end of the file, and excluding comments, will be included in the Psout output file. The following code shows a sample prolog that converts color assignments to line widths in a way similar to pen plotters:

```
*widthprolog
/ACADLayer { pop } def
/ACADColor { pop pop pop dup 0.5 mul setlinewidth pop} def
/ACADLtype { pop
userdict /Linedict known not { /Linedict 100 dict def } if
1 index cvn Linedict exch known not {
mark 1 index { dup 0 eq { pop 1 72.0 div } if abs } forall
counttomark 2 add ~-1 roll astore exch pop
1 index cvn exch Linedict begin def end }
{ pop } ifelse
Linedict begin cvx exec 0 setdash end } bind def
/bd{bind def}bind def /m{moveto}bd /l{lineto}bd /s{stroke}bd
/a{arc}bd /an{arcn}bd /gs{gsave}bd /gr{grestore}bd
/cp{closepath}bd /tr{translate}bd /sc{scale}bd /co{concat}bd
/ff{findfont}bd /sf{setfont}bd /sh{show}bd /np{newpath}bd
/sw{setlinewidth}bd /sj{setlinejoin}bd /sm{setmiterlimit}bd /cl{clip}bd
/fi{fill}bd
%%EndProlog
```

A complete discussion of Psout and Psin PostScript support is beyond the scope of this book. If you are interested in learning more, consult the PostScript section of the *AutoCAD Customization* manual. You can also learn a lot by looking at the PostScript files produced by Psout, browsing the Acad.psf file, and consulting the following publications:

- *Understanding PostScript* by David Holzgang (Sybex, 1992)
- *PostScript Language Program Design* by Adobe Systems Incorporated (Addison-Wesley Publishing Company, Inc.)
- *PostScript Language Reference Manual* by Adobe Systems Incorporated (Addison-Wesley Publishing Company, Inc.)
- *PostScript Language Tutorial and Cookbook* by Adobe Systems Incorporated (Addison-Wesley Publishing Company, Inc.)

You can also add a PostScript plotter to the plotter configuration. When you do this, AutoCAD plots the drawing as a series of vectors, just like any other plotter. If you have any filled areas in your drawing and you are plotting to a PostScript file, the vectors that are used to plot those filled areas can greatly increase plot-file size and the time it takes to plot your PostScript file.

Memory and AutoCAD Performance

Next to your computer's CPU, memory has the greatest impact on AutoCAD's speed. How much you have, and how you use it, can make a big difference to whether you finish that rush job on schedule or work late nights trying. In this section, we hope to clarify some basic points about memory and how AutoCAD uses it.

AutoCAD Release 14 is a virtual memory system. This means that when your RAM memory resources reach their limit, part of the data stored in RAM is temporarily moved to your hard disk to make more room in RAM. This temporary storage of RAM to your hard disk is called *memory paging*. Through memory paging, AutoCAD will continue to run, even though your work might exceed the capacity of your RAM.

AutoCAD uses memory in two ways. First, it stores its program code in RAM. The more programs you have open under Windows, the more RAM will be used. Windows controls the use of memory for program code, so if you start to reach the RAM limit, Windows will take care of memory paging. The second way AutoCAD uses memory is for storing drawing data. AutoCAD always attempts to store as much of your drawing in RAM as possible. Again, when the amount of RAM required for a drawing exceeds the actual RAM available, AutoCAD will page parts of the drawing data to the hard disk. The paging of drawing data is controlled strictly by AutoCAD. Since RAM is shared by program code and drawing data, your drawing size and the number of programs you have open under Windows will affect how much RAM you have available. For this reason, if you find your AutoCAD editing session is slowing down, try closing other applications you might have open. This will free up more memory for AutoCAD and the drawing file.

AutoCAD and Your Hard Disk

You will notice that AutoCAD slows down when paging occurs. If this happens frequently, the best thing you can do is add more RAM. But you can also improve the performance of AutoCAD under these conditions by insuring that you have adequate hard-disk space and that any free hard-disk space has been *defragmented* or *optimized*. A defragmented disk will offer faster access, thereby improving paging speed.

With previous versions of Windows, you were advised to set up a permanent swap file. With Windows 95 or NT 4, this is not necessary. Windows dynamically allocates swap-file space. However, you should make sure that there is enough free space on your hard disk to allow Windows to set up the space. A good guideline is to allow enough space for a swap file that is four times the size of your RAM capacity. If you have 32MB of RAM, for example, you need to allow space for a 128MB swap file (at a minimum). This will give your system 128MB of virtual memory.

What to Do for *Out of RAM* and *Out of Page Space* Errors

After you have used AutoCAD for some time, you may find some odd-looking files with an .ac$ extension in the \Windows\Temp directory. These files are the temporary files for storing unused portions of a drawing. They often appear if AutoCAD has been terminated abnormally. You can usually erase these files without any adverse effect.

If you've discarded all the old swap files and your disk is still unusually full, there may be some lost file clusters filling up your hard disk. Lost clusters are pieces of files that are not actually assigned to a specific file. Often they crop up when a program has terminated abnormally. To eliminate them and free up the disk space they're using, exit AutoCAD and run Scandisk, which is a standard Windows accessory.

Finally, if you are not in the habit of emptying your Recycle bin, you should do so now. Every file that you "delete" using the Windows Explorer is actually passed to the Recycle bin. You need to clear this out regularly.

When Things Go Wrong

AutoCAD is a complex program, and at times things won't go exactly right. If you run into a problem, chances are the problem is not insurmountable. Here are a few tips on what to do when things don't work.

Difficulty Starting Up or Opening a File

The most common reason that you'll have difficulty opening a file is the lack of free disk space. If you encounter errors while attempting to open files, check to see if you have adequate free disk space on all your drives.

If you've recently installed AutoCAD but you cannot get it started, you may have a configuration problem. Before you panic, try reinstalling AutoCAD from scratch. Particularly if you are installing the CD version, this does not take long (see Appendix B for installation instructions). Before you reinstall AutoCAD, use the Uninstall program to remove the current version of AutoCAD. Also make sure you have your Authorization code, serial number, and CD-Key handy. Make sure that you've closed all other programs when you run the AutoCAD installation, and as a final measure, restart your computer when you've completed the installation.

Restoring Corrupted Files

Hardware failures can result in data files becoming corrupted. When this happens, AutoCAD is unable to open the drawing file. Fortunately, there is hope for damaged files. In most cases, AutoCAD will run through a file-recovery routine automatically when it attempts to load a corrupted file. If you have a file you know is corrupted, you can start the file-recovery utility by clicking on File ➢ Drawing Utilites ➢ Recover. This opens the Select File dialog box, allowing you to select the file you want to recover. Once you enter the name, AutoCAD goes to work. You get a series of messages, most of which have little meaning to the average user. Then the recovered file is opened. You may lose some data, but a partial file is better than no file at all, especially when the file represents several days of work.

Another possibility is to attempt to recover your drawing from the BAK file—the last saved version before your drawing was corrupted. Rename the drawing BAK file to a DWG file with a different name, and then open it up. The drawing will contain only what was in your drawing when it was previously saved.

There may be situations when a file is so badly corrupted it cannot be restored. By backing up frequently, you can minimize the inconvenience of such an occurrence. You also might want to consider the Microsoft Office Plus package, which allows you to schedule backups, scan, and defragment your drives during off hours. Programs such as Scandisk can spot problem areas on your hard disk before they cause trouble.

Troubleshooting

AutoCAD is a large, complex program, so you are bound to encounter some difficulties from time to time. This section covers a few of the more common problems experienced while using AutoCAD.

You can see but cannot select objects in a drawing someone else has worked on   This may be happening because you have a Paper Space view instead of a Model Space view. To make sure you're in Model Space, type **tilemode**↵, and then type **1**↵. Or you can turn on the UCS icon (by typing **ucsicon**↵**on**↵, and if you see the triangular UCS icon in the lower-left corner, you are in Paper Space. You must go to Model Space before you can edit the drawing.

Another item to check is the layer lock setting. If a layer is locked, you won't be able to edit objects on that layer.

Grips do not appear when objects are selected   Make sure the Grips feature is enabled (Tools ➢ Grips). See Appendix B for details.

When you select objects, AutoCAD doesn't work as this book describes   Check the Selection settings to make sure they are set the same way as the exercise specifies (Tools ➢ Selection).

Text appears in the wrong font style, or an error message says AutoCAD cannot find font files   When you are working on files from another company, it's not uncommon to encounter a file that uses special third-party fonts that you do not have. You can usually substitute standard AutoCAD fonts for any fonts you don't have, without adverse effects. AutoCAD automatically presents a dialog box letting you select font files for the substitution. You can either choose a font file or press the Esc key to ignore the message (see Chapter 7 for more on font files). If you choose to ignore the error message, you might not see some of the text that would normally appear in the drawing.

DXF files do not import   Various problems can occur during the DXF import, the most common of which is that you are trying to import a DXF file into an existing drawing rather than a new drawing. Under some conditions, you can import a DXF file into an existing drawing using the Dxfin command, but AutoCAD may not import the entire file.

To ensure that your entire DXF file is safely imported, choose File ➢ Open and select *.dxf from the File Type drop-down list. Then import your DXF file.

If you know that the DXF file you are trying to import is in ASCII format and not a Binary DXF, take a look at the file with a text editor. If it contains odd-looking characters, chances are the file is damaged or contains extra data that AutoCAD cannot understand. Try deleting the lines of odd-looking characters, and then import the file again (be sure to make a backup copy of the file before you attempt this).

A file cannot be saved to disk Frequently, a hard drive will fill up quickly during an editing session. AutoCAD can generate temporary and swap files many times larger than the file you are editing. This can leave you with no room left to save your file. If this happens, you can empty the Recycle Bin to clear some space on your hard drive, or delete old AutoCAD BAK files you don't need. *Do not delete AutoCAD temporary files.*

AutoCAD does not display all the Paper Space viewports AutoCAD uses substantial memory to display Paper Space viewports. For this reason, it limits the number of viewports it will display at one time. Even though viewports don't display, they will still plot. Also, if you zoom in on a blank viewport while in Paper Space, you will be able to see its contents. The viewport regains visibility because you are reducing the number of viewports shown on the screen at one time.

You can increase the number of viewports AutoCAD will display at one time by resetting the *Maxactvp* system variable. (This is usually set to 48.) Be forewarned, however, that increasing the Maxactvp setting will cause AutoCAD to use more memory. If you have limited memory on your system, this will slow down AutoCAD considerably.

AutoCAD becomes impossibly slow when adding more Paper Space viewports As mentioned, AutoCAD consumes memory quickly when adding viewports. If your system resources are limited, you can reduce the Maxactvp system variable setting so that AutoCAD displays fewer viewports at one time. This will let you work on a file that has numerous viewports without decreasing your computer's performance. Try reducing Maxactvp to 8, and then reduce or increase the setting until you find the optimum value for your situation. Alternatively, you can use the `Mview OFF` option to turn off viewports when you are not working in them.

AutoCAD won't open a large file, and displays a "Page File Full" message AutoCAD will open drawing files larger than can fit into your system's RAM. In order to do this, however, AutoCAD attempts to store part of the drawing in a temporary file on your hard drive. If there isn't room on the hard drive, AutoCAD will give up. To remedy this problem, clear some space on your hard drive. It is not uncommon for AutoCAD to require as much as 10MB of hard-disk space for every 1MB of a drawing file.

The keyboard shortcuts for commands are not working If you are working on an unfamiliar computer, chances are the keyboard shortcuts (or command aliases) have been altered. The command aliases are stored in the Acad.pgp file. If you have installed the Release 14 Bonus Menu (see Appendix B for overall installation instructions), you can modify the keyboard shortcuts by using the Command Alias Editor option on the Bonus menu. Choose Bonus ➢ Tools ➢ Command Alias Editor, or type **aliasedit**↵ at the command prompt. Then you may add, remove, or edit the keyboard shortcut codes. See Chapter 19 for more on the Alias Editor.

Plots come out blank Check the scale factor you are using for your plot. Often, a blank plot means your scale factor is making the plot too large to fit on the sheet. Try plotting with the Scaled to Fit option. If you get a plot, then you know your scale factor is incorrect. See Chapter 14 for more on plotting options. Check your output before you plot by using the Full Preview option in the Print/Plot Configuration dialog box.

You cannot get your drawing to be properly oriented on the sheet If you want to change the orientation of your drawing on a plotted sheet and the Print/Plot Configuration orientation options don't seem to work, try rotating the UCS to align with your desired plot view, and then type **plan**↵. Adjust the view to display what you want to have plotted, and then use the View command (View ➢ Named Views) to save this view. When you are ready to plot, use the View option in the Print/Plot Configuration dialog box and plot the saved view, instead of rotating the plot.

Appendix B

Installing and Setting Up AutoCAD

Installing and Setting Up AutoCAD

Before Installing AutoCAD

Before you begin the installation process, be sure you have a drive with at least 100MB of free disk space. In addition, you should know your AutoCAD vendor's name and phone number.

You will also want to have at least an additional 50MB of free disk space for AutoCAD *temporary files* and *swap files*, plus another 20MB for the tutorial files you will create. (Temporary and swap files are system files AutoCAD creates as it works. You don't have to deal with these files directly, but you do have to allow room for them. If you want to know more about these files, see Appendix A.) If you are installing AutoCAD on a drive other than the one on which Windows is installed, make sure you have about 20MB free on the Windows drive. AutoCAD also stores temporary files there.

Finally, have your AutoCAD vendor's name and phone number ready. You will be asked to enter this information during the installation. You will also want to have your network and single-user authorization code ready. These can be obtained by calling the toll-free number listed in your AutoCAD package. Single-user systems have a 30-day grace period, so you can install and use AutoCAD without having to enter your authorization code right away.

Installing the AutoCAD Software

After you've made sure you've got enough disk space and you've closed all other programs, proceed with the following steps to install AutoCAD:

1. To begin your installation, be sure the AutoCAD CD is in your CD-ROM drive.
2. Click on the Windows Start button and choose Run.
3. At the Run dialog box, enter **D:setup** into the input box. Enter the drive letter of your CD-ROM in place of the *D* in this example. Click on OK when you are

ready. You'll see the Welcome dialog box warning you to make sure all other programs are closed. Click on Next to continue.

4. The next dialog box is the Software License Agreement. Read it, and if you accept the terms, click on Accept.

5. In the Serial Number dialog box, enter the serial number and CD Key, and then click on Next.

6. Next, you'll see the Personal Information dialog box. To personalize AutoCAD, you are asked for your name, company, and AutoCAD vendor's name and telephone number. This information will be displayed on the opening AutoCAD screen, so don't enter anything you'll regret later.

7. You are then asked to confirm the information you entered. If you have changed your mind, you can go back and change the information in the previous dialog box. Otherwise, click on Next.

8. Select the location for your AutoCAD files. The tutorial assumes you have AutoCAD on drive C and in a directory called \Program Files\AutoCAD R14— these are the defaults during the installation. When you're done, click on Next. If the installation location does not exist, you will see a message to this effect and asking if it is OK for the Installer to create it. Click on OK.

9. In the next dialog box, choose the type of setup you want. You have the choice of Typical, Full, Compact, or Custom. We encourage you to choose the Full installation, as this will enable you to take advantage of the Bonus utilities discussed in Chapter 19.

10. The installation will create a program folder with a set of AutoCAD-related programs and documents. Enter a name for this folder in the next dialog box or choose to accept the default, AutoCAD R14.

11. You'll finally get a Setup confirmation dialog box showing you a listing of program components the Installation will install. Click on Next to begin the actual file installation. This can take several minutes.

12. Once the installation is done, you will see the Setup Complete dialog box. You can have the install program automatically open the AutoCAD R14 Readme document by clicking on Yes I Want to View the Readme File. Click on Finish to exit the installation.

13. Start AutoCAD by double-clicking on the AutoCAD R14 shortcut on the Windows Desktop.

14. You will then see a message telling you that you have 30 days to authorize your copy of AutoCAD. The message will also give you the phone number for obtaining your Authorization code. Enter your Authorization code and click on

Authorize, or click on Defer to open AutoCAD without authorization. If you don't enter an Authorization code, you will see this message each time you open AutoCAD—until you finally enter it.

The Program Files

In the \Program Files\AcadR14 directory, you will see a number of subdirectories.

NOTE This book's tutorial assumes you are working from the directory where the program files are stored. However, if you prefer to work from a different directory, be sure you have set a path for the AutoCAD directory; otherwise, AutoCAD will not start properly. Consult your DOS manual for more information on Path statements.

Here are brief descriptions of each subdirectory's contents:

Adsrx contains sample ObjectARX code.

DRV contains the drivers used by AutoCAD to control input and output devices. These drivers include the display drivers for VGA and SVGA.

Fonts contains AutoCAD fonts.

Help contains AutoCAD Help documents.

Sample contains the menu compiler and the Shroom utility for creating a bigger shell.

Support contains the files for a variety of AutoCAD's functions, including sample title blocks, menus, hatch patterns, dialog box files, and so on.

Template contains AutoCAD template files.

Textures contains texture files for AutoCAD's rendering feature.

Vbakup contains Help files for AutoCAD VBA.

Vbasamp contains sample AutoCAD VBA applications.

Bonus contains the Bonus utilities and samples described in Chapter 19. You won't see this directory unless you've installed the Bonus utilities or performed the Full AutoCAD installation.

Configuring AutoCAD

In this section, you will learn how to *configure* AutoCAD. By configure, we mean to set up AutoCAD to work with the particular hardware you have connected to your computer. AutoCAD often relies on its own set of drivers to operate specialized equipment. By configuring AutoCAD, you tell it exactly what equipment it will be working with. With Release 14, you can configure AutoCAD at any time during an AutoCAD session through the Preferences dialog box.

The tutorials in this book assume that you are using the default Preferences settings. As you become more familiar with the workings of AutoCAD, you may want to make adjustments to the way AutoCAD works through the Preferences dialog box.

This dialog box can be accessed by clicking on Tools ➤ Preferences. It is further divided into sections shown as tabs across the top of the dialog box. The following sections describe the settings available on each of the tabs.

Files

The Files tab is where you tell AutoCAD where to place or find files it needs to operate. It uses a hierarchical list similar to the one presented by the Windows Explorer. You first see the general topics listed in the Search Path, File Names, and File Location list boxes. You can expand individual items in the list by clicking on the plus sign (+) shown to the left of the item.

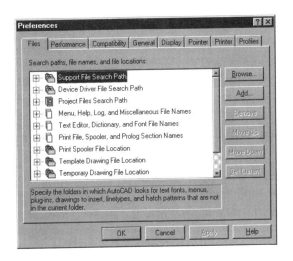

The following paragraphs explain what each item in the list box is for. Chances are you won't have to use most of them; others you may change occasionally.

Support File Search Path

AutoCAD relies on external files for many of its functions. Menus, text fonts, linetypes, and hatch patterns are a few examples of features that rely on external files. The Support File Search Path tells AutoCAD where to look for these files. You can add directory paths to this listing by clicking on the Add button and entering a new path, or by using the Browse button. It's probably not a good idea to delete any of the existing items under this heading unless you really know what you are doing.

Device Driver File Search Path

This item locates the device drivers for AutoCAD. Device drivers are applications that allow AutoCAD to communicate directly with the printers, plotters, and input devices. In most cases, you do not have to do anything with this setting.

Project Files Search Path

Eventually, you will receive a set of files from a consultant or other AutoCAD user that is dependent on Xref or raster images. Often, such files will expect the Xref or raster image to be in a particular directory. When such files are moved to another location with a different directory system, Xref-dependent files will not be able to find their Xrefs. The Project Files Search Path allows you to specify a directory where Xrefs or other dependent files are stored. If AutoCAD is unable to find an Xref or other file, it will look in the directory you specify in this listing.

To set it, highlight Project Files Search Path, and then click on the Add button. AutoCAD will suggest `Project1` as the directory name. You can change the name if you prefer. Click on the plus sign next to `Project1`, and then click on Browse to select a location for your Project support path. The search path is stored in a system variable called `Projectname`.

Menu, Help, Log, and Miscellaneous File Names

This item lets you set the location of a variety of support files including Menu, Help, Automatic Save, Log, and Configuration files. It also lets you set the default Internet address for the Launch Browser button on the AutoCAD Standard toolbar. If you have a network installation, you may also set the License Manager location on your network.

Text Editor, Dictionary, and Font File Names

Use this item to set the location of the Text Editor, custom and standard dictionaries, alternate font and font-mapping files. See Chapter 7 for more on this item.

Print File, Spooler, and Prolog Section Names

You can specify a print file name other than the default that is supplied by AutoCAD whenever you plot to a file. The Spooler option lets you specify an application intended to read and plot a plot file. The Prolog option is intended for PostScript export. It lets you specify the Prolog section from the Acad.psf file that you want AutoCAD to include with exported Encapsulate PostScript files. See Appendix A and Chapter 18 for more on exporting PostScript files and the Acad.psf file. The Prolog setting can also be controlled through the Psprolog system variable.

Print Spooler File Location

Print spooler applications will usually look in a specific directory for files that are to be plotted or printed. This item lets you set a directory location for print spool files.

Template Drawing File Location

When you select the Use a Template option in the Create New Drawing dialog box, AutoCAD looks at this setting for the location of template files. You can modify this setting, but chances are you won't need to.

Temporary Drawing File Location

AutoCAD creates temporary files to store portions of your drawing as you work on them. You usually don't have to think about these temporary files until they start crowding your hard disk, or unless you are working on a particularly large file on a system with little memory. This item lets you set the location for temporary files. The default location is the \Windows\Temp directory. If you have a hard drive that has lots of room and is very fast, you may want to change this setting to a location on that drive to improve performance.

Temporary External Reference File Location

If you are on a network and you foresee a situation where another user will want to open an Xref file of a file you are working on, you can set the Demand Load setting in the Performance tab to Enable with Copy. This causes AutoCAD to make and use a copy of any Xref that is currently loaded. This way, the original file can be opened by others. The Temporary External Reference File Location lets you specify the directory where AutoCAD will store this copy of an Xref.

Texture Maps Search Path

This item specifies the location for AutoCAD Render texture maps. In most cases, you won't have to change this setting. You can, however, add a directory name to this item for your own texture maps as you acquire or create them.

Performance

The options found on this tab affect the performance of AutoCAD. Generally, you can leave these settings alone, as they are tuned for a moderately powered computer. If you have a slow computer, you may want to change some of these settings. If you have a fast computer, you may alter some of these settings to improve display quality.

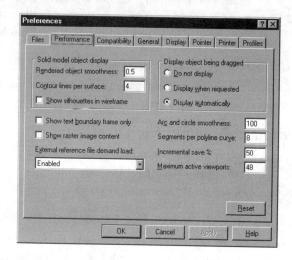

Solid Model Object Display

These settings affect the display of solid model wire-frame views. See Chapter 11 for details. These settings are also controlled by the Facetres, Isolines, and Dispsilh system variables.

Display Objects Being Dragged

These radio button options determine how objects are dragged as they are moved, copied, or rotated. Normally, with Display Automatically selected, you see the objects move with the cursor. The Display When Requested setting lets you determine whether or not to display objects as they are moving. When this option is selected, you will see a rubber-banding line indicating the move but you will not see an image of the object follow the cursor unless you type **drag**↵. Do Not Display turns off the dragging feature altogether. This setting is also controlled by the Dragmode system variable.

Show Text Boundary Frame Only

With this option selected, text will appear as a rectangle. See Chapter 7 for details. This setting is also controlled by the Qtextmode system variable.

Show Raster Image Content

With this option selected, AutoCAD will attempt to display the entire raster image as you perform pans and zooms; otherwise, only the bounding outline will be displayed. This setting is also controlled by the Rtdisplay system variable.

External Reference File Demand Load

Controls the Demand Load feature for external references. The Demand Load feature limits the amount of RAM used by an Xref, particularly if the Xref is clipped. See Chapter 9 for more on clipped Xrefs. This setting is also controlled by the Xloadctl system variable.

Arc and Circle Smoothness

This option controls the appearance of arcs and circles, particularly when you zoom in on them. In some instances, arcs and circles will appear to be octagons, even though they will plot as smooth arcs and circles. If you want arcs and circles to appear smoother, you can increase this setting. An increase will also increase memory use. This setting is also controlled by the Viewres system variable.

Segments per Polyline Curve

This setting controls the smoothness of polyline curves. A higher number will make a curved polyline appear smoother. This setting is also controlled by the Splinesegs system variable.

Incremental Save %

When you issue a Save, AutoCAD performs an incremental save until the file contains a 50 percent level of wasted space. When that level is reached, AutoCAD performs a full save, which takes more time. You can change the amount of wasted space allowed before a full save is performed. In general, you should leave this setting at 50 percent unless you are running into space limitations on your drive. This setting is also controlled by the Isavepercent system variable.

Maximum Active Viewports

AutoCAD limits to 48 the number of Paper Space viewports that display their contents. This limit keeps AutoCAD's memory consumption down. You may decrease this number to improve memory use while in Paper Space (prior versions of AutoCAD were set to 16). This setting is also controlled by the Maxactvp system variable.

The Reset button resets the values on this tab to the default settings. You'll find this button on several of the tabs in the Preferences dialog box.

Compatibility

These settings allow you to adjust AutoCAD to be more compatible with earlier versions. If you find you just cannot get used to the new way AutoCAD R14 works, you can change these settings to make AutoCAD14 a more familiar environment.

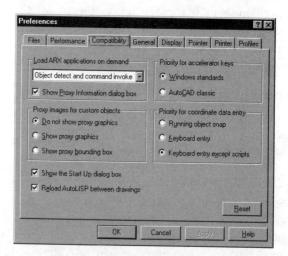

Load ARX Applications on Demand

AutoCAD R14 loads add-on applications, known as *ARX applications*, only when a command related to that application is invoked and a custom object associated with the application is detected. This is called *Demand Loading*. Two AutoCAD features, Render and Solids, are examples of such ARX add-on applications. With the settings in this group, you can control when Demand Loading is invoked. You usually don't have to change settings here, as they are set to optimize AutoCAD performance. This setting is also controlled by the Demandload system variable.

Proxy Images for Custom Objects

Custom objects are objects that are created by third-party add-on programs to AutoCAD. You can control the visibility of such objects through the options in this group. By default, the custom objects, called *proxy objects* when the add-on is not present, are represented as a wire frame. You have the option to not show them at all or to show them as a rectangle, known as a *bounding box*. This setting is also controlled by the Proxyshow system variable.

Priority for Accelerator Keys

Some keystroke combinations are reserved for Windows operations. Ctrl+c, for example, copies highlighted contents to the Clipboard. Older versions of AutoCAD, however, use these keystroke combinations differently; Ctrl+c executes the Cancel command in prior versions of AutoCAD. The settings in this group let you set which standard has priority: Windows or the prior AutoCAD versions.

Priority for Coordinate Data Entry

These options determine whether keyboard entry of coordinates overrides Running Object snaps.

Show the Start Up Dialog Box

This setting controls the appearance of the Start Up dialog box. When it is checked, you will see the Start Up dialog box whenever you create a new file or start AutoCAD.

Reload AutoLISP between Drawings

In prior versions of AutoCAD, AutoLISP programs had to be reloaded with each new file that was opened. This setting lets you maintain the prior method of handling AutoLISP programs or set AutoCAD to maintain the programs across file accesses. This setting is also controlled by the Lispinit system variable.

General

The options in this tab control a handful of miscellaneous operations. For the most part, you won't want to change these settings except perhaps the Automatic Save settings or the Log File feature.

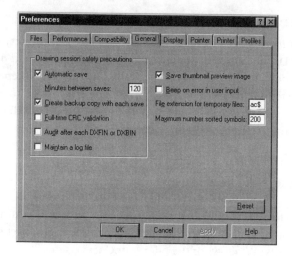

Drawing Session Safety Precaution

The settings in this group are concerned with file integrity. At the most basic level, you can turn the Automatic Save feature on or off and set the time interval between automatic saves. Other options affect whether or not files are checked as you work.

Automatic save lets you turn the Automatic Save feature on or off. It is highly recommended that you leave this feature on.

Minutes between saves lets you control the time interval at which AutoCAD performs automatic saves. This setting is also controlled by the Savetime system variable.

Create backup copy with each save lets you determine whether AutoCAD creates the BAK backup file each time you perform a save. This setting is also controlled by the Isavebak system variable.

Full-time CRC validation checks the integrity of each object as it is created in AutoCAD. (CRC stands for Cyclic Redundancy Check.) This can be used as a troubleshooting aid and can be turned on if you suspect there are problems with your hardware or with AutoCAD.

Audit after each DXFIN or DXBIN performs an audit on a file after a Dxfin or Dxbin is performed. The Audit command performs the same function (File ➢ Drawing Utilities ➢ Audit). This can be a helpful tool if you are importing files from questionable sources.

Maintain a log file turns on the Log File feature discussed in Chapter 18. This setting is also controlled by the Logfilemode system variable.

Save Thumbnail Preview Image

When you open files, you see a preview of the file in a preview window. This preview image is a small bitmap file that is saved with the AutoCAD file. This setting lets you determine whether that thumbnail is created or not. If you turn this feature off, you won't see those preview images for the files that you save from now on. This setting is also controlled by the Rasterpreview system variable.

Beep on Error in User Input

This option is a hold-over from the very early versions of AutoCAD. Just as the name implies, when you turn this option on, AutoCAD will beep whenever you do something AutoCAD doesn't understand. Chances are you won't want to turn this on.

File Extension for Temporary Files

Whenever you open a file, particularly a large one, AutoCAD creates temporary files that it uses to store drawing data when it starts to run out of RAM. Typically, these files have the .ac$ extension that AutoCAD is famous for leaving around. If you are on a network, you may want to change this extension so it doesn't conflict with files from another user, or so you can signal other users who may be opening a particular set of files.

Maximum Number Sorted Symbols

A few of AutoCAD's commands provide lists of items in text windows. This option determines how many items it will sort alphabetically in a list. If it is set to 0, listings are sorted by the order in which the item was created. This setting is also controlled by the Maxsort system variable.

Display

The settings on this tab let you control the appearance of AutoCAD. You can make AutoCAD look completely different with these settings if you choose. Scroll bars, fonts, and colors are all up for grabs.

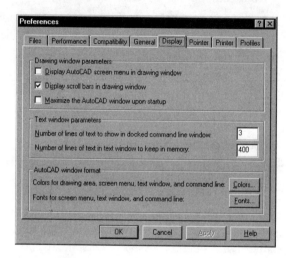

Drawing Window Parameters

These options control the general settings for the AutoCAD windows:

Display AutoCAD screen menu in drawing window turns on the old AutoCAD format screen menu that once appeared on the right side of the screen. If you really must have it, this is where you can turn it back on.

Display scroll bars in drawing window lets you turn the scroll bars on and off. If you've got a small monitor with low resolution, you may want to turn them off for a larger drawing area.

Maximize the AutoCAD window upon startup   determines whether the AutoCAD screen is maximized when you open AutoCAD. Again, if you have a small monitor, you may want to turn this option on.

Text Window Parameters

These settings control the display of the Command window.

Number of lines of text to show in docked command line window   lets you increase or decrease the number of text lines displayed. If you feel comfortable with AutoCAD and don't need to refer to the prompt too often, you may want to decrease this value to get a larger drawing area.

Number of lines of text in text window to keep in memory   determines how much of the Text window is retained in memory before it is discarded. You will be able to scroll back in the Text window to the number of lines indicated by this setting.

AutoCAD Window Format

Two buttons are offered here that open dialog boxes. These dialog boxes offer controls over the colors of the AutoCAD window and the fonts displayed on the menu bar, status bar, and screen menu.

Colors   opens a dialog box that lets you set the color for the various components of the AutoCAD window. This is where you can change the background color of the drawing area if you find that black doesn't work for you.

Fonts   opens a dialog box that lets you set the fonts of the AutoCAD window. You can select from the standard set of Windows fonts available in your system.

Pointer

There are two main functions on this tab. First, if you want to install a digitizing tablet, this is the place to install the AutoCAD drivers for it. Select the driver name from the list; then click on the Set Current button. You will be asked a series of questions regarding your hardware. When you are done, your tablet will be available for use with AutoCAD. You can always go back and change settings.

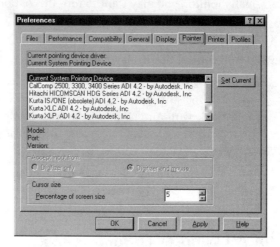

The other function is to allow control over the crosshair cursor in the AutoCAD drawing. Veteran AutoCAD users may prefer the larger cursor of previous versions. The Cursor Size input box can bet set to 100 to restore the full-screen cursor of previous versions.

Printer

The Printer tab lets you add, modify, or delete printers and plotters to your system. You can have multiple printers as well as raster file formats. See Chapter 14 for details on how to use this tab.

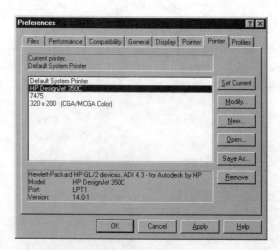

Profiles

If you are using Windows NT, you know that a user profile is saved for each log-in name. Depending on the log-in name you use, you will have a different Windows setup. The Profiles tab offers a similar function for AutoCAD users. You can store different Preferences settings and recall them at any time. You can also save them to a file with the .arg extension, then take that file to another system. It's a bit like being able to take your Preferences settings with you wherever you go.

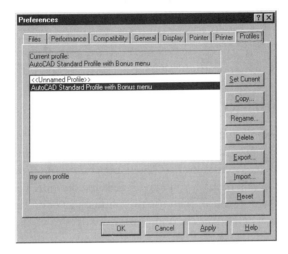

The main part of this tab displays a listing of profiles available. <<Unnamed Profile>> is the default profile. As you add more profiles, they will appear in the list.

To create a new profile, highlight a profile name from the list, and then click on Copy. The Copy Profile dialog box appears, allowing you to enter a profile name and a description of the profile. The description appears in the box below the list on the Profiles tab whenever that profile is selected.

Once you've created a new profile, you can modify the settings on the other tabs of the Preferences dialog box, and the new settings will be associated with the new profile. Profiles will store the way menus are set up, so profiles can be used as an aid to managing both your own customization schemes and third-party software. Profiles can also be a way to manage multiple users on the same computer. Each user can maintain his or her own profile, so users don't have to fight over how AutoCAD is set up. Here is a brief listing of the options on the Profiles tab:

Set Current installs the settings from the selected profile.

Copy creates a new profile from an existing one.

Rename allows you to rename a profile and change its description.

Delete removes the selected profile from the list.

Export lets you save a profile to a file.

Import imports a profile that has been saved to a file.

What Happened to the AutoCAD *.ini* File?

In prior Windows versions of AutoCAD, many of the AutoCAD settings were stored in the Acad.ini file. Release 14 now uses the Windows Registry to store information. Attempting to edit the Windows Registry is not recommended. Fortunately, the Preferences dialog box has been expanded to accommodate many of the settings that you might otherwise have accessed through the Acad.ini file.

Configuring Your Digitizing Tablet

If you are using a digitizer with AutoCAD for Windows, you will need to select some additional configuration options after you've installed the driver on the Pointer tab of the Preferences dialog box. These other options allow you to add a menu template.

As an alternative, you may choose to configure your tablet as the Windows pointing device. You can then use the Wintab driver on the Pointer tab of the Preferences dialog box to make the appropriate changes to your table's configuration.

Configuring the Tablet Menu Area

If you own a digitizing tablet and you would like to use it with the AutoCAD tablet menu template, you must configure your tablet menu.

1. First, securely fasten your tablet menu template to the tablet. Be sure the area covered by the template is completely within the tablet's active drawing area.
2. Choose Tools ➢ Tablet ➢ Configure. You will get this prompt:

   ```
   Digitize upper left corner of menu area 1:
   ```

   For the next series of prompts, you will be locating the four tablet menu areas, starting with menu area 1 (see Figure B.1).

FIGURE B.1

How to locate the tablet menu areas

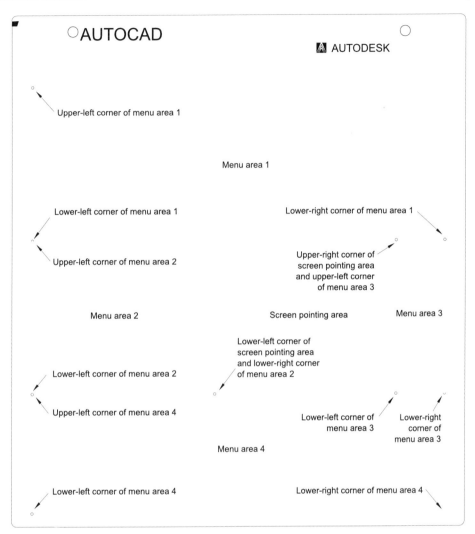

3. Locate the position indicated in Figure B.1 as the upper-left corner of menu area 1. Place your puck or stylus to pick that point. The prompt will change to

   `Digitize lower left corner of menu area 1:`

4. Again, locate the position indicated in Figure B.1 as the lower-left corner of menu area 1.

5. Continue this process until you have selected three corners for four menu areas.
6. When you are done selecting the menu areas, you will get this prompt:

   ```
   Do you want to respecify the Fixed Screen Pointing Area:
   ```

   Type **y**↵ and then pick the position indicated in Figure B.1.
7. Finally, you'll get this prompt:

   ```
   Digitize upper right corner of screen pointing area:
   ```

   Pick the position indicated in Figure B.1.

The three prompts that remain refer to a *Floating Screen Pointing Area*. This is an area on your tablet that allows you to select menu options and other areas on your screen outside the drawing area. This option is necessary because when you set up a digitizer for tracing, access to areas outside the drawing area is temporarily disabled. The Floating Screen Pointing Area lets you access pull-down menus and the status bar during tracing sessions (see Chapter 16).

8. If you never intend to trace drawings with your tablet, answer **n**↵ to all three prompts. Otherwise do the following three steps:
9. At the prompt:

   ```
   Do you want to specify the Floating Screen Pointing Area? <N>:
   ```

   type **y**↵.
10. At the prompt:

    ```
    Do you want the Floating Screen Pointing Area to be the same size as the
    Fixed Screen Pointing Area? <Y>:
    ```

    type **y**↵ if you want the Floating Screen Pointing Area to be the same as the Fixed Screen Pointing area, the area you specified in steps 6 and 7. Type **n**↵ if you want to use a separate area on your tablet for the Floating Screen Pointing Area.
11. The last prompt asks you if you want to use the F12 function key to toggle the Floating Screen Pointing Area on and off. (This is similar to the F10 key function of earlier releases of AutoCAD.) Enter **y**↵ or **n**↵, depending on whether or not you want to specify a different function key for the Floating Screen Pointing Area.

AutoCAD will remember this configuration until you change it again. Quit this file by selecting File ➢ Exit.

Turning On the Noun/Verb Option

If, for some reason, the Noun/Verb Selection method is not available, here are instructions on how to turn it back on.

1. Choose Tools ➤ Selection to display the Object Selection Settings dialog box.

2. In the Selection Modes button group, find the Noun/Verb Selection setting. Click on the check box to turn this option on.

3. Click on OK.

If it wasn't there before, you should now see a small square at the intersection of the crosshair cursor. This square is actually a pickbox superimposed on the cursor. It tells you that you can select objects, even while the command prompt appears at the bottom of the screen and no command is currently active. As you have seen earlier, the square will momentarily disappear when you are in a command that asks you to select points.

You can also turn on Noun/Verb Selection by entering ~**'pickfirst**↵ at the command prompt. When you are asked for New value for PICKFIRST <0>:, enter **1**↵ (entering **0** turns the Pickfirst function off). The Pickfirst system variable is stored in the AutoCAD configuration file. See Appendix D for more on system variables.

Other Selection Options

The Object Selection Settings dialog box lets you control the degree to which AutoCAD conforms to standard graphical user interface methods of operation. It also lets you adjust the size of the Object Selection pickbox.

In Chapter 2, you practiced selecting objects using the Noun/Verb Selection setting—one of several AutoCAD settings that make the program work more like other Windows programs. If you are used to working with other graphical environments, you may want to turn on some of the other options in the Selection Settings dialog box. Here are descriptions of them; in brackets you'll find the names of the system variables that control these features.

Use Shift to Add [Pickadd] With this option checked, you can use the standard GUI method of holding down the Shift key to pick multiple objects. When the Shift key is not held down, only the single object picked or the group of objects windowed will be selected. Previously selected objects are deselected, unless you hold down the Shift key during selection. To turn this feature on using system variables, set Pickadd to 0.

Press and Drag [Pickdrag] With this option checked, you can use the standard GUI method for placing windows: First click and hold down the pick button on the first corner of the window; then, while holding down the pick button, drag the other corner of the window into position. When the other corner is in place, let go of the pick button to finish the window. This setting applies to both Verb/Noun and Noun/Verb operations. In the system variables, set Pickdrag to 1 for this option.

Implied Window [Pickauto] When this option is checked, a window or crossing window will automatically start if no object is picked at the `Select objects:` prompt. This setting has no effect on the Noun/Verb setting. In the system variables, set Pickadd to 1 for this option.

Turning On the Grips Feature

If for some reason the Grips feature is not available, here are instructions for turning it back on:

1. Choose Tools ➣ Grips. The Grips dialog box appears.

2. At the top of the dialog box are the Select Settings check boxes. Click on the Enable Grips check box.

   3. Click on OK, and you are ready to proceed.

The Grips dialog box also lets you determine whether grips appear on objects that compose a block (see Chapter 4 for more on blocks), as well as set the grip color and size. These options can also be set using the system variables described in Appendix D.

You can also turn the Grips feature on and off by entering ~'**grips**↵. At the New value for GRIPS <0>: prompt, enter **1** to turn Grips on or **0** to turn Grips off. Grips is a system variable that is stored in the AutoCAD configuration file.

Setting Up AutoCAD to Use ODBC

AutoCAD can use the Open Database Connectivity (ODBC) driver that allows you to view Microsoft Access files. Unfortunately, the path to establishing an ODBC link from AutoCAD is somewhat tortuous. This section is intended to give you the basic information you need to know to establish a connection between AutoCAD and your Access 7 files.

To connect to an ODBC file, you need both the database file you want to link to and another database file called the *reference file*. In the following explanation, we will use a fictitious file named Mydb.mdb, which we will say is located in the \Acadr14\samples directory.

Create a Reference Database File

First, you will need to create a reference table in Access to emulate catalogs and schemas, as Access does not use catalogs and schemas directly. Take the following steps to create your reference file and tables:

   1. Open Access and choose File ≻ New Database.

   2. In the New dialog box, select the Blank Database icon from the General tab, and then click on OK.

   3. In the File New Database dialog box, enter a name for your new database file. For this example, use **acadlink**, although any name will work.

   4. Save the file in a place you can remember easily. For this example choose the \Acadr14 directory.

   5. Click on OK. Access will create a file called Acadlink.mdb.

   6. At the Acadlink:Database dialog box, click on the New button.

7. At the New Table dialog box, click on Design View. A view of a new database table appears.
8. Click on the cell just below the Field Name column, then enter **CATALOG_NAME**.
9. Press the Tab key and the word Text appears under the Data Type column.
10. Click on the cell below CATALOG_NAME in the Field Name column and enter **SCHEMA_NAME**.
11. Press Tab to see Text in the Data Type column.
12. Click on the Close button in the upper-right corner of the Table1 window.
13. Click on Yes when you see the message Do you want to save changes.
14. In the Save As dialog box, enter **SCHEMATA** for the table name.
15. You will see a Primary Key warning message. Click on No. You will return to the Acadlink:Database dialog box. The Schemata table now appears in the Tables tab.
16. Repeat steps 6 and 7.
17. Create the following four fields in the same way that you created the CATALOG_NAME and SCHEMA_NAME fields in steps 8–11. All fields should be of the Text data type.

    TABLE_CATALOG
    TABLE_SCHEMA
    TABLE_NAME
    TABLE_TYPE

18. Close the table window, click on Yes at the Save message, and enter the name **TABLES** in the Save As dialog box.
19. Click on No at the Primary Key warning message.

You've set up the tables. Now it's time to add some data to the tables. The data you add will be the name of the Access file to which you want to connect from AutoCAD, and the table names within that file.

1. At the Acadlink:Database dialog box, click on Schemata and click on Open. The Schemata table appears.
2. Under the CATALOG_NAME field, enter **NULL**.
3. Under the SCHEMA_NAME field, enter **d:\acadr14\sample\mydb** where *mydb* is the name of the actual database you want to access from AutoCAD (you needn't include the .mdb file extension). You can replace the file location with any location where you plan to store your database files.

4. Close the table.
5. Highlight TABLES in the Acadlink:Database dialog box; then click on Open.
6. Add a record for each table in your database that you wish to have access to. Under TABLE_CATALOG, enter **NULL**. Under TABLE_SCHEMA enter the name of the database file including the directory path. Leave off the .mdb extension, however. Under TABLE_NAME, enter the name of the table in the database file. Finally, under TABLE_TYPE, enter **Base Table**. Table B.1 shows a sample of what might be entered for the Mydb example, assuming that Mydb contains three tables named Computer, Employee, and Inventory.

TABLE B.1: ENTRIES FOR THE MYDB EXAMPLE			
TABLE_ CATALOG	TABLE_ SCHEMA	TABLE_ NAME	TABLE_ TYPE
NULL	D:\ACADR14\SAMPLE\MYDB	Computer	Base Table
NULL	D:\ACADR14\SAMPLE\MYDB	Employee	Base Table
NULL	D:\ACADR14\SAMPLE\MYDB	Inventory	Base Table

7. Close the table and save the file.
8. Exit Access.

Telling Windows Where the Reference File Resides

Next you need to tell Windows where to locate your database file. If you have Office 97 or Microsoft Access 97 installed, you should have an item in the Windows 95/NT Control Panel called 32bit ODBC. Make sure you have this item in the Control Panel, then do the following:

1. Double-click on 32bit ODBC.
2. At the ODBC Data Source Administrator, click on Add.
3. At the Create New Data Source dialog box, click on Microsoft Access Driver; then click on Finish.
4. In the ODBC Microsoft Access 97 Setup dialog box, enter **AutoCAD_ODBC** in the Data Source Name input box.
5. Click on the Select button in the Database button group.
6. In the Select Database dialog box, locate and select the Mydb.mdb file you created in the previous exercise.
7. In the ODBC Data Source Administrator dialog box, click on OK.

Telling AutoCAD Where the Reference File Resides

Now you need to tell AutoCAD the location of the reference file. Here are the steps to follow:

1. From the Windows Start menu, choose Programs ➢ AutoCAD R14 ➢ External Database Configuration.

2. In the External Database Configuration dialog box, click on Add.

3. For Select DBMS for New Environment, select ODBC; then enter **AutoCAD_ODBC** in the Environment Name input box. This is the same name you used when you configured the Windows Access driver. You may also enter a description in the Environment Description dialog box.

4. Click on OK. In the Environment dialog box, click on the Browse button and locate and select the reference database file. In our example we used the name Acadlink.

5. Make sure Not Supported is selected from the Set Schema drop-down list.

6. Enter **NULL** in the Default Catalog input box.

7. Click on OK; then click on OK again at the External Database Configuration dialog box.

Now you are ready to access your database from within AutoCAD. When you open the Administration dialog box within AutoCAD, you will see AUTOCAD_ODBC as one of the Environment options.

Appendix C

What's on the Companion CD-ROM

What's on the Companion CD-ROM

This appendix describes the materials supplied on the CD-ROM that comes with this book. The CD-ROM contains a number of useful utilities and resources that you can load and run at any time. Before you use them, however, it's best to get familiar with AutoCAD. Many of these utilities work within AutoCAD from the command line or from options on pull-down menus, and they offer prompts in a way similar to most other AutoCAD commands. Other utilities are stand-alone applications.

The Mastering AutoCAD Bonus Software

To help you get the most from AutoCAD and this book, we've included a set of programs and files. Once you load the CD, the Sybex interface will guide you to the following utilities:

Figures is a directory of the drawings used in this book. If you follow the tutorial chapter by chapter, you'll create these figures yourself. If you want to do the exercises out of sequence, use these files as needed.

Eye2eye is a directory of the Eye2eye add-on to AutoCAD, described later in this appendix. Eye2eye makes perspective viewing of your 3D models a snap.

Swlite is a portion of the SI Mechanical third-party add-on to AutoCAD. SWLite is a library of fasteners and utilities that will relieve you of the tedious job of creating and managing many standard parts.

The AutoCAD Instant Reference is now an online book, and it's on the companion CD-ROM. This is the definitive companion to *Mastering AutoCAD 14 for Mechanical Engineers*, and it's full of detailed descriptions of AutoCAD's commands and tools.

The ABCs of AutoLISP is a complete tutorial and reference book for AutoLISP, the AutoCAD programming language. This book has been a favorite of end users and developers alike, and it's now in an easy-to-use Web-browser format.

ActiveX samples help you get started with Automation in AutoCAD. You can use these examples along with Chapter 20 to explore the newest customization feature of AutoCAD.

Whip2 Netscape Communicator Plug-in allows you to view AutoCAD drawings over the Internet. It gives you full Pan and Zoom capabilities, as well as access to Web links that are embedded in AutoCAD files.

The Figures, Eye2eye, and SI Mechanical Utilities

The figures, Eye2eye, and SI Mechanical utilities are offered to you free as a part of this book. The SWLite software is a demo version of the software and will stop working six months after you install it, unless you decide to purchase the product.

If you plan to do any of the exercises in this book, you will want to install the figures to a Sample subdirectory on your hard disk. Whenever you are asked to open a file from the companion CD-ROM, you can then look in the Sample for the file.

Install the Eye2eye and SI-Mech utilities in their own separate subdirectories. You must also edit the AutoCAD Preferences dialog box to include these subdirectories in the AutoCAD_R14 path.

Opening the Installation Program

The software discussed in this appendix can be installed easily using the interface and installation program on the CD-ROM. Take the following steps to open the installation program:

1. Insert the *Mastering AutoCAD 14 for Mechanical Engineers* CD into your CD-ROM drive. This will start the installation application for the CD-ROM. You will first see the License Agreement screen.

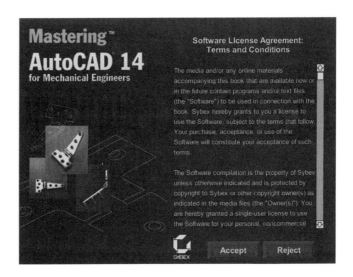

2. After you've read the agreement and accepted the terms, click on Accept. You'll see the *Mastering AutoCAD 14 for Mechanical Engineers* software installation interface.

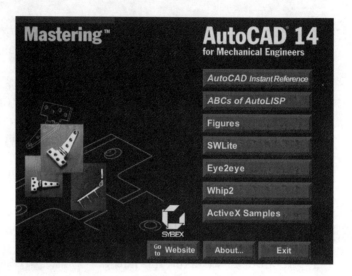

From here, you can select the software you wish to install. For details on each of these options, read the related sections that follow.

Installing and Using the Sample Drawing Files

The `Figures` directory contains sample drawing files for the exercises in this book. We have provided these drawings for you in case you decide to skip some of the book's tutorials. With these files, you can open the book to any chapter and start working, without having to construct the drawings from earlier chapters. A CD icon, like the one shown here in the margin, lets you know when a drawing file is available on the CD-ROM.

1. Start the *Mastering AutoCAD 14 for Mechanical Engineers* software installation program, as described earlier in this appendix.

2. Click on the Figures button.

3. Click on Continue. You will see the Winzip self-extractor dialog box. You have the option to accept the default directory shown in the input box, or you may enter a different location for the files.

4. When you've entered the location for the sample files, click on Winzip. The files will be installed on your computer.

5. Click on Close to return to the main installation program screen.

6. Proceed with another installation or click on Exit.

Once you've installed the sample files, you'll have free access to them during the exercises.

Installing the Bonus Add-On Packages

The companion CD-ROM also includes two add-on packages called Eye2eye and SWLite. Eye2eye is a utility that replaces the AutoCAD Dview command with an easy-to-use method for creating perspective views. Eye2eye uses a camera-target metaphor to let you place viewpoints and view directions in a drawing. SWLite is a utility that demonstrates the power of the AutoCAD open environment. It is a library of fasteners, other hardware, and utilities specific to Mechanical Engineering. SWLite was created by Design Pacifica and is available from your local vendor or directly from Design Pacifica.

Eye2eye

Eye2eye is a set of AutoLISP utilities that aid the viewing of 3D models in AutoCAD. If you find you are struggling with the AutoCAD Dview command, these utilities will be of benefit to you.

Installing Eye2eye

Here's how to install Eye2eye:

1. Start the *Mastering AutoCAD 14 for Mechanical Engineers* software installation program, as described earlier in this appendix.

2. Click on the Eye2Eye button.

3. Click on Continue. You see the Winzip self-extractor dialog box. You have the option to accept the default directory shown in the input box, or you may enter a different location for the files. We recommend that you install the Eye2Eye files in a subdirectory called Eye2Eye under the \Program Files\AutoCAD R14\ directory where AutoCAD is installed.

4. When you've entered the location for the Eye2Eye files, click on Unzip. The files will be installed on your computer.

5. Click on Close to return to the main installation screen.

6. Click on Exit.
7. Start AutoCAD and choose Tools ➤ Preferences.
8. Click on the Files tab at the top of the dialog box.
9. Click on Support Files Search Path.
10. Click on the Add button, and then click on Browse.
11. Locate and select the \Program Files\AutoCAD R14\Eye2eye directory, and then click on OK.
12. Click on OK in the Preferences dialog box.

Eye2eye is now available for loading and use. The next section describes how to access the Eye2eye features.

Using Eye2eye

Eye2eye is based on the idea of using a camera and target object to control your Perspective views. The camera and target objects are called *eyes*, hence the name Eye2eye. To set up a view, simply place the camera, or Eyeto block (see Figure C.1), where you want your point of view. Then place the target, or Eyefrom block, where you want your center of attention. Use Eye2eye's Showeye command to display the perspective.

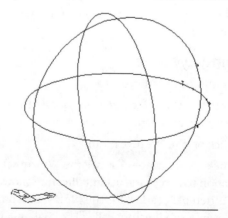

FIGURE C.1
The Eyeto block used with Eye2eye

The following list gives a brief description of all the Eye2eye commands:

Showeye displays the current eye-target perspective view in the current viewport. Be sure to have the desired viewport active before using this command.

Findeye draws a temporary vector between the camera and target points to help locate these points.

Crosseye displays a view that reverses the camera and target locations.

Mtarg moves the target location. You must be in an Orthogonal view for this command.

Meye moves the camera location. You must be in an Orthogonal view for this command.

Seteye allows you to turn the Perspective mode on or off, set camera "focal length," turn the camera and target objects on or off, pan the Perspective view, or set up multiple viewports for Eye2eye.

Paneye lets you pan the Perspective view. This is useful for fine-tuning your Perspective view.

Matcheye sets the camera and target objects to the current Perspective view. You must be in a Perspective viewport to use this command. This is useful when you've changed your view using methods outside of the Eye2eye command set, such as the Dview command.

Eye2eye assumes that you have a fairly good grasp of AutoCAD's 3D functions and the use of Paper Space viewports. It is best suited as a tool for viewing your model after you've created the basic massing.

Once you've built a 3D model, load the Eye2eye utilities by entering the following command at the command prompt:

`(load"eye2eye")`↵

You can also choose Tools ≻ Load Applications. At the dialog box, click on the File button and locate and load the `Eye2eye.lsp` file in the `\Program Files\AutoCAD R14\Eye2eye` directory.

After you've loaded Eye2eye, set up multiple viewports with the following steps:

1. Open a file containing a 3D model. Set up your display so that you can see all of the model plus some room for the camera location. Make sure this is a Plan or Top-down view.

2. Go to the World Coordinate System and enter **eye2eye**↵ at the command prompt.

3. Since this is the first time you are using Eye2eye, you are prompted to pick a camera point. Do so. Don't worry about the exact placement of your camera just yet; you will get a chance to adjust its location later. The camera object, a block called Eyeto, appears.

4. Next, you are prompted to select a target location. Do so. As with the camera location, it isn't important to place the target in an exact location at this time. You can easily adjust it later. Eye2eye will switch to Paper Space and set up four viewports: one large viewport (to the right) for your Perspective view and three smaller viewports (to the left) for your top, front, and left-side Orthogonal views. You will use the Orthogonal views to manipulate the camera and target points (see Figure C.2). You are now ready to use the Eye2eye utilities.

5. You'll also see the prompt:

   Perspctv Off/ON/Focal lngth/Pan/cLose eyes/Show eyes/fInd eyes/move Camera/move Target/Hide/Match prspctv/eXit:

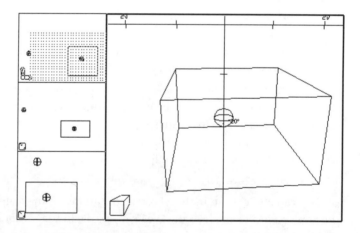

FIGURE C.2

A sample of the viewport arrangement created with Eye2eye's Seteye command

Because your views are in Paper Space, you can easily enlarge or resize a view for easy editing. Just remember that while using the Eye2eye utilities, you must be in Tiled Model Space.

You are now ready to use the Eye2eye utilities. The options shown in the Eye2eye prompt perform some of the same functions that are described in the beginning of this section.

Perspctv Off/ON turns the Perspective mode off or on in the Perspective viewport.

Focal length changes the focal length of the "camera."

Pan (Paneye) lets you pan the Perspective view. This is useful for fine-tuning your Perspective view.

Close eyes turns off the layer on which the camera and target "eyes" reside. This turns off the visibility of the eyes.

Show eyes turns on the layer on which the camera and target eyes reside. This turns on the visibility of the eyes.

Find eyes helps you locate the camera and target eyes by zooming into an area that just includes the eyes. You must be in a non-Perspective viewport before this option will take effect.

Move Camera (Meye) moves the camera location. You must be in an Orthogonal view for this command.

Move Target (Mtarg) moves the target location. You must be in an Orthogonal view for this command.

Hide performs a hidden-line removal on the Perspective view.

Match prspctv (Matcheye) sets the camera and target objects to the current Perspective view. You must be in a Perspective viewport to use this command. This is useful when you've changed your view using methods outside of the Eye2eye command set, such as the Dview command.

Exit terminates the Eye2eye command.

Using Eye2eye for Model Construction

If you want to use Eye2eye as a tool to help you construct a 3D model, be sure you set up your Model Space work area *before* you proceed with the previous steps. You can use the Ddsetup utility provided with Eye2eye to accomplish this. To use Ddsetup, enter

    (load "ddsetup")↵

at the command prompt, and then enter **ddsetup**↵. A dialog box appears from which you can choose a unit style, sheet size, and scale. Once you've selected the appropriate options, click on OK. The drawing is then set up according to your selections. You will also see the grid dots. They are set to represent 1-inch intervals in your final plot size.

You can now set up Eye2eye as described above.

Setting the Eyes

Try moving the camera and target objects in the Orthogonal views. You can use the Meye and Mtarg commands to help you locate the camera and target objects. When you use these commands, a temporary red-and-green vector is drawn to show you the location of the camera and target. The red portion of the vector shows the direction of the camera; the green portion shows the direction of the target. Use the Redraw command to remove the vectors. They are not true AutoCAD objects and will not plot.

Mtarg and Meye work just like the Move command, but you don't have to select an object: The camera (Meye) or target (Mtarg) is automatically selected. You only need to select the Base point and Second point for the move. You can also use the standard AutoCAD Move command to move the camera and target.

As you move the eye and target locations with Mtarg and Meye, the Perspective viewport will automatically update to display the new eye-to-target orientation. If you happen to move the eye or target blocks using the Move command, you can update the Perspective viewport by clicking on the Perspective viewport and entering **showeye**↵. A Perspective view based on the camera-target locations is displayed. You can then use the AutoCAD View command to save your view or go on to make minor adjustments. For example, you can use the Paneye command to adjust your Perspective view before saving it.

Controlling the Eyes

You will notice some lines and numbers on the Perspective view. These are parts of the camera object showing you the angle below or above horizontal in 10-degree increments. You can turn off the display of camera and target objects by using the Hide Eyes option of the Seteye command. To turn them back on again, use the Show Eyes option. Alternately, you can simply turn the Eyes layer on or off. The Eyes layer is the layer on which the Eyeto and Eyefrom blocks were constructed.

SWLite

SWLite enriches the mechanical design and engineering environment by providing an extensive array of fasteners as both 2D and 3D blocks; a shaft generator including tapered and keyed termination; segments with gear, cam, and pulley detail; true automatic ordinate dimensioning; dimensioning aids; drawing formats; weld, finish, and GD&T symbols; Bill of Material formatter; and much more. SWLite is a third-party add-on to AutoCAD, and it is yours to use for a period of six months.

Installing SWLite

Here's how to install SWLite:

1. Start the *Mastering AutoCAD 14 for Mechanical Engineers* software installation program, as described earlier in this appendix.
2. Click on the SWLite button.
3. Click on Continue. You will see the Winzip self-extractor dialog box. You have the option to accept the default directory shown in the input box, or you may enter a different location for the files.
4. When you've entered the location for the SWLite sample files, click on Unzip. The files will be installed on your computer.
5. Click on Close to return to the main installation program screen.
6. Proceed with another installation or click on Exit.
7. Go to the file you just created and click on SWLite.
8. Follow the prompts on the screen to complete the installation.

The ActiveX Automation Samples

With AutoCAD Release 14, you have a new way to create custom applications in AutoCAD. ActiveX Automation is a new feature that allows AutoCAD users to take control of AutoCAD by creating Visual Basic applications or through other applications that support Automation.

Chapter 20 provides an introduction to ActiveX Automation as it relates to AutoCAD. There are some sample applications included on the CD-ROM that you can experiment with when you're reading Chapter 20. The following steps show you how to access those examples.

1. Start the *Mastering AutoCAD 14 for Mechanical Engineers* software installation program, as described earlier in this appendix.
2. Click on the ActiveX Samples button.
3. Click on Continue. You will see the Winzip self-extractor dialog box. You have the option to accept the default directory shown in the input box, or you may enter a different location for the files.
4. When you've entered the location for the ActiveX sample files, click on Unzip. The files will be installed on your computer.
5. Click on Close to return to the main installation program screen.
6. Proceed with another installation or click on Exit.

The AutoCAD Instant Reference

Mastering AutoCAD was designed to demonstrate commands in the context of everyday activities that you might encounter in your work. It shows you what commands to use in a given situation. As a result, we can't always show every option or permutation of a command. This is where the *AutoCAD 14 Instant Reference* comes in.

The *AutoCAD 14 Instant Reference* is the perfect companion to *Mastering AutoCAD 14 for Mechanical Engineers*. You can think of it as a dictionary of AutoCAD commands that describes each command in detail, including all of the command options.

The CD-ROM contains a Modern Age Books electronic version of the *AutoCAD 14 Instant Reference*. You will need to install the Modern Age Books viewer to use this online book.

1. Start the *Mastering AutoCAD 14 for Mechanical Engineers* software installation program, as described earlier in this appendix.

2. Click on the *AutoCAD Instant Reference* button. You'll see a message telling you that you are about to install the Modern Age Books viewer. You have the option to go back to the previous screen or to install the viewer.

3. Click on Install, and then follow the instructions.

The ABCs of AutoLISP

In previous editions of *Mastering AutoCAD*, an introduction to AutoLISP appeared in Chapter 20. In this edition, the AutoLISP information was replaced by a chapter on ActiveX automation. Now, the companion CD-ROM includes a complete AutoLISP tutorial and source book. *The ABCs of AutoLISP* offers a detailed look at AutoCAD's macro-programming language.

The *ABCs of AutoLISP* is in a format that can be viewed using any Internet Web browser, such as Netscape Navigator or Microsoft Internet Explorer. To use the *ABCs of AutoLISP*, simply install it on your hard drive and double-click on the `Contents.htm` or on the `ABC's_of_AutoLISP.htm` file. The following steps show you how to install the *ABCs of AutoLISP*:

1. Start the *Mastering AutoCAD 14 for Mechanical Engineers* software installation program, as described earlier in this appendix.

2. Click on the *ABCs of AutoLISP* button. You will see a brief description of the *ABCs of AutoLISP* and the default location where it will be installed.

3. Click on Continue. You will see the Winzip self-extractor dialog box. You have the option to accept the default directory shown in the input box, or you may enter a different location for the files.

4. When you've entered the location for the *ABCs of AutoLISP* files, click on Unzip. The files will be installed on your computer.

5. Click on Close to return to the main installation program screen.

6. Proceed with another installation or click on Exit.

If you want to uninstall the *ABCs of AutoLISP*, simply delete the folder containing the files.

The Whip2 Netscape Communicator Plug-in

In Chapter 19 we discuss how you can add drawings to a Web page. Before you can view such drawings, you need to install the Whip2 plug-in for your Web browser. You can obtain this plug-in from the AutoCAD Web site, but as a convenience, we've included it on the CD-ROM—so you won't have to spend your time downloading this fairly large file.

WARNING

If you have installed the AutoCAD Internet Publishing Kit for Release 13, you will have to uninstall it before installing the Whip2 plug-in. All of the features of the Internet Publishing Kit are duplicated in Release 14, so you won't be losing any features.

To install the Whip2 plug-in, first make sure that you have Netscape Communicator 3.0 or later. Next, perform the following steps.

1. Make sure all other programs are closed.

2. Start the *Mastering AutoCAD 14 for Mechanical Engineers* software installation program, as described earlier in this appendix.

3. Click on the Whip2 button. The Whip2 installation will start.

4. Follow the instructions to complete the Whip2 installation.

If you are using Microsoft Internet Explorer, perform the following steps:

1. If your system requires that you log on to your Internet Service Provider before opening your Web browser, do so now.

2. Start the *Mastering AutoCAD 14 for Mechanical Engineers* software installation program as described earlier in this appendix.

3. Click on the Go To Website button. Your default Web browser will open a Web page with links to various other Web pages. If your default browser is Microsoft Internet Explorer, click on the word *here*. This will take you to the Whip2 ActiveX Control automatic download page.

The Whip2 ActiveX Control software will take about 20 minutes to download. When it is done, your Internet Explorer will be able to view AutoCAD DWF files online.

As an alternative, you might visit the Whip2 ActiveX Control Web site on your own. Here is the URL:

```
http://pilot1.autodesk.com/products/acadr14/features/whpdwnie.htm
```

Appendix D

System and Dimension Variables

System and Dimension Variables

This appendix discusses AutoCAD's system variables. It is divided into two sections: *system variables* and *dimension variables*. The general system variables let you fine-tune your AutoCAD environment. Dimension variables govern the specific dimensioning functions of AutoCAD.

System variables are accessible directly from the command prompt, and transparently (while in another command), by entering the variable name preceded by an apostrophe. These variables are also accessible through the AutoLISP interpreter by using the Getvar and Setvar functions.

We've divided this appendix into two main sections, *Setting System Variables* and *Setting Dimension Variables*. This division is somewhat artificial, because as far as AutoCAD is concerned, there is no difference between system variables and dimension variables—you use both types of variables the same way. But because the set of dimension variables is quite extensive, this set is in a separate table for clarity.

Setting System Variables

Table D.1 lists the system variables and notes whether each is read-only or adjustable. Most of these variables have counterparts in other commands, as listed in the table. For example, Angdir and Angbase can be adjusted using the Ddunits command (Format ➤ Units). Many, such as Highlight and Expert, do not have equivalent commands. These must be adjusted at the command line (or through AutoLISP).

TABLE D.1: SYSTEM VARIABLES

Variable Name	Associated Command	Where Saved	Use
Acadprefix	Preferences	NA	The ACAD environment setting.
Acadver	NA	NA	The AutoCAD version number.
Acisoutver	Acisout	With drawing	Controls ACIS version of SAT files.
Aflags	Units	NA	Controls attribute mode settings: 1 = invisible; 2 = constant; 4 = verify; 8 = preset.
Angbase	Units	With drawing	Controls direction of 0 angle, relative to the current UCS.
Angdir	Units	With drawing	Controls positive direction of angles: 0 = counterclockwise; 1 = clockwise.
Apbox	Draw and Edit commands	Registry	Displays Autosnap aperture box when Autosnap is activated: 0 = off; 1 = on.
Aperture	Draw and Edit commands	Registry	Sets size of Osnap cursor in pixels.
Area (read-only)	Area	NA	Displays last area calculation; use with Setvar or with AutoLISP's Getvar function.
Attdia	Insert/Attribute	With drawing	Controls the Attribute dialog box: 0 = no dialog box; 1 = dialog box.
Attmode	Attdisp	With drawing	Controls attribute display mode: 0 = off; 1 = normal; 2 = on.
Attreq	Insert	With drawing	Controls the prompt for attributes: 0 = no prompt or dialog box for attributes (attributes use default values); 1 = normal prompt or dialog box upon attribute insertion.
Auditctl	Config	Registry	Controls whether an audit file is created: 0 = disable; 1 = enable creation of ADT file.

Continued

TABLE D.1: SYSTEM VARIABLES (CONTINUED)

Variable Name	Associated Command	Where Saved	Use
Aunits	Units	With drawing	Controls angular units: 0 = decimal degrees; 1 = degrees-minutes-seconds; 2 = grads; 3 = radians; 4 = surveyor's units.
Auprec	Units	With drawing	Controls the precision of angular units determined by decimal place.
Autosnap	Draw/Edit	Registry	Controls Autosnap display and features: 0 = everything off; 1 = marker on; 2 = snaptip on; 4 = magnet on.
Backz (read-only)	Dview	With drawing	Displays distance from Dview target to back clipping plane.
Blipmode	NA	With drawing	Controls appearance of blips: 0 = off; 1 = on.
Cdate (read-only)	Time	NA	Displays calendar date/time read from system date (YYYY MMDD .HHMMSSMSEC).
Cecolor	Color	With drawing	Controls current default color assigned to new objects.
Celtscale	NA	With drawing	Controls current linetype scale for individual objects.
Celtype	Linetype	With drawing	Controls current default linetype assigned to new objects.
Chamfera	Chamfer	With drawing	Controls first chamfer distance.
Chamferb	Chamfer	With drawing	Controls second chamfer distance.
Chamferc	Chamfer	With drawing	Controls chamfer distance for Angle option.
Chamferd	Chamfer	With drawing	Controls chamfer angle for Angle option.
Chammode	Chamfer	NA	Controls method of chamfer: 0 = use 2 distances; 1 = use distance and angle.

Continued

TABLE D.1: SYSTEM VARIABLES (CONTINUED)

Variable Name	Associated Command	Where Saved	Use
Circlerad	Circle	NA	Controls the default circle radius: 0 = no default.
Clayer	Layer	With drawing	Sets the current layer.
Cmdactive (read-only)	NA	NA	Displays whether a command, script, or dialog box is active: 1 = command active; 2 = transparent command active; 4 = script active; 8 = dialog box active (values are cumulative, so 3 = command and transparent command are active).
Cmddia	NA	Registry	Controls use of dialog boxes for some commands: 0 = don't use dialog box; 1 = use dialog box.
Cmdecho	AutoLISP	NA	With AutoLISP, controls display of prompts from embedded AutoCAD commands: 0 = no display of prompt; 1 = display prompts.
Cmdnames (read-only)	NA	NA	Displays the English name of the currently active command.
Cmljust	Mline	With drawing	Sets method of justification for Multilines: 0 = top; 1 = middle; 2 = bottom.
Cmlscale	Mline	With drawing	Sets scale factor for Multiline widths: a 0 value collapes the Multiline to a single line; a negative value reverses the justification.
Cmlstyle (read-only)	Mline	With drawing	Displays current Multiline style by name.

Continued

TABLE D.1: SYSTEM VARIABLES (CONTINUED)

Variable Name	Associated Command	Where Saved	Use
Coords	F6, Ctrl+D	With drawing	Controls coordinate readout: 0 = coordinates displayed only when points are picked; 1 = absolute coordinates dynamically displayed as cursor moves; 2 = distance and angle displayed during commands that accept relative distance input.
Cursorsize	NA	Registry	Determines size of crosshairs as a percentage of the screen size (1–100).
Cvport (read-only)	Vports	With drawing	Displays ID number of current viewport.
Date (read-only)	Time	NA	Displays date and time in Julian format.
Dbmod (read-only)	NA	NA	Displays drawing modification status: 1 = object database modified; 2 = symbol table modified; 4 = database variable modified; 8 = window modified; 16 = view modified.
Dctcust	Spell	Registry	Sets default custom spelling dictionary file name, including path.
Dctmain	Spell	Registry	Sets default main spelling dictionary file name; requires specific keywords for each language; use AutoCAD Help for complete list of keywords.
Delobj	NA	With drawing	Controls whether source objects used to create new objects are retained: 0 = delete objects; 1 = retain objects.
Demandload	NA	Registry	Controls loading of third-party applications required for custom objects in drawing (0–3).

Continued

TABLE D.1: SYSTEM VARIABLES (CONTINUED)

Variable Name	Associated Command	Where Saved	Use
Diastat (read-only)	NA	NA	Displays how last dialog box was exited: 0 = Cancel; 1 = OK.
Dispsilh	All curved solids	With drawing	Controls silhouette display of curved 3D solids: 0 = no silhouette; 1 = silhouette curved solids.
Distance (read-only)	Dist	NA	Displays last distance calculated by Dist command.
Donutid	Donut	NA	Stores default inside diameter of a donut.
Donutod	Donut	NA	Stores default outside diameter of a donut.
Dragmode	NA	With drawing	Controls dragging: 0 = no dragging; 1 = if requested; 2 = automatic drag.
Dragp1	NA	Registry	Controls regeneration-drag input sampling rate.
Dragp2	NA	Registry	Controls fast-drag input sampling rate.
Dwgcodepage (read-only)	NA	With drawing	Displays code page of drawing (see Syscodepage).
Dwgname (read-only)	Open	NA	Displays drawing name specified by user.
Dwgprefix (read-only)	NA	NA	Displays drive and directory of current file.
Dwgtitled (read-only)	NA	NA	Displays whether a drawing has been named: 0 = untitled; 1 = named by user.
Edgemode	Trim/Extend	Registry	Controls how trim and extend boundaries are determined: 0 = boundaries defined by object only; 1 = boundaries defined by objects and their extension.
Elevation	Elev	With drawing	Controls current 3D elevation relative to current UCS.

Continued

TABLE D.1: SYSTEM VARIABLES (CONTINUED)

Variable Name	Associated Command	Where Saved	Use
Expert	NA	NA	Controls prompts, depending on level of user's expertise: 0 = normal prompts; 1 = suppresses About to regen and Really want to turn the current layer off prompts; 2 = suppresses Block already defined and A drawing with this name already exists. prompt for Block command; 3 = suppresses An item with this name already exists prompt for the Linetype command; 4 = suppresses An item with this name already exists for the UCS/Save and Vports/Save options; 5 = suppresses An item with this name already exists for Dim/Save and Dim/Override commands.
Explmode	Explode	With drawing	Controls whether blocks inserted with different X, Y, and Z values are exploded: 0 = blocks are not exploded; 1 = blocks are exploded.
Extmax (read-only)	Zoom	With drawing	Displays upper-right corner coordinate of Extents view.
Extmin (read-only)	Zoom	With drawing	Displays lower-left corner coordinate of Extents view.
Facetratio	Shade/Hide	NA	Controls aspect ratio of faceting of curved 3D surfaces: value is 0 or 1, where 1 increases the density of the mesh.

Continued

TABLE D.1: SYSTEM VARIABLES (CONTINUED)

Variable Name	Associated Command	Where Saved	Use
Facetres	Shade/Hide	With drawing	Controls appearance of smooth, curved 3D surfaces when shaded or hidden: value can be between 0.01 and 10. The higher the number, the more faceted (and smoother) the curved surface, and the longer the time needed for shade and hidden-line removal.
Filedia	Dialog box	Registry	Sets whether a file dialog box is used by default: 0 = don't use unless requested with a tilde (~); 1 = use whenever possible.
Filletrad	Fillet	With drawing	Stores fillet radius.
Fillmode	Fill	With drawing	Controls fill status: 0 = off; 1 = on.
Fontalt	Open, Dxfin, other File ➢ Import options	Registry	Lets you specify an alternate font when AutoCAD cannot find the font associated with a file. If no font is specified for Fontalt, AutoCAD displays warning message and dialog box where you manually select a font.
Fontmap	Open, Dxfin, other Import functions	Registry	Similar to Fontalt, but lets you designate a set of font substitutions through a font-mapping file. Example line from mapping file: romans:c:\Acadr13\fonts\times.ttf. This replaces Romans font with Times TrueType font. Font-mapping file can be any name, with the extension .fmp.

Continued ▶

TABLE D.1: SYSTEM VARIABLES (CONTINUED)

Variable Name	Associated Command	Where Saved	Use
Frontz (read-only)	DVIEW	With drawing	Stores front clipping plane for current viewport; use with Viewmode system variable.
Gridmode	Grid	With drawing	Controls grid: 0 = off; 1 = on.
Gridunit	Grid	With drawing	Controls grid spacing.
Gripblock	Grips	Registry	Controls display of grips in blocks: 0 = show insertion point grip only; 1 = show grips of all objects in block.
Gripcolor	Grips	Registry	Controls color of unselected grips: Choices are integers from 1 to 255; default is 5.
Griphot	Grips	Registry	Controls color of hot grips: Choices are integers from 1 to 255; default is 1.
Grips	Grips	Registry	Controls use of grips: 0 = grips disabled; 1 = grips enabled (default).
Gripsize	Grips	Registry	Controls grip size (in pixels), from 1 to 255 (default is 3).
Handles (read-only)	NA	With drawing	Displays status of object handles: 0 = off; 1 = on.
Hideprecision	Hide/Shade	NA	Controls the Hide/Shade precision accuracy: 0 = single-level precision; 1 = double-level precision.
Highlight	Select	NA	Controls whether objects are highlighted when selected: 0 = no highlighting; 1 = highlighting.
Hpang	Hatch	NA	Sets default hatch pattern angle.
Hpbound	Hatch	Registry	Controls type of object created by the Hatch and Boundary commands: 0 = region; 1 = polyline.
Hpdouble	Hatch	NA	Sets default hatch doubling for user-defined hatch pattern: 0 = no doubling; 1 = doubling at 90° offset.

Continued

TABLE D.1: SYSTEM VARIABLES (CONTINUED)

Variable Name	Associated Command	Where Saved	Use
Hpname	Hatch	NA	Sets default hatch pattern name; use a period (.) to set to no default.
Hpscale	Hatch	NA	Sets default hatch pattern scale factor.
Hpspace	Hatch	NA	Sets default line spacing for user-defined hatch pattern; cannot be 0.
Indexctl	N/A	With drawing	Controls whether layer and spatial indexes are created and saved in drawings: 0 = no index; 1 = layer index; 2 = spatial index; 3 = both.
Inetlocation	Browser	Registry	Stores the Internet location used by the Browser command.
Insbase	Base	With drawing	Stores insertion base point set by user with Base command.
Insname	Insert	NA	Sets default block or file name for the Insert command; enter a period (.) to set to no default.
Isavebak	Save	Registry	Controls the creation of BAK files: 0 = no BAK file created; 1 = BAK file created.
Isavepercent	Save	Registry	Determines whether to do a full or an incremental save based on the amount of wasted space tolerated in a drawing file: (0–100).
Isolines	Curved solids	With drawing	Specifies the number of lines on a solid's surface to help visualize its shape.
Lastangle (read-only)	Arc	NA	Displays ending angle for last arc drawn.
Lastpoint	NA	NA	Stores last point entered by user; can be accessed with @.

Continued

TABLE D.1: SYSTEM VARIABLES (CONTINUED)

Variable Name	Associated Command	Where Saved	Use
Lastprompt	NA	NA	Stores last command issued at command prompt.
Lenslength (read-only)	Dview	With drawing	Displays focal length of lens used for perspective display.
Limcheck	Limits	With drawing	Controls object creation outside drawing limits: 0 = OK; 1 = not.
Limmax	Limits	With drawing	Controls coordinate of drawing's upper-right limit.
Limmin	Limits	With drawing	Controls coordinate of drawing's lower-left limit.
Lispinit	Load		Preserves AutoLISP-defined functions and variables beyond current: 0 = AutoLISP variables preserved; 1 = AutoLISP functions valid for current session only.
Locale (read-only)	NA	NA	Displays ISO language code used by your version of AutoCAD.
Logfilemode	NA	Registry	Determines whether text screen is recorded to log file: 0 = no 1 = yes.
Logfilename	NA	Registry	Specifies name/path of Logfile.
Loginname (read-only)	NA	NA	Displays user's login name.
Ltscale	Ltscale	With drawing	Controls the global linetype scale factor.
Lunits	Units	With drawing	Controls unit styles: 1 = scientific; 2 = decimal; 3 = engineering; 4 = architectural; 5 = fractional.
Luprec	Units	With drawing	Controls unit accuracy by decimal place or size of denominator.
Maxactvp	Viewports/Vports	With drawing	Controls maximum number of viewports to regenerate at one time.

Continued

TABLE D.1: SYSTEM VARIABLES (CONTINUED)

Variable Name	Associated Command	Where Saved	Use
Maxsort	NA	Registry	Controls maximum number of items to be sorted when a command displays a list.
Measurement	Bhatch, Linetype	With drawing	Sets drawing units as English or metric: 0 = English; 1 = metric.
Menuctl	NA	Registry	Controls whether side menu changes in response to a command name entered from the keyboard: 0 = no response; 1 = menu response.
Menuecho	NA	NA	Controls messages and command prompt display from commands embedded in menu: 0 = display all messages; 1 = suppress menu item name; 2 = suppress command prompts; 4 = Disable ^P toggle of menu echo; 8 = debugging aid for DIESEL expressions.
Menuname (read-only)	Menu	With drawing	Displays name of current menu file.
Mirrtext	Mirror	With drawing	Controls mirroring of text: 0 = disabled; 1 = enabled.
Modemacro	NA	NA	Controls display of user-defined text in status line using DIESEL programming commands.
Mtexted	Mtext	Registry	Controls name of program used for editing Mtext objects.
Offsetdist	Offset	NA	Controls default offset distance.
Olehide	NA	Registry	Controls display of OLE objects.
Orthomode	F8, Ortho	With drawing	Controls Ortho mode: 0 = off; 1 = on.

Continued

TABLE D.1: SYSTEM VARIABLES (CONTINUED)

Variable Name	Associated Command	Where Saved	Use
Osmode	Osnap	With drawing	Sets current default Osnap mode: 0 = none; 1 = endpoint; 2 = midpoint; 4 = center; 8 = node; 16 = quadrant; 32 = intersection; 64 = insert; 128 = perpendicular; .256 = tangent; 512 = nearest; 1024 = quick; 2048 = apparent intersection. If more than one mode is required, enter the sum of those modes.
Osnapcoord	Osnap	Registry	Controls coordinates entered at the command line : 0 = Running Osnap settings override keyboard coordinate entry; 1 = keyboard entry overrides Osnap settings; 2 = keyboard entry overrides Osnap settings except in scripts.
Pdmode	Ddptype	With drawing	Controls type of symbol used as a point during Point command.
Pdsize	Point	With drawing	Controls size of symbol set by Pdmode.
Pellipse	Ellipse	With drawing	Controls type of object created with Ellipse command: 0 = true NURBS ellipse; 1 = polyline representation of ellipse.
Perimeter (read-only)	Area/List	NA	Displays last perimeter value derived from Area, Dblist, and List commands.
Pfacevmax (read-only)	Pface	NA	Displays maximum number of vertices per face. (Pfaces are 3D surfaces designed for use by third-party software producers and are not designed for end users.)

Continued

TABLE D.1: SYSTEM VARIABLES (CONTINUED)

Variable Name	Associated Command	Where Saved	Use
Pickadd	Select	Registry	Determines how items are added to a selection set: 0 = only most recently selected item(s) become selection set (to accumulate objects in a selection set, hold down Shift while selecting); 1 = selected objects accumulate in a selection set as you select them (hold down Shift while selecting items to remove those items from the selection set).
Pickauto	Select	Registry	Controls automatic window at Select objects prompt: 0 = window is disabled; 1 = window is enabled.
Pickbox	Select	Registry	Controls size of object-selection pickbox (in pixels).
Pickdrag	Select	Registry	Controls how selection windows are used: 0 = click on each corner of the window; 1 = click and hold on first corner, then drag and release for the second corner.
Pickfirst	Select	Registry	Controls whether you can pick object(s) before you select a command: 0 = disabled; 1 = enabled.
Pickstyle	Group/Hatch	With drawing	Controls whether groups and/or associative hatches are selectable: 0 = neither is selectable; 1 = groups only; 2 = associative hatches only; 3 = both groups and associative hatches.
Platform (read-only)	NA	NA	Indicates Windows platform being used.

Continued

TABLE D.1: SYSTEM VARIABLES (CONTINUED)

Variable Name	Associated Command	Where Saved	Use
Plinegen	Pline/Pedit	With drawing	Controls how polylines generate linetypes around vertices: 0 = linetype pattern begins and ends at vertices; 1 = linetype patterns ignore vertices and begin and end at polyline beginning and ending.
Plinetype	Pline	Registry	Controls whether AutoCAD creates optimized 2D polylines and/or converts existing plines to optimized plines: 0 = plines in existing drawings are not converted *and* new plines are not optimized; 1 = plines in existing drawings are not converted *but* new plines are optimized; 2 = plines in existing drawings are not converted *and* new plines are optimized.
Plinewid	Pline	With drawing	Controls default polyline width.
Plotid	Plot	Registry	Sets default plotter based on its description.
Plotrotmode	Plot	Registry	Controls orientation of your plotter output.
Plotter	Plot	Registry	Sets default plotter, based on its integer ID.
Polysides	Polygon	NA	Controls default number of sides for a polygon.
Popups (read-only)	NA	NA	Displays whether the current system supports dialog boxes, menu bars, and pull-down menus: 0 = no; 1 = yes.

Continued

TABLE D.1: SYSTEM VARIABLES (CONTINUED)

Variable Name	Associated Command	Where Saved	Use
Projectname	Preferences	Registry	Assigns a project name to a drawing; the project name can be associated with one or more folders.
Projmode	Trim/Extend	Registry	Controls how Trim and Extend affect objects in 3D: 0 = objects must be coplanar; 1 = trims/extends based on a plane parallel to the current UCS; 2 = trims/extends based on a plane parallel to the current view plane.
Proxygraphics	NA	With drawing	Controls whether images of proxy objects are stored in a drawing: 0 = images not stored; 1 = images saved.
Proxynotice	NA	Registry	Issues a warning to the user when a proxy object is created (that is, when the user opens a drawing containing custom objects created using an application which is not loaded): 0 = no warning; 1 = warning displayed.
Proxyshow	NA	Registry	Specifies if and how proxy objects are displayed: 0 = No display; 1 = graphic display of all proxy objects; 2 = only bounding box shown.
Psltscale	Pspace	With drawing	Controls Paper Space linetype scaling.
Psprolog	Psout	Registry	Controls what portion of the Acad.psf file is used for the prologue section of a Psout output file. Set this to the name of the section you want to use.

Continued

TABLE D.1: SYSTEM VARIABLES (CONTINUED)

Variable Name	Associated Command	Where Saved	Use
Psquality	Psin	Registry	Controls how images are generated in AutoCAD with the Psin command. Value is an integer: 0 = only bounding box is drawn; >0 = number of pixels per AutoCAD drawing unit; <0 = outline with no fills, and absolute value of setting determines pixels per drawing units.
Qtextmode	Qtext	With drawing	Controls the quick text mode: 0 = off; 1 = on.
Rasterpreview	Save	Registry	Controls whether raster preview images are saved with the drawing and sets the format type: 0 = no preview image created; 1 = BMP preview image.
Regenmode	Regenauto	With drawing	Controls Regenauto mode: 0 = off; 1 = on.
Re-init	Reinit	NA	Reinitializes I/O ports, digitizers, display, plotter, and Acad.pgp: 1 = digitizer port; 2 = plotter port; 4 = digitizer; 8 = display; 16 = PGP file reload.
Rtdisplay	Rtpan, Rtzoom	Registry	Controls display of raster images during realtime Pan and Zoom.
Savefile (read-only)	Autosave	Registry	Displays filename that is autosaved.
Savename (read-only)	Save	NA	Displays user file name under which file is saved.
Savetime	Autosave	Registry	Controls time interval between Automatic Saves, in minutes: 0 = disable Automatic Save.
Screenboxes (read-only)	Menu	Registry	Displays number of slots or boxes available in side menu.

Continued

TABLE D.1: SYSTEM VARIABLES (CONTINUED)

Variable Name	Associated Command	Where Saved	Use
Screenmode (read-only)	NA	Registry	Displays current display mode: 0 = text; 1 = graphics; 2 = dual screen.
Screensize (read-only)	NA	NA	Displays current viewport size in pixels.
Shadedge	Shade	With drawing	Controls how drawing is shaded: 0 = faces shaded, no edge highlighting; 1 = faces shaded, edge drawn in background color; 2 = faces not filled, edges in object color; 3 = faces in object color, edges in background color.
Shadedif	Shade	With drawing	Sets difference between diffuse, reflective, and ambient light. Value represents percentage of diffuse reflective light.
Shpname	Shape	NA	Controls default shape name.
Sketchinc	Sketch	With drawing	Controls sketch record increment.
Skpoly	Sketch	With drawing	Controls whether Sketch uses regular lines or polylines: 0 = line; 1 = polyline.
Snapang	Snap	With drawing	Controls snap and grid angle.
Snapbase	Snap	With drawing	Controls snap, grid, and hatch pattern origin.
Snapisopair	Snap	With drawing	Controls isometric plane: 0 = left; 1 = top; 2 = right.
Snapmode	F9/Snap	With drawing	Controls Snap toggle: 0 = off; 1 = on.
Snapstyl	Snap	With drawing	Controls snap style: 0 = standard; 1 = isometric.
Snapunit	Snap	With drawing	Controls snap spacing given in X and Y values.

Continued

TABLE D.1: SYSTEM VARIABLES (CONTINUED)

Variable Name	Associated Command	Where Saved	Use
Sortents	NA	Registry	Controls whether objects are sorted based on their order in database: 0 = disabled; 1 = sort for object selection; 2 = sort for Osnap; 4 = sort for Redraw; 8 = sort for Mslide; 16 = sort for Regen; 32 = sort for Plot; 64 = sort for Psout.
Splframe	Pline/Pedit/ 3D Face	With drawing	Controls display of spline vertices, defining mesh of a surface-fit mesh, and display of "invisible" edges of 3D faces: 0 = no display of spline vertices, display only fit surface of a smoothed 3D mesh, and no display of "invisible" edges of 3D face; 1 = spline vertices are displayed, only defining mesh of a smoothed 3D mesh is displayed, "invisible" edges of 3D face are displayed.
Splinesegs	Pline/Pedit	With drawing	Controls number of line segments used for each spline patch.
Splinetype	Pline/Pedit	With drawing	Controls type of spline curve generated by Pedit spline: 5 = quadratic B-spline; 6 = cubic B-spline.
Surftab1	Rulesurf/Tabsurf/ Revsurf/Edgesurf	With drawing	Controls number of facets in the m direction of meshes.
Surftab2	Revsurf/Edgesurf	With drawing	Controls number of facets in the n direction of meshes.
Surftype	Pedit	With drawing	Controls type of surface fitting used by Pedit's Smooth option: 5 = quadratic B-spline surface; 6 = cubic B- spline surface; 8 = Bezier surface.
Surfu	3D Mesh	With drawing	Controls surface density in the m direction.

Continued

TABLE D.1: SYSTEM VARIABLES (CONTINUED)

Variable Name	Associated Command	Where Saved	Use
Syscodepage (read-only)	NA	NA	Displays system code page specified in Acad.xmx.
Tabmode	Tablet	NA	Controls tablet mode: 0 = off; 1 = on.
Target (read-only)	Dview	With drawing	Displays coordinate of perspective target point.
Tdcreate (read-only)	Time	With drawing	Displays time and date of file creation in Julian format.
Tdindwg (read-only)	Time	With drawing	Displays total editing time in days and decimal days.
Tdupdate (read-only)	Time	With drawing	Displays time and date of last file update, in Julian format.
Tdusrtimer (read-only)	Time	With drawing	Displays user-controlled elapsed time in days and decimal days.
Tempprefix (read-only)	NA	NA	Displays location for temporary files.
Texteval	NA	NA	Controls interpretation of text input: 0 = AutoCAD takes all text input literally; 1 = AutoCAD interprets "(" and "!" as part of an AutoLISP expression, unless either the Text or Dtext command is active.
Textfill	Text	Registry	Controls display of Bitstream, TrueType, and PostScript Type 1 fonts: 0 = outlines; 1 = filled.
Textqlty	Text	With drawing	Controls resolution of Bitstream, TrueType, and PostScript Type 1 fonts: values from 1.0 to 100.0. The lower the value, the lower the output resolution. Higher resolutions improve font quality but decrease display and plot speeds.
Textsize	Text/Dtext	With drawing	Controls default text height.
Textstyle	Text/Dtext	With drawing	Controls default text style.

Continued

TABLE D.1: SYSTEM VARIABLES (CONTINUED)

Variable Name	Associated Command	Where Saved	Use
Thickness	Elev	With drawing	Controls default 3D thickness of object being drawn.
Tilemode	Mspace/Pspace	With drawing	Controls Paper Space and Model Space access: 0 = Paper Space enabled; 1 = strictly TiledModel Space.
Tooltips	Icon tool palettes	Registry	Controls display of tool tips: 0 = off; 1 = on.
Tracewid	Trace	With drawing	Controls trace width.
Treedepth	Treestat	With drawing	Controls depth of tree-structured spatial index affecting speed of AutoCAD database search. First two digits are for Model Space nodes; second two digits are for Paper Space nodes. Use positive integers for 3D drawings and negative integers for 2D drawings. Negative values can improve speed of 2D operation.
Treemax	Regen/Treedepth	Registry	Limits memory use during regens by limiting maximum number of nodes in spatial index created with Treedepth.
Trimmode	Chamfer/Fillet	Registry	Controls whether lines are trimmed during Chamfer and Fillet commands: 0 = no trim; 1 = trim (as with pre-Release 13 versions of AutoCAD).
Ucsfollow	UCS	With drawing	Controls whether AutoCAD automatically changes to plan view of UCS while in Model Space: 0 = UCS change does not affect view; 1 = UCS change causes view to change with UCS.

Continued

TABLE D.1: SYSTEM VARIABLES (CONTINUED)

Variable Name	Associated Command	Where Saved	Use
Ucsicon	Ucsicon	With drawing	Controls UCS icon: 0 = off; 1 = on; 2 = UCS icon appears at origin.
Ucsname (read-only)	UCS	With drawing	Displays name of current UCS.
Ucsorg (read-only)	UCS	With drawing	Displays origin coordinate for current UCS relative to World Coordinate System.
Ucsxdir (read-only)	UCS	With drawing	Displays X direction of current UCS relative to World Coordinate System.
Ucsydir (read-only)	UCS	With drawing	Displays Y direction of current UCS relative to World Coordinate System.
Undoctl (read-only)	Undo	NA	Displays current state of Undo feature: 0 = off; 1 = Undo enabled; 2 = only one command can be undone; 4 = Autogroup mode enabled; 8 = group is currently active.
Undomarks (read-only)	Undo	NA	Displays number of marks placed by Undo Marks option command.
Unitmode	Units	With drawing	Controls how AutoCAD displays fractional, foot-and-inch, and surveyors angles: 0 = industry standard; 1 = AutoCAD input format.
Useri1–Useri5	AutoLISP	With drawing	Five user variables capable of storing integer values.
Userr1–Userr5	AutoLISP	With drawing	Five user variables capable of storing real values.
Users1–Users5	AutoLISP	With drawing	Five user variables capable of storing string values.
Viewctr (read-only)	NA	With drawing	Displays center of current view in coordinates.
Viewdir (read-only)	Dview	With drawing	Displays camera viewing direction in coordinates.

Continued

TABLE D.1: SYSTEM VARIABLES (CONTINUED)

Variable Name	Associated Command	Where Saved	Use
Viewmode (read-only)	Dview	With drawing	Displays view-related settings for current viewport: 1 = perspective on; 2 = front clipping on; 4 = back clipping on; 8 = UCS follow on; 16 = front clip not at a point directly in front of the viewer's eye.
Viewsize (read-only)	NA	With drawing	Displays height of current view in drawing units.
Viewtwist (read-only)	Dview	With drawing	Displays twist angle for current viewport.
Visretain	Layer	With drawing	Controls whether layer setting for Xrefs is retained: 0 = current layer color; linetype and visibility settings retained when drawing is closed; 1 = layer settings of Xref drawing always renewed when file is opened.
Vsmax (read-only)	NA	With drawing	Displays coordinates of upper-right corner of virtual screen.
Vsmin (read-only)	NA	With drawing	Displays coordinates for lower-left corner of virtual screen.
Worlducs (read-only)	UCS	NA	Displays status of WCS: 0 = current UCS is not WCS; 1 = current UCS is WCS.
Worldview	Dview/Vpoint	With drawing	Controls whether Dview and Viewpoint operate relative to UCS or WCS: 0 = current UCS is used; 1 = WCS is used.
Xclipframe	Xref	With drawing	Controls visibility of Xref clipping boundaries: 0 = clipping boundary is not visible; 1 boundary is visible.

Continued

TABLE D.1: SYSTEM VARIABLES (CONTINUED)			
Variable Name	Associated Command	Where Saved	Use
Xloadctl	Xref/Xclip	Registry	Controls Xref demand loading and creation of copies of original Xref: 0 = no demand loading allowed, entire Xref drawing is loaded; 1 = demand loading allowed, and original Xref file is kept open; 2 = demand loading allowed, using a copy of Xref file stored in AutoCAD temp files folder.
Xloadpath	Xref	Registry	Creates a path for storing temporary copies of demand-loaded Xref files.
Xrefctl	Xref	Registry	Controls whether Xref log files are written: 0 = no log files; 1 = log files written.

Setting Dimension Variables

In Chapter 8, nearly all of the system variables related to dimensioning are shown with their associated options in the Dimension Styles dialog box. Later in this appendix, you'll find a complete discussion of all elements of the Dimension Styles dialog box and how to use them.

This section provides further information about the dimension variables. For starters, Table D.2 lists the variables, default status, and a brief description of what each does. You can get a similar listing by entering **dimstyle**↵ at the command prompt, then typing **st** to select the Status option. Alternatively, you can use the AutoCAD Help system. This section also discusses a few system variables that do not show up in the Dimension Styles dialog box.

TABLE D.2: DIMENSION VARIABLES

General Dimension Controls

Dimension Variable	Default Setting	Description
Dimaso	On	Turns associative dimensions on and off.
Dimsho	On	Updates dimensions dynamically while dragging.
Dimstyle	Standard	Name of current dimension style.
Dimupt	Off	Controls user positioning of text during dimension input: 0 = automatic text positioning; 1 = user-defined text positioning allowed.

Scale

Dimension Variable	Default Setting	Description
Dimscale	1.0000	Overall scale factor of dimensions.
Dimtxt	.18 (approx. 3/16")	Text height.
Dimasz	.18 (approx. 3/16")	Arrow size.
Dimtsz	0"	Tick size.
Dimcen	.09 (approx. 3/32")	Center mark size.
Dimlfac	1.0000	Multiplies measured distance by a specified scale factor. This factor should be the reciprocal of the part's scale relative to full size.

Offsets

Dimension Variable	Default Setting	Description
Dimexo	.0625 or 1/16"	Extension line origin offset.
Dimexe	.18 (approx. 3/16")	Amount extension line extends beyond dimension line.
Dimdli	.38 (approx. 3/8")	Dimension line offset for continuation or base.
Dimdle	0"	Amount dimension line extends beyond extension line.

Continued

TABLE D.2: DIMENSION VARIABLES (CONTINUED)

Tolerances

Dimension Variable	Default Setting	Description
Dimalttz	0	Controls zero suppression of tolerance values: 0 = leaves out zero feet and inches; 1 = includes zero feet and inches; 2 = includes zero feet; 3 = includes zero inches; 4 = suppresses leading zeros in decimal dimensions; 8= suppresses lead-ing zeros in decimal dimensions.
Dimdec	4	Sets decimal place for primary tolerance values.
Dimtdec	4	Sets decimal place for tolerance values.
Dimtp	0"	Plus tolerance.
Dimtm	0"	Minus tolerance.
Dimtol	Off	When on, shows dimension tolerances.
Dimtolj	1	Controls vertical location of tolerance values relative to nominal dimension: 0 = bottom; 1 = middle; 2 = top.
Dimtzin	0	Controls zero suppression in tolerance values: 0 = leaves out zero feet and inches; 1 = includes zero feet and inches; 2 = includes zero feet; 3 = includes zero inches; 4 = suppresses leading zeros in decimal dimensions; 8= suppresses lead-ing zeros in decimal dimensions; 12 = suppresses leading and trailing zeros in decimal dimensions.
Dimlim	Off	When on, shows dimension limits.

Rounding

Dimension Variable	Default Setting	Description
Dimrnd	0"	Rounding value.
Dimzin	0	Controls zero suppression dimension text: 0 = leaves out zero feet and inches; 1 = includes zero feet and inches; 2 = includes zero feet; 3 = includes zero inches; 4 = suppresses leading zeros in decimal dimensions; 8= suppresses lead-ing zeros in decimal dimensions; 12 = suppresses leading and trailing zeros in decimal dimensions.

Continued

TABLE D.2: DIMENSION VARIABLES (CONTINUED)

Dimension Arrow & Text Control

Dimension Variable	Default Setting	Description
Dimadec	-1	Controls the number of decimal places shown for angular dimension text: -1 = uses the value set by Dimdec dimension variable; 0–8 = specifies the actual number of decimal places to be shown.
Dimaunit	0	Controls angle format for angular dimensions; settings are the same as for Aunits system variable.
Dimblk	" "	User-defined arrow block name.
Dimblk1	" "	User-defined arrow block name for first end of dimension line used with Dimsah.
Dimblk2	" "	User-defined arrow block name for second end of dimension line used with Dimsah.
Dimfit	3	Controls location of text and arrows for extension lines. If space is not available for both: 0 = text and arrows placed outside; 1 = text has priority, arrows are placed outside extension lines; 2 = arrows have priority; 3 = AutoCAD chooses between text and arrows, based on best fit; 4 = a leader is drawn from dimension line to dimension text when space for text not available; 5 = No leader.
Dimgap	1/16" or 0.09"	Controls distance between dimension text and dimension line.
Dimjust	0	Controls horizontal dimension text position: 0 = centered between extension lines; 1 = next to first extension line; 2 = next to second extension line; 3 = above and aligned with the first extension line; 4 = above and aligned with second extension line.
Dimsah	Off	Allows use of two different arrowheads on a dimension line. See Dimblk1 and Dimblk2.
Dimtfac	1.0"	Controls scale factor for dimension tolerance text.
Dimtih	On	When on, text inside extensions is horizontal.
Dimtoh	On	When on, text outside extensions is horizontal.

Continued

TABLE D.2: DIMENSION VARIABLES (CONTINUED)

Dimension Arrow & Text Control (continued)

Dimension Variable	Default Setting	Description
Dimtad	0	When on, places text above the dimension line.
Dimtix	Off	Forces text between extensions.
Dimtvp	0	Controls text's vertical position based on numeric value.
Dimtxsty	Standard	Controls text style for dimension text.
Dimunit	2	Controls unit style for all dimension style groups except angular. Settings are same as for Lunit system variable.

Dimension & Extension Line Control

Dimension Variable	Default Setting	Description
Dimsd1	Off	Suppresses the first dimension line.
Dimsd2	Off	Suppresses the second dimension line.
Dimse1	Off	When on, suppresses the first extension line.
Dimse2	Off	When on, suppresses the second extension line.
Dimtofl	Off	Forces a dimension line between extension lines.
Dimsoxd	Off	Suppresses dimension lines outside extension lines.

Alternate Dimension Options

Dimension Variable	Default Setting	Description
Dimalt	Off	When on, alternate units selected are shown.
Dimaltf	25.4000	Alternate unit scale factor.
Dimaltd	2	Alternate unit decimal places.
Dimalttd	2	Alternate unit tolerance decimal places.
Dimaltu	2	Alternate unit style. See Lunits system variable for values.

Continued

TABLE D.2: DIMENSION VARIABLES (CONTINUED)

Alternate Dimension Options (continued)

Dimension Variable	Default Setting	Description
Dimaltz	0	Controls the suppression zeros for alternate dimension values.
Dimpost	" "	Adds suffix to dimension text.
Dimapost	" "	Adds suffix to alternate dimension text.

Colors

Dimension Variable	Default Setting	Description
Dimclrd	0 or *byblock*	Controls color of dimension lines and arrows.
Dimclre	0 or *byblock*	Controls color of dimension extension lines.
Dimclrt	0 or *byblock*	Controls color of dimension text.

Finally, if you want to write macros, scripts, or AutoLISP programs to control dimension styles, we'll talk about using two options of the Dimstyle command to set and recall dimension styles from the command line: **dimstyle↵s↵** and **dimstyle↵r↵**.

If you want to change a setting through the command line instead of through the Dimension Styles dialog box, you can enter the system variable name at the command prompt.

Controlling Associative Dimensioning

As discussed in Chapter 8, you can turn off AutoCAD's *associative dimensioning* by changing the Dimaso setting. The default for Dimaso is On.

The Dimsho setting controls whether the dimension value is dynamically updated while a dimension line is being dragged. The default for this setting is On.

Storing Dimension Styles through the Command Line

Once you have set the dimension variables as you like them, you can save the settings by using the Dimstyle command. The Dimstyle/Save command records all of the current dimension variable settings (except Dimaso) with a name you specify.

1. At the command prompt, enter **dimstyle**↵.
2. At the Save/Restore/Status/Variables/Apply/?: prompt, type **s**↵.
3. When the ?/Name for new dimension style: prompt appears, you can enter a question mark (**?**) to get a listing of any dimension styles currently saved, or you can enter a name under which you want the current settings saved.

For example, suppose you change some of your dimension settings through dimension variables instead of through the Dimension Styles dialog box, as shown in the following list:

Dimtsz	0.120
Dimtad	Off
Dimtih	On
Dimtoh	On

These settings are typical for an ANSI standard of dimensioning; you might save them under the name Standard, as you did in an exercise Chapter 8. When you want to return to these settings, use the Restore option of the Dimstyle command, described in the next section.

Restoring a Dimension Style from the Command Line

To restore a dimension style you've saved using the Dimstyle Save option, take these steps:

1. At the command prompt, enter **dimstyle**↵.
2. At the Save/Restore/STatus/Variable/Apply/?: prompt, type **r**↵.
3. At the following prompt

    ?/Enter dimension style name or RETURN to select dimension:

    you have three options: Enter a question mark (**?**) to get a listing of saved dimension styles; enter the name of a style, such as Stan, if you know the name of the style you want; or use the cursor to select a dimension on the screen whose style you want to match.

Notes on Metric Dimensioning

This book assumes you are using feet and inches as units of measure. The AutoCAD user community is worldwide, however, and you may be using the metric system in your work. As long as you are not mixing US (feet and inches) and metric measurements, using the English version of AutoCAD is fairly easy. With the Units command, set your measurement system to decimal, then draw distances in millimeters or centimeters. At plot time, select the MM radio button (millimeters) under Paper Size and Orientation in the Print/Plot Configuration dialog box.

If your drawings are to be in both foot-and-inch and metric measurements, you will be concerned with several settings:

Dimlfac sets the scale factor for dimension values. The dimension value will be the measured distance in AutoCAD units times this scale factor. Set Dimlfac to 25.4 if you have drawn in inches but want to dimension in millimeters. The default is 1.00.

Dimalt turns the display of alternate dimensions on or off. Alternate dimensions are dimension text added to your drawing in addition to the standard dimension text.

Dimaltf sets the scale factor for alternate dimensions (i.e., metric). The default is 25.4, which is the millimeter equivalent of 1".

Dimaltd sets the number of decimal places displayed in the alternate dimensions.

Dimapost adds suffix to alternate dimensions, as in 4.5mm.

If you prefer, you can use the metric template drawing supplied by AutoCAD.

1. Start a new drawing.
2. In the Create New Drawing dialog box, select the Use a Template option.
3. Click on ACADISO.DWT, and click on OK to open the template.

This drawing is set up for metric/ISO standard drawings, except you will need to set plot to MM in the New Drawing dialog box.

A Closer Look at the Dimension Styles Dialog Box

As you saw in Chapter 8, you can control the appearance and format of dimensions through dimension styles. To get the Dimension Styles dialog box, click on the Dimension Style button on the Dimension toolbar or enter **ddim**↵ at the command line.

The three buttons in the Dimension Styles dialog box—Geometry, Format, and Annotation—open related dialog boxes that control the variables associated with these three aspects of AutoCAD's dimensioning system. You'll get a closer look at these dialog boxes in the next section.

Within the Dimension Styles dialog box, dimensions are divided into "families" as a way of classifying the different types of dimensions available in AutoCAD. The dimension families are Angular, Diameter, Linear, Leader, Ordinate, and Radial. The Parent family affects all the dimension families globally. You can fine-tune your dimension styles by making settings to each family independently. If you don't set a family, its settings default to the Parent settings. To change the settings of a family, click on the family name's radio button before making changes in the Geometry, Format, or Annotation dialog boxes.

The following paragraphs describe the options in the Geometry, Format, and Annotation subdialogs. Each description specifies the dimension variables (in parentheses) that are related to the dialog box option. As you work through this appendix, you may want to refer back to the figures in Chapter 8 that illustrate these subdialogs.

The Geometry Dialog Box

This dialog box lets you control the placement and appearance of dimension lines, arrowheads, extension lines, and center marks. You can also set a scale factor for these dimension components.

The Dimension Line Group Refer to Figure D.1 for examples of the effects of these options.

> **Suppress (Dimsd1, Dimsd2)** suppresses the dimension line to the left or right of the dimension text.
>
> **Extension (Dimdle)** sets the distance that dimension lines are drawn beyond extension lines, when using the standard AutoCAD dimension tick for arrows. This option is unavailable (grayed out) when the filled arrow is selected in the Arrowheads group.
>
> **Spacing (Dimdli)** determines the distance between dimension lines from a common extension line generated by the Baseline Dimension or Continue Dimension options on the Dimension toolbar.
>
> **Color (Dimclrd)** sets the color of dimension lines. The standard AutoCAD Color dialog box appears, allowing you to visually select a color.

FIGURE D.1

Examples of how the Dimension Line options affect dimensions

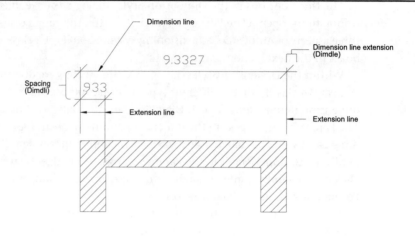

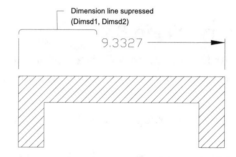

The Extension Line Group Refer to Figure D.2 for examples of the effects of these options.

Suppress (Dimse1, Dimse2) suppresses the first or second dimension extension line.

Extensions (Dimsexe) sets the distance that extension lines extend beyond the dimension line.

Origin Offset (Dimexo) sets the distance the extension line is offset from its point of origin on the object being dimensioned.

Color (Dimclre) sets the color of extension lines. The standard AutoCAD Color dialog box appears, allowing you to visually select a color.

FIGURE D.2

Examples of how the Extension Line options affect extension lines

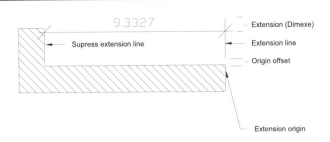

The Arrowheads Group These options let you control the type of arrowhead AutoCAD applies to dimensions.

1st shows you a drop-down list of choices to set the arrowheads at both ends of the dimension line. When you select an arrowhead, the graphic above the group shows you how the arrow will look. You can also click on the graphic to cycle through the selections.

2nd shows you a drop-down list of choices to set different arrowheads for each side of a dimension. This option works like the 1st option, but only sets one arrowhead.

Size (Dimasz) sets the size of the arrowhead.

In the 1st and 2nd drop-down lists is a choice called User; this lets you use a block in place of the standard arrows. A dialog box opens, in which you enter the name of the block you wish to use for the arrow. The block must already exist in the drawing before you can add it, and it must follow the guidelines described in the *Drawing Blocks for Your Own Dimension Arrows and Tick Marks* section of this appendix.

The Center Group These options let you determine what center mark is drawn when using the Dimcenter command. Center marks are also drawn when dimension lines are placed outside a circle or arc using the Diameter or Dimradius commands. All these settings are controlled by the Dimcen system variable.

Mark adds a center mark.

Line creates a center mark and lines.

None suppresses the creation of center marks and lines.

Size controls the size of the center marks.

The Scale Group These options let you control the overall scaling of dimensions.

Overall Scale (Dimscale) sets the scale factor for the size of dimension components, text and arrow size, and text location. This setting has no effect on actual dimension text or the distances being dimensioned.

Scale to Paper Space is meaningful only if you dimension objects in a Model Space viewport while you're in Paper Space. When this option is enabled, AutoCAD adjusts the scaling of dimension components to Paper Space.

The Format Dialog Box

This subdialog contains the following general settings:

User Defined (Dimupt) overrides the dimension text location settings and lets you place the text manually when the dimensions are drawn.

Force Line Inside (Dimtofl) forces a dimension line to be drawn between extension lines under all conditions.

Fit (Dimfit) lets you determine how dimension text and arrows are placed between extension lines. In the drop-down list, Text and Arrows causes both arrows and text to be placed outside extensions if space isn't available. Text Only gives text priority, so that if space is available for text only, arrows will be placed outside extension lines. With the Best Fit option, AutoCAD determines whether text or arrows fit better, and draws the dimension accordingly. Leader draws a leader line from the dimension line to the dimension text when space for text is not available.

The settings in the Format subdialog control the location of dimension text. The Text, Horizontal Justification, and Vertical Justification groups include a graphic that demonstrates the effect of your selected option on the dimension text. You can also click on the graphic to scroll through the options.

The Text Group These options let you control the text location.

Inside Horizontal (Dimtih) orients text horizontally when it occurs between extension lines, regardless of the dimension lines' orientations.

Outside Horizontal (Dimtoh) orients text horizontally when it occurs outside the extension lines, regardless of the dimension lines' orientations.

The Horizontal Justification Group   These settings can also be controlled using the Dimjust system variable.

>   **Centered**   centers the dimension text between the extension lines.
>
>   **1st Extension Line**   places the text next to the first extension line.
>
>   **2nd Extension Line**   places the text next to the second extension line.
>
>   **Over 1st Extension**   places the text over the first extension line, aligned with the extension line.
>
>   **Over 2nd Extension**   places the text over the second extension line, aligned with the extension line.

Vertical Justification   These options can also be set using the Dimtad system variable.

>   **Centered**   places the dimension text in line with the dimension line.
>
>   **Above**   places the text above the dimension line, as is typical in architectural dimensioning. The distance from the text to the dimension line can be set with the Gap option in the Annotation subdialog.
>
>   **Outside**   places the text outside the dimension line at a point farthest away from the origin point of the first extension line. This effect is similar to the Above option, but is more apparent in circular dimensions.
>
>   **JIS**   places text in conformance with the Japanese Industrial Standards.

The Annotation Dialog Box

This dialog box controls the dimension text. You can determine the style, color, and size of text, as well as the unit style, tolerance, and alternate dimensions.

Primary Units and Alternate Units   These two groups offer the same options. The Primary Units options affect only the main dimension text, and the Alternate Units options control alternate units when they are enabled. Alternate units are dimension values that are shown in brackets next to the standard dimension value, and are helpful when two unit systems, such as U.S. (feet and inches) and metric, are used in the same drawing. The Enable Units checkbox turns on the alternate units (see Figure D.3).

FIGURE D.3

An example of alternate units

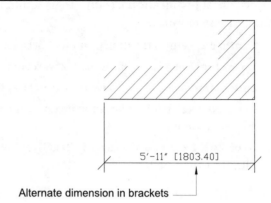

Alternate dimension in brackets

Prefix lets you include a prefix in dimension text, and **Suffix** lets you include a suffix in dimension text. For the primary dimension text, Prefix and Suffix are controlled by the same Dimpost system variable. For alternate dimension text, Prefix and Suffix share the Dimapost system variable. See Table D.2 for details on Dimpost and Dimapost.

Both the Primary Units and Alternate Units groups have a Units button, which you click to bring up the Primary Units and Alternate Units dialog boxes. These dialog boxes contain the same options; when enabled, the Primary Units options affect only the main dimension text, and the Alternate Units options affect only alternate dimension text. Here are descriptions of these options:

Units and **Angles** let you select the unit and angle styles (which are the same styles as for the Units command described in Chapter 3). For Primary Units, the Units option is controlled by the Dimunit system variable, and the Angles option is controlled by the Dimaunit system variable. For Alternate Units, the Units option is controlled by the Dimaltu.

Dimension Precision and **Tolerance Precision (Dimdec, Dimtdec, Dimaltd, Dimalttd)** set the number of decimal places you want to use for these values. Dimension precision can also be set using the Dimdec system variable, and tolerance precision using Dimtdec. For Alternate Units, dimension precision can be set using Dimaltd, and tolerance precision using Dimalttd.

Dimension Zero Suppression (Dimzin, Dimaltz) controls how AutoCAD handles zeros in dimensions. The Leading option suppresses leading zeros in a decimal dimension (for example, 0.3000 becomes .3000). The Trailing option suppresses trailing zeros (so that 8.8000 becomes 8.8, or 45.0000 becomes 45). The 0 Feet option suppresses zero feet values in a dimension so that 0'–4" becomes 4".

0 Inches suppresses zero inch values in a dimension, so that 12' –0" becomes 12'. The Dimzin system variable controls this option for Primary Units; the Dimaltz system variable controls this option for Alternate Units.

Tolerance Zero Suppression (Dimtzin, Dimalttz) controls how AutoCAD handles zeros in tolerance dimensions. (See the description of Dimension Zero Suppression, just above.) The Dimtzin system variable controls this option for Primary Units; Dimalttz controls this option for Alternate Units.

Scale lets you specify a scale factor to linear dimensions. This setting affects the dimension text value. For example, say you have drawn an object to one-half its actual size. To have your dimensions reflect the true size of the object, you would set the Linear input box to 2. When you place dimensions in the drawing, AutoCAD multiplies the drawing distances by 2 to derive the dimension text value. By checking the Paper Space Only check box, you tell AutoCAD to apply the scale factor only when dimensioning in Paper Space. These two options in Alternate Units perform the same function for alternate dimension text. In Primary Units, the Scale options are controlled by the Dimlfac system variable. When Paper Space Only is enabled, Dimlfac becomes a negative value. For Alternate Units, the scale options are controlled by the Dimaltf system variable.

The Tolerance Group   These options affect both primary and alternate units.

Method sets the type of tolerance displayed in a dimension. Choose from the following: The Symmetrical option adds a single tolerance value with a plus-minus (±) sign; this is the same as Dimtol set to 1 and Dimlim set to 0. The Deviation option adds two stacked values: one a plus value and the other a minus value. The Limits option places two stacked dimension values, showing the allowable range for the dimension instead of the single dimension (this is the same as Dimtol set to 0 and Dimlim set to 1). The Basic option draws a box around the dimension text; the Dimgap system variable set to a negative value produces the same result.

Upper Value   sets the maximum tolerance limit. This is stored in the Dimtp system variable.

Lower Value   sets the minimum tolerance limit. This is stored in the Dimtm system variable.

Justification   sets the vertical location of stacked tolerance values.

Height   sets the height for tolerance values. This is stored in the Dimtfac system variable as a ratio of the tolerance height to the default text height used for dimension text.

The Text Group   These options let you control the appearance of text in a dimension.

Style (Dimtxsty)   sets the text style used for dimension text.

Height (Dimtxt)   sets the current text height for dimension text.

Gap (Dimgap)   sets a margin around the dimension text, within which margin the dimension line is broken.

Color (Dimclrt)   sets the color of the dimension text.

Round Off (Dimrnd)   sets the amount that dimensions are rounded off to the nearest value. This setting works in conjunction with the Dimtol system variable.

Importing Dimension Styles from Other Drawings

Dimension styles are saved within the current drawing file only. You don't have to recreate the dimension style for each new drawing, however. If you have loaded the Release 14 Bonus menu (see Appendix B for overall installation instructions), you can import a dimension style that you created in another drawing.

First, save the dimension style using the Bonus ➢ Tools ➢ Dimstyle Export option. The dimension style is saved to a user-specified DIM file. For example, you might call it `Designer.dim` or `mydim1.dim`.

To restore an exported dimension style, choose Bonus ➢ Tools ➢ Dimstyle Import. The selected DIM file is loaded into your current drawing.

Drawing Blocks for Your Own Dimension Arrows and Tick Marks

If you don't want to use the arrowheads supplied by AutoCAD for your dimension lines, you can create a block of the arrowheads or tick marks you like, to be used in the Arrowheads group of the Dimension Styles/Geometry dialog box.

A CLOSER LOOK AT THE DIMENSION STYLES DIALOG BOX

TIP

To get to the Arrowhead options, click on the Dimension Styles button on the Dimension toolbar, or type **ddim**↵. At the Dimension Styles dialog box, click on Geometry, and choose User from the drop-down list for the 1st arrowhead.

For example, say you want to have a tick mark that is thicker than the dimension lines and extensions. You can create a block of the tick mark using a color to which you assign a thick pen weight in the Plot command's Pen Parameters, and then assign that block to the Arrowhead setting. This is done by first opening the Geometry dialog box from the Dimension Styles dialog box (click on the Dimension Styles button on the Dimension toolbar to open the Dimension Styles dialog box). Next, choose User Arrow from the 1st drop-down list in the Arrowheads group. At the User Arrow dialog box, enter the name of your arrow block.

When you draw the arrow block, make it one unit long. The block's insertion point will be used to determine the point of the arrow that meets the extension line, so make sure you place the insertion point at the tip of the arrow. Because the arrow on the right side of the dimension line will be inserted with a zero rotation value, create the arrow block so that it is pointing to the right (see Figure D.4). The arrow block is rotated 180° for the left side of the dimension line.

FIGURE D.4

The orientation and size of a block used in place of the default arrow

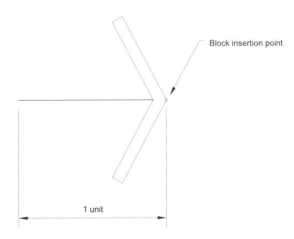

To have a different type of arrow at each end of the dimension line, create a block for each arrow. Then, in the Dimension Styles/Geometry dialog box, choose User in the drop-down list for the 1st arrowhead, and enter the name of one block. Then choose User in the drop-down list for the 2nd arrowhead, and enter the name of the other block.

Index

Note to the Reader: Throughout this index first level entries are in **bold**. **Boldface** pae numbers indicate primary discussions of a topic. *Italic* page numbers indicate illustrations.

Symbols

< > (angle brackets) in prompts, 57
' (apostrophe), system variables and, 824
* (asterisk) as a wildcard character, 173
@ (at sign) in distance specifications, 49–50, 51
\ (backslash)
    in attributes, 782
    in custom menus, 960–961, *961*
, (comma) for entering Cartesian coordinates, 49–50, 52
… (ellipsis) in menus, 15
= (equal to)
    in filtering selections, 407
    and saving blocks, 150
> (greater than) in filtering selections, 407
– (hyphen)
    copying objects with, 203
    in distance entries, 95
    double-hyphen symbol (—) in custom menus, 962
< (less than symbol)
    in Block Definition dialog box, 149
    for entering polar coordinates, 49–50, 52
    in filtering selections, 407
!= (not equal to) in filtering selections, 407
+ (plus sign) in custom menus, 961–962
? (question mark) as a wildcard character, 173
~ (tilde) in custom menus, 962
_ (underscore) in custom menus, 961

A

ABCs of AutoLISP on book's CD-ROM, 1032, 1042–1043
absolute coordinates, finding in drawings, 814
.ac$ files, 1000
acad.ini file, 1022
acad.lin file, 966, 970, 972
acad.log file, 184, 826
acad.lsp file, 879–880
acad.mnc file, 955
acad.mnr file, 955
acad.mns file, 945, 955
acad.mnu file, 955
accelerator keys
    creating for custom menus, 965–966
    entering commands with, 26
    Priority for Accelerator Keys option in Preferences dialog box, 1015
    troubleshooting, 1004
accessing. *See also* displaying; opening; viewing
    Edit Vertex mode for polylines, 428
    external databases, **789–798**
        adding rows to database tables, 797–798
        AutoCAD SQL Extension (ASE) defined, **789–790**
        configuring AutoCAD for external databases, 791–793
        determining database structure, 791
        displaying External Database toolbar, 793
        finding records in databases, 795–797
        opening databases from AutoCAD, 794–795
        setting up AutoCAD SQL Extension (ASE) to locate database files, 790, *790*
    Snap mode, 102–103
ACIS solid-modeling engine, 576
Active application, 936–937, *936*
ActiveX Automation, 877, 905–937
    AutoCAD and, **906–907**, 909–910
    automation techniques, **931–937**
        creating drawings, 932–933
        drawing lines, 933–935, *934*
        overview of, 931
        setting active properties, 936–937, *936*
        user interaction, 935–936
    clients, servers and, 907–908, *907*
    defined, **877, 905–906**
    example, **910–913**, *910*
    examples on book's CD-ROM

accessing, 1041
Active application, 936–937, *936*
defined, **1032**
GridMaker application, 932–935, *934*
Highlight application, 935–936
and integrating AutoCAD with other applications, 906–907
objects, properties, and methods, **908–909**, **913–931**
    Application object, 915–918
    Block object, 921–922, *921*
    classes and, **908–909**
    collection objects, **913**
    defined, **908–909**
    Dictionary object, 923
    DimStyle object, 924
    Document object, 919–920, 933
    entity properties, 922–923
    Group collection object, 924
    Layer object, 924–926
    LineType object, 926
    object hierarchy, **914–915**, *914*
    Plot object, 929–930
    Preferences object, 918–919
    RegisteredApplication collection object, 926
    SelectionSet collection object, 924, 936
    TextStyle object, 927
    UserCoordinateSystem object, 927–928
    Utility object, 930–931
    View object, 928
    Viewport object, 928–929
  variables, 911
add object selection command, 70
Add a Printer dialog box, 650, *650,* 651
adding
  3D faces to 3D surfaces, 596–598, *596, 597, 599, 600*
  attributes to blocks, 763–765, *765*
  drawing labels with database links, 800–802
  DWF files to Web pages, **889–891**
    overview of, 889
    specifying filenames and opening views, 890–891
    what your Internet service provider needs to know, 891
  feet and inch distances, 411–412
  Help messages to custom menus, 963–964
  linetypes to drawings, 176–177
  notes with arrows, **350–352**
    overview of, 350–351, *350, 351*
    using multiline text with leaders, 351–352
  plotters and printers, 649–651
  rows to database tables, 797–798
  text objects to drawings, 291–293, *292*

text to drawings, 266–269, *268*
tolerance notation to dimensions, 356–358
volumes of two solids, 475–477, *476*
Additional Parameters options, 655–664. *See also* Print/Plot Configuration dialog box
  Adjust Area Fill option, 656, *657*
  Autospool option, 658
  Display option, 658–659, *659,* 664
  Extents option, 660, *660,* 664
  Hide Lines option, 656
  Limits option, 660, *661*
  Plot to File option, 657–658
  Text Fill option, 655
  Text Resolution option, 655
  View option, 661–662, *662*
  Window option, 662–663, *663*
add-ons. *See also* CD-ROM (book's)
  Eye2eye add-on, **877, 1032, 1035–1040**
    creating 3D models, 1039
    defined, **877, 1032, 1035**
    installing, 1035–1036
    moving and controlling camera and target objects, 1040
    overview of, 1036–1039, *1036*
    setting up viewports, 1037–1038, *1038*
  installing, 1035
  SI-Mechanical add-on, 885, 1032, 1033
Adjust Area Fill option in Print/Plot Configuration dialog box, 656, *657*
Adjust Planar Coordinates dialog box, 718, *718*
adjusting. *See also* changing; editing
  appearance of materials for 3D rendering, 699–700, *699, 701*
  camera and target positions in perspective views, 629–631, *630, 631*
  dimensions in traced drawings, 738
  distances between camera and target in perspective views, 627–629, *628, 629*
  raster image brightness, contrast, and strength, 748–749, *748, 749*
  raster image quality and transparency, 750–751, *750*
Administration dialog box, 794–795, *794,* 799, 802
ADS (AutoCAD Development System)
  defined, **877**
  Load AutoLISP, ADS, and ARX Files dialog box, 878–879
  Xdata Attachment command, 876
Aerial View window, 9, 32, 255–257, *256*
Alias Editor dialog box, 871–873, *871*
Align option of Dtext command, 297, *297*
Aligned Dimension tool, 342–343, *343*
aligning
  lines in traced drawings, 735–737, *736, 737*

objects in 3D space, **622–623**, *623*
text, **275–278**
    Osnaps and, 277–278, *277*
    overview of, 275–276, *275*, *276*
two 3D objects, **508–521**
    finishing up, 514–521
    overview of, 508–513, *509*, *510*, *511*, *513*
all object selection command, 69
alternate dimension options variables, 1073–1074
America Online Autodesk forum, 886
angle brackets (< >) in prompts, 57
angles
    Angle option of Xline command, **221**
    dimensioning, 345–346, *346*
    entering
        measurement system and, 95–96
        overview of, 50, *51*
    setting angle display style, 94
    setting direction of 0 base angle, 95
Angular Dimension tool, 345–346, *346*
animation, linking to AutoCAD, 844
Annotation dialog box, 314, 315, 1081–1084, *1082*
ANSI-style dimensions, 312, *312*
answering attribute prompts with Enter Attributes dialog box, 772
Anti-Aliasing option in Render dialog box, 724–725, *724*
AOL Autodesk forum, 886
apostrophe ('), system variables and, 824
appending data to dimension text, 328–329
applications. *See also* utilities
    Active application, 936–937, *936*
    Application object in ActiveX Automation, **915–918**
    GridMaker application, 932–935, *934*
    Highlight application, 935–936
    RegisteredApplication collection object in ActiveX Automation, **926**
applying ordinate dimensions, 354–355, *354*, *355*
architectural scale factors, calculating, 318
Architectural unit style, 94, *95*
arcs. *See also* circles; lines
    arc (a) command, 55–60, *56*, *59*, *60*, 120–122, *121*
    Arc Aligned Text tool, **858**
    Arc and Circle Smoothness option in Preferences dialog box, 255, 1013
    Arc option of Polyline tool, 419
    Arc tool, 55–60, *56*, *59*, *60*, **120–122**, *121*
    dimensioning radii, diameters, and arcs, 347–348, *348*
    display of in enlarged drawings, 218
    finding distances along, **228–230**

    finding an exact distance along an arc, 229–230, *231*
    finding a point a particular distance from another point, 228–229, *229*
    UCS orientation and, 487
areas
    of complex shapes, 817–820, *818*, *819*
    of objects, 814–817, *815*, *816*
    recalling last area calculation, 817
    saving area data as attributes in drawing files, **820**
arrays
    array cells, **208**, *208*
    Array tool, **206–208**, *207*, *208*, 598, *600*, **602**
    defined, **199**
    polar arrays
        creating, **200–204**, *202*, *204*
        defined, **199**
        Polar Array tool and 3D solid models, 533–534, *534*
    rectangular arrays
        creating at an angle, 207
        defined, **199**
arrows
    adding notes with, **350–352**
        overview of, 350–351, *350*, *351*
        using multiline text with leaders, 351–352
    arrow and text control dimension variables, 1072–1073
    creating blocks for dimension arrows, **1084–1086**, *1085*
    selecting arrow styles, 317–319
ARX. *See* ObjectARX programming environment
ASE. *See* AutoCAD SQL Extension
assigning
    layers to objects, 163–166
    linetypes to layers, **174–181**
        adding linetypes to drawings, 176–177
        overview of, 174, *175*
        setting linetype scale, 177–181, *179*
associative dimensioning feature. *See also* dimensions
    controlling, 1074
    defined, **311**
    definition points and, 331, *331*
    and editing dimensions with other objects, 338–339, *338*, *339*
    turning off, 341
asterisk (*) as a wildcard character, 173
at sign (@) in distance specifications, 49–50, *51*
Attach Image dialog box, 742, *742*, *743*
Attach option in External Reference dialog box, 376
Attach Xref dialog box, 377–378, 385

attaching drawings as external references, 372–375, *374*
attdisp command, 778–779, *779*
attredef command, 779–780
attributes, 487, 761–789, **808**
   answering attribute prompts with Enter Attributes dialog box, 772
   attdisp command, 778–779, *779*
   attredef command, 779–780
   Attribute Definition dialog box
      attribute definition modes, 768
      overview of, 763–764, *763*
   attribute tags, **764**
   backslash (\) in, 782
   balloons, leaders and, 762
   blocks and
      adding attributes to blocks, 763–765, *765*
      extracting block information using attributes, 783–784
      inserting blocks containing attributes, 768–771, *770*
      redefining blocks containing attributes, 779–780
   Bonus Text Tools toolbar Global Attribute Edit tool, 778
   case sensitivity of, 769
   creating, **762–772**
      adding attributes to blocks, 763–765, *765*
      changing attribute specifications, 765–768, *767*
      inserting blocks containing attributes, 768–771, *770*
      linking attributes to parts, 762
   ddattext command, 785, 787
   defined, **761–762**
   Draw menu Block submenu Define Attributes command, 763–764
   editing, **772–780**
      editing individual attributes, 773
      editing several attributes in succession, 773–774, *774*
      Global Attribute Edit tool, **778**, **858**
      global editing, 776–778
      making invisible attributes visible, 778–779, *779*
      making minor changes to attributes' appearance, 774–776, *775*
      redefining blocks containing attributes, 779–780
   exercises, **808**, *809*
   Explode Attributes to Text tool, **858**
   extracting and exporting attribute information, **781–789**
      Attribute Extraction dialog box, 785–788, *785*
      attribute template files and, 781, 784
      to comma-delimited format (CDF), 785–787
      to data exchange format (DXF), 788
      determining what to extract, 781–783
      with Export Data dialog box, 788
      exporting to Microsoft Access, 789
      exporting to Microsoft Excel, 788
      extracting block information using attributes, 783–784
      performing the extraction, 785–788
      to space-delimited format (SDF), 787–788
      text editor line endings, 783
   Make Displayable Attribute dialog box, 801–802, *801*
   Modify menu Object submenu Attribute submenu
      Global command, 774–778, *775*
      Single command, 773–774, *774*
   saving area data as attributes in drawing files, **820**
   spaces in attribute values, 782
   UCS orientation and, 487
   uses for, 780
Audit after Each DXFIN or DXBIN option in Preferences dialog box, 1017
Audit utility, 828
auto object selection command, 70
Auto option of Undo command, 200
AutoCAD 14 Instant Reference, **1032**, **1042**
AutoCAD
   ActiveX Automation and, **906–907**, **909–910**
   advantages of, **989–990**
   configuring for external databases, 791–793
   defined, **5–6**
   displaying screen menu in Drawing window, 8, 1018
   exiting, **35**
   installing, **1006–1008**
      AutoCAD directory contents, 1008
      the installation process, 1006–1008
      preparing to install, 1006
   on networks, **987–988**
   newsgroups, 886
   opening databases from, 794–795
   program window, **9–13**
      Command window, 9, *12*, 13, 18–19
      drawing area, 9, 11–12
      menu bar, 9, *11*
      overview of, 9–13, *10*
      selecting coordinate points in drawing area, 12
      status bar, 9, *12*, 93
      User Coordinate System (UCS) icon, 12–13
   projects
      managing, 983–984
      Project Files Search Path option in Preferences dialog box, 1010

when to stop using AutoCAD, 990
setting up for Open Database Connectivity (ODBC) drivers, **1027–1030**
    creating reference database files, 1027–1029
    telling AutoCAD the location of reference files, 1030
    telling Windows the location of reference files, 1029
starting, 7
supporting, **979–984**
    getting outside support, 979–980
    learning AutoCAD, 982
    making AutoCAD easier to use, 982–983
    managing AutoCAD projects, 983–984
    preparing the staff, 981–982
    selecting in-house experts, 980–981
tips for DOS version users, 8
troubleshooting, **1002–1004**
VBA Edition, **938–939**, *938*
AutoCAD Alias Editor dialog box, 871–873, *871*
AutoCAD Development System (ADS)
    defined, **877**
    Load AutoLISP, ADS, and ARX Files dialog box, 878–879
    Xdata Attachment command, 876
AutoCAD Drawing Explorer utility, 917
AutoCAD Extended Batch Plot utility. *See* Batch Plot utility
AutoCAD SQL Extension (ASE)
    defined, **789–790**
    setting up to locate database files, 790, *790*
AutoCAD Text window. *See also* windows
    capturing and saving text data from, 827–828
    displaying data in, 84–85, *85*
    Number of Lines of Text in Text Window to Keep in Memory option in Preferences dialog box, 1019
    overview of, 268
    Status screen, 820–822, *821*
    Text Window Parameters options in Preferences dialog box, 1019
AutoCAD Window Format options in Preferences dialog box, 1019
AutoLISP language, **877–884**
    ABCs of AutoLISP on book's CD-ROM, **1032**, **1042–1043**
    defined, **877**
    Load AutoLISP, ADS, and ARX Files dialog box, 878–879
    loading programs automatically, 879–880
    loading and running programs, 878–879
    macros, **881–884**, 963
        creating, 881–883, *881*, *882*
        loading into custom menus, 963
    saving as files, 883–884, *883*
    Reload AutoLISP between Drawings option in Preferences dialog box, 1015
    Xdata Attachment command, 876
Automatic save feature, 32–33, 128, **1016**
automation. *See* ActiveX Automation
Autoselect feature, **72–74**, *73*, *74*
AutoSnap settings in Object Snap Settings dialog box, 142
Autospool option in Print/Plot Configuration dialog box, 658

B

Back option of Undo command, 200
backgrounds in 3D rendering, 702–704, *703*, *704*
backing up files, **985–987**
    Create Backup Copy with Each Save option in Preferences dialog box, 1016
backslash (\)
    in attributes, 782
    in custom menus, 960–961, *961*
Backspace key, **48**
.bak files, 985
balloons, attributes and, 762
base points, selecting objects with, 64–68, *66*, *67*, *68*
Baseline Dimension tool, 326–327, *327*
Batch Plot utility, **675–678**. *See also* labels; plotting
    File tab, 675–676, *675*
    Layers tab, 676–677, *677*
    Logging tab, 678, *678*
    Plot Area tab, 677, *677*
    Plot Stamping tab, 678, *678*
BC option of Dtext command, 296
Beep on Error in User Input option in Preferences dialog box, 1017
Begin option of Undo command, 200
Begin OR and Begin AND operators in filtering selections, 408–409
Bind option in External Reference dialog box, 377
Bisect option of Xline command, 221
bitmap images
    in 3D rendering, **714–718**, *715*, *716*, *719*
    BMP files, 651
BL option of Dtext command, 296
blips
    blipmode command, **54**
    clearing screen of, 54
    defined, 30, **54**
blocks, **134–152**, **166**, **181–182**. *See also* grouping objects; symbols
    attributes and
        adding to blocks, 763–765, *765*

extracting block information using attributes, 783–784
inserting blocks containing attributes, 768–771, *770*
redefining blocks containing attributes, 779–780
Block Definition dialog box, 134–136, *134*, 149
Block object in ActiveX Automation, **921–922**, *921*
block-related tools in Bonus Standard toolbar, **869**
clipping, 380
converting external reference files to, **376**
creating for dimension arrows and tick marks, **1084–1086**, *1085*
creating symbols from, **134–136**
defined, **134**
deleting, **239–241**
    deleting all unused blocks, 240–241
    selectively deleting unused blocks, 239–240
versus external reference files (Xrefs), **372–376**
frozen layers and, 261
versus grouping objects, 152
importing from external reference files, **374**, **382–384**
layers and, 166
Make Block tool or Block (B) command, **134–136**
marking intervals on curves with, **453**
replacing existing files with, 151
retaining objects used to create, 136
saving as drawing files, **150**
selecting objects for, 135
UCS orientation and, 487
unblocking, exploding, or redefining, **147–149**, *149*
uses for, **151–152**
viewing, 181–182
wblock command, 151
BMP files. *See* bitmap images
Bonus tools, 850–876. *See also* utilities
    Bonus Layer Tools toolbar, 852–853
    Bonus menu Cookie Cutter Trim command, 867–868
    Bonus menu Tools submenu, **870–876**
        Command Alias Editor command, 871–873
        Dimstyle Export and Dimstyle Import commands, 874–875
        Get Selection Set command, 876
        List Entity Xdata command, 876
        Pline Converter command, 876
        Popup Menu command, 870–871
        System Variable Editor command, 873–874
        Xdata Attachment command, 876
    Bonus Standard toolbar, **859–870**
        block-related tools, 869
        Copy Nested Entities tool, 869
        Extend to Block Entities tool, 869
        Extended Change Properties tool, 865–866
        Extended Clip tool, 868, *869*
        Extended Trim tool, 867–868
        List Xref/Block Entities tool, 870
        Move Copy Rotate tool, 870
        Multiple Entity Stretch tool, 867
        Multiple Pedit tool, 866–867
        Pack 'n Go tool, 381, 863–865
        Quick Leader tool, 870
        Revision Cloud tool, 862–863, *863*
        Trim to Block Entities tool, 869
        Wipeout tool, 748, 859–861, *860*, *861*, *862*
    Bonus Text Tools toolbar, **306, 853–858**
        Arc Aligned Text tool, 858
        Change Multiple Text Items tool, 856–857
        Explode Attributes to Text tool, 858
        Explode Text tool, 858
        Find and Replace Text tool, 857–858
        Global Attribute Edit tool, 778, 858
        overview of, 306
        Text Fit tool, 858
        Text Mask tool, 853–855, *855*
    installing, 850
    Layer Manager
        saving layer settings of external reference files, 382
        saving and recalling layer settings, 851–852
Boolean operations, regions and, **555–558**, *557*, *558*
Boolean submenu, Modify menu
    Subtract command, 471–472, *472*, 516, *517*, 820
    Union command, 476, 526–528, *526*, *527*
Boundary Creation dialog box, 815–817, *816*
boundary windows, sizing, 266
box primitives, creating solid models from, 477–478
Box tool, 477–478, 552–553
BR option of Dtext command, 296
Break option of polyline Edit Vertex mode, 429–430, *430*
breaking lines, **224–226**, *225*
brightness of raster images, 748–749, *748*, *749*
Browse option in External Reference dialog box, 377, 381
Browse/Search dialog box, 184–186, *185*, *186*
bubble-jet printers, **996**
Button Editor dialog box, 951–953
Button Properties dialog box, 949–950
buttons. *See also* toolbars
    creating custom buttons, 949–950
    creating custom icons, 951–953
    editing existing buttons, 955
Bylayer property, 180–181, 367, *368*

C

Cadence and *Cadalyst* utilities and information, 876–877, 886
calculating
 architectural scale factors, 318
 areas
  of complex shapes, 817–820, *818*, *819*
  of objects, 814–817, *815*, *816*
  recalling last area calculation, 817
  saving area data as attributes in drawing files, 820
Calculator, 409–413
 adding feet and inch distances, 411–412
 finding the midpoint between two points, 409
 finding a point relative to another point, 411
 guidelines, 412
 operators and functions listed, 413
 Osnap modes in Calculator expressions, 409–411
calibrating tablet digitizers, 732–734, *734*
capturing text data from AutoCAD Text window, 827–828
Cartesian coordinates. *See also* coordinates
 entering distances with, 51–53, *51*, *52*, *53*
cascading menus, creating, 962
case sensitivity of attributes, 769
CDF files (comma-delimited format), exporting attribute information to, 785–787
CD-ROM (book's), 877, 885, 917, 932–937, 1032–1044
 ABCs of AutoLISP, 1032, 1042–1043
 ActiveX Automation examples
  accessing, 1041
  Active application, 936–937, *936*
  defined, **1032**
  GridMaker application, 932–935, *934*
  Highlight application, 935–936
 AutoCAD 14 Instant Reference, 1032, 1042
 AutoCAD Drawing Explorer utility, 917
 Eye2eye add-on, **877, 1032, 1035–1040**
  creating 3D models, 1039
  defined, **877, 1032, 1035**
  installing, 1035–1036
  moving and controlling camera and target objects, 1040
  overview of, 1036–1039, *1036*
  setting up viewports, 1037–1038, *1038*
 Figures directory, 1032, 1033, 1034–1035
 installing, **1033–1035**
  add-ons, 1035
  opening the installation program, 1033–1034
  and using sample drawing files, 1034–1035
 Nozzle3d file, 27–28, *28*
 SI-Mechanical add-on, 885, 1032, 1033
 SWLite software, 1032, 1033, 1040–1041
 Whip2 Netscape Plug-in, 900–901, 1033, 1043–1044
Celtscale system variable, 180
Center option of Dtext command, 295
Chamfer tool, 222, 474–475, *475*, **531–532**, *532*
Change Dictionaries dialog box, 301–302, *301*
Change Group options in Object Grouping dialog box, 157
Change Multiple Text Items tool, 856–857
Change Properties dialog box, 166
changing. *See also* adjusting; editing
 attribute specifications, 765–768, *767*
 fonts, 269–272, *271*
 layer property of lines, 168–169
 layers, 167–168, *168*
 layers of objects, 367, *368*
 length of objects, 231–232
 mesh control settings, 610, *611*
 polyline width, 424, *424*
 shape of viewports in Paper Space, 396, 399–400
 spline control points with Refine option, 447–450, *448*, *449*
 spline tolerance setting, 446, *446*
 style settings for groups of dimensions, 337
 style settings of individual dimensions, 333–337
 text boundary window width, 274, *274*
 text height, 269–272, *271*
Character Map dialog box, 289–290, 298
checking interference in 3D solid models, 559–561, *560*
checking spelling, 299–302
 Check Spelling dialog box, 299–301
 selecting dictionaries, 301–302
child dimension styles, 349
circles. *See also* arcs
 Arc and Circle Smoothness option in Preferences dialog box, 255, 1013
 creating filled circles with Donut command, 456, *457*
 dimensioning radii, diameters, and arcs, 347–348, *348*
 display of in enlarged drawings, 218
 extruding circular surfaces, 617–619, *618*
 UCS orientation and, 487
classes, ActiveX Automation objects and, **908–909**
cleaning up
 line work, **222–226**, *222*
 traced drawings, **735–738**
  adjusting dimensions, 738
  aligning lines, 735–737, *736*, *737*
  straightening lines, 735
clearing screen of blips, 54
click and drag actions, 22, 43–44

clicking on a point, 12
clients
    ActiveX Automation and, 907–908, *907*
    client/server networks, **988**
Clip command in Modify menu Object submenu, 378–380, *379*
clip panes in perspective views of 3D surfaces, 636–638, *636, 637*
Clipboard
    copying 3D solid models with, 562–564, *564*
    copying information to dialog boxes from, 828
    exporting drawings from, 845–846
clipping
    blocks, 380
    raster images, 745–748, *747*
    Xref views with Xclip command, **378–380**, *379*
clockwise, copying objects in a clockwise direction, 203
Close option
    of Multiline command, 232
    of Spline command, 442
closing
    AutoCAD, 35
    Draw toolbar, 33
    menus, 14
    Modify toolbar, 33
co (copy) command, 109, *110*, 212–213, 368–369, *369*, 602
collection objects in ActiveX Automation, 913
colors
    Bylayer property, **180–181**, 367, *368*
    changing color properties, **180–181**
    color dimension variables, **1074**
    coloring text, 272–274
    Colors option in Preferences dialog box, 1019
    layers and, 166
column copies, 204–213
comma (,) for entering Cartesian coordinates, 49–50, 52
comma-delimited format (CDF), exporting attribute information to, 785–787
Command Alias Editor command in Bonus menu Tools submenu, 871–873
Command window
    Angle/Length of chord/<End point>: prompt, 57–58, *59*
    Center/<Start point>: prompt, 56–58, *59*
    overview of, 9, *12*, 13, **18–19**
command-line commands. *See also* menus; toolbars
    arc (a), 120–122, *121*
    area, 814–815, *815*
    attdisp, 778–779, *779*
    attredef, 779–780
    blipmode, 54

block (b), 134–136
command prompts, **55–60**, *59*, *60*
copy (co), 109, *110*, 212–213, 368–369, *369*, 602
copyclip, 562–564, *564*
ddattext, 785, 787
ddedit, 276, 278–279
ddim, 1076
ddrename, 286
dimalt, 1076
dimaltd, 1076
dimaltf, 1076
dimapost, 1076
dimedit, 330, 336
dimlfac, 1076
dimoverride, 337
dimstyle, 1069, 1074
displaying data in AutoCAD Text window, **84–85**, *85*
dist (distance), 94
draworder command, 371
dtext, 291–293, *292*, 295–296, *296*
dxfin, 831–832
dxfout, 831
ellipse (el), 106–107, *107*
entering
    with accelerator keys, 26
    from the keyboard, 8, 15, 26
    from pull-down menus and toolbars, 18–19, 26
explode (x), 147–149, *149*
export (exp), 150–151
fillet (f)
    editing 3D solid models, 502–505, *503*, *505*, 528–529, *529*
    overview of, 117–118, *118*, 127–128, *128*, 222–224
    polylines and, 421
gridunit, 101
hot grip commands, 80
hpconfig, 669
keyboard equivalents for menu commands, 15
layer, 162–163
logfileon and logfileoff, 825
massprop, 574
move (m), 68, 71–72
mslide, 547–548
Noun/Verb selection method: selecting objects before commands, **70–72**, **74–75**, **1025**
Number of Lines of Text to Show in Docked Command Line Window option in Preferences dialog box, 1019
object selection commands, **69–70**
    add, 70
    all, 69
    auto, 70

crossing, 69
crossing polygon, 69
fence, 69
last, 69
multiple, 69
previous, 69
remove, 70
single, 70
window, 69
window polygon, 70
offset (o), 124–125, *124, 125,* 127–128, *128,* 421–422, *422*
pasteclip, 562–564, *564*
qtext (Quick Text), 305, *305*
redo, 49
redraw (r), 54, 253–254, 288, 714
regen, 54, 147, 218, **253–254**, 288, 714
reinit, 673
repeating, 79
rm, 101, 102
savetime, 128
selecting, **55–60**, *56, 59, 60*
sketch, 454–455
soldraw, 568–569
solprof, 569–571, *571*
solview, 565–568, *566, 568*
time, 823, *823*
toolbar, 947–948
undo (u), 49
viewres, 254–255
wblock, 151
xclip, 378–380, *379*
xline, 220–221
xref, 240
zoom (z), 108
common base extension lines, drawing dimensions from, 325–327, *326, 327*
companion CD-ROM. *See* CD-ROM (book's)
company-wide layering conventions, 984–985
Compatibility tab in Preferences dialog box, 1014–1015
   Load ARX Applications on Demand option, 1014–1015
   overview of, 1014, *1014*
   Priority for Accelerator Keys option, 1015
   Priority for Coordinate Data Entry option, 1015
   Proxy Images for Custom Objects option, 1015
   Reload AutoLISP between Drawings option, 1015
   Show the Start Up Dialog Box option, 1015
compiling PostScript fonts into AutoCAD font format, 272
CompuServe Autodesk forum, 886
Cone tool, 550, *550*

configuring. *See also* Preferences dialog box; setting up
   AutoCAD for external databases, 791–793
   digitizing tablets, 1022
   tablet digitizer menu templates, 995, 1022–1024, *1023*
   tablet digitizers for tracing, 731, *731*
Connect option of Sketch command, 455
Constant attribute definition mode, 768
construction lines as tools, 226–228, *227, 228*
Continue Dimension tool, 321–327
continuing dimensions, 321–327. *See also* dimensions
   drawing dimensions from common base extension lines, 325–327, *326, 327*
   from older dimensions, 327
   from previous dimensions, 325
   using Osnaps while dimensioning, 322–325, *324*
contrast of raster images, 748–749, *748, 749*
Control option of Undo command, 200
control points of spline curves
   adding with Fit Data option, 444–445, *445*
   changing with Splinedit command Refine option, 447–450, *448, 449*
   defined, **442**
controlling
   associative dimensioning feature, 1074
   camera and target objects in Eye2eye add-on, 1040
   display smoothness with Viewres command, 254–255
   layer visibility, 169–170, *170*
   layers with Layer command, 162–163
   object visibility and overlap with raster images, 742–743
   tangency at the beginning and end points of splines, 444
   UCS icon display, 479
   Xref memory use, 380–381, *381*
converting
   3D solid models to 2D drawings, **561–564**
      copying solid models using the Clipboard, 562–564, *564*
      overview of, 561–562
   external reference files to blocks, 376
   flyouts to custom toolbars, 43
   objects to polylines, 427–428
   polylines
      Pline Converter command in Bonus menu Tools submenu, 876
      to solids, 470–471
      to splines, 428
Cookie Cutter Trim command in Bonus menu, 867–868

coordinates, 12, 49–53. *See also* User Coordinate System; World Coordinate System
    coordinate readout, **104–105**, *105*
    creating 3D meshes by specifying coordinates, 611
    entering distances
        with relative Cartesian coordinates, 51–53, *51, 52, 53*
        with relative polar coordinates, 49–50, *50, 51*
    finding absolute coordinates in drawings, 814
    Priority for Coordinate Data Entry option in Preferences dialog box, 1015
    selecting coordinate points in drawing area, 12, 486
    spherical and cylindrical coordinate formats, 593–595, *594, 595*
copy (co) command, 109, *110*, 212–213, 368–369, *369*, 602
Copy Link command In Edit menu, 846
Copy Nested Entities tool, **869**
Copy tool, 109, *110*, **112–113**, 368–369, *369*, 602
copyclip command, 562–564, *564*
copying
    3D solid models using the Clipboard, 562–564, *564*
    drawings
        from existing drawings, 118–123, *119, 121, 122, 123*
        with Offset tool, 124–125, *124, 125*, 127–128, *128*, 421–422, *422*
    information to dialog boxes from Clipboard, 828
    objects
        on any layer, 368–369, *369*
        with base points, 64–68, *66, 67, 68*
        with minus sign (-), 203
    objects multiple times, **199–213**
        creating polar arrays, 200–204, *202, 204*
        creating random multiple copies, 212–213
        creating row and column copies, 204–213
corrupted files, recovering, 828, 1001–1002
CRC Validation option in Preferences dialog box, 1017
Create Backup Copy with Each Save option in Preferences dialog box, 1016
Create Group options in Object Grouping dialog box, 157
Create New Drawing dialog box, 198–199, 215
Create Slide File dialog box, 547–548, *547*
creating. *See also* customizing; defining
    3D solid models, **468–478, 500–501, 506–508, 529–531**
        from 2D drawings, 506–507, *506, 508*
        adding volumes of two solids, 475–477, *476*
        from box primitives, 477–478

        converting polylines to solids, 470–471
        drawing solid cylinders, 469–470
        overview of, 500–501, *501, 502*, 529–531, *530*
        preparation, 469
        removing the volume of one solid from another, 471–472, *472*
        verifying designs with the Hide tool, 472–474, *473*
    3D surface models, **584–591**
        from 2D polylines, 586–587, *588*
        elevation and, 584–585, *585*, 586–587, *588*
        making horizontal surfaces opaque, 590, *590, 591*
        overview of, 584–585, *584, 585*
        setting elevation and thickness before drawing, 588–589, *589*
        setting layers, 591, *591*
        thickness property and, 584, *584*, 586, *588*
    3D surfaces, **592–603, 611–619**
        adding 3D faces, 596–598, *596, 597, 599, 600*
        defining surfaces using two objects, 612–614, *613, 614*
        extruding circular surfaces, 617–619, *618*
        extruding objects along a straight line, 615–616, *615, 616*
        hiding unwanted surface edges, 600, *601*
        laying out 3D form objects, 593–595
        spherical and cylindrical coordinate formats, 593–595, *594, 595*
        using point filters, 592
        using pre-defined 3D surface shapes, 602
accelerator keys for custom menus, 965–966
attributes, **762–772**
    adding attributes to blocks, 763–765, *765*
    changing attribute specifications, 765–768, *767*
    inserting blocks containing attributes, 768–771, *770*
    linking attributes to parts, 762
AutoLISP macros, 881–883, *881, 882*
blocks for dimension arrows and tick marks, **1084–1086**, *1085*
cascading menus, 962
company-wide layering conventions, **984–985**
complex 3D surfaces, **603–611**
    changing mesh control settings, 610, *611*
    creating 3D meshes by specifying coordinates, 611
    curved 3D surfaces, 603–610, *604*
dimension styles, **312–319, 349**
    Annotation dialog box and, 314, 315, 1081–1084, *1082*

changing style settings for groups of dimensions, 337
changing style settings of individual dimensions, 333–337
dimension styles versus ANSI-style dimensions, 312, *312*
Dimension Styles dialog box and, **1076–1084**
Format dialog box and, 316, 1080–1081
Geometry dialog box and, 317–318, 1077–1080, *1078, 1079*
importing dimension styles from other drawings, **1084**
overview of, **312–313**, *312*
parent and child styles in dimension families, 349
Primary Units dialog box and, 314–315
selecting arrow style and setting dimension scale, 317–319
setting dimension unit style, 313–315
setting height for dimension text, 315
setting location of dimension text, 316
using dimension families to fine-tune dimension styles, 349
division lines and dimmed text in custom menus, 962
drawings
    with ActiveX Automation, 932–933
    exercises, 85–86, *86*, 129, *129*
    from existing drawings, 118–123, *119, 121, 122, 123*
    overview of, 44–48, *47*
DWF files, 888–889
filled circles with Donut command, 456, *457*
flyouts, 953–954
hatch patterns, 974–979
layers, 159–161, *159, 160*, 165–166
multilines, 232–233, *232*
multiple views, 262
object links to databases, 799–800, *800*
polar arrays, 200–204, *202, 204*
polyline spline curves, 438–440
polylines, 417–420, *418, 419*
random multiple copies, 212–213
reference database files for ODBC drivers, 1027–1029
row and column copies, 204–213
shortcut to acad.log file, 826
solid filled areas, 455–456
spline curves, 441–442, *441, 443*
styles, 281–282
symbols
    from blocks, **134–136**
    from existing drawings, 140–147, *140*
    from external reference files (Xrefs), 152
    from shapes, 152
    from Windows Character Map dialog box, 289–290, 298
templates, 196–198
viewports in Paper Space, 392, *392*
Web-compatible drawings, **887–889**
    creating DWF files, 888–889
    installing Internet Utilities, 888
    opening Internet Utilities toolbar, 888
Crop Window option in Render dialog box, 722–723, *722*
crosshair cursor, 48
crossing object selection command, 69
crossing polygon object selection command, 69
crossing windows
    defined, **72**, *73*
    of Stretch command, 338–339, *338*
cross-referenced files. *See* external reference files
Ctrl key, selecting objects with, 111, 369–370
cursors
    crosshair cursor, **48**, *55*
    Object Selection cursor, **54–55**, *55*, 61
    object snap (Osnap) marker, **55**, *55*, 65–67, *66, 67*
    Percent of Screen Size option in Preferences dialog box, 55
    Point Selection cursor, **48**, **54–55**, *55*
    sizing, 55
    Standard cursor, **54–55**, *55*, **71**
    types of, **54–55**, *55*
    Zoom Realtime cursor, 211
curved 3D surfaces, 603–610, *604*
curves. *See* polylines; spline curves
customizing, 944–979. *See also* ActiveX Automation; creating
    hatch patterns, **974–979**
    linetypes, **966–974**
        acad.lin file and, 966, 970, 972
        creating complex linetypes, 971–974, *972, 973*
        creating linetypes, 969–970
        linetype codes and, 970–971, *970, 971*
        viewing available linetypes, 966–967, *967, 968*
    menus, **956–966**
        adding Help messages, 963–964
        calling commands, 960
        creating accelerator keys, 965–966
        creating cascading menus, 962
        creating division lines and dimmed text, 962
        creating pull-down menus, 956–957, 959
        double-hyphen symbol (—) in, 962
        loading AutoLISP macros in menus, 963
        loading custom menus, 957–959
        Menu Customization dialog box and, 957–959

pausing for user input with backslash (\),
960–961, *961*
tilde (~) in, 962
underscore (_) in, 961
using plus sign (+) for long lines, 961–962
toolbars, **944–955**
creating custom buttons, 949–950
creating custom icons, 951–953
creating flyouts, 953–954
creating toolbars, 945–947
Customize Toolbars dialog box and, 946–947
editing existing buttons, 955
New Toolbar dialog box and, 946
Toolbars dialog box and, 944–945
Windows menu files and, 955
cutting and pasting AutoCAD images into other applications, 845–846
Cylinder tool, 469, **475–477**, *476*, 544
cylindrical coordinate formats, 593–595, *594*, *595*

D

data
displaying in AutoCAD Text window, 84–85, *85*
translation tools, 575–576
data exchange format. *See* DXF files
databases, 789–808. *See also* Microsoft Access
accessing external databases, **789–798**
adding rows to database tables, 797–798
AutoCAD SQL Extension (ASE) defined, **789–790**
configuring AutoCAD for external databases, 791–793
determining database structure, 791
displaying External Database toolbar, 793
finding records in databases, 795–797
opening databases from AutoCAD, 794–795
setting up AutoCAD SQL Extension (ASE) to locate database files, 790, *790*
External Database toolbar
Administration button, 794, 799
defined, **25**
Export Links button, 806
Rows button, 795, 800, 802
Select Objects button, 805
SQL Editor tool, 807
linking objects to databases, **798–808**
adding labels with links, 800–802
creating links, 799–800, *800*
deleting links, 804–805
filtering selections and exporting links, 805–807
finding and selecting objects through databases, 803–804

updating rows and labels, 802–803
using SQL statements, 807–808
setting up AutoCAD for Open Database Connectivity (ODBC) drivers, **1027–1030**
creating reference database files, 1027–1029
telling AutoCAD the location of reference files, 1030
telling Windows the location of reference files, 1029
ddattext command, 785, 787
ddedit command, 276, **278–279**
ddim command, 1076
ddrename command, 286
Decimal unit style, 91, **94**
dedicated file servers, 987–988
defaults
defined, **56**
setting default text style, 283
Define Attributes command in Draw menu, 763–764
defining. *See also* creating
blocks, 147–149, *149*
blocks containing attributes, 779–780
User Coordinate System
in the current view plane, 491–492, *491*
with three points, 484–486, *485*
UCS planes with objects, 486–487, *486*
definition points, 331–333, *331*, *333*
deleting
external reference files (Xrefs), 240
named User Coordinate Systems, 492–493
object links to databases, 804–805
styles, **239–241**
deleting all unused styles, 240–241
selectively deleting unused styles, 239–240
the volume of one solid from another, 471–472, *472*
Web page links, 896
DElta option of Lengthen command, 231
designing 3D solid models, 578
desktop publishing, 833–839
exporting raster files, 651, 833–836
exporting vector files, **836–839**
.EPS (encapsulated PostScript) files, 837
overview of, 836
PostScript font substitution, 837–839
Destination option in Render dialog box, 720, 722
Detach option in External Reference dialog box, 376
Device and Default Selection dialog box, 648–649, *648*
Device Driver File Search Path option in Preferences dialog box, 1010

dialog boxes
copying information to dialog boxes from Clipboard, 828
defined, **19**
Show the Start Up Dialog Box option in Preferences dialog box, 1015
diameters, dimensioning, 347–348, *348*
dictionaries for checking spelling, 301–302
Dictionary object in ActiveX Automation, 923
digitizing tablets
calibrating, 732–734, *734*
configuring, 1022
defined, **995–996**
menu templates, 995, 1022–1024, *1023*
reconfiguring for tracing, 731, *731*
dimalt command, 1076
dimaltd command, 1076
dimaltf command, 1076
dimapost command, 1076
Dimaso system variable, 341
dimensions, 25, 311–359, 1069–1086
adding notes with arrows, **350–352**
overview of, 350–351, *350*, *351*
using multiline text with leaders, 351–352
adding tolerance notation, 356–358
applying ordinate dimensions, 354–355, *354*, *355*
associative dimensioning feature
controlling, 1074
defined, **311**
definition points and, 331, *331*
and editing dimensions with other objects, 338–339, *338*, *339*
turning off, 341
creating blocks for dimension arrows and tick marks, **1084–1086**, *1085*
creating dimension styles, **312–319**, 349
Annotation dialog box and, 314, 315, 1081–1084, *1082*
changing style settings for groups of dimensions, 337
changing style settings of individual dimensions, 333–337
dimension styles versus ANSI-style dimensions, 312, *312*
Dimension Styles dialog box and, **1076–1084**
Format dialog box and, 316, 1080–1081
Geometry dialog box and, 317–318, 1077–1080, *1078*, *1079*
importing dimension styles from other drawings, **1084**
overview of, **312–313**, *312*
parent and child styles in dimension families, 349
Primary Units dialog box and, 314–315
restoring dimension styles, 1075
saving dimension styles, 1075
selecting arrow style and setting dimension scale, 317–319
setting dimension unit style, 313–315
setting height for dimension text, 315
setting location of dimension text, 316
using dimension families to fine-tune dimension styles, 349
ddim command, 1076
defined, **311**
dimalt command, 1076
dimaltd command, 1076
dimaltf command, 1076
dimapost command, 1076
Dimension menu
Aligned command, 342–343, *343*
Angular command, 345–346, *346*
Diameter command, 347, *348*
Dimedit command, 330, 336
Dimlin command, 320–321, *321*
Leader command, 350–351, *350*, *351*
Oblique command, 353, *353*
Override command, 337
Tolerance command, 356–358
Dimension Style Export dialog box, 874–875
Dimension Style Import dialog box, 875
Dimension Styles dialog box, **1076–1084**
Annotation subdialog box, 314, 315, 1081–1084, *1082*
Format subdialog box, 316, 1080–1081
Geometry subdialog box, 317–318, 1077–1080, *1078*, *1079*
overview of, 1076–1077
dimension text
appending data to, 328–329
moving fixed dimension text, 334–335, *335*
rotating dimension text, 336–337
setting height, 315
setting location of, 316
Dimension toolbar
Aligned Dimension tool, 342–343, *343*
Angular Dimension tool, 345–346, *346*
Baseline Dimension tool, 326–327, *327*
Continue Dimension tool, 321–327
defined, **25**
Diameter Dimension tool, 347, *348*
Dimension Edit tool, 330, 336, 353, *353*
Dimension Style tool, 1076
Dimension Update tool, 335
Leader tool, 350–351, *350*, *351*
Linear Dimension tool, 320–321, *321*
opening, 319
Ordinate Dimension tool, 354–355, *354*, 356

Radius Dimension tool, 348
Tolerance tool, 356–358
dimension variables, **341**, **348**, **1069–1074**
　alternate dimension options variables, 1073–1074
　color variables, 1074
　Dimaso, 341, 1070
　dimension arrow and text control variables, 1072–1073
　dimension and extension line control variables, 1073
　Dimtix and Dimtofl, 348
　general dimension control variables, 1070
　offset variables, 1070
　rounding variables, 1071
　scale variables, 1070
　tolerance variables, 1071
dimensioning nonorthogonal objects, **342–349**
　dimensioning angles, 345–346, *346*
　dimensioning nonorthogonal linear distances, 342–344, *343*, *344*, *345*
　dimensioning radii, diameters, and arcs, 347–348, *348*
dimlfac command, 1076
dimstyle command, 1069, 1074
Dimstyle Export and Dimstyle Import commands in Bonus menu Tools submenu, 874–875
DimStyle object in ActiveX Automation, **924**
Dimtix and Dimtofl system variables, 348
editing, **327–341**
　appending data to dimension text, 328–329
　changing style settings for groups of dimensions, 337
　changing style settings of individual dimensions, 333–337
　with Dimension Edit tool, 330, 336
　dimensions with other objects, 338–341, *338*, *339*
　with grips, 332–333, *333*, 337
　locating definition points, 331–333, *331*, *333*
　Modify Dimension dialog box, 328–329
　moving fixed dimension text, 334–335, *335*
　multiple dimensions, 330
　rotating dimension text, 336–337
　with Stretch tool, 338–341, *338*, *339*
leaders, **350–352**
　attributes and, 762
　overview of, 350–351, *350*, *351*
　Quick Leader tool, **870**
　using multiline text with, 351–352
linear dimensions, **319–327**
　continuing dimensions, 321–327
　continuing dimensions from older dimensions, 327
　continuing dimensions from previous dimensions, 325

　defined, **319**
　Dimlin command and, 320–321, *321*
　drawing dimensions from common base extension lines, 325–327, *326*, *327*
　placing horizontal ad vertical dimensions, 320–321, *321*
　using Osnaps while dimensioning, 322–325, *324*
metric dimensioning, 1076
in Paper Space, **404**
Setup View tools and adding dimensions, 572–573, *573*
skewing dimension lines, 330, **352–353**, *353*
in traced drawings, 738
UCS orientation and, 487
dimmed text in custom menus, 962
Direct Distance method, **53**
directions, copying objects in a clockwise direction, 203
directories
　AutoCAD directories, 1008
　Figures directory on book's CD-ROM, 1032, 1033, 1034–1035
disk drives
　AutoCAD and, 1000
　lost clusters on, 1000
　space requirements for 3D rendering, 684
Display Objects Being Dragged option in Preferences dialog box, 1012
Display option in Print/Plot Configuration dialog box, 658–659, *659*, 664
Display Order submenu in Tools menu, 371, 744–745
Display tab in Preferences dialog box, 8, 212, **1018–1019**
　AutoCAD Window Format options, 1019
　Colors option, 1019
　Display AutoCAD Screen Menu in Drawing Window option, 8, 1018
　Display Scroll Bars in Drawing Window option, 212, 1018
　Fonts option, 1019
　Maximize the AutoCAD Window upon Startup option, 1019
　Number of Lines of Text in Text Window to Keep in Memory option, 1019
　Number of Lines of Text to Show in Docked Command Line Window option, 1019
　overview of, 1018, *1018*
　Text Window Parameters options, 1019
displaying. *See also* accessing; opening; viewing
　AutoCAD screen menu in Drawing window, 8, 1018
　data in AutoCAD Text window, 84–85, *85*
　drawings in Paper Space viewports, 392–393
　External Database toolbar, 793

scroll bars in drawing window, 212, 1018
Surfaces toolbar, 595
Distance tool, 205–206
distances. *See also* coordinates
    adding feet and inch distances, 411–412
    dimensioning nonorthogonal linear distances, 342–344, *343, 344, 345*
    Dist (Distance) command, 94
    entering, **49–53**
        from the keyboard with Direct Distance method, 53
        measurement system and, 95–96
        with relative Cartesian coordinates, 51–53, *51, 52, 53*
        with relative polar coordinates, 49–50, *50, 51*
    finding along arcs, **228–230**
        finding an exact distance along an arc, 229–230, *231*
        finding a point a particular distance from another point, 228–229, *229*
    measuring distances as you draw lines, 105
    measuring system and measuring distances between points, 94
Distant Light option in Lights dialog box, 687, *687*
Divide command in Draw menu Point submenu, 230, 450–452, *452,* 453
division lines in custom menus, 962
docked toolbars, 9, 21–24
Document object in ActiveX Automation, 919–920, 933
Donut command, 456, *457*
double-hyphen symbol (—) in custom menus, 962
downloading DWG files from the Web, 896–899
drag and drop feature, inserting symbols with, 187–188
Draw menu
    Block submenu Define Attributes command, 763–764
    Boundary command, 815–817, *816*
    Donut command, 456, *457*
    Hatch command, 455–456
    Multiline command, 232
    Point submenu
        Divide command, 230, 450–452, *452,* 453
        Measure command, 230, *231,* 452–453
    Polygon command, 470
    Ray command, 226–228, *227,* 228
    Region command, 555–558, *557, 558,* 820
    Solids submenu
        Box command, 477–478, 552–553
        Cone command, 550, *550*
        Cylinder command, 469, 475–477, *476,* 544
        Extrude command, 470–471, 516, *516*
        Interference command, 559–561, *560*
        Revolve command, 536–538, *539*
        Section command, 553–554, *555*
        Setup submenu Drawing command (soldraw), 568–569
        Setup submenu Profile command (solprof), 569–571, *571*
        Setup submenu View command (solview), 565–568, *566, 568*
        Slice command, 551–553, *552, 553*
        Sphere command, 542–545, *545*
        Torus command, 539–542, *539, 541, 542*
        Wedge command, 549, *549*
    Surfaces submenu
        3D Face command, 596–600, *596, 597, 599, 600, 601*
        Revolved Surface command, 617–619, *618*
        Ruled Surface command, 612–614, *613, 614,* 616
        Tabulated Surface command, 615–616, *615, 616*
    Trace command, 455
Draw toolbar, 25, 33, 40–44. *See also* toolbars
    Arc tool, 120–122, *121*
    closing, 33
    defined, **25**
    Ellipse tool, 106–107, *107*
    Make Block tool, 134–136
    Multiline Text tool, 266–270, *268*
    overview of, **40–44**, *41*
    Polygon tool, 470
    Polyline tool, 418–421, *418, 419*
    Rectangle tool, 105
    Region tool, 555–558, *557, 558,* 820
    Spline tool, 441–442, *441, 443*
drawing aids, 100–105
    coordinate readout, 104–105, *105*
    Drawing Aids dialog box, 101, *101,* 103, 458
    Grid mode
        defined, **100**
        gridunit command and system variable, 101
        setting grid spacing, 100–102
        toggling on or off, 100
        using Snap mode with, 103–104
    Ortho mode, **104,** 247–248
    Qtext, **305,** *305*
    Snap mode
        accessing, 102–103
        coordinate readout and, 105
        defined, **102**
        snapunit command and system variable, 03
        toggling on or off, 102–103
        using Grid mode with, 103–104
drawing area. *See also* measurement system
    clearing screen of blips, 54
    defined, **9, 11–12**

selecting coordinate points in, 12
setting size of, 96–98
drawing files. *See* files
drawing information, 182–184, 814–828
   capturing and saving text data from AutoCAD Text window, 827–828
   copying information to dialog boxes from Clipboard, 828
   determining drawing status, 820–823, *821*
   finding, **814–820**
      absolute coordinates in drawings, 814
      areas of complex shapes, 817–820, *818, 819*
      areas or location of objects, 814–817, *815, 816*
   logging AutoCAD session activity, 182–184, 824–826
   recalling last area calculation, 817
   recovering corrupted files, 828, 1001–1002
   saving area data as attributes in drawing files, 820
   in system variables, 824
   tracking time spent on drawings, 823, *823*
drawing labels. *See also* Batch Plot utility
   adding with database links, 800–802
   updating, 802–803
Drawing Limits command in Format menu, 97–98, *99*
Drawing Session Safety Precaution options in Preferences dialog box, 1016–1017
drawing template files, 196–199
   creating, **196–198**
   defined, **196**
   Drawing Template File option in Save Drawing As dialog box, 197
   naming, 197
   Template Drawing File Location option in Preferences dialog box, 1011
   using, 198–199
Drawing Units option in Print/Plot Configuration dialog box, 664–665
Drawing Utilities, 828
Drawing window, displaying AutoCAD screen menu in, 8, 1018
drawings. *See also* files; objects; tracing
   adding linetypes to, **176–177**
   adding text objects to, **291–293**, *292*
   adding text to, **266–269**, *268*
   attaching as external references, **372–375**, *374*
   clearing screen of blips, 54
   copying
      from existing drawings, 118–123, *119, 121, 122, 123*
      with Offset tool, **124–125**, *124, 125,* 127–128, *128,* 421–422, *422*
   Create New Drawing dialog box, 198–199, 215
   creating
      with ActiveX Automation, 932–933

exercises, 85–86, *86,* 129, *129*
from existing drawings, 118–123, *119, 121, 122, 123*
overview of, **44–48**, *47*
creating symbols from, 140–147, *140*
displaying in Paper Space viewports, 392–393
editing, **33–34**, *34*
editing with grips, **75–81**
   defined, **75**, 81
   moving and rotating drawings, 78–80, *80*
   stretching lines, 75–78, *77*
Figures directory on book's CD-ROM, **1032**, **1033**, **1034–1035**
importing drawing settings, **213–217**
managing, **988–989**
measurement system and accuracy of, 92
moving, 78–80, *80*
naming, 46
opening, *27–28, 28,* 35–36
orienting, 1004
planning and laying out, **124–128**
   finishing touches, 127–128, *128*
   setting up layouts, 124–125, *124, 125*
   using layouts, 126–127, *126, 127*
rotating, 78–80, *80*
scaling, **730**, **738–739**
scanning, **730**, **739**
tracking time spent on, 823, *823*
undoing parts of with Undo command options, 200–201
draworder command, 371
drives
   AutoCAD and, 1000
   lost clusters on, 1000
   space requirements for 3D rendering, 684
dtext command, 291–293, *292,* **295–296,** *296*
Dview command. *See* perspective
DWF files
   adding to Web pages, **889–891**
      overview of, 889
      specifying filenames and opening views, 890–891
      what your Internet service provider needs to know, 891
   creating, 888–889
   DWF Export Options dialog box, 889, *889*
DWG files
   defined, **46**
   downloading from the Web, **896–899**
DXB files
   Audit after Each DXFIN or DXBIN option in Preferences dialog box, 1017
   plotting, 651
DXF files (data exchange format)
   Audit after Each DXFIN or DXBIN option in Preferences dialog box, 1017

defined, **829**
DXF translators, **576**
dxfin command, 831–832
dxfout command, 831
exporting, 829–831, 836
exporting attribute information to, 788
importing, 831–832, 1002–1003
.dxx files, 788
DYnamic option of Lengthen command, 232

E

Edge option for Trim tool, 115, *116*
Edge Surface tool, 601, 608–609
Edit menu
    Copy Link command, 846
    Paste Special command, 840, 845–846
Edit Polyline tool, 622
Edit Row dialog box, 797–798, *797*
editing. *See also* adjusting; changing
    3D models from previous versions of AutoCAD, **468**
    3D solid models
        with Chamfer tool, 474–475, *475*, 531–532, *532*
        with Fillet tool, 502–505, *503*, *505*, 528–529, *529*
        with Slice tool, 551–553, *552*, *553*
    3D surfaces, **602**
    attributes, **772–780**
        editing individual attributes, 773
        editing several attributes in succession, 773–774, *774*
        global editing, 776–778
        making invisible attributes visible, 778–779, *779*
        making minor changes to attributes' appearance, 774–776, *775*
        redefining blocks containing attributes, 779–780
    with Backspace key, 48
    dimensions, **327–341**
        appending data to dimension text, 328–329
        changing style settings for groups of dimensions, 337
        changing style settings of individual dimensions, 333–337
        with Dimension Edit tool, 330, 336
        dimensions with other objects, 338–341, *338*, *339*
        with grips, 332–333, *333*, 337
        locating definition points, 331–333, *331*, *333*
        Modify Dimension dialog box, 328–329
        moving fixed dimension text, 334–335, *335*
        multiple dimensions, 330
        rotating dimension text, 336–337
        with Stretch tool, 338–341, *338*, *339*
    drawings, **33–34**, *34*
    drawings with grips, **75–81**
        grips defined, **75**, **81**
        moving and rotating drawings, 78–80, *80*
        stretching lines, 75–78, *77*
    with Escape key, 48
    group members, 154–156, *154*, *155*, *156*
    meshes, 619–622, *620*
    multilines, 237–239, *238*
    objects, 33–34, *34*, 109, *110*
    polyline vertices, **428–438**
        accessing, 428
        Break option, 429–430, *430*
        Insert option, 430–431, *431*, *432*
        Move option, 433, *434*
        Next option, 429
        Previous option, 429
        Straighten option, 433, *435*
        Tangent option, 435, *436*
        Undo option, 438
        Width option, 437, *437*
    polylines, **421–428**
        changing polyline width, 424, *424*
        joining polylines, 423, *423*
        options, 425
        overview of, **421–424**, *422*, *423*, *424*
        smoothing polylines, 425–427, *426*
    text, 278–279
    text objects, 293–294
    toolbar buttons, 955
Effects options in Text Style dialog box, 284–285, *284*, *285*
Element Properties option in Multiline Styles dialog box, 235–236, *236*
elements. *See* blocks; layers; linetypes; shapes; styles
elevation and creating 3D surface models, **584–589**, *585*, *588*, *589*
Ellipse tool, 106–107, *107*
ellipsis (…) in menus, 15
embedding data, 844–845
encapsulated PostScript (.EPS) files, 837
End option of Undo command, 200
End OR and End AND operators in filtering selections, 408–409
Engineering unit style, 91, **94**, 95
Enter Attributes dialog box, 772
Enter option for Trim tool, 116, *116*
entering. *See also* selecting
    angles
        measurement system and, 95–96
        overview of, 50, *51*

commands
    with accelerator keys, 26
    from the keyboard, 8, 15, 26
    from pull-down menus and toolbars, 18–19, 26
distances, **49–53**
    from the keyboard with Direct Distance method, 53
    measurement system and, 95–96
    with relative Cartesian coordinates, 51–53, *51, 52, 53*
    with relative polar coordinates, 49–50, *50, 51*
    Model Space from Paper Space, 395
    Paper Space, 391
entity properties in ActiveX Automation, 922–923
Environment dialog box, 792–793, *792*
.EPS (encapsulated PostScript) files, 837
equal to (=)
    in filtering selections, 407
    and saving blocks, 150
Erase option of Sketch command, 455
Erase tool in Modify toolbar, 34, *34*
Escape key, 48, 367
exchanging data. *See* exporting; importing; Object Linking and Embedding
Exit option of Sketch command, 455
exiting. *See also* closing
    AutoCAD, **35**
Explode Attributes to Text tool, 858
Explode Text tool, 858
exploding
    blocks, 147–149, *149*
    polylines, 421
Export Data dialog box, 150–151, 788, 829–830
Export Links dialog box, 806–807, *806*
Export Options dialog box, 830–831, *830*
exporting. *See also* importing; Object Linking and Embedding
    attribute information
        to comma-delimited format (CDF), 785–787
        to data exchange format (DXF), 788
        to Microsoft Access, 789
        to Microsoft Excel, 788
        to space-delimited format (SDF), 787–788
    AutoCAD files to previous Release formats, 832
    Dimension Style Export dialog box, 874–875
    drawings using the Clipboard, 845–846
    DWF Export Options dialog box, 889, *889*
    DXF files, 829–831, 836
    information about links between objects and databases, 805–807
    PostScript files, 752–754, *753*, 837, 997–999
    raster files, 651, 833–836
    vector files, **836–839**

DXF files, 829–831, 836
.EPS (encapsulated PostScript) files, 837
HPGL plot file format, 839
PostScript font substitution, 837–839
Extend to Block Entities tool, 869
Extended Change Properties tool, 865–866
Extended Clip tool, 868, *869*
Extended Trim tool, 867–868
Extents option in Print/Plot Configuration dialog box, 660, *660*, 664
External Database Configuration dialog box, 791–793, *791*
External Database toolbar. *See also* databases
    Administration button, 794, 799
    defined, **25**
    Export Links button, 806
    Rows button, 795, 800, 802
    Select Objects button, 805
    SQL Editor tool, 807
external reference files (Xrefs), 371–385
    Attach Xref dialog box, 377–378
        Overlay option, 377, 385
    attaching drawings as external references, **372–375,** *374*
    versus blocks, **372–376**
    clipping Xref views with Xclip command, **378–380,** *379*
    controlling Xref memory use, **380–381,** *381*
    converting to blocks, 376
    creating symbols from, 152
    defined, **371–372**
    deleting, 240
    External Reference dialog box, 376–377
    External Reference File Demand Load option in Preferences dialog box, 380–381, *381*, 1013
    importing named elements from, **374, 382–384**
    importing settings from, 216–217
    Insert toolbar and, 375
    List Xref/Block Entities tool, **870**
    moving, 381
    nesting, **384–385**
    Reference toolbar, External Reference Bind tool, 374, 383
    saving layer settings of, 382
    Temporary External Reference File Location option in Preferences dialog box, 1011
    Xbind dialog box, 383–384
    xref command, **240**
extracting and exporting attribute information, 781–789. *See also* attributes
    Attribute Extraction dialog box, 785–788, *785*
    attribute template files and, 781, 784
    to comma-delimited format (CDF), 785–787
    to data exchange format (DXF), 788

determining what to extract, 781–783
with Export Data dialog box, 788
exporting to Microsoft Access, 789
exporting to Microsoft Excel, 788
extracting block information using attributes, 783–784
performing the extraction, 785–788
to space-delimited format (SDF), 787–788
text editor line endings, 783
Extrude tool, 470–471, 516, *516*
extruding
circular surfaces, 617–619, *618*
objects along a straight line, 615–616, *615*, *616*
Eye2eye add-on, 877, 1032, 1035–1040. *See also* CD-ROM (book's); perspective views
creating 3D models, 1039
defined, **877, 1032, 1035**
installing, 1035–1036
moving and controlling camera and target objects, 1040
overview of, 1036–1039, *1036*
setting up viewports, 1037–1038, *1038*

F

F2 key, drawing editor/AutoCAD Text window toggle, 85, 827
F7 key, Grid mode toggle, 100
F9 key, Snap mode toggle, 102–103
fabrication shops and 3D solid models, 577
families of dimensions, 349
feature control symbols, 356–358
fence object selection command, 69
Fence option for Trim tool, 365, *366*
Figures directory on book's CD-ROM, 1032, 1033, 1034–1035
file extensions. *See also* DWF files; DWG files; DXF files
.ac$, 1000
.bak, 985
.dwg, 46
.dxb, 651, 1017
.dxx, 788
File Extension for Temporary Files option in Preferences dialog box, 1017
.lin, 966
.sld, 546
File menu
Drawing Utilities command, 828
Export command, 150–151, 788
New command, 90, 91
Open command, 27–28, *28*
Print Preview command, 646

Save As command, 33, 380
Save command, 32–33
File Output Configuration dialog box, 721–722, *721*
File tab in Batch Plot utility, 675–676, *675*
files, 985–987. *See also* drawings; external reference files
3D rendering to files, 721–722
.ac$ files, 1000
acad.ini file, 1022
acad.lin file, 966, 970, 972
acad.log file, 184, 826
acad.lsp file, 879–880
acad.mnc file, 955
acad.mnr file, 955
acad.mns file, 945, 955
acad.mnu file, 955
attribute template files, 781, 784
Automatic save feature, **32–33, 128, 1016**
backing up, 985–987
Create Backup Copy with Each Save option in Preferences dialog box, 1016
BMP files, 651
drawing.dwg files, 199
drawing template files, **196–199**
creating, **196–198**
defined, **196**
Drawing Template File option in Save Drawing As dialog box, 197
naming, 197
Template Drawing File Location option in Preferences dialog box, 1011
using, 198–199
DXB files, 651, 1017
DXF translators, **576**
file formats available for drag and drop, 187
finding, **184–187**
graphics file formats, 651
IGES (International Graphics Exchange Standard), 576, 829
importing drawing settings, **213–217**
labeling drawings with filenames, **987**
maintaining, **985**
menu files in Windows, **955**
naming, 46
New File Setup Wizard, **214–216**
Nozzle3d file, 27–28, *28*
opening, 27–28, *28*, 35–36
in a particular view, 260
as read-only, 240
troubleshooting, 1003
Pack 'n Go tool, **381, 863–865**
PC2 and PCP files, 652, 673–674
PCX files, 651

Plot to File option, 657–658
PostScript files
    exporting, 752–754, *753*, 837, 997–999
    importing, 751–752
    plotting, 651
Raster File Export option in Add a Printer dialog box, 651
recovering corrupted files, 828, 1001–1002
replacing existing files with blocks, 151
SAT file translation, **576**
saving
    AutoLISP macros as files, 883–884, *883*
    automatically, 32–33, 128, 1016
    blocks as drawing files, 150
    Create Backup Copy with Each Save option in Preferences dialog box, 1016
    Incremental Save option in Preferences dialog box, 1013
    with preview images, 150
    Release 14 files to previous Release formats, 832
    Save Thumbnail Preview Image option in Preferences dialog box, 1017
    troubleshooting, 1003
    to Web or FTP sites, 899–900
stereolithography (STL) translation, **577**
Targa files, 651
temporary files
    .ac$ files, 1000
    File Extension for Temporary Files option in Preferences dialog box, 1017
    Temporary Drawing File Location option in Preferences dialog box, 1011
    Temporary External Reference File Location option in Preferences dialog box, 1011
TIFF files, 651
translation tools, 575–576
unnamed.dwg files, 199
vector file exporting, **836–839**
    DXF files, 836
    .EPS (encapsulated PostScript) files, 837
    HPGL plot file format, 839
    PostScript font substitution, 837–839
Files tab in Preferences dialog box, 1009–1011
    Device Driver File Search Path option, 1010
    Menu, Help, Log, and Miscellaneous File Names option, 1010
    overview of, 1009–1010, *1009*
    Print File, Spooler, and Prolog Section Names option, 1011
    Print Spooler File Location option, 1011
    Project Files Search Path option, 1010
    Support File Search Path option, 1010
    Template Drawing File Location option, 1011
    Temporary Drawing File Location option, 1011
    Temporary External Reference File Location option, 1011
    Text Editor, Dictionary, and Font File Names option, 1010
    Texture Maps Search Path option, 1011
Fillet tool
    editing 3D solid models, 502–505, *503*, *505*, 528–529, *529*
    overview of, **117–118**, *118*, 127–128, *128*, 222–224
    polylines and, 421
filling solid areas, 455–459. *See also* polylines
    creating, 455–456
    creating filled circles with Donut command, 456, *457*
    toggling solid fills on and off, 421, 458–459, *458*
filtering
    layers, 172–174, *173*
    point filters, **592**
    Point Filters command in Object Snap (Osnap) menu, 219–220
    selections, **405–409, 805–807**
        creating complex selection sets, 408–409
        filtering objects by location, 407
        Object Selection Filters dialog box, 406–407
        of objects linked to external databases, 805–807
        overview of, 405–407
        saving filter criteria, 407
Find and Replace Text tool, 857–858
finding
    distances along arcs, **228–230**
        finding an exact distance along an arc, 229–230, *231*
        finding a point a particular distance from another point, 228–229, *229*
    drawing information, **814–820**
        absolute coordinates in drawings, 814
        areas of complex shapes, 817–820, *818*, *819*
        areas or location of objects, 814–817, *815*, *816*
    layers, 171–174
    the midpoint between two points, 409
    objects through databases, 803–804
    a point relative to another point, 411
    records in databases, 795–797
Fit Data options of Splinedit command, 444–447
    adding control points, 444–445, *445*
    changing spline tolerance setting, 446, *446*
    controlling tangency at the beginning and end points, 444
    when to use, 447
Fit option of Dtext command, 297, *297*

Fit Tolerance option of Spline command, 442
Floating Model command in View menu, 393
floating toolbars, 9, 21–24
flyouts. *See also* toolbars
   converting to custom toolbars, 43
   creating, **953–954**
   defined, **20–21, 43–44**
   Flyout Properties dialog box, 954
   most recently selected option as default for toolbar tools, 44
   opening, 43
fonts. *See also* text
   changing, 269–272, *271*
   examples of, **287–289,** *287, 288*
   Font options in Text Style dialog box, 284
   font substitution, **302–304,** 837–839
   Fontmap system variable, **303–304**
   Fonts option in Preferences dialog box, 1019
   PostScript fonts
      compiling into AutoCAD font format, 272
      font substitution, 837–839
   Text Editor, Dictionary, and Font File Names option in Preferences dialog box, 1010
   Textfill system variable, **288**
   troubleshooting, 1002
   types of, 267
Format dialog box, 316, 1080–1081
Format menu
   Drawing Limits command, 97–98, 99
   Multiline Style command, 234
   Point Style command, 229–230, *229*
   Rename command, 286
   Text Style command, 281
   Thickness command, 588–589, *589*, 593
formatting text, 269–274
   adding color, stacked fractions, and symbols, 272–274
   changing text height and font, 269–272, *271*
   changing width of text boundary window, 274, *274*
Fractional unit style, 94, 95
fractions, stacked, **272–274**
frames
   Show Text Boundary Frame Only option in Preferences dialog box, 1013
   turning off in raster images, 750–751, *750*
freehand sketching, 454–455
Freeze Object Layer tool, 853
freezing layers, 171–172, 261
FTP sites, saving files to, 899–900
Full-time CRC Validation option in Preferences dialog box, 1017
functions in the Calculator, 413

G

general dimension control variables, 1070
General tab in Preferences dialog box, 182–184, 824–826, 1016–1018
   Audit after Each DXFIN or DXBIN option, 1017
   Automatic Save option, 32–33, 128, 1016
   Beep on Error in User Input option, 1017
   Create Backup Copy with Each Save option, 1016
   Drawing Session Safety Precaution options, 1016–1017
   File Extension for Temporary Files option, 1017
   Full-time CRC Validation option, 1017
   Maintain a Log File option, 182–184, 824–826, *825*, 1017
   Maximum Number Sorted Symbols option, 1017–1018
   Minutes between Saves option, 1016
   overview of, 1016, *1016*
   Save Thumbnail Preview Image option, 1017
Geographic Location dialog box, 688–689
Geometric Tolerance dialog box, 358
Geometry Calculator. *See* Calculator
Geometry dialog box, 317–318, 1077–1080, *1078, 1079*
Get Selection Set command in Bonus menu Tools submenu, 876
Global Attribute Edit tool, 778, 858
global attribute editing, 776–778
graphics
   graphics file formats, **651**
   linking to AutoCAD, 844
graphics display resolution and performance, 254–255, 994–995
greater than (>) in filtering selections, 407
Grid mode. *See also* drawing aids
   defined, **100**
   gridunit command, 101
   gridunit command and system variable, 101
   setting grid spacing, 100–102
   toggling on or off, 100
   using Snap mode with, 103–104
Grid option on status bar, 245–246
GridMaker application, 932–935, *934*
gridunit command, 101
grips, 75–81, 1025–1026
   defined, **75, 81**
   editing dimensions with, 332–333, *333*, 337
   Esc key and, **367**
   hot grip commands, 80
   hot grips, **76**
   moving and rotating drawings, 78–80, *80*
   stretching lines, 75–78, *77*
   troubleshooting, 1002
   turning on, 1026–1027

Group collection object in ActiveX Automation, 924
grouping objects, 152–157
    versus blocks, 152
    editing group members, 154–156, *154, 155, 156*
    Object Grouping dialog box, 156–157, 246–247, *247*
    overview of, **152–154**, *153*
grouping operators in Object Selection Filters dialog box, 408

H

hard clip limits of plotters and printers, 652–653, *652*
hard disk drives
    AutoCAD and, 1000
    lost clusters on, 1000
    space requirements for 3D rendering, 684
hatch patterns. *See also* texture maps
    creating, **974–979**
    Hatch command, **455–456**
    text in hatched areas, 269, 341
height of text, changing, 269–272, *271*
Help
    adding Help messages to custom menus, 963–964
    Help window, **81–84**
    Menu, Help, Log, and Miscellaneous File Names option in Preferences dialog box, 1010
    for World Wide Web features, 900
Hewlett-Packard
    Hpconfig plotter command, 669
    HPGL plot file format, 839
Hide Lines option in Print/Plot Configuration dialog box, 656
Hide tool, 472–474, *473*, 520, *521*, 591, *591*
hiding
    parts of perspective views of 3D surfaces, 636–638, *636, 637*
    unwanted 3D surface edges, 600, *601*
Highlight application, 935–936
highlighting objects, 61–62, *62*
Hor option of Xline command, 221
horizontal dimensions, 320–321, *321*
hot grip commands, 80
hot grips, 76
hotkeys. *See* accelerator keys
hpconfig command, 669
HPGL plot file format, 839
hyphen (-)
    copying objects with, 203
    in distance entries, 95
    double-hyphen symbol (—) in custom menus, 962

I

IGES (International Graphics Exchange Standard), 576, 829
Image Adjust dialog box, 748–749, *748, 749*
Image dialog box, 741, *741*, 742–743
importing. *See also* exporting; linking; Object Linking and Embedding; opening
    AutoCAD files from previous versions, 832
    Dimension Style Import dialog box, 875
    dimension styles from other drawings, 1084
    DXF files, 831–832, 1002–1003
    named elements from external reference files, **374, 382–384**
    PostScript files, 751–752
    raster images, 740–742, *741, 742*
    settings
        from external reference files (Xrefs), 216–217
        overview of, 213–216
    text files, 291
Include Path option in Attach Xref dialog box, 378
Incremental Save option in Preferences dialog box, 1013
information. *See* drawing information
initializing plotter ports, 673
ink-jet printers, 996
Inquiry submenu in Tools menu
    Area command, 814–815, *815*, 817–819, *818, 819*
    ID Point command, 814
    Set Variables command, 824
    Setup submenu Mass Properties command, 574
    Status command, 820–822, *821*
    Time command, 823, *823*
Inquiry toolbar, 25
Insert menu Raster Image command, 741, 742
Insert option of polyline Edit Vertex mode, 430–431, *431, 432*
Insert toolbar
    defined, **25**
    external reference files and, 375
    Insert Block tool, 137
inserting
    blocks containing attributes, 768–771, *770*
    symbols, **136–139**, *137, 139*, 273–274
    with drag and drop feature, 187–188
installing
    AutoCAD, **1006–1008**
        AutoCAD directory contents, 1008
        the installation process, 1006–1008
        preparing to install, 1006
    Bonus tools, 850
    Eye2eye add-on, 1035–1036
    Internet Utilities, 888

INSTALLING • LAYERS

software on book's CD-ROM, **1033–1035**
    add-ons, 1035
    opening the installation program, 1033–1034
    and using sample drawing files, 1034–1035
    SWLite software, 1041
interference checking for 3D solid models,
    559–561, *560*
International Graphics Exchange Standard (IGES),
    576, 829
Internet. *See* World Wide Web
Internet Utilities toolbar, 888
Intersect tool, 510–511, *511*
invisible attributes
    Invisible attribute definition mode, **768**
    making visible, 778–779, *779*
ISO Pen Widths option, 668
Isolate Object Layer tool, 853
Isometric view options, 496–498, *496*, *497*

J

joining
    3D solid models with Union command, 526–528,
    526, *527*
    multilines, 237–239, *238*
    polylines, 423, *423*
Justification option of Multiline command, 232
justifying text objects, 295–296, *296*

K

keyboard. *See also* command-line commands
    entering commands from, 8, 15, 26
    entering distances with Direct Distance method, 53
    keyboard equivalents for menu commands, 15
keyboard macros. *See* macros
keyboard shortcuts
    entering commands with, 26
    Priority for Accelerator Keys option in Preferences
    dialog box, 1015
    troubleshooting, 1004

L

labels. *See also* Batch Plot utility
    adding with database links, 800–802
    adding notes with arrows, **350–352**
        overview of, 350–351, *350*, *351*
        using multiline text with leaders, 351–352
    updating, 802–803
laser printers, 996
last object selection command, 69

last used tool feature on Standard toolbar, 209
Layer & Linetype Properties dialog box, 159–163,
    159, *160*, 177–180, 181
    Linetype tab, ISO Pen Widths option, 668
layer command, 162–163
Layer Manager
    saving layer settings of external reference files, 382
    saving and recalling layer settings, 851–852
layers, 9, **158–184**
    3D surface models and, 591, *591*
    assigning linetypes to, **174–181**
        adding linetypes to drawings, 176–177
        overview of, 174, *175*
        setting linetype scale, 177–181, *179*
    assigning to objects, 163–166
    blocks and, 166
    Bonus Layer Tools toolbar, **852–853**
    changing, **167–168**, *168*
    changing layer property of lines, 168–169
    changing layers of objects, 367, *368*
    colors, linetypes and, 166
    controlling with Layer command, 162–163
    controlling layer visibility, 169–170, *170*
    copying objects on any layer, **368–369**, *369*
    creating, **159–161**, *159*, *160*, 165–166
    creating company-wide layering conventions,
    984–985
    defined, **9**, **158**, *158*
    deleting, **239–241**
        deleting all unused layers, 240–241
        selectively deleting unused layers, 239–240
    filtering, 172–174, *173*
    finding, **171–174**
    Freeze Object Layer tool, **853**
    freezing and thawing, 171–172, 261
    importing from external reference files, 374,
    382–384
    Layer & Linetype Properties dialog box, 159–163,
    159, *160*, 177–180, 181, 668
    Layer Filter dialog box, 172–173
    layer name or number display in Object Properties
    toolbar, 9
    Layer object in ActiveX Automation, **924–926**
    locking and unlocking, 172
    Log File option, **182–184**
    managing lists of, 172–174
    Paper Space viewports and, 396
    Release 14 Layer & Linetype Properties dialog box,
    162
    renaming, 165
    saving layer settings of external reference files, 382
    Set Layer Filters dialog box, 173–174, *173*
    setting in individual Paper Space viewports,
    400–403, *402*

Setup View tools and, 571, *572*
turning off, 171
viewing, 181–182, 382
Layers tab in Batch Plot utility, 676–677, *677*
layouts. *See* planning and laying out drawings
leaders, 350–352. *See also* dimensions
    attributes and, 762
    overview of, 350–351, *350, 351*
    Quick Leader tool, **870**
    using multiline text with, 351–352
learning AutoCAD, 982
Lengthen tool, 231–232
less than symbol (<)
    in Block Definition dialog box, 149
    for entering polar coordinates, 49–50, 52
    in filtering selections, 407
lighting effects in 3D rendering, 704–711
    named scenes, 707–711, *709, 710, 711, 712*
    spotlights, 705–707, *706, 707*
Lights dialog box. *See also* 3D rendering in Numbers section of index
    Distant Light option, 687, *687*
    Modify option, 692
    North Location option, 690, *690*
    Spotlight option, 705
    sunlight angle settings and, 687, *687*
Lights tool, 686–687, 691–692
Limits option in Print/Plot Configuration dialog box, 660, *661*
.lin files, 966
Linear Dimension tool, 320–321, *321*
linear dimensions, 319–327. *See also* dimensions
    continuing dimensions, 321–327
    continuing dimensions from older dimensions, 327
    continuing dimensions from previous dimensions, 325
    defined, 319
    Dimlin command and, 320–321, *321*
    drawing dimensions from common base extension lines, 325–327, *326, 327*
    placing horizontal and vertical dimensions, 320–321, *321*
    using Osnaps while dimensioning, 322–325, *324*
lines, 217–228. *See also* arcs; multilines; polylines; spline curves
    aligning in traced drawings, 735–737, *736, 737*
    breaking, 224–226, *225*
    changing layer property of, 168–169
    cleaning up line work, **222–226,** *222*
    dimension and extension line control variables, 1073
    drawing with ActiveX Automation, 933–935, *934*

extruding objects along a straight line, 615–616, *615, 616*
measuring distances as you draw lines, 105
rays, **226–228,** *227, 228*
roughing in line work, 217–221, *217*
rubber-banding, **47,** *47*
Space property, 182
straightening in traced drawings, 735
stretching, 75–78, *77*
UCS orientation and, 487
using construction lines as tools, 226–228, *227, 228*
linetypes
    assigning to layers, **174–181**
        adding linetypes to drawings, 176–177
        overview of, 174, *175*
        setting linetype scale, 177–181, *179*
    Bylayer and Byblock properties, **180–181,** 367, 368
    Celtscale system variable, 180
    customizing, **966–974**
        acad.lin file and, 966, 970, 972
        creating complex linetypes, 971–974, *972, 973*
        creating linetypes, 969–970
        linetype codes and, 970–971, *970, 971*
        viewing available linetypes, 966–967, *967, 968*
    deleting, **239–241**
        deleting all unused linetypes, 240–241
        selectively deleting unused linetypes, 239–240
    importing from external reference files, **374,** **382–384**
    Layer & Linetype Properties dialog box, 159–163, *159, 160,* 177–180, 181, 668
    layers and, 166
    LineType object in ActiveX Automation, **926**
    Ltscale system variable, 179
    Ltype setting in Print/Plot Configuration dialog box, 669
    Release 14 Layer & Linetype Properties dialog box, 162
    setting hardware linetypes, 669
    setting linetype scales in Paper Space, 403
    setting for objects individually, **180–181**
Link Path Names dialog box, 799–800, *799, 800*
linking. *See also* Object Linking and Embedding
    attributes to parts, 762
    editing links, 843–844, *843*
    Excel spreadsheets to AutoCAD, 839–843, *841, 842*
    objects to databases, **798–808**
        adding labels with links, 800–802
        creating links, 799–800, *800*

deleting links, 804–805
filtering selections and exporting links, 805–807
finding and selecting objects through databases, 803–804
updating rows and labels, 802–803
using SQL statements, 807–808
sound, video, graphics, and animation, 844
Links dialog box, 801, *801*, 804, *804*
List button on Object Properties toolbar, 181–182
List Entity Xdata command in Bonus menu Tools submenu, 876
List tool Area option, 814–815, *815*
List View option in External Reference dialog box, 377
List Xref/Block Entities tool, 870
Load ARX Applications on Demand option in Preferences dialog box, 1014–1015
Load AutoLISP, ADS, and ARX Files dialog box, 878–879
loading. *See* displaying; opening; running; starting
locating
definition points, 331–333, *331*, *333*
objects in reference to other objects, 106–107, *107*
Lock Object Layer tool, 853
locking layers, 172
logging
AutoCAD session activity, 824–826
block and layer information, 182–184
logfileon and logfileoff commands, 825
Maintain a Log File option in Preferences dialog box, 182–184, 824–826, *825*, 1017
Menu, Help, Log, and Miscellaneous File Names option in Preferences dialog box, 1010
Logging tab in Batch Plot utility, 678, *678*
lost clusters, 1000
Ltscale system variable, 179
Ltype setting in Print/Plot Configuration dialog box, 669

---M---

macros, 881–884, 963. *See also* AutoLISP
creating, 881–883, *881*, *882*
loading into custom menus, 963
saving as files, 883–884, *883*
Maintain a Log File option in Preferences dialog box, 182–184, 824–826, *825*, 1017
maintaining files, 985
Make Block tool, 134–136
Make Displayable Attribute dialog box, 801–802, *801*

managing
AutoCAD projects, **983–984**
files, **985**
Mapping dialog box, 717–718
Mapping tool, 717
Mark option of Undo command, 200
marking
curves into specified lengths, **452–453**
intervals on curves with blocks, **453**
segments of equal length on curves, **450–452**, *452*
mass properties of 3D solid models, 573–574
Match Properties tool, 368
materials in 3D rendering, 696–701
adjusting appearance of, 699–700, *699*, *701*
assigning to objects, 698, *699*
Materials dialog box, 696–697, *696*, 699
Materials Library dialog box, 697, *697*
Modify Granite Material dialog box, 699–700, *699*, *701*
selecting, 696–697
Maxactvp system variable, 398
Maximize the AutoCAD Window upon Startup option in Preferences dialog box, 1019
Maximum Active Viewports option in Preferences dialog box, 1014
Maximum Number Sorted Symbols option in Preferences dialog box, 1017–1018
MC option of Dtext command, 296
Measure command in Draw menu Point submenu, 230, *231*, 452–453
measurement system, 90–100
Architectural unit style, **94**, 95
Decimal unit style, 91, **94**
defined, **90**, 94
drawing accuracy and, 92
Engineering unit style, **91**, **94**, 95
and entering distances and angles, 95–96
Fractional unit style, **94**, 95
and measuring distances between points, 94
metric system and scale factors, 99–100
Precision option, 94
scales, sheet sizes, and setting drawing area size limits, 96–98
setting angle display style, 94
setting direction of 0 base angle, 95
setting unit display in status bar, 94
setting unit style, **90–94**
text size and scale factors, **98–100**
Units Control dialog box, 92, *92*
Mechanical toolbar, 40
memory
AutoCAD performance and, 999–1000
controlling Xref memory use, **380–381**, *381*

Number of Lines of Text in Text Window to Keep in Memory option in Preferences dialog box, 1019
Out of RAM and Out of Page Space errors, **1000**, **1003**
video RAM, 994
menu bar, 9, *11*
menu files in Windows, 955
Menu, Help, Log, and Miscellaneous File Names option in Preferences dialog box, 1010
menus, 9, 13–19, **956–966**
  closing, 14
  customizing, **956–966**
    adding Help messages, 963–964
    calling commands, 960
    creating accelerator keys, 965–966
    creating cascading menus, 962
    creating division lines and dimmed text, 962
    creating pull-down menus, 956–957, 959
    double-hyphen symbol (—) in, 962
    loading AutoLISP macros in menus, 963
    loading custom menus, 957–959
    Menu Customization dialog box and, *957–959*
    pausing for user input with backslash (\\), 960–961, *961*
    tilde (~) in, 962
    underscore (_) in, 961
    using plus sign (+) for long lines, 961–962
  ellipsis (…) in, 15
  versus entering commands from the keyboard, 8, 26
  keyboard equivalents for menu commands, 15
  versus keyboard-entry of commands, 8, 26
  Menu, Help, Log, and Miscellaneous File Names option in Preferences dialog box, 1010
  Object Snap (Osnap) menu, 65–67, *66*, *67*
  pop-up menus
    Pan pop-up menu, 210–211
    Popup Menu command in Bonus menu Tools submenu, 870–871
  selecting options, 14–17, *17*, *18*, *23*
  tablet digitizer menu templates, 995, 1022–1024, *1023*
  Windows menu files and customizing toolbars, 955
meshes. *See also* 3D surfaces in Numbers section of index
  changing mesh control settings, 610, *611*
  creating 3D meshes by specifying coordinates, 611
  editing, 619–622, *620*
messages
  adding Help messages to custom menus, 963–964

Out of RAM and Out of Page Space messages, 1000, 1003
Outside limits, 99
metric system
  metric dimensioning, 1076
  scale factors and, 99–100
Microsoft Access. *See also* ActiveX Automation; databases
  exporting attribute information to, 789
  setting up AutoCAD for Open Database Connectivity (ODBC) drivers, **1027–1030**
    creating reference database files, 1027–1029
    telling AutoCAD the location of reference files, 1030
    telling Windows the location of reference files, 1029
Microsoft Excel. *See also* ActiveX Automation
  exporting attribute information to, 788
  linking spreadsheets to AutoCAD, 839–843, *841*, *842*
Microsoft Windows 95
  Character Map dialog box, 289–290, 298
  creating shortcut to acad.log file, 826
  emptying Recycle Bin, 1000
  menu files, 955
  telling Windows the location of ODBC reference files, 1029
Middle option of Dtext command, 295
minus sign (-), copying objects with, 203
Minutes between Saves option in Preferences dialog box, 1016
Mirror tool, 606, *606*
mirroring text, 299
ML option of Dtext command, 296
Model Space. *See also* Paper Space
  defined, **385**
  re-entering Model Space from Paper Space, 395
  Tiled Model Space, **385**, **388**
  tiled viewports in, **385–388**, *387*, *389*
  User Coordinate System (UCS) icon and, 13
  View menu Floating Model Space command, 391, 393, 395, 398
  zooming or panning in, 572
Modify Dimension dialog box, 328–329
Modify Distant Light dialog box, 692, *692*
Modify Granite Material dialog box, 699–700, *699*, *701*
Modify II toolbar
  defined, **25**
  Edit Polyline tool, 622
  Intersect tool, 510–511, *511*
  Subtract tool, 471–472, *472*, 516, *517*, 820
  Union tool, 476, 526–528, *526*, *527*

Modify menu
   3D Operations submenu
      Align command, 508, *509*, 622–623, *623*
      Rotate 3D command, 512–514, *513*, 519, *520*, 624
   Boolean submenu
      Intersect command, 510–511, *511*
      Subtract command, 471–472, *472*, 516, *517*, 820
      Union command, 476, 526–528, *526*, *527*
   Chamfer command, 222, 474–475, *475*, 531–532, *532*
   Copy command, 109, *110*, 212–213, 368–369, *369*, 602
   Fillet command
      editing 3D solid models, 502–505, *503*, *505*, 528–529, *529*
      overview of, 117–118, *118*, 127–128, *128*, 222–224
      polylines and, 421
   Move command, 68, 71–72, 602
   Object submenu
      Clip command, 378–380, *379*
      Polyline command, 621
      Spline command, 444–450
   Scale command, 534–535, *536*, 602
Modify MText dialog box, 275–276, *275*, *276*, 282–283
Modify option in Lights dialog box, 692
Modify Properties dialog box, 166, 358
Modify Text dialog box, 294, *294*
Modify toolbar. *See also* toolbars
   Array tool, 206–208, *207*, *208*, 598, *600*, 602
   Chamfer tool, 222, 474–475, *475*, 531–532, *532*
   closing, 33
   Copy tool, 109, *110*, 112–113, 368–369, *369*, 602
   defined, **25**
   Erase tool, 34, *34*
   Explode tool, 147, 421
   Fillet tool
      editing solid models, 502–505, *503*, *505*
      overview of, 117–118, *118*, 127–128, *128*, 222–224
      polylines and, 421
   Lengthen tool, 231–232
   Mirror tool, 606, *606*
   Move tool, 68, 71–72, 602
   Offset tool, 124–125, *124*, *125*, 127–128, *128*, 421–422, *422*
   overview of, **41–44**, *41*
   Scale tool, 534–535, *536*, 602
   Stretch tool, 338–341, *338*, *339*, 602
      Multiple Entity Stretch tool, **867**
   Trim tool, 111–116, *113*

monitor resolution and performance, 254–255, 994–995
mouse. *See also* digitizing tablets
   button functions, 63
   click and drag actions, **22**, **43–44**
   right-clicking, 65
mouse pointers. *See* cursors
Move command, 68, 71–72, 602
Move Copy Rotate tool, 870
move (m) command, 68, 71–72
Move option of polyline Edit Vertex mode, 433, *434*
moving
   camera and target objects in Eye2eye add-on, 1040
   drawings, 78–80, *80*
   external reference files, 381
   fixed dimension text, 334–335, *335*
   objects, 68, 71–72
      with base points, 64–68, *66*, *67*, *68*
   toolbars, 21–24
   viewports in Paper Space, 396, *397*
moving between viewports in Paper Space, 393, *394*
MR option of Dtext command, 296
mslide command, 547–548
Multiline Text tool, 266–270, *268*
multilines, 232–239. *See also* lines
   creating, **232–233**, *232*
   defined, **232**, **237**
   joining and editing, 237–239, *238*
   Multiline Styles dialog box, **234–237**
      Element Properties option, 235–236, *236*
      Multiline Properties option, 237
      overview of, 234–235
Multiple Entity Stretch tool, 867
multiple object selection command, 69
Multiple Pedit tool, 866–867

N

named elements. *See* blocks; layers; linetypes; shapes; styles
named scenes in 3D rendering, 707–711, *709*, *710*, *711*, *712*
named User Coordinate Systems
   restoring and deleting, 492–493
   saving, 492
naming
   drawing files, 46
   renaming layers, 165
   renaming styles, 285–286
   styles, 285–286
   templates, 197
nesting external reference files, 384–385

Netscape Whip2 Plug-in, 900–901, 1033, 1043–1044
networks, AutoCAD on, **987–988**
New Command Alias dialog box, 872, *872*
New command in File menu, 90, 91
New Distant Light dialog box, 687–688
New File Setup Wizard, 214–216
New Toolbar dialog box, 946
newsgroups, 886
Next option of polyline Edit Vertex mode, 429
nonorthogonal objects, 342–349
    dimensioning angles, 345–346, *346*
    dimensioning nonorthogonal linear distances, 342–344, *343*, *344*, *345*
    dimensioning radii, diameters, and arcs, 347–348, *348*
Non-Uniform Rational B-Splines. *See* NURBS
North Location dialog box, 690, *690*
not equal to (!=) in filtering selections, 407
notes with arrows, 350–352. *See also* labels
    overview of, 350–351, *350*, *351*
    using multiline text with leaders, 351–352
Noun/Verb selection method: selecting objects before commands, 70–72, 74–75, **1025**
Nozzle3d file, 27–28, *28*
Number of Lines of Text in Text Window to Keep in Memory option in Preferences dialog box, 1019
Number of Lines of Text to Show in Docked Command Line Window option in Preferences dialog box, 1019
NURBS (Non-Uniform Rational B-Splines) curves, defined, **107**

—————O—————

Object Linking and Embedding (OLE), 839–846. *See also* exporting; importing; linking
    cutting and pasting AutoCAD images into other applications, 845–846
    embedding data, 844–845
    linking
        editing links, 843–844, *843*
        Excel spreadsheets to AutoCAD, 839–843, *841*, *842*
        sound, video, graphics, and animation, 844
    overview of, 839
Object Properties toolbar. *See also* toolbars
    changing layer visibility, 170, *170*
    color and linetype property settings, 180–181
    defined, **25**
    layer name or number display, 9
    List button, 181–182
    moving, 22–23
    Properties tool, 166, 181, 275, **358–359**
    Sun tool, 261

Object Selection cursor, **54–55**, *55*, 61
Object Selection Filters dialog box, 406–407
Object Selection Settings dialog box, 369–371, 1025–1026
Object Snaps. *See* Osnaps
Object Sort Method dialog box, 370–371
Object submenu, Modify menu
    Clip command, 378–380, *379*
    Polyline command, 621
    Spline command, 444–450
Object UCS tool, **486–487**, *486*
ObjectARX programming environment
    defined, **877**
    Load ARX Applications on Demand option in Preferences dialog box, 1014–1015
    Load AutoLISP, ADS, and ARX Files dialog box, 878–879
objects, **61–75**, **106–118**. *See also* ActiveX Automation; drawings; text objects
  assigning layers to, 163–166
  Bylayer and Byblock properties, 180–181, 367, 368
  changing layers of objects, 367, 368
  changing length of, **231–232**
  controlling visibility and overlap with raster images, 742–743
  converting to polylines, **427–428**
  copying
    on any layer, 368–369, *369*
    with base points, 64–68, *66*, *67*, *68*
    with minus sign (-), 203
  copying multiple times, **199–213**
    creating polar arrays, 200–204, *202*, *204*
    creating random multiple copies, 212–213
    creating row and column copies, 204–213
  defining UCS planes with, **486–487**, *486*
  dividing into segments of equal length, **450–452**, *452*
  dividing into specified lengths, **452–453**
  editing, 33–34, *34*, 109, *110*
  filtering selections, **405–409**
    creating complex selection sets, 408–409
    filtering objects by location, 407
    Object Selection Filters dialog box, 406–407
    overview of, 405–407
    saving filter criteria, 407
  grouping, **152–157**
    versus blocks, 152
    editing group members, 154–156, *154*, *155*, *156*
    Object Grouping dialog box, 156–157, 246–247, *247*
    overview of, **152–154**, *153*
  highlighting, **61–62**, *62*
  linking to databases, **798–808**

adding labels with links, 800–802
creating links, 799–800, *800*
deleting links, 804–805
filtering selections and exporting links, 805–807
finding and selecting objects through databases, 803–804
updating rows and labels, 802–803
using SQL statements, 807–808
locating in reference to other objects, 106–107, *107*
moving, 68, 71–72
moving with base points, 64–68, *66, 67, 68*
selecting, **61–75, 111, 369–371, 935–936**
    with ActiveX Automation, **935–936**
    Autoselect feature, 72–74, *73, 74*
    for blocks, 135
    by windowing them, 63–64, *64*
    close or overlapping objects, 111
    with Ctrl key, 111, 369–370
    Noun/Verb method: selecting objects before commands, 70–72, 74–75, 1025
    overview of, **61–64**, *62, 64*
    selection commands listed, 69–70
    with selection cycling and Object Selection Settings dialog box, 369–371, 1025–1026
    through databases, 803–804
    troubleshooting, 1002
    undoing selections, 62–63
    using base points, 64–68, *66, 67, 68*
text objects, **291–298**
    adding to drawings, 291–293, *292*
    defined, **291**
    editing, 293–294
    justifying, 295–296, *296*
    using special characters with, 297–298
trimming, **110–118, 867–868**
    Cookie Cutter Trim command in Bonus menu, 867–868
    Edge, Project, Undo, and Enter options, 115–116, *116*
    Extended Trim tool, **867–868**
    Fence option, 365, *366*
    overview of, 110–115, *113, 114, 115*
    Trim to Block Entities tool, **869**
Wipeout objects, 854–855
zooming on, 107–108, *108*
Oblique command in Dimension menu, 353, *353*
offset dimension variables, 1070
Offset tool, 124–125, *124, 125,* 127–128, *128,* 421–422, *422*
online services, utilities on, 886
opaque, making horizontal surfaces opaque, 590, *590, 591*

Open Database Connectivity (ODBC) driver setup, 1027–1030
creating reference database files, 1027–1029
telling AutoCAD the location of reference files, 1030
telling Windows the location of reference files, 1029
Open Drawing dialog box, Select Initial View option, 260
opening. *See also* accessing; displaying; importing
databases from AutoCAD, 794–795
Dimension toolbar, 319
DXF files, 831–832
files, 27–28, *28,* 35–36
    in a particular view, 260
    as read-only, 240
    troubleshooting, 1003
flyouts, 43
Internet Utilities toolbar, 888
toolbars from the command-line, 947–948
operators
in the Calculator, 413
grouping operators in Object Selection Filters dialog box, 408
relational operators in filtering selections, 407
optical drives for backups, 987
Optimizing Pen Motion dialog box, 671–672, *671*
ordinate dimensions, 354–355, *354, 355*
Orientation options in Print/Plot Configuration dialog box, 654–655, *654*
orienting drawings, 1004
origin point in User Coordinate System, setting, 482–483, *484*
Ortho mode, 104, 247–248, 354–355, *356*
Orthogonal view options, 496–498, *496, 497*
Osnaps
Osnap menu
    overview of, **65–67,** *66, 67*
    Point Filters command, 219–220
Osnap modes in Calculator expressions, 409–411
Osnap Settings dialog box, 141–144, *141,* 322–323
Osnap toolbar, **25**
Running Osnaps, 322–323, 427, 452
and selecting points, **486**
using while dimensioning, 322–325, *324*
Out of RAM and Out of Page Space errors, 1000, 1003
output devices, 648–651. *See also* plotting; printers
adding, 649–651
selecting, 648–649, *648*
Outside limits message, 99
overlapping

controlling object overlap with raster images, 742–743
overlapping viewports in Paper Space, 400
Overlay option in Attach Xref dialog box, 377, 385
Override command in Dimension menu, 337

P

Pack 'n Go tool, 381, 863–865
Pan pop-up menu, 210–211
Pan Realtime tool
   defined, **208–212**, *210*, *211*
   drawing regeneration and, 253
   in Model Space, 572
   in Paper Space, 394, 572
Paper Size options in Print/Plot Configuration dialog box, 654–655, *654*
Paper Space. *See also* Model Space
   defined, **388–390**, *390*
   dimensioning in, 404
   entering, 391
   Maxactvp system variable, 398
   panning in, 394, 572
   and plotting multiple views in different scales, 388
   Psltscale system variable, 403
   re-entering Model Space, 395
   scaling views in, 398–400, *399*
   setting linetype scales, 403
   setting up, 391
   Setup View tools and, 566–568, *568*
   Tilemode system variable and, 391
   User Coordinate System (UCS) icon and, 13
   uses for, 405
   viewports
      changing shape of, 396, 399–400
      creating, 392, *392*
      displaying drawings in, 392–393
      floating Model Space viewports, 400
      layers and, 396
      moving, 396, *397*
      moving between, 393, *394*
      overlapping, 400
      saving viewport views, 400
      setting layers in individual viewports, 400–403, *402*
      setting maximum number of active viewports, 398
      troubleshooting, 1003
      zooming in, 394, *395*, 572
parallel lines. *See* multilines
parallel projection, 624
parent dimension styles, 349
Paste Special dialog box command, 840–841, *840*, 845–846

pasteclip command, 562–564, *564*
patterns. *See* hatch patterns; texture maps
pausing for user input in custom menus, 960–961, *961*
PC2 and PCP files, 652, 673–674
PCI bus, 994
PCX files, 651
Pdmode system variable, 451
peer-to-peer networks, 988
Pen Assignments options, 667–670. *See also*
   Print/Plot Configuration dialog box
      ISO Pen Widths option and, 668
      Ltype setting, 669
      overview of, 667–668, *667*
      Pen No. setting, 668–669
      Speed setting, 669
      Width option, 670, *670*
Percent option of Lengthen command, 232
Percent of Screen Size option in Preferences dialog box, 55
performance, 253–262
   Aerial view and, 255–257, *256*
   controlling display smoothness with Viewres command, 254–255
   memory and, 999–1000
   of monitors, 994–995
   Regen and Redraw commands and, 253–254, 288
   saving views, 257–260, *259*
Performance tab in Preferences dialog box, 1012–1014
   Arc and Circle Smoothness option, 255, 1013
   Display Objects Being Dragged option, 1012
   External Reference File Demand Load option, 380–381, *381*, 1013
   Incremental Save option, 1013
   Maximum Active Viewports option, 1014
   overview of, 1012, *1012*
   Rendered Object Smoothness option, 691
   Segments per Polyline Curve option, 1013
   Show Raster Image Content option, 1013
   Show Text Boundary Frame Only option, 1013
   Solid Model Object Display option, 1012
Period (.) option of Sketch command, 455
perspective views of 3D surfaces, 624–638. *See also*
   Eye2eye add-on
      adjusting camera and target positions, 629–631, *630*, *631*
      adjusting distances between camera and target, 627–629, *628*, *629*
      changing your point of view, 631–634, *632*, *633*, *634*
      hiding parts of the view with clip panes, 636–638, *636*, *637*
      overview of, 624–626, *625*

versus parallel projection, 624
setting up perspective views, 626–627, *627*
twisting the camera, 635, *635*
using Dview command Zoom option as a telephoto lens, 634–635
Photo Real Render Options dialog box, 694, *694, 695*
Pickstyle system variable, 156
placing horizontal and vertical dimensions, 320–321, *321*
Plan view, 49
planning and laying out drawings, 124–128
finishing touches, 127–128, *128*
setting up layouts, 124–125, *124, 125*
using layouts, 126–127, *126, 127*
Pline Converter command in Bonus menu Tools submenu, 876
plines. *See* polylines
Plot object in ActiveX Automation, 929–930
Plot Rotation and Origin dialog box, 666, *666*
Plot to File option in Print/Plot Configuration dialog box, 657–658
Plotted Inches option in Print/Plot Configuration dialog box, 664–665
plotters, 997
plotting. *See also* Print/Plot Configuration dialog box; printers
3D wire-frame models, 651
Add a Printer dialog box, 650, *650*, 651
Batch Plot utility, **675–678**. *See also* labels
File tab, 675–676, *675*
Layers tab, 676–677, *677*
Logging tab, 678, *678*
Plot Area tab, 677, *677*
Plot Stamping tab, 678, *678*
determining plotter's origins, **653–654**, *653*
graphics file formats, **651**
hard clip limits of plotters, 652–653, *652*
Hewlett-Packard plotter Hpconfig command, 669
multiple views in different scales in Paper Space, 388
output devices, **648–651**
adding, 649–651
selecting, 648–649, *648*
plotting service bureaus, **679**
previewing plots, **646–647, 672–673**
Reconfigure a Printer dialog box, 650–651, *650*
reinitializing plotter ports, 673
rotating plots, 666
saving plotter settings, 652, 673–674
troubleshooting, **664, 1004**
plus sign (+) in custom menus, 961–962
point filters, 592
Point Filters command in Object Snap (Osnap) menu, 219–220

Point Selection mode of the cursor, 48
Point Style dialog box, 229–230, *229*, 451, *451*
Point submenu, Draw menu
Divide command, 230, 450–452, *452*, 453
Measure command, 230, *231*, 452–453
Pointer tab in Preferences dialog box
overview of, **1019–1020**, *1020*
Percent of Screen Size option, 55
pointers. *See* cursors
polar arrays
creating, 200–204, *202, 204*
defined, **199**
Polar Array tool and 3D solid models, 533–534, *534*
polar coordinates, entering distances with, 49–50, *50, 51*
Polygon tool, 470
Polyline command in Modify menu Object submenu, 621
polylines, 105, **417–440, 450–460**. *See also* spline curves
converting
Pline Converter command in Bonus menu Tools submenu, 876
to solids, 470–471
to splines, 428
converting objects to polylines, 427–428
creating 3D surface models from 2D polylines, **586–587**, *588*
creating, **417–420**, *418, 419*
creating polyline spline curves, **438–440**
defined, **105, 417**
dividing into segments of equal length, **450–452**, *452*
dividing into specified lengths, **452–453**
Edit Polyline tool, **622**
Edit Vertex options, **428–438**
accessing, 428
Break option, 429–430, *430*
Insert option, 430–431, *431, 432*
Move option, 433, *434*
Next option, 429
Previous option, 429
Straighten option, 433, *435*
Tangent option, 435, *436*
Undo option, 438
Width option, 437, *437*
editing, **421–428**
changing polyline width, 424, *424*
joining polylines, 423, *423*
options, 425
overview of, **421–424**, *422, 423, 424*
smoothing polylines, 425–427, *426*
exercises, **459**, *460*
marking intervals on with blocks, **453**

Multiple Pedit tool, **866–867**
Pline Converter command in Bonus menu Tools submenu, 876
polyline options, 420–421, *420*
Polyline tool, 418–421, *418*, *419*
Segments per Polyline Curve option in Preferences dialog box, 1013
Sketch command and, **454–455**
solid filled areas, **455–459**
    creating, **455–456**
    creating filled circles with Donut command, 456, *457*
    toggling solid fills on and off, 421, 458–459, *458*
versus spline curves, **441**
UCS orientation and, 487
pop-up menus. *See also* menus
    Pan pop-up menu, 210–211
    Popup Menu command in Bonus menu Tools submenu, 870–871
ports, reinitializing plotter ports, 673
PostScript files
    exporting, 752–754, *753*, 837, 997–999
    importing, 751–752
    plotting, 651
PostScript fonts
    compiling into AutoCAD font format, 272
    font substitution, 837–839
PostScript printers, 997
Precision measurement system option, 94
pre-defined 3D surface shapes, 602
Preferences dialog box, 1009–1022
    Compatibility tab, **1014–1015**
        Load ARX Applications on Demand option, 1014–1015
        overview of, 1014, *1014*
        Priority for Accelerator Keys option, 1015
        Priority for Coordinate Data Entry option, 1015
        Proxy Images for Custom Objects option, 1015
        Reload AutoLISP between Drawings option, 1015
        Show the Start Up Dialog Box option, 1015
    Display tab, **8**, **212**, **1018–1019**
        AutoCAD Window Format options, 1019
        Colors option, 1019
        Display AutoCAD Screen Menu in Drawing Window option, 8, 1018
        Display Scroll Bars in Drawing Window option, 212, 1018
        Fonts option, 1019
        Maximize the AutoCAD Window upon Startup option, 1019
        Number of Lines of Text in Text Window to Keep in Memory option, 1019
        Number of Lines of Text to Show in Docked Command Line Window option, 1019
        overview of, 1018, *1018*
        Text Window Parameters options, 1019
    Files tab, **1009–1011**
        Device Driver File Search Path option, 1010
        Menu, Help, Log, and Miscellaneous File Names option, 1010
        overview of, 1009–1010, *1009*
        Print File, Spooler, and Prolog Section Names option, 1011
        Print Spooler File Location option, 1011
        Project Files Search Path option, 1010
        Support File Search Path option, 1010
        Template Drawing File Location option, 1011
        Temporary Drawing File Location option, 1011
        Temporary External Reference File Location option, 1011
        Text Editor, Dictionary, and Font File Names option, 1010
        Texture Maps Search Path option, 1011
    General tab, **182–184**, **824–826**, **1016–1018**
        Audit after Each DXFIN or DXBIN option, 1017
        Automatic Save option, 32–33, 128, 1016
        Beep on Error in User Input option, 1017
        Create Backup Copy with Each Save option, 1016
        Drawing Session Safety Precaution options, 1016–1017
        File Extension for Temporary Files option, 1017
        Full-time CRC Validation option, 1017
        Maintain a Log File option, 182–184, 824–826, *825*, 1017
        Maximum Number Sorted Symbols option, 1017–1018
        Minutes between Saves option, 1016
        overview of, 1016, *1016*
        Save Thumbnail Preview Image option, 1017
    Performance tab, **1012–1014**
        Arc and Circle Smoothness option, 255, 1013
        Display Objects Being Dragged option, 1012
        External Reference File Demand Load option, 380–381, *381*, 1013
        Incremental Save option, 1013
        Maximum Active Viewports option, 1014
        overview of, 1012, *1012*
        Rendered Object Smoothness option, 691
        Segments per Polyline Curve option, 1013
        Show Raster Image Content option, 1013
        Show Text Boundary Frame Only option, 1013
        Solid Model Object Display option, 1012

Pointer tab
    overview of, **1019–1020**, *1020*
    Percent of Screen Size option, 55
Printer tab
    adding output devices, 649–651, *649*
    overview of, **1020**, *1020*
    Profiles tab, **24**, **1021–1022**, *1021*
Preferences object in ActiveX Automation, 918–919
Preset attribute definition mode, 768
Preset UCS tool, 482, 493–495, *495*
previewing plots, 646–647, 672–673
Previous command in View menu Zoom submenu, 30
previous object selection command, 69
Previous option of polyline Edit Vertex mode, 429
Primary Units dialog box, 314–315
Print File, Spooler, and Prolog Section Names option in Preferences dialog box, 1011
Print Preview command in File menu, 646
Print Spooler File Location option in Preferences dialog box, 1011
Print/Plot Configuration dialog box, 648–649, 654–674. *See also* plotting
    Additional Parameters options, **655–664**
        Adjust Area Fill option, 656, *657*
        Autospool option, 658
        Display option, 658–659, *659*, 664
        Extents option, 660, *660*, 664
        Hide Lines option, 656
        Limits option, 660, *661*
        Plot to File option, 657–658
        Text Fill option, 655
        Text Resolution option, 655
        View option, 661–662, *662*
        Window option, 662–663, *663*
    Device and Default Selection dialog box, 648–649, *648*
    Optimization settings, **671–672**
    overview of, 647–648, *647*
    Paper Size and Orientation options, 654–655, *654*
    Pen Assignments options, **667–670**
        ISO Pen Widths option and, 668
        Ltype setting, 669
        overview of, 667–668, *667*
        Pen No. setting, 668–669
        Speed setting, 669
        Width option, 670, *670*
    Plot Preview option, 672–673
    Plot Rotation and Origin dialog box, 666, *666*
    saving settings, 652, 673–674
    Scale, Rotation, and Origin options, **664–666**
        overview of, 664
        Plotted Inches and Drawing Units options, 664–665

Rotation and Origin option, 665–666, *665*
Scaled to Fit option, 664
Printer tab in Preferences dialog box
    adding output devices, 649–651, *649*
    overview of, **1020**, *1020*
printers. *See also* plotting
    Add a Printer dialog box, 650, *650*, 651
    adding, 649–651
    hard clip limits of, 652–653, *652*
    Reconfigure a Printer dialog box, 650–651, *650*
    selecting, 648–649, *648*
    types of, **996–997**
Priority for Accelerator Keys option in Preferences dialog box, 1015
Priority for Coordinate Data Entry option in Preferences dialog box, 1015
Profiles tab in Preferences dialog box, 24, 1021–1022, *1021*
program window, 9–13. *See also* windows
    Command window, 9, *12*, 13, 18–19
    drawing area, 9, 11–12
    menu bar, 9, *11*
    overview of, 9–13, *10*
    selecting coordinate points in drawing area, 12
    status bar, 9, *12*, 93
    User Coordinate System (UCS) icon, 12–13
Project option for Trim tool, 115–116, *116*
projects
    managing, 983–984
    Project Files Search Path option in Preferences dialog box, 1010
    when to stop using AutoCAD, 990
prompts
    answering attribute prompts with Enter Attributes dialog box, 772
    command-line prompts, **55–60,** *59, 60*
properties. *See also* ActiveX Automation
    Button Properties dialog box, 949–950
    Bylayer and Byblock, 180–181, 367, *368*
    Change Properties dialog box, 166
    changing
        color properties, 180–181
        layer property of lines, 168–169
    Element Properties option in Multiline Styles dialog box, 235–236, *236*
    Extended Change Properties tool, **865–866**
    Flyout Properties dialog box, 954
    Layer & Linetype Properties dialog box, 159–163, *159, 160*, 177–180, 181
        Linetype tab, ISO Pen Widths option, 668
    mass properties of 3D solid models, **573–574**
    Match Properties tool, **368**
    Modify Properties dialog box, 166, 358
    in Multiline Styles dialog box

Element Properties option, 235–236, *236*
Multiline Properties option, 237
Object Properties toolbar
  changing layer visibility, 170, *170*
  color and linetype property settings, 180–181
  defined, **25**
  layer name or number display, 9
  List button, 181–182
  moving, 22–23
  Properties tool, 166, 181, 275, **358–359**
  Sun tool, 261
Properties tool, 166, 181, 275, **358–359**
Release 14 Layer & Linetype Properties dialog box, 162
setting active properties with ActiveX Automation, **936–937**, *936*
Space property of lines, 182
thickness property of 3D surface models, 584, *584*, 586, 588–589, *588*, *589*
prototyping 3D solid models, 575–577
  ACIS solid-modeling engine and SAT file translation, 576
  data translation tools and, 575–576
  DXF translators, 576
  IGES translation tools, 576
  stereolithography (STL) translation, 577
Proxy Images for Custom Objects option in Preferences dialog box, 1015
pull-down menus. *See* menus

Q

qtext (Quick Text) command, 305, *305*
question mark (?) as a wildcard character, 173
Quick Leader tool, 870
Quit option of Sketch command, 455
quote ('), system variables and, 824

R

r (redraw) command, 54, 253–254, 288, 714
radii, dimensioning, 347–348, *348*
RAM. *See* memory
random multiple copies, 212–213
rapid prototyping of 3D solid models, 575–577
  ACIS solid-modeling engine and SAT file translation, 576
  data translation tools and, 575–576
  DXF translators, 576
  IGES translation tools, 576
  stereolithography (STL) translation, 577
raster images, 651, 740–751
  adjusting brightness, contrast, and strength, 748–749, *748*, *749*

adjusting overall quality and transparency, 750–751, *750*
Attach Image dialog box, 742, *742*, 743
clipping, 745–748, *747*
controlling object visibility and overlap with, 742–743
exporting, 651, 833–836
importing, 740–742, *741*, *742*
Raster File Export option in Add a Printer dialog box, 651
reordering, 744–745, *744*
Show Raster Image Content option in Preferences dialog box, 1013
turning off frames, 750–751, *750*
Wipeout tool and, 748
ray tracing in 3D rendering, 712–713, *714*
rays, 226–228, *227*, *228*
Realtime command in View menu Zoom submenu, 30–32, *31*
recalling
  last area calculation, 817
  layer settings, 851–852
Reconfigure a Printer dialog box, 650–651, *650*
reconfiguring tablet digitizers for tracing, 731, *731*
Record option of Sketch command, 455
records, finding in databases, 795–797
recovering. *See* restoring
Rectangle tool, 105
rectangular arrays, defined, **199**
redefining. *See also* creating; defining
  blocks, 147–149, *149*
  blocks containing attributes, 779–780
Redo command or tool, 49
Redraw tool, 54, 253–254, 288, 714
Redraw view, 54
Reference toolbar
  defined, **25**
  External Reference Bind tool, 374, 383
Refine option of Splinedit command, 447–450, *448*, *449*
regen command, 54, 147, 218, 253–254, 288, 714
Regenmode system variable, 261–262
regions in 3D solid models, 555–558, *557*, *558*
RegisteredApplication collection object in ActiveX Automation, 926
reinitializing plotter ports, 673
relational operators in filtering selections, 407
relative Cartesian coordinates, entering distances with, 51–53, *51*, *52*, *53*
relative polar coordinates, entering distances with, 49–50, *50*, *51*
Release 14 Layer & Linetype Properties dialog box, 162
Reload AutoLISP between Drawings option in Preferences dialog box, 1015

Reload option in External Reference dialog box, 377
remove object selection command, 70
removing. *See* deleting
renaming layers, 165
Render toolbar. *See also* 3D rendering in Numbers section of index
    defined, 25
    Hide tool, 472–474, *473*, 520, *521*, 591, *591*
    Shade tool, 520, *521*
Rendered Object Smoothness option in Preferences dialog box, 691
reordering raster images, 744–745, *744*
repeating commands, 79
replacing existing files with blocks, 151
reports, Pack 'n Go reports, 865
resolution
    of monitors, 994–995
    Text Resolution option in Print/Plot Configuration dialog box, 655
restoring
    corrupted files, 828, 1001–1002
    dimension styles, 1075
    named User Coordinate Systems, 492–493
Retain Objects option for Make Block tool or Block command, 136
Revision Cloud tool, 862–863, *863*
Revolve tool, 536–538, *539*
Revolved Surface tool, 617–619, *618*
Right option of Dtext command, 295
right-clicking mouse, 65
rm command, 101, 102
rotating
    3D models, 512–514, *513*, 519, *520*
    dimension text, 336–337
    drawings, 78–80, *80*
    objects in 3D space, 624
    plots, 666
    the UCS plane about the X, Y, or Z axis, 488–489, *489*
Rotation and Origin option in Print/Plot Configuration dialog box, 665–666, *665*
roughing in line work, 217–221, *217*
rounding dimension variables, **1071**
row copies, 204–213
rows, adding to database tables, 797–798
Rows dialog box, 796, *796*, 800, 802–803
rubber-banding lines, 47, *47*
Ruled Surface tool, 612–614, *613*, *614*, 616
running AutoLISP programs, 878–879
Running Osnaps, 322–323, 427, 452

—S—

SAT file translation, 576
Save As command, 33, 380
Save command, 32–33
Save Drawing As dialog box
    Drawing Template File option, 197
    Save as Type option, 832
Save Path option in External Reference dialog box, 377
Save Thumbnail Preview Image option in Preferences dialog box, 1017
savetime command, 128
saving
    area data as attributes in drawing files, 820
    AutoLISP macros as files, 883–884, *883*
    blocks as drawing files, 150
    dimension styles, 1075
    files
        AutoLISP macros as files, 883–884, *883*
        automatically, 32–33, 128, 1016
        blocks as drawing files, 150
        Create Backup Copy with Each Save option in Preferences dialog box, 1016
        Incremental Save option in Preferences dialog box, 1013
        with preview images, 150
        Release 14 files to previous Release formats, 832
        Save Thumbnail Preview Image option in Preferences dialog box, 1017
        troubleshooting, 1003
        to Web or FTP sites, 899–900
    layer settings, 851–852
    layer settings of external reference files, 382
    named User Coordinate Systems, 492
    object selection filter criteria, 407
    Paper Space viewport views, 400
    plotter settings, 652, 673–674
    text data from AutoCAD Text window, 827–828
    views, 257–260, *259*
Scale option of Multiline command, 232
Scale, Rotation, and Origin options, 664–666. *See also* Print/Plot Configuration dialog box
    overview of, 664
    Plotted Inches and Drawing Units options, 664–665
    Rotation and Origin option, 665–666, *665*
Scale tool, 534–535, *536*, 602
Scaled to Fit option in Print/Plot Configuration dialog box, 664
scales
    calculating architectural scale factors, 318

plotting multiple views in different scales in Paper Space, 388
scale dimension variables, **1070**
scale factors and size of text, **98–100, 279–280**
scaling views in Paper Space, 398–400, *399*
setting dimension scale, 317–319
and setting drawing area size limits, 96–98
setting linetype scales in Paper Space, 403
scaling drawings, 730, 738–739
Scandisk utility, 1000
scanning drawings, 730, 739
scenes in 3D rendering, 707–711, *709, 710, 711, 712*
screen menu, displaying in Drawing window, 8, 1018
screens
    clearing screen of blips, 54
    resolution and performance, 254–255, 994–995
scroll bars, displaying in Drawing window, 212, 1018
SDF files (space-delimited format), exporting attribute information to, 787–788
searching. *See* finding
Section tool, 553–554, *555*
Segments per Polyline Curve option in Preferences dialog box, 1013
Select DBMS for New Environment dialog box, 792, *792*
Select File dialog box, 27–28, *27, 28*
Select Initial View option in Open Drawing dialog box, 260
Select Objects dialog box, 805–806, *805*
Select Objects toolbar, 25
Select Slide File dialog box, 548, *548*
selecting. *See also* entering
    arrow styles, 317–319
    commands, 55–60, *56, 59, 60*
    coordinate points in drawing area, 12, 486
    in-house AutoCAD experts, 980–981
    materials in 3D rendering, 696–697
    menu options, 14–17, *17, 18,* 23
    objects, **61–75, 111, 369–371, 935–936**
        with ActiveX Automation, **935–936**
        Autoselect feature, 72–74, *73, 74*
        for blocks, 135
        by windowing them, 63–64, *64*
        close or overlapping objects, 111
        with Ctrl key, 111, 369–370
        Noun/Verb method: selecting objects before commands, 70–72, 74–75, 1025
        overview of, **61–64,** *62, 64*
        selection commands listed, 69–70
        with selection cycling and Object Selection Settings dialog box, 369–371, 1025–1026
    through databases, 803–804
    troubleshooting, 1002

Undoing selections, 62–63
using base points, 64–68, *66, 67, 68*
output devices, 648–649, *648*
plotters and printers, 648–649, *648*
selection filtering, **405–409, 805–807**
    creating complex selection sets, 408–409
    filtering objects by location, 407
    Object Selection Filters dialog box, 406–407
    of objects linked to external databases, 805–807
    overview of, 405–407
    saving filter criteria, 407
spell checking dictionaries, 301–302
Selection command in Tools menu, 369–371, 1025–1026
SelectionSet collection object in ActiveX Automation, 924, 936
servers
    ActiveX Automation and, 907–908, *907*
    dedicated file servers, **987–988**
service bureaus, 679
Set Layer Filters dialog box, 173–174, *173*
setting
    active properties with ActiveX Automation, 936–937, *936*
    angle display style, 94
    default text style, 283
    dimension scale, 317–319
    dimension unit style, 313–315
    direction of 0 base angle, 95
    drawing area size, 96–98
    drawing area size limits, 96–98
    grid spacing, 100–102
    hardware linetypes, 669
    height for dimension text, 315
    layers in individual Paper Space viewports, 400–403, *402*
    layers when creating 3D surface models, 591, *591*
    linetype scale, 177–181, *179*
    linetype scales in Paper Space, 403
    location of dimension text, 316
    maximum number of active viewports in Paper Space, 398
    time interval between automatic saves, 128, 1016
    UCS origin, 482–483, *484*
    units of measurement display in status bar, 94
    units of measurement style, 90–94
setting up. *See also* configuring; Preferences dialog box
    AutoCAD for Open Database Connectivity (ODBC) drivers, **1027–1030**
        creating reference database files, 1027–1029
        telling AutoCAD the location of reference files, 1030

telling Windows the location of reference files, 1029
AutoCAD SQL Extension (ASE) to locate database files, 790, *790*
drawing layouts, 124–125, *124, 125*
Paper Space, 391
perspective views, 626–627, *627*
viewports in Eye2eye add-on, 1037–1038, *1038*
Setup View tools, 565–573
and adding dimensions, 572–573, *573*
layers and, 571, *572*
Paper Space and, 566–568, *568*
Setup Drawing tool (soldraw), 568–569
Setup Profile tool (solprof), 569–571, *571*
Setup View tool (solview), 565–568, *566, 568*
Shade tool, 520, *521*
shadows in 3D rendering, 691–696
Modify Distant Light dialog box, 692, *692*
Photo Real Render Options dialog box, 694, *694, 695*
ray tracing and, 713, *714*
Shadow Options dialog box, 692–693, *692, 693,* 694–696, *695*
shapes
calculating areas of complex shapes, 817–820, *818, 819*
creating symbols from, 152
deleting, **239–241**
deleting all unused shapes, 240–241
selectively deleting unused shapes, 239–240
UCS orientation and, 487
using pre-defined 3D surface shapes, 602
shortcut keys
entering commands with, 26
Priority for Accelerator Keys option in Preferences dialog box, 1015
troubleshooting, 1004
shortcut to acad.log file, 826
Show Raster Image Content option in Preferences dialog box, 1013
Show the Start Up Dialog Box option in Preferences dialog box, 1015
Show Text Boundary Frame Only option in Preferences dialog box, 1013
SI-Mechanical add-on, 885, 1032, 1033
single object selection command, 70
single quote ('), system variables and, 824
single-line text objects, 291–298
adding to drawings, 291–293, *292*
defined, **291**
editing, 293–294
justifying, 295–296, *296*
using special characters with, 297–298

sizing
boundary windows, 266
text, 98–100, 279–280
sketch command, 454–455
skewing dimension lines, 330, **352–353,** *353*
.sld files, 546
Slice tool, 551–553, *552, 553*
slides, 546–548
creating, **547–548**
defined, **546**
Mslide and Vslide tools, 546, *547*
smoothing polylines, 425–427, *426*
smoothness of display, controlling, 254–255
Snap mode. *See also* drawing aids
accessing, 102–103
coordinate readout and, 105
defined, **102**
snapunit command and system variable, 03
toggling on or off, 102–103
using Grid mode with, 103–104
Snap option on status bar, 245–246
soldraw command, 568–569
solid filled areas, 455–459. *See also* polylines
creating, 455–456
creating filled circles with Donut command, 456, *457*
Solid command, 455
toggling solid fills on and off, 421, 458–459, *458*
Solid Model Object Display option in Preferences dialog box, 1012
solid models. *See* 3D solid models in Numbers section of index
Solids submenu, Draw menu
Box command, 477–478, 552–553
Cone command, 550, *550*
Cylinder command, 469, 475–477, *476,* 544
Extrude command, 470–471, 516, *516*
Interference command, 559–561, *560*
Revolve command, 536–538, *539*
Section command, 553–554, *555*
Setup submenu Drawing command (soldraw), 568–569
Setup submenu Profile command (solprof), 569–571, *571*
Setup submenu View command (solview), 565–568, *566, 568*
Slice command, 551–553, *552, 553*
Sphere command, 542–545, *545*
Torus command, 539–542, *539, 541, 542*
Wedge command, 549, *549*
Solids toolbar. *See also* toolbars
Box tool, 477–478, 552–553
Cone tool, 550, *550*
Cylinder tool, 469, 475–477, *476,* 544

defined, **25**
Extrude tool, 470–471, 516, *516*
Interference tool, 559–561, *560*
Revolve tool, 536–538, *539*
Section tool, 553–554, *555*
Setup Drawing tool (soldraw), 568–569
Setup Profile tool (solprof), 569–571, *571*
Setup View tool (solview), 565–568, *566, 568*
Slice tool, 551–553, *552, 553*
Sphere tool, 542–545, *545*
Torus tool, 539–542, *539, 541, 542*
Wedge tool, 549, *549*
solprof command, 569–571, *571*
solview command, 565–568, *566, 568*
sound, linking to AutoCAD, 844
Space property of lines, 182
space-delimited format (SDF), exporting attribute information to, 787–788
special characters. *See* symbols
Specify On-screen option in Attach Xref dialog box, 378
specifying. *See* entering; selecting
Speed setting in Print/Plot Configuration dialog box, 669
spelling checker, 299–302
Check Spelling dialog box, 299–301
selecting dictionaries, 301–302
Sphere tool, 542–545, *545*
spherical coordinate formats, 593–595, *594, 595*
spline curves, 438–450. *See also* polylines
Close option, 442
control points
adding with Fit Data option, 444–445, *445*
changing with Splinedit command Refine option, 447–450, *448, 449*
defined, **442**
converting polylines to, **428**
creating, **441**–**442**, *441, 443*
creating polyline spline curves, **438–440**
Fit Data options, **444**–**447**
adding control points, 444–445, *445*
changing spline tolerance setting, 446, *446*
controlling tangency at the beginning and end points, 444
when to use, 447
Fit Tolerance option, 442
versus polylines, **441**
Splinedit command
Fit Data options, 444–447, *445, 446*
Refine option, 447–450, *448, 449*
text on spline curves, 859
spotlights in 3D rendering, 705–707, *706, 707*
spreadsheets. *See also* Microsoft Excel
linking to AutoCAD, 839–843, *841, 842*

SQL Editor dialog box, 807–808, *807*
stacked fractions, 272–274
Standard toolbar. *See also* toolbars
Bonus Standard toolbar, **859–870**
block-related tools, 869
Copy Nested Entities tool, 869
Extend to Block Entities tool, 869
Extended Change Properties tool, 865–866
Extended Clip tool, 868, *869*
Extended Trim tool, 867–868
List Xref/Block Entities tool, 870
Move Copy Rotate tool, 870
Multiple Entity Stretch tool, 867
Multiple Pedit tool, 866–867
Pack 'n Go tool, 381, 863–865
Quick Leader tool, 870
Revision Cloud tool, 862–863, *863*
Trim to Block Entities tool, 869
Wipeout tool, 748, 859–861, *860, 861, 862*
defined, **25**
Distance tool, 94, 205–206
last used tool feature, 209
List tool Area option, 814–815, *815*
Mass Properties tool, 574
Match Properties tool, 368
Pan Realtime tool
defined, **208**–**212**, *210, 211*
drawing regeneration and, 253
in Model Space, 572
in Paper Space, 394, 572
Redo button, 49
Redraw button, 54, 253–254, 288, 714
Undo button, 49
Zoom flyout, 209
Zoom Previous button, 30
Zoom Realtime button, 30–32, *31*, 108, 253
Zoom Window button, 29–30, *29*, 108
standard windows, 72, *73*
starting
AutoCAD, 7
AutoLISP programs, 878–879
AutoLISP programs automatically, 879–880
status bar
defined, **9**, *12*, **93**
Grid and Snap options on, 245–246
status of drawings, 820–823, *821*
stereolithography (STL) translation, 577
Straighten option of polyline Edit Vertex mode, 433, *435*
straightening lines in traced drawings, 735
strength of raster images, 748–749, *748, 749*
stretching
dimensions with Stretch tool, **338**–**341**, *338, 339,* **602**

lines, 75–78, *77*
Multiple Entity Stretch tool, **867**
Style Name options in Text Style dialog box, 284
Style option of Multiline command, 232
styles
   defined, **280**
   dimension styles, **312–319**, **349**
      Annotation dialog box and, 314, 315, 1081–1084, *1082*
      changing style settings for groups of dimensions, 337
      changing style settings of individual dimensions, 333–337
      Dimension Style Export dialog box, 874–875
      Dimension Style Import dialog box, 875
      dimension styles versus ANSI-style dimensions, 312, *312*
      Dimension Styles dialog box and, **1076–1084**
      Dimstyle Export and Dimstyle Import commands in Bonus menu Tools submenu, 874–875
      DimStyle object in ActiveX Automation, **924**
      Format dialog box and, 316, 1080–1081
      Geometry dialog box and, 317–318, 1077–1080, *1078*, *1079*
      overview of, **312–313**, *312*
      parent and child styles in dimension families, 349
      Primary Units dialog box and, 314–315
      selecting arrow style and setting dimension scale, 317–319
      setting dimension unit style, 313–315
      setting height for dimension text, 315
      setting location of dimension text, 316
      using dimension families to fine-tune dimension styles, 349
   Text Style dialog box, **283–285**
      Effects options, 284–285, *284*, *285*
      Font options, 284
      Style Name options, 284
   text styles, **239–241**, **280–286**
      creating, **281–282**
      deleting all unused styles, 240–241
      importing from external reference files, **374**, **382–384**
      renaming, 285–286
      selectively deleting unused styles, 239–240
      setting default style, 283
      TextStyle object in ActiveX Automation, **927**
      using, **282–283**, *282*
Sub Sampling option in Render dialog box, 723, *723*
substituting fonts, 302–304, 837–839
Subtract tool, 471–472, *472*, 516, *517*, 820

Sun tool, 261
sunlight angle settings in 3D rendering, 686–691, *689*, *690*
   Geographic Location dialog box, 688–689
   Lights dialog box, 687, *687*
   New Distant Light dialog box, 687–688
   North Location dialog box, 690, *690*
   Sun Angle Calculator dialog box, 688
Support File Search Path option in Preferences dialog box, 1010
supporting AutoCAD, 979–984
   getting outside support, 979–980
   learning AutoCAD, 982
   making AutoCAD easier to use, 982–983
   managing AutoCAD projects, 983–984
   preparing the staff, 981–982
   selecting in-house experts, 980–981
surface models. *See* 3D surfaces in Numbers section of index
Surfaces submenu, Draw menu
   3D Face command, 596–600, *596*, *597*, *599*, *600*, *601*
   Revolved Surface command, 617–619, *618*
   Ruled Surface command, 612–614, *613*, *614*, *616*
   Tabulated Surface command, 615–616, *615*, *616*
Surfaces toolbar. *See also* toolbars
   3D Face tool, 596–600, *596*, *597*, *599*, *600*, *601*
   displaying, 595
   Edge Surface tool, 608–609
   Edge tool, 601
   Revolved Surface tool, 617–619, *618*
   Ruled Surface tool, 612–614, *613*, *614*, *616*
   Tabulated Surface tool, 615–616, *615*, *616*
SWLite software, **1032**, **1033**, **1040–1041**
Symbol document, 357
symbols, **134–157**. *See also* blocks; Symbols section of the index
   creating
      from blocks, **134–136**
      from existing drawings, 140–147, *140*
      from external reference files (Xrefs), 152
      from shapes, 152
      from Windows Character Map dialog box, 289–290, 298
   exercises, **188–189**, *188*, *189*
   feature control symbols, 356–358
   inserting, **136–139**, *137*, *139*, 273–274
      with drag and drop feature, 187–188
   Maximum Number Sorted Symbols option in Preferences dialog box, 1017–1018
   tolerance symbols, 357, *357*
   using with text objects, 297–298
system variables, **1046–1069**. *See also* dimensions, dimension variables

apostrophe (') and, 824
Celtscale, 180
Dimaso, 341
Dimtix and Dimtofl, 348
drawing information in, **824**
Fontmap, 303–304
Gridunit, 101
listed, **1046–1069**
Ltscale, 179
Maxactvp, 398
Pdmode, 451
Pickstyle, 156
Psltscale, 403
Regenmode, 261–262
Snapunit, 103
System Variable Editor command in Bonus menu Tools submenu, 873–8744
Textfill, 288
Tilemode, 391
Visretain, 382

T

tablet digitizers
    calibrating, 732–734, *734*
    configuring, 1022
    defined, **995–996**
    menu templates, 995, 1022–1024, *1023*
    reconfiguring for tracing, 731, *731*
Tabulated Surface tool, 615–616, *615*, *616*
Tangent option of polyline Edit Vertex mode, 435, *436*
tape backup systems, 986
Targa files, 651
TC option of Dtext command, 295
technical support. *See* supporting AutoCAD
templates
    attribute template files, 781, 784
    drawing template files, **196–199**
        creating, **196–198**
        defined, **196**
        Drawing Template File option in Save Drawing As dialog box, 197
        naming, 197
        Template Drawing File Location option in Preferences dialog box, 1011
        using, 198–199
    menu templates, 995
Temporary External Reference File Location option in Preferences dialog box, 1011
temporary files
    .ac$ files, 1000
    File Extension for Temporary Files option in Preferences dialog box, 1017

Temporary Drawing File Location option in Preferences dialog box, 1011
Temporary External Reference File Location option in Preferences dialog box, 1011
text, 98–100, 265–280. *See also* attributes; fonts; symbols; text styles
    adding to drawings, 266–269, *268*
    aligning, 275–278
        Osnaps and, 277–278, *277*
        overview of, 275–276, *275*, *276*
    arrow and text control dimension variables, 1072–1073
    Bonus Text Tools toolbar, **306**, **853–858**
        Arc Aligned Text tool, 858
        Change Multiple Text Items tool, 856–857
        Explode Attributes to Text tool, 858
        Explode Text tool, 858
        Find and Replace Text tool, 857–858
        Global Attribute Edit tool, 778, 858
        overview of, 306
        Text Fit tool, 858
        Text Mask tool, 853–855, *855*
    changing text height and font, 269–272, *271*
    ddedit command, **276**, **278–279**
    dimension text
        appending data to, 328–329
        moving fixed dimension text, 334–335, *335*
        rotating dimension text, 336–337
        setting height, 315
        setting location of, 316
    editing, **278–279**
    formatting, **269–274**
        adding color, stacked fractions, and symbols, 272–274
        changing text height and font, 269–272, *271*
        changing width of text boundary window, 274, *274*
    in hatched areas, 269, 341
    mirroring, **299**
    scale factors and size of, **98–100**, **279–280**
    Show Text Boundary Frame Only option in Preferences dialog box, 1013
    spelling checker, **299–302**
        Check Spelling dialog box, 299–301
        selecting dictionaries, 301–302
    on spline curves, 859
    UCS orientation and, 487
    uses for, **306**
    using multiline text with leaders, 351–352
text boundary windows, changing width of, 274, *274*
Text Editor, Dictionary, and Font File Names option in Preferences dialog box, 1010
text editor line endings in attribute files, 783

text files, importing, 291
Text Fill option in Print/Plot Configuration dialog box, 655
text objects, 291–298. *See also* objects
adding to drawings, 291–293, *292*
defined, **291**
editing, 293–294
justifying, 295–296, *296*
using special characters with, 297–298
Text Resolution option in Print/Plot Configuration dialog box, 655
Text Style dialog box, 283–285
Effects options, 284–285, *284*, *285*
Font options, 284
Style Name options, 284
text styles, 239–241, 280–286
creating, **281–282**
deleting all unused styles, 240–241
importing from external reference files, **374, 382–384**
renaming, 285–286
selectively deleting unused styles, 239–240
setting default style, 283
TextStyle object in ActiveX Automation, **927**
using, 282–283, *282*
Text window. *See also* windows
capturing and saving text data from, 827–828
displaying data in, 84–85, *85*
Number of Lines of Text in Text Window to Keep in Memory option in Preferences dialog box, 1019
overview of, 268
Status screen, 820–822, *821*
Text Window Parameters options in Preferences dialog box, 1019
Textfill system variable, **288**
texture maps in 3D rendering, **714–718**. *See also* hatch patterns
Adjust Planar Coordinates dialog box, 718, *718*
Mapping dialog box, 717–718
overview of, 714–717, *715*, *716*, *719*
Texture Maps Search Path option in Preferences dialog box, 1011
thawing layers, 171–172, 261
thickness property of 3D surface models, 584, *584*, 586, 588–589, *588*, *589*
third-party software, 884–885
3D Face tool, 596–600, *596*, *597*, *599*, *600*, *601*
3D models. *See also* User Coordinate System; World Coordinate System
aligning two 3D objects, **508–521**
finishing up, 514–521
overview of, 508–513, *509*, *510*, *511*, *513*
creating with Eye2eye add-on, **1039**

editing models from previous versions of AutoCAD, **468**
rotating, **512–514**, *513*, 519, *520*
viewing, **495–500**
Isometric and Orthogonal view options, 496–498, *496*, *497*
overview of, 495–496
with Viewpoint Presets dialog box, 498–500, *499*
3D Operations submenu, Modify menu
Align command, 508, *509*, 622–623, *623*
Rotate 3D command, 512–514, *513*, 519, *520*, 624
3D rendering, 683–725
backgrounds, **702–704**, *703*, *704*
bitmap images in, **714–718**, *715*, *716*, *719*
disk space requirements, 684
example, **684–686**, *685*
lighting effects, **704–711**
named scenes, 707–711, *709*, *710*, *711*, *712*
spotlights, 705–707, *706*, *707*
Lights dialog box
Distant Light option, 687, *687*
Modify option, 692
North Location option, 690, *690*
Spotlight option, 705
sunlight angle settings and, 687, *687*
materials, **696–701**
adjusting appearance of, 699–700, *699*, *701*
assigning to objects, 698, *699*
Materials dialog box, 696–697, *696*, *699*
Materials Library dialog box, 697, *697*
Modify Granite Material dialog box, 699–700, *699*, *701*
selecting, 696–697
output options, **719–722**
File Output Configuration dialog box, 721–722, *721*
rendering to a file, 721–722
rendering to Render window, 719–720, *720*
Windows Render Options dialog box, 720, *720*
ray tracing, **712–713**, *714*
Render dialog box, **685**, 722–725
Anti-Aliasing option, 724–725, *724*
Background option, 702
Crop Window option, 722–723, *722*
Destination option, 720, 722
More Options option, 694
overview of, 685, *685*
Raytrace Rendering Options dialog box, 724
Shadow option, 692
Sub Sampling option, 723, *723*

Render toolbar
  defined, **25**
  Lights tool, 686–687, 691–692
  Mapping tool, 717
  Materials tool, 696
  Render tool, 689–690
  Scenes tool, 707–708, 711, *712*
Rendered Object Smoothness option in Preferences dialog box, 691
shadows, **691–696**
  Modify Distant Light dialog box, 692, *692*
  Photo Real Render Options dialog box, 694, *694*, *695*
  ray tracing and, 713, *714*
  Shadow Options dialog box, 692–693, *692*, *693*, 694–696, *695*
sunlight angle settings, **686–691**, *689*, *690*
  Geographic Location dialog box, 688–689
  Lights dialog box and, 687, *687*
  New Distant Light dialog box, 687–688
  North Location dialog box, 690, *690*
  Sun Angle Calculator dialog box, 688
texture maps, **714–718**, **1011**
  Adjust Planar Coordinates dialog box, 718, *718*
  Mapping dialog box, 717–718
  overview of, 714–717, *715*, *716*, *719*
  Texture Maps Search Path option in Preferences dialog box, 1011
View menu Render submenu
  Lights command, 686–687, 691–692
  Mapping command, 717
  Materials command, 696
  Render command, 689–690
  Scenes command, 707–708, 711, *712*
Z buffer shaded models, 686
3D solid models, 468–478, 500–505, 507–508, 525–579
  converting to 2D drawings, **561–564**
    copying solid models using the Clipboard, 562–564, *564*
    overview of, 561–562
  creating, **468–478, 500–501, 506–507, 529–531**
    from 2D drawings, 506–507, *506*, *508*
    adding volumes of two solids, 475–477, *476*
    from box primitives, 477–478
    converting polylines to solids, 470–471
    drawing solid cylinders, 469–470
    overview of, 500–501, *501*, *502*, 529–531, *530*
    preparation, 469
    removing the volume of one solid from another, 471–472, *472*
    verifying designs with the Hide tool, 472–474, *473*
  defined, **468**
  design tips, **578**
  Draw menu Solids submenu
    Box command, 477–478, 552–553
    Cone command, 550, *550*
    Cylinder command, 469, 475–477, *476*, 544
    Extrude command, 470–471, 516, *516*
    Interference command, 559–561, *560*
    Revolve command, 536–538, *539*
    Section command, 553–554, *555*
    Setup submenu Drawing command (soldraw), 568–569
    Setup submenu Profile command (solprof), 569–571, *571*
    Setup submenu View command (solview), 565–568, *566*, *568*
    Slice command, 551–553, *552*, *553*
    Sphere command, 542–545, *545*
    Torus command, 539–542, *539*, *541*, *542*
    Wedge command, 549, *549*
  editing
    with Chamfer tool, 474–475, *475*, 531–532, *532*
    with Fillet tool, 502–505, *503*, *505*, 528–529, *529*
    with Slice tool, 551–553, *552*, *553*
  fabrication shops and, **577**
  interference checking, **559–561**, *560*
  joining with Union command, **526–528**, *526*, *527*
  mass properties, **573–574**
  Polar Array tool and, 533–534, *534*
  rapid prototyping, **575–577**
    ACIS solid-modeling engine and SAT file translation, 576
    data translation tools and, 575–576
    DXF translators, 576
    IGES translation tools, 576
    stereolithography (STL) translation, 577
  regions, **555–558**, *557*, *558*
  Scale tool and, 534–535, *536*, 602
  Setup View tools, **565–573**
    and adding dimensions, 572–573, *573*
    layers and, 571, *572*
    Paper Space and, 566–568, *568*
    Setup Drawing tool (soldraw), 568–569
    Setup Profile tool (solprof), 569–571, *571*
    Setup View tool (solview), 565–568, *566*, *568*
  Solid Model Object Display option in Preferences dialog box, 1012
  Solids toolbar
    Box tool, 477–478, 552–553
    Cone tool, 550, *550*
    Cylinder tool, 469, 475–477, *476*, 544
    defined, **25**
    Extrude tool, 470–471, 516, *516*

Interference tool, 559–561, *560*
Revolve tool, 536–538, *539*
Section tool, 553–554, *555*
Setup Drawing tool (soldraw), 568–569
Setup Profile tool (solprof), 569–571, *571*
Setup View tool (solview), 565–568, *566, 568*
Slice tool, 551–553, *552, 553*
Sphere tool, 542–545, *545*
Torus tool, 539–542, *539, 541, 542*
Wedge tool, 549, *549*
UCS orientation and, 487
3D surfaces, 468, 583–638
aligning objects in 3D space, **622–623**, *623*
creating 3D surface models, **584–591**
from 2D polylines, 586–587, *588*
elevation and, 584–585, *585*, 586–587, *588*
making horizontal surfaces opaque, 590, *590, 591*
overview of, 584–585, *584, 585*
setting elevation and thickness before drawing, 588–589, *589*
setting layers, 591, *591*
thickness property and, 584, *584*, 586, *588*
creating 3D surfaces, **592–603, 611–619**
adding 3D faces, 596–598, *596, 597, 599, 600*
defining surfaces using two objects, 612–614, *613, 614*
extruding circular surfaces, 617–619, *618*
extruding objects along a straight line, 615–616, *615, 616*
hiding unwanted surface edges, 600, *601*
laying out 3D form objects, 593–595
spherical and cylindrical coordinate formats, 593–595, *594, 595*
using point filters, 592
using pre-defined 3D surface shapes, 602
creating complex 3D surfaces, **603–611**
changing mesh control settings, 610, *611*
creating 3D meshes by specifying coordinates, 611
curved 3D surfaces, 603–610, *604*
defined, **468, 583–584**
Draw menu Surfaces submenu
3D Face command, 596–600, *596, 597, 599, 600, 601*
Revolved Surface command, 617–619, *618*
Ruled Surface command, 612–614, *613, 614, 616*
Tabulated Surface command, 615–616, *615, 616*
editing, **602**
meshes
changing mesh control settings, 610, *611*
creating 3D meshes by specifying coordinates, 611
editing, 619–622, *620*
rotating objects in 3D space, **624**
Surfaces toolbar
3D Face tool, 596–600, *596, 597, 599, 600, 601*
defined, **25**
displaying, 595
Edge Surface tool, 601, 608–609
Revolved Surface tool, 617–619, *618*
Ruled Surface tool, 612–614, *613, 614, 616*
Tabulated Surface tool, 615–616, *615, 616*
viewing in perspective, **624–638**. See also Eye2eye add-on
adjusting camera and target positions, 629–631, *630, 631*
adjusting distances between camera and target, 627–629, *628, 629*
changing your point of view, 631–634, *632, 633, 634*
hiding parts of the view with clip panes, 636–638, *636, 637*
overview of, 624–626, *625*
versus parallel projection, 624
setting up perspective views, 626–627, *627*
twisting the camera, 635, *635*
using Dview command Zoom option as a telephoto lens, 634–635
3D wire-frame models
defined, **467**
example, 586, *588*
plotting, 651
tick marks, creating blocks for dimension tick marks, **1084–1086**, *1085*
TIFF files, 651
tilde (~) in custom menus, 962
Tiled Model Space, 385, 388
Tiled Viewport Layout dialog box, 386–388
tiled viewports in Model Space, 385–388, *387, 389*
Tilemode system variable, 391
time
Minutes between Saves option in Preferences dialog box, 1016
tracking time spent on drawings, 823, *823*
TL option of Dtext command, 295
toggling
Grid mode, 100
layers, 171
Snap mode, 102–103
solid fills on and off, 421, 458–459, *458*
tolerance
adding tolerance notation to dimensions, 356–358

changing spline tolerance setting, 446, *446*
Fit Tolerance option of Spline command, 442
tolerance dimension variables, **1071**
tool tips, 19, *20*
toolbar command, 947–948
toolbars, 19–26, 944–955
    Bonus Standard toolbar, **859–870**
        block-related tools, 869
        Copy Nested Entities tool, 869
        Extend to Block Entities tool, 869
        Extended Change Properties tool, 865–866
        Extended Clip tool, 868, *869*
        Extended Trim tool, 867–868
        List Xref/Block Entities tool, 870
        Move Copy Rotate tool, 870
        Multiple Entity Stretch tool, 867
        Multiple Pedit tool, 866–867
        Pack 'n Go tool, 381, 863–865
        Quick Leader tool, 870
        Revision Cloud tool, 862–863, *863*
        Trim to Block Entities tool, 869
        Wipeout tool, 748, 859–861, *860, 861, 862*
    book's versus screen images of, 44
    customizing, **944–955**
        creating custom buttons, 949–950
        creating custom icons, 951–953
        creating flyouts, 953–954
        creating toolbars, 945–947
        Customize Toolbars dialog box and, 946–947
        editing existing buttons, 955
        New Toolbar dialog box and, 946
        Toolbars dialog box and, 944–945
        Windows menu files and, 955
    Dimension toolbar
        Aligned Dimension tool, 342–343, *343*
        Angular Dimension tool, 345–346, *346*
        Baseline Dimension tool, 326–327, *327*
        Continue Dimension tool, 321–327
        defined, **25**
        Diameter Dimension tool, 347, *348*
        Dimension Edit tool, 330, 336, 353, *353*
        Dimension Style tool, 1076
        Dimension Update tool, 335
        Leader tool, 350–351, *350, 351*
        Linear Dimension tool, 320–321, *321*
        opening, 319
        Ordinate Dimension tool, 354–355, *354, 356*
        Radius Dimension tool, 348
        Tolerance tool, 356–358
    docked versus floating toolbars, **9**, **21–24**
    Draw toolbar, **25, 33, 40–44**
        Arc tool, 120–122, *121*
        closing, 33
        defined, **25**

Ellipse tool, 106–107, *107*
Make Block tool, 134–136
Multiline Text tool, 266–270, *268*
overview of, **40–44**, *41*
Polygon tool, 470
Polyline tool, 418–421, *418, 419*
Rectangle tool, 105
Spline tool, 441–442, *441, 443*
External Database toolbar, 25
flyouts
    converting to custom toolbars, 43
    creating, **953–954**
    defined, **20–21, 43–44**
    Flyout Properties dialog box, 954
    most recently selected option as default for toolbar tools, 44
    opening, 43
Inquiry toolbar, 25
Insert toolbar
    defined, **25**
    external reference files and, 375
    Insert Block tool, 137
Internet Utilities toolbar, 888
listed, **24–26**
location of, *9, 11*
Mechanical toolbar, **40**
Modify II toolbar
    defined, **25**
    Edit Polyline tool, 622
    Intersect tool, 510–511, *511*
    Subtract tool, 471–472, *472*, 516, *517*, 820
    Union tool, 476, 526–528, *526, 527*
Modify toolbar
    Array tool, 206–208, *207, 208*, 598, *600*, 602
    Chamfer tool, 222, 474–475, *475*, 531–532, *532*
    closing, 33
    Copy tool, 109, *110*, 212–213, 368–369, *369*, 602
    defined, **25**
    Erase tool, 34, *34*
    Explode tool, 147, 421
    Fillet tool, 117–118, *118*, 127–128, *128*, 222–224, 421, 502–505, *503, 505*, 528–529, *529*
    Lengthen tool, 231–232
    Mirror tool, 606, *606*
    Move tool, 68, 71–72, 602
    Offset tool, 124–125, *124, 125*, 127–128, *128*, 421–422, *422*
    overview of, **41–44**, *41*
    Scale tool, 534–535, *536*, 602
    Stretch tool, 338–341, *338, 339*, 602
    Trim tool, 111–116, *113*

moving, 21–24
Object Properties toolbar
    changing layer visibility, 170, *170*
    color and linetype property settings, 180–181
    defined, **25**
    layer name or number display, 9
    List button, 181–182
    moving, 22–23
    Properties button, 166, 181
Object Snap toolbar, 25
    opening from the command-line, **947–948**
Reference toolbar
    defined, **25**
    External Reference Bind tool, 374, 383
Render toolbar
    defined, **25**
    Lights tool, 686–687, 691–692
    Mapping tool, 717
    Materials tool, 696
    Render tool, 689–690
    Scenes tool, 707–708, 711, *712*
Select Objects toolbar, 25
Solids toolbar
    Box tool, 477–478, 552–553
    Cylinder tool, 469, 475–477, *476*, 544
    defined, **25**
    Extrude tool, 470–471, 516, *516*
Standard toolbar
    defined, **25**
    Distance tool, 94, 205–206
    last used tool feature, 209
    List tool Area option, 814–815, *815*
    Mass Properties tool, 574
    Match Properties tool, 368
    Pan Realtime tool, 208–212, *210*, *211*, 253, 394, 572
    Redo button, 49
    Redraw button, 54, 253–254, 288, 714
    Undo button, 49
    Zoom flyout, 209
    Zoom Previous button, 30
    Zoom Realtime button, 30–32, *31*, 108, 253
    Zoom Window button, 29–30, *29*
Surfaces toolbar
    3D Face tool, 596–600, *596*, *597*, *599*, *600*, *601*
    defined, **25**
    displaying, *595*
    Edge Surface tool, 608–609
    Edge tool, 601
    Revolved Surface tool, 617–619, *618*
    Ruled Surface tool, 612–614, *613*, *614*, 616
    Tabulated Surface tool, 615–616, *615*, *616*
tool tips, **19**, *20*

UCS toolbar
    defined, **25**
    Object UCS tool, 486–487, *486*
    overview of tools, **481–482**
    Preset UCS tool, 482, 493–495, *495*
    World UCS tool, 488
    X Axis Rotate UCS tool, 488–489, *489*
    Y Axis Rotate UCS tool, 489
    Z Axis Rotate UCS tool, 487–488, *488*
    Z Axis Vector UCS tool, 490, *490*
Viewpoint toolbar
    defined, **26**
    SE Isometric button, 470
Zoom toolbar, 26
Tools menu. *See also* Preferences dialog box
    Display Order submenu, 371, 744–745
    Drawing Aids command, 101, 102
    Grips command, 1026–1027
    Inquiry submenu
        Area command, 814–815, *815*, 817–819, *818*, *819*
        ID Point command, 814
        Set Variables command, 824
        Setup submenu Mass Properties command, 574
        Status command, 820–822, *821*
        Time command, 823, *823*
    Load Application command, 878
    Object Snap Settings command, 141–144
    Selection command, 369–371, 1025–1026
    UCS Save command, 492
    UCS submenu
        Origin command, 482–484, *484*
        overview of, **481**, *481*
Tools submenu, Bonus menu, 870–876
    Command Alias Editor command, 871–873
    Dimstyle Export and Dimstyle Import commands, 874–875
    Get Selection Set command, 876
    List Entity Xdata command, 876
    Pline Converter command, 876
    Popup Menu command, 870–871
    System Variable Editor command, 873–874
    Xdata Attachment command, 876
Torus tool, 539–542, *539*, *541*, *542*
Total option of Lengthen command, 232
TR option of Dtext command, 295
tracing, 455, 487, 730–738
    cleaning up traced drawings, **735–738**
        adjusting dimensions, 738
        aligning lines, 735–737, *736*, *737*
        straightening lines, 735
    defined, **730**
    tablet digitizers

calibrating, 732–734, *734*
configuring, 1022
defined, **995–996**
menu templates, 995, 1022–1024, *1023*
reconfiguring for tracing, 731, *731*
Trace command, 455
tracing lines from drawings, 735
UCS orientation and, 487
tracking time spent on drawings, 823, *823*
transparency, adjusting in raster images, 750–751, *750*
Tree View option in External Reference dialog box, 377
triangle. *See* Ortho mode
trimming objects, 110–118, 867–868
Cookie Cutter Trim command in Bonus menu, 867–868
Edge, Project, Undo, and Enter options, 115–116, *116*
Extended Trim tool, 867–868
Fence option, 365, *366*
overview of, 110–115, *113, 114, 115*
Trim to Block Entities tool, **869**
troubleshooting, 664, 1001–1004
AutoCAD, 1002–1004
fonts, 1002
grips, 1002
keyboard shortcuts, 1004
object selection, 1002
opening files, 1003
orienting drawings, 1004
Paper Space viewports, 1003
plotting, 664, 1004
restoring corrupted files, 828, 1001–1002
saving files, 1003
starting up or opening files, 1001
T-square. *See* Ortho mode
Turn Object Layer Off tool, 853
turning on grips, 1026–1027
turning off
associative dimensioning feature, 341
frames in raster images, 750–751, *750*
layers, 171, 853
UCS icon, 13
twisting the camera in perspective views, 635, *635*
2D drawings
converting 3D solid models to, 561–564
copying solid models using the Clipboard, 562–564, *564*
overview of, 561–562
creating 3D surface models from 2D polylines, 586–587, *588*
creating solid models from, 506–507, *506, 508*
Txtpath.lsp utility, 859

U

U or Undo command, 49
unblocking blocks, 147–149, *149*
underscore (_) in custom menus, 961
undoing
Edit Vertex options, 438
object selections, 62–63
parts of drawings with Undo command options, 200–201
Undo command or button overview, 49
Undo option for Trim tool, 116, *116*
Union tool, 476, 526–528, *526, 527*
Units Control dialog box, 92, *92*
units of measurement. *See* measurement system
Unload option in External Reference dialog box, 377
unlocking layers, 172, 853
unnamed.dwg files, 199
updating database rows and drawing labels, 802–803
User Coordinate System (UCS), 12–13, 478–495. *See also* coordinates; World Coordinate System
defined, **478**, **480–481**
defining
in the current view plane, 491–492, *491*
with three points, 484–486, *485*
UCS planes with objects, 486–487, *486*
named UCSs
restoring and deleting, 492–493
saving, 492
returning to the World Coordinate System, **488**
rotating the UCS plane about the X, Y, or Z axis, 488–489, *489*
setting UCS origin, 482–483, *484*
Tools menu UCS submenu
Origin command, 482–484, *484*
overview of, **481**, *481*
UCS icon
controlling display of, 479
defined, **12–13, 478**
Paper Space versus Model Space mode and, 13
turning on or off, 13
World Coordinate System and, 12
World UCS icon, 478, *478*
UCS Orientation dialog box, 494–495, *494, 495*
UCS Save command, 492
UCS toolbar
defined, **25**
Object UCS tool, 486–487, *486*
overview of tools, **481–482**
Preset UCS tool, 482, 493–495, *495*
World UCS tool, 488

X Axis Rotate UCS tool, 488–489, *489*
Y Axis Rotate UCS tool, 489
Z Axis Rotate UCS tool, 487–488, *488*
Z Axis Vector UCS tool, 490, *490*
UserCoordinateSystem object in ActiveX Automation, **927–928**
Z axis
   defined, **480**
   defining a UCS origin and plane with, 489–490, *490*
   rotating the UCS plane about, 487–489, *488, 489*
user interaction using ActiveX Automation, 935–936
utilities. *See also* Bonus tools; CD-ROM (book's)
   AutoCAD Drawing Explorer, 917
   Batch Plot utility, **675–678**
      File tab, 675–676, *675*
      Layers tab, 676–677, *677*
      Logging tab, 678, *678*
      Plot Area tab, 677, *677*
      Plot Stamping tab, 678, *678*
   Cadence and *Cadalyst* utilities, 876–877, 886
   Eye2eye, 877
   on online services, 886
   Scandisk, 1000
   third-party software, **884–885**
   Txtpath.lsp, 859
   on the World Wide Web, 885
Utility object in ActiveX Automation, 930–931

V

variables. *See also* dimensions, dimension variables; system variables
   in ActiveX Automation, 911
vector file exporting, 836–839
   DXF files, 836
   .EPS (encapsulated PostScript) files, 837
   HPGL plot file format, 839
   PostScript font substitution, 837–839
Ver option of Xline command, 221
Verify attribute definition mode, 768
verifying 3D designs with the Hide tool, 472–474, *473*
vertical dimensions, 320–321, *321*
video display resolution and performance, 254–255, 994–995
video images, linking to AutoCAD, 844
video RAM, 994
View menu
   3D Viewpoint submenu, 496–498, *496, 497, 498*
   Floating Model Space command, 391, 393, 395, 398

Paper Space command, 391, 394, 395, 398
Redraw View command, 54
Render submenu
   Lights command, 686–687, 691–692
   Mapping command, 717
   Materials command, 696
   Render command, 689–690
   Scenes command, 707–708, 711, *712*
Tiled Model Space command, 395
Tiled Viewports submenu, Layout command, 386
Zoom submenu
   Previous command, 30
   Realtime command, 30–32, *31*
   Window command, 29–30, *29*
View object in ActiveX Automation, 928
View option in Print/Plot Configuration dialog box, 661–662, *662*
viewing. *See also* accessing; displaying; opening
   3D models, **495–500**
      Isometric and Orthogonal view options, 496–498, *496, 497*
      overview of, 495–496
      with Viewpoint Presets dialog box, 498–500, *499*
   3D surfaces in perspective, **624–638**. *See also* Eye2eye add-on
      adjusting camera and target positions, 629–631, *630, 631*
      adjusting distances between camera and target, 627–629, *628, 629*
      changing your point of view, 631–634, *632, 633, 634*
      hiding parts of the view with clip panes, 636–638, *636, 637*
      overview of, 624–626, *625*
      versus parallel projection, 624
      setting up perspective views, 626–627, *627*
      twisting the camera, 635, *635*
      using Dview command Zoom option as a telephoto lens, 634–635
   available linetypes, 966–967, *967, 968*
   blocks, 181–182
   layers, 181–182, 382
   Web page links, 896
   Web pages, 891–893, *892, 894*
Viewpoint Presets dialog box, 498–500, *499*
Viewpoint toolbar
   defined, **26**
   SE Isometric button, 470
viewports. *See also* windows
   adding dimensions in, 572–573, *573*
   Maximum Active Viewports option in Preferences dialog box, 1014

in Paper Space
    changing shape of, 396, 399–400
    creating, 392, *392*
    displaying drawings in, 392–393
    floating Model Space viewports, 400
    layers and, 396
    moving, 396, *397*
    moving between, 393, *394*
    overlapping, 400
    saving viewport views, 400
    setting layers in individual viewports, 400–403, *402*
    setting maximum number of active viewports, 398
    troubleshooting, 1003
  setting up in Eye2eye add-on, 1037–1038, *1038*
  tiled viewports in Model Space, 385–388, *387*, *389*
  Viewport object in ActiveX Automation, **928–929**
viewres command, 254–255
views. *See also* perspective views; windows
    clipping Xref views with Xclip command, 378–380, *379*
    creating multiple views, 262
    Hidden view, 472–474, *473*
    opening files in a particular view, 260
    Plan view, **49**
    Redraw view, 54
    saving, 257–260, *259*
    saving Paper Space viewport views, 400
    scaling in Paper Space, 398–400, *399*
visibility
    controlling layer visibility, 169–170, *170*
    controlling object visibility with raster images, 742–743
Visretain system variable, 382
Visual Basic, AutoCAD VBA Edition, **938–939**, *938*

W

wblock command, 151
WCS. *See* World Coordinate System
Wedge tool, 549, *549*
Whip2 Netscape Plug-in, 900–901, 1033, 1043–1044
Width option of polyline Edit Vertex mode, 437, *437*
Width option in Print/Plot Configuration dialog box, 670, *670*
width of text boundary windows, 274, *274*
Window command in View menu Zoom submenu, 29–30, *29*
window object selection command, 69

Window option in Print/Plot Configuration dialog box, 662–663, *663*
window polygon object selection command, 70
windows. *See also* viewports; views
    Aerial View window, 9, 32, 255–257, *256*
    AutoCAD Text window
        capturing and saving text data from, 827–828
        displaying data in, 84–85, *85*
        Number of Lines of Text in Text Window to Keep in Memory option in Preferences dialog box, 1019
        overview of, 268
        Status screen, 820–822, *821*
        Text Window Parameters options in Preferences dialog box, 1019
    changing width of text boundary windows, 274, *274*
    Command window, 9, *12*, 13, **18–19**
    crossing windows
        defined, **72**, *73*
        of Stretch command, 338–339, *338*
    Help window, **81–84**
    program window, **9–13**
        Command window, 9, *12*, 13, 18–19
        drawing area, 9, 11–12
        menu bar, 9, *11*
        overview of, 9–13, *10*
        selecting coordinate points in drawing area, 12
        status bar, 9, *12*, 93
        User Coordinate System (UCS) icon, 12–13
    selecting objects by windowing them, 63–64, *64*
    standard windows, **72**, *73*
Windows Render Options dialog box, 720, *720*
Wintab-compatible digitizers, 995–996
Wipeout objects, 854–855
Wipeout tool, 748, 859–861, *860*, *861*, *862*
wire-frame models. *See also* 3D wire-frame models in Numbers section of index
    defined, **467**
Wizards, New File Setup Wizard, 214–216
work area. *See* drawing area; measurement system
World Coordinate System (WCS). *See also* coordinates; User Coordinate System
    defined, **478**, 480
    User Coordinate System (UCS) icon and, 12
    World UCS icon, 478, *478*
    World UCS tool, **488**
World Wide Web, 885–900
    adding DWF files to Web pages, **889–891**
        overview of, 889
        specifying filenames and opening views, 890–891

what your Internet service provider needs to know, 891
Autodesk and third-party information, 885–886
creating Web-compatible drawings, **887–889**
creating DWF files, 888–889
installing Internet Utilities, 888
opening Internet Utilities toolbar, 888
downloading DWG files from the Web, **896–899**
establishing Web space, 887
getting Help, 900
saving files to Web or FTP sites, 899–900
Web pages
adding links to other drawings or Web pages, 894–896, *895*
viewing, 891–893, *892*, *894*
viewing and deleting links, 896
Whip2 Netscape Plug-in, 900–901, 1033, 1043–1044

—X—

X axis. *See also* User Coordinate System
rotating the UCS plane about, 488–489, *489*
X Axis Rotate UCS tool, 488–489, *489*
X (Explode) command, 147–149, *149*
Xbind dialog box, 383–384
xclip command, 378–380, *379*
Xdata Attachment command in Bonus menu Tools submenu, 876
Xrefs. *See* external reference files

—Y—

Y axis. *See also* User Coordinate System
rotating the UCS plane about, 488–489, *489*
Y Axis Rotate UCS tool, 489

—Z—

Z axis. *See also* User Coordinate System
defined, **480**
defining a UCS origin and plane with, 489–490, *490*
rotating the UCS plane about, 487–489, *488*, *489*
Z Axis Rotate UCS tool, 487–488, *488*
Z Axis Vector UCS tool, 490, *490*
Z buffer shaded models, 686
zooming
in Model Space, 572
on objects, 107–108, *108*
in Paper Space, 394, *395*, 572
Qtext command and, 305, *305*

using Dview command Zoom option as a telephoto lens, 634–635
Zoom commands, 28–32, *29*, *31*, 108
Zoom flyout on Standard toolbar, 209
Zoom Realtime button in Standard toolbar, 30–32, *31*, 108, 253
Zoom Realtime command and cursor, 211–212, 253
Zoom toolbar, **26**
Zoom Window button in Standard toolbar, 29–30, *29*, 108

What's on the CD

This CD-ROM is packed with valuable resources, including electronic versions of two Sybex books, add-ons, utilities, references, and drawing files for the exercises in the book. More specifically, the CD-ROM includes:

AutoCAD 14 Instant Reference, a Sybex best-seller, now an electronic book. This is the definitive companion to *Mastering AutoCAD 14 for Mechanical Engineers*, providing you with detailed descriptions of AutoCAD's commands and tools.

The ABCs of AutoLISP is a complete tutorial and reference book for AutoLISP, the AutoCAD macro-programming language. This book has been a favorite of end users and developers alike, and it's now in an easy-to-use Web-browser format.

Eye2eye is an add-on utility that greatly simplifies the process for obtaining 3D perspective views of your 3D models. No more lost views and unwieldy commands. Set your views using an easy-to-use camera and target metaphor.

ActiveX samples help you get started with Automation in AutoCAD. You can use these examples along with Chapter 20 to explore the newest customization feature of AutoCAD.

Whip2 Netscape Communicator plug-in allows you to view AutoCAD drawings over the Internet. It gives you full Pan and Zoom capabilities as well as access to Web links embedded in AutoCAD files.

SWLite is a demo version of a library of fasteners and utilities that you can use, free, for six months. SWLite is an abbreviated version of another Design Pacifica's product, SI Mechanical. Purchase information is available in the readme.txt file on the CD-ROM.

All the drawing files from the exercises in the book are included so you can easily study any topic at any time. You can experiment with the files on your own without worrying about losing or corrupting them.